Optoelectronics and Optical Communication

OPTOELECTRONICS AND OPTICAL COMMUNICATION

By

ARIJIT SAHA
B.Sc., B.Tech., M.Tech.
Assistant Professor
Electronics & Communication Engineering
B P Poddar Institute of Management & Technology
Poddar Vihar, 137 VIP Road, Kolkata
West Bengal

NILOTPAL MANNA
B.E., M.Tech.
Senior Assistant Professor
Electronics & Instrumentation Engineering
JIS College of Engineering
Kalyani, Nadia
West Bengal

UNIVERSITY SCIENCE PRESS
(An Imprint of Laxmi Publications Pvt. Ltd.)
BANGALORE • CHENNAI • COCHIN • GUWAHATI • HYDERABAD
JALANDHAR • KOLKATA • LUCKNOW • MUMBAI • PATNA
RANCHI • NEW DELHI

Published by :
UNIVERSITY SCIENCE PRESS
(An Imprint of Laxmi Publications Pvt. Ltd.)
113, Golden House, Daryaganj,
New Delhi-110002
Phone : 011-43 53 25 00
Fax : 011-43 53 25 28
info@laxmipublications.com
www.laxmipublications.com

Price: ₹ **395.00** *Only.* *First Edition* : 2011

OFFICES

Bangalore	080-26 75 69 30	**Chennai**	044-24 34 47 26
Cochin	0484-237 70 04, 405 13 03	**Guwahati**	0361-251 36 69, 251 38 81
Hyderabad	040-24 65 23 33	**Jalandhar**	0181-222 12 72
Kolkata	033-22 27 43 84	**Lucknow**	0522-220 99 16
Mumbai	022-24 91 54 15, 24 92 78 69	**Patna**	0612-230 00 97
Ranchi	0651-221 47 64		

U00-9628-395-OPTOELECTRO OPTICA COMMU-SAH C—3719/011/06
Typeset at : Shubham Composer, New Delhi. *Printed at :* Ajit Printers, Delhi.

Dedication

To our family for their constant support and encouragement to complete this work

CONTENTS

PREFACE

Information transfer is one of the basic needs of mankind. Speech or narration is one kind of information transfer within a limited distance. This distance barrier is overcome by incorporation of some processing techniques and with the use of transmission media. Information is not confined to the speech signal only. There are various types of information signals, to mention a few — video signal, digital data, parameters related to industrial instruments, computer communication, remote teaching and learning, remote conference, radio signal etc., which are required to be transmitted from one part of the world to other part.

Superiority of transmission link is determined by how fast and how much error-free data or information can be transmitted. Obviously this leads to the bandwidth limitations. Till the last century, the transmission was mainly relied on metal wires. RF link has also been used, but this suffers the distance barrier and some other constraints. With the advent of optical fiber, there came a revolution in communication process and technology. Information in form of light energy can be guided through optical fiber at faster rate with less error and loss. Optical transmission made great impact on communication and also leads to the development of optical and electro-optical devices. This book intends to provide a comprehensive knowledge of optical sources, detectors, associated electronics and their application in the field of communication.

This book is divided into two parts consisting of fifteen chapters. The first part consisting of seven chapters lay the foundations for optical sources, detectors, optical devices and associated electronic circuitry. The first chapter gives the basic idea how light may be generated by semiconductors with the brief description of bandstructures of material. Chapter 2 describes LED or Light Emitting Diode, its structures, materials, spectral response, efficiency, characteristics etc., in detail. Realizations of different colors of LEDs and various types of drive circuits for LED are also included. Laser— another optical source has been discussed in Chapter 3. Generation of different types of laser— solid state laser, liquid dye laser, gas laser, semiconductor laser and others have been described here with their efficiency, modulation techniques and drive circuits.

Chapter 4 is devoted to various types of photodetectors, their structures, performance as well as characteristics, and optical detecting schemes. Various modulation schemes and electro-optical modulators are described at Chapter 5. Holography applications of optical sources and photodetectors are discussed at Chapter 6 and various types of display devices— LED display, LCD display, Plasma display etc., are described at Chapter 7.

The second part consisting of eight chapters is devoted to application of optics in the field of transmission including optical fiber. Chapter 8 introduces the application of optics in the field of communication mentioning its manifolds advantages. The essential optical communication media— waveguides and optical fibers are discussed at Chapter 9. The optical fiber characteristics are described at Chapter 10.

Various types of optical modulation techniques and schemes for analog and digital types are elaborated at Chapter 11. Optical multiplexing concepts, schemes and protocols are explained at Chapter 12. The hardware and interconnecting devices for optical communication are mentioned at Chapter 13. Optoelectronic integrated circuits that are useful for optical communication and other optoelectronic applications are discussed at Chapter 14. Finally Chapter 15 describes various aspects of optical transmission for short to long distance communication and their protocols.

Simple language is used throughout the book. Clear diagrams and numerous examples are provided for all the topics to facilitate easy understanding of the concepts. The principles of devices and circuits are presented with appropriate analyses and derivations. This book presents materials for both undergraduate and postgraduate students.

Suggestions and corrections are welcome for the improvement of the book.

—Authors

1 INTRODUCTION TO OPTOELECTRONICS

1.1 INTRODUCTION

The interaction of light and optical processes with electronic processes is dealt within a separate category known as '*Optoelectronics*'. At the same time the devices in which such interactions take place, are called optoelectronic devices. Generally the optoelectronic devices are made with semiconductors. For the development of microelectronics the elemental semiconductors like Silicon (Si) and Germanium (Ge) have been extensively used. But for the development of optoelectronic devices these elemental semiconductors were not at all efficient. The question arises why? To understand this, let us have a short detour on basic of solid state device theory.

According to quantum theory, electrons are located near the minimum of conduction band whereas holes are located near the maximum of valence band. Figure 1.1(*a*) and Fig. 1.1(*b*) represent variations of electron energy with momentum. In figure 1.1(*a*) the minima of conduction band and maxima of valence band does not occur at same momentum, thus forming indirect bandgap. On the other hand, the maxima of valence band and minima of conduction band occur at same momentum at Fig. 1.1(*b*) forming direct bandgap. The electrons of the conduction band have the tendency to recombine with holes of valence band and in this process, energy is released in the form of electromagnetic waves of different wavelengths.

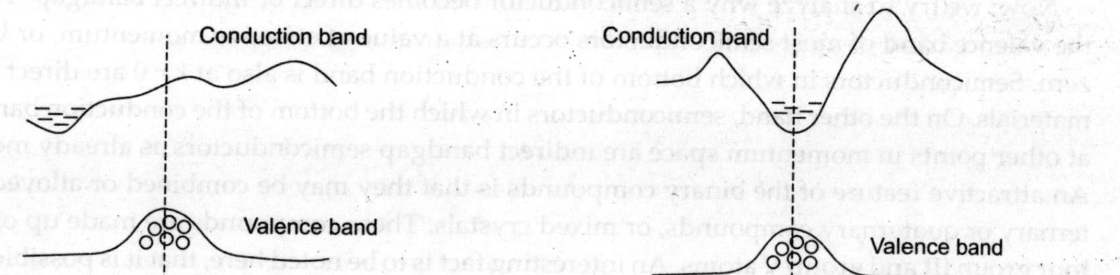

Fig. 1.1(*a*) Fig. 1.1(*b*)

The relation is given as $E = hc/\lambda$, where E is the emitted energy equivalent to bandgap energy, h is the Plank's constant, c is the velocity of light and λ is the wavelength. It may be noted that under suitable condition, certain wavelength may be generated. Our matter of interest lies in the wavelengths of infrared to visible to ultraviolet ranges, *i.e.*, 0.3 mm to 0.03 μm.

The fundamental bandgap of the elemental semiconductors like Silicon and Germenium is indirect. This implies that the light emission efficiency of these semiconductors is very poor. In fact most of the energy emitted from these semiconductors is in the form of heat energy, rather than light energy. Thus, it became clear that Si, which was considered by many as an almost universal semiconductor, was unable to perform optoelectronic applications. Naturally, for optoelectronic applications, it was the necessity to find out semiconductor materials which will be efficient in emitting the energy in the form of light. It turned out that compound semiconductors offered many of the desired properties and could also be synthesized easily. The compound semiconductors are the direct bandgap semiconductors, where the release of the energy is mostly in the form of light. Compound semiconductors are made from elements of different columns of periodic table. Some examples are III-V, IV-VI, IV-IV, or II-VI compounds. The III-V compounds were the first and most widely used type of compound semiconductors for optoelectronic applications. An interesting feature for compound semiconductors is to be noted that they have same average number of valence electrons per atom as of Si. In other words, Si has four valence electrons and in III-IV or II-VI compounds, the sum of valence electrons is eight.

In 1950, Indium antimonide (InSb) was the first III-V compound semiconductor to be discovered. The compound could be very easily synthesized. The electron mobility and the ionic component in the crystal binding also attracted interest in this particular compound. As a matter of fact, these properties are still of immense interest. Moreover, because of its low bandgap, $E_g = 0.17$ eV, InSb has become important for the development of far infrared detector technology. But the development of semiconductor lasers turned the interest to other two very important III-V semiconductors, *e.g.*, GaAs ($E_g = 1.43$ eV) and InP ($E_g = 1.35$ eV). GaP ($E_g = 2.1$ eV) is also a very important compound semiconductor used in the visible region of the spectrum. But an interesting fact is that GaP is an indirect bandgap semiconductor. However by certain doping techniques, it is possible to improve the radiative efficiency. Compared to Ge and Si, InP and GaAs have much high electron mobilities and velocities, which is a very important property utilized for the development of high-speed electronic devices.

Now, we try to analyze why a semiconductor becomes direct or indirect bandgap. The top of the valence band of most semiconductors occurs at a value of effective momentum, or k equal to zero. Semiconductors in which bottom of the conduction band is also at $k = 0$ are direct bandgap materials. On the other hand, semiconductors in which the bottom of the conduction band occurs at other points in momentum space are indirect bandgap semiconductors as already mentioned. An attractive feature of the binary compounds is that they may be combined or alloyed to form ternary or quaternary compounds, or mixed crystals. These compounds are made up of three or four group III and group V atoms. An interesting fact is to be noted here, that it is possible to select different binary compounds to select different bandgaps, and therefore light sources of varying emission energies may be obtained. It is possible to vary the bandgap continuously and monotonically by the alloying process. Hence, the band-structures, the electronic and optical properties of the compound semiconductors may also be varied by the alloying process.

The formation of heterojunctions, which have become essential for the design of high performance electronic and optoelectronic devices, have become possible by the formation of ternary and quaternary compounds of varying bandgaps. Some examples of ternary and quaternary compounds are $Al_xGa_{1-x}As$, $GaAs_{1-x}P_x$, $In_{1-x}Ga_xP$, $In_xGa_yAl_{1-x-y}As$, and $In_{1-x}Ga_xAs_yP_{1-y}$. The quaternary compounds mentioned above are extremely important for optical communication, as their bandgaps correspond to the spectral window in which silica fibers have their lowest loss and dispersion.

1.2 CHOICE OF MATERIALS

Optical part of the electromagnetic frequency spectrum may be divided into three basic bands: infrared, visible and ultraviolet. The frequency and wavelength are related by $c = \nu\lambda$, where c is the velocity of light (3×10^8 m/s), ν is the frequency in hertz (cycles/sec), and λ is the wavelength of the electromagnetic wave (in meters). In fact the optical spectrum spans the wavelength range of 0.005 to 20 microns (μm).

Optoelectronic devices are used in various applications. Depending on these applications light is transmitted through different media. In free-space communication light is transmitted through air, whereas in guided wave communication light transmits through optical fibers or waveguides. The choice of materials depends largely on the application and the transmission medium. Optical fibers have their lowest loss at 1.55 μm. The lowest dispersion and optical amplification in fibers have been demonstrated at 1.3 μm. Hence, the materials of choice for the fabrication of fibers used for long distance communication are InP-based materials. On the contrary for short haul communication GaAs/AlGaAs is used. As semiconductor sources are developed for the visible spectrum, using II-VI and III-V compounds, appropriate guiding materials are also being developed. In this way, different materials are chosen depending on the applications and the advantages they offer.

The most important III-V compounds for the manufacture of visible LEDs are GaP and $GaAs_xP_{1-x}$. These alloys and heterostructures with different compositions are grown on GaAs or GaP substrates. Such an important heterojunction is formed by GaAs-$Ga_{0.51}In_{0.49}P$ which is used in bipolar devices and may be important for phototransistors. Some large bandgap nitrides like GaN, AlN, and InN have gained much importance as optoelectronic materials for emitting blue light. The ternary derivatives of these binary compounds, like AlGaAsN, are also important for heterojunctions.

The most important semiconductors in the near infrared region are GaAs, AlAs, and their ternary derivatives $Al_xGa_{1-x}As$. The GaAs-Al_xGa_{1-x} as heterojunction, being lattice-matched, is widely used for waveguiding and active optoelectronic devices in the near-infrared wavelength region. Though this wavelength range does not overlap with the desirable wavelengths of optical fibers, the heterojunction still is used for sources and detectors due to availability of high quality GaAs substrates and the ability to epitaxially grow high-quality heterojunctions. Heterostructure lasers made of GaAs-AlGaAs have very low threshold currents. Detectors made of these materials have high responsivity.

For longer wavelengths, the most important materials used for optoelectronic device applications are InP and the ternary and quaternary semiconductors lattice-matched to InP. Some examples of ternary compounds are $Al_{0.48}In_{0.52}As$, $Ga_{0.47}In_{0.53}As$ and quaternary compounds are InGaAlAs and InGaAsP. The $Ga_xIn_{1-x}As_yP_{1-y}$/InP and the $Al_xGa_yIn_{(1-x-y)}As/Al_{0.48}In_{0.52}$ as heterojunctions are extensively used for LEDs, Lasers, detectors, solar cells, modulators etc. The bandgaps of the above mentioned materials are compatible with the region of low-loss and low-dispersion of optical fibers, and hence are important for optical fiber communication. Moreover some additional advantage lead to applications of these heterojunctions to optoelectronic integrated circuits (OEICs).

Till now we have confined our discussion only to lattice-matched materials and heterojunctions. But there is another degree of freedom in bandgap tailoring is obtained if lattice-matching requirements are forgone, or if thin, coherently stained active regions are utilized. Usage of quantum wells provides additional tunability of the bandgap and hence the wavelength of use. Again, we have restricted our discussion to the III-V compounds. The II-VI compounds are also of immense importance for application to optoelectronic devices.

1.3 EFFECT OF TEMPERATURE AND PRESSURE ON BANDGAP

Several important effects occur in the lattice with the variation of temperature of the semiconductor. There will be expansion or contraction in the lattice and the oscillations of the lattice atoms around their mean positions will increase or decrease. In fact the electron-lattice interaction changes with temperature. Experimentally it is observed that the bandgap changes with temperature. Though the dependence varies in different temperature ranges, but approximately obeys the empirical relation, first proposed by Varshni:

$$E_g(T) = E_g(0) - \frac{AT^2}{T+B} \tag{1.1}$$

where E(0) is the energy bandgap at 0°K and A and B are constants.

Comprehensive hydrostatic pressure decreases the interatomic spacing, which increases the fundamental energy bandgap. Hydrostatic pressure can be applied by immersing the crystal in a suitable pressure-transmitting medium, which itself is compressed with a piston or anvil arrangement. Due to this pressure, in general, all energy levels in a semiconductor are shifted. This change can be expressed by the equation

$$E(P) = E(0) + E_1 \Delta a \tag{1.2}$$

which denotes a linear change in the energy level with change in lattice constant Δa. This relation is valid only for small changes in Δa. E(0) is the energy level without the application of hydrostatic pressure, and E_1 is a pressure coefficient or deformation potential. The change in the energy gap of a semiconductor can then be expressed by the equation:

$$E_g(P) = E_g(0) \pm (E_{1C} + E_{1V})\,\Delta a \tag{1.3}$$

where E_{1V} and E_{1C} are the respective pressure coefficients for the valence and conduction bands. Different valleys have different pressure coefficients and some may even have negative values. It can be shown that the processes of alloying and application of hydrostatic pressure are equivalent. A change of the bandgap due to hydrostatic pressure causes changes in other properties of the materials. A change in the bandgap will cause a change in the dielectric constant, which indirectly alters the ionization energies of impurity atoms and the binding energies of *excitons.*

Other than temperature and pressure, electric field also has a profound effect on the optical processes. A magnetic field changes the energy levels, due to its effect on carrier motion. It usually splits the energy levels. Depending on the magnitude of the field at which the splitting takes place, the effects are known as *Zeeman* splitting or *Landau* splitting.

1.4 FORMATION AND RECOMBINATION OF ELECTRON-HOLE PAIR

The creation and annihilation of electron-hole pairs is the basis of operation of almost all optoelectronic devices. An electron is raised in energy from the valence band to the conduction band, thus creating a hole in the valence band. In principle, any energetic particle incident on a semiconductor which can impart energy, at least equal to the bandgap energy of the valence band electron can create electron-hole pairs. The simplest way to create an electro-hole pair is to irradiate the semiconductor. In this way photons with sufficient energy are absorbed and hence the electrons are raised from the valence band to the conduction band. This process is termed as *absorption*. The reverse process associated with emission of excess energy is called *recombination*. Recombination may be either *radiative* or *nonradiative*. In the nonradiative recombination, the excess energy is usually imparted to phonons and dissipated as heat. In the radiative recombination, the excess energy is dissipated as photons. This is the *luminescent* process, which may be classified according to the method of creation of electron-hole pairs. Radiative recombination of electron-hole pairs created by injection of photons is termed as *photoluminescence*. *Electroluminescence* is the process of radiative recombination of electron-hole pairs following injection with a *p-n* junction or similar device. Similarly radiative recombination of electron-hole pairs created by electron bombardment is known as *cathodoluminescence*.

In the equilibrium condition of a semiconductor, the carrier densities can be calculated from an equilibrium Fermi level by using Fermi-Dirac or Boltzmann statistics. If excess carriers are created within the semiconductor, nonequilibrium conditions are generated and the concept of Fermi level becomes invalid. The nonequilibrium distribution functions for electrons and holes are defined as

$$f_e(E) = \frac{1}{1 + \exp\left(\frac{E - E_{fe}}{k_B T}\right)} \tag{1.4}$$

and

$$1 - f_h(E) = \frac{1}{1 + \exp\left(\frac{E - E_{fe}}{k_B T}\right)} \tag{1.5}$$

E_{fe} and E_{fh} are the quasi-Fermi levels for electrons and holes, respectively. When there is no excitation, $E_{fe} = E_{fh} = E_F$. The difference $(E_{fe} - E_{fh})$ is a measure of the deviation from equilibrium. For the nondegenerate case we obtain

$$f_e(E) \cong \exp\left(\frac{E_{fe} - E}{k_B T}\right) \tag{1.6}$$

$$f_h(E) \cong \exp\left(\frac{E - E_{fh}}{k_B T}\right) \tag{1.7}$$

The nonequilibrium carrier concentrations are given by

$$e = N_C \exp\left(\frac{E_{f_e} - E_C}{k_B T}\right) \tag{1.8}$$

$$h = N_V \exp\left(\frac{E_V - E_{f_h}}{k_B T}\right) \tag{1.9}$$

where N_C and N_V are the effective density of states in the conduction and valence bands, respectively and E_C and E_V are the Fermi-levels for the conduction and valence bands respectively. If an *n*-type semiconductor with an equilibrium electron density n_o is uniformly irradiated with intrinsic photoexcitation so as to produce Δn electron-hole pairs with a generation rate G, the nonequilibrium electron and hole concentrations will be given by

$$n = \Delta n + n_o \tag{1.10}$$

$$p = \Delta n + n_i^2 / n_o \tag{1.11}$$

The excess carriers created in the semiconductor must recombine. Actually in the steady state condition the recombination rate is equal to the generation rate. These processes involve the transition of carriers across the energy bandgap and hence are different for direct and indirect bandgap semiconductors. In a direct bandgap semiconductor an electron raised to the conduction band dwells there for a very short time and recombines with a valence band hole thereby emitting light of energy equal to the bandgap. Hence, the probability of *radiative recombination* in quite high in direct bandgap semiconductors. In indirect bandgap semiconductors the *nonradiative recombination* process predominates. These will be dealt in detail in the next chapter.

1.5 FRANZ-KELDYSH AND STARK EFFECTS

As already discussed, semiconductors emit light of energy equal to the bandgap. Since the light emitted has equal energy that of the bandgap, hence the same semiconductor absorbs that same wavelength of light emitted by it. The process of absorption in semiconductors is affected in presence of strong electric field. This effect is called *Franz-Keldysh effect*, which results in the absorption of photons with energies less than the bandgap of the semiconductor.

The *Stark effect* is the change in atomic energy due to the application of an electric field. The electric field affects the higher order orbits of the electrons so that the center of gravity of the elliptical orbit and the focus are displaced with respect to each other and linearly aligned in the direction of the electric field. Due to this there is a splitting of the energy of the outer 2*s* or 2*p* states, and the energy shift is simply given as $\Delta E = qdE$, where d is the eccentricity of the orbit. This is termed as *linear Stark effect*. The effect of the electric field on ground state orbits also results in an energy shift of the state. This is known as *second-order* or *quadratic Stark effect*. Both these effects are very useful in designing the optoelectronic devices.

1.6 SEMICONDUCTOR HETEROJUNCTIONS

A heterojunction is formed when two semiconductors of different bandgaps and lattice constants are brought together, usually be epitaxy. These heterojunctions are very essential components for forming optoelectronic devices. For example in semiconductor lasers these heterojunctions provide both optical as well as carrier confinement. These will be discussed in detail in Chapter 3.

A heterojunction forms a two-dimensional channel of carriers at the interface with superior transport properties.

Heterojunction semiconductors belong to one of two categories: (*a*) ***iso-type***, or (*b*) *aniso-type*. An iso-type semiconductor may be in the form of (*n-n*) or (*p-p*), whereas an aniso-type semiconductor may in the form of (*n-p*) or (*p-n*).

1.7 OPTOELECTRONIC DEVICES

Optoelectronic devices can be broadly classified as optoelectronic sources and detectors. An optoelectronic source generates an electromagnetic wave in the optical range. This serves as the information carrier. An ideal optoelectronic source should generate a stable single-frequency electromagnetic wave. But in practice, LEDs and even semiconductor laser diodes emit a range of frequencies. These LEDs and ILDs are compact, light weight and consume moderate power. Moreover they are easy to modulate.

The optoelectronic detector converts the optical signal back into electrical signal. Generally *p-i-n* diodes and avalanche photodiodes are employed for this purpose. Normally the photocurrent generated by these detectors is proportional to the incident optical power and hence to the information input. For practical purposes these should be of small size and low power consumption. Furthermore they are required to be of linear characteristics and fast response compatible to optical signals.

1.8 CONCLUSION

Apart from the optical sources and detectors, with the advent of technology many more optoelectronic devices have developed. Optoelectronic modulation devices are one of the most important parts of optical signal processing. There are different techniques of modulating optical signal, *e.g.*, electro-optic modulation, acousto-optic modulation, magneto-optic modulation, etc. Different modulators like electro-optic (pockels) modulators, longitudinal phase modulators, transverse phase modulators, intensity modulators, kerr modulators have been designed based on these techniques. Moreover Raman-Nath acouto-optic modulator, bragg acouto-optic modulator, acousto-optic deflector, acousto-optic spectrum analyzer, tunable acousto-optic filters, frequency shifters, optical isolators are some of the most important modulation devices.

Different display devices are used for optoelectronic devices. Liquid crystal display, Alphanumeric display, Plasma display are to name a few. The LCDs are the most used display devices now a day, though plasma display devices have taken a major share of the market.

While more and more storage of data in small place is the most important concern, optical storage may be able to solve the problem to a quite extent. And hologram is the solution to the optical storage. In the early days of holography, it could not be thought of how much data we can store there. With the advent of computer generated hologram, this new storage concept came. In fact holographic memories are one of the most researched areas of late.

2

LIGHT EMITTING DIODES

2.1 INTRODUCTION

The light emitting diode or LED was discovered in 1904. The only semiconductor diode in existence at that time was the Silicon Carbide (SiC) point contact rectifier. Needless to mention, the level of light emission and luminance from such devices were very small. Since then, the light emission from semiconductor devices has attracted the interest of researchers all over the world. It is only recently, with the development of wider bandgap III-IV compounds, the visible light emission from a biased *pn* junction has been demonstrated. There is wide application of LED in the form of indicator lamps, display devices and sensors.

In a junction LED, photons of near bandgap energy are generated by the process of *injection luminescence* or *electro-luminescence*, where a large population of electrons are injected into normally empty conduction band and recombine with holes in the valence band. The device emits light by spontaneous emission, which is different from the stipulated emission as observed in junction lasers. Since the spontaneous emission requires a smaller forward bias, the LEDs operate at lower current densities than the lasers. Also LEDs do not need any optical cavity or mirror facets to provide the feedback of photons. The emitted photons have random phases and therefore LED produces incoherent light. The linewidth of the spontaneous emission is approximately equal to the photo-luminescence linewidth, which is a few times K_BT and is typically 30-50 nm at room temperature Thus, many optical modes are supported and therefore the LED performs as a convenient multimode optical source, suitable to use with multimode fibers.

Some of the distinct advantages of LED as light source are that it possesses simple fabrication procedure, it is of low cost and uses simpler drive circuitry. Light emitting diodes are reliable in operation because they operate at lower powers. Therefore they do not suffer catastrophic degradation or even gradual degradation as observed in case of lasers. Ideally, the LED exhibits a linear light output-current characteristics, unlike that of an injection laser and hence it is suitable for analog modulation. The light-current characteristics is also less sensitive to temperature. The main disadvantages of LEDs are the low power output, harmonic distortion due to the multimodal output and a smaller modulation bandwidth. However, superradiant LEDs have been developed which have output powers comparable to that of lasers.

2.2 THE ELECTROLUMINESCENCE PROCESS

The LED converts input electrical energy to output optical radiation in the visible or infrared portion of the spectrum, depending on the semiconductor material. The energy conversion takes place in two stages: first, the energy of the carriers in the semiconductors is raised above their equilibrium value by electrical input energy, and second, most of these carriers after spending a mean lifetime in their higher energy state, give up their energy as spontaneous emission of photons with the energy nearly equal to the bandgap E_g of the semiconductor. If the semiconductor is doped with *p* and *n* type, then the energy of the current carriers can be enhanced with the application of a forward bias to *p-n* junction. Under forward bias, the minority carriers are injected on both sides of the junction and these excess minority carriers defuse away from the junction, recombining with the majority carriers. This process is illustrated at Fig. 2.1. The excess hole concentration at the *n*-side and excess electron concentration at *p*-side away from the depletion region are expressed by the following equations:

$$\delta p(x) = \delta p_0 e^{-x/L_h} \tag{2.1}$$

and

$$\delta n(x) = \delta n_0 e^{-x/L_e} \tag{2.2}$$

where $\delta p_0 = \delta p(0)$ and $\delta n_0 = \delta n(0)$

p_0 and n_0 are the equilibrium hole and electron concentrations respectively. $L_h = (\tau_h D_h)^{1/2}$ and $L_e = (\tau_e D_e)^{1/2}$ are the diffusion lengths. τ_h and τ_e are recombination lifetimes for holes and electrons. D_h and D_e are the diffusion coefficients.

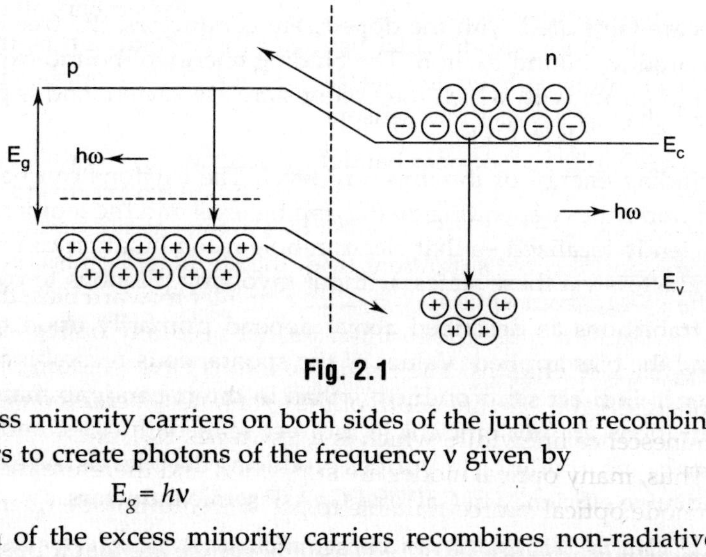

Fig. 2.1

Most of the excess minority carriers on both sides of the junction recombine radiatively with the majority carriers to create photons of the frequency ν given by

$$E_g = h\nu \tag{2.3}$$

A small fraction of the excess minority carriers recombines non-radiatively, where excess energy of these carriers is dissipated in form of heat in the lattice. Normally the rate of recombination is proportional to the forward bias injection rate and hence the diode current under such condition may expressed as

$$J = J_h(0^+) + J_e(0^-) = q(D_h p_{N0}/L_h + D_e n_{P0}/L_e)\,(e^{q/kBT} - 1) \tag{2.4}$$

Most of the recombination occurs close to the junction, however some of the minority carriers diffuse away and recombine takes place in regions further away from the junction. The *internal quantum efficiency* of the device is defined as the rate of emission of photons divided by the rate of supply of electrons. Since all the injected electrons do not recombine radiatively with holes, the quantum efficiency may be less than 100%. Under reverse bias condition of the diode, no injection takes place across the depleted junction and therefore, no light is emitted.

For light emission in a semiconductor, the wavelength or the energy of emission close to the bandgap is important. The transitions at band to band, impurity band, donor to acceptor and excitonic transitions are the matter of interest. The peak in the electron density distribution in the conduction band occurs at an energy of $k_BT/2$ above the band minimum. Therefore, the band to band transition energy is slightly higher than E_g, and some of the emitted photons are reabsorbed. This self-absorption does not exist in the indirect bandgap semiconductors due to the involvement of phonons. The radiative efficiency of phonons, participating in the radiative transitions in indirect bandgap semiconductor is low. However, most radiative transitions involve energy levels within the bandgap for which the emission energy is smaller than the bandgap.

The impurity induced radiative transitions and donor to acceptor transitions have the emission energy close to the bandgap energy. However, there is an interesting aspect of impurity levels, which plays important role in the operation of LEDs. A carrier in an impurity state is trapped in the potential of the impurity, and so becomes localized in space. As the impurity level becomes deeper, its binding energy increases and the localization of carriers is enhanced. This leads to a spatial localization of the carriers in the impurity levels causing a spread in their momentum (or K) values. The spread of the impurity in k-space is sufficient to allow a significant rate of transitions between states and the bandedges without involvement of phonons. This, in effect, increases the radiative efficiency for the indirect bandgap semiconductors.

Since the LEDs are fabricated with the doped semiconductors, the free exciton couples with impurity atom to produce bound exciton. The binding energy of bound exciton depends on the binding energy of impurity with which the excitons are associated, and is given by

$$E_{ex} = 0.1\, E_i$$

where E_i is the binding energy of the impurity level. The excitons can be associated with the neutral or ionized impurities depending on the doping level and the temperature. Bound exciton may also be sufficiently localized so that electron-hole recombination can take place in indirect bandgap semiconductors via these states, without involving phonons tc conserve momentum.

The radiative transitions as described above depend primarily upon the impurities in the semiconductor and the bias applied. Values of the spontaneous recombination coefficient, B_r is significantly lower in indirect semiconductors than in direct bandgap materials. Therefore, the band-to-band transitions are not efficient in indirect semiconductors, and generally, radiative efficiency increases by incorporating impurities producing deep levels. As example, presence of O or N enhances radiative efficiency in GaP and GaAsP semiconductors.

Example 2.1. *Calculate the intrinsic carrier concentration in a semiconductor GaAs at room temperature (300°K) from the following data-$m_e = 0.07\, m$, $m_h = 0.56\, m$, $E_g = 1.43$ eV, where m is mass of an electron at free space $= 9.11 \times 10^{-31}$ kg.*

Solution:

$k = 1.38 \times 10^{23}$, $h = 6.626 \times 10^{-34}$, $1\text{eV} = 1.6 \times 10^{-19}$

The number of free electrons per unit volume $= n = 2(2\pi m_e kT/h^2)^{3/2}\, e^{-E_g/2kT}$

The number of free holes per unit volume $= p = 2(2\pi m_h kT/h^2)^{3/2}\, e^{-E_g/2kT}$

Intrinsic carrier concentration $= n_i = (np)^{1/2} = 2(2\pi kT/h^2)^{3/2}\, (m_e m_h)^{3/4}\, e^{-E_g/2kT}$

$$= 2\left(\frac{2\pi \times 1.38 \times 10^{-23} \times 300 \Big/ \left(6.626 \times 10^{-34}\right)^2}{\left[0.07 \times 0.56 \times \left(9.11 \times 10^{-31}\right)^2\right]^{3/4}}\right) \exp\left[-1.43 \times 1.6 \times 10^{-19} / 2 \times 1.38 \times 10^{23} \times 300\right]$$

$$= 2.2 \times 10^{12}\ \text{m}^{-3}$$

Example 2.2. *Calculate the diffusion potential of GaAs p-n junction in equilibrium at room temperature (300°K). Assume the acceptor and donor impurity concentrations are $5 \times 10^{23} m^{-3}$ and $5 \times 10^{21} m^{-3}$.*

Solution:

$$\text{Diffusion potential } V_D = \left(\frac{kT}{e}\right) \ln\left(\frac{N_a N_d}{n_i^2}\right)$$

$$= \left(\frac{1.38 \times 10^{23} \times 300}{1.6 \times 10^{-19}}\right) \ln\left(\frac{5 \times 10^{23} \times 5 \times 20^{21}}{\left(2.2 \times 10^{12}\right)^2}\right) \left(\text{assuming } n_i = 2.2 \times 10^{12}\text{m}^{-3} \text{ from example 2.1}\right)$$

$$= 1.234\ \text{V}$$

2.3 LED MATERIALS

The wavelength of light emission depends on the factors of the semiconductor materials in terms of their bandgap energy, as well as heavily doping capability with *n*-type and *p*-type impurities and thereby forming a junction diode. Lower bandgap is required for infrared and far-infrared applications, and larger bandgap materials (around 2eV) are needed for a visible light source. With the increase of the bandgap, melting point of the material increases and its structural stability decreases. Also, the higher bandgap materials have higher resistivity and are difficult to dope to high levels that may be required for LED fabrication.

The binary compounds GaAs and Gap and their ternary derivative $GaAs_{1-x}P_x$ are the most important of III-V compounds in application of radiative emission. GaAs has direct bandgap of 1.43 eV at room temperature and emits in the near infra-red region (λ = 860 nm). Zn-diffusion at *n*-GaAs or Si doping of solution grown GaAs is normally used for Light emitting diodes. Si, for its amphoteric nature, can incorporate into GaAs crystal as donor or acceptor, depending on growth temperature. The main radiative transition in a Zn-doped material is due to band-to-band process and there is large re-absorption of the emitted radiation. On the other hand, re-absorption is less severe for Si-doped diode and its emission wavelength is about 940 nm. The binding energy is in order of 0.1 eV. However, Si-doped GaAs light emitting diodes are more expensive to fabricate than the Zn-doped devices.

GaP is an indirect bandgap semiconductor material with bandgap of 2.26 eV (λ = 549) and impurity levels in the bandgap are required to be used as an efficient light emitter. GaP is commonly used for red and green emission. Nitrogen as doping material acts as donor-like isoelectronic trap in GaP with binding energy of 8meV. The isoelectronic trap forms a bound exciton, emitting a near bandgap wavelength of 550 nm (green). The quantum efficiency is smaller than 1%. But higher level of N doping ($\geq 10^{14}$ per cm^3) raises the defect levels in the bandgap due to the formation of N-N molecular complexes, and emits yellow light at 590 nm. Red emission occurs if GaP is simultaneously doped with Zn and O, which occupy, respectively, substitutional Ga and P sites in the lattice. O impurity in GaP raises a deep donor with a binding energy of 0.8 eV. However, an isoelectronic trap, with binding energy of approximately 0.3 eV, is formed when Zn and O atoms occur as nearest neighbors in the lattice. The bound exciton associated with this trap level recombines to produce emission of wavelength of $\lambda = 640$ nm.

The ternary $GaAs_{1-x}P_x$ changes from direct to indirect bandgap material for $x = 0.45$, with $E_g = 2.1$ eV. Thus, diodes may be fabricated using the direct bandgap and indirect bandgap materials with N doping to emit red ($GaAs_{0.6}P_{0.4}$), orange ($GaAs_{0.35}P_{0.65}$:N) and yellow ($GaAs_{0.15}P_{0.85}$:N) lights. Table 2.1 provides the characteristics of some visible light emitting diodes.

Table 2.1

Structure	*Material*	*Bandgap Type*	*Peak Wavelength (nm)*	*Colour*	*Typical Performance (lm/W)*
Homojunction	GaAsP	Direct	650	Red	0.15
	GaP:Zn,O	Indirect	700	Red	0.4
	GaAsP:N	Indirect	630	Red	1.0
			585	Yellow	1.0
	GaP:N	Indirect	565	Yellow-green	2.6
	GaP	Indirect	555	Green	0.6
	SiC	Indirect	480	Blue	0.04
Single heterojunction	AlGaAs	Direct	650	Red	2.0
Double heterojunction	AlGa As	Direct	650	Red	4.0
	AlGaP	Direct	620	Orange	20.0
	AlInGaP	Direct	595	Amber	20.0
	AlInGap	Direct	570	Yellow-green	6.0
	GaN	Direct	450	Blue	0.6
Double heterojunction with transparent substrate	AlGaAs	Direct	650	Red	8.0

In the recent development, the InP based compounds such as the ternary $In_{0.53}Ga_{0.47}As$ and the quanternaries $InGa_{1-x}As_{1-y}$ are found to be most effective to produce emission of longer wavelengths in the near infrared region of the spectrum. Devices made of these materials have the wavelengths of 1.1 to 1.6 μm, at the spectral region of low loss and minimum dispersion in optical fibers. Therefore, they are important for optical fiber communication. For the far infrared region of the spectrum the antimonides like GaSb, InSb and their alloys are important. However, use of these compounds is still restricted due to lack of suitable lattice-matched substrates. For the emission of blue light, higher bandgap material like binary Gallium Nitrate (GaN) is extremely important ($E_g = 3.5$ eV), which would be suitable for satellite communication. But *p*-type doping of the material is a problem. Silicon Carbide (SiC) of IV-V compound and the III-V compounds are the promising semiconductor materials under development for making blue LEDs. High melting point of SiC as well as its mechanical and chemical properties make difficult to develop its device technology. Yellow LEDs are currently commercially available by doping with Al, B, Be and Sc boosting up the emission in various parts of the spectrum. Other promising materials and compounds for light emission, which are under study, are Diamond, ZnS, ZnSe, CdTe, HgTe, ZnTe and their ternary derivatives.

2.4 DEVICE CONSTRUCTION AND EFFICIENCY

Figure 2.2 presents the schematic of a typical surface-emitting LED. The whole structure is developed by epitaxial techniques, while the junction is usually formed by diffusion. The device is designed such a way that the most of the radiative recombination takes place in the side of the junction nearest to the surface, greatly reducing the probability of reabsoption. Thus it is ensured that the most of the current flowing across the diode is carried by the injected carriers into the surface layer.

In a junction LED, three stages of processes are observed. First is the *excitation* or *the injection* process, where the energy of carriers is raised by forward bias injection. Next, most of these carriers give up their excess energy as photons which is termed as *recombination* process. Finally, the generated photons leave the semiconductor producing photocurrent in a detector or providing optical effects. This is called the *extraction* process. The overall device efficiency η_o can be expressed in terms of efficiency of each of the processes as

$$\eta_o = \eta_{in}\,\eta_r\,\eta_e \tag{2.5}$$

where η_{in}, η_r and η_e are the injection efficiency, radiative recombination efficiency and the extraction efficiency respectively. η_o is also termed as the external conversion efficiency.

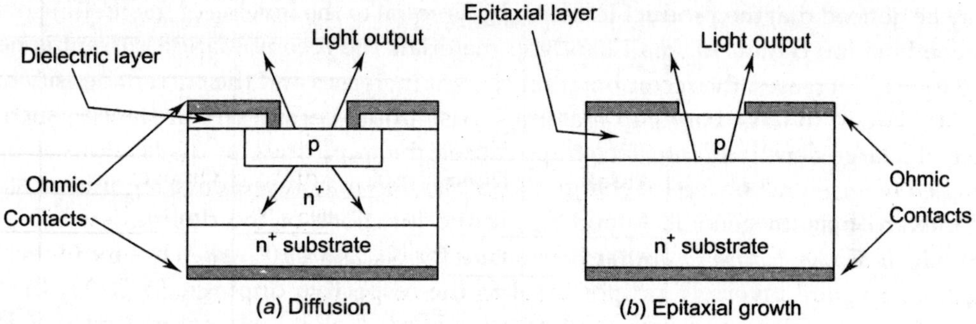

Fig. 2.2

2.4.1 Injection Efficiency

In the device, as shown in Fig. 2.2, which is formed to an asymmetric n^+–p junction diode, electrons injection is much more dominant than the hole injection. For small carrier diffusion lengths, as normally found in the most of III-V compounds, it is almost impossible to determine the injection efficiency experimentally. However the injection efficiency η_{in} can be calculated from the junction theory. The fraction of total diode current due to diffusion of electrons that are injected into p-side of the junction may be expressed as

$$\eta_{in} = \frac{D_e n_{po}/L_e}{D_e n_{po}/L_e + D_h p_{no}/L_h} \tag{2.6}$$

or,
$$\eta_{in} = \left(1 + \frac{D_h L_e p_{no}}{D_e L_h n_{po}}\right)^{-1} \tag{2.7}$$

Using the Einstein relation and the equality relationship $n_o p_o = n_i^2$, equation (2.7) becomes

$$\eta_{in} = \left(1 + \frac{\mu_h p_{no} L_e}{\mu_e n_{po} L_h}\right)^{-1} \tag{2.8}$$

In most III-V compounds, $\mu_e >> \mu_h$ and $L_e \approx L_h$. For full ionization condition, $p_{po} = N_A$ and $n_{no} = N_D$. Then the injection efficiency $\eta_{in} \approx 1$. A high injection efficiency can be ensured by making $N_D >> N_A$, forming an asymmetric junction. The calculation of the injection efficiency is made neglecting the recombination in the space charge layer or the depletion layer of the junction. From the Shockley-Read-Hall theory, the current due to space charge recombination is estimated by the relationship,

$$I_{rec} = \frac{qW}{2} s V_{th} N_T n_i A \left(e^{qVf/2kBT} - 1\right) \tag{2.9}$$

where, s is the capture cross-section,

N_T is the density of the recombination center or traps,

$V_{th} = (3k_BT/m)^{1/2}$ is the thermal velocity of the carriers,

n_i is the intrinsic carrier concentration,

A is the area of the depletion layer of the diode,

W is the width of the depletion layer, and

V_f is the applied forward bias to the diode.

It may be noticed that the product term $sV_{th}N_T$ is equal to the inverse of the lifetime of the deep level recombination center. In small bandgap materials the recombination current is negligible. As the bandgap increases, the recombination current increases and the current density may be as high as 10^{-2} A/cm^2 in large bandgap materials. Also under certain circumstances, such as n the presence of a large density of interface traps caused the impurities, or dislocations as in a lattice mismatched hetero-junction, space charge recombination may represent a significant shunt path at low forward bias. In general, L_e and L_h are the functions of the doping levels N_A and N_D respectively. In GaAs, L_e and L_h are fairly constant for N_A, $N_D \leq 10^{18}$ cm^{-3}, but for higher levels of doping L_e and L_h are inversely proportional to the respective dopings. In GaAs, the injection efficiency η_{in} is possible to be achieved up to 0.8, while in GaP it ranges from 0.6 to 0.8. However, to determine the device efficiency η_o, the injection efficiency η_{in} is not of much significance, as its value is higher than other two types of efficiencies of equation (2.5) and moderate variation of it does not affect much on the overall efficiency.

2.4.2 Recombination Efficiency

The junction LED emits light by the spontaneous emission of photons, and any optical amplification process through stimulated radiation as observed in laser is not present here. Thus limits the internal quantum efficiency (η_i) or radiative efficiency (η_r). The radiative efficiency η_r is defined as the ratio of the number of photons generated to the number of injected electrons. It is obvious that the elimination of any possible nonradiative processes improves the radiative efficiency. The main nonradiative processes occurred in a LED structure are Shockley-Read-Hall recombination, band to band and trap-assisted Auger recombination, and recombination via surface states.

In direct bandgap semiconductors, for homojunction devices, the radiative efficiency is usually in the range of 0.5. This is due to that some of the injected electrons penetrate deeply into the active region before recombining and many of the photons generated are absorbed. Moreover the photons produced downward in the device will also be absorbed. In the indirect bandgap materials, the radiative efficiency is normally very low, unless, recombination can take place via an impurity level. To achieve a reasonable level of radiative efficiency for an indirect bandgap semiconductor with a substitutional impurity, carriers must be tightly bound to the core of the impurity atoms. A deep level in energy gap is required for this and the transition wavelength much larger than the corresponding energy gap. Thus, for III-V compounds like GaP or GaAsP, near infrared or infrared light is observed. For example, the typical radiative efficiencies of GaP:Zn,O and GaP:N are 0.3 and 0.03 respectively.

Nonradiative surface recombination is more important in direct bandgap semiconductor than in indirect bandgap materials. The radiative efficiency or internal quantum efficiency can be improved with the use of double heterojunction structures to as high as 0.6 to 0.8.

2.4.3 Extraction Efficiency

The extraction efficiency or the external conversion efficiency ultimately determines the amount of light emerging from the LED device. If the light is not extracted efficiently, the overall efficiency becomes low, in spite of higher values of injection and radiative efficiencies. The extraction efficiency depends on several factors. One of the factors is the absorption coefficient which is dictated to a large extent by the type of bandgap, — whether it is direct or indirect. The extraction efficiency is higher, if the light-emitting junction is constructed very close to the surface in a direct bandgap device. This can be explained by considering the examples of GaAs and GaP:Zn,O, whose absorption and luminescence spectra are shown in Figs. 2.3(*a*) and 2.3(*b*) respectively.

In GaAs, the absorption coefficient at the peak of the output is ~ 10^3 cm^{-1}. Since $\Phi_x = \Phi_0 e^{-\alpha x}$, the half of the emitted radiation is reabsorbed while traveling as little as 2 μm as $\Phi_x = \Phi_0 e^{-0.6} \approx \Phi_0/2$, where $\alpha = 3 \times 10^3$ cm^{-1}. In case of GaP:Zn,O material, where $\alpha = 3$ cm^{-1} at the peak of the emission spectrum and reabsorption in a few microns of material is negligible. But, unfortunately, while the junction approaches to the surface, radiative recombination efficiency reduces. The use of heterojunction eliminates this problem to a great extent.

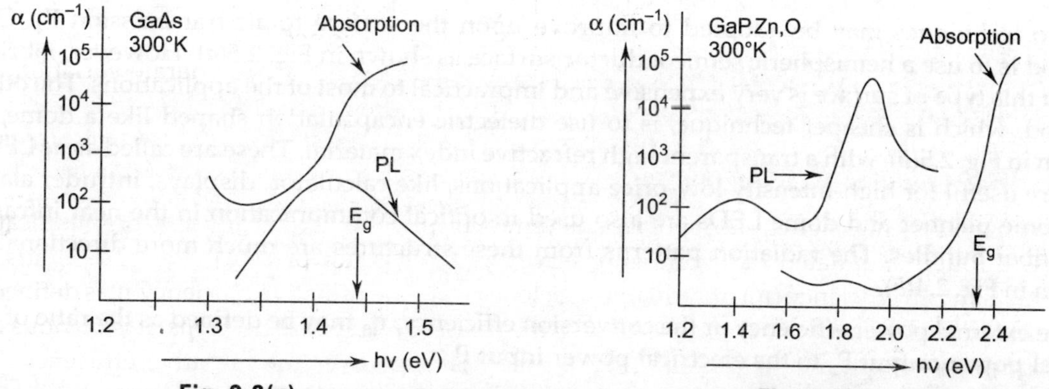

Fig. 2.3(*a*) **Fig. 2.3(*b*)**

Another factor to play the important role on the extraction efficiency is the radiation geometry of the LED, which emits through the planner surface. The intensity distribution of the radiation is Lambertian, in which the power radiated from a unit area into a unit solid angle, expressed in units of Wsr^{-1}m^{-2}, is constant in all direction. This is shown in Fig. 2.4(*a*). However, most of the emitted radiation that strikes the semiconductor-air interface at angles greater than the critical angle, θ_c remains trapped by total internal reflection.

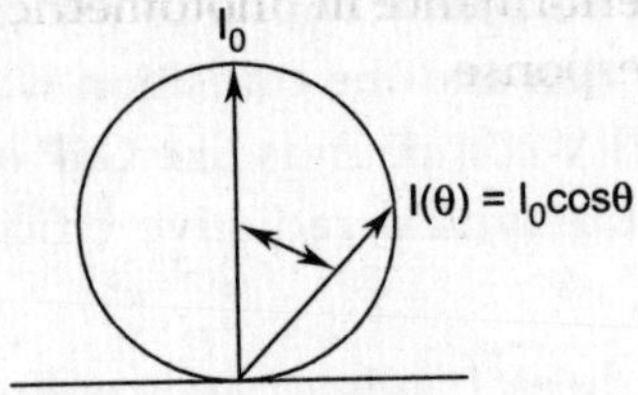

Fig. 2.4(a)

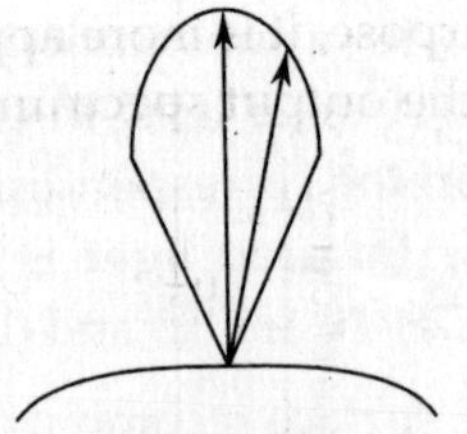
Fig. 2.4(b)

The critical angle is given by $\theta_c = \sin^{-1}(n_{r2}/n_{r1})$ (2.10)

where n_{r2} and n_{r1} are refractive indexes of air and semiconductor respectively. The high dielectric constant or refractive index of semiconductor compounds makes the critical angle very small. The transmission F_T, light radiated to the total light generated, at the semiconductor-air interface may be related by the following expression,

$$F_T = \frac{1}{4}\left(\frac{n_{r2}}{n_{r1}}\right)^2\left(1 - \left(\frac{n_{r1} - n_{r2}}{n_{r1} + n_{r2}}\right)^2\right) \tag{2.11}$$

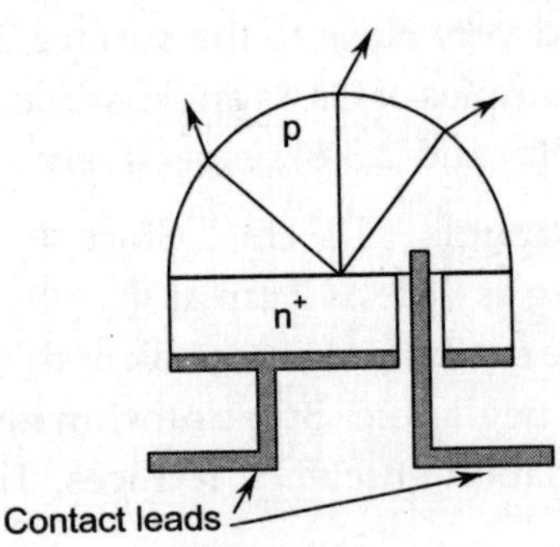

Fig. 2.5(a)

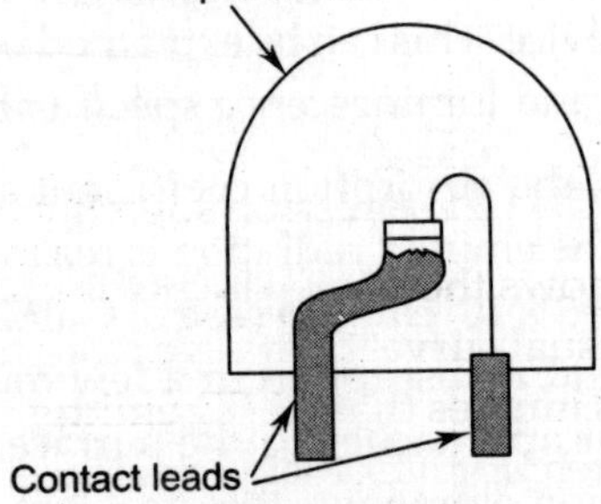

Fig. 2.5(b)

Two techniques may be adopted to improve upon the surface to air transmission F_T. One method is to use a hemispheric semiconductor surface as shown in Fig. 2.5(*a*). However, fabrication of this type of surface is very expensive and impractical to most of the applications. The other method, which is cheaper technique, is to use dielectric encapsulation shaped like a dome, as shown in Fig. 2.5(*b*), with a transparent high refractive index material. These are called *dome* LEDs, and are useful for high-intensity low-price applications, like calculator, displays, intruder alarm etc. Some planner and dome LEDs are also used in optical communication in the near infrared with fiber bundles. The radiation patterns from these structures are much more directional as shown in Fig. 2.4(*b*).

The external power efficiency or the conversion efficiency, η_0, may be defined as the ratio of the optical power output P_o to the electrical power input P_e,

$$\eta_0 = \frac{P_o}{P_e} \times 100\% \tag{2.12}$$

Typical values for η_0 in common LEDs are in the range of 1% to 5%.

2.4.4 Luminous Efficiency

Finally, it is interesting to note that there is physiological impact or eye response on different colors of lights. Hence for daily and common applications of LEDs, this factor must not be neglected. For this purpose, it is more appropriate to express the performance in photometric units, which normalizes the output spectrum of the device with eye response.

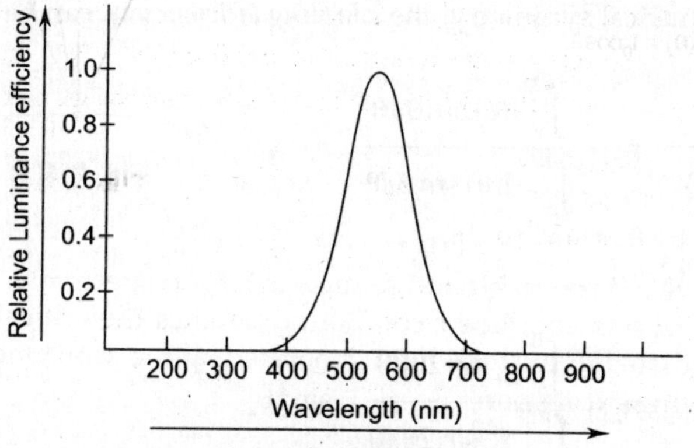

Fig. 2.6

Thus the *luminous efficiency* or *luminosity* is expressed by

$$\eta_L = \frac{\int_\lambda V(\lambda) P_o(\lambda) d\lambda}{\int_\lambda P_o(\lambda) d\lambda} \tag{2.13}$$

where, $P_o(\lambda)$ is the LED emission spectrum and $V(\lambda)$ is the eye response.

Figure 2.6 shows the approximate normalized spectral response of human eye, also known as the photopic visual curve. In the eye response curve, the peak is obseved at the wavelength of 500 to 600 nm. This implies that a LED emitting at the peak of the eye response in the green or yellow will appear much brighter than a device emitting equal amount of energy in red or blue. Considering the eye response into account the overall efficiency may be express as

$$\eta_o = \eta_{in} \eta_r \eta_e \eta_L \tag{2.14}$$

2.4.5 Coupling Loss

In most practical applications, light emitting devices are coupled with the optical fiber cables, which introduces some amount of loss, called *coupling loss*. Figure 2.7 explains how coupling loss is introduced.

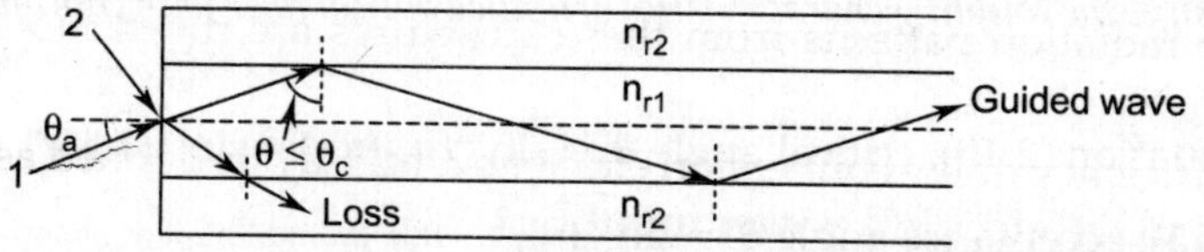

Fig. 2.7

Consider, refractive indexes are n_{r1} and n_{r2} for core and the cladding of the optical fiber ($n_{r1} > n_{r2}$) and θ_a is the acceptance angle. Then,

$$\theta_a = \sin^{-1}(n_{r1}^2 - n_{r2}^2)^{1/2} = \sin^{-1}(A_n) \tag{2.15}$$

A_n is the numerical aperture of the fiber. It is assumed that the entire light incident on the exposed end of the fiber core is within the angle θ_a, and is coupled by the internal reflection. The radiated intensity for a Lambertian distribution is expressed as

$$I(\theta) = I_0 \cos\theta \tag{2.16}$$

With the assumption that the device is smaller than and in close proximity to the fiber core, and also assuming a cylindrical symmetry, the *coupling efficiency* η_c can be expressed as

$$\eta_c = \frac{\int_0^{\theta a} I(\theta)\sin\theta\, d\theta}{\int_0^{\pi/2} I(\theta)\sin\theta\, d\theta} \tag{2.17}$$

Substituting equation (2.16),

$$\eta_c = \frac{\int_0^{\theta a} I_o \sin 2\theta\, d\theta}{\int_0^{\pi/2} I_o \sin 2\theta\, d\theta} = \sin^2\theta_a \tag{2.18}$$

Example 2.3. *Calculate the injection efficiency of a GaAs diode in which $N_a = 10^{23}\ m^{-3}$ and $N_d = 10^{21}\ m^{-3}$. Assume that at room temperature (300°K) $\mu_e = 0.85\ m^2v^{-1}s^{-1}$, $\mu_h = 0.04\ m^2v^{-1}s^{-1}$ and $L_e = L_h$.*

Solution: From equation (2.7)

$$\eta_{in} = \left(1 + \frac{D_h L_e p_{no}}{D_e L_h n_{po}}\right)^{-1}$$

or,

$$\eta_{in} = \left(1 + \frac{D_h L_e N_a}{D_e L_h N_d}\right)^{-1}$$

$$= \left(1 + \frac{\mu_h L_e N_a}{\mu_e L_h N_d}\right)^{-1}$$

$$= \left(1 + \frac{0.04 \times 10^{23}}{0.85 \times 10^{21}}\right)^{-1} = 0.1752$$

Example 2.4. *Determine the coupling characteristics of a silica optical fiber with core and cladding refractive indices of 1.45 and 1.50.*

Solution: From equation (2.10), critical angle $\theta_c = \sin^{-1}(n_{r2}/n_{r1}) = \sin^{-1}(1.45/1.50) = 75.16°$

From equation (2.15), acceptance angle $\theta_a = \sin^{-1}(n_{r1}{}^2 - n_{r2}{}^2)^{1/2}$

$= \sin^{-1}(1.5^2 - 1.45^2)^{1/2} = 8.48°$

2.5 LIGHT OUTPUT FROM LED

The amount of light emitted due to forward bias injection can be calculated by the following process. Consider Fig. 2.8. The photons are emitted due to the recombination of an electron at energy $E(k'') = E_2$ with a hole at energy $E(k') = E_1$.

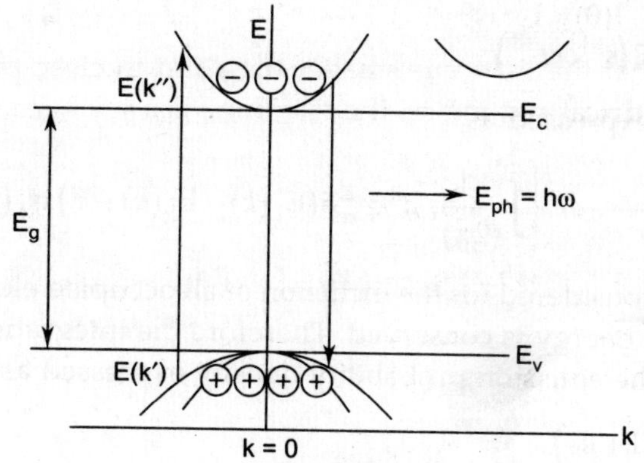

Fig. 2.8

E(k″) and E(k′) are given as

$$E(k'') = E_g + \frac{h^2 k''^2}{2m_e^*} \tag{2.19}$$

$$E(k') = \frac{h^2 k'^2}{2m_h^*} \tag{2.20}$$

Assuming $k'' = k' = k$, and $E_2 - E_1 = E = h\nu$, is the emitted photon.

$$k^2 = \frac{2m_r^*}{h^2}\left(E - E_g\right) \tag{2.21}$$

where m_r^* is the reduced mass.

$$m_r^* = \frac{m_e^* m_h^*}{m_e^* + m_h^*} \tag{2.22}$$

Therefore, near the bottom of the conduction band

$$E_2 = E_g + \frac{m_r^*}{m_e^*}\left(E - E_g\right) \tag{2.23}$$

Emission of photons depends on the density of states in both the conduction and valence bands. Conservation of energy and momentum forces the photon to interact with electrons and holes of specific energy and momentum. This interaction is described by the *joint density of states*, $N_J(E)$. Hence for an one-to-one correspondence between E_2 and the photon energy E, the following incremental relationship can be written,

$$N(E_2)\, dE_2 = N_J(E)\, dE \tag{2.24}$$

or,

$$N_J(E) = \frac{m_r^*}{m_e^*} N(E_2) \tag{2.25}$$

For parabolic electron-hole bands, $N_J(E)$ per unit volume can be expressed as

$$N_J(E) = \frac{\left(2m_r^*\right)^{3/2}}{2\pi^2 h^3} (E - E_g)^{1/2} (eV)^{-1} cm^{-3} \tag{2.26}$$

The spontaneous emission rate $r_{sp}(E)$ and total spontaneous emission rate per unit volume R_{sp} are defined by

$$r_{sp}(E) = P_{em} N_J(E) f_n(E_2)[1 - f_p(E_1)] \quad (s^{-1}(eV)^{-1}cm^{-3}) \tag{2.27}$$

And $$R_{sp} = \int_E r_{sp}(E)dE\left(s^{-1}cm^{-3}\right) \tag{2.28}$$

Alternately R_{sp} may be expressed as

$$R_{sp} = \int \frac{n_r q^2 E}{3\pi\varepsilon_0 m_0^2 h^2 c^3} dE\left\{\int \frac{1}{(2\pi)^3} d^3k p_{cv}^2 \delta(E_2(k) - E_1(k) - E) x f_n(E_2)\left[1 - f_p(E_1)\right]\right\} \tag{2.29}$$

The integral over d^3k is considered for the inclusion of all occupied electron and hole states and the δ-function assures that energy is conserved. Therefore the integration gives the joint density of states. The term P_{em} is the emission probability density expressed as

$$P_{em} = \frac{n_r q^2 E p_{cv}^2 (1 + u_E)}{3\pi\varepsilon_0 m_0^2 h^2 c^3} \tag{2.30}$$

P_{em} also includes the factor n_E, which is the photon mode density. It should be noticed that the emission probability has a factor $(1 + u_E)$, where u_E is the number of photons per mode or the photon occupation. For spontaneous emission, due to the fact that photons are created by recombination escape, u_E is very small and the factor becomes unity. For example, in GaAs, $2p_{cv}^2/m_0$ = 23 eV. Assuming E = 1.5 eV, P_{em} is calculated to $1.8 \times 10^9 s^{-1}$. This implies that the recombination lifetime $\tau_r = 1/P_{em}$ = 0.6 ns.

The Fermi function in equations 2.27 and 2.29 is used to ensure that a state in conduction band is filled and a state in valence band is empty. The distribution functions $f(E_2)$ and $f(E_1)$involve electron and hole at quasi-Fermi levels E_{fn} and E_{fp} respectively. Under thermal equilibrium condition, the quasi-Fermi levels merge into a single Fermi level within the bandgap. Then equilibrium distribution can be used. With Boltzman approximation, $f_n(E_2)\,[1 - f_p(E_1)] = e^{-E/k_BT}$. Under these conditions,

$$r_{sp}(E) = \frac{\left(2m_r^*\right)^{3/2}}{2\pi^2 h^3 \tau_r}\left(E - E_g\right)^{1/2} e^{-E_g/k_BT} e^{-(E-E_g)/k_BT} \tag{2.31}$$

In case of weak injection in an LED, where quasi-Fermi levels are at several k_BT away from the band-edges, but still within the band gap, the spontaneous emission rate can be written as

$$r_{sp}(E) = \frac{\left(2m_r^*\right)^{3/2}}{2\pi^2 h^3 \tau_r}\left(E - E_g\right)^{1/2} e^{(E_{fn} - E_{fp} - E_g)/k_BT} e^{-(E-E_g)/k_BT} \tag{2.32}$$

Spectral distribution of the injection electroluminescent rate function $r_{sp}(E)$ is plotted in Fig. 2.9.

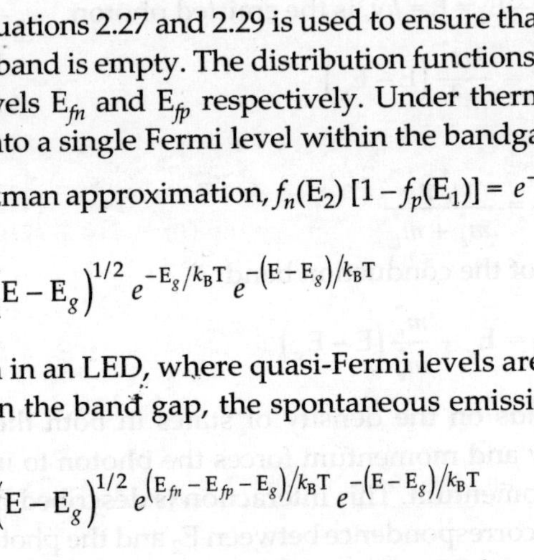

Fig. 2.9

The photon flux is derived by the integration of equation (2.32).

$$\Phi_0 = V_1 \int_0^{\alpha} r_{sp}(E)dE$$

$$= \frac{V_1}{\sqrt{2h^3\tau_r}} \left\{\frac{m_r^*}{\pi}\right\}^{3/2} e^{(E_{fn} - E_{fp} - E_g)/k_BT} \left(s^{-1}\right) \quad (2.33)$$

V_l is the active region volume. In this integration operation P_{em} (a function of photon energy) has been substituted by $1/\tau_r$, due to that the effect of the value of E is small compared to the change in $(E - E_g)$. When the injection level is increased, the quasi-Fermi levels move into the bands and probability function $f_n(E_2)\,[1 - f_p(E_1)]$ increases. As a consequence the light output increases.

The responsivity, R of an LED is defined as the ratio of the emitted optical power $P_0 = \Phi_0 h\nu$ to the injection current I. This means

$$R = \frac{\Phi_0 h\nu}{I} = \eta_0 \frac{h\nu}{q} = \frac{1.24\,\eta_0}{\lambda}\,(W/A) \quad (2.34)$$

(λ is in μm)

Example 2.5. *Determine the responsivity of an LED with $\eta_0 = 5\,\%$ at $\lambda = 0.5\ \mu m$.*

Solution: The responsivity $R = \frac{1.24\,\eta_0}{\lambda}\,(W/A) = \frac{1.24 \times 0.05}{0.5}\,(W/A) = 123\,(mW/A)$

Example 2.6. *Calculate the output power for a $n^+\!-\!p$ GaAs LED with electron current of 1.0 mA and efficiency $\eta_0 = 50\,\%$.*

Solution: It is assumed that the total current in the asymmetric diode is equal to the electron injection current, *i.e.*, $\eta_{in} = 1$.

From equation (2.14), it follows that $\eta_r\,\eta_e = 50\,\% = 0.5$ (η_L is considered as unity)

From equation (2.34), $R = \frac{1.24\,\eta_0}{\lambda}\,(W/A)$

Therefore output power $= \frac{1.24\,I\eta_0}{\lambda}\,(W)$

$$= \frac{1.24 \times 1 \times 10^{-3} \times 0.5}{0.87}\,(W) \qquad [\text{For GaAs LED } \lambda = 0.87\ \mu m]$$

$= 0.713$ mW

2.6 LED STRUCTURES

LED behavior discussed so far and the structure as shown in Fig. 2.2 is of simple homojunction type. There are two main disadvantages with this structure, which limit the internal quantum efficiency. First, the surface states on *p*-layer close to the light-emitting junction result in rising of nonradiative recombination. If the surface is far from the junction, then the probability of reabsorption increases. Secondly, the electrons injected from the n^+-region into the *p*-region as minority carriers diffuse away from the junction and gradually recombine with the majority carriers. Therefore same number of photons are created in a greater volume of material, resulting in the reduction of internal quantum efficiency. Overcoming these disadvantages, various different types of LED structures have been developed. Their uses depend on their light emitting efficiency, brightness and applications.

2.6.1 Heterojunction LED

The problems stated above are removed by use of double heterostructure diode, whose schematic and band diagram are shown at Fig. 2.10. In this structure, the GaAs layer, into which electrons are injected from the top n^+ AlGaAs layer, can be made thin (within 1000 to 2000°A). these electrons are contained within this thickness by the potential barrier of the *p*-GaAs/*p*-AlGaAs heterojunction. The top AlGaAs layer serves several important functions. The nonradiative recombination is reduced because the interface state density at this lattice-matched junction is orders of magnitude lower than the free surface state density. Secondly, due to its larger bandgap, it acts as a window layer in which the emitted photons, with energy close to the GaAs bandgap, are not reabsorbed. Finally, the injection efficiency is also higher. Thus , both the internal as well as the external quantum efficiencies are enhanced. Such heterostructure LEDs are generally realized by epitaxial growth and InP-based materials such as InGaAsP/InP or InGaAs/InGaAsP/InP are also used to realize emission of higher wavelengths.

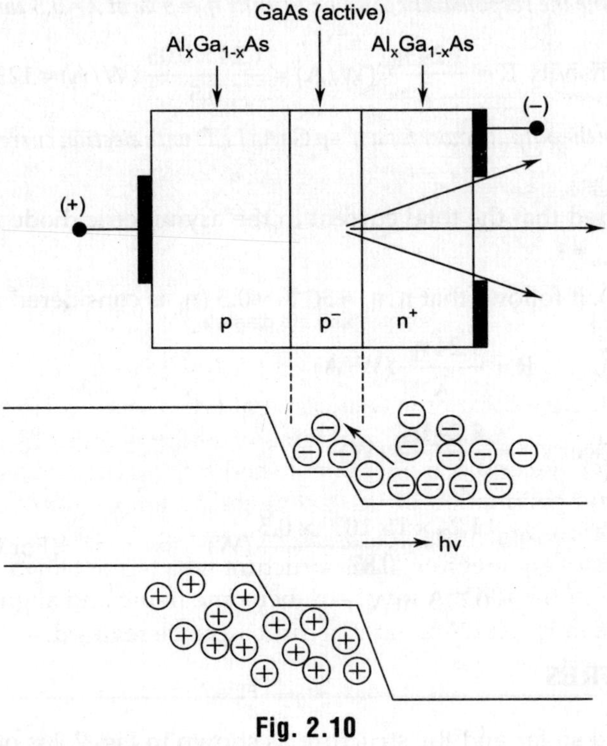

Fig. 2.10

2.6.2 Burrus Surface-emitting LED

This double heterostructure LED are specially developed in order to high radiance and efficient coupling of the emitted light to a fiber. This structure was conceived and realized by Burrus and Dawson. The schematic diagram of the structure is shown in Fig. 2.11. The bottom p^+ GaAs and top n^+ GaAs layers are provided to achieve low-resistance ohmic contacts. A deep well in the top *n*-GaAs layer is etched up to the top of *n*-AlGaAs layer to avoid reabsorption of the emitted radiation. The well can be conveniently realized by selective etchant that etches GaAs but not AlGaAs. The well can also be made use of supporting the fiber, which is butt-coupled to the device and held in place with epoxy resin of appropriate refractive index to enhance external power efficiency of the device. A thin layer of SiO_2 is provided to isolate the contact from the gold heat sink. Photons are generated in the thin layer of *p*-GaAs, and emission is ensured from top surface by the heterostructure and reflection from the back crystal face. Thus, very high forward radiance is obtained in this device. The top *n*-GaAs contact layer ensures low contact resistance and thermal resistance, and therefore high current densities and high radiation intensity are achieved. The fiber is to be properly aligned to optimize coupling of emitted radiation. However, there is some loss due to Lambertian distribution of the radiation intensity. The coupling efficiency is typically 1% to 2%.

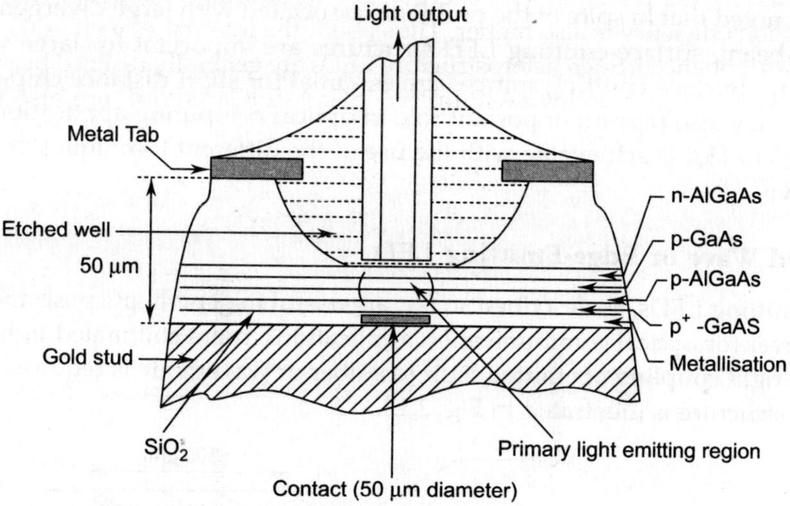

Fig. 2.11

In order to improve the coupling efficiency, several types of lens coupling are used with surface-emitting heterostructures, schematic diagrams of which are shown in Figs. 2.12(*a*), 2.12(*b*) and 2.12(*c*). In Fig. 2.12(*a*), the end of a fiber is polished into a spherical lens. This is used for the purpose of collection and collimation of the divergent radiation from LED. A glass microlens inserted between the device and the fiber, as shown in Fig. 2.12(*b*), serves the same purpose. Figure 2.12(*c*) demonstrates an integrated lens structure which is obtained by etching and polishing the top surface layer of the LED. A lot of cumbersome fixing and alignment work is avoided with use of the structure in Fig. 2.12(*c*), but it is difficult to be realized.

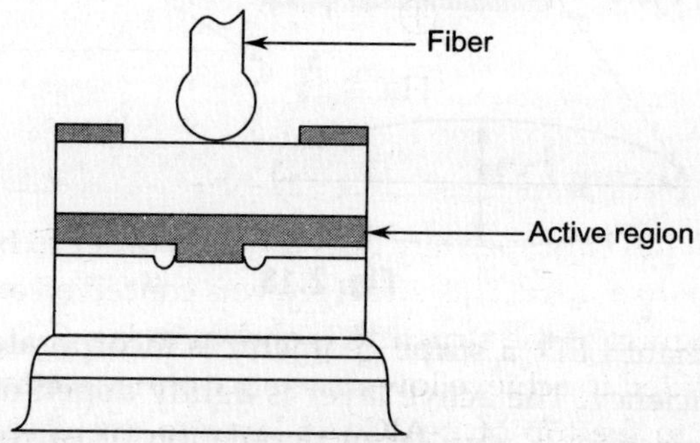

Fig. 2.12(*a*)

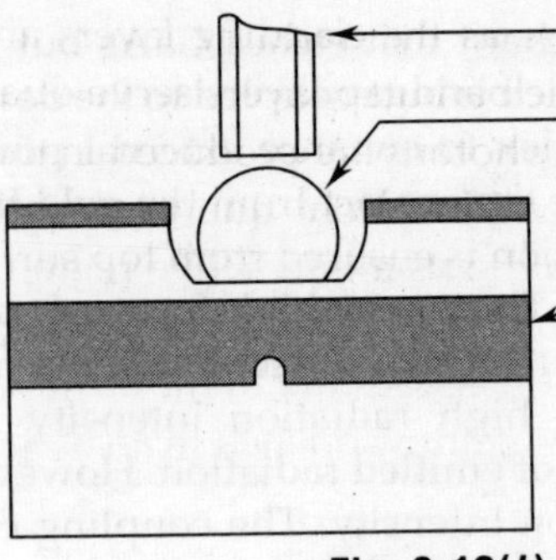

Fig. 2.12(*b*)

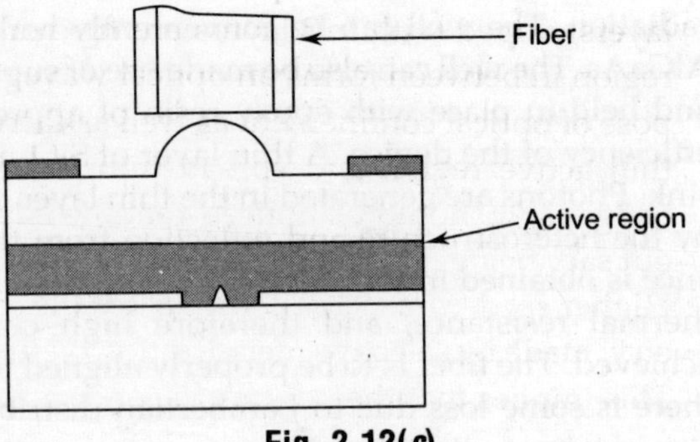

Fig. 2.12(*c*)

It should be noted that in spite of the problems associated with large divergence and coupling of the radiant beam, surface-emitting LED structures are important for large-volume and low cost applications. Surface-emitting sources are essential for short distance chip-to-chip communication. They may also play an important role in optical computing applications. The coupling efficiency of 5% to 15% is achievable with the use of the different techniques of lens coupling as discussed above.

2.6.3 Guided Wave or Edge-Emitting LED

The surface-emitting LEDs, as described so far, are useful in general purpose and low cost applications. However for optical communication applications, more collimated light emission is desirable, where tight coupling of emitted light to a fiber or waveguide is required. The basic edge-emitting LED structure is illustrated in Fig. 2.13.

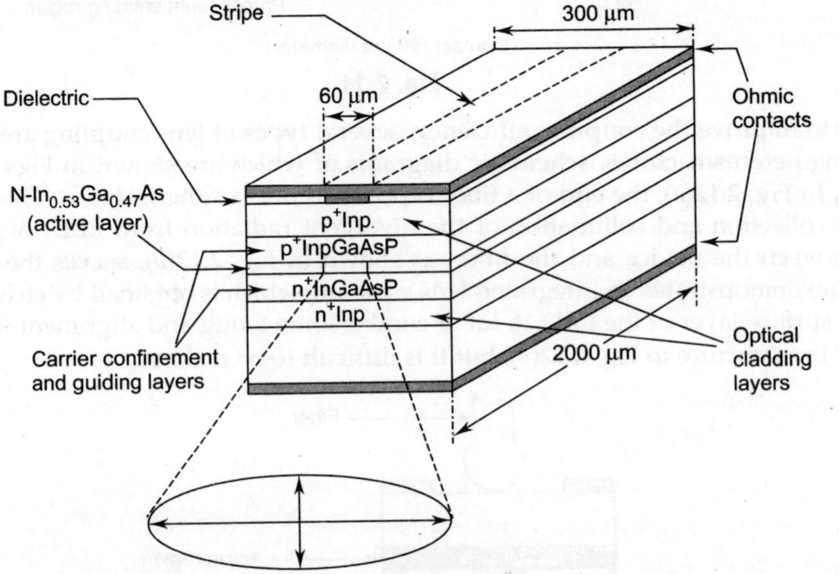

Fig. 2.13

Generally for superradiant LED, a stripe geometry is incorporated, as in a junction laser, to improve the injection efficiency. The active layer is lightly doped or undoped of $In_{0.53}Ga_{0.47}As$. With the forward bias injection, a very large population of carriers for recombination takes place in this region. On the both sides of the active region, heavily doped InGaAs layers are incorporated for the purpose of the carrier confinement. Outer sides of the carrier confinement layers, doped with InP layers are further incorporated to serve as the cladding layers and the region in between forms an optical waveguide. Thus these wide bandgap layers serve dual purpose of optical confinement as well as carrier confinement. The photons are produced in the very thin active region and spread into guiding layers without reabsorption, due to their larger badgaps. It should be noticed that the edge-emitting structure allows a thin active region without sacrificing the radiative efficiency, in comparison with a homojunction surface-emitting LED. Higher carrier injection densities for the same drive current is observed due to the stripe geometry, made by selective metallization on the top surface through window opened in an SiO_2 layer. This implies that the power efficiency is improved.

Edge-emitting LED structure is very similar to the junction laser. However, it does not have the feedback cavity and is still an incoherent light source. Most of the light is guided to come out of one of the edges of the structure. This is ensured by cleaving and providing a reflective coating at one of the edges to make it non-emitting side, and providing an antireflective coating at the other edge to make it emitting end. The divergence of the emitted radiation is greatly reduced by waveguide. Half power width is approximately 30° in the direction perpendicular to the plane of the layers. However in the plane of layers the output is still Lambertian and the half-power width is observed at approximately 120°. Optical confinement can be achieved by selective etching and regrowth of a higher bandgap material. This technique is more common in laser fabrication and will be discussed chapter 3. The waveguide and reduced beam divergence improve coupling efficiency of the radiated beam into fibers. The larger operating current density in a smaller structure can create the problem of heat generation. However, the edge-emitting LEDs with InP based material have large potential to play important role as high power sources in short distant optical communication, where dispersion in fibers is not too severe.

2.6.4 Drive Circuitry

The LED is operated under forward bias injection conditions. The amount of forward bias depends on the forward cit-in voltage, which in turn, is determined by the bandgap of the semiconductor. Typical forward cut-in voltage for LED is between 1 to 2 V, and current drawn is 20 to 100 mA. An LED can be operated by Direct Current (DC) or Alternating Current (AC) drive circuit.

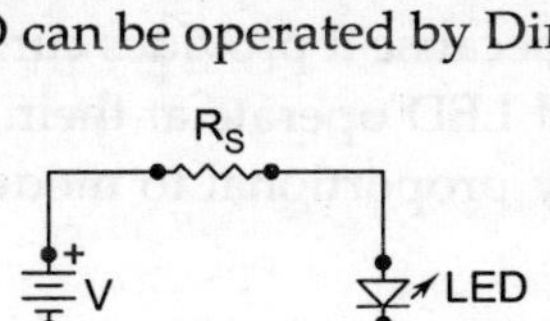

Fig. 2.14(*a*)

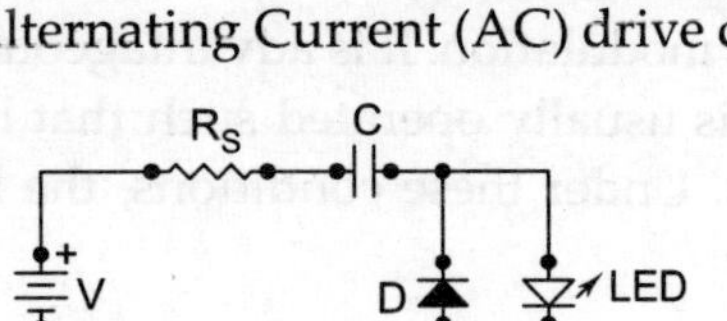

Fig. 2.14(*b*)

A simple DC drive circuit is shown in Fig. 2.14(*a*). The current through LED is provided by the voltage source V is limited by the series resistance R_S. Under operating conditions, the voltage drop across the LED is V_L, which is the operating voltage of the device. If the operating current is I_L, then the circuit equation can be written as

$$V - V_L = I_L R_S \tag{2.35}$$

It may often be desirable to operate the LED in an AC circuit. Depending on the applications, several types of circuits can be used. A common form LED circuit in AC operation is shown in

Fig. 2.14(*b*). Here a diode with reverse polarity is placed in parallel with the LED to prevent it from high reverse bias voltage. A capacitor and a current limiting resistor are placed in series with the two diodes. The voltage drop across the diodes at both the polarities is limited to very small value equivalent to the cut-in voltages of the diodes. Then the current flowing through the circuit is approximately equal to the supply voltage divided by the impedance due to the resistance and capacitance. The series resistance also prevents the LED from high turn-on transients.

The two circuits schematic described above are for continuous operation. If it is desired to switch the LED on or off, or to modulate the output, the circuits shown in Figs. 2.15(*a*) and 2.15(*b*) respectively may be used. In Fig. 2.15(*a*), a switch is placed at the base of a transistor. When switch is open no base voltage is available at the transistor and hence it remains off, providing no current through LED. The LED, in turn remains off. When switch is closed, sufficient voltage is available the base of the transistor and the transistor becomes on. Collector current then passes through the LED to make in on.

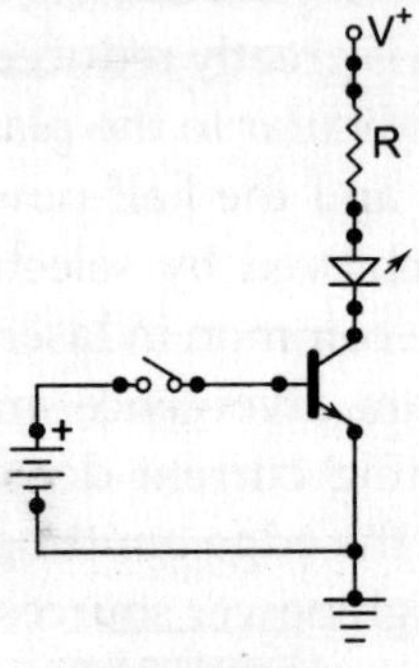

Fig. 2.15(*a*)

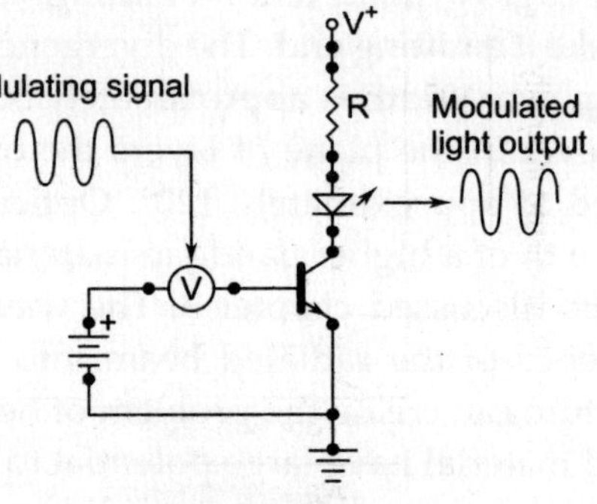

Fig. 2.15(*b*)

The circuit schematic of Fig. 2.15(*b*) is the basic circuit for communication application, where high frequency signal is impressed upon the continuous wave output of the light source. Here base voltage of the transistor is varied with the application of modulating signal, which in turn, varies the collector current Thus the drive current of the LED is varied in respect of modulating signal, and so the light output. The circuit is the example of internal modulation and also called current modulation. It is advantageous to use the transistor because it provides current gain. The circuit is usually operated such that both the transistor and LED operate at their linear output regions. Under these conditions, the light output is directly proportional to modulating signal voltage.

2.7 DEVICE PERFORMANCE CHARACTERISTICS

Performance of LED is determined by its electrical and optical characteristics. For commercial LEDs, the performance parameters are generally provided by the manufacturers. A typical set of performance characteristics of AlGaAs and InGaAsP are discussed here.

2.7.1 Spectral Response

The spectral response is the parameter that indicates the relative light output intensity at different wavelength. As example, the relative output intensity against the wavelength in nanometers of AlGaAs is plotted in Fig. 2.16, where the peak intensity is observed approximately at 835 nm. The spectral linewidth, measured in units of wavelength between the half maximum intensity points is termed as the *Full Width at Half Maximum* (FWHM). The FWHM of AlGaAs is around 40 nm, which is typical of LEDs operating at 800 nm. In LEDs emitting at longer wavelengths of 1100 to 1600 nm, the material used is typically InGaAsP/InP, and the linewidth of spectral response is broadened by alloy scattering. The linewidth is also shifted and broadened by heavy doping in the active region. Therefore, in surface emitting LEDs, where the active region is heavily doped, the emission wavelengths are longer but of lower energies and the spectral response is broadened. Variation in temperature also affects the response characteristics. The bandgap of most III-IV materials decrease at the rate of 5 meV/°K with the increase of temperature. So linewidth of the spectrum also increases with the increase of temperature because of larger energy spread of carrier distribution and phonon scattering. Therefore, proper heat-sinking and cooling arrangements must be incorporated with the device for the applications where the emission wavelengths and linewidths are desired to be invariant.

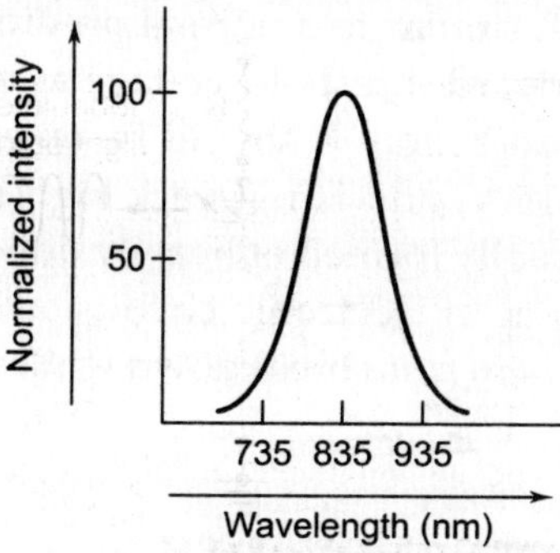

Fig. 2.16

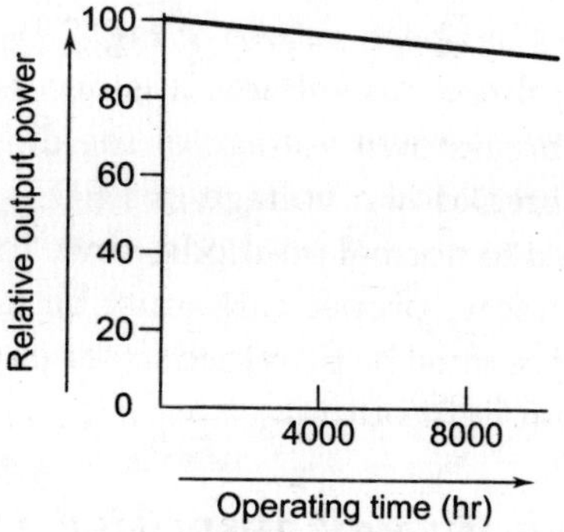

Fig. 2.17

2.7.2 Output Power-Time Characteristics

The output power of an electroluminescent device or a source decreases with the time. Therefore the lifetime of the device in communication applications is an important parameter. A typical plot for relative output power against operating time of AlGaAs LED is shown in Fig. 2.17. It should be noted that the power output has been decreased by a few percents after operating for 8000 hrs.

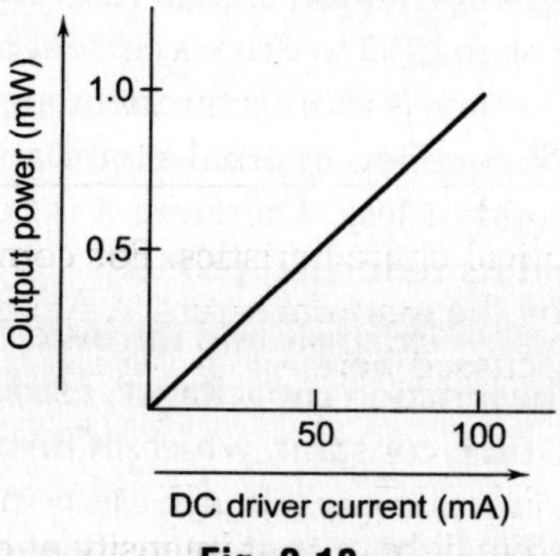

Fig. 2.18

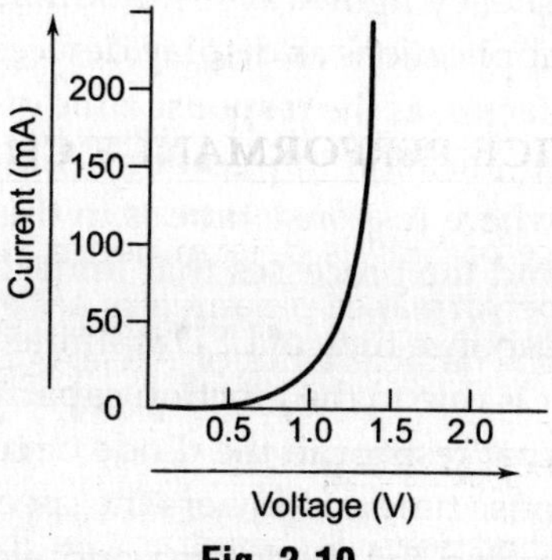

Fig. 2.19

2.7.3 Light (Power) Current Characteristics

The light power output with the current is the most important parameter of an LED. The characteristic of an edge-emitting device is shown at Fig. 2.18. Linearity in this curve is also important for the application of modulation in analog communication. Nonlinearity in some LEDs may be observed due to material properties and device configuration. It should be noted that edge-emitting devices radiate less optical power into air than the surface-emitting devices, although the radiation intensity of the edge-emitting devices can be very high at the emitting face. This is due to the dominant role of interfacial recombination and reabsorption in edge-emitting devices. However, higher radiation intensity from them can be efficiently utilized at fibers by suitably coupling with small apertures.

2.7.4 Diode Current-Voltage Characteristics

This parameter describes the electrical performance of the device. Typical current voltage characteristics has been shown at Fig. 2.19, which is very much similar to a normal pn-diode, but of different threshold voltage. It is necessary to know the forward threshold voltage as well as the reverse breakdown voltage of the device for an efficient operation. It should be noted that the reverse breakdown voltage of LED is in range of 5 V to 15 V, and is not very high enough as compared to normal pn-diode, since the LED junction is usually formed of heavily doped materials. Therefore, proper care must be taken while using them in electronic circuits, and suitable impedance must be provided in series of them, to prevent them from breakdown while extracting maximum light output.

2.8 FREQUENCY RESPONSE AND MODULATION BANDWIDTH

In communication applications, the transition time plays an important role, which must be minimized for an efficient utilization. Since the maximum amount of information is desired to be transmitted in least possible time. Higher transmission speed of signal may be achieved if signal changes rapidly with time. Signal speed is described by the bandwidth, which is width of the signal spectrum in frequency domain. High frequency is involved in high speed communication, in other word, high speed systems require wideband signals to represent the information as well as wideband devices and systems to accommodate the signals. In optical communication system high frequency light wave is modulated by the wideband information signal. In everyday commercial applications as display devices, the response time of an LED to drive a current pulse is not limiting factor, as the response time of most LEDs is $\leq 1\mu s$, which is more than adequate. However for communication applications, LEDs sources of high frequency internal modulation are required, where response time is in the order of a nanosecond or less. Therefore it is necessary to understand the processes that limit the response time and its reduction process.

The response time of LED is limited by two main factors, — extrinsic and intrinsic. The extrinsic factor is due to the junction capacitance of the LED. The junction capacitance, associated with the external resistor in the diode circuit increases the RC time constant, which in turn, increases the response time and lowers the speed of operation. The junction capacitance can be made lower by decreasing the diode area and also by decreasing the bias. The change in bias is related to doping. The charge storage and diffusion capacitance of *pn*-junction of LED under forward bias condition affect the intrinsic response time. When the applied bias to the diode is modulated at high speed, the stored charge cannot respond so rapidly. The excess carriers injected during forward bias are removed by diffusion and recombination with minority carriers. Therefore, minority carriers recombination life time in the junction material plays an important role to determine the modulation bandwidth.

For analysis and derivation of frequency response, Fig. 2.20 is considered, where electrons are injected into *p*-type active region of width *d*. It is assumed that *p*-*n* junction is at $x = 0$ (junction might be heterojunction also) and excess carriers recombine with a lifetime τ throughout thickness *d*, the narrow depletion region in a forward biased junction is neglected. The excess electron density $n(x)$ in the active region is lost by spontaneous recombination and diffusion. Therefore, when injection is stopped, neglecting the drift current, the rate of change of the excess carriers may be expressed as

$$\frac{\partial n(x)}{\partial t} = -R_{sp} + D_e \frac{\partial^2 n}{\partial x^2} \tag{2.36}$$

or,

$$\frac{\partial n(x)}{\partial t} = -\frac{n(x)}{\tau} + D_e \frac{\partial^2 n}{\partial x^2} \tag{2.37}$$

(Assuming $Rsp = n(x)/\tau$, where $\tau_0 = 0.5$ ns)

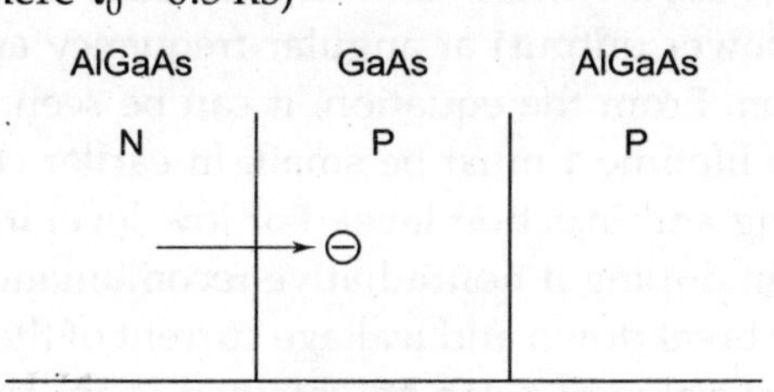

Fig. 2.20

The one dimensional geometry of heterojunction LED is shown in Fig. 2.20. In case of small signal modulation operation of the diode, a time dependent current is superimposed on a dc bias. So the excess carriers will have a steady state and a frequency dependent component. The excess carriers can be written as

$$n(x, t) = n_0(x) + n_j(x)\, e^{jwt} \tag{2.38}$$

From equations (2.37) and (2.38), we can write for dc parts and frequency dependent part respectively as

$$D_e \frac{\partial^2 n_0}{\partial t} - \frac{n_0}{\tau} = 0 \tag{2.39}$$

and

$$D_e \frac{\partial^2 n_j}{\partial x^2} - \frac{n_j(1 + j\omega\tau)}{\tau} = 0 \tag{2.40}$$

By analogy with the definition $L_e = (D_e\tau)^{1/2}$, we may define $L_e(\omega) = [D_e\tau/(1 + j\omega\tau)]^{1/2}$. Therefore, equation (2.40) can be rewritten as

$$\frac{\partial^2 n_j(x)}{\partial x^2} - \frac{n_j(x)}{L_e^2(\omega)} = 0 \tag{2.41}$$

The frequency response of the LED $r(w)$ is defined as

$$r(w) = \frac{\Phi_j(w)}{J_j(w)/q} \tag{2.42}$$

where the numerator is the ac photon flux and the denominator is the electron flux. J_1 is the ac current density. At the *p-n* junction, where it is assumed that only electrons cross the boundary, $n_1(x) = n_{10}$. it is assumed that $d >> L_e$, so that $n_1(d) = 0$ under these boundary conditions the solution of equation 2.41 is given by

$$n_1(x) = n_{10}\, e^{-x/L_e(w)} \tag{2.43}$$

Now
$$\Phi_j(w) = \frac{1}{\tau}\int_0^d n_j(x)\,dx = \frac{n_{10}L_e(w)}{\tau} \tag{2.44}$$

and
$$J_j(w) = qD_e \frac{\partial n_j(x)}{dx} = \frac{n_{10}}{L_e(w)} \tag{2.45}$$

Therefore,
$$r(w) = \frac{|L_e(w)|^2}{T} = \frac{1}{\left[1 + w^2\tau^2\right]^{1/2}} \tag{2.46}$$

The same result can be obtained when d is finite, but there is heterojunction barrier at $x = d$. Here $r(w)$ is the response (power output) at angular frequency ω and τ is the minority carrier lifetime in the injected region. From the equation, it can be seen that for good high frequency response, the recombination lifetime τ must be small. In earlier chapter, we have seen that the value of τ is related to doping and injection level. For low level injection into *p*-type material, τ decreases with the increase in doping if nonradiative recombination is considered to be negligible, as $\tau = \tau_r$. Considering the breakdown and leakage current of the diode and the solubility limit of the dopant specie, the best value of τ can be obtained is approximately 1 ns. An alternative approach is to operate in the bi-molecular recombination regime in lightly doped material. For high level injection into a thin active region, the injected carrier density Δn is much greater than the equilibrium value. Under this condition, if J is entirely a recombination current, then the number of recombinations per unit volume per second in an active region of width d is J/qd. This must be equal to $\Delta n/\tau_r$. Therefore the τ_r may be derived as

$$\tau_r = \left(\frac{qd}{JB_r}\right)^{1/2} \tag{2.47}$$

Thus, τ_r may be reduced by decreasing d and increasing J. But very high value of J leads to heat generation developing heat-sinking problems and distortion in modulated signal. It should be noted that, the recombination lifetime is limited by the surface recombination rate in conventional surface-emitting LEDs with thin active regions, while edge-emitting LEDs with thin undoped active region operating in the bi-molecular recombination region have better frequency response. It is also to be noted that all the material and device parameters are interrelated and they are to be optimized to achieve the best frequency response.

Modulation bandwidth is the alternate way to express the frequency response of a LED. This is defined by the frequency at which the optical power received at a detector reduced to half or 3 db with respect to the transmitted power. With the increase in modulation frequency, LED does not emit the total amount of light, since all the injected carriers cannot recombine. Assuming a linear relationship of light-power against current for both the LED and detector, the bandwidth may be defined as

$$\Delta f(db) = 10 \log_{10} \frac{P_{out}(f)}{P_{out}(dc)} \quad (2.48)$$

$$= 10 \log_{10} \frac{I_{out}(f)}{I_{out}(dc)} \quad (2.49)$$

where I_{out} is the current at the detector. The 3db point occurs when the ratio of currents is equal to ½. This is shown in Fig. 2.21. Note that electrical bandwidth occurs at current ratio is 0.707, as electrical power is proportional to square of the current (when $[I_{out}(f)/I_{out}(dc)]^2 = ½$). Therefore electrical bandwidth is smaller than the optical bandwidth.

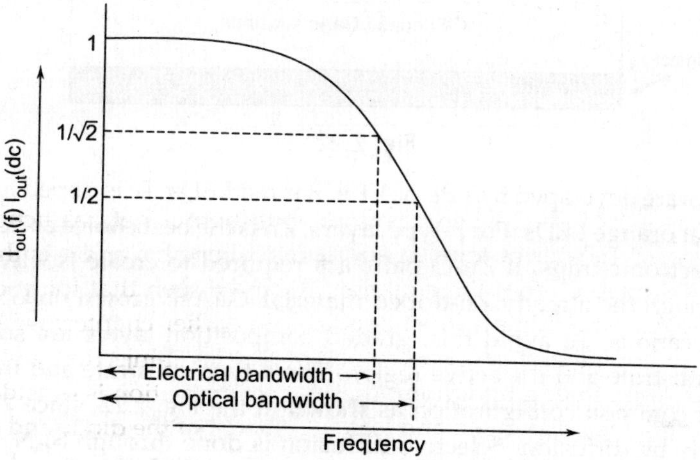

Fig. 2.21

As summary, the modulation bandwidth of an LED depends on the device configuration (surface-emitting or edge-emitting type), doping level in the active layer, the lifetime of the injected carriers, and the parasitic capacitance and resistance in the circuit. It may be noted that in LEDs the photons are generated by spontaneous recombination, and the radiative lifetime under these conditions are always larger than that under stimulated emission, as in a laser. Therefore, lasers have higher modulation bandwidths. However bandwidths over 1 GHz are now available in practical LEDs. The intensity modulation by variation of the drive current is possible by both analog and digital signals. But the analog signal modulation requires higher signal to noise ratio and so it is suitable for short distance applications. Digital modulation has better noise immunity and is applied for long distance communication.

2.9 MANUFACTURING PROCESS AND APPLICATIONS

LED is the simplest and most robust type of electroluminescent devices. The cross-section of a standard red-emitting LED is shown in Fig. 2.22. the device consists of GaAsP layers on GaAs substrate. GaP substrates are also used for some LEDs (orange-emitting). The substrate material is similar to other electronic and optoelectronic devices and grown in ingots by the Horizontal-Bridgemann or Czochralski techniques. For P-cotaining compounds, the resulting pressure at the melting point is very high (*e.g.*, 40 atmospheric pressures for GaP). B_2O_3 is also used for encapsulation introdcing B impurities in the single crystal. For high quality substrates the defect density varies within 10^2 to 10^5 per sq cm.

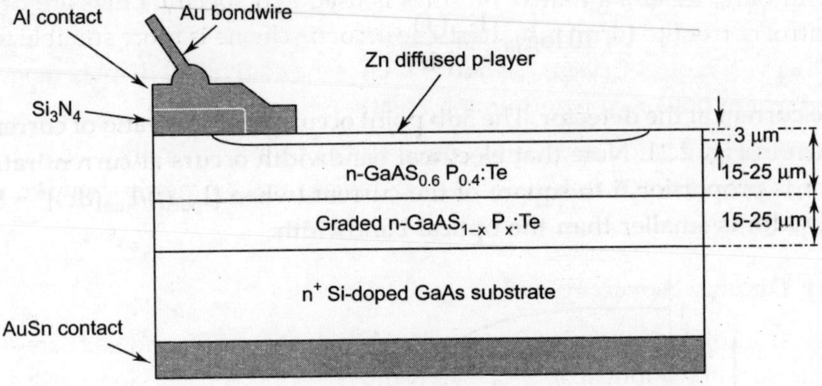

Fig. 2.22

The epitaxial layers are developed by LPE or VPE. For red LEDs, Te is doped for *n*-type layer and S is used as dopant for orange LEDs. For *p*-type layers, Zn is the best choice as dopant. NH_3 is used for creating N isoelectronic traps. If Zn-O pairs are required to create isoelectronic traps, then Ga_2O_3 is passed through the already Zn-doped material. GaAsP grown on GaAs substrate may produce misfit dislocations. To avoid this, graded composition layers are sometimes incorporated between the substrate and the active region. The *n*-type substrate and the top Zn-diffused *p*-layer are the most common configuration as shown in the Fig. 2.22, since Zn can be incorporated relatively easily by diffusion. Selective diffusion is done through Si_3N_4 mask to form *p-n* junction. Contact layers of Al and AuSn are generally formed on *p* and *n* sides respectively. Alternatively, AuBe and AuGe are also used to form the contact layers on *p* and *n* sides respectively. Gold bond wire is used to make contact to bias source. To realize scattering light and uniform radiation from the junction, the metallization is made on it, such that the current distribution over the active region is uniform.

The completed LED is encapsulated in a dome, as shown in Fig. 2.5. For visible LEDs, to eliminate any possible spurious reflections from bottom surfaces, special coating is applied. Lamps and displays are the most common applications of visible LEDs. The luminous intensity, colour, uniform visibility from all directions and contrasting background are the deciding factors to make choice of an LED for a particular application.

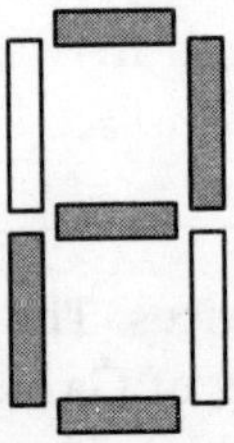

Fig. 2.23(*a*)

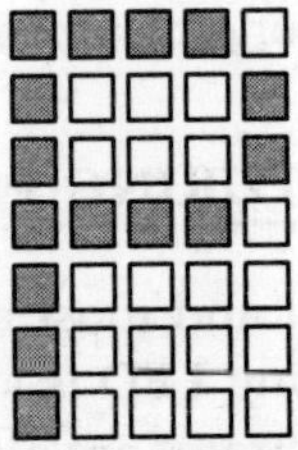

Fig. 2.23(*b*)

In the display applications like calculator etc., several individual LED chips are clustered to make capable of forming the numerals. Two schemes are illustrated here. In the first scheme as in

Fig. 2.23(*a*), a seven segment system is used, where each of the segments may consist of several LEDs. A digital logic control system is used to feed current to the desired segment. In the second scheme, as shown in Fig. 2.23(*b*), a matrix of LEDs is used and specific LEDs are driven by the external logic control current to form numerals. The second scheme is more suitable for alphanumeric display. However, Liquid Crystal Display (LCD) is the equivalent display device and finds wide applications where power consumption is greatly reduced.

For sensing and communication applications, infrared LEDs are generally used. GaAs LEDs are suitable for sensing applications, whereas InP-based devices are commonly used for communication purposes.

2.9.1 LEDs for Display Applications

A part from the application in the field of optical communication, LEDs and Led arrays are widely used in display application, optical interconnect and sensors. As mention earlier, in display application, Liquid Crystal Displays (LCD) are found to be of competing technology. Active matrix liquid crystal displays or AMLCDs have offered the best opportunity for wide scale implementation. Thin film transistor LCD or TFT-LCD, which is the most versatile form of technology, has several advantages like low drive voltage, low cost, full colour display capability and easy availability. However, these displays posses major disadvantages of power consumption, limited contrast, narrow viewing angle, slow speed, high cost of polarizers and colour filters, manufacturing complexity, the need of backplane illumination and bulkiness for some applications. This technology has the limitations of amorphous Si electronic properties, low drive current, high parasitic capacitance, non-ideal gate dielectric materials and structures of TFTs, improvement over contrast ratio, lack of electrical stability of TFTs, non-optimal driver circuit configurations, architecture and voltage requirements, complex technology of liquid crystals, barriers in liquid crystal cell technology to improve visual angles, lack of implementation of system partitioning and packaging techniques.

Field Emission Displays (FEDs) are the most promising and attractive alternatives to AMLCDs. They have reduced manufacturing complexities and cost, increased brightness, improved viewing angle. FEDs produce their own light using colour phosphors and are faster than the liquid crystals. Therefore video display is possible using simple matrix addressing scheme. The small dimension of the FED may include dense arrays, which eliminates the requirement of large arrays AMLCDs.

Electroluminescent Displays (ELDs) and Light Emitting Diodes (LEDs) also provides similar advantages in terms of performance and manufacturability. Crystalline III-V compound based light emitting diodes are very reliable devices. However the realization blue light emitting devices was still elusive in last decade. A serious study in the recent years led to realize reliable and cost effective blue emitting LEDs. Commercially viable blue LEDs based on the wide bandgap semiconductor gallium nitride and indium gallium nitride were invented by Shuji Nakamura while working in Japan at Nichia Corporation in 1993 and became widely available in the late 1990s. They can be added to existing red and green LEDs to produce white LED lamps and full-colour displays, though white LEDs today rarely use this principle. The materials that are generally used are mentioned in section 2.3 and table 2.1. Figure 2.24 gives the scheme for potential material system.

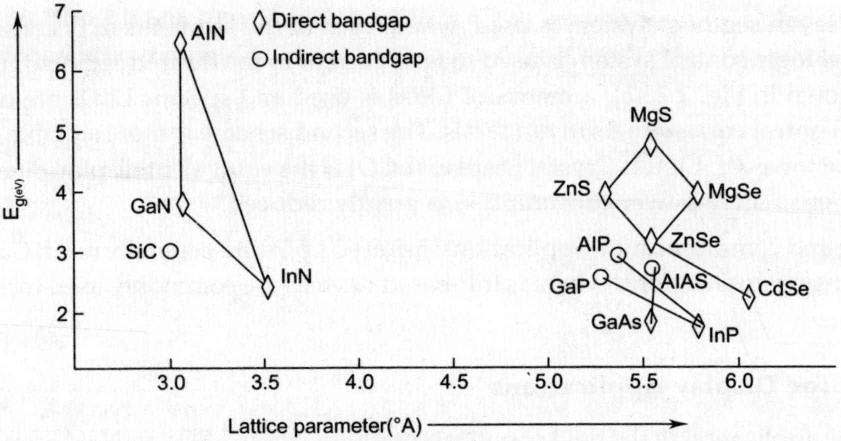

Fig. 2.24

Using III-V compounds, red LEDs are made with direct bandgap AlGaAs and indirect bandgap GaAsP isoelectronically doped with N. Green-emitting LEDs are made with indirect gap GaP:N. Blue LEDs from Silicon Carbide (SiC), having an indirect bandgap of 2.86 eV possess very low luminous efficiency (≈ 0.5 lumens/ watt) and external quantum efficiency (0.02%). Nitrides like GaN and the ternaries In GaN and AlGaN are the suitable alternatives. However, due to low sticking coefficient of N atoms on the surface, epitaxial layers of the compounds usually have a large density of N vacancies, which leads to a very high background donor density. Consequently, *p*-type doping is a problem. This problem has been solved by Zn-doping or Mg-doping and subsequent thermal annealing to form *p*-type material. Non-availability of substrate material that is lattice-matched to GaN is the other problem. Even the closest, sapphire (Al_2O_3) is so poorly matched that the epitaxial layers have a high dislocation density. Polycrystalline GaN is now used as a buffer between the sapphire substrate and GaN layer to reduce the defect density. Figure 2.25 shows the schematic structure of a typical blue-emitting nitride based LED and some operating characteristics are shown in Fig. 2.26.

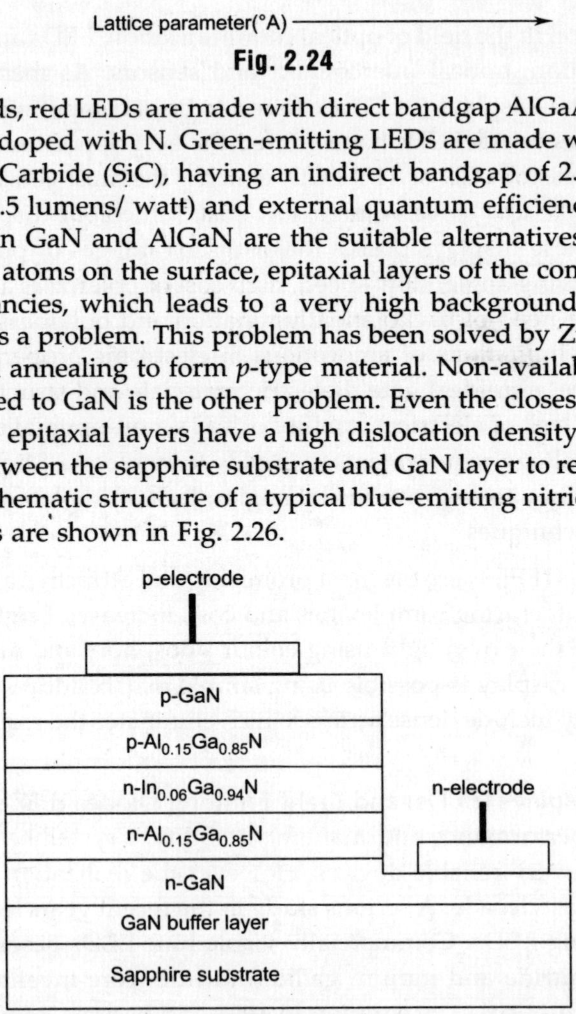

Fig. 2.25

From the material scheme in Fig. 2.24, it is apparent that the ZnMgSSe and ZnCdSe are also important materials for blue-green emitters, and they can be grown lattice-matched to GaAs substrates which is technologically useful. High brightness has been observed in the light emitting diodes with materials. Output powers of 1.3 mW (at 512nm-green) and 0.3 mW (at 489nm-blue) have been demonstrated with these II-IV compounds based LEDs with typical lifetime 100 hrs.

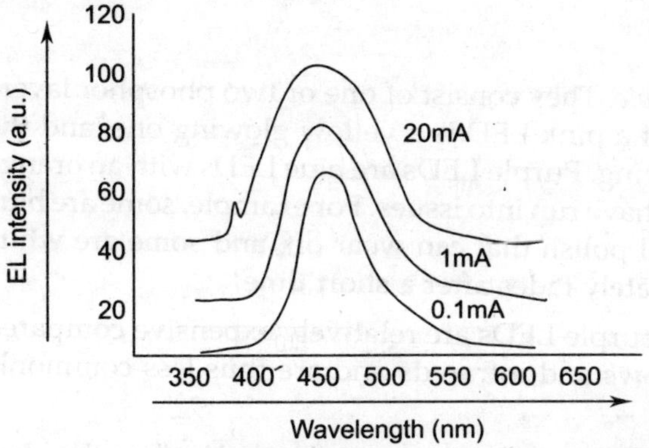

Fig. 2.26(*a*)

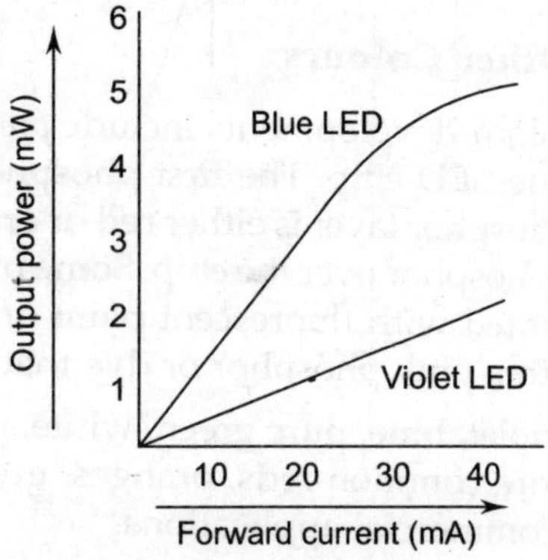

Fig. 2.26(*b*)

With the success of commercially viable blue LED, the inventors and manufacturers turned their attention towards white LED, as white LED can be produced with the combination of red, green and blue LED. However the principle of producing white LED is different. Most white LEDs in production today use a 450 nm – 470 nm blue GaN (gallium nitride) LED covered by a yellowish phosphor coating usually made of cerium-doped yttrium aluminium garnet (Ce^{3+}:YAG) crystals which have been powdered and bound in a type of viscous adhesive. The LED chip emits blue light, part of which is efficiently converted to a broad spectrum centered at about 580 nm (yellow) by the Ce^{3+}:YAG. The single crystal form of Ce^{3+}:YAG is actually considered a scintillator rather than a phosphor. Since yellow light stimulates the red and green receptors of the eye, the resulting mix of blue and yellow light gives the appearance of white, the resulting shade often called *lunar white*. This approach was developed by Nichia and was used by them from 1996 for manufacturing of white LEDs.

The pale yellow emission of the Ce^{3+}: YAG can be tuned by substituting the cerium with other rare earth elements such as terbium and gadolinium and can even be further adjusted by substituting some or all of the aluminium in the YAG with gallium. Due to the spectral characteristics of the diode, the red and green colors of objects in its blue plus yellow light are not as vivid as in broad-spectrum light. Manufacturing variations make the LEDs produce light with different color temperatures, from warm yellowish to cold bluish; the LEDs have to be sorted during manufacture by their actual characteristics.

White LEDs can also be made by coating near ultraviolet (NUV) emitting LEDs with a mixture of high efficiency europium based red and blue emitting phosphors plus green emitting copper and aluminium doped zinc sulfide (ZnS:Cu,Al). This is a method analogous to the way fluorescent lamps work. However the ultraviolet light causes photodegradation to the epoxy resin and many other materials used in LED packaging, causing manufacturing challenges and shorter lifetimes. This method is less efficient than the blue LED with YAG:Ce phosphor, as the Stokes shift is larger and more energy is therefore converted to heat, but yields light with better spectral characteristics, which renders colors better. Due to the higher radiative output of the ultraviolet LEDs than of the blue ones, both approaches offer comparable brightness. The newest method used to produce white light LEDs uses no phosphors at all and is based on *homoepitaxially* grown zinc selenide (ZnSe) on a ZnSe substrate which simultaneously emits blue light from its active region and yellow light from the substrate.

A new technique just developed by Michael Bowers, a graduate student at Vanderbilt University in Nashville, involves coating a blue LED with quantum dots that glow white in response to the blue light from the LED. This technique produces a warm, yellowish-white light similar to that produced by incandescent bulb.

2.9.2 Other Colours

Recent colour developments include *pink* and *purple*. They consist of one or two phosphor layers over a blue LED chip. The first phosphor layer of a pink LED is a yellow glowing one, and the second phosphor layer is either red or orange glowing. Purple LEDs are blue LEDs with an orange glowing phosphor over the chip. Some pink LEDs have run into issues. For example, some are blue LEDs painted with fluorescent paint or fingernail polish that can wear off, and some are white LEDs with a pink phosphor or dye that unfortunately fades after a short time.

Ultraviolet, blue, pure green, white, pink and purple LEDs are relatively expensive compared to the more common reds, oranges, greens, yellows and infrareds and are thus less commonly used in commercial applications.

The semiconductor chip is encased in a solid plastic lens, which is much tougher than the glass envelope of a traditional light bulb or tube. The plastic may be colored, but this is only for cosmetic reasons or to improve the *contrast ratio*; the color of the packaging does not substantially affect the color of the light emitted.

Light emitting diode arrays are gradually finding large scale commercial application in electronic billboards, traffic lights, annunciators, automobile lights, etc. In most of these applications, crystalline materials on crystalline substrates are utilized to make the arrays. Very high degrees of brightness and reliability have been achieved in producing LEDs. It is interesting to note that almost three-quarters of worldwide sales of III-IV compounds is made up of visible and infrared LEDs.

2.10 DEFECTS AND DEVICE RELIABILITY

For both general and specialized applications of any item, one of the foremost requirements is the reliability over a long period of time. Unlike junction lasers, whch operate under much larger drive currents, the LEDs generally do not suffer catastrophic degradation. But their performance characteristics gradually deteriorates with the time. There are two classes of defects that are known to degrade the performance of these devices. First is the dislocations, which may propagate from the substrate during epitaxy into active region. Currently the dislocation in doped substrate is approximately $10–10^2$ per sqcm and the density in the active layer can be reduced by a small geometry and incorporation of appropriate buffer layers that act as dislocation filters. The second is the deep level defects in the forbidden energy gap. Deep level defects are due to point defects such as vacancies, interstitial defects, impurity atoms or conformed from combination of these.

Dislocation in the active area results in *dark line defects* and *dark spots*, which cause a more rapid degradation of the device. The degradation depends on operating temperature, injection current density and impurity concentration in the active region.

Deeps levels reduce the radiative efficiency of the LED. In addition, point defects and impurity migration into active region may be created or enhanced by the recombination process. This results to a slow or long-term degradation. The degradation rate is related to the activation energy E_d of the degradation process by the Arrhenius equation-

$$\gamma_d = C\exp(-E_d/k_BT) \tag{2.50}$$

where γ_d is the degradation rate (s^{-1}) and C is the proportionality constant. The activation energy E_d represents an energy barrier to impurity migration or defect generation. Consequently, larger the E_d, the smaller is the degradation rate. E_d is dependent upon the material parameters, epitaxial techniques, device processing and the geometry. The radiant output power as a function of time P(t) may be expressed as

$$P(t) = P(0)\exp(-\gamma_d t) \qquad (2.51)$$

Based on these equations, it is expected that the operation lifetime may be in the order of 10^6 to 10^9 hours of continuous wave (*cw*) operation, if values of E_d is in the range of 0.6 to 1.0 eV. Operation lifetime also depends the material system, material synthesis, device fabrication and efficient heat-sinking technique. Without proper heat-sinking, the junction temperature can quickly increase to much higher than the ambient temperature.

Example 2.5. *A LED has an activation barrier energy $E_d = 0.6$ ev to long term degradation. At room temperature operation, find the time after which the radiant output power will fall to half its initial value if the value of $C = 2\times10^3.hr^{-1}$.*

Solution: Using equation 2.50, $\gamma_d = C\exp(-E_d/k_BT)$

Hence, $\gamma_d = 2\times10^3\exp(-0.6/k_BT) = 7.6\times10^{-9}\ hr^{-1}$

Using equation (2.51), $P(t) = P(0)\exp(-\gamma_d t)$

Therefore, $P(t)/P(0) = \exp(-\gamma_d t) = 0.5$ *i.e.,* $\exp(\gamma_d t) = 2$

So, $t = \ln 2/\gamma_d = \ln 2/7.6\times10^{-9} = 91.2\times10^6$ hr

2.11 LED DRIVE CIRCUITS

Proper operation of LED in the electronic circuits involves certain considerations and they are to be taken care of while designing the drive circuits. The current limiting resistor as shown at Fig. 2.27(*a*) is a universal method used in simple LED circuits for driving 5mm LEDs. This primitive method is extremely simple and very cost efficient. However, the current limiting resistor is not suitable for all simple LED circuits. For example, high power LED circuits always require a constant current source. Employing a simple LED driver circuit such as the current limiting resistor in a high power LED application results in an unstable and potentially dangerous circuit design. This is primary because the forward junction voltage will continuously vary, and can lead to thermal runaway.

(*a*) **Current limitation :** Drive current for LED driver circuit is to be selected by first referring to the LED manufacture's datasheet. Most often, the typical forward current rating is chosen as their specified LED drive current. The luminous output vs. forward current graph is considered for reference to determine how much light output this will produce. It may be noted that under-driving an LED light is more efficient and allows for a safety margin if ambient temperature should increase. Next, the forward current vs. forward voltage graph in the manufacture's datasheet is referred, which indicates the forward voltage drop according to the drive current as selected for the LED electric circuits. This information is critical during the process of determining the resistor to be utilized for current limiting.

Fig. 2.27(*a*)

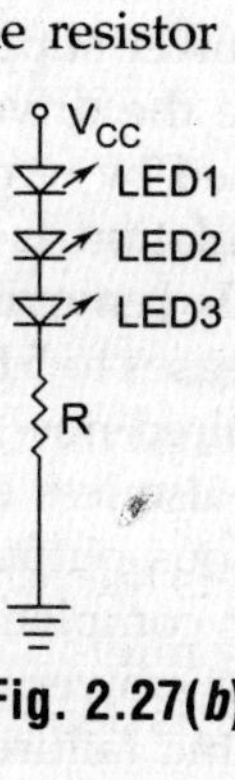

Fig. 2.27(*b*)

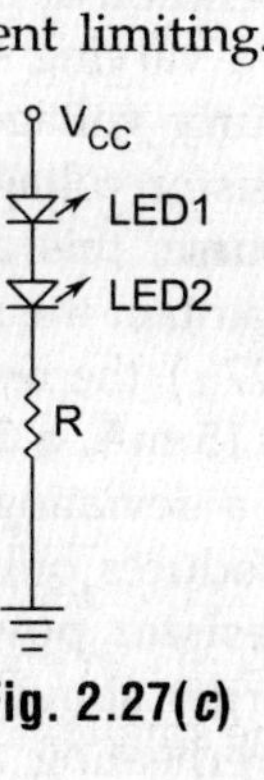

Fig. 2.27(*c*)

(*b*) **LED circuit efficiency and stability :** In the world of custom LED lighting, it is extremely important to understand the relationship between circuit stability and circuit efficiency. A stable LED circuit is less responsive to variations in source voltage. However, the circuit dissipates more heat from the resistor and is therefore less efficient. An unstable circuit is more efficient, but variations in source voltage will cause dramatic fluctuations in current. This can lead to overdriven light emitting diodes causing damage, or if underdriven, the LED lights may not light at all. When designing an LED circuit with current liming resistor circuit, the goal is to find the best compromise between stability and efficiency. To accomplish this, it is best to consider a few different variations of the same LED driver circuit in an attempt to determine the most feasible solution. This becomes extremely critical when designing for LED tail lights.

(*c*) **LED clusters :** The LED forward voltage drop is determined by matching selected drive current with a voltage value from the LED datasheet. Continue by dividing the measured source voltage by the LED forward drop, then round down to the nearest whole number. This calculation determines the maximum number of LEDs suitable per single LED cluster. For example, suppose our measured source voltage V_{CC} is 12 volts and our selected drive current is 15 mA. According to the LED datasheets corresponding with this example problem, the LED forward voltage drop should equal 3.1 volts. We divide the 12-volt source by 3.1 volts to determine the LED forward voltage drop goes into the source voltage 3.8 times. Rounding down to the nearest whole number, we can determine that an LED cluster of three will be the maximum number of LEDs suitable per a single resistor, as shown in Fig. 2.27(*b*).

(*d*) **Calculating LED resistor values :** In previous steps, it has been determined how many LEDs will fit into circuit based on the source voltage and LED forward voltage drop. In the next step, the proper resistor value is to be calculated. The total number of LEDs is multiplied by the LED forward voltage drop to calculate for total LED voltage drop. Now the total LED voltage drop is subtracted from the source voltage to calculate for the resistor voltage drop. Finally, the voltage across the resistor is divided by selected drive current as estimated earlier to determine the exact resistor value. As example, considering the maximum number of three LEDs per resistor, the LED forward voltage drop, say 3.1 volts, is multiplied by 3 for a total LED forward voltage drop of 9.3 volts. This is subtracted from the source voltage, 12 volts, to determine a voltage drop of 2.7 volts across the current limiting resistor. Finally, when the resistor voltage drop of 2.7 volts is divided by the selected drive current of 15 mA (0.015 amps), the resistor value of R is obtained to be 180 ohm.

(*e*) **Power supply voltage variations :** The above estimation for the most efficient resistor values, source voltage, drive current, and LED forward voltage drop is not adequate for a stable operation of LED. This circuit design is susceptible to becoming unstable under the condition of a varying source voltage. If the power supply of the circuit is not voltage regulated, the source voltage can change and cause the drive current to vary. To stabilize the circuit, the resistor voltage drop must be increased by removing one or more LEDs from the circuit. Of course, this also means that recalculation is required for the resistor value in order to maintain the selected drive current. By removing one LED from the circuit as shown in Fig. 2.27(*c*), the resistor voltage drop increases to 5.8 volts. To maintain the selected drive current of 15 mA, a 387-ohm resistor is required now. This overall circuit design is less susceptible to a deviating source voltage, but is also less efficient. Although the circuit is very stable, it produces only two-thirds the luminous output of our previous unstable circuit design.

(*f*) **Resistor power dissipation :** After completing the circuit design, it is important not to forget about calculating for resistor power dissipation. Failing to do so may result in overheating and possibly catastrophic failure. Simply multiply the drive current by the resistor voltage drop to calculate for resistor power. Resistor power should not exceed the resistor's maximum power rating, in watts. In the previous circuit example, the resistor voltage drop was estimated as 5.8 volts. Multiply this by drive current of 15 mA to obtain the resistor power dissipation as 87 mW or 0.087 watts. In this example, a resistor rated at one-quarter watt is more than sufficient to dissipate the power.

(*g*) **LED forward voltage drop variation :** The LED forward voltage drop will vary between manufacturing lots. Each time the LED manufacture produces an LED, the forward voltage drop can be slightly different from the last. Refer to the LED manufacture datasheets to determine the minimum and maximum possible forward voltage drops. Using these values along with the calculated resistor value, the circuit current is recalculated. The drive current may be changed drastically due to the voltage variation across the LED junction. The best solution is usually to reduce the total number of LEDs per series circuit. This will minimize total variations in forward voltage across the junctions. A resistor driven circuit may not be feasible in the application if a reduction in LED per series circuit is not possible, or is possible but total current remains unstable. In this case, the recommended constant current source circuit configuration may be used as a substitute for the resistor driven circuit.

2.11.1 LED Drive Circuits for Digital Transmission

The operation of the LED for binary transmission requires the switching on and off of a current in the range of several tens to several hundreds of milliamperes. This must be performed at high speed in response to logic voltage levels at the driving circuit input. A common method of achieving this current switching operation of LED is shown in Fig. 2.28. The circuit uses a single stage bipolar transistor Q in common emitter mode. When V_{in} is low, the transistor Q is off and very small current passes through the LED to keep it at off condition. When V_{in} is high, the transistor Q is at saturation and sufficient current is passed through the LED to make it illuminated. The single stage circuit provides current gain as well as low saturation voltage (collector-emitter saturation voltage V_{CE} is approximately 0.3 V).

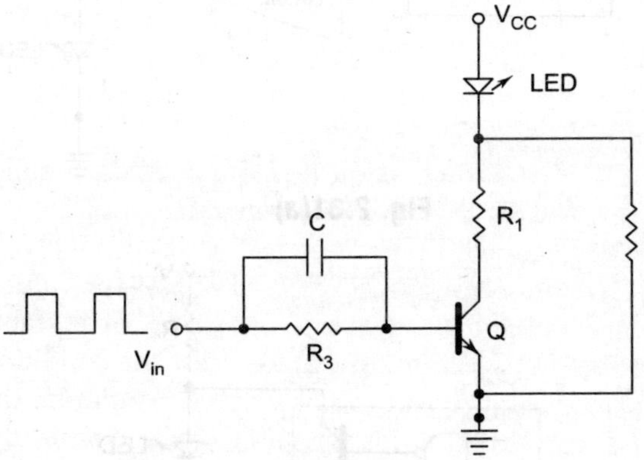

Fig. 2.28

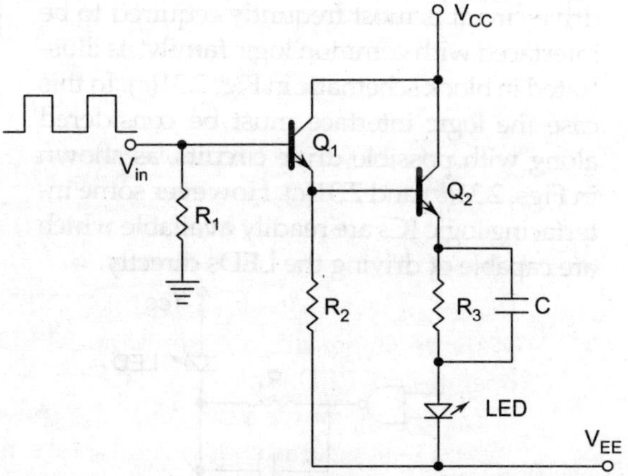

Fig. 2.29

The resistor R1 limits the maximum current through the LED. However independent bias to the device may be provided by the incorporation of resistor R_2. The value of R_2 is sufficiently high such that the illumination of the LED very low or nearly zero. Therefore, when Q is at cut-off, LED does not glow or offers very illumination. The switching speed of the common emitter configuration is limited by space charge and diffusion capacitance, thus limiting the bandwidth. A speed-up capacitor C may be used at the base drive as shown in Fig. 2.28 to compensate the bandwidth requirement, which overdrives or pre-emphasizes the base current during the switch-on period.

Increased switching speed may be obtained from an LED without pulse shaping or speed-up element by the use of a low impedance driving circuit, whereby charging of the space charge and diffusion capacitance occur as rapidly as possible. This may be achieved with the use of an emitter follower driving circuit as demonstrated in Fig. 2.29. When driving pulse at V_{in} is high, both the transistors Q_1 and Q_2 are on, providing the required current to the LED to glow. The use of a compensating matching network R_3C in Fig. 2.29 provides fast direct modulation of LEDs with relatively low drive power. This circuit, with optimum values for matching network, is capable of providing optical rise time of 2.5 ns for LEDs with capacitance of 180 pF, allowing 100 Mbits/sec operations.

Another type of low impedance shunt configuration LED driver circuit is shown at Fig. 2.30. In this configuration the switching transistor is placed in parallel with the LED, providing low impedance path for switching off the LED by shunting current around it. When the driving pulse is at high condition, the transistor switches on providing low impedance path and therefore LED becomes off due unavailability of required current to glow. When input is low, transistor is at cut-off and so LED illuminates as sufficient current passes through it. The switch-on performance of the circuit is determined by the combination of resistor and internal capacitance of LED. Stored space charge may be removed by slightly reverse biasing the LED while switching off. This may be achieved by placing the transistor emitter at negative potential (V_{EE} is negative) with respect to ground. A schottky diode D may be incorporated parallel and reverse to the LED as shown in Fig. 2.31 to limit the extent of reverse bias without introducing any extra minority carrier stored charge into the circuit.

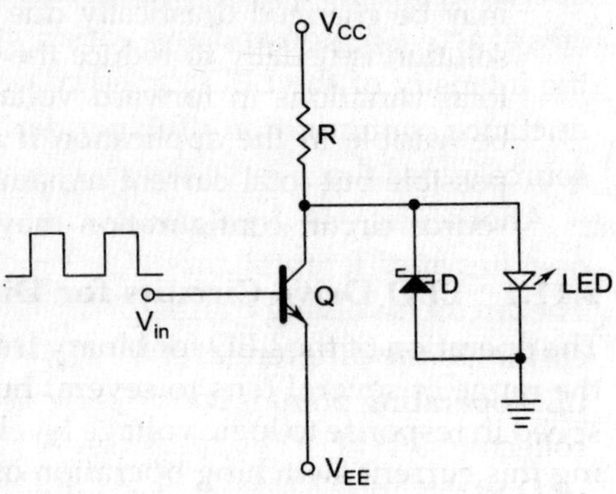

Fig. 2.30

In digital tramission application, the LED drive circuit is most frequntly required to be interfaced with common logic family, as illustrated in block schematic in Fig. 2.31(*a*). In this case the logic interface must be considered along with possible drive circuits, as shown in Figs. 2.31(*b*) and 2.31(*c*). However some interfacing logic ICs are readily available which are capable of driving the LEDs directly.

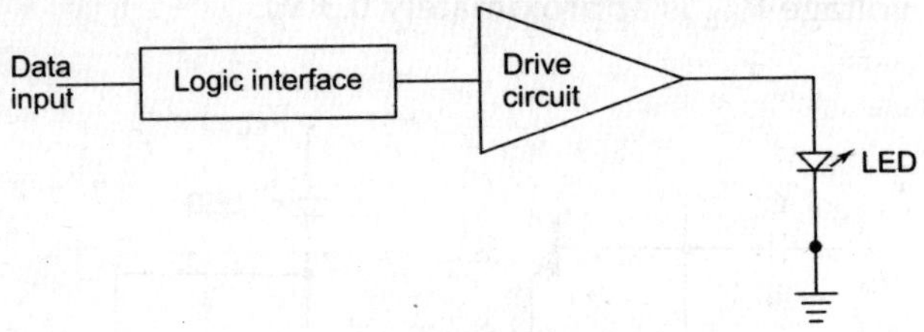

Fig. 2.31(*a*)

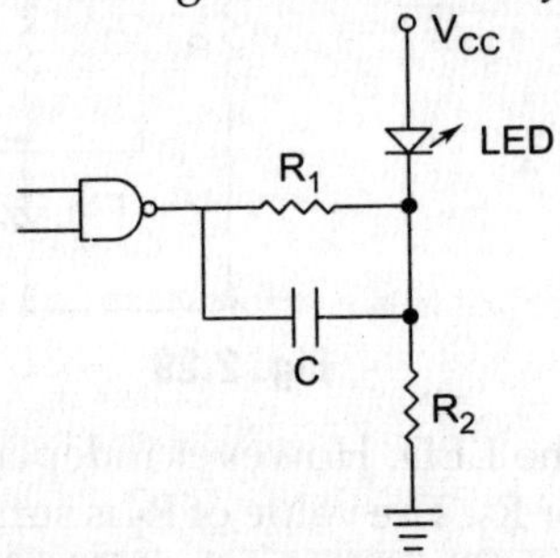

Fig. 2.31(*b*)

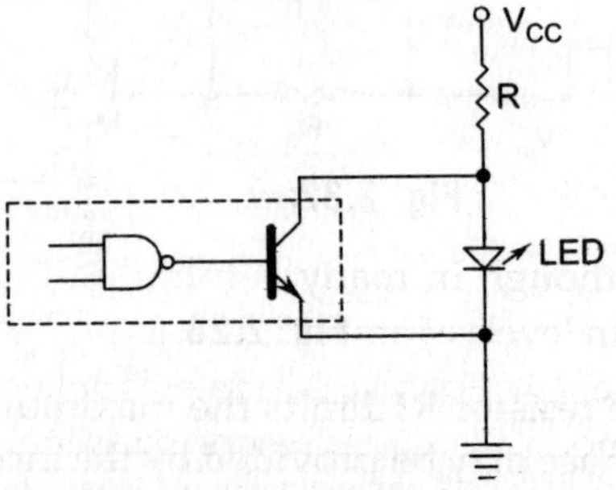

Fig. 2.31(*c*)

2.11.2 LED Drive Circuits for Analog Transmission

For analog transmission the drive circuit must be able to deliver the light output from a LED source accurately following a time-varying input voltage waveform in both amplitude and phase.

Therefore, it is important that the LED output power responds linearly to the input voltage or current. Unfortunately, the inherent nonlinearity within the LED itself sometimes creates distortion and nonlinearity on the signal. Thus the performance of analog transmission systems is limited by the LED characteristics unless suitable compensation is incorporated in the drive circuit. However, if the distortion level is not critical, simple transistor driver circuit may be utilized.

Two possible high speed drive circuit configurations are illustrated in Figs. 2.32(*a*) and 2.32(*b*). The drive circuit of Fig. 2.32(*a*) consists of a common emitter transconductance amplifier which converts an input base voltage into collector current. The transistor is biased at class A mode of operation and its quiescent collector current is about half of the peak value. The transistor operates at active region and any variation in base voltage causes variation at collector current and so the intensity of the LED varies in accordance with the input base voltage. A similar trans-conductance configuration utilizing darlington transistor pair as shown in Fig. 2.32(*b*) reduces the source impedance. This circuit is capable of driving high radiance LEDs at frequencies of 70 MHz.

Another simple drive circuit configuration for analog transmission is presented in Fig. 2.33. It operates at differential amplifier mode over its linear region, which directly modulates the LED. The circuit involving transistor Q_3 and resistors R_3, R_4, R_5 forms a current generator and feeds the current to the differential stage (Q_1 and Q_2), this is also limits the maximum current of the device. LED operating point is controlled by the voltage V_{ref}. Any variation of input voltage V_{in} varies the collector current of Q_1, affecting the variation in collector current of Q_2. Therefore, the intensity of LED varies in accordance with input voltage V_{in}. The transimpedance of the driver is reduced through current series feedback provided by two resistors R_1 and R_2, which are normally assigned equal values. Variation of these feedback resistors can be used to compensate for the transfer function of both the drive circuit as well as the LED.

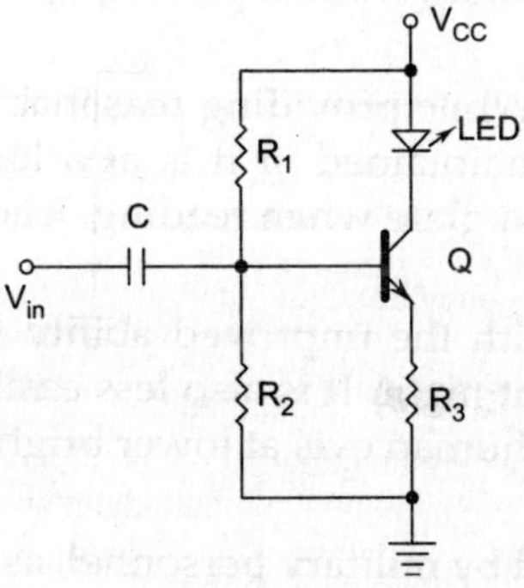

Fig. 2.32(*a*)

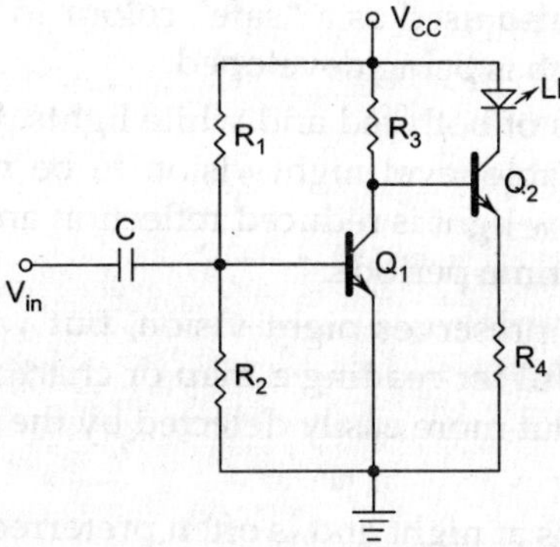

Fig. 2.32(*b*)

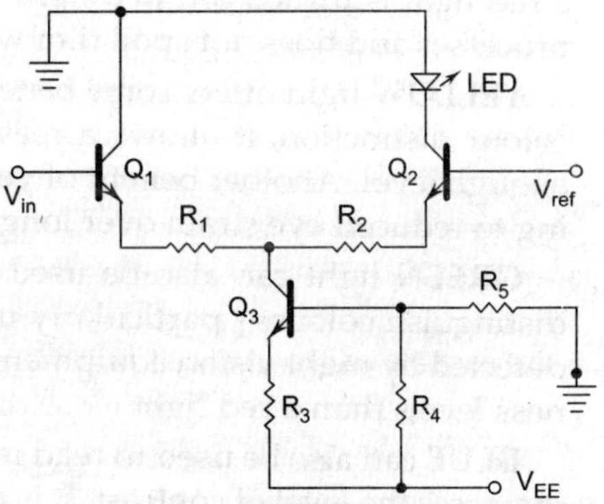

Fig. 2.33

Although in many communication applications where a single analog signal is transmitted certain levels of amplitude and phase distortion can be tolerated, the same may not be applicable in frequency multiplexed systems where a high degree of linearity is required in order to minimize intermodulation interference between individual channels. Furthermore, extremely low levels of amplitude and phase distortion are required to be maintained for baseband video transmission of TV signals. For such applications simple drive circuit as described here may not be adequate without some linearization to compensate for both LED and drive circuit nonlinearities. Several compensation circuits like negative feedback technique, phase shift modulation for selective harmonics are employed depending on the application.

2.12 CONCLUSION

In this chapter, various types of LED, their structures, materials, parameters, characteristics and applications have been discussed. The most recent innovation in InGaN/SiC technology is "True Blue" LEDs. The die improvement to Cotco's High Brightness LED family greatly enhances the performance and the ability to obtain a "True Blue" wavelength. The process combines highly efficient InGaN materials with a Silicon Carbide (SiC) substrate in a geometrically enhanced vertical chip structure, to maximize total light extraction. This material also incorporates a current distribution layer that requires only a single wire bond for increased reliability. These LEDs are available in 3mm and 5mm packages, with 15°A and 25°A viewing angles and luminous intensities up to 8,200mcd.

"True Blue" LEDs find enormous application potential in the field of medical electronics. Advancements in skin and pain relief treatments are turning more and more medical manufacturers to Light Emitting Diode (LED) technology. According to clinical studies, packets of light energy from LED's can benefit skin health in a variety of ways including non-invasive therapies for skin rejuvenation and healing. Blue LED light is also believed to be a contractor and restrictor that can aid in pain relief and the reduction of swelling. Blue LED therapies can be used where traditional treatments such as ice are currently being used. As light therapies evolve, the uses of LEDs and their unique properties will continue to be expanded within the medical community.

This chapter may be concluded with a brief guide to the most common colours available for LEDs and the practical uses to which they are most often put.

WHITE light allows perfect colour distinction, but can spoil dark-adapted vision, which then requires a period of re-adjustment, once the light source is extinguished.

RED light is traditionally used for preserving night vision. Red light does not cause your pupils to contract so much and as a result your eyes do not need to re-adjust to the darkness once a red light is turned off. Red light is also used as a "safe" colour in monochrome photographic processes and does not spoil film which is being developed.

YELLOW light offers some benefits of both red and white lights. While providing reasonable colour distinction, it allows a reasonable level night vision to be maintained, if it is at a low enough level. Another benefit of yellow light is reduced reflection and glare when reading, leading to reduced eye strain over longer time periods.

GREEN light can also be used to preserves night vision, but with the improved ability to distinguish colours - particularly useful for reading a map or chart at night. It is also less easily detected by night vision equipment, but more easily detected by the human eye, at lower brightness levels than a red light.

BLUE can also be used to read maps at night and is often preferred by military personnel, as it increases the level of contrast. It is also the classic backstage working light colour for theatres and shows. Blue LED light has the same effect as UV light on most fluorescent materials.

BLUE-GREEN brings similar night vision benefits to the green light and the blue light, but with improved colour distinction. Some users prefer the blue-green light for this reason.

INFRARED is for use in conjunction with night vision equipment. Otherwise it is invisible to the human eye.

ULTRAVIOLET is most commonly used to detect forged banknotes, which glow under UV light. Some UV LED torches have become popular at clubs and parties, where they are used to make fluorescent materials shine brightly.

The colour of the light we see is determined by its wavelength, which in turn is often used to denote the colour of light. The wavelengths associated with light are tiny and measured in nanometres (nm) - billionths of a metre. LEDs produce light that is almost *coherent*, meaning that it is almost all on one wavelength, giving very pure colours. Here is an approximate guide to the colours of light and their wavelengths.

MID INFRARED:	4600 nm – 1600 nm – Invisible
LOWER INFRARED:	1300 nm – 870 nm – Invisible
	850 nm – 810 nm – Nearly invisible, very dull red
NEAR INFRARED:	780 nm – Very dim cherry red when viewed directly
	770 nm – Deep cherry red when viewed directly
	740 nm – Deep cherry red
RED:	700 nm – Deep red
	660 nm – Red
	645 nm – Bright red
	630 nm – "He-Ne Laser" red
	620 nm – Orange-red
ORANGE:	615 nm – Reddish orange
	610 nm – Orange
	605 nm – Amber
YELLOW:	590 nm – "Sodium" yellow
	585 nm – Yellow
	575 nm – Lemon yellow / greenish
GREEN:	570 nm – Blueish green
	565 nm – Blue-green
	555 nm – Blueish lime green
	550 nm – Emerald green
	525 nm – Pure green
BLUE-GREEN:	505 nm – Greenish blue / turquoise
	500 nm – Greenish cyan
	495 nm – Sky blue
BLUE:	475 nm – Azure blue
	470 nm – 460nm-Bright blue
	450 nm – Pure blue
BLUE-VIOLET:	444 nm – Deep blue
	430 nm – Bluish-violet
VIOLET:	405 nm – Pure violet
	400 nm – Deep violet
NEAR ULTRAVIOLET:	395 nm – Deep purple with a slight red tinge
ULTRAVIOLET (UV/A):	370 nm – Nearly invisible. May appear a very dim purple when viewed through Woods Glass.
WHITE:	White LEDs come in a wide range of shades from yellowish to purplish white. The most noticeable colour cast is generally in the centre of the beam and is usually slightly blue, surrounded by "pure" white. The size and noticeability of the blue centre is often a guide to the quality of the LED: the whiter the beam, the better the quality.

REVIEW QUESTIONS

1. In an LED, which of the following factors affects most severely the efficiency of the diode and cannot be eliminated even in principle?
   (*a*) Fresnel reflection (*b*) Back emission
   (*c*) Total internal reflection (*d*) Absorption

2. Which of the following materials is not suitable for making a LED?
   (*a*) GaAs (*b*) Silicon
   (*c*) InGaAs (*d*) GaAlAs
3. In which of the following the densities of electrons and holes are same?
   (*a*) Intrinsic semiconductor (*b*) Extrinsic semiconductor
   (*c*) *p-n* diode (*d*) Forward biased *p-n* junction
4. At what temperature, the majority carrier concentrations in *p-n* homojunctionare are almost equal to the dopant concentration?
   (*a*) Absolute zero temperature (*b*) Normal temperature
   (*c*) High temperature (*d*) All temperatures.
5. What type of material makes an efficient LED?
   (*a*) Metal (*b*) Insulator
   (*c*) Indirect bandgap type semiconductor (*d*) Direct bandgap type semiconductor
6. Which of the following pairs are suitable for making a heterojunction?
   (*a*) Si and Ge (*b*) Si and GaAs (*c*) GaAs and AlAs (*d*) GaAs and GaAlAs
7. A LED with external quantum efficiency of 0.012 is coupled with an optical fiber of NA of 0.15 in air medium. What is the overall source-fiber coupling efficiency?
   (*a*) 1.8×10^{-4} (*b*) 2.7×10^{-4} (*c*) 3.2×10^{-4} (*d*) 7.8×10^{-4}
8. Which combination of materials is most suitable for 'True Blue' LED?
   (*a*) GaAs and AlAs (*b*) GaAs and GaAlAs
   (*c*) InGaN and SiC (*d*) InGaN and GaAs
9. What is the reason that limits the internal quantum efficiency in homojunction LED?
   (*a*) Minority carriers diffuse away from junction and recombine
   (*b*) Increase of nonradiative recombination
   (*c*) Both of these
   (*d*) None of these
10. Which of the following statement is true in LED?
   (*a*) Optical bandwidth is greater than electrical bandwidth
   (*b*) Optical bandwidth is equal to electrical bandwidth
   (*c*) Optical bandwidth is lower than electrical bandwidth
   (*d*) None of these
11. What are the direct bandgap and indirect bandgap type of semiconductors? Give at least two examples of each. Which of these are more suitable for fabricating LEDs? Mention reasons.
12. Schematically illustrate the variation of carrier concentration across an $\overset{+}{n}-p$ homojunction, in equilibrium and under forward bias.
13. Assuming that a $\overset{+}{n}-p$ homojunction is forward biased and net rate of recombination per unit volume in the *p*-side is equal to $\Delta n/\tau_p$, where Δn is the local excess minority concentration in the *p*-side and τ_p is a constant. Prove that τ_p is the mean lifetime of the excess minority carriers.
14. Calculate the injection efficiency of a GaAs diode where $N_a = 10^{21}/m^3$ and $N_d = 10^{23}/m^3$. Assuming at room temperature of 300°K $\mu_e = 0.85$ m^2/vsec and $\mu_h = 0.04$ m^2/vsec and $L_e = L_h$, what is the mean lifetime of the excess minority carriers.
15. (*a*) What are homojunctions and heterojunctions?
   (*b*) Discuss the unique properties of *p-n-n* double heterostructure LED with its energy level diagram.
16. In a forward biased *p-n* homojunction, the probability that an electron in the conduction band at energy level E_2 will recombine with a hole in the valance band at energy E_1 is proportional

to the concentration of electrons $n(E_2)$ at E_2, and proportional to the concentration of holes $p(E_1)$ at E_1. The probability of a photon energy E_{ph} {= $(E_2 - E_1)$}, being radiated by the diode may be obtained by integrating the product $n(E_2)\ p(E_1)$ over all values of E_1 or E_2.

(*a*) Calculate the power P radiated by the *p-n* diode in respect of photon energy

(*b*) What is the photon energy when the maximum power occurs?

(c) What is the power at this photon energy?

Assume that $n(E_2) = A \exp[-(E_2 - E_c)/kT]$

$p(E_1) = B \exp[-(E_v - E_1)/kT]$, A and B are the proportionality constants.

17. Describe the LED structures commonly used for optical fiber communication and discuss their merits and demerits.
18. What are the parameters required to be optimized to maximize the external efficiency of LED? Discuss the techniques to optimize these parameters.
19. Which of the two types of LEDs made of Si and GaAs with identical dimensions and doping levels is expected to exhibit a larger modulation bandwidth and why?
20. What are the different classes of defects that degrade the performance of LED?
21. Explain why edge-emitting LEDs is expected to be of high modulation bandwidth.
22. Describe the construction idea for LED of white colour.

3

LASERS

3.1 INTRODUCTION

The word 'LASER' is an acronym for *Light Amplification by Stimulated Emission of Radiation*. Albert Einstein first predicted in 1917, that an atom can emit radiation by two different processes—spontaneous emission and stimulated emission. Basov and Prochorov in the USSR and Townes and Schawlow in the USA first put forward the fact that the stimulated emission process could be used in the construction of coherent optical sources. Finally in 1960 T.H.Maiman first achieved laser action at optical frequencies in ruby. Since then the development of lasers has been extremely rapid and laser action has been demonstrated with gases, solids, liquids, free electrons, semiconductors, etc.

The three main components of any laser system are the optical resonator, the amplifying medium and the pump. The optical resonator consists of a pair of mirrors facing each other and provides optical feedback to the amplifier so that it can act as a source of radiation. We will discuss the optical resonators in detail in a later part of the chapter. The amplifying medium consists of a collection of atoms, ions or molecules which act as an amplifier for light waves. Under normal conditions, the number of atoms in the lower energy state is always larger than the number in the excited states. In such a collection of atoms if an optical wave passes through, then more absorption will take place than emission and hence the wave will get attenuated. Hence in order to have amplifications, it is necessary to have more number of atoms in the excited state than in the lower state—a situation called 'population inversion'. In such a condition, an optical wave passing through the collection will induce more emissions than absorption, and hence gets amplified. The pump is the source of energy which maintains the medium in its population inverted state.

3.2 EMISSION AND ABSORPTION OF RADIATION

If an electron in an atom undergoes transition between two energy states or levels it either emits or absorbs a photon, which can be described in terms of a wave of frequency ν where $\nu = \Delta E/h$, ΔE is the energy difference between the two levels concerned, and $h = 6.626 \times 10^{-34}$ J s is Planck's constant. We now consider the electron transitions that may take place between the two energy levels of the hypothetical atomic system shown in Fig. 3.1.

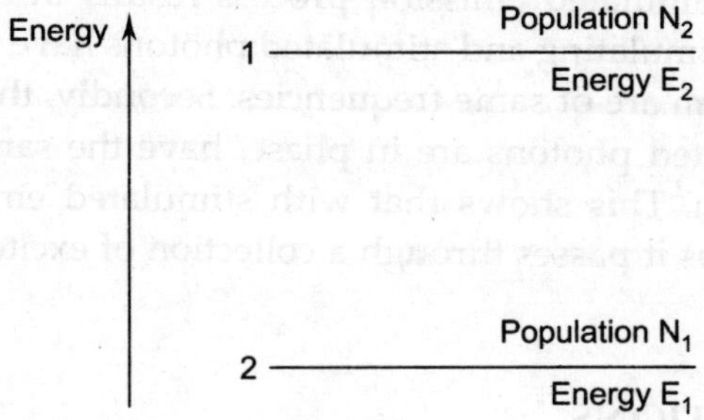

Fig. 3.1

If an electron is in the lower energy level E_1 then in the presence of photons of energy $(E_2 - E_1)$ it may be excited to the upper level E_2 by absorbing a photon. As the absorption transition can only occur in the presence of photons of appropriate energy, it is often referred to as *stimulated absorption*. Alternatively if the electron is in the level E_2 it may return to the ground state with the emission of a photon. The emission process may occur in two distinct ways. They are: (*a*) the *spontaneous emission* process in which the electron drops to the lower level in an entirely random manner and (*b*) the *stimulated emission* process in which the electron is 'triggered' to undergo the transition by the presence of photons of energy $(E_2 - E_1)$. The emission and absorption processes are illustrated in Figs. 3.2(*a*), (*b*) and (*c*). The black dot indicates the state of the atom before and after the transition.

Since, under normal circumstances the probability of occurrence of the spontaneous process is much higher, we do not observe the stimulated process normally. The average time the electron exists in the excited state before making a spontaneous transition is called the lifetime τ_{21} of the excited state. The '21' here indicates the two energy levels involved. Since the spontaneous emission process is random in nature, the radiation emitted will be incoherent in nature. A similar emission process in semiconductors provides the basic mechanism for light generation within the Light Emitting Diode (LED).

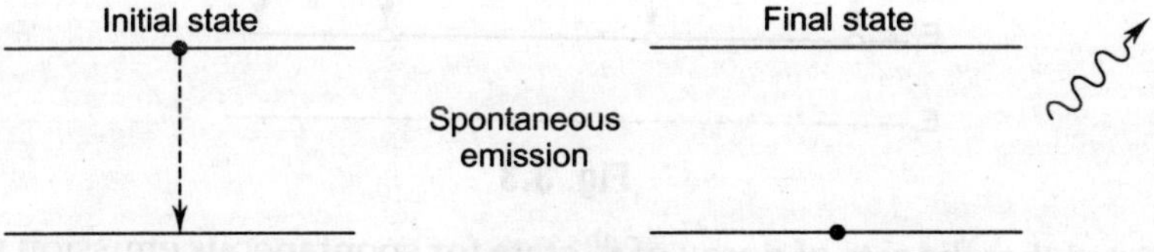

Fig. 3.2(*a*)

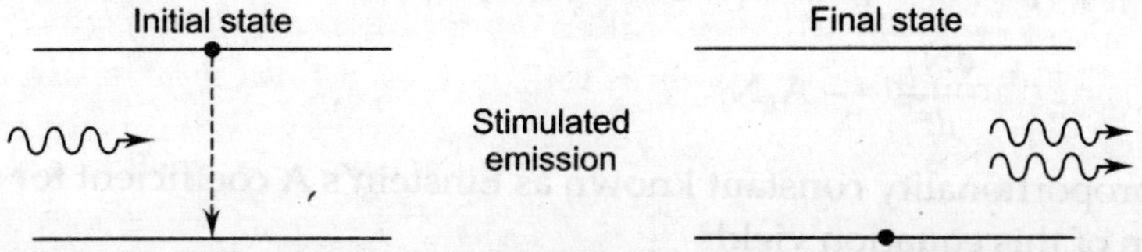

Fig. 3.2(*b*)

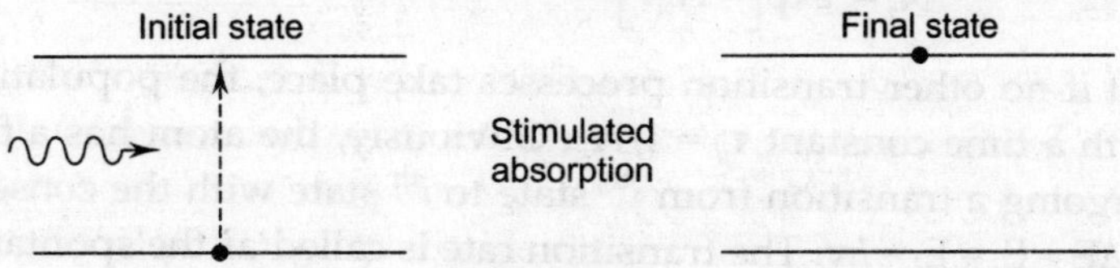

Fig. 3.2(*c*)

In contrast to this, the stimulated emission process results in coherent radiation since the waves associated with the stimulating and stimulated photons have identical energies and hence the light associated with them are of same frequencies. Secondly, the waves associated with the stimulating and the stimulated photons are in phase, have the same state of polarization and travel in the same direction. This shows that with stimulated emission the amplitude of an incident wave can increase as it passes through a collection of excited atoms and hence leads to amplification.

3.3 EINSTEIN'S RELATIONS

In 1917 Einstein demonstrated that the rates of the three transition processes of stimulated absorption, spontaneous emission and stimulated emission are related mathematically. Einstein's concept is based on the quantum hypothesis which states that the radiation energy comes in discrete packages of $\Delta E = h\nu$. An obvious conclusion is that there must be at least two distinctly separated energy states E_i and E_j as shown in Fig. 3.3. The number of atoms per unit volume in the respective states is called as the population density of those states and is denoted by N_i and N_j.

3.3.1 Spontaneous Emission

When an excited atom in an ensemble undergoes transition from a higher energy level 'E_j' to any lower energy level 'E_i' by emitting a photon having energy $\Delta E = E_j - E_i = h\nu$ without being stimulated by any external field, the transition is called spontaneous transition and the process is termed as spontaneous emission as already mentioned in the previous section. ν is called the characteristic frequency of any transition between the concerned states.

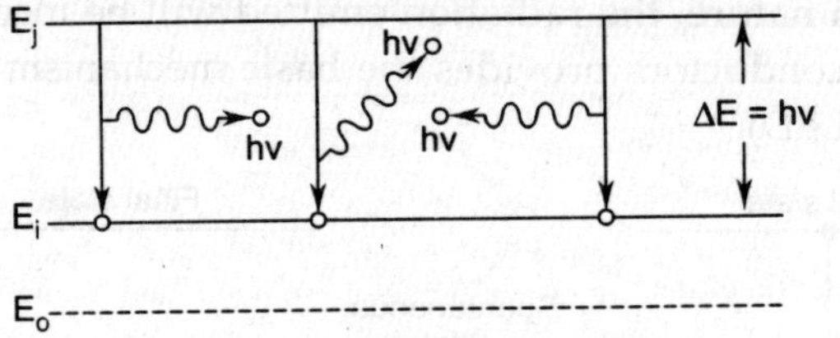

Fig. 3.3

Einstein predicted that the rate of decay of jth state for spontaneous emission process is directly proportional to the population per unit volume (population density) of jth state, that is,

$$\frac{dN_j}{dt} = -A_{ji}N_j \tag{3.1}$$

where A_{ji} is the proportionality constant known as Einstein's A coefficient for spontaneous emission. The solution of this equation yields

$$N_j = \exp\left[-A_{ji}t\right] \tag{3.2}$$

This means that if no other transition processes take place, the population density of the jth state decreases with a time constant $\tau_{ji} = 1/A_{ji}$. Obviously, the atom has a finite probability per unit time of undergoing a transition from j^{th} state to i^{th} state with the consequent emission of a photon of energy $\Delta E = E_j - E_i = h\nu$. The transition rate is called as the spontaneous transition rate and is often expressed as A_{ji}. To restate, A_{ji} is the probability per unit time that an excited atom will spontaneously emit a photon in a random direction and pass from an upper energy state to a lower one. It is apparent that $i \rightarrow j$ transition cannot be spontaneous.

The stimulated transition, on the other hand, may be $i \rightarrow j$ transition or $j \rightarrow i$ transition. As the name suggests, stimulation of the constituent atoms by external photons (radiation) is necessary for such transitions. For stimulation to be physically effective, the frequency of the stimulating photons should be ideally ν, the characteristic frequency between the concerned states of the atom. During stimulated transitions, decrease in population of one level means the increase in population of a specific level, since it depends on the characteristic frequency between the specific levels. Whereas in case of spontaneous absorption, the excited particle may jump to any higher level usually determined by the Pauli exclusion principle. When the transition is from higher state to lower state, energy will be emitted, while during lower to higher state transition, absorption of energy will take place.

3.3.2 Stimulated Absorption

In this process the atom absorbs a photon having energy $\Delta E = h\nu$ from the radiative field and thus lifts the atom from i^{th} state to j^{th} state. The rate of excitation is proportional to the population of the ith state and the spectral energy density '$\rho(\nu)$' of the radiative field. Since the process ends at the jth level, a decrease in population of ith state means an increase of population of jth state. Hence, the rate equation may be written as:

$$\frac{dN_j}{dt} = +B_{ij}N_i\rho(\nu) = -\frac{dN_i}{dt} \tag{3.3}$$

B_{ij} is the Einstein's B coefficient for stimulated absorption. The schematic of stimulated absorption is shown at Fig. 3.4

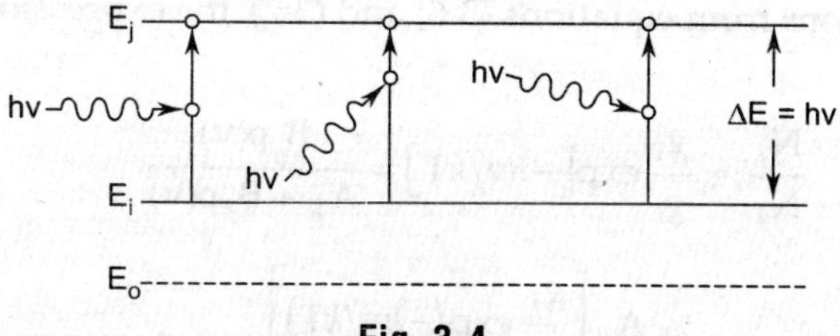

Fig. 3.4

3.3.3 Stimulated Emission

This process is just the reverse of absorption and the excited atom of jth state when stimulated by the external field releases an energy $\Delta E = h\nu$ to the field and comes down to the ith level. This is shown in Fig. 3.5. The decay rate of jth state depends on population of the excited state and the spectral energy density of the stimulating field. Hence the rate equation may be written as:

$$\frac{dN_j}{dt} = -B_{ji}N_j\rho(\nu) = -\frac{dN_i}{dt} \tag{3.4}$$

B_{ji} is the Einstein's B coefficient for stimulated emission. It should be noted that the Einstein's B coefficients are fundamental parameters that determine the interaction of the atom with the external radiation field. It is to be noted that the Einstein's coefficient, for a given transition, are independent of external conditions, such as, temperature, pressure, power of radiation, etc.

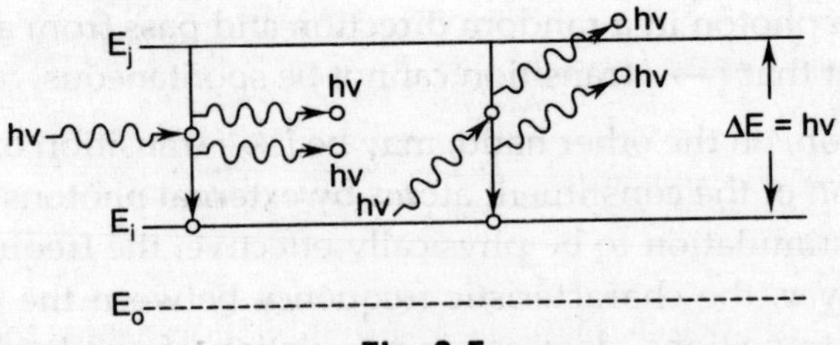

Fig. 3.5

At thermodynamic equilibrium, all these processes must balance each other to maintain the population of both the states. If all other unidentified transitions have been neglected for the further analysis, we may write

$$\frac{dN_j}{dt} = -A_{ji}N_j + B_{ij}N_i\rho(\nu) - B_{ji}N_j\rho(\nu) = -\frac{dN_i}{dt} \tag{3.5}$$

Since at equilibrium the rate of change of population of both the states is zero,

$$\frac{N_j}{N_i} = \frac{B_{ij}\rho(\nu)}{A_{ji} + B_{ji}\rho(\nu)} \tag{3.6}$$

But Boltzmann distribution statistics provides a different expression relating the populations between the two states, which is given below.

$$\frac{N_j}{N_i} = \frac{g_j}{g_i}\exp\left[-h\nu/kT\right] \tag{3.7}$$

where, g_i and g_j are the degeneracy of the levels concerned, and $k = 1.381 \times 10^{-23}$ is Boltzmann's constant.

Comparing both relations from equations (3.6) and (3.7), the expression for the spectral energy density becomes

$$\frac{N_j}{N_i} = \frac{g_j}{g_i}\exp\left[-h\nu/kT\right] = \frac{B_{ij}\rho(\nu)}{A_{ji} + B_{ji}\rho(\nu)} \tag{3.8}$$

or,

$$\rho(\nu) = \frac{A_{ji}\left[\dfrac{g_j}{g_i}\exp(-h\nu/kT)\right]}{B_{ij} - B_{ji}\left[\dfrac{g_j}{g_i}\exp(-h\nu/kT)\right]} = \frac{A_{ji}}{B_{ji}}\frac{1}{\left[\dfrac{B_{ij}g_i}{B_{ji}g_j}\exp(h\nu/kT)\right] - 1} \tag{3.9}$$

This is the expression for energy density per frequency interval in terms of Einstein's A and B coefficients.

Again according to Planck the spectral energy density $\rho(\nu)$ in frequency interval $d\nu$ for black body radiation is given by:

$$\rho(\nu) = \frac{8\pi h\nu^3}{c^3\left[\exp(h\nu/kT) - 1\right]} = \frac{8\pi h\nu^3}{c^3}\frac{1}{\left[\exp(h\nu/kT) - 1\right]} \tag{3.10}$$

A comparison between equations (3.9) and (3.10) connects all the three basic transitions as shown as follows:

$$\frac{B_{ij}g_i}{B_{ji}g_j} = 1 \tag{3.11}$$

and
$$\frac{A_{ji}}{B_{ji}} = \frac{8\pi h\nu^3}{c^3} \tag{3.12}$$

These are the well-known Einstein's relations. The first relation relates the probabilities of stimulated absorption and emission, while the second relation gives the ratio between the probabilities of spontaneous emission and stimulated emission. It may be observed from equation (3.11) that when the degeneracies of the two levels are equal ($g_i = g_j$), then the probabilities of absorption and stimulated emission are equal. Furthermore, the ratio of the stimulated emission rate to the spontaneous emission rate is given by:

$$\frac{\text{Stimulated emission rate}}{\text{Spontaneous emission rate}} = \frac{B_{ji}\rho(\nu)}{A_{ji}} = \frac{1}{\exp(h\nu/kT)-1} \tag{3.12a}$$

It is interesting to note from equation (3.12) that the ratio of the probabilities of spontaneous emission and stimulated emission is proportional to the cube of the transition frequency. This is the basic reason for which laser transition at shorter wavelengths is difficult to obtain.

The above discussion indicates that the process of stimulated emission competes with the process of spontaneous emission and absorption. Hence, if we wish to amplify a beam of light by stimulated emission then we must increase the rate of this process in relation to the other two processes. To achieve this for a given pair of energy levels we must increase both the population density N_j of the upper level in relation to the population density N_i of the lower level and the radiation density. We shall show that to produce laser action we must create a condition in which $N_j > \left(\frac{g_j}{g_i}\right) N_i$ even though $E_j > E_i$, *i.e.*, we must create a *population inversion*. Before describing this situation in detail we want to look more closely at the process of absorption and light amplification by an atomic system.

Example 3.1. *Estimate the ratio of the stimulated emission rate to the spontaneous emission rate for an incandescent lamp at a temperature of 900 K. It may be assumed that the average operating wavelength is 0.6 μm.*

Solution:

The average operating frequency is given by:

$$\nu = \frac{c}{\lambda} = \frac{3\times 10^8}{0.6\times 10^{-6}} = 5\times 10^{14}\ \text{Hz}$$

Applying equation (3.12 a) the ratio is obtained as:

$$\frac{\text{Stimulated emission rate}}{\text{Spontaneous emission rate}} = \frac{1}{\exp\left(\dfrac{6.626\times 10^{-34}\times 5\times 10^{14}}{1.381\times 10^{-23}\times 900}\right)-1}$$

$$= \frac{1}{\exp(26.65)-1}$$

$$= 2.7\times 10^{-12}$$

3.4 LIGHT AMPLIFICATION BY AN ATOMIC SYSTEM

Let us consider a beam of light having irradiance I(*z*) at central frequency ν and frequency width Δν passing along direction *z* through a collection of atoms confined in an elementary volume *dxdydz*, as shown in Fig. 3.6. If any of the characteristic frequencies of the concerned atoms is the same as that of the light beam, the beam will lose some of its energy due to stimulated absorption. The irradiance of the output beam which is coming out of this elementary space is I(*z* + *dz*). Then the net amount of energy absorbed by this collection of atoms is given as

$$\left[I(z)-I(z+dz)\right]dxdy = \left[I(z)-I(z)-(dI/dz)dz\right]dxdy = -(dI/dz)dxdydz \qquad (3.13)$$

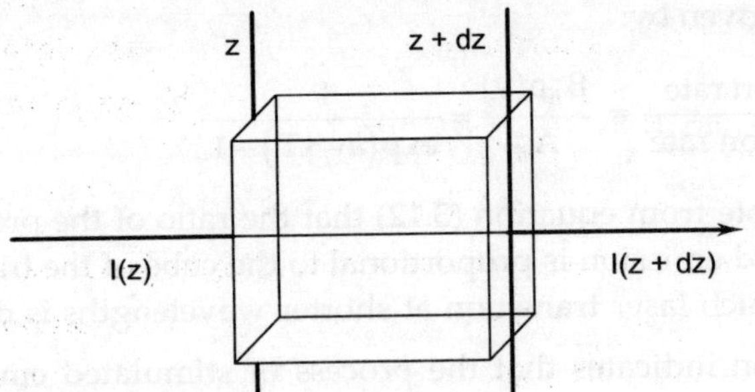

Fig. 3.6

During spontaneous transition energy released by the excited atoms is radiated in all directions. Obviously, only a negligible fraction of it contributes to the beam traveling along z-direction and is neglected for further analysis. Only the energies released and absorbed due to stimulated emission and absorption will affect the beam under consideration. The amount of energy absorbed by the atoms in going from lower to higher level will be ξ_{ij} *hνdxdydz*, where ξ_{ij} is the number of atoms undergoing stimulated absorptions per unit time per unit volume. Similarly the energy released through stimulated emissions for transitions from higher to lower level will be ξ_{ij} *hνdxdydz*.

Now, using equation (3.12), the number of stimulated emissions per unit volume per unit time when the cluster of atoms is radiated by a wave having energy density ρ(ν) will be:

$$\xi_{ij} = \int B_{ji}\rho(\nu)N_j g(\nu)d\nu = \frac{A_{ji}N_j c^3}{8\pi h}\int \frac{\rho(\nu)g(\nu)}{\nu^3}d\nu \qquad (3.14)$$

where, *g*(ν)*d*ν is the line shape function. In most of the cases, while dealing with laser radiation only finite amount of energy is released in a bandwidth that is much smaller than the same expressed by the line shape transition. Hence, if it is assumed that the active particles are interacting with near monochromatic radiation of frequency ν′, then ρ(ν) has a sharp peak at ρ_ν when compared to *g*(ν). The factor $g(\nu)/\nu^3$ can be taken out of the integral in such condition. So the integral may be simplified to

$$\xi_{ij} = \frac{A_{ji}N_j c^3}{8\pi h}\frac{g(\nu)}{\nu^3}\int \rho(\nu)\,d\nu = \frac{A_{ji}N_j c^3}{8\pi h}\frac{g(\nu)}{\nu^3}\rho_\nu \qquad (3.15)$$

where, $\int \rho(\nu)d\nu = \rho_\nu$. Similarly using equations (3.11) and (3.12), we may write

$$\xi_{ij} = \frac{A_{ji}c^3}{8\pi h}\frac{N_i g_j}{g_i}\frac{g(\nu)}{\nu^3}\rho_\nu \tag{3.16}$$

Thus, the net rate of energy absorbed by the active particles within the specified volume will be:

$$(\xi_{ij} - \xi_{ji})\, h\nu dxdydz = \frac{A_{ji}c^3}{8\pi h}\left(\frac{g_j}{g_i}N_i - N_j\right)\frac{g(\nu)}{\nu^3}\rho_\nu h\nu dxdydz \tag{3.17}$$

Now, the irradiance I_ν which is a function of z, and the spectral energy density, ρ_ν, *i.e.*, the energy per unit volume are related through $I_\nu(z) = c\rho_\nu$. Using this relation along with Equations (3.9) and (3.10), we get

$$\frac{A_{ji}c^3}{8\pi}\left(\frac{g_j}{g_i}N_i - N_j\right)\frac{g(\nu)}{\nu^2}\rho_\nu\, dxdydz = -\frac{dI_\nu(z)}{dz}dxdydz$$

or,
$$\frac{dI_\nu(z)}{dz} = \frac{A_{ji}c^2}{8\pi\nu^2}\left(N_j - \frac{g_j}{g_i}N_i\right)g(\nu)\, I_\nu(z) = -\alpha(\nu)I_\nu(z) \tag{3.18}$$

or,
$$\frac{dI_\nu(z)}{dz} = \left[\frac{A_{ji}\lambda_0^2}{8\pi}\left(N_j - \frac{g_j}{g_i}N_i\right)g(\nu)\right]I_\nu(z) = -\alpha(\nu)I_\nu(z) \tag{3.19}$$

where
$$\alpha(\nu) = \frac{A_{ji}c^2}{8\pi\nu^2}g(\nu)\left(\frac{g_j}{g_i}N_i - N_j\right) = \frac{A_{ji}c^2}{8\pi\nu^2}g(\nu)\Delta N \tag{3.20}$$

This is known as the central laser equation and $\alpha(\nu)$is the absorption coefficient and ΔN stands for the population difference per unit volume. The solution of this differential equation leads to:

$$I_\nu(z) = I_\nu(0)\exp[-\alpha(\nu)z] \tag{3.21}$$

The relation is the well-known Beer's law where the irradiance of the input beam is denoted by $I_\nu(0)$. According to the Boltzmann's distribution law, at thermal equilibrium $N_i(g_j / g_i) > N_j$, which indicates a positive value of $\alpha(\nu)$. The above relation predicts that the intensity decreases with z causing an attenuation of the beam. This phenomenon was well known to the scientists for quite a long time and is the basic reason behind the absorption lines in the solar spectra.

On the contrary, if $N_i(g_j / g_i) < N_j$, $\alpha(\nu)$ is negative which implies an amplification of the beam with z, such amplification was first predicted by Einstein in 1917. But it had never been observed before the invention of NH_3-maser in 1953 and then in Ruby laser in 1959. This building up of the light energy, which is also called light amplification by stimulated emission, is the underlying principle for laser operation. The condition $N_i(g_j / g_i) < N_j$ is known as the population inversion and only under this condition one may expect amplification of the beam. Obviously when $N_i(g_j / g_i) = N_j$ there will be no amplification or attenuation of the beam and it will pass through the system without gain or loss of energy.

3.5 POPULATION INVERSION

Under the conditions of thermal equilibrium according to Boltzmann distribution statistics (equation 3.7), the numbers of atoms are more in the lower energy level E_1 than in the higher energy level E_2 of a two level atomic system. This situation, which is normal for structures at room temperature, is shown in Fig. 3.7(*a*). We note that if the energy difference between E_1 and E_2

were nearly equal to kT (≅ 2.0 eV at room temperature) then the population of the upper level would be $1/e$ or 0.37 of that of the lower level.

However in order to achieve light amplification, it is necessary to create a non-equilibrium distribution of atoms such that the population of the upper energy level is greater than the population of the lower energy level (*i.e.*, $N_2 > N_1$). This condition is known as population inversion and is shown in Fig. 3.7(*b*).

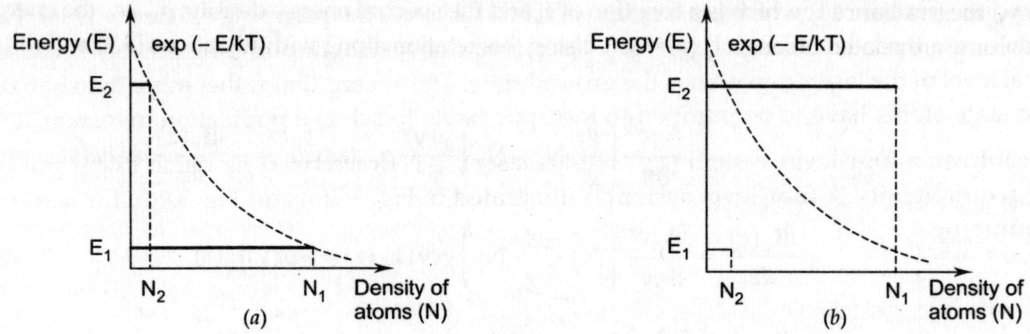

Fig. 3.7

Clearly if we are to create a population inversion, we must supply a large amount of energy to excite atoms into upper level E_2. This excitation process is called 'pumping'. Pumping produces a non-thermal equilibrium situation. A common method used for pumping is the stimulated absorption, *i.e.*, the energy levels which one hopes to use for laser action are pumped by intense irradiation of the system. Now as B_{12} and B_{21} are equal (assuming $g_1 = g_2$), once atoms are excited into upper level the probabilities of further stimulated absorption and emission are equal so that even with very intense pumping the best that can be achieved with the two level system, is equality of the population of the two levels.

Hence, we must look for materials with either three or four energy levels. To aid attainment of population inversion both systems display a central *metastable* state in which the atoms spend an unusually long time.

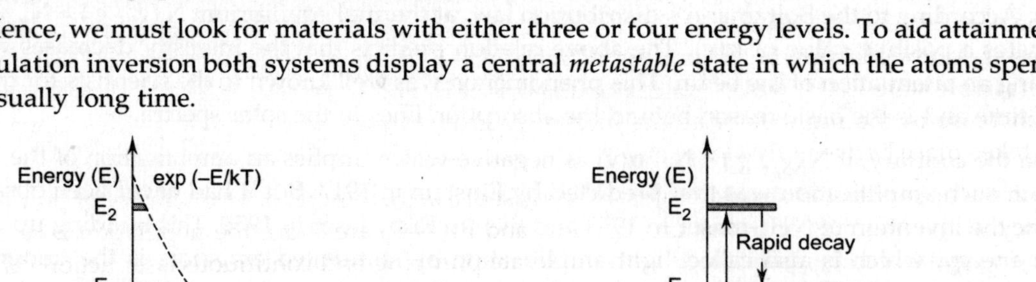

Fig. 3.8

The three-level system first proposed by Bloembergen is shown in Fig. 3.8 (Fig. 3.8(*a*) represents the distribution before pumping and Fig. 3.8(*b*) for distribution after pumping and the transitions). In this system of atoms, if the collection of atoms is intensely illuminated the electrons can be excited (*i.e.*, pumped) into the level E_2 from the ground state E_0. From E_2 the electrons decay by non-radiative processes to the metastable level E_1 and a population inversion can be created between E_1 and E_0. Ideally, the transition from E_2 to E_1 should be very fast, to ensure that there are always vacant states at E_2. On the contrary, the transition from E_1 to E_0 should be very slow. This allows the number of atoms at E_1 to increase significantly, as the probability of spontaneous emission is relatively small. Eventually N_1 may become greater than N_0 and this leads to the population inversion.

The level E_2 should preferably consist of a large number of closely spaced levels so that pumping uses a wide spectral range of the pumping radiation. Hence, the pumping efficiency is increased. Even then, the three-level lasers (*e.g.*, ruby laser), require very high pumping powers because the terminal level of the laser transition is the ground state. This means that rather more than half of the ground state atoms have to be pumped to the upper state to achieve population inversion.

By contrast, a four-level system (*e.g.*, He-Ne laser), is characterized by much lower pumping power requirements. A four-level system is illustrated in Fig. 3.9(*a*) and Fig. 3.9(*b*) for before and after pumping.

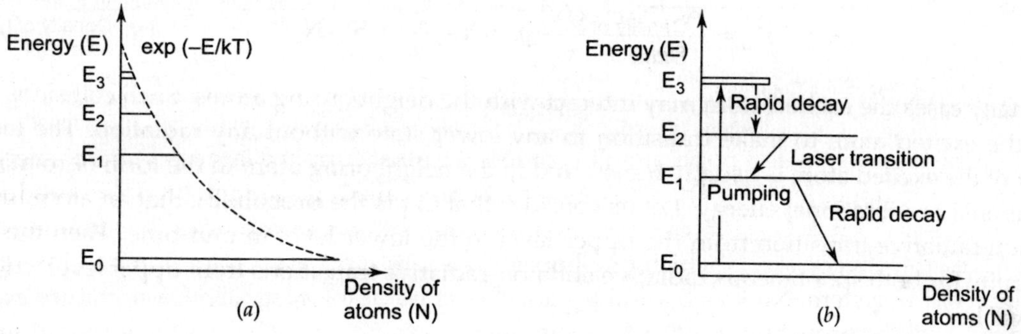

Fig. 3.9

If $(E_1 - E_0)$ is quite large compared to kT (*i.e.*, the thermal energy at the temperature of operation), then the populations of E_1, E_2 and E_3 are all very small in conditions of thermal equilibrium. Hence, if atoms are pumped from the ground state to the level E_3 from which they decay very rapidly to the metastable state E_2. However, since populations of E_3 and E_1 remain essentially unchanged a small increase in the number of atoms in energy level E_2 creates a population inversion, and lasing takes place between this level and E_1.

Again the upper level E_3 should preferably consist of a large number of levels for better pumping efficiency. If the lifetimes of transitions $E_3 \rightarrow E_2$ and $E_1 \rightarrow E_0$ are short, the population inversion between E_2 and E_1 can be maintained with moderate pumping and continuous laser action can be achieved more readily. The details of the mechanisms used for pumping lasers can be quite complex and, in addition to optical pumping, pumping can occur by electron bombardment or by an electrical discharge, the passage of a current, the release of chemical energy, etc. The energy level schemes of the media used in lasers are often complex, but they can usually be approximated by either three or four level schemes.

3.6 LASER RATE EQUATIONS

Since the variation of populations of different energy states play a major role in laser operation, in this section, we will study the rate at which the populations of different atomic states are changing in the presence of external pumping.

3.6.1 Two-Level System

Let us first consider a two level system having E_1 and E_2 energy levels with population N_1 and N_2 respectively. The system of the active particles is radiated by a wave of frequency ν with energy density ρ. According to equation (3.16), the number of particles per unit volume that will be excited to the upper state due to the absorption of this radiation will be:

$$\Gamma_{12} = \frac{A_{21}c^3}{8\pi hn^3}\frac{N_1 g_2}{g_1}\frac{g(\nu)}{\nu_3}\rho_\nu = \Psi_{12}\frac{N_1 g_2}{g_1} \tag{3.22}$$

where, $$\Psi_{12} = \frac{A_{21}c^3}{8\pi hn^3}\frac{g(\nu)}{\nu_3}\rho_\nu \tag{3.23}$$

Similarly the number of particles that will escape from upper state to lower state because of stimulated emission:

$$\Gamma_{21} = \frac{A_{21}N_2 c^3}{8\pi hn^3}\frac{g(\nu)}{\nu_3}\rho_\nu = \Psi_{21}N_2 = \Psi_{12}N_2 \tag{3.24}$$

In many cases the excited atom may interact with the neighbouring atoms. Such collisions may cause the excited atom to make transition to any lower state without any radiation. The loss of energy of the excited atom is usually transferred to the neighboring atom in the form of rotational, translational or vibrational energy. Let us consider that C_{21} is the probability that an atom undergoes non-radiative transition from the upper level to the lower level in unit time. Then the total probability for both spontaneous radiative and non-radiative transitions from upper level to lower level will be:

$$\Phi_{21} = A_{21} + C_{21} \tag{3.25}$$

Hence, the rate of change of population of both the levels may be written as:

$$\frac{dN_1}{dt} = -\Psi_{21}\left(N_1\frac{g_2}{g_1} - N_2\right) + \Phi_{21}N_2 \tag{3.26}$$

$$\frac{dN_2}{dt} = \Psi_{12}\left(N_1\frac{g_2}{g_1} - N_2\right) - \Phi_{21}N_2 \tag{3.27}$$

Now at non-thermal equilibrium steady state:

$$\frac{dN_1}{dt} = \frac{dN_2}{dt} = 0 \tag{3.28}$$

that is,

$$\frac{N_2}{N_1\dfrac{g_2}{g_1}} = \frac{\Psi_{12}}{\Psi_{12} + \Phi_{21}} \tag{3.29}$$

In such a situation the expression for the population difference will be:

$$\frac{N_2 - N_1\dfrac{g_2}{g_1}}{N_2 + N_1\dfrac{g_2}{g_1}} = \frac{\Phi_{21}}{2\Psi_{12} + \Phi_{21}} \tag{3.30}$$

The numerator of the above expression is the population difference and the denominator gives the total number of active particles, *i.e.*,

$$\Delta N = N_2 - N_1 \frac{g_2}{g_1} \quad (3.31)$$

and

$$N = N_2 + N_1 \frac{g_2}{g_1} \quad (3.32)$$

Substituting equations (3.31) and (3.32) in equation (3.30), we get:

$$\Delta N = -\frac{\Phi_{21}}{2\Psi_{12} + \Phi_{21}} N = -\frac{1}{1 + 2\Psi_{12} / \Phi_{21}} N \quad (3.33)$$

Since Φ_{21} and Ψ_{12} are positive quantities ΔN is always less than zero. Or it may be concluded that by external pumping alone it is not possible to obtain a steady state population inversion in a two level system.

If radiative transition dominates the non-radiative transitions from upper level to lower level $A_{21} >> C_{21}$ and $\Phi_{21} \approx A_{21}$. Substituting this and Ψ_{21} from equation (3.23), we get:

$$\Delta N = -\frac{1}{1 + \dfrac{c^2}{4\pi h n^2} \dfrac{g(\nu)}{\nu^3} I_\nu} N = -\frac{1}{1 + \dfrac{I_\nu}{I_{sat}} g(\nu)} N \quad (3.34)$$

where,

$$I_{sat} = \frac{4\pi h n^2}{c^2} \nu^3 \quad (3.35)$$

I_{sat} is called as *saturation intensity*.

3.6.2 Three-Level System

Let us consider a three-level system. E_1, E_2 and E_3 are the energy levels and their corresponding population densities are N_1, N_2 and N_3. Further we assume that the transitions will take place only between these three levels. If N represents the number of active particles per unit volume, then

$$N = N_1 + N_2 + N_3 \quad (3.36)$$

Now the rate of change in population density for the three levels can be written as:

$$\frac{dN_3}{dt} = \Omega(N_1 - N_3) - \Phi_{32} N_3 \quad (3.37a)$$

$$\frac{dN_2}{dt} = \Psi(N_1 - N_2) + \Phi_{32} N_3 - \Phi_{21} N_2 \quad (3.37b)$$

$$\frac{dN_1}{dt} = \Omega(N_3 - N_1) - \Psi(N_2 - N_1) + \Phi_{21} N_2 \quad (3.37c)$$

where, Ω stands for the rate of pumping for $1 \rightarrow 3$ transition and Φ_{ij} denotes rate of both radiative and non-radiative transition from i^{th} level to j^{th} level and Ψ represents transition rate for stimulated emission from level $2 \rightarrow 1$.

At steady state, rate of change of population density for all the states must be zero. Hence

$$\frac{dN_1}{dt} = \frac{dN_2}{dt} = \frac{dN_3}{dt} = 0 \quad (3.38)$$

From these three equations we get,

$$\frac{N_2 - N_1}{N} = \frac{\left[\Omega(\Phi_{32} - \Phi_{21}) - \Phi_{32}\Phi_{21}\right]}{\left[3\Omega\Psi + 2\Omega\Phi_{21} + 2\Phi_{32}\Psi + \Phi_{32}\Omega + \Phi_{32}\Phi_{21}\right]}$$

$$\frac{N_2 - N_1}{N} = \frac{(\Phi_{32} - \Phi_{21})\left[\Omega - \dfrac{\Phi_{32}\Phi_{21}}{(\Phi_{32} - \Phi_{21})}\right]}{\left[3\Omega\Psi + 2\Omega\Phi_{21} + 2\Phi_{32}\Psi + \Phi_{32}\Omega + \Phi_{32}\Phi_{21}\right]} \tag{3.39}$$

From the above relation, it may be concluded that to obtain population inversion between levels 2 and 1, that is for $(N_2 - N_1)$ to be positive, a necessary but not sufficient condition is $\Phi_{32} > \Phi_{21}$. If Ψ is very small which is true below the threshold of laser oscillation and if $\Phi_{32} >> \Phi_{21}$ equation (3.39) may be written as:

$$\frac{N_2 - N_1}{N} = \frac{\Omega - \Phi_{21}}{\Omega + \Phi_{21}} \tag{3.40}$$

As the laser starts oscillating, Ψ becomes large and that in turn reduces the population inversion $(N_2 - N_1)$. This obviously results in a reduced amplification. When the laser oscillates under steady state, the intensity of the output radiation increases to such a value that $(N_2 - N_1)$ remains same at the threshold. Similarly laser rate equations for four level system can also be found out. It is left as an exercise.

3.7 OPTICAL FEEDBACK

We know in an electronic oscillator, an amplifier tuned to a particular frequency is provided with positive feedback and, when switched on, any electrical signal of the appropriate frequency appearing at the input will be amplified. The amplified output is fed back to the input and amplified yet again and so on.

Light amplification occurs in a laser by oscillation. Hence it is necessary to obtain positive feedback in a laser to amplify the light. In a laser positive feedback may be obtained by placing the gain medium between a pair of mirrors which, in fact form an optical cavity. The optical cavity provides positive feedback of the photons by reflection at the mirrors at either end of the cavity. The structure is known as a Fabry-Perot resonator. The initial stimulus is provided by any spontaneous transitions between appropriate energy levels in which the emitted photon travels along the axis of the system. Although the amplification of the signal from a single pass through the medium is quite small, after multiple passes the net gain can be large. Furthermore, if one mirror is made partially transmitting, useful radiation may escape from the cavity. Saturation is reached when the gain provided by the medium exactly matches the loses incurred during a complete round trip. The major loses are resulted due to scattering and absorption in the amplifying medium, scattering, absorption and diffraction at the mirrors and non-useful transmission through the mirrors.

In practice a number of different mirror curvatures and configurations are used depending on the applications and the type of laser being used. A detailed analysis of the effects of different mirror systems requires a rigorous application of the diffraction theory, which is beyond the scope of this book. Using simple ray tracing techniques, however, it is quite easy to anticipate the results of such an analysis in that, mirror configurations which retain a ray of light, initially inclined at a small angle to the axis within the optical cavity after several reflections are likely to be useful. Such cavities are said to be stable.

Some commonly used mirror configurations for laser cavity are shown in Fig. 3.10. Each of them has various advantages and disadvantages. For example, the plane-parallel configuration is difficult to align. If the mirrors are not perfectly parallel (to within about 1 sec of arc) the optical beam will "walk-off" the mirrors after a few reflections. Again, the mirrors need to be flat within λ/100. But the advantage of such a configuration is that the radiation beam makes maximum use of the laser medium, as there is no focusing of the beam within the laser cavity. In contrast the confocal arrangement is relatively easy to align (to within about 1.5 minutes of arc). The disadvantage of such configuration is that the use of the active medium is restricted. Hence the mode volume is small. If maximum beam coherence is required (*i.e.*, uniphase operation) we generally use the hemispherical system, whereas if maximum power output is required (*e.g.*, gas lasers) large radius resonator is used.

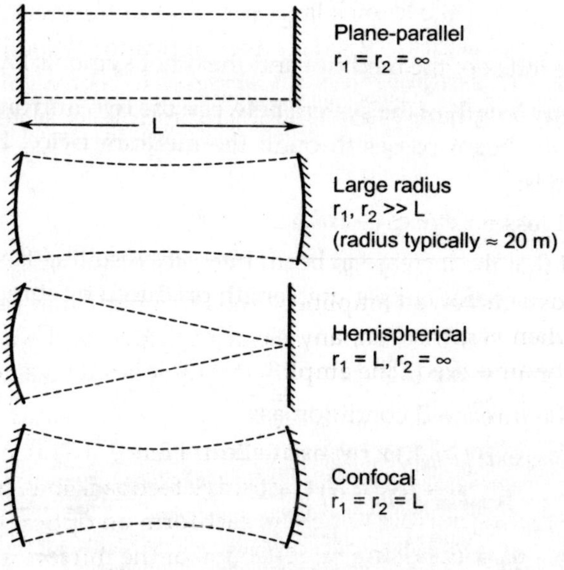

Fig. 3.10

In some cases the mirror configurations give rise to unstable cavities, where a ray which is initially traveling with a small angle of incidence to the cavity axis will diverge away from the axis after a number of reflections. Though such resonators have high loss, but they have some useful properties. They can efficiently use the mode volume. Since these unstable resonators have high losses, they may be used with high-gain media such as carbon dioxide.

3.8 THRESHOLD CONDITION FOR LASER OSCILLATION

The population inversion is the necessary condition for the laser oscillations but not a sufficient condition. The steady state conditions for the lasing to take place are achieved when the total gain in the amplifying medium exactly balances the total losses. This minimum value of the gain required to initiate and sustain the laser oscillations within the amplifying medium is known as the threshold gain. This applies to a CW laser which gives continuous output, rather than pulsed lasers where slightly different conditions exist.

The loss of the system is due to a number of different processes. Some of the most important ones are mentioned below:

1. Absorption and scattering at the mirrors.
2. Transmission at the mirrors—one of the mirrors is made as reflective as possible to minimize losses, whereas the other mirror is partially transmissive for the useful output to come out.
3. Diffraction losses at the mirrors.
4. Scattering at the optical inhomogeneities in the laser medium.
5. Absorption in the laser medium due to transitions other than the desired transitions.

To simplify the matter all losses except those due to transmissions through the mirrors may be included in a single loss coefficient per unit length α cm^{-1}. The value of α is given as:

$$\alpha = \left(\frac{g_2}{g_1} N_1 - N_2\right) \frac{B_{21} h\nu_{21} n}{c} \tag{3.41}$$

where n is the refractive index of the medium and the other symbols have their usual meanings.

We assume that the cavity length of the system between the two mirrors of reflectivities r_1 and r_2 is L. On each round trip the beam passes through the medium twice. Hence the fractional loss incurred by the light beam is:

$$\text{Fractional loss} = r_1 r_2 \exp(-2\,\alpha L) \tag{3.42}$$

Furthermore it is found that the increase in beam intensity resulting from stimulated emission is exponential. Hence if the gain coefficient per unit length produced by stimulated emission is γ cm^{-1}, the fractional round trip gain is given by:

$$\text{Fractional gain} = \exp(2\,\gamma L) \tag{3.43}$$

Hence we may write the threshold condition as:

$$\exp(2\gamma_{th} L) \times r_1 r_2 \exp(-2\alpha L) = 1$$

or,

$$r_1 r_2 \exp 2(\gamma_{th} - \alpha)\, L = 1$$

or,

$$\gamma_{th} = \alpha + \frac{1}{L} \ln \frac{1}{r_1 r_2} \tag{3.44}$$

If the reflectivities of the two mirrors are identical, we may write:

$$\gamma_{th} = \alpha + \frac{1}{L} \ln \frac{1}{r} \tag{3.44a}$$

The first term on the right hand side of the equation (3.44) represents the volume losses and the second term the transmission loss through the mirrors in the form of the useful output. If γ is high then it is relatively easy to achieve laser action, mirror alignment is not critical and dust can be tolerated on the mirrors. On the other hand, with low gain media, such losses cannot be tolerated and the mirrors must have high reflectance, and be carefully aligned without any dust.

It has to be noted that a laser with a high gain medium will not necessarily have a high efficiency. The efficiency is the output optical power to the input pumping power. Hence, it depends on how effectively the pumping power is utilized to produce population inversion, which in turn depends on the probabilities of different kinds of transitions from the upper level and on the losses in the system. However, it must be noted that the parameters displayed in equation (3.44) are totally dependent on the laser type.

Example 3.2. *For a GaAs laser, the reflectivities of the two facets are 0.3 for uncoated facets (i.e., 30% of the radiation is reflected at the facets) and $\alpha = 10\ cm^{-1}$. If the laser diode is 600 μm long, estimate the threshold gain for the laser.*

Solution:

We may obtain the threshold gain for the laser from equation (3.44 a) as:

$$\gamma_{th} = \alpha + \frac{1}{L} \ln \frac{1}{r}$$

$$= 10 + \frac{1}{600 \times 10^{-4}} \ln \frac{1}{0.3} = 30\ cm^{-1}$$

3.9 LINESHAPE FUNCTION

We have already mentioned that the radiation comes out of a collection of atoms when they make transitions between two energy levels. But they are never perfectly monochromatic, *i.e.,* they have a spectral width. This can be seen in both emission and absorption. If we were to measure the transmission as a function of frequency for the transition between the two energy states E_1 and E_2 we would obtain the bell-shape curve shown in Fig. 3.11(*a*). The emission curve would be the inverse of this, as illustrated in Fig. 3.11(*b*).

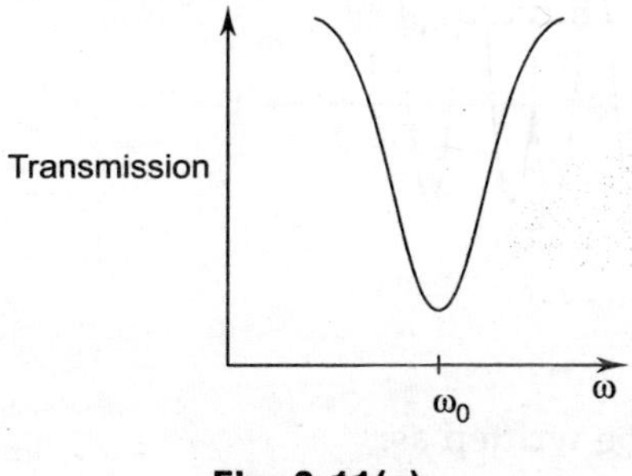

Fig. 3.11(*a*)

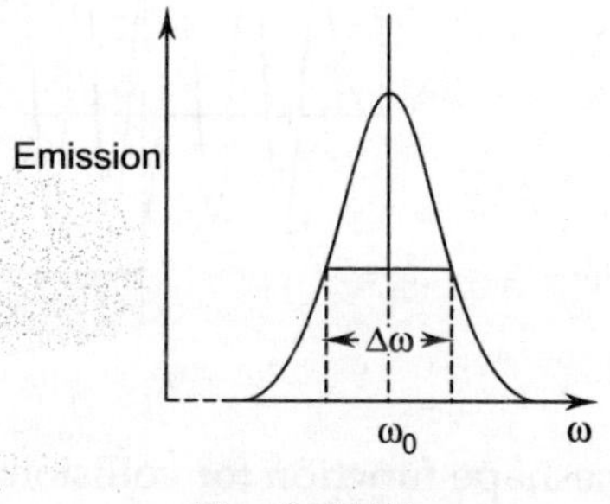

Fig. 3.11(*b*)

This line broadening is described in terms of the lineshape function $g(\omega)$. This can also be used to describe a frequency probability curve. Hence, we may define $g(\omega)d\omega$ as the probability that a given transition between the two energy levels will result in absorption (or emission) of a photon whose frequency lies between ω and $(\omega + d\omega)$. Hence, we see that a photon of energy $h\omega$ may not necessarily stimulate another photon of energy $h\omega$. Thus $g(\omega)\ d\omega$ is the probability that the stimulated photon will have the energy between $h\omega$ and $h(\omega + d\omega)$.

In this section we will discuss some important line broadening mechanisms and obtain the corresponding $g(\omega)$. A study of line broadening is extremely important since it determines the operational characteristics of the laser such as the number of oscillating modes, the threshold populational inversion etc. The various broadening mechanisms can be broadly classified as homogeneous or inhomogeneous broadening. In the case of homogeneous broadening, like collision (or pressure) and natural broadening, the mechanism acts to broaden the response of each atom in an identical manner and for such a case the probability of emission or absorption of radiation at a certain frequency is the same for all the atoms in the collection. Homogeneous broadening leads to a Lorentzian lineshape function. In the case of inhomogeneous broadening, like Doppler broadening, different groups of atoms are distinguished by different frequency responses or lineshapes for the same transition. Local variations of temperature, pressure, applied magnetic field as well as local variations due to crystal imperfections also lead to inhomogeneous broadening of the emission or absorption lineshapes. Inhomogeneous broadening leads to a Gaussian frequency distribution.

3.9.1 Collision Broadening

Random collision takes place between the atoms in a gas. If the atoms are very close the energy levels of the atoms change in such a collision, due to their mutual interaction. If T_c represents the time between collisions and ΔT_c the collision time, then the order of magnitudes of those quantities are 10^{-6} *s* and 10^{-13} *s* respectively. Thus the collision time is very small compared to the time between collisions and thus the collisions may be taken to be almost instantaneous. Since the collision time ΔT_c is random, the phase of the wave after the collision is arbitrary with respect to the phase before the collision. Thus each collision may be assumed to lead to random phase changes as shown in Fig. 3.12. Hence the wave shown in Fig. 3.12 is no longer monochromatic and this broadening is referred to as collision (or pressure) broadening. Clearly the higher the pressure (and temperature) of the gas the more frequently will the atoms suffer collision and the greater will be the spectral broadening.

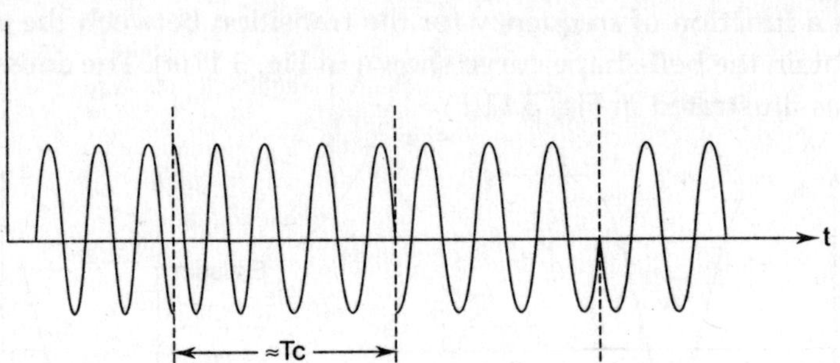

Fig. 3.12

The lineshape function for collision broadening can be written as:

$$g(\omega) = \left(\frac{\varepsilon_0}{\pi}\right)^2 \frac{1}{2} \frac{1}{(\omega - \omega_0)^2 + 1/T_0^2} \tag{3.45}$$

where ε_0 is the electric field with the wave at $t = 0$. The above functional form is a Lorentzian. We can show the normalized lineshape function for collision broadening to be:

$$g(\omega)\, d\omega = \frac{T_0}{\pi} \frac{1}{1 + (\omega - \omega_0)^2 T_0^2}\, d\omega \tag{3.46}$$

and the FWHM will be:

$$\Delta\omega_c = \frac{2}{T_0} \tag{3.47}$$

where T_0 represents the mean time between two collisions, ω_0 is the characteristic frequency. Thus a mean collision time of ~ 10^{-6} s corresponds to a $\Delta\nu$ of 0.3 MHz.

3.9.2 Natural Broadening

It has already been seen that an excited atom can emit its energy in the form of spontaneous emission. In order to investigate the spectral distribution of this spontaneous radiation, we recall that the rate of decrease of the number of atoms in level 2 due to transitions from level 2 to level 1 is given by equation (3.1)

$$\frac{dN_2}{dt} = -A_{21}N_2 \tag{3.48}$$

We may write the electric field associated with the spontaneous radiation as:

$$\varepsilon(t) = \varepsilon_0 \exp(i\omega t)\exp(-t/2t_{SP}) \tag{3.49}$$

where $t_{SP} = 1/A_{21}$ and we have used the relation that intensity is proportional to the square of the electric field. Thus the electric field associated with spontaneous emission decrease exponentially. The effect of this is similar to collision broadening in that it effectively shortens the wave trains and produces a broadened spectral line.

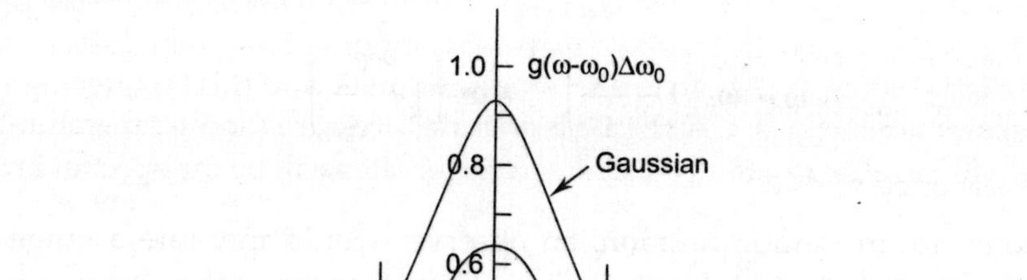

Fig. 3.13

The lineshape function associated with the spontaneously emitted radiation can be written as:

$$g(\omega) = K\frac{1}{(\omega-\omega_0)^2 + 1/4t_{SP}^2} \tag{3.50}$$

where K is a constant of proportionality which is determined such that $g(\omega)$ is normalized. The normalized lineshape function is written as:

$$g(\omega) = \frac{2t_{SP}}{\pi}\frac{1}{1+4(\omega-\omega_0)^2 t_{SP}^2} \tag{3.51}$$

The above functional form is again a Lorentzian and is plotted in Fig. 3.13. The FWHM of the Lorentzian is

$$\Delta\omega_N = \frac{1}{t_{SP}} \tag{3.52}$$

3.9.3 Doppler Broadening

Atoms move randomly in a gas. When a moving atom interacts with electromagnetic radiation, the apparent frequency of the wave is different from that seen from a stationary atom; this is called the Doppler effect and the broadening caused by this is termed Doppler broadening. The frequency as measured by the observer increases if the source and observer approach one another and decreases as they recede.

To obtain $g(\omega)$ for Doppler broadening, we consider a radiation of frequency ω passing through a collection of atoms which have a resonant frequency ω_0 and which move randomly. If the radiation is assumed to propagate along the x-direction, then the apparent frequency seen by the atom having a x-component of velocity v_x will be:

$$\tilde{\omega} = \omega\left(1 - \frac{v_x}{c}\right) \tag{3.53}$$

For a strong interaction, the frequency of the incident radiation must be such that $\tilde{\omega} = \omega_0$. Thus

$$\omega = \omega_0\left(1 - \frac{v_x}{c}\right)^{-1} \approx \omega_0\left(1 + \frac{v_x}{c}\right) \tag{3.54}$$

where we have assumed $v_x << c$.

Since the atoms are in random motion, an observer would measure a range of frequencies depending on the magnitude and direction of v_x. That is, as far as the observer is concerned the collection of atoms would be emitting at a range of different resonant frequencies resulting in a broadening of the emission lineshape. The individual Doppler shifted resonant frequencies contribute to the smooth Doppler broadened lineshape.

The lineshape function associated with Doppler broadening can be expressed as:

$$g(\omega)d\omega = \frac{c}{\omega_0}\left(\frac{M}{2\pi k_B T}\right)^{\frac{1}{2}} \exp\left[-\frac{Mc^2}{2k_B T}\frac{(\omega - \omega_0)^2}{\omega_0^2}\right] d\omega \tag{3.55}$$

where M is the mass of the atom, T is the absolute temperature of the gas and k_B is the Boltzmann's constant. Equation (3.55) corresponds to a Gaussian distribution. The lineshape function is peaked at ω_0, and the FWHM is given by:

$$\Delta\omega_D = 2\omega_0\left(\frac{2k_B T}{Mc^2}\ln 2\right)^{\frac{1}{2}} \tag{3.56}$$

Equation (3.55) may be written in terms of $\Delta\omega_D$ as:

$$g(\omega)d\omega = \frac{2}{\Delta\omega_D}\left(\frac{\ln 2}{\pi}\right)^{\frac{1}{2}} \exp\left[-4\ln 2\frac{(\omega - \omega_0)^2}{(\Delta\omega_D)^2}\right] d\omega \tag{3.57}$$

Figure 3.13 shows a comparative plot of a Lorentzian and a Gaussian line having the same FWHM. It can be seen from the plot, that the peak value of the Gaussian is more and that the Lorentzian has a much longer tail. Doppler broadening is the predominant mechanism in most gas lasers emitting in the visible region.

3.10 LASER MODES

If we examine a laser output with a spectrometer of very high resolving power, we will find that it consists of a number of discrete frequency components. To understand how these discrete lines are formed and how they are related to the laser transition lineshape we need to examine the effects of the mirrors on the light within the laser cavity.

3.10.1 Axial Modes

The operating modes of a physical system are formed by physical laws and boundary conditions. Consider a guitar string that is plucked. The physical law is the restoring force that tends to bring the string to its initial, straight position. The boundary conditions are that on either end of the string, the amplitude of the vibration must be zero. The superimposed waveforms that form the mode move with a fixed velocity is dependant on the properties of the string, and have wavelengths equal to $2L/p$, where L is the length of the string, and p is a positive integer. These are the *modes* of the system.

Similarly since the Fabry-Perot cavity forms a resonant cavity, when sufficient population inversion exists in the amplifying medium the radiation builds up and becomes established as standing waves between the mirrors. These standing waves exist only at frequencies for which the distance between the mirrors is an integral multiple of half wavelengths. Thus when optical spacing between the mirrors is L the resonance condition along the axis of the cavity is given by:

$$p\frac{\lambda}{2n} = L \tag{3.58}$$

where p is an integer, n is the refractive index of the amplifying medium and λ is the emission wavelength. Alternatively, discrete emission frequencies ν are defined by:

$$\nu = \frac{pc}{2Ln} \tag{3.59}$$

Generally p has a very large value ($\approx 10^6$). As p has such a large value, many different values of p are possible for only a small change in wavelength. Each value of p which satisfies equation (3.59) defines a *mode* of the cavity. Since equations (3.58) and (3.59) apply for the case when L is along the longitudinal axis of the cavity, the frequencies given by equation (3.59) are known as *longitudinal* or *axial modes*. Furthermore, from equation (3.59) it may be observed that these modes are separated by a frequency interval $\delta\nu$ given by:

$$\delta\nu = \frac{c}{2nL} \tag{3.60}$$

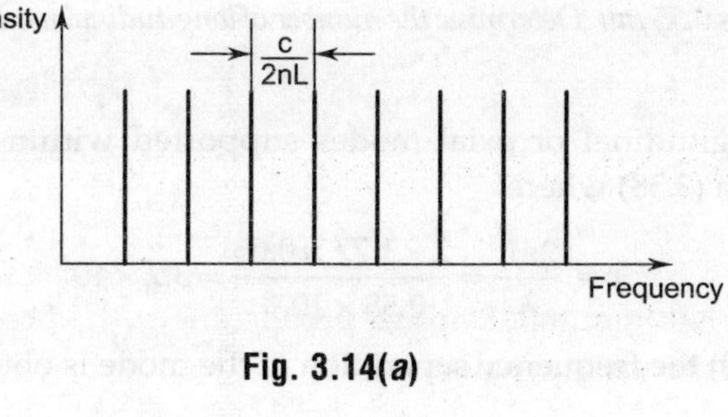

Fig. 3.14(*a*)

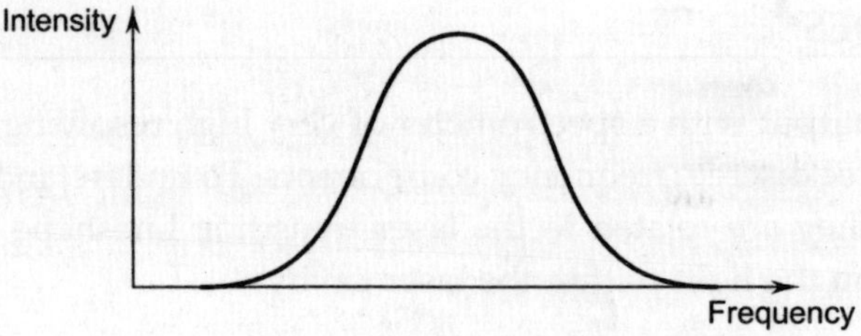

Fig. 3.14(*b*)

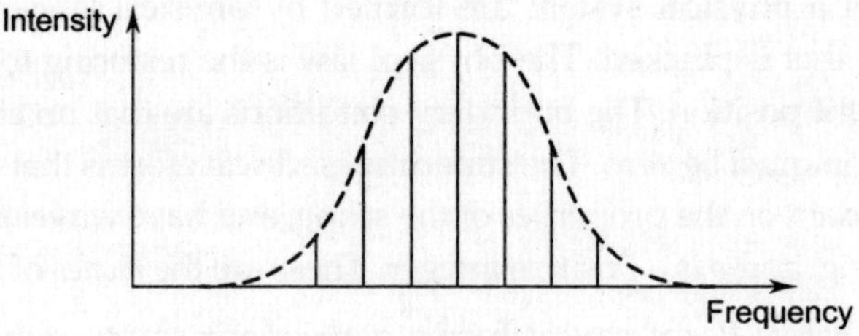

Fig. 3.14(*c*)

The mode separation in terms of the free space wavelength, assuming $\delta\nu << \nu$ and $\nu = c/\lambda$, is given by:

$$\delta\lambda = \frac{\lambda\delta\nu}{\nu} = \frac{\lambda^2}{c}\delta\nu \qquad (3.61)$$

Hence substituting $\delta\nu$ from equation (3.60) gives:

$$\delta\lambda = \frac{\lambda^2}{2nL} \qquad (3.62)$$

In addition it is to be noted that equation (3.61) can be used to determine the device spectral linewidth as a function of wavelength when it is quoted in Hertz, or vice versa.

As equation (3.60) is independent of *p*, the frequency separation of adjacent modes is the same irrespective of their actual frequencies. The modes of oscillation of the laser cavity will therefore consist of a large number of frequencies, given by equation (3.59) and separated by $c/2n$L, as shown in Fig. 3.14 (*a*). Although the figure shown in 3.14 (*a*) indicates that a large number of modes may be generated within the laser cavity, the spectral output from the device is defined by the gain curve as shown in Fig. 3.14 (*b*). This situation is shown in Fig. 3.14 (*c*). The axial modes all contribute to a single 'spot' of light in the laser output, whereas the transverse modes may give rise to a pattern of spots in the output.

Example 3.3. *A ruby laser contains a crystal length 5 cm with a refractive index of 1.77. The peak emission wavelength from the laser is 0.55 μm. Determine the number of longitudinal modes and their frequency separation.*

Solution:

The number of longitudinal or axial modes supported within the laser structure may be obtained from equation (3.58) where:

$$p = \frac{2nL}{\lambda} = \frac{2 \times 1.77 \times 0.05}{0.55 \times 10^{-6}} = 3.2 \times 10^5$$

Using equation (3.60) the frequency separation of the mode is obtained as:

$$\delta\nu = \frac{3\times 10^8}{2\times 1.77\times 0.05} = 1.7 \text{ GHz}$$

Example 3.4. *Estimate the spectral output (number of modes) of an AlGaAs laser supported by the gain spectrum which has a bandwidth of 7 nm. The laser has a cavity length of 230 μm and the emission wavelength is 820 nm. The refractive index of the material is 3.3.*

Solution:

The number of longitudinal modes supported within the laser structure may be obtained from equation (3.58) as:

$$p = \frac{2nL}{\lambda} = \frac{2\times 3.3\times 230\times 10^{-6}}{820\times 10^{-9}} = 1851$$

However all these modes are not within the gain curve. Using equation (3.60) the frequency separation of the mode is obtained as:

$$\delta\nu = \frac{3\times 10^8}{2\times 3.3\times 230\times 10^{-6}} = 197.6 \text{ GHz}$$

The corresponding wavelength separation is given by equation (3.61) as:

$$\delta\lambda = \frac{\lambda^2}{c}\delta\nu = \frac{\left(820\times 10^{-9}\right)^2 \times 197.6\times 10^9}{3\times 10^8} = 0.44 \text{ nm}.$$

Thus, the number of modes supported by the gain spectrum $= \frac{7}{0.44} \cong 16$.

If the mode spacing can be increased, say by reducing the cavity length, then the number of modes will decrease and the laser will come closer to being a single-mode device.

Example 3.5. *Determine the length of the optical cavity and the number of longitudinal modes emitted by an AlGaAs laser emitting at wavelength of 0.95 μm. The modes are separated by 200 GHz. The refractive index of AlGaAs is 3.3.*

Solution:

From equation (3.60) we have:

$$L = \frac{c}{2n\delta\nu} = \frac{3\times 10^8}{2\times 3.3\times 200\times 10^9}$$

$$= 227.3 \text{ μm}.$$

The number of modes p is given by equation (3.58) as:

$$p = \frac{2nL}{\lambda} = \frac{2\times 3.3\times 227.3}{0.95} \cong 1579$$

Of these longitudinal modes, only those within the spontaneous emission spectrum will actually participate in lasing.

3.10.2 Transverse Modes

Axial modes are formed by plane waves traveling axially along the laser cavity on a line joining the centers of the mirrors. For any real laser cavity there will probably be waves traveling just off-axis that are able to replicate themselves after covering a closed path as illustrated in Fig. 3.15. These will also give rise to resonant modes. But since these modes have the components of their electromagnetic fields which are transverse to the direction of propagation and they are termed as transverse electromagnetic (or TEM) modes.

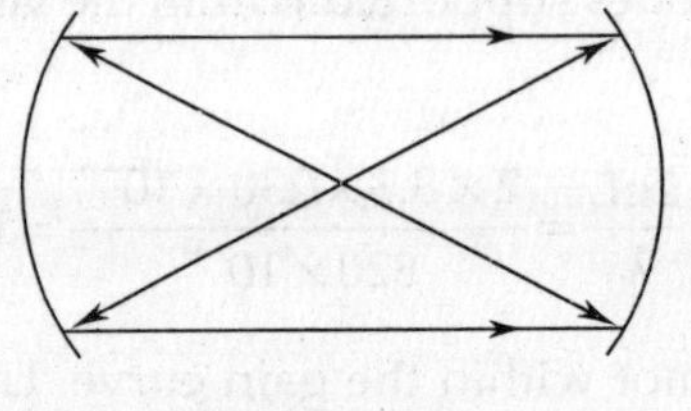

Fig. 3.15

The transverse modes determine the intensity distributions on the cross-sections of the beam. In a laser with cylindrical symmetry, the transverse mode patterns are described by a combination of a Gaussian beam profile with a Laguerre polynomial. The simplest mode is the Gaussian mode, which has complex amplitude. It appears as a central bright dot, as illustrated in Fig. 3.16 (*a*). Gaussian beams are usually the preferred output of most lasers, since they are easy to manipulate, are circularly symmetric, and usually have the greatest overall and concentrated intensity of all the transverse modes. They are stable as well, which means they retain their shape as they propagate.

The Gaussian mode is a specific case of the more generalized **Hermite-Gaussian** (HG) modes. The HG modes are also referred to as **Transverse Electro-Magnetic**, or **TEM**. A TEM mode is described as TEM_{mn}, where m and n are the **indices** of the mode. Here *m* refers to the number of intensity minima in the direction of the electric field oscillation, and *n* refers to the number of minima in the direction of the magnetic field oscillation. An HG mode appears to be a grid of dots, or blobs. In Fig. 3.16 (*b*), a CCD image of an HG_{02} mode is shown. The Gaussian mode is a TEM_{00}, or HG_{00} mode.

Laguerre-Gaussian modes or LG modes, like the Gaussian mode, are circularly symmetric. However, all LG modes except LG_{00} are hollow. Their key feature is the presence of a screw phase dislocation, which means that it has orbital angular momentum. One cool application of this is the transfer of this momentum to a particle, making it spin. This screw phase dislocation is also the origin of the hollow center of an LG beam, since that type of phase dislocation appears as a dark spot. Figure 3.16 (*c*) illustrates such a mode.

Fig. 3.16(*a*)

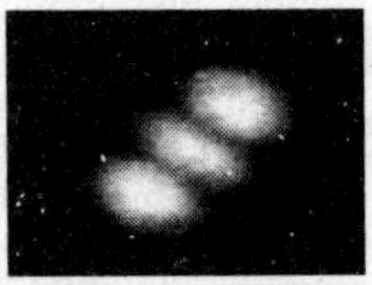

Fig. 3.16(*b*)

Fig. 3.16(*c*)

The intensity at a point r,φ (in polar coordinates) from the centre of the mode is given by:

$$I_{pl}(r,\varphi) = I_0 \rho^l \left[L_p^l(\rho)\right]^2 \cos^2(l\varphi) e^{-\rho} \tag{3.63}$$

where $\rho = 2r^2/\omega^2$, and $L_p{}^l$ is the associate Laguerre polynomial of order p and index l. ω is the spot size of the mode corresponding to the Gaussian beam radius.

The overall size of the mode is determined by the Gaussian beam radius W, and this may increase or decrease with the propagation of the beam, however the modes preserve their general shape during propagation. Higher order modes are relatively larger compared to the TEM_{00} mode, and thus the fundamental Gaussian mode of a laser may be selected by placing an appropriately sized aperture in the laser cavity.

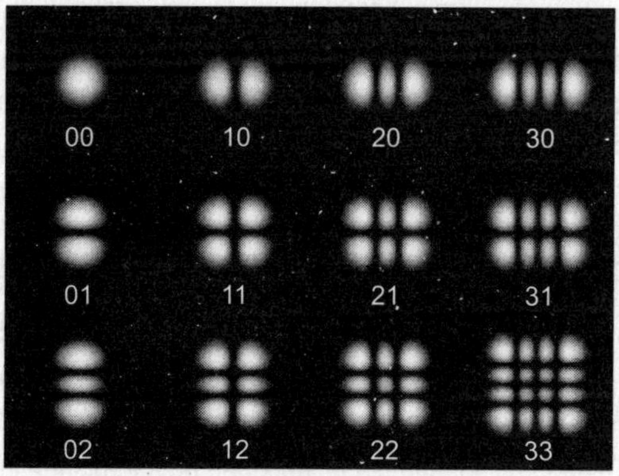

Fig. 3.17

In many lasers, the symmetry of the optical resonator is restricted by polarizing elements such as Brewster's angle windows. In these lasers, transverse modes with rectangular symmetry are formed. These modes are designated TEM_{mn} with m and n being the horizontal and vertical orders of the pattern. The intensity at point x,y is given by:

$$I_{mn}(x,y) = I_0 \left[H_m\left(\frac{\sqrt{2}x}{W}\right)\exp\left(\frac{-x^2}{W^2}\right)\right]^2 \left[H_n\left(\frac{\sqrt{2}y}{W}\right)\exp\left(\frac{-y^2}{W^2}\right)\right]^2 \tag{3.64}$$

where $H_m(x)$ is the mth order Hermite polynomial. Rectangular transverse mode patterns TEM(mn) is shown at Fig. 3.17.

The TEM_{00} mode corresponds to exactly the same fundamental mode as in the cylindrical geometry. Modes with increasing m and n show lobes appearing in the horizontal and vertical directions, with in general $(m+1)\,(n+1)$ lobes present in the pattern. Higher-order modes have a larger spatial extent than the 00 mode.

The phase of each lobe of a TEM_{mn} is offset by π radians with respect to its horizontal or vertical neighbours. This is equivalent to the polarization of each lobe being flipped in direction.

The overall intensity profile of a laser's output may be made up from the superposition of any of the allowed transverse modes of the laser's cavity, though often it is desirable to operate only on the fundamental mode, *i.e.*, TEM_{00} mode. The greatest degree of coherence, together with the highest level of spectral purity, may be obtained from a laser which operates in only TEM_{00} mode. Higher order transverse modes only occur when the width of the cavity is sufficient for them to oscillate. Consequently, they may be eliminated by narrowing the laser cavity suitably.

3.10.3 Multimodes

While proper misalignment of a laser cavity (with a Brewster window) can produce certain HG modes by geometric selection of portions of the gain material, there is a much easier way to control this for use in an astigmatic mode converter. In general, there is competition between different transverse and longitudinal modes for use of the gain material in a laser cavity. Therefore, in general, there is more than one simultaneously oscillating transverse mode. The superposition of all these simultaneous modes is called a multimode.

By selectively disabling parts of the gain medium, specific diagonal HG modes can be selected. This disabling is achieved by mounting a thin wire or human hair (about 25 microns across) at 45 degrees to the axis of the cylindrical lenses, and placing it on a translation stage between the Brewster window and the output coupler of the laser cavity. By translating it across the beam, an entire multimode, a smaller multimode, a single diagonal mode, or no mode at all can be selected from a single initial multimode. Intuitively, it would seem that a multimode with the greatest power output would contain the most single modes.

3.11 MODE LOCKING

Mode locking is a technique in optics by which a laser can be made to produce pulses of light of extremely short duration, on the order of picoseconds (10^{-12}s) or femtoseconds (10^{-15}s) or attoseconds (10^{-18}s). The basis of the technique is to induce a fixed phase relationship between the modes of the laser's resonant cavity. The laser is then said to be *phase-locked* or *mode-locked*. Interference between these modes causes the laser light to be produced as a train of pulses. Depending on the properties of the laser, these pulses may be of extremely brief duration, as short as a few femtoseconds.

In a simple laser, each of these modes will oscillate independently, with no fixed relationship between each other, in essence like a set of independent lasers all emitting light at slightly different frequencies. The individual phase of the light waves in each mode is not fixed, and may vary randomly due to such things as thermal changes in materials of the laser. The output of such a laser depends on the relative phases, frequencies and amplitudes of the modes. The total electric field as a function of time may be written as:

$$E(t) = \sum_{n=0}^{N-1} (E_0)_n \exp\{i(\omega_n t + \delta_n)\} \tag{3.65}$$

where $(E_0)_n$, ω_n and δ_n are the amplitude, angular frequency and phase of the nth mode. Usually all these parameters are time varying, so that the modes are incoherent and the total irradiance is simply the sum of the irradiances of the individual modes. Hence, we may write:

$$I = NE_0^2 \tag{3.66}$$

where for simplicity we have assumed that all N modes have the same amplitude E_0. The irradiance may exhibit small fluctuations if two or three modes happen to be in phase at any given time.

In lasers with only a few oscillating modes, interference between the modes can cause beating effects in the laser output, leading to random fluctuations in intensity; in lasers with many thousands of modes, these interference effects tend to average to a near-constant output intensity, and the laser operation is known as a *c.w.* or *continuous wave*.

If instead of oscillating independently, each mode operates with a fixed phase between it and the other modes, the laser output behaves quite differently. Instead of a random or constant output intensity, the modes of the laser will all constructively interfere with one another periodically, producing an intense burst or pulse of light. Such a laser is said to be *mode-locked* or *phase-locked*. The total irradiance must now be found by adding the individual electric fields rather than the irradiances. Using equation (3.65), the resultant electric field can now be written as:

$$E(t) = E_0 \exp(i\delta) \sum_{n=0}^{N-1} \exp(i\omega_n t) \tag{3.67}$$

where we forced the various modes to maintain the same relative phase δ to one another, *i.e.*, $\delta_n = \delta$.

For convenience, let us write the angular frequency ω_n as $\omega n = \omega - n\delta\omega$, where ω is the angular frequency of the highest frequency mode and δω is the angular frequncy separation between modes, which may be written as

$$\delta\omega = \pi \frac{c}{L} \tag{3.68}$$

Hence, these pulses occur separated in time by $\tau = 2L/c$, which is the time taken for the light to make exactly one round trip of the laser cavity. Equation (3.67) for E(*t*) can then be written as:

$$E(t) = E_0 \exp(i\delta) \sum_{n=0}^{N-1} \exp\{i(\omega - n\delta\omega)t\}$$

$$= E_0 \exp\{i(\omega t + \delta)\} \sum_{n=0}^{N-1} \exp\left(-\frac{i\pi cnt}{L}\right)$$

or

$$E(t) = E_0 \exp\{i(\omega t + \delta)\}\,(1 + \exp(-i\psi) + \exp(-i2\psi) + K + \exp\{-i(N-1)\psi) \tag{3.69}$$

where $\psi = \dfrac{\pi ct}{L}$. From equation (3.69) we may write:

$$E(t) = E_0 \exp\{i(\omega t + \delta)\} \frac{\sin(N\psi/2)}{\sin(\psi/2)} \tag{3.70}$$

The irradiance I is then $I = E(t) . E^*(t)$

or

$$I(t) = E_0^2 \frac{\sin^2(N\psi/2)}{\sin^2(\psi/2)} \tag{3.71}$$

Thus we see that the irradiance I(*t*) is periodic ($\Delta\Psi = 2\pi$) in the time interval $\tau = 2L/c$. The maximum value of the irradiance is $N^2E_0^2$. This occurs for $\Psi = 0$ or $p\pi$, *p* being an integer, where the value of the fraction $\left[\dfrac{\sin^2(N\psi/2)}{\sin^2(\psi/2)}\right]$ equals N^2. Similarly, the irradiance has a minimum value of zero when $N\Psi/2 = p\pi$, *p* being an integer which is not zero, *i.e.*, when $\Psi = 2p\pi/N$ or, $t = (1/N)(2L/c)\,p$. Thus the time duration of the maxima, which is the time taken for the irradiance to fall from the maximum value to the adjacent zero ($p = 1$) is $(1/N)(2L/c)$. Hence, we find that the output of a mode locked laser consists of a sequence of short pulses separated in time by $\tau = 2L/c$, each of a peak power equal to N times the average power.

The duration of each pulse of light is determined by the number of modes which are oscillating in phase (in a real laser, it is not necessarily true that *all* of the laser's modes will be phase-locked). If there are N modes locked with a frequency separation $\Delta\nu$, the overall modelocked bandwidth is $N\,\Delta\nu$, and the wider this bandwidth, the shorter the pulse duration from the laser. In practice, the actual pulse duration is determined by the shape of each pulse, which is in turn determined by the exact amplitude and phase relationship of each longitudinal mode. For example, for a laser producing pulses with a Gaussian temporal shape, the minimum possible pulse duration Δt is given by

$$\Delta t = \frac{0.44}{N\Delta\nu} \tag{3.72}$$

The value 0.44 is known as the *time-bandwidth product* of the pulse, and varies depending on the pulse shape. For ultrashort pulse lasers, a hyperbolic-secant-squared (sech^2) pulse shape is often assumed, giving a time-bandwidth product of 0.315.

Using this equation, we can calculate the minimum pulse duration which can be produced by a laser. For the HeNe laser with a 1.5 GHz bandwidth, the shortest Gaussian pulse which can be produced would be around 300 picoseconds; for the 128 THz bandwidth Ti:sapphire laser, this duration would be only 3.4 femtoseconds. These values represent the shortest possible Gaussian pulses supported by the laser's bandwidth; in a real mode-locked laser, the actual pulse duration depends on many other factors, such as the actual pulse shape, and the overall dispersion of the cavity.

3.11.1 Mode Locking Methods

Methods for producing mode locking in a laser may be classified as either *active* or *passive*. Active methods typically involve using an external signal to induce a modulation of the intra-cavity light. Passive methods do not use an external signal, but rely on placing some element into the laser cavity which causes self-modulation of the light.

Active Mode Locking

Mode locking is achieved by forcing the longitudinal modes to maintain fixed phase relationships. This can be accomplished by modulating the gain (or loss) of the laser cavity at a frequency equal to the intermode frequency separation $\delta\nu = c/2L$. Let us imagine that the loss modulation is provided by a shutter placed near one of the mirrors. The shutter is closed most of the time, which corresponds to very high losses. This shutter is only opened very briefly every 2L/c seconds, which corresponds to the cavity round trip time of the wave packet. If the wave packet is exactly as long in time, as the shutter stays open, and if it arrives exactly when the shutter is open, it will be unaffected by the presence of the shutter. Any parts of the wave packet, which arrive after the shutter closes or before the shutter opens is eliminated. Thus the phase relationships of the oscillating modes are continuously restored by the periodic operation of the shutter.

The most common active modelocking technique places a standing wave acousto-optic modulator into the laser cavity. When driven with an electrical signal, this produces a sinusoidal amplitude modulation of the light in the cavity. Considering this in the frequency domain, if a mode has optical frequency ν, and is amplitude modulated at a frequency f, the resulting signal has sidebands at optical frequencies $(\nu - f)$ and $(\nu + f)$. If the modulator is driven at the same frequency as the cavity mode spacing Δν, then these sidebands correspond to the two cavity modes adjacent to the original mode. Since the sidebands are driven in-phase, the central mode and the adjacent modes will be phase-locked together. Further operation of the modulator on the sidebands produces phase-locking of the $(\nu - 2f)$ and $(\nu + 2f)$ modes, and so on until all modes in the gain bandwidth are locked.

This process can also be considered in the time domain. The amplitude modulator acts as a weak *shutter* to the light bouncing between the mirrors of the cavity, attenuating the light when it is "closed", and letting it through when it is "open". If the modulation rate f is synchronised to the cavity round-trip time τ, then a single pulse of light will bounce back and forth in the cavity. The actual strength of the modulation does not have to be large; a modulator that attenuates 1% of the light when "closed" will mode-lock a laser, since the same part of the light is repeatedly attenuated as it traverses the cavity.

Related to this Amplitude Modulation (AM) active modelocking is Frequency Modulation (FM) modelocking, which uses a modulator device based on the electro-optic effect. This device, when placed in a laser cavity and driven with an electrical signal, induces a small, sinusoidally varying frequency shift in the light passing through it. If the frequency of modulation is matched to the round-trip time of the cavity, then some light in the cavity sees repeated up-shifts in frequency, and some repeated down-shifts. After many repetitions, the up-shifted and down-shifted light is swept out of the gain bandwidth of the laser. The only light which is unaffected is that which passes through the modulator when the induced frequency shift is zero, which forms a narrow pulse of light.

The third method of active modelocking is synchronous modelocking, or synchronous pumping. In this, the pump source (energy source) for the laser is itself modulated, effectively turning the laser on and off to produce pulses. Typically, the pump source is itself another mode-locked laser. This technique requires accurately matching the cavity lengths of the pump laser and the driven laser.

Passive Mode Locking

Passive mode locking techniques are those that do not require a signal external to the laser (such as the driving signal of a modulator) to produce pulses. Rather, they use the light in the cavity to cause a change in some intracavity element, which will then itself produce a change in the intracavity light. The most common type of device which will do this is a saturable absorber.

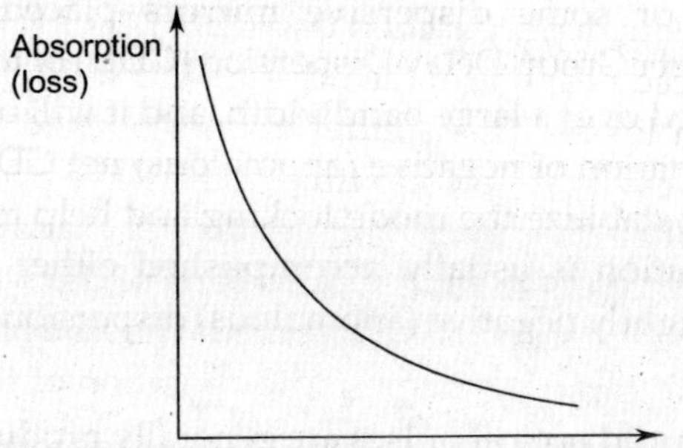

Fig. 3.18

A saturable absorber is an optical device that exhibits an intensity-dependent transmission. What this means is that the device behaves differently depending on the intensity of the light passing through it. For passive modelocking, ideally a saturable absorber will selectively absorb low-intensity light, and transmit light which is of sufficiently high intensity, as shown in Fig. 3.18.

When placed in a laser cavity, a saturable absorber will attenuate low-intensity constant wave light. However, because of the somewhat random intensity fluctuations experienced by an un-modelocked laser, any random, intense spike will be transmitted preferentially by the saturable absorber. As the light in the cavity oscillates, this process repeats, leading to the selective amplification of the high-intensity spikes, and the absorption of the low intensity light. After many round-trips, this leads to a train of pulses and mode locking of the laser.

Considering this in the frequency domain, if a mode has optical frequency ν, and is amplitude modulated at a frequency $n f$, the resulting signal has sidebands at optical frequencies $(\nu - n f)$ and $(\nu + n f)$ and enables much stronger modelocking for shorter pulses and more stability than active modelocking, but has startup problems.

Saturable absorbers are commonly liquid organic dyes, but they can also be made from doped crystals and semiconductors. Semiconductor absorbers tend to exhibit very fast response times (~100 *fs*), which is one of the factors that determines the final duration of the pulses in a passively mode-locked laser. In a *colliding pulse mode locked laser* the absorber steepens the leading edge while the lasing medium steepens the trailing edge of the pulse.

There are also passive mode locking schemes that do not rely on materials that directly display an intensity dependent absorption. In these methods, nonlinear optical effects in intra-cavity components are used to provide a method of selectively amplifying high-intensity light in the cavity, and attenuation of low-intensity light. One of the most successful schemes is called **Kerr-lens mode locking (KLM)**. This uses a nonlinear optical process, the optical Kerr effect, which results in high-intensity light being focussed differently than low-intensity light. By careful arrangement of an aperture in the laser cavity, this effect can be exploited to produce the equivalent of an ultra-fast response time saturable absorber.

Practical Mode Locked Lasers

In practice, a number of design considerations affect the performance of a modelocked laser. The most important are the overall dispersion of the laser's optical resonator, which can be controlled with a prism compressor or some dispersive mirrors placed in the cavity, and optical nonlinearities. For excessive net Group Delay Dispersion (GDD) of the laser cavity, the phase of the cavity modes cannot be locked over a large bandwidth, and it will be difficult to obtain very short pulses. For a suitable combination of negative (anomalous) net GDD with the Kerr nonlinearity, soliton like interactions may stabilize the mode locking and help to generate shorter pulses. The shortest possible pulse duration is usually accomplished either for zero dispersion (without nonlinearities) or for some slightly negative (anomalous) dispersion (exploiting the soliton mechanism).

The shortest directly produced optical pulses are generally produced by Kerr-lens mode locked Ti-sapphire lasers, and are around 5 femtoseconds long. Some advanced techniques (involving high harmonic generation with amplified femtosecond laser pulses) can be used to produce optical features with durations as short as 100 attoseconds. Other achievements, important particularly for laser applications, concern the development of modelocked lasers which can be pumped with laser diodes, can generate very high average output powers (tens of watts) in sub-picosecond pulses, or generate pulse trains with extremely high repetition rates of many GHz.

Pulse durations less than approximately 100 fs are too short to be directly measured using optoelectronic techniques *i.e.*, photodiodes, and so indirect methods such as autocorrelation or Frequency-resolved optical grating are used.

Modelocking in Nd-lasers either uses the established techniques of Ti:Sa lasers, or use an etalon to select a single 1 nm wide line and a Gires-Tournois-Interferometer to compensate the GVD. Alternatively a bulky prism compressor compensates the GVD for all Nd lines. Active modelocking is fast enough for this laser material (not for Nd:glass), but the round-trip frequency of the cavity must be adjusted with 1 in e^6 precision onto the AOM standing wave frequency. For this the diffracted light from the AOM is detected by a photo diode, and minimized. Depending on the sign of the detuning the relative phase between the diffracted signal and the reflection on the front side of the AOM changes and can be detected by an interference of the two on a second photodiode.

Applications

The short pulses generated by mode locking can be used to probe the process of the reaction at a very high temporal resolution, allowing the detection of short-lived intermediate molecules. This method is particularly useful in biochemistry, where it is used to analyse details of protein folding and function. The same can be used in nuclear fusion (inertial confinement fusion).

3.12 Q-FACTOR

Since the mirrors of a laser system form an open resonator cavity there are always losses associated with any mode. The main types of loss mechanisms are the finite reflectivities of the mirrors, scattering and absorption in the medium filling the resonator cavity and the diffraction spill-over when the field undergoes reflection from the mirrors. This dissipation in energy is described in terms of the quality factor Q of the mode, which may be defined as

$$Q = \frac{\omega_0 \times \text{energy stored in the mode}}{\text{energy dissipated per second in that mode}} \tag{3.73}$$

where $\omega_0 = 2\pi\nu_0$, corresponds to the oscillating frequency of a mode. Thus if W(t) represents the energy stored in the mode at time t, then from the definition of Q.

$$\frac{dW}{dt} = -\frac{\omega_0}{Q} W \tag{3.74}$$

The negative sign indicates that there is usually a loss in energy. We can integrate equation (3.74) to obtain

$$W(t) = W_0 \exp(-\omega_0 t/Q) \tag{3.75}$$

where $W_0 = W$ at $t = 0$. Equation (3.75) corresponds to an exponential decay of energy with time. It follows from equation (3.75) that the energy deacys to $1/e$ times of its value at $t = 0$ in a time

$$t_c = \frac{Q}{\omega_0} = \frac{Q}{2\pi\nu_0} \tag{3.76}$$

which is also referred to as the passive cavity lifetime.

In order to calculate the Q of a passive resonator, we first find the energy left in the cavity after one complete cycle of oscillation. Let W_0 be the total energy contained within the cavity at $t = 0$. One complete cycle of oscillation in the cavity corresponds to a pair of reflections from mirrors M_1 and M_2 (with power reflection coefficients R_1 and R_2 respectively) and two traversals through the laser medium, which is assumed to have a net power absorption coefficient α_c per unit length. Thus the energy remaining within the cavity after one complete cycle would be

$$W(t) = W_0 R_1 R_2 \exp(-2a_c L) \tag{3.77}$$

where L is the length of the cavity. Also, one complete cycle correspond to

$$t = \frac{2n_0 L}{c} \tag{3.78}$$

where n_0 is the refractive index of the medium inside the cavity. Thus substituting the value of t from equation (3.78) in equation (3.75), we obtain for the energy inside the cavity

$$W(t) = W_0 \exp\left(-\frac{2\pi v_0}{Q}\frac{2n_0 L}{c}\right) \tag{3.79}$$

Comparing equations (3.77) and (3.79), we obtain

$$2a_c L - \ln R_1 R_2 = \frac{4\pi v_0 n_0 L}{Qc}$$

or,

$$Q = \frac{4\pi v_0 n_0 L}{c}\frac{1}{2\alpha_c L - \ln R_1 R_2} \tag{3.80}$$

3.13 Q-SWITCHING

Q-switching, sometimes known as **giant pulse formation**, is a technique by which a laser can be made to produce a pulsed output beam. The technique allows the production of light pulses with extremely high (gigawatt) peak power, much higher than would be produced by the same laser if it were operating in a continuous wave (constant output) mode. Compared to modelocking, another technique for pulse generation with lasers, Q-switching leads to much lower pulse repetition rates, much higher pulse energies, and much longer pulse durations. Both techniques are sometimes applied at once.

Q-switching was first proposed in 1958 by Gordon Gould, and independently discovered and demonstrated in 1961 or 1962 by R.W. Hellwarth and F.J. McClung using electrically switched Kerr cell shutters in a ruby laser.

3.13.1 Principle of Q-Switching

Q-switching is achieved by putting some type of variable attenuator inside the laser's optical resonator. When the attenuator is functioning, light which leaves the gain medium does not return, and lasing cannot begin. This attenuation inside the cavity corresponds to a decrease in the ***Q-factor*** or ***quality factor*** of the optical resonator. A high Q-factor corresponds to low resonator losses per roundtrip, and vice versa. The variable attenuator is commonly called a "Q-switch", when used for this purpose.

Initially the laser medium is pumped while the Q-switch is set to prevent feedback of light into the gain medium (producing an optical resonator with low Q). This produces a population inversion, but laser operation cannot yet occur since there is no feedback from the resonator. Since the rate of stimulated emission is dependent on the amount of light entering the medium, the amount of energy stored in the gain medium increases as the medium is pumped. Due to losses from spontaneous emission and other processes, after a certain time the stored energy will reach some maximum level; the medium is said to be *gain saturated*. At this point, the Q-switch device is quickly changed from low to high Q, allowing feedback and the process of optical amplification by stimulated emission to begin. Because of the large amount of energy already stored in the gain medium, the intensity of light in the laser resonator builds up very quickly; this also causes the energy stored in the medium to be depleted almost as quickly. The net result is a short pulse of light output from the laser, known as a *giant pulse*, which may have a very high peak intensity.

In the ordinary pulsed mode, the output of an insulating crystal laser such as the Nd:YAG consists of many random 'spikes' of about 1 μs duration with a separation of about 1 μs; the length of the train of spikes depends principally on the duration of the exciting flashtube source, which may be about 1 ms. Peak powers within the 'spikes' are typically of the order of kilowatts. When the laser is Q-switched, however, the result is a single 'spike' of great power, as discussed above, typically in the megawatt range, with a duration of 10—100 ns. Although there is a vast increase in the peak power of the Q-switched laser, the total energy emitted is less than in the non-Q-switched operation due to losses associated with the Q-switching mechanism.

There are two important requirements for effective Q-switching:

- The Q-switch must switch rapidly in comparison to the buildup of laser oscillations, otherwise the latter will build up gradually and a longer pulse will be obtained thus reducing the peak power. In practice, the Q-switch should operate in a time less than 1 ns; and
- The rate of pumping must be faster than the spontaneous decay rate of the upper lasing level, otherwise the upper level will empty more quickly than it can be filled so that a sufficiently large population inversion will not be achieved.

3.13.2 Methods of Q-Switching

There are two main types of Q-switching: (*i*) active Q-switching, and (*ii*) passive Q-switching.

Active Q-Switching

Here, the Q-switch is an externally-controlled variable attenuator. This may be a mechanical device such as a shutter, chopper wheel or spinning mirror placed inside the cavity, or (more commonly) it may be some form of modulator such as an acousto-optic device or an electro-optic device—a Pockels cell or Kerr cell. The reduction of losses (increase of Q) is triggered by an external event, typically an electrical signal. The pulse repetition rate can therefore be externally controlled.

Modulators generally allow a faster transition from low to high Q, and provide better control. An additional advantage of modulators is that the rejected light may be coupled out of the cavity and can be used for something else. Alternatively, when the modulator is in its low-Q state, an externally-generated beam can be coupled *into* the cavity through the modulator. This can be used to "seed" the cavity with a beam that has desired characteristics (such as transverse mode or wavelength). When the Q is raised, lasing builds up from the initial seed, producing a Q-switched pulse that has characteristics inherited from the seed. We discuss here the rotating mirror method of Q-switching and the electro-optic Q-switching.

Rotating Mirror Method

This is the first method to be developed for Q-switching. In this method one of the mirrors is rotated at a very high angular velocity so that the optical losses are large except for the brief interval in each rotation cycle when the mirrors are very nearly parallel. Just before this point is reached a trigger mechanism initiates the flashlamp discharge to pump the laser (we assume here an insulating crystal type of laser). Since the mirrors are not yet parallel, population inversion can build up without any laser action starting. When the mirrors become parallel, Q-switching occurs allowing the Q-switched pulse to develop as shown in Fig. 3.19.

The repetition rate of the laser firing is determined by controlling the flashlamp and not by the speed of rotation of the mirror, which may be as high as 60000 rev min^{-1}. If the laser fired every revolution, the repetition rate would be ≈ 1000 times per second, a rate which is prohibitive in insulating crystals due to the excessive heating of the laser rod which would occur. Although these type of Q-switches are cheap, reliable and rugged, the method suffers from the major disadvantage of being slow. This results in an inefficient production of Q-switched pulses with lower peak power than can be produced by other methods.

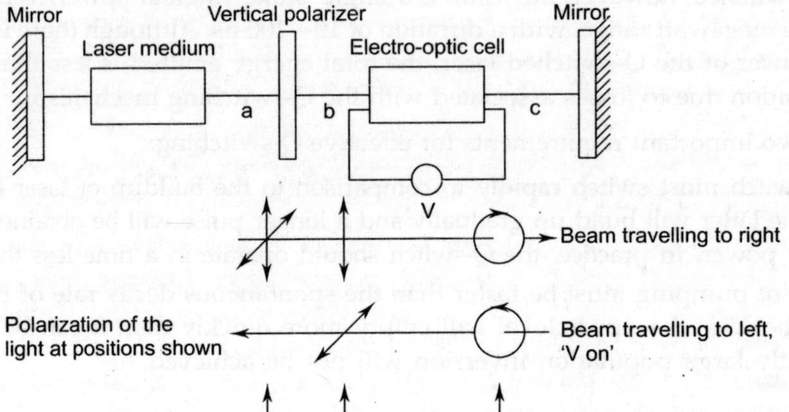

Fig. 3.19

It can be shown that the peak power, *i.e.*, the greatest rate of change of population inversion and hence of photon emission, occurs when the population inversion drops to the threshold inversion N_{th}. To estimate the average power in the pulse, we need to evaluate the pulse duration. We imagine that the Q-switched pulse oscillates to and fro between the laser mirrors and that each time it strikes one of the mirrors a fraction $(1 - r)$ of its energy is lost due to transmission. The pulse will then make $1/(1 - r)$ passes along the length of the cavity. This happens in a time $(1/(1 - r))(L/c)$. This is often referred to as cavity lifetime T_c and may be taken as the duration of the pulse. The power of the pulse is then approximately $P = E/T_c$, which, as $E = \frac{1}{2} h\nu_{21} (N_i - N_f)V$, we can write as

$$P = \frac{(N_i - N_f) h\nu_{21} Vc(1 - R)}{2L} \tag{3.81}$$

Electro-Optic and Acousto-Optic Q-Switching

A variety of electro-optic techniques may be used including the Pockels cell or Kerr cell (high voltage driven) which affects the polarization and the acousto-optic type (RF driven) which deflects part of the beam out of the cavity thus reducing gain.

There are three main differences from the optical side between Electro-Optic (EO) Q-switches (*e.g.*, Pockels Cell) and Acousto-Optic (AO) Q-switches:

1. With EO, a much higher loss can be achieved. If tuned properly, the single-pass loss can approach 100%. AOMs can get up to ~80%, but that is either in materials with low damage threshold and with a small aperture (~1 mm) or in high-power devices which need 50 to 100 W of RF power into the AOM and water cooling. Even than, 50 to 60% is more typical.
2. EOs can be switched off faster. In AOMs the switching time is determined by the time the acoustic wave needs to cross the beam. 6 km/s (*i.e.*, 6 mm/μs) is a typical acoustic velocity, so that the switching time is about 170 ns per mm beam diameter. That can be a problem especially on large systems.
3. AOMs can be modulated much faster. While the switching time is slow, you can switch on and off as much and fast as you want only limited by this switching time. 100s of kHz repetition rate is not a problem for the AOM. EOs need a certain recovery time. An additional advantage of AOMs is that they require only low voltages.

The electro-optic, acousto-optic and magneto-optic modulators discussed in Chapter 5 can be used as fast Q-switches. For example, if a Pockels cell is used and the laser output is not naturally polarized, then a polarizer may be placed in the cavity along with the Pockels cell as shown in Fig. 3.19.

A voltage is applied to the cell to produce the effect of a quarter-wave plate which converts the linearly polarized light incident on it into circularly polarized light. The laser mirror reflects the light and in doing so reverses its direction of rotation so that on re-passing through the electro-optic cell it emerges as a plane polarized light, but at 90° to its original direction of polarization. This light is therefore not transmitted by the polarizer and the cavity is 'switched off'. When the applied voltage is reduced to zero, there is no rotation of the plane of polarization and the Q-switching occurs. This change of voltage can be accomplished in less than 10 ns and hence very effective Q-switching occurs.

Alternative arrangements involving Kerr cells and acousto-optic modulators are available. In the case of acousto-optic modulator, application of an acoustic signal to the modulator deflects some of the beam out of the cavity, and hence a high loss takes place. When the sound wave is shut off, Q-switching occurs as before. Acousto-optic devices are often used when the laser medium is continuously pumped and repetitively Q-switched, as is frequently the case with CO_2 and Nd:YAG lasers.

Passive Q-Switching

In this case, the Q-switch is a saturable absorber , a material whose transmission increases when the intensity of light exceeds some threshold. The material may be an ion-doped crystal like Cr:YAG which is used for Q-switching of Nd:YAG lasers, a bleachable dye, or a passive semiconductor device. Initially, the loss of the absorber is high, but still low enough to permit some lasing once a large amount of energy is stored in the gain medium. As the laser power increases, it saturates the absorber, *i.e.*, rapidly reduces the resonator loss, so that the power can increase even faster. Ideally, this brings the absorber into a state with low losses to allow efficient extraction of the stored energy by the laser pulse. After the pulse, the absorber recovers to its high-loss state before the gain recovers, so that the next pulse is delayed until the energy in the gain medium is fully replenished. The pulse repetition rate can only indirectly be controlled, *e.g.* by varying the laser's pump power and the amount of saturable absorber in the cavity. Direct control of the repetition rate can be achieved by using a pulsed pump source as well as passive Q-switching.

3.13.3 Typical Performance

A typical Q-switched laser (*e.g.*, a Nd:YAG laser) with a resonator length of *e.g.*, 10 cm can produce light pulses of several tens of nanoseconds duration. Even when the average power is well below 1 W, the peak power can be many kilowatts. Large-scale laser systems can produce Q-switched pulses with energies of many joules and peak powers in the gigawatt region. On the other hand, passively Q-switched microchip lasers (with very short resonators) have generated pulses with durations far below one nanosecond and pulse repetition rates from hundreds of hertz to several megahertz (MHz).

Applications

Q-switched lasers are often used in applications which demand high laser intensities in nanosecond pulses, such as dentistry, metal cutting or pulsed holography. However, Q-switched lasers can also be used for measurement purposes, *e.g.*, for distance measurements (range finding) by measuring the time it takes for the pulse to get to some target and the reflected light to get back to the sender.

By altering the pulse rate and duration a time structure can be developed resulting in new branches of chemistry/physics-namely ultra-fast photo-chemistry, which deals with femtosecond (10^{-15} *s*) time scale reactions.

3.14 CLASSES OF LASER

The Helium-Neon laser used in the Optics Laboratory produces an intense beam of light. The power level of the laser used in the Optics Laboratory is approximately 1.5 mW. This is a low-power laser. The beam will not harm the skin, even upon direct exposure. The human eye, however, is much more susceptible to injury by laser light. The lens of the eye focuses the beam to a very small spot on the retina, the light sensitive layer at the back of the eye. The resulting high intensity may lead to a local burn, which may cause degradation of vision, depending on the exact location and size of the spot.

The components of the eye are vulnerable to different wavelengths of electromagnetic radiation. Each wavelength region must be considered separately when evaluating the potential hazards. The far ultraviolet and the far infrared wavelengths are entirely absorbed by the cornea and the lens. Near ultraviolet light is partially absorbed in the cornea, lens and the retina, and can damage all three. Light with visible or near infrared wavelengths poses the greatest danger to the retina because the lens focuses these wavelengths on the retina.

Tissue damage by heating occurs when the heat energy deposited by light exceeds the capacity of the tissues to safely carry the heat away. Lasers hazards or the laser's potential for causing immediate injury to the eye or skin and/or potential for causing fires from direct exposure to the beam or from reflections from diffuse reflective surfaces are characterized by Classes. The class of a laser depends on the laser power or energy and on the emission wavelength. Since August 1, 1976, commercially produced lasers have been classified and identified by labels affixed to the laser. In cases where the laser has been fabricated on campus or is otherwise not labeled, Health Physics should be consulted on the appropriate laser classification and labeling. Lasers are classified using physical parameters of the laser, power, wavelength and exposure duration. The Class of a laser is defined by the American National Standards Institute (ANSI) according to the degree of hazard presented. A qualitative description of laser classes follows:

3.14.1 Class I Lasers

Class I lasers are low-powered and do not emit hazardous radiation under normal operating conditions because they are completely enclosed from any possible viewing by a person and the laser system is interlocked to prevent the laser from being on when exposed. Class I lasers are exempt from any control measures. Equipment, such as laser printers and laser disc players, are examples of this class.

3.14.2 Class II Lasers

Class II lasers emit accessible laser light in the visible portion of the spectrum with power levels less than 1 mW radiant power and are capable of creating eye damage through chronic exposure. The human eye blink reflex, which occurs within 0.25 seconds of exposure to the Class II laser beam, provides adequate protection. It is possible to overcome the blink response and stare into the Class II laser long enough to damage the eye. Class II lasers are exempt from any control measures. Equipment, such as some visible continuous wave Helium-Neon lasers and some laser pointers, are examples of Class II lasers. Caution labels (yellow) should be placed on the laser equipment. No known skin exposure hazard exists and no fire hazard exists in these types of lasers.

3.14.3 Class IIA Lasers

Class IIa lasers are special purpose lasers that emit accessible visible laser light with power levels less than 1 mW radiant power and are not intended for viewing. This class of lasers causes injury when viewed directly for more than 1,000 seconds. Class IIA lasers are exempt from any control measures. Equipments such as some bar code readers are examples of Class IIA lasers.

3.14.4 Class IIIA Lasers

Class IIIA lasers are systems with power levels of 1 to 5 mW that normally would not produce a hazard if viewed for only momentary periods with the unaided eye. They pose severe eye hazards when viewed through optical instruments (*e.g.*, microscopes, binoculars, or other collecting optics). Class IIIA lasers must be labeled. Products that have a Class IIIA laser should have a laser emission indicator to tell when the laser is in operation. A warning label shall be placed on or near the laser in a conspicuous location and caution users to avoid staring into the beam or directing the beam toward the eye of individuals. A key operated power switch should be used to prevent unauthorized use. No known skin hazard of fire hazard exists in this type. Equipment, such as some visible continuous wave Helium-Neon lasers and some solid state laser pointers, are examples of Class IIIA lasers.

3.14.5 Class IIIB Lasers

Class IIIB lasers are systems with power levels of 5 mW to 500 mW for continuous wave lasers or less than 10 J/cm² for a 0.25 s pulsed laser. These lasers will produce an eye hazard if viewed directly. This includes intrabeam viewing or specular reflections. Higher power lasers in this class will also produce hazardous diffuse reflections. Specific control measures covered in Class IIIB lasers shall be used in areas where entry by unauthorized personnel can be controlled. These lasers must have a key switch to prevent unauthorized use. A laser emission indicator, along with a 3 to 5 second time delay after power is applied to allow the operator to move away from the beam path and a mechanical shutter to turn the beam off during use is provided. Skin may be burned at the higher levels of power output as well as the flash point of some materials which could catch fire. Example is an argon ion laser. A red *DANGER* label and aperture label must be affixed to the laser.

3.14.6 Class IV Lasers

Class IV lasers are systems with power levels greater than 500 mW for continuous wave lasers or greater than 10 J/cm^2 for a 0.25 s pulsed laser. These can and will cause eye damage. The Class IV range can and will cause materials to burn on contact as well as skin and clothing to burn. This includes intrabeam viewing, specular reflections or diffuse reflections. These laser systems must have a key lockout switch to prevent unauthorized use, inter-locks to prevent the system from being used with the protective covers off and emission indicators to show that the laser is in use. Mechanical shutters are provided to block the beam and *Red* DANGER labels and aperture labels are affixed to the laser. The reflected beam should be considered as dangerous as the primary beam. Example is CO_2 laser.

3.14.7 Embedded Lasers

Embedded lasers are found in laser products with lower class ratings. Laser printers, CD players, and laser welders may have Class III or Class IV lasers in their protective and interlocked housings. When such a laser system is used as intended, the lower laser class applies. When such a system is opened (*e.g.*, for service or alignment) and the embedded laser beam is accessible, the requirements for the higher class of the embedded laser must be implemented.

3.14.8 Registration of Laser Systems

Any laser system that has a power output of greater than 5 mW must be registered with the FDA and Center for Devices and Radiological Health if it has an exposed beam, such as for entertainment (*i.e.*, Laser light shows) or for medical use (such as surgery) where someone other than the operator may come in contact with it.

3.15 LASER SAFETY GUIDELINES

Follow these guidelines at all times to ensure the safety of yourself and others working in the optics laboratory.

- Be aware of the laser hazards present.
- Know the appropriate laser Class and take the necessary precautions.
- Follow operation instructions carefully.
- Never point a laser beam at a person.
- Never look directly into the laser beam.
- Never look directly into specular, mirror-like reflections of the laser beam.
- Do not wear any objects on your hands and wrists that may cause specular reflection, such as watches or jewelry.

- Do not let the laser beam or any of its reflections leave the experiment table unless the experiment requires it.
- Keep all the beams in a plane parallel to the experiment table.
- Do not bring your eye level down to the plane of the laser beam, keep the beam axis below typical standing and sitting positions.
- Close your eyes if you bend down to pick something up off the floor.
- Use beam stops and carefully plan the placement and movement of optical elements.
- Laser light scattered by rough surfaces, such as paper is harmless to the eye at low power levels. However, it is still a good idea not to stare at bright diffuse spots for a long time.
- Leave the room lights **ON** when possible. The eye's pupils open wide in a dark room and present a larger target for dangerous laser light.
- Use common sense and be alert at all times.

3.16 DIFFERENT TYPES OF LASER SYSTEMS

A list of laser types, their operational wavelengths, and their applications is being presented here. There exist various of kinds of lasers, but most of them are not used beyond specialised research.

1. Gas lasers
2. Chemical lasers
3. Dye lasers
4. Metal-vapor lasers
5. Solid-state lasers
6. Semiconductor lasers
7. Other types of lasers

Gas lasers

Laser gain medium and type	Operation wavelength(s)	Pump source	Applications and notes
Helium-neon laser	632.8 nm (543.5 nm, 593.9 nm, 611.8 nm, 1.1523 μm, 1.52 μm, 3.3913 μm)	Electrical discharge	Interferometry, holography, spectroscopy, barcode scanning, alignment, optical demonstrations.
Argon laser	454.6 nm, 488.0 nm, 514.5 nm (351 nm,457.9 nm, 465.8 nm, 476.5 nm, 472.7 nm, 528.7 nm)	Electrical discharge	Retinal phototherapy (for diabetes), lithography, confocal microscopy, pumping other lasers.
Krypton laser	416 nm, 530.9 nm, 568.2 nm, 647.1 nm, 676.4 nm, 752.5 nm, 799.3 nm	Electrical discharge	Scientific research, mixed with argon to create "white-light" lasers, light shows.
Xenon ion laser	Many lines throughout visible spectrum extending into the UV and IR.	Electrical discharge	Scientific research.

Nitrogen laser	337.1 nm	Electrical discharge	Pumping of dye lasers, measuring air pollution, scientific research. Nitrogen lasers can operate super-radiantly (without a resonator cavity). Amateur laser construction.
Carbon dioxide laser	10.6 μm, (9.4 μm)	Transverse (high power) or longitudinal (low power) electrical discharge	Material processing (cutting, welding, etc.), surgery.
Carbon monoxide laser	2.6 to 4 μm, 4.8 to 8.3 μm	Electrical discharge	Material processing (engraving, welding, etc.), photo-acoustic spectroscopy.
Excimer laser	193 nm (ArF), 248 nm (KrF), 308 nm (XeCl), 353 nm (XeF)	Excimer recombination via electrical discharge	Ultraviolet lithography for semiconductor manufacturing, laser surgery, LASIK.

Chemical lasers

Laser gain medium and type	Operation wavelength(s)	Pump source	Applications and notes
Hydrogen fluoride laser	2.7 to 2.9 μm for Hydrogen fluoride (<80% Atmospheric transmittance)	Chemical reaction in a burning jet of ethylene and nitrogen trifluoride (NF_3)	Used in research for laser weaponry by the U.S. DOD, operated in continuous wave mode, can have power in the megawatt range.
Deuterium fluoride laser	~3800 nm (3.6 to 4.2 μm) (~90% Atm. transmittance)	chemical reaction	MIRACL, Pulsed Energy Projectile & Tactical High Energy Laser
COIL (Chemical oxygen-iodine laser)	1.315 μm (<70% Atmospheric transmittance)	Chemical reaction in a jet of singlet delta oxygen and iodine	Laser weaponry, scientific and materials research, laser used in the U.S. military's Airborne laser, operated in continuous wave mode, can have power in the megawatt range.

Dye lasers

Laser gain medium and type	Operation wavelength(s)	Pump source	Applications and notes
Dye lasers	390–435 nm (stilbene), 460–515 nm (coumarin 102), 570–640 nm (rhodamine 6G), many others	Other laser, flash lamp	Research, spectroscopy, birthmark removal, isotope separation. The tuning range of the laser depends on which dye is used.

Metal-vapour lasers

Laser gain medium and type	Operation wavelength(s)	Pump source	Applications and notes
Helium-cadmium (HeCd) metal-vapor laser	441.563 nm, 325 nm	Electrical discharge in metal vapour mixed with helium buffer gas.	Printing and typesetting applications, fluorescence excitation examination (*i.e.*, in U.S. paper currency printing), scientific research.
Helium-mercury (HeHg) metal-vapor laser	567 nm, 615 nm	Electrical discharge in metal vapour mixed with helium buffer gas.	Rare, scientific research, amateur laser construction.
Helium-selenium (HeSe) metal-vapor laser	up to 24 wavelengths between red and UV	Electrical discharge in metal vapour mixed with helium buffer gas.	Rare, scientific research, amateur laser construction.
Copper vapor laser	510.6 nm, 578.2 nm	Electrical discharge	Dermatological uses, high speed photography, pump for dye lasers.
Gold vapor laser	627 nm	Electrical discharge	Rare, dermatological and photodynamic therapy uses.

Solid-state lasers

Laser gain medium and type	Operation wavelength(s)	Pump source	Applications and notes
Ruby laser	694.3 nm	Flashlamp	Holography, tattoo removal. The first type of visible light laser invented; May 1960.
Nd:YAG laser	1.064 μm, (1.32 μm)	Flashlamp, laser diode	Material processing, rangefinding, laser target designation, surgery, research, pumping other lasers (combined with frequency doubling to produce a green 532 nm beam). One of the most common high power lasers. Usually pulsed (down to fractions of a nanosecond)
Er:YAG laser	2.94 μm	Flashlamp, laser diode	Periodontal scaling, Dentistry

Neodymium YLF (Nd:YLF) solid-state laser	1.047 and 1.053 μm	Flashlamp, laser diode	Mostly used for pulsed pumping of certain types of pulsed Ti:sapphire lasers, combined with frequency doubling.
Neodymium doped Yttrium orthovanadate (Nd: YVO4) laser	1.064 mm	Laser diode	Mostly used for continuous pumping of mode-locked Ti:sapphire or dye lasers, in combination with frequency doubling. Also used pulsed for making and micromachining.
Neodymium doped yttrium calcium oxoborate $Nd{:}YCa_4O(BO_3)_3$ or simply Nd:YCOB	~1.060 μm (~530 nm at second harmonic)	laser diode	Nd:YCOB is a so called "self-frequency doubling" or SFD laser material which is both capable of lasing and which has nonlinear characteristics suitable for second harmonic generation. Such materials have the potential to simplify the design of high brightness green lasers.
Neodymium glass (Nd:Glass) laser	~1.062 μm (Silicate glasses), ~1.054 μm (Phosphate glasses)	Flashlamp, laser diode	Used in extremely high power (terawatt scale), high energy (megajoules) multiple beam systems for inertial confinement fusion. Nd:Glass lasers are usually frequency tripled to the third harmonic at 351 nm in laser fusion devices.
Titanium sapphire (Ti:sapphire) laser	650–1100 nm	Other laser	Spectroscopy, LIDAR, research. This material is often used in highly-tunable mode-locked infrared lasers to produce ultrashort pulses and in amplifier lasers to produce ultrashort and ultra-intense pulses.
Thulium YAG (Tm:YAG) laser	2.0 μm	Laser diode	LIDAR.
Ytterbium YAG (Yb:YAG) laser	1.03 μm	Laser diode, flashlamp	Optical refrigeration, materials processing, ultrashort pulse research, multiphoton microscopy, LIDAR.
Ytterbium doped glass laser (rod, plate/chip, and fiber)	1.0 μm	Laser diode.	Fiber version is capable of producing several-kilowatt continuous power, having ~70-80% optical-to-optical and ~25% electrical-to-optical efficiency. Material processing: cutting, welding, marking; nonlinear fiber optics:

			broadband fiber-nonlinearity based sources, pump for fiber Raman lasers; distributed Raman amplification pump for telecommunications.
Holmium YAG (Ho:YAG) laser	2.1 μm	Laser diode	Tissue ablation, kidney stone removal, dentistry.
Cerium doped lithium strontium(or calcium) aluminum fluoride (Ce:LiSAF, Ce:LiCAF)	~280 to 316 nm	Frequency quadrupled Nd:YAG laser pumped, excimer laser pumped, copper vapor laser pumped.	Remote atmospheric sensing, LIDAR, optics research.
Chromium doped chrysoberyl (alexandrite) laser	Typically tuned in the range of 700 to 820 nm	Flashlamp, laser diode, mercury arc (for CW mode operation)	Dermatological uses, LIDAR, laser machining.
Erbium doped and erbium-ytterbium codoped glass lasers	1.53–1.56 μm	Laser diode	These are made in rod, plate/chip, and optical fiber form. Erbium doped fibers are commonly used as optical amplifiers for telecommunications.
Trivalent uranium doped calcium fluoride ($U:CaF_2$) solid-state laser	2.5 μm	Flashlamp	First 4-level solid state laser (November 1960) developed by Peter Sorokin and Mirek Stevenson at IBM research labs, second laser invented overall (after Maiman's ruby laser), liquid helium cooled, unused today.
Divalent samarium doped calcium fluoride ($Sm:CaF_2$) laser	708.5 nm	Flashlamp	Also invented by Peter Sorokin and Mirek Stevenson at IBM research labs, early 1961. Liquid helium cooled, unused today.
F-center laser.	2.3–3.3 μm	Ion laser	Spectroscopy

Semiconductor lasers

Laser gain medium and type	Operation wavelength(s)	Pump source	Applications and notes
Semiconductor laser diode (general information)	0.4–20 μm, depending on active region material.	Electrical current	Telecommunications, holography, printing, weapons, machining, welding, pump sources for other lasers.
GaN	0.4 μm	Electrical current	Optical discs.
AlGaAs	0.63–0.9 μm	Electrical current	Optical discs, laser pointers, data communications. 780 nm

			Compact Disc player laser is the most common laser type in the world. Solid-state laser pumping, machining, medical.
InGaAsP	1.0–2.1 μm	Electrical current	Telecommunications, solid-state laser pumping, machining, medical.
lead salt	3–20 μm	Electrical current	
Vertical Cavity Surface Emitting Laser (VCSEL)	850–1500 nm, depending on material	Electrical current	Telecommunications
Quantum cascade laser	Mid-infrared to far-infrared.	Electrical current	Research, Future applications may include collision-avoidance radar, industrial-process control and medical diagnostics such as breath analyzers.
Hybrid silicon laser	Mid-infrared	Electrical current	Research

Other types of lasers

Laser gain medium and type	Operation wavelength(s)	Pump source	Applications and notes
Free electron laser	A broad wavelength range (about 100 nm - several mm); one free electron laser may be tunable over a wavelength range	relativistic electron beam	Atmospheric research, material science, medical applications.
"Nickel-like" Samarium laser	X-rays at 7.3 nm wavelength	Lasing in ultra-hot samarium plasma formed by double pulse terawatt scale irradiation fluences created by Rutherford Appleton Laboratory's Nd:glass Vulcan laser.	First demonstration of efficient "saturated" operation of a sub–10 nm X-ray laser, possible applications in high resolution microscopy and holography, operation is close to the "water window" at 2.2 to 4.4 nm where observation of DNA structure and the action of viruses and drugs on cells can be examined.
Raman laser, uses inelastic stimulated Raman scattering in a nonlinear media, mostly fiber, for amplification	1–2 μm for fiber version	Other laser, mostly Yb-glass fiber lasers	Complete 1–2 μm wavelength coverage; disitributed optical signal amplification for telecommunications; optical solitons generation and amplification
Nuclear pumped laser	See gas lasers	Nuclear fission	Research

We will discuss some specific laser systems and their important operating characteristics. The systems that will be considered are some of the important lasers of widespread use of today for different applications. The lasers considered here are:

1. solid state lasers: ruby, Nd:YAG, Nd:glass;
2. gas lasers: He-Ne, argon ion, CO_2 and N_2;
3. excimer lasers;
4. liquid lasers: dyes;
5. semiconductor lasers.

3.16.1 Solid State Lasers

Solid-state laser is a laser that uses a gain medium that is a solid, rather than a liquid such as dye lasers or a gas such as gas lasers. Semiconductor based lasers are also in the solid state, but are generally considered separately from solid-state lasers (see semiconductor laser).

Generally, the active medium of a solid-state laser consists of a glass or crystalline *host* material to which is added a *dopant* such as neodymium, chromium, erbium, or other ions. Many of the common dopants are rare earth elements, because the excited states of such ions are not strongly coupled with thermal vibrations of the crystalline lattice (phonons), and the lasing threshold can be reached at relatively low brightness of pump.

There are many hundreds of solid-state media in which laser action has been achieved, but relatively few types are in widespread use. Of these, probably the most common type is neodymium doped YAG . Neodymium-doped glass (Nd:glass) and Ytterbium doped glasses and ceramics are used in solid-state lasers at extremely high power (terawatt scale), high energy (megajoules multiple beam systems for inertial confinement fusion. Titanium doped sapphire is also widely used for its broad tunability. The first solid-state material used for lasing was ruby. Ruby lasers are still used for some applications, but are not common due to their low efficiency.

Solid-state lasers are used in all sorts of applications including materials processing (cutting, drilling, welding, marking, heat treating, etc.), semiconductor fabrication (wafer cutting, IC trimming), the graphic arts (high-end printing and copying), medical and surgical, rangefinders and other types of measurement, scientific research, entertainment, and many others where high peak power and/or high continuous power are required. A high energy pulsed YAG laser has even been used in rocket propulsion experiments. The largest lasers (with the highest peak power) in the world are solid state lasers. Many of the laser projectors for light shows and for other laser displays use solid state rather than gas lasers like argon or krypton ion. And, that green laser pointer is a Diode Pumped Solid State (DPSS) laser.

- Solid-state lasers can be pulsed, CW (Continuous Wave), or quasi-CW.
- Pulsed solid-state lasers generally use various versions of xenon flash-lamps like the original ruby laser. Q-switching is used to stabilize and boost peak power output by preventing the laser cavity from resonating (*e.g.*, one of the mirrors is blocked or forced to be misaligned) until the population inversion has built up fully.
- Energy output is measured in joules (Watt-seconds) per pulse. Multiply this by the number of pulses/second to calculate average power output. To determine the peak power in each pulse requires knowledge of the pulse shape.

- Flash-lamp pumped solid-state lasers are used where high peak power is required as most other pumping methods can't even come close. However, the average power and efficiency may be quite low compared to approaches using high power laser diode pumping.
- CW solid-state lasers may use xenon or krypton arc lamps or other sources of intense broad spectrum light. However, the trend today is toward the use of arrays of high power laser diodes to do the pumping. These can be designed to have a wavelength that matches an absorption band in neodymium (around 800 nm) making for very efficient excitation. The diode pumped approaches are rapidly taking over due to their efficiency, resulting lower power consumption and heat dissipation, much more compact size, as well as much higher reliability and lower (virtually no) maintenance. In fact, the relatively common, but still expensive green laser pointers currently available (prior to the introduction of direct injection green laser diodes) are diode pumped solid-state lasers with a frequency doubler as part of the laser cavity. Power output is measured the same way as for other CW lasers.
- Quasi-CW solid-state lasers are actually pulsed lasers but operating with a Pulse Repetition Rate (PRR) high enough to appear (for the intended application) to be continuous. Generally, this means a PRR of at least 50 pps but may be 10s or 100s of kHz if electro-optically Q-switched.

Depending on the application, the average power output or peak pulse energy or power may be the relevant measurement of performance. Note that the output would appear as CW to the human eye, if frequency is doubled to 532 nm (green) and it would not be suitable for laser TV or light show scanning since it really isn't continuous.

It may be noticed that while the *peak* power of the output pulses for a Q-switched laser may be many orders of magnitude greater than with the Q-switch disabled or bypassed, for the fundamental output wavelength (*e.g.*, 1,064 nm for a Nd:YAG laser), the *total* energy (pulsed) or *average* power (quasi-CW) is generally lower. However, when frequency multiplication is used (*e.g.*, 532 nm green from a Nd:YAG laser), total energy or average power may actually be greater with the Q-switch because frequency multiplication is a non-linear process and is much more efficient as the peak power increases. Depending on the application, the high peak power may more than offset the lower total energy or average power or vice-versa.

Efficiency of Solid State Lasers

Wall plug efficiency can vary from well under 1 percent for flash-lamp and arc lamp pumped solid-state lasers to 25 percent or more for those pumped with laser diode pumped.

A (laser) diode pumped Nd:YAG may have a 40% efficiency (operating multimode with good thermal control of the diodes), and the pump diodes themselves have about a 45% efficiency, resulting in a net 18% of efficiency from electrical power to the diodes to output beam power. However, at increased pump powers, thermal issues may cause the efficiency to decrease after a certain point. This decrease is power dependent, as well as resonator and pump assembly design dependent.

Ruby Lasers

The ruby laser was the first laser invented in 1960 by Maiman (shown in Fig. 3.20). It consisted of a ruby crystal formed into a cylinder surrounded by a spiral flash tube to provide a flash of white light that triggers the laser action. The ruby crystal was enclosed within a polished aluminum cylindrical cavity cooled by forced air. The ruby cylinder forms a Fabry-Perot cavity by optically polishing the ends to be parallel to within a third of a wavelength of light. Each end was coated with evaporated silver; one end was made less reflective to allow some radiation to escape as a beam.

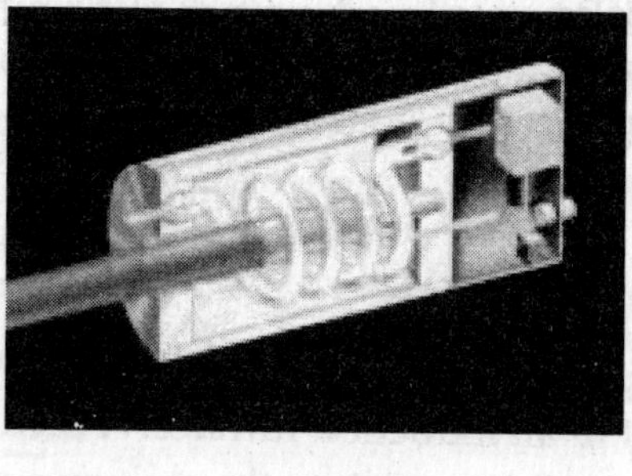

Fig. 3.20

Ruby is a synthetically grown aluminum oxide crystal in which some of the aluminum ions have been replaced with chromium ions (Cr^{3+}). Chromium ions replace aluminum ions in the lattice and the crystal field partially removes the degeneracy of the isolated ions to provide levels for pumping and the laser transitions. In this case some of the energy levels of the Cr^{3+} ions are almost independent of the crystal field and hence they remain sharp. Other levels, on the contrary, are strongly dependent on the crystal field so that lattice vibrations, which cause fluctuations in the crystal field, broaden these levels quite substantially. The ^2E and $^4A_2$ levels remain sharp while the $^4T_1$ and $^4T_2$ levels become broad as shown in Fig. 3.22. Thus the pump transitions are spectrally broad while the transitions R_1 and R_2 are narrow. The wavelenghts of the R_1 and R_2 laser lines are temperature dependent. The energy level diagram shows that ruby is basically a three level system. Thus more than half of the total number of ions has to be pumped to level 2 via level 3 to create a population inversion. Thus the laser has a very low efficiency compared to a four level system discussed later.

Chromium gives ruby its characteristic red color and is responsible for the lasing behavior of the crystal. Typical construction of a Ruby laser is shown in Fig. 3.21. The green and blue wavelengths in the flash excite electrons in the chromium atoms to a higher energy level. Upon returning to their normal state, the electrons emit their characteristic ruby-red light. The mirrors reflect some of this light back and forth inside the ruby crystal, stimulating other excited chromium atoms to produce more red lights, until the light pulse builds up to high power and drain the energy stored in the crystal.

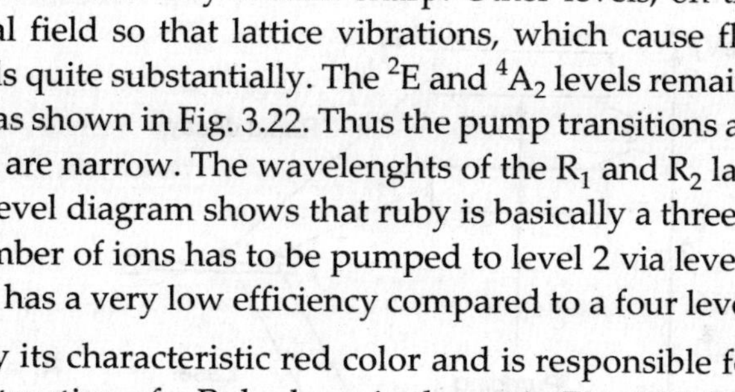

Fig. 3.21

Photo-pumped by a fast discharge flash-lamp, the first ruby lasers operated in pulsed mode for reasons of heat dissipation and the need for high pumping powers. Nelson and Boyle (1962) constructed a continuous lasing ruby by replacing the flash lamp with an arc lamp. The ruby laser produces pulses of visible light at a wavelength of 694.3 nm, which appears as deep red to human eyes. Typical ruby laser pulse lengths are on the order of a millisecond. These short pulses of red light are visible to the human eye, if the viewer carefully watches the target area where the pulse will fire.

Manufacture of Synthetic Ruby Crystals

Ruby rods for lasers are made synthetically. Aluminum Oxide (Al_2O_3) with a very small amount of chromium impurity is melted in an induction furnace. A seed crystal (perhaps a natural ruby or a chip of another synthetic crystal) is stuck into the melt on a rod and then slowly withdrawn. A cylindrical rod of "ruby" crystal is formed and is slowly pulled out of the melt. This rod is then cut up and ground with diamond machining equipment to form the precisely shaped laser rod. The ends are polished to extreme levels and then treated with whatever optical coatings are desired, depending on the design of the laser (*i.e.*, mirrors directly coated on the rod or external mirrors are used).

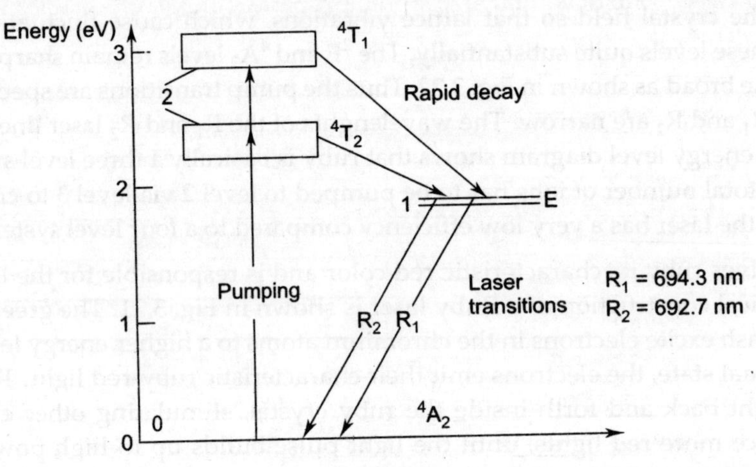

Fig. 3.22

There are several ways to do this. The first is the easiest, to pull from the melt. Al_2O_3 can be molten in molybdenum crucibles and a crystal can be pulled directly from the melt. Even single crystal tubes and other shapes having a fixed cross section can be pulled using a technique called "edge defined growth."

If colored sapphire or ruby is required there are two more methods used. The first, Vernuile, uses a hydrogen-oxygen flame and drops powdered Al_2O_3 plus dopant through the flame. The flame melts the powder, which falls on the seed crystal and crystallizes. Because only the surface of the crystal is molten the dopant gets incorporated into the bulk. The crystals are called boules, and look vaguely like a pop bottle, with a small neck, opening up into a cylindrical crystal. Most colored sapphire and ruby is made this way, including the watch jewels.

The last commonly used method is flux growth. The Al_2O_3 is dissolved in a molten salt, usually lead oxide plus cryolite, in a platinum crucible. The crystals come out of solution as the melt is cooled just like sugar in hot water. These are the most desirable of the synthetic stones because they look more like natural stones after cutting, and the process is the most expensive.

Applications

Ruby lasers have declined in use with the discovery of better lasing media. They are still used in a number of applications where short pulses of red light are required. Holographers around the world produce holographic portraits with ruby lasers, in sizes up to a metre squared. The red 694.3 nm laser light is preferred to the 532 nm green light of frequency-doubled Nd:YAG. Many non-destructive testing labs use ruby lasers to create holograms of large objects such as aircraft tires to look for weaknesses in the lining. Ruby lasers were used extensively in tattoo and hair removal, but are being replaced by alexandrite lasers and Nd:YAG lasers in this application.

Nd:YAG Lasers

Laser operation of Nd:YAG was first demonstrated by Geusic *et al.* at Bell Laboratories in 1964. Nd:YAG (neodymium-doped yttrium aluminium garnet; $Nd{:}Y_3Al_5O_{12}$) is a crystal that is used as a lasing medium for solid-state lasers. This material possesses good optical, thermal, and mechanical properties and is likely the most used solid-state laser material. It has a relatively low threshold and high efficiency. The dopant, triply ionized neodymium (Nd^{3+}), typically replaces yttrium in the crystal structure, since they are of similar size. Some properties of Nd:YAG is shown in Table 3.1.

Table 3.1

Property	*Value*
Chemical formula	$Nd^{3+}{:}Y_3Al_5O_{12}$
Crystal structure	Cubic
Mass density	4.56 g/cm³
Moh hardness	8 to 8.5
Young's modulus	280 gpa
Tensile strength	200 mpa
Melting point	1970°C
Thermal conductivity	13 to 14 w / (m k)
Thermal expansion coefficient	$7 \text{ to } 8 \cdot 10^{-6}$/k
Thermal shock resistance parameter	790 w/m
Birefringence	None (only thermally induced)
Refractive index at 1064 nm	1.82
Temperature dependence of refractive index	$7 \text{ to } 10 \cdot 10^{-6}$/k
Nd density for 1% atm. Doping	$1.36 \cdot 10^{20}$ cm^{-3}
Fluorescence lifetime	230 µs

Its basic properties are: strongest transition at room temperature = 1064 nm, gain bandwidth = 257 GHz, spontaneous fluorescence lifetime = 230 us, melting point = 1,970 °C, change in line center of –1.4 GHz per °C. YAG is an extremely hard material and has good thermal conductivity. The main absorption bands of Nd:YAG are located at 570 to 600 nm, 730 to 760 nm, 790 to 820 nm, and 860 to 890 nm. The lasing output of YAG is randomly polarized so the crystal must be cut at the Brewster angle or there must be some additional optics inside the cavity if a linearly polarized beam is required.

Other than the main transition at 1064 nm there are also transitions near 940, 1120, 1320, and 1440 nm. Nd:YAG lasers operate in both pulsed and continuous mode. Pulsed Nd:YAG lasers are typically operated in the so called Q-switching mode: An optical switch is inserted in the laser cavity waiting for a maximum population inversion in the neodymium ions before it opens. Then the light wave can run through the cavity, depopulating the excited laser medium at maximum population inversion. In this Q-switched mode output powers of 20 megawatts and pulse durations of less than 10 nanoseconds are achieved.

Generally the crystalline host is doped with around 1% neodymium by weight but the amount of the neodymium dopant in the material varies according to its use. For continual wave output, the doping is significantly lower than for pulsed lasers. The lightly doped CW rods can be optically distinguished by being less colored, almost white, while higher-doped rods are pink-purplish.

The Nd^{3+} ions provide the energy levels for both the pumping and the laser transitions. Though the YAG host itself does not participate directly in the lasing action, it does have two important roles. When an Nd^{3+} ion is placed in a host crystal lattice it is subject to the electrostatic field of the surrounding ions, the so called *crystal field*. This crystal field of the host interacts with the electron energy levels in a variety of ways depending on such factors as its symmetry and strength, and on the electron configuration of the impurity.

A Nd^{3+} ion, which is free to move in a gaseous discharge, have many degenerate energy states (*i.e.*, many energy states with the same energy). When the ion is placed in the host the crystal field splits some of the energy levels, thereby partly removing the degeneracy. A rather more important effect in the case of the Nd:YAG laser is that the crystal field modifies the transition probabilities between the various energy levels of the Nd^{3+} ion so that some transitions, which are forbidden in the free ion, become allowed.

Figure 3.23 illustrates the simplified energy level diagram for the Nd^{3+} ion in YAG showing the principal laser transitions. Laser emission also results from transitions between the $^4F_{3/2}$ levels and the $^4I_{15/2}$ and $^4I_{13/2}$ levels, but at only one tenth of the intensity of the transition shown. The net result is that the ground and the first excited state energy levels of the Nd^{3+} ion split into the groups of levels. The symbols used to describe the energy levels in this case arise from Russel-Saunders or LS coupling. The symbol for an energy level is written $^{2S+1}X_J$. Here S is the vector sum of the electron spins of the ion. X gives the vector sum L of the orbital angular momentum quantum numbers, where values of L = 0, 1, 2, 3, 4, ...are designated by S, P, D, F, G ... Finally J is the vector sum of S and L. Thus in the $^4F_{3/2}$, S = 3/2 (there are three unpaired electrons in the 4f subshell of the Nd^{3+} ion), J = 3/2 and L =3.

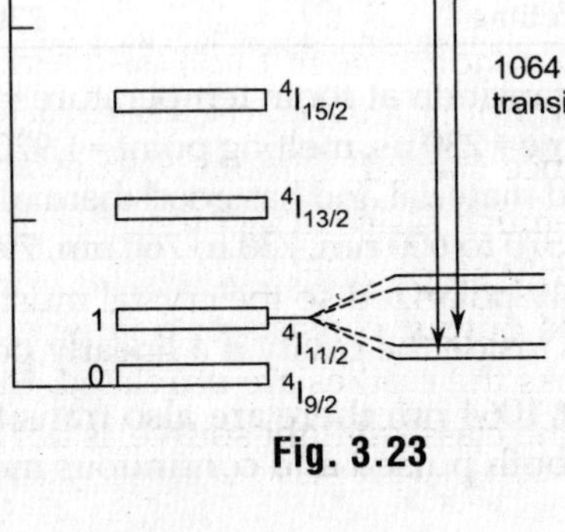

Fig. 3.23

As can be seen from Fig. 3.23 the N*d*:Yag laser is essentially a four level system, *i.e.*, the terminal laser level $^4I_{11/2}$ is sufficiently far removed from the ground state $^4I_{9/2}$. Whilst a number of laser transitions may occur between some of the pairs of levels shown to the right of the figure, the most intense line at 1604 nm arises from the superposition of the two transitions shown.

N*d*:YAG has been effectively pumped by various sources including flash lamps (xenon and krypton), krypton CW arc lamps, tungsten-halogen lamps, and high power laser diodes. At current densities of less than 4,000 A/cm^2, both xenon and krypton have a good match with the absorption curve of N*d*:YAG laser material. Even some more exotic methods have been used, such as sun-pumped, flashbulb-pumped, and explosively-pumped. The pumping excites the Nd^{3+} ions from the ground state to the various energy states above the $^4F_{3/2}$ state. In fact there are many more states at higher energies than are shown in Fig. 3.23. The presence of several possible pumping transitions contributes to the efficiency of the laser when using a pumping source with a broad spectral output. To ensure that the maximum possible radiation from the flashlamp is absorbed in the laser medium, close optical coupling is required. The usual arrangement is shown in the Fig. 3.24. A linear flashlamp and the lasing medium in the form of a rod are placed inside a highly reflecting elliptical cavity. If the flashlamp is along one focal axis and the laser rod along the other, then the properties of the ellipse ensures that most of the radiations from the flashlamp passes through the laser. The flashlamp is fired by discharging a capacitor bank through the lamp. The discharge is often initiated by using a secondary high voltage (≈ 20 *k*V) trigger pulse.

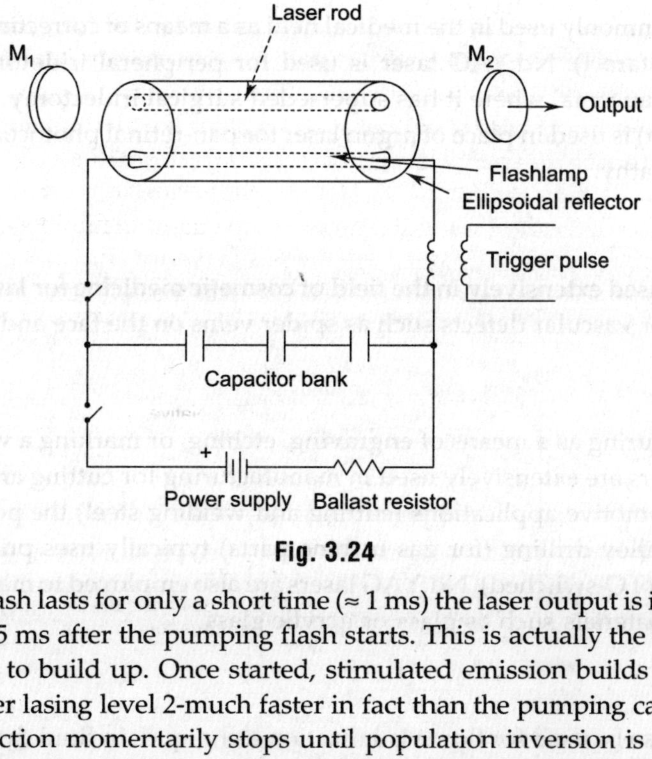

Fig. 3.24

As the pumping flash lasts for only a short time (≈ 1 ms) the laser output is in the form of a pulse, which starts about 0.5 ms after the pumping flash starts. This is actually the time required for the population inversion to build up. Once started, stimulated emission builds up rapidly and thus depopulates the upper lasing level 2-much faster in fact than the pumping can replace the excited atoms so that laser action momentarily stops until population inversion is achieved again. This process then repeats itself so that the output consists of a large number of spikes of about 1 μs duration with about 1μs separation. As the s pikes are unrelated, the coherence of the laser pulse, though is much greater than that of the classical light source, is not as great as might be expected.

The optical cavity may be formed by grinding the ends of the ND:YAG rod flat and parallel and then silvering them. More usually, external mirrors are used as shown in Fig. 3.24. As discussed for ruby laser, one mirror is made totally reflecting while the other is usually is about 10% transmitting to give an output. A large amount of heat is dissipated by the flashlamp and consequently the laser rod quickly becomes very hot. To avoid damage resulting from this, and to allow a resonable pulse repetition rate, cooling has to be provided by forcing air over the crystal using the reflector as a container. For higher power lasers it is required to use water cooling. Provided sufficient cooling is available it is possible to replace the flashlamp with quartz-halogen lamps and to operate the laser continuously.

Applications

Neodymium-YAG lasers have become very important because they can be used to produce high powers. Such lasers have been constructed to produce over a kilowatt of continuous laser power at 1065 nm and can achieve extremely high powers in a pulsed mode. Neodymium-YAG lasers are used in pulse mode in laser oscillators for the production of a series of very short pulses for research with femtosecond time resolution. These lasers are used in several applications in everyday life. Here we discuss a few of them.

Ophthalmology

Nd:YAG lasers are commonly used in the medical field as a means of correcting posterior capsular opacification (after-cataract). Nd:YAG laser is used for peripheral iridotomy in patients with acute angle closure glaucoma, where it has superseded surgical iridectomy. Frequency-doubled Nd:YAG laser (532 nm) is used in place of argon laser for pan-retinal photocoagulation in patients with diabetic retinopathy.

Cosmetic Medicine

These lasers are also used extensively in the field of cosmetic medicine for laser hair removal and the treatment of minor vascular defects such as spider veins on the face and legs.

Manufacturing

It is used in manufacturing as a means of engraving, etching, or marking a variety of metals and plastics. Nd:YAG lasers are extensively used in manufacturing for cutting and welding steel and super alloys. For automotive applications (cutting and welding steel) the power levels are typically 1–5 *k*W. Super alloy drilling (for gas turbine parts) typically uses pulsed Nd:YAG lasers (millisecond pulses, not Q-switched). Nd:YAG lasers are also employed to make subsurface markings in transparent materials such as glass or acrylic glass.

Fluid Dynamics

Nd:YAG lasers can also be used for flow visualization techniques in fluid dynamics (for example particle image velocimetry or induced fluorescence).

Dentistry

Nd:YAG lasers are used for soft tissue surgeries in the oral cavity, such as gingivectomy, periodontal sulcular debridement, frenectomy, biopsy, and coagulation of graft donor sites.

Additional Frequencies

For many applications, the infrared light is frequency-doubled or -tripled using nonlinear optical materials such as lithium triborate to obtain visible (532 nm, green) or ultraviolet light. A green laser pointer is a frequency doubled Nd:YVO_4 DPSS laser. Nd:YAG can be also made to lase at its non-principal wavelength. The line at 946 nm is typically employed in "blue laser pointer" DPSS lasers, where it is doubled to 473 nm.

Nd:Glass Lasers

Instead of a crystal, an amorphous glass, with its high optical homogeneity is used as the host material for neodymium. Its lasing characteristics are similar to Nd:YAG but since glass has a lower thermal conductivity, cooling becomes a problem for high power applications. The main reason Nd:Glass is used for these lasers is that it is the lasing medium that can be manufactured in sizes larger than a foot in diameter (as required to keep the fluence-energy density within acceptable limits to avoid damage to the optics in these multi-kilojoule pulsed lasers). Not to mention that it is also less expensive to prepare glass rods than to grow YAG crystals. Neodymium glass lasers have emerged as the design choice for research in laser-initiated thermonuclear fusion.

Local electric fields within the glass modify the Nd^{3+} ion energy levels in much the same way as the crystal field in YAG. The Nd^{3+} ions, however, may be in a variety of slightly different environments causing the linewidth to be much broader than in YAG and thereby raising the threshold pumping power required for laser action, as the threshold power level is proportional to Δv. In consequence, Nd:Glass lasers are operated in the pulsed mode and the output spectral line width is greater than in Nd:YAG. On the contrary, glass can be doped more heavily than YAG (6% as opposed to 1.5%) and up to three times as much energy can therefore be produced by Nd:Glass lasers. In fact these pulsed lasers generate pulses as short as 10^{-12} seconds with peak powers of 10^9 kilowatts.

Types of Nd:Glass

There are slight variations in the peak wavelengths for different types of Nd doped glasses. These differences are only very slight and should not be of great concern. The following are some glass types and peak emission wavelengths:

- LG-670 (silicate): 1,061 nm.
- Q-246 (silicate): 1,062 nm.
- LG-760 (phosphate): 1,054 nm.
- LGH-5 (phosphate): 1,054 nm.
- Q-88 (phosphate): 1,054 nm.

Only a few types of Nd:doped glass is commercially available. Most common are silicates and phosphates. Doping concentration varies from 0.5% to 6%.

Diode Pumped Lasers

CW solid state lasers may use xenon or krypton arc lamps or other sources of intense broad spectrum light. However, the trend today is toward the use of arrays of high power laser diodes to do the pumping. These can be designed to have a wavelength that matches an absorption band in neodymium (around 807 nm) making for very efficient excitation. The diode pumped approaches are rapidly taking over due to their efficiency, resulting lower power consumption and heat dissipation, much more compact size, as well as much higher reliability and lower (virtually no) maintenance. In fact, the relatively common, but still expensive green laser pointers currently available (prior to the introduction of direct injection green laser diodes) are diode pumped solid state lasers with a frequency doubler as part of the laser cavity. Power output is measured the same way as for other CW lasers.

Wall plug efficiency can vary from well under 1 percent for flashlamp and arc lamp pumped solid state lasers to 25 percent or more for those pumped with laser diode.

Diode pumped Nd:YAG may have a 40% efficiency (operating multimode with good thermal control of the diodes), and the pump diodes themselves have about a 45% efficiency, resulting in a net 18% of efficiency from electrical power to the diodes to output beam power. However, at increased pump powers, thermal issues may cause the efficiency to decrease after a certain point. This decrease is power dependent, as well as resonator and pump assembly design dependent.

Pumping is usually carried out longitudinally, *i.e.*, the pumping radiation enters through one end of the laser rod. This end has a dielectric coating so that it reflects 1.06 μm radiation (and hence acts as one of the cavity mirrors) whilst being transparent at the pumping wavelength. The cavity output coupling mirror is then situated some distance beyond the other end of the rod.

Microchip Diode Pumped Solid State Lasers

While what comes to mind when the term "solid state laser" is mentioned will likely be that of a ruby or YAG rod pumped by a linear or helical lamp, modern CW lasers are more likely to be diode pumped and small to medium power ones will use a small piece of Nd:YAG or Nd:YVO_4 as the lasing medium. For example, typical green laser pointers and a 100 mW green diode pumped solid state (DPSS) laser, both of these use a chip of Nd:YVO_4 (abbreviated as vanadate) to produce the 1,064 nm IR intra-cavity beam which is frequency doubled generating green light at 532 nm.

- For the green laser pointer, the vanadate is typically about 0.5 × 2 × 2 mm and the actual active portion is probably less than 0.5 mm long × 0.1 mm in diameter as determined by the lasing mode volume and the pump beam waist.
- For the 100 mW green laser, the vanadate is typically about 1.2 × 3 × 3 mm around 1.2 mm long × 0.15 mm in diameter.
- For a 2 W green DPSS laser, the vanadate may be 3 × 3 × 3 mm with 3 mm long × 0.3 mm diameter active.

Since microchip lasers can use very little actual lasing material and the pump diodes are also very small, they can be very compact, and potentially mass produced and inexpensive. In addition to green laser pointers and low to medium power DPSS IR and green lasers based on YAG or vanadate, all sorts of other solid state lasing materials can be used. Of particular interest for communications are erbium (Er) doped materials which lase around 1,530 nm, a wavelength which is optimal for fiber-optic cable.

Microchip lasers also don't necessarily need high pump power. Depending on type, cavity design, and pump beam shape, a few mW of pump beam may be enough to exceed the lasing threshold and they have very high slope efficiency (percent increase in laser output versus increase in pump input) as well.

Fiber Lasers

An interesting extension of the idea of an end pumped solid rod laser is the fiber laser. These lasers use optical fibers instead of solid rods. The basic structure of a glass fiber laser is shown in Fig. 3.25. An optical fiber, whose core is doped with rare earth ions, is positioned between two mirrors adjacent to its end faces which form the laser cavity. Light from a pumping laser source is launched through one mirror into the fiber core which is a wave-guiding resonant structure and forms a Fabry-Perot cavity. The optical output from the device is coupled through the mirror on the other fiber end face, as shown in Fig. 3.25. Thus the fiber laser is effectively an optical wavelength converter in which the photons at the pumping wavelength are absorbed to produce the required population inversion and stimulated emission. Hence a lasing output is provided at a wavelength which is characterized by the dopant in the fiber.

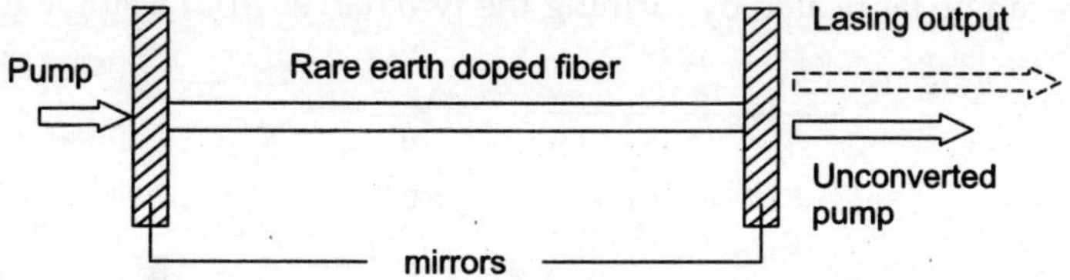

Fig. 3.25

There are fifteen rare earth elements or lanthanides and they occupy the penultimate row of the periodic table. They range from lanthanum (La), with an atomic number of 57, to lutetium with an atomic number of 71. Ionization of the rare earths normally takes place to form a trivalent state and the two major dopants currently used for fiber lasers are neodymium (Nd^{3+}) and erbium (Er^{3+}). In common with the Nd:YAG laser, the former element provides a four level scheme with significant lasing outputs at wavelengths of 0.90, 1.06 and 1.32 μm. The latter element gives a three level scheme with major useful lasing transitions at 0.80, 0.98 and 1.55 μm. If the imperfection losses in a fiber laser are low, then in a four level system the threshold power decreases inversely with the length of the fiber gain medium. In a three level system, however, there is an optimum length that gives the minimum threshold power which is independent of the value of the imperfection losses.

To date, silica-based glasses have provided the major host materials although fluorozirconate fibers doped with both neodymium and erbium ions have produced lasers emitting at wavelengths of 1.05 and 1.35 μm, and 1.55 μm respectively. In addition, fluoride glasses with other dopants give lasing outputs in the mid-infrared wavelength range.

Both neodymium and erbium doped silica fiber lasers employ co-dopants such as phosphorous pentoxide (P_2O_5), germania (*e.g.*, GeO_2, $GeCl_4$) or alumina (Al_2O_5). Dopant levels are generally low (at 400 ppm) in order to avoid concentration quenching which causes a reduction in the population of the upper lasing levels as well as crystallization within the glass matrix.

The basic Fabry-Perot cavity fiber laser shown in Fig. 3.25 can be easily constructed from standard optical components but it suffers from several limitations. In particular, the launching of light from the pump laser through one of the mirrored fiber ends can cause damage to the mirror coating as well as a substantial reduction in the launch efficiency. Furthermore, the gain spectrum of most rare earth ions extends over a wavelength range of some 50 nm. Unless the dielectric coatings on the mirrors are specially designed for broadband performance, however, the lasing output will be restricted to between 5 and 10 nm. Such a linewidth is too narrow for the provision of a broadband optical source but too wide to be used in single frequency laser applications such as coherent transmission. Different alternative fiber laser structures have therefore been fabricated which do not require dielectric or metallic mirrors. Two of such structures are the fiber ring resonator and the fiber loop reflector made from a series concatenation of distributed reflectors using loops of fiber. They are shown in Fig. 3.26.

The fiber ring resonator may employ the coherent beam splitting properties of the single-mode fiber fused directional coupler. In this case two of the arms of the coupler are spliced together as shown in Fig. 3.26 (*a*) to form a circulating pathway in which the light can travel. Hence an optical cavity without mirrors is formed. The finesse of this structure is determined by the splitting ratio of the coupler. (The finesse of the Fabry-Perot cavity provides a measure of its filtering properties and can be defined as the free spectral range divided by the full width half maximum permitted by the cavity). High performance fiber ring resonators can be fabricated by forming the two halves from a single fiber length.

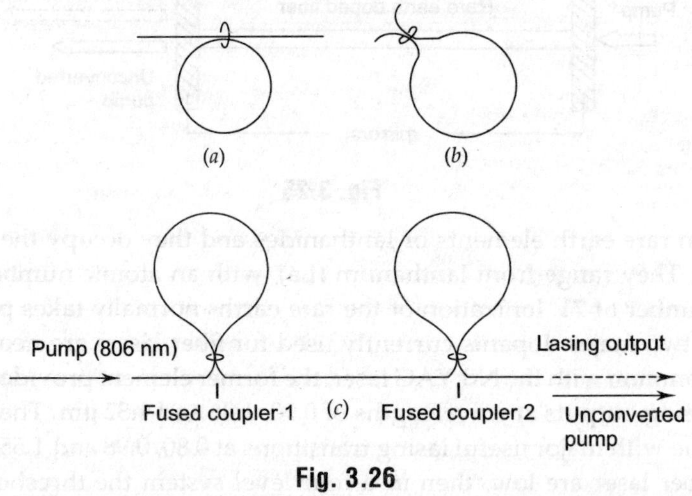

Fig. 3.26

The structure of a fiber loop reflector is shown in Fig. 3.26 (*b*). In contrast to the fiber ring where there is energy storage within the resonant structure, the fiber loop is a non-resonant interferometer (it constitutes a Sagnac interferometer). Light which enters the loop through the input fiber end forms forward and backward (reflected) waves which are counter propagating, providing a coherent superposition of the clockwise and counterclockwise propagating fields. Thus the single fiber loop performs as a distributed all-fiber reflector which may be used to form a fiber laser. Again, when two such loops are joined together in series, as shown in Fig. 3.26 (*c*), a resonator is obtained. This two loop structure provides the all-fiber laser.

3.16.2 Gas Lasers

The most widely used type of lasers are the gas lasers. They range from the low power He-Ne to very high power CO_2 lasers. According to whether the transitions are between the electronic energy levels of atoms or ions, or between vibrational/rotational levels of molecules, gas lasers can be classified into three different types. In general, the energy levels involved in the lasing process are well defined and the absence of broad bands effectively eliminates the possibility of optical pumping. Though other methods can be used, most gas lasers are excited by electron collisions in a gas discharge. We now consider typical examples from each of those classes mentioned above.

Atomic Laser-the Helium-Neon Laser

A helium-neon laser, usually called a **He-Ne laser**, is a type of small gas laser. The He-Ne laser was the first gas laser to be invented, by Ali Javan, William Bennet Jr. and Donald Herriot at Bell Labs, who in 1961 achieved continuous wave emission of the laser on the 1.15 μm wavelength line. He -Ne lasers have many industrial and scientific uses, and are often used in laboratory demonstrations of optics. Its usual operation wavelength is 632.8 nm, in the red portion of the visible spectrum.

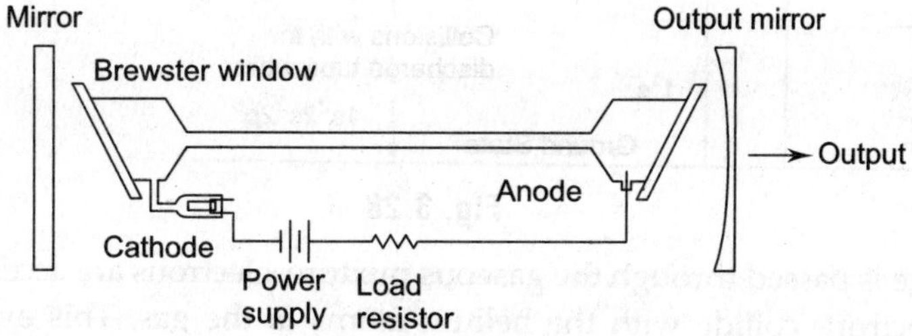

Fig. 3.27

A schematic diagram of the He-Ne laser is shown in the Fig. 3.27. The He-Ne laser consists of a long and narrow discharge tube (of diameter of about 1 cm and about 80 cm long). The tube contains helium gas at a pressure of about 1 torr and neon at a pressure of about 0.1 torr (a torr is a unit of pressure equivalent to 1/760 of an atmosphere). The main purpose of the helium is to act as a continuous reservoir of energy (supplied with electrical discharge) for the neon.

Optical pumping methods as used in solid state lasers are inefficient for gas lasers, because the gas atoms are characterized by sharp energy levels compared to solids, which are characterized by energy bands rather than energy levels. The energy or pump source of the He-Ne laser is provided by an electrical discharge of around 1000V through an anode and cathode at each end of the glass tube. A current of 5 to 100 mA is typical for CW operation. The optical cavity of the laser typically consists of a plane, high-reflecting mirror at one end of the laser tube, and a concave output coupler mirror of approximately 1% transmission at the other end. The optical output power ranges from 1 m to 100 mW.

The red He-Ne laser wavelength is usually reported as 632 nm. However, the true wavelength in air is 632.816 nm, so 633 nm is actually closer to the true value. For the purposes of calculating the photon energy, the vacuum wavelength of 632.991 nm should be used. The precise operating wavelength lies within about 0.002 nm of this value, and fluctuates within this range due to thermal expansion of the cavity.

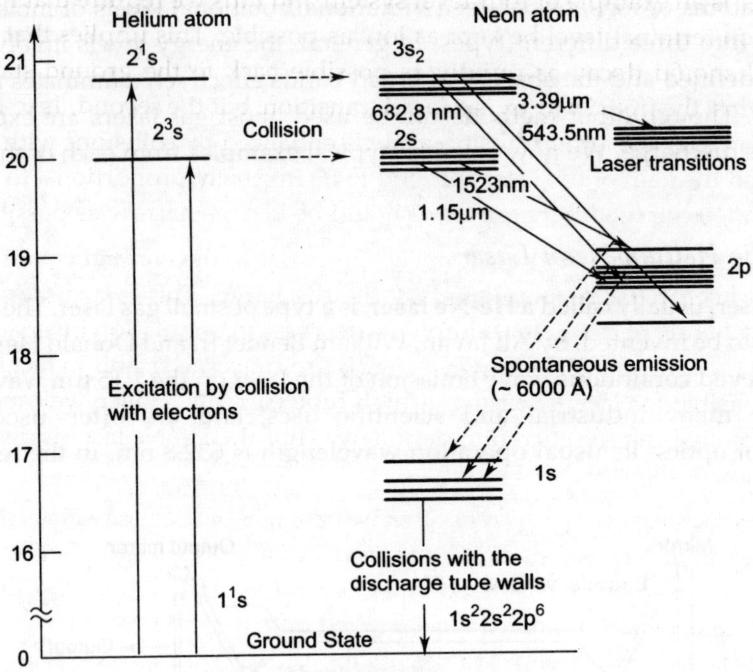

Fig. 3.28

When a discharge is passed through the gaseous mixture, electrons are accelerated down the tube. The accelerated electrons collide with the helium atoms in the gas. This excites helium from the ground state to one of the two metastable states designated 2^3s and 2^1s (as illustarted in Fig. 3.28). Collision of the excited helium atoms with the ground-state neon atoms results in transfer of energy to the neon atoms, exciting neon electrons into the $3s_2$ level. This is due to a coincidence of energy levels between the helium and neon atoms. The probability of this resonant transfer of energy is proportional to , where ΔE is the energy difference between the excited states of the two atoms involved. The energy level diagram for helium and neon atoms is shown in Fig. 3.28. Here, the neon states are labeled by the so called Paschen notation (the numerical subscripts have no spectroscopic significance and are used only as labels) while the helium states are labeled according to the L-S representation. The Fig. 3.28 shows that there is a group of four neon levels at almost the same energies as each of the two excited helium states, and resonant transfer thus occurs quite readily. This process is given by the reaction equation:

$$\text{He}(2^1s)^* + \text{Ne} + \Delta\text{E} \rightarrow \text{He}(1^1s) + \text{Ne}3s_2^* \quad (3.82)$$

where (*) represents an excited state, and ΔE is the small energy difference between the energy states of the two atoms, of the order of 0.05 *e*V or 387 cm^{-1}, which is supplied by kinetic energy. A population inversion is thus created between the 3*s* and (3*p*, 2*p*) groups of levels and also between the 2*s* and 2*p* levels. Transitions between the 3*s* and 2*s* levels and between the 3*p* and 2*p* levels are forbidden by quantum mechanical rules.The number of neon atoms entering the excited states builds up as further collisions between helium and neon atoms occur, causing a population inversion. Spontaneous and stimulated emission between the 3*s* and 2*p* states results in emission of 632.82 nm wavelength light, the typical operating wavelength of a He-Ne laser. Because the neon upper level saturates with higher current and the lower level varies linearly with current, the He-Ne laser is restricted to low power operation to maintain population inversion.

The He-Ne laser is an example of four level system and thus we require that the population of the lasing transition terminal level be kept as low as possible. This implies that the electrons in the terminal level should decay as rapidly as possible back to the ground state. In neon this happens in two steps; the first, $2p$ to $1s$, is a rapid transition, but the second, 1s to the ground state is not so rapid. However the latter transition rate is enhanced by collisions with the walls of the discharge tube, and the gain of the laser is found to be inversely proportional to the tube radius. So for this the diameter of the discharge tube should be kept as narrow as possible.

The transition $2p$ to 1s is also of interest since it gives rise to the familiar colour of 'neon lights'. Thus the $2p$ level itself must be populated by the discharge. This is a disadvantage since an increase in the population of the 2p level implies a decrease in population inversion (at least as far as the 1.15 μm, 632.8 nm and 543.5 nm wavelengths are concerned), and in fact this effect is to a large extent responsible for lasing ceasing at high tube currents. Hence we cannot increase the output simply by increasing the current indefinitely. Thus the He-Ne laser is destined to remain a relatively low-power device.

With the correct selection of cavity mirrors, other wavelengths of laser emission of the HeNe laser are possible. There are infrared transitions at 3.39 μm and 1.15 μm wavelengths, and a variety of visible transitions, including a green (543.5 nm, the so-called GreenNe laser), a yellow (594 nm) and an orange (612 nm) transition. The typical 633 nm wavelength red output of a HeNe laser actually has a much lower gain compared to other wavelengths such as the 1.15 μm and 3.39 μm lines, but these can be suppressed by choosing cavity mirrors with optical coatings that reflect only the desired wavelengths.

Each of the four major laser transitions (at 3.39 μm, 1.15 μm, 632.8 nm and 543.5 nm) shares with one of the others either a common starting or terminating level. Thus the transitions are always competing with each other and precautions must be taken to prevent the two unwanted wavelengths from lasing. This can be achieved quite easily by using the multilayer coated mirrors as mentioned in the previous paragraph. The very low absorption loss of such mirrors is an essential feature as the gain in the He-Ne medium is rather small.

The basic structure of the He-Ne laser is relatively simple. The essential elements are already illustrated in Fig. 3.27. The discharge is usually initiated by a high voltage 'trigger' pulse and then maintained at a current of 10 to 20 mA. The mirrors forming the resonant cavity are sometimes cemented to the ends of the discharge tube, thereby forming a gas-tight seal. On the other hand the mirrors can be external to the tube which is then sealed with glass windows oriented at the Brewster angle to the axis of the tube. This arrangement allows 100% transmission for the radiation which has its electric vector vibrating parallel to the plane of incidence, thereby ensuring the maximum possible gain in each round trip. The Brewster windows result in plane polarized output beam. Though this arrangement is complicated compared to the previous one, it enables us to insert mode selecting, frequency stabilizing and other devices in the cavity. The mirrors can also be altered to allow operation with other output characteristics and at other wavelengths.

The gain bandwidth of the laser is dominated by Doppler broadening, and is quite narrow at around 1.5 GHz for the 633 nm transition lasing on a single longitudinal mode. The visible output of the HeNe laser, and it's excellent spatial quality, makes the HeNe a useful source for holography and as a reference for spectroscopy. It is also one of the benchmark systems for the definition of the meter. Prior to the invention of cheap, abundant diode lasers, HeNe lasers were used in barcode scanners.

Ion Laser the Argon Laser

An **ion laser** is a gas laser which uses an ionized gas as its lasing medium. Like other gas lasers, ion lasers feature a sealed cavity containing the laser medium and mirrors forming a Fabry-Perot resonator. Unlike He-Ne lasers, the energy level transitions that contribute to laser action come from ions. Because of the large amount of energy required to excite the ionic transitions used in ion lasers, the required current is much greater, and as a result all but the smallest ion lasers are water cooled. The most powerful CW lasers operating in the visible region are the inert gas ion lasers such as the argon and krypton ion lasers. CW outputs of several watts can readily be obtained while, if the laser is pulsed, powers up to a kilowatt in microsecond pulses can be generated.

Argon Laser

The Argon laser was invented in 1964 by William Bridges at Hughes Aircraft and is one of a family of Ion lasers that use a noble gas as the active medium. Argon lasers are used for retinal phototherapy (for diabetes), lithography, and pumping other lasers. Argon lasers emit at several wavelengths through the visible and ultraviolet spectrum (see Table 3.2)

Common argon and krypton lasers are capable of emitting continual wave output of several milliwatts to tens of watts continually. In comparison with the helium-neon lasers requiring just a few milliamps, the current used for pumping the krypton laser ranges in several amperes, as the gas has to be ionized. The ion laser tube produces a lot of waste heat and requires active cooling.

Argon as an Excitation Medium

The neutral argon atom is pumped to the 4*p* energy level-the origin of the lasing transition-by two collisions with electrons. The gas atoms are ionized by the first electron collision in a high current discharge ($\approx$ 15 to 50 A). The second excites it from the ground state E_1 either directly to the 4*p* level (E3), some 35 eV above the atomic ground state or to E_4, from which it cascades almost immediately to 4*p*. As the electron energies are only a few electron volts, the excitation must be a result of multiple collisions. A population inversion forms between the 4*p* levels and the 4*s* (E_2), level which is about 33.5 eV above the ground state. The 4*p* ions eventually decay to 4*s*, either spontaneously or when stimulated to do so by a photon of appropriate energy. The wavelength of the photon depends on the specific energy levels involved, but will be between 351 nm and 520 nm. The ion decays spontaneously from 4s to the ground state emitting an ultraviolet photon, about 72 nm. The energy level diagram form Argon ion laser is illustrated in Fig. 3.29.

The tube design of the argon laser is much more complicated compared to that of the He-Ne laser. This is mainly because of the much higher energy required to pump the ionic levels and the need to dissipate the heat energy released. The current density can be increased by concentrating the discharge with a magnetic field applied along the axis of the tube. A magnetic field produced by a solenoid envelopes the plasma and enhances population inversion. It tends to force the free electrons toward the center of the tube, increasing the probability of a pumping collision and thus reducing the number of ions which collide with and damage the walls of the tube. The tubes are usually made of kovar, beryllium oxide, graphite, ceramics, or copper.

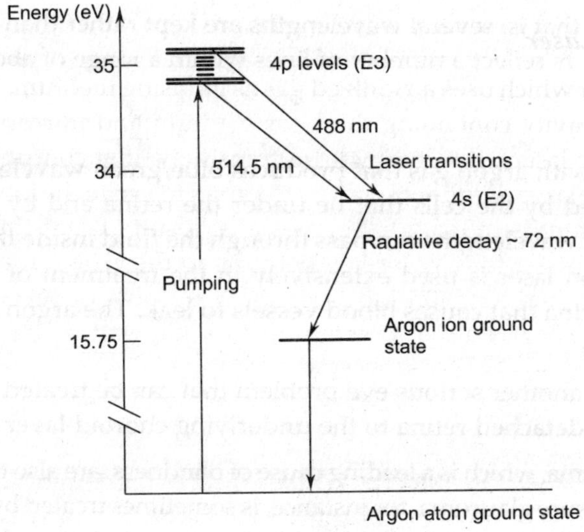

Fig. 3.29

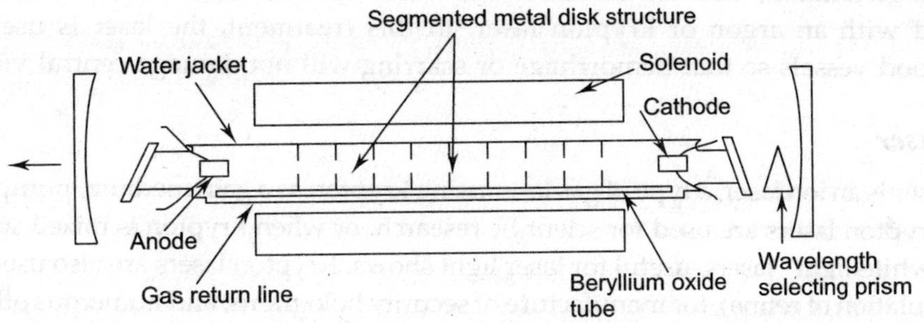

Fig. 3.30

As already mentioned, to dissipate the large amount of heat generated, most ion lasers are water cooled. They often include a series of metal disks to act as heat exchangers, as shown in Fig. 3.30. Holes in the center of these disks define the active region of the laser. Because of the high current involved, the cathode must be an excellent electron emitter and a getter is often incorporated to clean up any impurities which may otherwise affect the cathode.

The discharge is initiated by a high voltage pulse and then maintained by a d.c. voltage of 200 V during operation, the positive ions tend to collect at the cathode and may eventually cause the discharge to be extinguished. Proper pressure inside the cavity must be maintained to optimize the 'gain.' Varying the number of argon atoms in the volume will change the time between collisions with free electrons, which varies the average electron energy. To maintain the proper pressure, a gas return path is provided between the cathode and anode. The pressure-balanced design of the plasma tube brings stability to this argon ion laser. Pulsed ion lasers tend to be simpler and, with a low duty cycle, the heat generated is small enough to be dissipated by convective cooling.

There are many competing energy bands. These can be preferentially selected using a small prism in front of one end mirror. This prism selects a specific wavelength to send back through the cavity to stimulate identical emissions, which stimulates more, and so on. This describes single line operation. Hence the laser can be tuned to different wavelengths. Removal of the prism allows for broadband operation, that is, several wavelengths are kept rather than keeping only a particular wavelength. The mirrors reflect a number of lines within a range of about 70 nm maximum.

Application

The argon laser is filled with argon gas that produces blue/green wavelengths. These particular wavelengths are absorbed by the cells that lie under the retina and by the red hemoglobin in blood, but the blue-green wavelengths can pass through the fluid inside the eye without damage. For this reason, the argon laser is used extensively in the treatment of diabetic retinopathy, a severe disorder of the retina that causes blood vessels to leak. The argon laser can burn and seal these blood vessels.

Retinal detachment is another serious eye problem that can be treated by the argon laser. The laser is used to weld the detached retina to the underlying choroid layer of the eye.

Several forms of glaucoma, which is a leading cause of blindness, are also treated with argon lasers. The very serious angle closure glaucoma, for instance, is sometimes treated by using the laser to create a tiny opening in the iris, allowing excess fluid inside the eye to drain to reduce pressure.

Macular degeneration, a severe condition that affects central vision in older adults, is sometimes treated with an argon or krypton laser. In this treatment, the laser is used to destroy abnormal blood vessels so that hemorrhage or scarring will not damage central vision.

Krypton Laser

A krypton laser is an ion laser, a type of gas laser using krypton as a gain medium, pumped by electric discharge. Krypton lasers are used for scientific research, or when krypton is mixed with argon, for creation of "white-light" lasers, useful for laser light shows. Krypton lasers are also used in medicine (*e.g.*, for coagulation of retina), for manufacture of security holograms, and numerous other purposes.

Krypton lasers emit at several wavelengths through the visible spectrum (see Table 3.2). The most intense line is at 647.1 nm.

Table 3.2 Wavelengths of laser lines for Argon and Krypton (TEM_{00} lines)

Ar^+-ion Laser λ (nm)	*Kr^+-ion Laser λ (nm)*
351.1–363.8 multiple lines	337.4–356.4 multiple lines
457.9–514.5 multiple lines	406.7–415.4 multiple lines
514.5	647.1–676.4 multiple lines
501.7	799.3
496.5	752.5
488.0	676.4
476.5	647.1
472.7	568.2
465.8	530.9
457.9	520.8
454.5	482.5
	476.5
	413.1

Molecular Laser: the Carbon Dioxide Laser

The possibility of using molecular vibrations for laser action was clearly pointed out by Polanyi in 1961. More recently, Patel has described in some detail the laser action on the vibrational-rotational (V–R) transitions of CO_2 in an electrical discharge. Shortly afterwards, progress towards high power and efficiency was achieved by Patel with the addition of N_2 and by Moeller and Rigden with the addition of He. Other molecular systems have been made to lase, but so far CO_2 is the most important power producer.

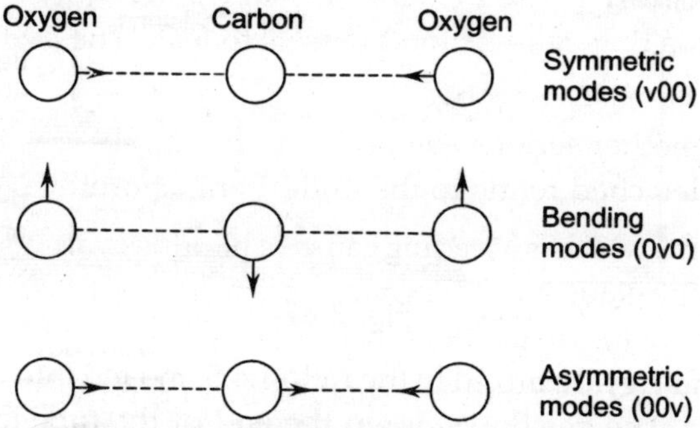

Fig. 3.31

The carbon dioxide laser is the most important molecular laser and indeed it is arguably the most important of all lasers from the standpoint of technological applications. In molecular lasers only the electronic ground state is of interest. Thus, molecular translation, vibration and rotation are the principal forms of energy storage and interchange. Carbon dioxide is basically a linear, symmetric molecule with the carbon atom balanced against the two oxygen atoms, O-C-O. Therefore, three characteristic vibrational modes exist (as shown in Fig. 3.31): the *symmetric* mode, where the two oxygen atoms vibrate against each other; the *bending* mode, where the carbon atom moves out of the molecular axis, thus bending the molecule; and the *asymmetric* mode, where the two oxygen atoms oscillate against the carbon atom. Each of these vibrating modes is quantized, and the particular vibrational state of the molecule is designated by three integral numbers (v_1, v_2, v_3). The first number is the quantum level or excitation number of the symmetric mode; the second number is the excitation number of the bending mode; and the third number is the excitation number of the asymmetric mode. Only the energy levels of the pure modes are given, but all the mixed vibrational modes exist as well.

In addition to these vibrational modes, the molecules can also rotate. Each and every vibrational state is degenerate or further subdivided into a whole series of levels brought about by gross rotation of the vibrating molecule. These levels are also quantized and designated by J, the rotational quantum number. The energy separation between these molecular levels is small and the laser output is therefore in the infrared. The important energy levels of the molecule are shown in Fig. 3.32, which also shows the ground state and the first excited state of the vibrational modes of nitrogen.

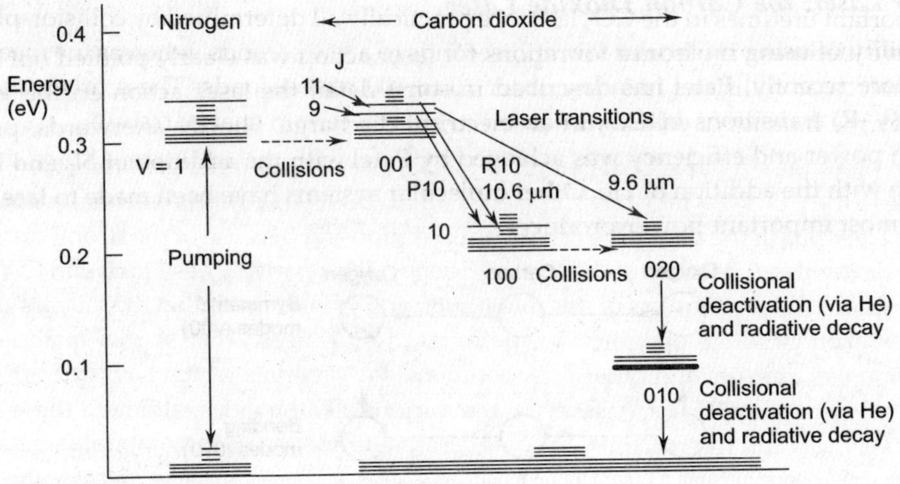

Fig. 3.32

Three gases (CO_2, N_2 and He) are mixed in the ratio 1:4:5.and fed into one end of a discharge tube at a pressure of a few torr. The gas flows down the end of the tube in about one second and is pumped out the far end with a mechanical forepump. An electrical discharge is maintained between the metallic end flanges of the tube. The ballast resistance is required because of the negative dynamic resistance of the discharge. With a fully reflecting mirror on the left and a partially transmitting mirror on the right, the device becomes a laser which radiates in the far infrared at 10.6 microns.

Population inversion is not a result of direct electronic excitation but is caused by vibrational energy interchange between N_2 and CO_2. The lowest vibrational level of N_2 is essentially metastable since the molecule is symmetric and cannot radiate. In a nitrogen discharge, about 20–30 per cent of the molecules are vibrationally excited. The first energy level of N_2, *i.e.*, $n = 1$ matches the level of the 001 state of CO_2 within 2 mV, (well within the thermal energy spread of about 25 mV). The (100) CO_2 levels have a lower energy and cannot be populated in this way, so that population inversion is created between the (001) and (100) levels giving stimulated emission at about 10.6 microns. When the two gases are mixed, effective resonance collision transfer occurs between the two states. Also, the second level of N_2 matches the 002 mode of CO_2 and so on up the ladder. The net result is that the asymmetric ladder, 00*n*, becomes highly excited; or it has a high vibrational temperature.

However, when the laser is operative, the 100 mode and its chain become heavily populated. This chain cannot radiate to the ground state and can only radiate weakly to the lower bending modes. However, there is a close "Fermi" resonance of the 100 mode with the 020 mode, and collisions can cause this transition.

The helium gas has a dual role. Firstly, it increases the thermal conductivity to the walls of the tube, thereby decreasing the temperature and Doppler broadening, which in turn increases the gain. Secondly, it increases laser efficiency by indirectly depleting the population of the (100) level, which is linked by resonant collisions to the (020) and (010) levels, the latter being depleted via collisions with the helium atoms.

The important lifetimes in the CO_2 laser are practically all determined by collision phenomena. The radiative lifetimes vary from a few milliseconds to a few seconds, whereas the mean free time between molecular collisions is of the order 10 to 100 nanoseconds. These times must then be scaled up by the number of collisions needed to effect a certain energy transfer.

Nitrogen plays the similar role of helium in He-Ne laser. While other gas lasers have efficiencies of 0.1% or less, the CO_2 laser may have efficiency up to about 30%. This is essentially due to the ease with which electrons in the discharge can cause excitation and because of the strong coupling of the various levels involved. Because of this high efficiency, it is relatively easy to obtain CW outputs of 100 W for a laser 1 m long. Powers of this magnitude and greater mean that the mirrors should have very low absorption, while operation in the infrared also means that special materials must be used for windows, mirrors and other laser components. Materials which have been used successfully include germanium, GaAs, ZnS, ZnSe and various alkali halides, although these suffer from being relatively soft and hygroscopic. In some cases, a diffraction grating mounted on a piezoelectric transducer is used instead of the high reflectance mirror. The grating permits tuning of the laser output over the large range of distinct lines.

The lasing transition at 10.6 microns is a simultaneous change of the vibrational and rotational quantum states. However, this may occur on any one of a number of rotational transitions. The selection rules for the vibrational-rotational transition are that J must change by ±1. Those transitions where J increases by 1 are called P-branch transitions (for example, **P10**), and those that decrease are called R-branch transitions (for example, R10). The wavelengths associated with the dominant transitions are **P18-10.57** microns, **P20-10.59** microns, **P22-10.61** microns, etc. When the inversion in the total vibrational level occurs because of the discharge, the P branch transitions have more gain than their R counterparts. Since the P and R transitions compete with each other, the medium lases only on the P branches unless special precautions are taken. Those P branches which have the highest gain are most likely to show lasing action.

Discharge Properties

Although the CO_2, N_2, and He discharge can be operated either with dc, ac, RF, or pulses, it appears that the maximum average output power can be achieved either with dc or low-frequency ac applied directly to the electrodes. The power is less with RF, probably because it is hard to keep a long length of the discharge uniformly excited. Where the discharge must be pulsed, the average power may be only 1/10 to 1/8 of the dc value; also the optimum pressure and output coupling conditions change.

Electrical characteristics of the CO_2 laser are high voltage and low current with a negative dynamic resistance. Except for the large negative resistance, the discharges are similar to cold-cathode, glow discharges. In flowing gas systems such as these, almost any material may be used for the electrodes. For glass tubes which are bolted together with O rings, metal flanges between the sections are most convenient. For glass-to-metal seal systems, the Kovar seals themselves can be used for the electrodes.

CO_2 Laser Characteristics

Important CO_2 laser characteristics are high unsaturated gain, high-power output and good efficiency. Gain is a strong function of tube diameter and that output power is essentially independent of diameter. The gain curves for larger diameter tubes can be substantially changed simply by adding some turbulence to the gas flow. The gain reaches its maximum at low currents, falling off as current increases. This fall-off is apparently caused in part by the heating of the gas on the tube axis.

So far, the optimum transmission coefficient of the output mirror has been determined by trial and error. Ideally, to achieve the maximum power, the transmission coefficient for a tube with a given diameter and length should be known. Fortunately, the power curve is quite broad with variation in transmission coefficient, so exact values are not needed. Based on tube diameter and discharge length, a rough estimate of the required transmission coefficients can be obtained.

The discharge tubes should normally have water jackets for cooling, but in the flowing gas system the effect of cooling is only marginal. For example, the output power from the 50 mm bore tube may decrease by only 10 per cent as the wall temperature increases 30° to 40°C. Sealed-off lasers, however, do not dissipate their power by dumping the stored energy into the flowing gas, but must dissipate their heat to the walls. Thus, water jackets are a must in order to cool the central gas and thereby avoid loss in output power.

Applications

Many applications have been suggested for CO_2 lasers, but only a few of these have been given serious consideration and analysis. This is partly because little engineering has been done at 10 microns.

Because most materials absorb at 10 microns, considerable interest has been shown in CO_2 lasers for cutting applications. This means cutting such things as paper, cardboard, plastics, glass, quartz, wood (tree surgery), meat, flesh (bloodless surgery), metal, and rock. Except for the last two items, the cutting process is just what one might imagine if a small intense flame were to burn through the material. In a sense, the process is very similar to electron-beam cutting and welding, but it is much simpler to use. For metal, the initial absorption depends on the surface condition, and the cutting depends strongly on the size and thermal conductivity of the material. Thin-sheet stock can be cut, but molten metal tends to freeze out on the edges giving a rough cut. Thick metal, however, tends to form a molten pool which must be removed by some means if further penetration is to occur. It was discovered that irradiation of hard rocks with power densities of the order of 200 W per square cm caused the surface to become white hot. The resulting mechanical stress would cause a general weakening of the rock; in some cases, it caused the rock to crumble. At higher power densities, the rock would melt, relieving the stress locally. Further research into the comminution of hard materials is under way. Most materials are opaque at 10 microns, and any problem which requires controlled surface heating or burning might find a potential solution with the CO_2 laser.

The CO_2 laser is a specialized laser that is filled with carbon dioxide gas and uses an infrared emission for cutting tissue through heat absorption. It is one of the most common lasers used in surgery and is good for precise cutting and vaporization of tissue, such as that needed in the treatment of superficial lesions or removing small volumes of tissue.

The CO_2 laser is used by ophthalmic plastic surgeons to remove fine wrinkles from around the eyes. This laser precisely removes the outermost layer of skin and the underlying dermis, allowing the regrowth of wrinkle-free new skin.

Types of CO_2 Lasers

Although there are many types of structures for CO_2 lasers. Followings are the most common.

- **Axial gas flow:** The gas mixture is pumped in one end of the tube and taken out from the other. This provides fresh gas to replace the (CO_2) depleted due to dissociation of the gas molecules. He and N_2 are added to the mixture to boost efficiency. Power output of this type of laser is typically 40 to 80 W per meter of tube length (more or less independent of tube diameter). Folded optical systems may be used to reduce the total physical length of the laser. This approach is practical for output powers of up to a couple of *k*W.
- **Transverse gas flow:** Instead of flowing down the tube, the gas flows across it providing the highest power ratings for continuous CO_2 laser operation. Power outputs of 10 *k*W per meter are possible with Transverse Excitation Atmospheric (TEA) designs.
- **Sealed tube:** The structure of these is very similar to that of sealed He-Ne and Ar/Kr ion lasers but with features (bore diameter, etc.) geared to much longer CO_2 wavelength. Power output of sealed CO_2 lasers ranges from a few watts to perhaps 100 W (maybe more).

The most common types of excitation are a direct electrical discharge (usually DC) and Radio Frequency (RF). Either current control or pulse width modulation can be used for power control (since this does make sense for a CO_2 laser as output power is directly related to tube current). To boost power output and provide some redundancy, some CO_2 lasers have multiple separately excited discharge tubes which are optically combined.

Flowing Gas CO_2 Lasers

This was the way the earliest CO_2 lasers were constructed. The tube was not sealed but required an active pumping system and gas supply to operate. Such lasers are very easy to construct. These have maximum power outputs in the 20 to 100 W range.

More modern CO_2 lasers in this power class, and up to about 500 W feature a low or zero maintenance sealed tube. However, very large output power lasers (*e.g.*, *k*Ws to tens of *k*Ws) still use a flowing gas design. Controlling the output power of a flowing gas CO_2 laser is done with a combination of discharge voltage and gas flow. As the voltage increases, the gas flow must be increased and vice-versa.

- A Variac may be used for voltage control. Even many commercial CO_2 lasers have used a manual or motorized Variac and neon sign transformer or phase control of a neon sign transformer for this.
- A control valve (manual or electrically operated) is used to adjust the rate of gas flow into the laser.

In commercial lasers, the tube voltage is either mechanically or electronically coupled to the flow valve or ('witty valve' as it is known). As you increase the voltage, the flow valve is opened in proportion to the power setting.

Sealed CO_2 Lasers

Commercial sealed CO_2 lasers have much in common with sealed He-Ne lasers. However, you can't just take an axial flow CO_2 design, seal it up, and expect the laser to work for more than a few minutes. The discharge process breaks down the CO_2 to produce CO and O_2 which quickly poison the lasing process. There are a number of solutions to this problem including the addition of other gases like H_2 or H_2O (water vapor) to the gas mix to react with the CO and O_2 to regenerate CO_2 or the use of a high temperature (300°C) cathode to act as a catalyst to stimulate recombination.

- The typical sealed CO_2 tube has an operating voltage of between 3 and 12 *k*V at 2 to 15 mA DC. Some may require 20 *k*V at 20 mA or more. (The corresponding values for He-Ne tubes are: .6 to 5 *k*V at 3 to 8 mA DC.
- A high starting voltage is required to initiate the discharge. For He-Ne laser tubes, the actual value is typically 3 to 5 times the operating voltage with the manufacturer's specifications being somewhat higher to assure reliable starting.
- The value of the negative resistance of the CO_2 discharge is between – 200 KΩ and – 600 KΩ. (The corresponding value for He-Ne tubes is about – 50 KΩ). The effective ballast resistor value is selected to be 30 to 50% higher than this to assure stability.

Therefore, requirements for the power supply to be used with sealed CO_2 and He-Ne lasers are very similar. However, due to the significantly larger negative resistance characteristic of the CO_2 laser, there is more incentive to push part of the effective ballast resistance into the control of a switch-mode inverter rather than a simple power wasting resistor.

One thing that is significantly different for a CO_2 laser compared to the He-Ne variety is efficiency: The electrical to optical efficiency of a typical small sealed CO_2 laser is around 5 to 8 percent compared to less than .1 percent for a He-Ne laser. However, the efficiency of large (flowing gas) CO_2 lasers can exceed 20 percent. The best efficiency is only achieved when tube diameter and gas pressure are optimal. For example, the optimum gas pressure for a sealed CW CO_2 laser using an 8 mm inside diameter glass tube (cooled to room temperature by a surrounding water jacket) is about 14 torr.

For tubes of different diameters, the plasma scaling laws for CO_2 laser operation work as follows, where D is the tube diameter or waveguide cross section:

- Optimum gas pressure varies as 1/D.
- Optimum current density varies as 1/D and thus the total current varies as D.
- Discharge voltage per unit length then varies as 1/D.
- Power input and laser output per unit length are then independent of D (ignoring waveguide losses, in the case of waveguide lasers).
- Tunability about each individual transition varies as 1/D

Transverse Excited Atmospheric (TEA) CO_2 Lasers

One problem with electrically excited axial CO_2 lasers is that there is an upper limit to the gas pressure at which a discharge can be maintained. This is well under 100 torr. Unlike helium-neon and many other gas lasers, the CO_2 laser will operate up to atmospheric pressure (and probably beyond) with gain/power output proportional to pressure. Thus, 10 to 20 times more power would be possible running at 1 atmospheric pressure.

Most high power (*k*Ws) CO_2 lasers are constructed using TEA (Transverse Excited Atmospheric) designs. Rather than having a long tube with a electrical discharge along its length and the CO_2 gas mixture flowing from end to end, a series of electrodes and gas inlets are spaced along the tube as illustrated in the Fig. 3.33. In this manner, higher pressures can be used since the electrical discharge doesn't need to go the full length of the tube—only across it. And, fresh gas can flow to all parts of the tube and doesn't get depleted along the way.

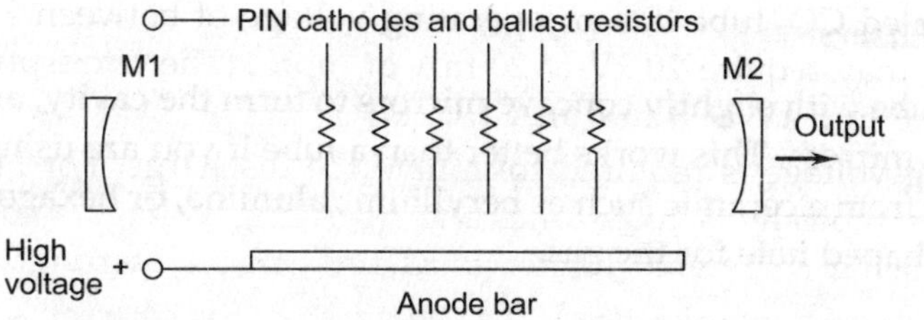

Fig. 3.33

TEA lasers must operate pulsed (a continuous discharge can't be maintained at higher pressures) and due to the shape of the cavity, the output beam has a rectangular cross-section. Gas flow is a major design consideration since the intense electrical pulses are very effective at destroying and rearranging chemical bonds.

Basically, there is a chamber with CO_2 laser gas in it. The gas is at atmospheric pressure (hence the "A" in "TEA"). Two electrodes run down the length of the chamber, one on the left side, one on the right. These electrodes are connected to a capacitor bank which, when discharged, sends a spark/pulse across the gap inside the chamber. The pulse excites the CO_2 laser gas, which then begins spontaneous emission. A reflector and an output coupler outside the chamber form the resonating cavity. The "TE" in "TEA" comes from the fact that the excitation is performed transverse to the direction of the laser output.

The problem with TEA lasers is that the high voltage pulse causes the CO_2 gas in the laser gas mixture to dissociate into CO and O_2. If new gas is not flown into the chamber or the dissociated gas is not re-circulated through a catalyst to recombine the molecules, the pulse rate of the laser is limited. For high repetition rate TEA lasers, it is not uncommon for the gas to flow at many meters per second through the chamber.

CO_2 TEA lasers operated with a normal gas mix typically produce a 'spike and tail' output, with ~1/3 of the energy in the ~100nS spike, and 2/3 in the few hundred ns to 1 μs 'tail'. The 'spike' is a gain switched feature — the 'tail' from *v-v* transfer from the N_2. If the laser is operated with reduced nitrogen the tail goes away. If the laser has a short cavity, and is fairly heavily coupled, the spike shortens. The exact numbers are very system dependent of course. It is quite easy to get approximately 50 ns pulses with no tail.

Slab Type CO_2 Lasers

The big problem with getting power out of CO_2 lasers is that when the gas mixture gets hot, the power goes away. The fundamental concept of a slab CO_2 laser is that you can extract heat from the laser gas without any moving parts by having a pair of planar water-cooled electrodes separated by a small gap. The laser gas is RF-excited by these electrodes.

To get around this, the Rofin-Sinar lasers use a Tulip resonator (named for its inventor, not its shape) which has an unstable resonator in one axis and a waveguide resonator in the other. Power is extracted off of one side of the slab. The actual output that comes off of this is strangely shaped, but external correcting optics and knife edges produce something that looks and works amazingly like a Gaussian beam.

Waveguide CO_2 Lasers

Instead of using a tube with slightly concave mirrors to form the cavity, an alternative is to use a waveguide and flat mirrors. This works better than a tube if you are using small diameters. The waveguide is made from a ceramic such as beryllium , alumina, or hexagonal boron nitride, with a round or square shaped hole for the gas.

RF Excited CO_2 Lasers

CO_2 lasers can be operated using radio frequency (including microwave) excitation instead of a direct electrical discharge but this results in more complex resonator/electrode configurations, more complex driving electronics and additional safety issues. Depending on the size and type of laser, maximum achievable power output and efficiency may be lower as well.

However, there are many efficient, compact RF excited lasers in the market. For example, the Coherent Gem Select 100 is water cooled sealed tube RF excited waveguide CO_2 laser using a folded resonator. Its rated output power is 100 W but typically produces 120 W. The laser is only about 7" (H) × 8" (W) × 31" (L) and weighs under 52 pounds including power supply, and runs on 200 to 240 VAC at 14 A max.

It is possible to make an RF-excited CO_2 laser with the electrodes in the form of broad plates that are closely spaced. With such a configuration, the gas mixture can be efficiently cooled by the electrodes. This is diffusion cooling. Of course, the discharge area is poorly shaped for power extraction by stable resonators, so these lasers use complex resonator designs.

Although the CO_2, N_2, and He discharge can be operated either with DC, AC, RF, or pulses, it appears that the maximum average output power can be achieved either with DC or low frequency AC applied directly to the electrodes. The power is less with RF, probably because it is hard to keep a long length of the discharge uniformly excited. Where the discharge must be pulsed, the average power may be only 1/10th to 1/8th of the DC value. Also, the optimum pressure and output coupling conditions change.

Gas Dynamic CO_2 Laser

A gas dynamic CO_2 laser essentially uses a rocket engine as the exciter. The propellants are burned to create a very high velocity, high pressure, and high temperature stream of gases (nitrogen and CO_2) in a lase-able ratio. As the exhaust expands through a nozzle, the temperature of the gases drop very rapidly. The lower energy levels are rapidly and selectively depopulated (aided by trace chemicals introduced into the combustion chamber exclusively for that purpose) while the upper energy levels (populated because of the high temperature of the gases in the chamber) remain energized, resulting in an instant, complete population inversion. After a time delay determined by the relaxation time of those upper energy states, the stimulated emission begins. The optical resonator cavity is positioned a certain distance beyond the throat of the nozzle (determined by exhaust gas velocity), so as to initiate the lasing process at the proper time. These lasers are quite difficult and expensive, but produce tremendous beam power. Gas dynamic lasers suffer from the disadvantage of large size and the rocket-like roar as the gas expands.

Other Molecular Laser: the Nitrogen Laser

As the number of molecular systems with energy levels which may be suitable for laser action is very large, it is not surprising that there are other molecular lasers and it is likely that many more will be discovered. One such molecular laser, pumped by electrical discharge and emitting in the ultraviolet range is nitrogen laser. The nitrogen laser differs markedly from the CO_2 laser. In the latter the transitions are between molecular rotational/vibrational energy levels, while in nitrogen the laser transitions are between electronic energy states.

Gain Medium

The gain medium of nitrogen laser is nitrogen molecules in the gas phase. The nitrogen laser is a 3-level laser. In contrast to a ruby laser or to other more typical 4-level lasers, the upper laser level of nitrogen is directly pumped, imposing no speed limits on the pump. Pumping is normally provided by direct electron impact; the electrons must have sufficient energy, or they will fail to excite the upper laser level. Typically reported optimum values are in the range of 80 to 100 eV per Torr-cm pressure of nitrogen gas. There is a 40 ns upper limit of laser lifetime at low pressures and the lifetime becomes shorter as the pressure increases. The lifetime is only 1 to 2 ns at 1 atmosphere. For various reasons the efficiency of the nitrogen laser is low, typically 0.1% or less, though nitrogen lasers with efficiency of up to 3% have been reported in the literature.

The strongest lines are at 337.1 nm wavelength in the ultraviolet. Other lines have been reported at 357.6 nm, also ultraviolet. This information refers to the second positive system of molecular nitrogen, which is by far the most common. There are also lines in the far-red and infrared from the first positive system, and a visible blue laser line from ionized molecular nitrogen positive ion.

The metastable lower level lifetime is 40 μs. Thus, the laser self-terminates, typically in less than 20 ns. This type of self-termination is sometimes referred to as "bottlenecking in the lower level". Thus, in contrast to the helium-neon laser, the nitrogen laser can only operate in pulsed mode. Many nitrogen lasers terminate even more rapidly, but this is usually because the drive circuitry cannot provide sufficiently energetic electrons for more than a few nanoseconds. Very fast rise time pumping mechanisms similar to those used for TEA lasers must be used. It might be thought that the short pulse-width of the nitrogen laser is caused by the short lifetime of the upper laser level, but that is demonstrably false. Several organic dyes with upper level lifetimes of less than 10 ns have been used in continuous mode. The upper level lifetime does limit the attainable pulse-width in the case of the nitrogen laser, but this is caused by the bottlenecking issue and is not an inherent feature of short upper-state lifetime.

Repetition rates can range as high as a few kHz, provided adequate gas flow and cooling of the structure are provided. Cold nitrogen is a better medium than hot nitrogen, and this appears to be part of the reason that the pulse energy and power drop as the repetition rate increases to more than a few pulses per second. There are also, apparently, issues involving ions remaining in the laser channel.

Gas pressure is generally between some mbar to several bars. Air (*i.e.*, 78%). Nitrogen can be used, but more than 0.5% oxygen poisons the laser, and air provides much less output than pure nitrogen or a mixture of nitrogen and helium.

Pulse length of the laser system is between a few hundred picoseconds (typically at 1 atmosphere partial pressure of nitrogen) and a maximum of approximately 30 nanoseconds at reduced pressure, typically some dozens of Torr, though FWHM (Full Width at Half Maximum) pulse-widths of approximately 6 or 8 ns are typical.

The gain in nitrogen is so large that it can be used as a simple amplifier; *i.e.*, in many applications it is not necessary to provide positive feedback (such high gain is termed super-radiant). Though nitrogen lasers can operate super-fluorescently (without a resonator cavity), it is quite common to put a mirror at one end so that all output can be collected from the other end. Commercial nitrogen lasers are capable of producing some kW to more than 3 MW peak power pulses.

Application

The nitrogen lasers are often used in photochemical investigations, for pumping of other lasers *e.g.*, dye lasers and measurement of air pollution.

3.16.3 Excimer Lasers

The first excimer laser was invented in 1971 by Nikolai Basov , V. A. Danilychev and Yu. M. Popov, at the P. N. Lebedev Physical Institute, using a xenon dimer (Xe_2) excited by an electron beam to give stimulated emission at 172 nm wavelength. A later improvement was the use of noble gas halides (originally Xe), invented (and patented) in 1975 by George Hart and Stuart Searles of the United States Government's Naval Research Laboratory. Since 1972, a large number of lasers based on excimers have been developed; which cover the wavelength range from about 120 nm to 500 nm. Rare gas halide excimer lasers are especially efficient with XeF and XrF giving the highest efficiencies (some 10 –15%).

An excimer laser is a powerful laser which is basically always operated in the ultraviolet (UV) spectral region and generates nanosecond pulses. The excimer gain medium is a gas mixture, typically containing a noble gas (rare gas) (*e.g.*, argon, krypton, or xenon) and a halogen (*e.g.*, fluorine or chlorine), apart from helium and/or neon as buffer gas. An excimer gain medium is pumped with short (nanosecond) current pulses in an electric discharge (or sometimes with an electron beam), which create so-called *excimers* (excited dimers).

Laser action in an excimer molecule occurs because it has a bound (associative) excited state, but a repulsive (disassociative) ground state. This is because noble gases such as xenon and krypton are highly inert and do not usually form chemical compounds. However, when in an excited state, they can form temporarily-bound molecules with themselves (dimers) or with halides (complexes) such as fluorine and chlorine. The excited compound can give up its excess energy by undergoing spontaneous or stimulated emission, resulting in a strongly-repulsive ground state molecule which very quickly (on the order of a picosecond) disassociates back into two unbound atoms. This forms a population inversion between the two states. Most "excimer" lasers are of the noble gas halide type, for which the term *excimer* is strictly speaking a misnomer (since a dimer refers to a molecule of two identical or similar parts). The correct but less commonly used name for such is **exciplex laser**. The wavelength of an excimer laser depends on the molecules used, and is usually in the ultraviolet. Some of the excimers with their typical wavelengths is shown in Table 3.3.

Table 3.3

Excimer	*Wavelength*
Ar_2^*	126 nm
Kr_2^*	146 nm
F_2	157 nm
Xe_2^*	172 & 175 nm
ArF	193 nm
KrF	248 nm
XeBr	282 nm
XeCl	308 nm
XeF	351 nm
CaF_2	193 nm
KrCl	222 nm
Cl_2	259 nm
N_2	337 nm

Excimer lasers are usually pumped by an intense electron beam source or by a fast discharge. The electrons in the electron beam are accelerated until they have energies of 1 MeV and then transmitted to the laser chamber in pulses giving beam currents of about 100 kA. Electron beam generators with this sort of capability are rather large and alternative pumping mechanisms are being investigated.

Typical excimer lasers emit pulses with a repetition rate up to a few kilohertz and average output powers between a few watts and hundreds of watts, which makes them the most powerful laser sources in the ultraviolet region, particularly for wavelengths below 300 nm. The power efficiency varies between 0.2% and 2%.

Early excimer lasers have had rather limited lifetimes due to different problems arising *e.g.*, from the corrosive nature of the used gases and from contamination of the gas with chemical byproducts and dust created by the electric discharge, and from the high load on the thyratrons in the power supply. Other challenges are the ablation of material from the electrodes and the high peak power of the required current pulses, which often allowed the thyratron switches to last only for a couple of weeks or months. However, a lot of engineering, involving *e.g.*, the use of corrosion-resistant materials, advanced gas re-circulating and purification systems, and solid state high voltage switches, has mitigated challenges of the excimer laser concept to a significant degree. The lifetime of modern excimer lasers is now limited by that of the UV optics, which has to stand high fluxes of short-wavelength radiation, to something of the order of a few billion pulses.

The UV light from an excimer laser is well absorbed by biological matter and organic compounds. Rather than burning or cutting material, the excimer laser adds enough energy to disrupt the molecular bonds of the surface tissue, which effectively disintegrates into the air in a tightly controlled manner through ablation rather than burning. Thus excimer lasers have the useful property that they can remove exceptionally fine layers of surface material with almost no heating or change to the remainder of the material which is left intact.These properties make excimer lasers well suited to precision micromachining organic material (including certain polymers and plastics), or delicate surgeries such as eye surgery (LASIK).

Formation and Decay

Simplified energy level diagram of excimer laser is shown at Fig. 3.34. Under the molecular orbital formalism, a typical ground-state molecule has electrons in the lowest possible energy levels. According to Hund's rule, at most two electrons can occupy a given orbital, and if an orbital contains two electrons they must be in opposite spin states. The highest occupied molecular orbital is called the HOMO and the lowest unoccupied molecular orbital is called the LUMO; the energy gap between these two states is known as the HOMO/LUMO gap. If the molecule absorbs light whose energy is larger than this gap, an electron in the HOMO may be excited to the LUMO. This is called the molecule's excited state.

Excimers are only formed when one of the dimer components is in the excited state. When the excimer returns to the ground state, its components dissociate and often repel each other. The wavelength of an excimer's emission is longer (smaller energy) than that of the excited monomer's emission. An excimer can thus be measured by fluorescent emissions.

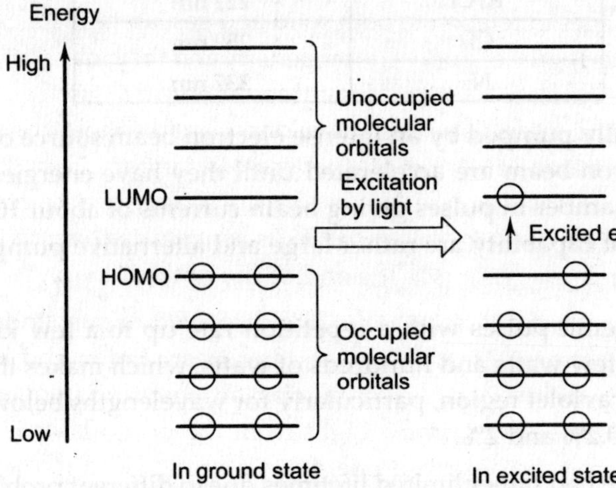

Fig. 3.34

Because excimer formation is dependent on a bimolecular interaction, it is promoted by high monomer density. Low-density conditions produce excited monomers that decay to the ground state before they interact with an unexcited monomer to form an excimer.

Applications

The short wavelengths of excimer lasers in the ultraviolet spectral region make possible a number of applications:

- The generation of very fine patterns with photolithographic methods, *e.g.*, in semiconductor chip production;
- Material processing with laser ablation, exploiting the very short absorption lengths of the order of a few micrometers in many materials;
- Pulsed laser deposition;
- Laser marking and micro-structuring of glasses and plastics;
- Fabrication of fiber Bragg gratings;
- Ophthalmology (eye surgery), particularly for vision correction with ArF lasers at 193 nm;

- Psoriasis treatment with XeCl lasers at 308 nm;
- Pumping other lasers, *e.g.*, certain dye lasers.

The beam output from an excimer is in general multimode and not of good quality when compared to other lasers. In laser drilling systems the excimer is employed similar to a conventional light source. The accuracy comes from the imaging system and the fact that UV light has a short wavelength.

Excimer lasers are quite large and bulky devices, which is a disadvantage in their medical applications, although their size is rapidly decreasing with ongoing development.

Excimer lasers raise a variety of safety issues, related to the use of high voltages, the handling of poisonous gases (halogens), and the risk of causing skin cancer and eye damage by irradiation with ultraviolet light.

3.16.4 Liquid Lasers: Dye

Liquids have useful advantages in relation to both solid and gas laser media. Solids are very difficult to prepare with the requisite degree of optical homogeneity and they may suffer permanent damage if overheated. Gases do not suffer from these difficulties but have a much smaller density of active atoms. Several different liquid lasers have been developed, but the most important is the dye laser. Dye laser was first demonstrated in 1965 at IBM laboratories in the US, by Peter P. Sorokin and J. R. Lankard. They discovered the dye laser action during a fluorescence research of organic dye molecules, which were excited by Ruby laser. In 1967 scientists discovered the possibility to tune the emitted wavelength, using a grating at the end of the optical cavity.

The laser medium in dye lasers is an organic dye dissolved or suspended in a liquid carrier. These lasers are optically pumped, typically by another laser. Argon and krypton ion lasers are some of the most commonly used CW dye laser pumps; nitrogen and frequency doubled Nd:YAG lasers are common pulsed dye laser pumps. Light from the pump laser is absorbed by the dye molecules, which then re-emit the absorbed energy as a broad-band fluorescence. The dye laser optics includes a wavelength selection device that allows only a single color of the fluorescence light to resonate within the laser cavity. Readjusting this wavelength selector produces a different wavelength for the laser light. So, dye lasers are continuously tunable, usually within the full range of wavelengths produced by the fluorescent dye.

Tunability of a dye laser depends on the type of dye used in it because of the characteristic fluorescence of a given dye molecule. Also, since the optimum reflectivity of mirrors is wavelength dependent, some of the laser's optics may need to be changed when running it with a new dye. Still, by using different dyes and optics these lasers can produce continually variable laser light over a large range of wavelengths, typically a 50–100 nm wide range. This makes dye lasers very useful in any applications that require a single pure color of light that can be scanned over a range of different colors, such as scientific or diagnostic spectroscopy. Since energy must be conserved, however, the dye laser's wavelength output is always longer than the wavelength of its pump laser, *i.e.* it fluorescence. The energy difference between the absorbed and emitted photons ultimately appears as heat. Figure 3.35 shows the fluorescence ranges (tuning curves) of some commonly used dyes pumped with an ion laser.

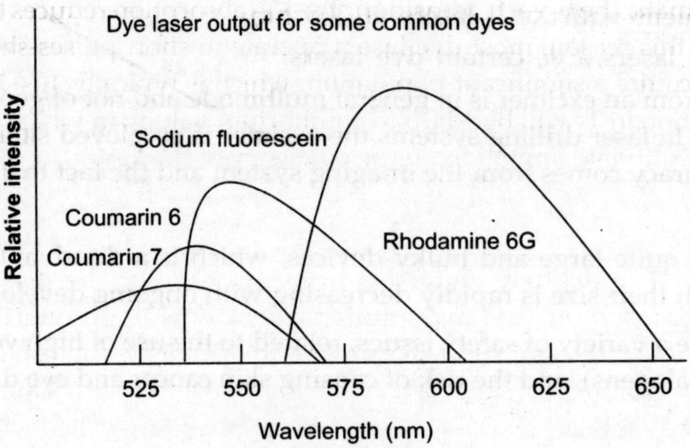

Fig. 3.35

The broad fluorescence spectrum can be explained by the energy level diagram of a typical dye molecule as shown at Fig. 3.36.The molecule has two groups of closely spaced electronic energy levels; the singlet states (S_0, S_1 and S_2) and the triplet states (T_1 and T_2). The singlet states occur when the total spin of the excited electrons in each molecule is zero (the value of 2S + 1 is thus unity). The triplet states occur when the total spin is unity (2S + 1 = 3). Each electronic energy level is broadened into a near continuum of levels by the effects of the vibration and rotation of the dye molecule and also by the effects of the solvent molecules. Pumping results in the excitation of the molecule from the ground state S_0 to the first excited state S_1. This is followed by the rapid nonradiative decay processes to the lower energy levels in S_1. The laser transition is then from these levels to a level in S_0. Since there are many such rotational/vibrational levels within S_0 and S_1, there are many transitions resulting in an emission line which is very broad. As the termination of the laser transition in S_0 is at an energy much larger than kT above the bottom of S_0, the dye laser is a four level system and threshold is reached with a very small population inversion.

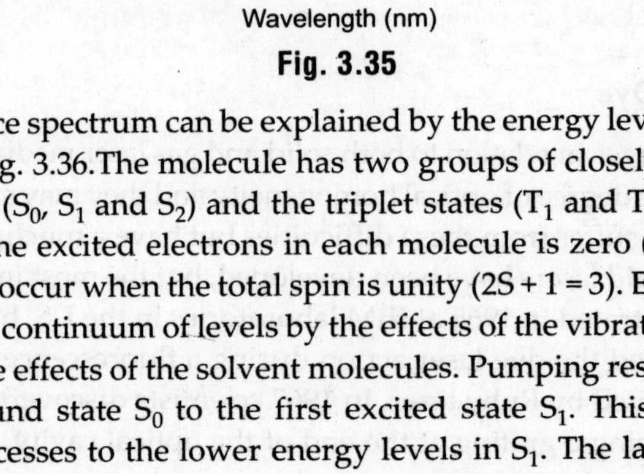

Fig. 3.36

Although the triplet states are not directly involved in the laser action, they have a profound effect as there is a small probability of a transition $S_1 \rightarrow T_1$, even though this is forbidden by quantum mechanical selection rules. Since the transition $T_1 \rightarrow S_0$ is also forbidden, molecules 'pile up' in T_1. The transition $T_1 \rightarrow T_2$ is allowed, and unfortunately the range of frequencies required for this transition is almost exactly the same as the laser transition frequencies. Thus often a significant number of molecules make the $S_1 \rightarrow T_1$ transition, $T_1 \rightarrow T_2$ absorption reduces the gain and may stop the lasing action. For this reason, most dye lasers operate in short pulses-shorter in fact, than the time taken for T_1 to acquire a significant population, which is typically 1μs. For long pulse or CW operation, the population in T_1 will build up to equilibrium values, in which case the absorption is high and becomes the ultimate limitation on the efficiency of the laser.

Construction

Since organic dyes tend to degrade under the influence of light, the dye solution is normally circulated from a large reservoir. The dye solution can be flowing through a cuvette, *i.e.*, a glass container, or be as a *dye jet, i.e.*, as a sheet-like stream in open air from a specially-shaped nozzle. With a dye jet, one avoids reflection losses from the glass surfaces and contamination of the walls of the cuvette. These advantages come at the cost of a more-complicated alignment.

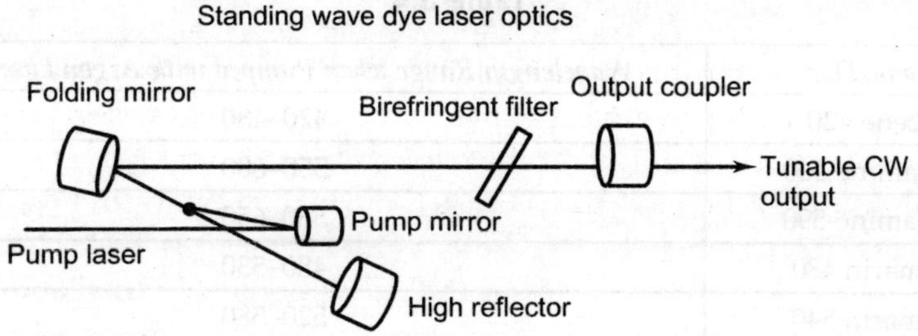

Fig. 3.37

Not all of the absorbed pump energy is re-emitted as the fluorescent light. Some of this energy goes into the collisional excitation of the molecules in the dye and the suspending fluid. This wasted energy heats up the fluid and can adversely affect laser operation through evaporation of the fluid or by altering dye properties. So, in most dye lasers, especially the CW operated ones, the fluid is circulated over cooling radiator coils with the help of a circulating pump motor. In some dye lasers the circulating fluid dye enters a transparent container where the pump laser is focused onto the dye. A standing wave type dye laser is shown schematically in Fig. 3.37.

In the dye laser schematic shown above, the dye jet is flowing out of the diagram. The pump laser is focused onto the jet using a small mirror called the pump mirror. The folded optical cavity of this laser involves three mirrors: the high reflector, the folding mirror, and the output coupler. This arrangement allows for having a long optical cavity in a smaller space. In addition to the mirrors, this laser has a three-plate birefringent filter that selects the desired wavelength to resonate within the cavity. As the polarization axes of the plates rotate with respect to one another different wavelengths can pass through. In this way the output wavelength of this dye laser can be continuously altered. In other dye laser designs the high reflector is a diffraction grating, instead of a mirror. Alignment of the grating with respect to the lasing axis selects the wavelength of the laser.

Dye lasers come in a variety of optical configurations. The standing wave type depicted above is one of the simplest. Other common forms have a closed loop type optical cavity and are called ring-dye lasers. Depending on its pump laser, some dye lasers operate CW while others produce pulsed output. In general, dye lasers are difficult to operate and cannot deliver very high powers. Some of the organic dyes used in these lasers are known carcinogens while the health risks of many of the other commonly used dyes are unknown. As a result, fluid leaks from the circulating dye or airborne dyes from evaporating fluids could cause environmental risk factors that some applications may not be able to afford. Today, in many of these applications dye lasers are often replaced with the more robust solid-state (Ti:sapphire) or semiconductor lasers.

Chemicals Used

Some of the dyes used are Rhodamine 6G, fluorescein, coumarin, stilbene, umbelliferone, tetracene, malachite green, and others. Adamantane is added to some dyes to prolong their life. Cycloheptatriene and cyclooctatetraene (COT) can be added as triplet quenchers for rhodamine G, increasing the laser output power. Output power of 1.4 kilowatt at 585 nm was achieved using Rhodamine 6G with COT in methanol-water solution. Table 3.4 shows the range of wavelengths accessible with typical commercial dyes.

Table 3.4

Organic Dye	*Wavelength Range when Pumped with Argon Laser (nm)*
Stilbene 420	420–480
Rhodamine 560	550–600
Rhodamine 590	560–650
Coumarin 480	480–530
Coumarin 540	520–580
LD 700	710–850
DCM	610–730
IR140	920–980
LDS 821	800–940

3.16.5 Semiconductor Lasers

A semiconductor laser diode is a laser where the active medium is a semiconductor similar to that found in a light-emitting diodes. The most common and practical type of laser diode is formed from a *p-n* junction and powered by injected electrical current. The *p-n* junction provides the active medium. Thus to obtain laser action we need only to meet the other necessary requirements of population inversion and optical feedback. These devices are sometimes referred to as injection laser diodes to distinguish them from optically pumped laser diodes, which are more easily produced in the laboratory. To obtain stimulated emmision, there must be a region of the device where there are many excited electrons and vacant states (*i.e.*, holes) present together. This is achieved by forward biasing a junction formed from very heavily doped *p* and *n* materials.

In n^+-type material, the Fermi level lies within the conduction band. Similarly for the p^+-type material the Fermi level lies in the valence band. The equilibrium and forward biased energy level diagrams for a junction formed from such so-called degenerate materials are illustrated in Fig. 3.38 (the broken lines represent the Fermi level). When the junction is forward biased with a voltage that is nearly equal to the energy gap voltage E_g/e, electrons and holes are injected across the junction in sufficient numbers to create a population inversion in a narrow zone called the *active region* (shown in Fig. 3.39).

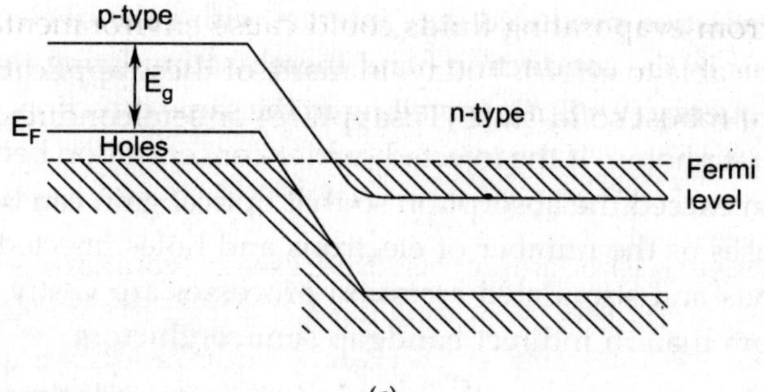

(a)

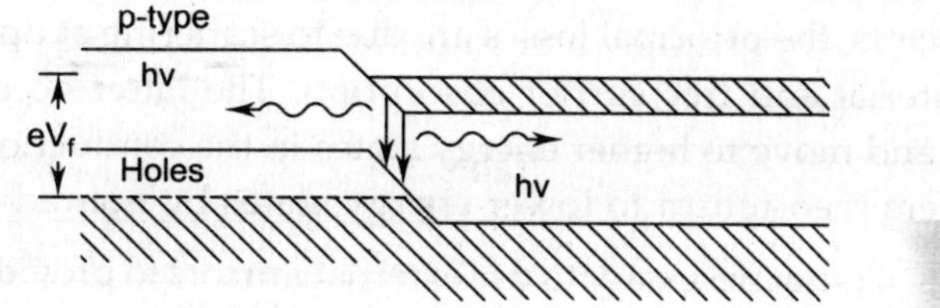

(b)

Fig. 3.38

The thickness t of the active region can be approximated by the diffusion length L_e of the electrons injected into the *p*-region, assuming that the doping level of the *p*-region is less than that of the *n*-region so that the junction current is carried substantially by electrons. For heavily doped GaAs at room temperature L_e is 1–3 μm.

As mentioned earlier, when this structure is forward biased, holes from the *p*-region are injected into the *n*-region, where electrons are the majority carriers. Similarly, electrons from the n-region are injected into the *p*-region, where holes are the majority carriers. In the case of direct bandgap materials *e.g.*, GaAs, the electrons and the holes have a high probability of recombining radiatively. When such radiative recombination happens spontaneous emission may take place—that is, the electron may re-occupy the energy state of the hole, emitting a photon with energy equal to the difference between the electron and hole states involved. These injected electrons and holes represent the injection current of the diode, and spontaneous emission gives the laser diode below lasing threshold similar properties to an LED. Spontaneous emission is necessary to initiate laser oscillation, but it is a source of inefficiency once the laser is oscillating.

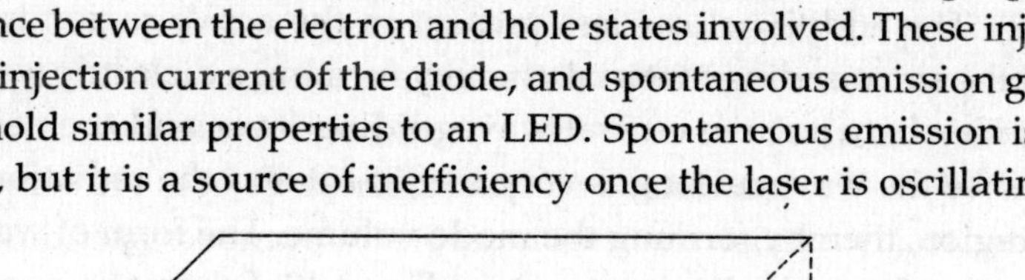

Fig. 3.39

The recombination radiation produced may interact with valence electrons and be absorbed, or interact with electrons in the conduction band thereby stimulating the production of further photons of the same frequency ($\nu = E_g/h$), travelling in the same direction, with the same polarization and phase as the first photon. If the injected carrier concentration becomes large enough, the stimulated emission can exceed the absorption so that optical gain can be achieved in the active region. The gain increases as the number of electrons and holes injected across the junction increases. The spontaneous and stimulated emission processes are vastly more efficient in direct bandgap semiconductors than in indirect bandgap semiconductors.

Laser oscillations occur, as usual, when the round trip gain exceeds the total losses over the same distance. In semiconductors, the principal losses are due to scattering at optical inhomogeneities in the semiconductor material and free carrier absorption. The latter occurs when electrons and holes absorb a photon and move to higher energy states in the conduction band or valence band respectively. The carriers then return to lower energy states by nonradiative processes.

In case of diode lasers, it is not necessary to use external mirrors to provide positive feedback. The high refractive index of the semiconductor material ensures that the reflectance of the material/air interface is sufficiently large even though it is only about 0.32. In the simplest form of laser diode, an optical waveguide is made on that crystal surface, such that the light is confined to a relatively narrow line. The two ends of the crystal are cleaved along natural crystal planes normal to the plane of the junction so that the end faces are parallel, to form perfectly smooth, parallel edges, forming a Fabry-Perot resonator.

For GaAs the junction plane is (100) and the cleaved faces are (110) planes. However, compared to other types of lasers, this cavity is much smaller, being approximately 250–500 μm long, 5–15 μm wide, and 0.1–0.2 μm thick. These dimensions are commonly referred to as the longitudinal, lateral, and transverse dimensions of the cavity, respectively.

The radiation generated within the active region spreads out into the surrounding lossy GaAs, although there is, in fact, some confinement of the radiation within a region called the *mode volume* (shown in Fig. 3.39). The additional carriers present in the active region increase its refractive index above that of the surrounding material, thereby forming a dielectric waveguide. The difference in refractive index between the central waveguiding layer and the neighboring regions is only about 0.02. Hence the waveguiding is very inefficient and the radiation extends some way beyond the active region, thereby forming the mode volume. The form of waveguiding achieved in simple homojunction laser diodes is shown in Fig. 3.40. Schematic construction of a GaAs homojunction semiconductor laser having side lengths 200–400 μm has been shown at Fig. 3.40(*a*). The emission is confined to the junction region. The narrow thickness *d* of this region causes a large beam divergence. The very small change in refractive index in the junction region is shown in Fig. 3.40(*b*) and the resulting poor confinement of the optical radiation to the gain region is shown at Fig. 3.40(*c*).

This type of waveguiding only works just well enough to allow laser action to occur as a result of very vigorous pumping. In fact homojunction lasers can usually only be operated in the pulsed mode at room temperature because of the threshold pumping current density required is so high, being typically of the order of 400 A μm^{-2}.

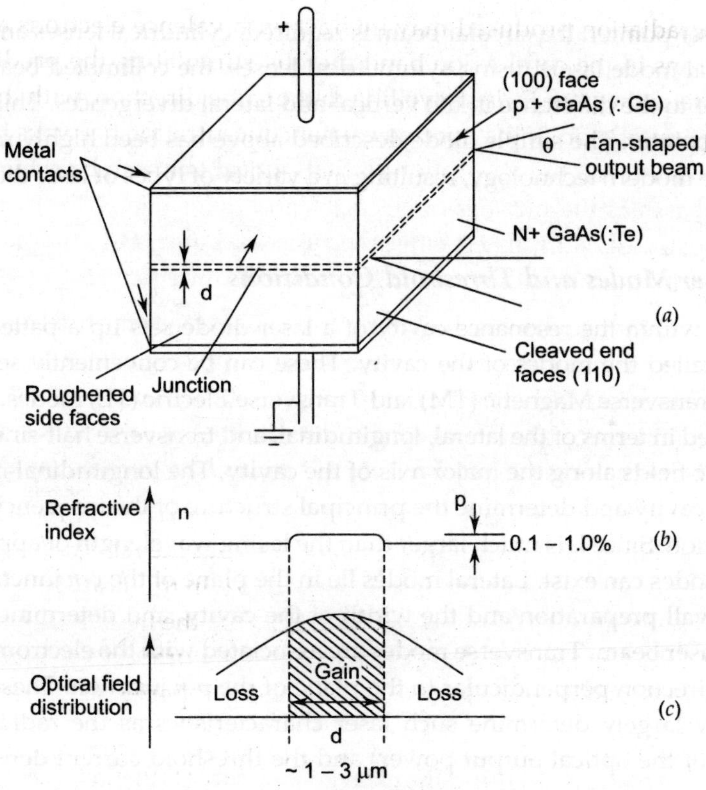

Fig. 3.40

Some important properties of laser diodes are determined by the geometry of the optical cavity. Generally, in the vertical direction, the light is contained in a very thin layer, and the structure supports only a single optical mode in the direction perpendicular to the layers. In the lateral direction, if the waveguide is wide compared to the wavelength of light, then the waveguide can support multiple lateral optical modes, and the laser is known as "multi-mode". These laterally multi-mode lasers are adequate in cases where one needs a very large amount of power, but not a small diffraction-limited beam; for example in printing, activating chemicals, or pumping other types of lasers.

In applications where a small focused beam is needed, the waveguide must be made narrow, on the order of the optical wavelength. This way, only a single lateral mode is supported and one ends up with a diffraction limited beam. Such single spatial mode devices are used for optical storage, laser pointers, and fiber optics. Note that these lasers may still support multiple longitudinal modes, and thus can lase at multiple wavelengths simultaneously.

The wavelength emitted is a function of the band-gap of the semiconductor and the modes of the optical cavity. In general, the maximum gain will occur for photons with energy slightly above the band-gap energy, and the modes nearest the gain peak will lase most strongly. If the diode is driven strongly enough, additional *side modes* may also lase. Some laser diodes, such as most visible lasers, operate at a single wavelength, but that wavelength is unstable and changes due to fluctuations in current or temperature.

Due to diffraction, the beam diverges (expands) rapidly after leaving the chip, typically at 30 degrees vertically by 10 degrees laterally. A lens must be used in order to form a collimated beam like that produced by a laser pointer. If a circular beam is required, cylindrical lenses and other optics are used. For single spatial mode lasers, using symmetrical lenses, the collimated beam ends up being elliptical in shape, due to the difference in the vertical and lateral divergences. This is easily observable with a red laser pointer. The simple diode described above has been highly modified in recent years to accommodate modern technology, resulting in a variety of types of laser diodes, as described later.

Semiconductor Laser Modes and Threshold Conditions

The optical radiation within the resonance cavity of a laser diode sets up a pattern of electric and magnetic field lines called the modes of the cavity. These can be conveniently separated into two independent sets of Transverse Magnetic (TM) and Transverse Electric (TE) modes. Again each set of modes can be described in terms of the lateral, longitudinal, and transverse half-sinusoidal variations of the electromagnetic fields along the major axis of the cavity. The longitudinal modes are related to the length L of the cavity and determine the principal structure of the frequency spectrum of the emitted optical radiation. Since L is much larger than the lasing wavelength of approximately 1 μm, many longitudinal modes can exist. Lateral modes lie in the plane of the *p-n* junction. These modes depend on the side wall preparation and the width of the cavity, and determine the shape of the lateral profile of the laser beam. Transverse modes are associated with the electromagnetic field and beam profile in the direction perpendicular to the plane of the *p-n* junction. These modes are very important, since they largely determine such laser characteristics as the radiation pattern (the angular distribution of the optical output power) and the threshold current density.

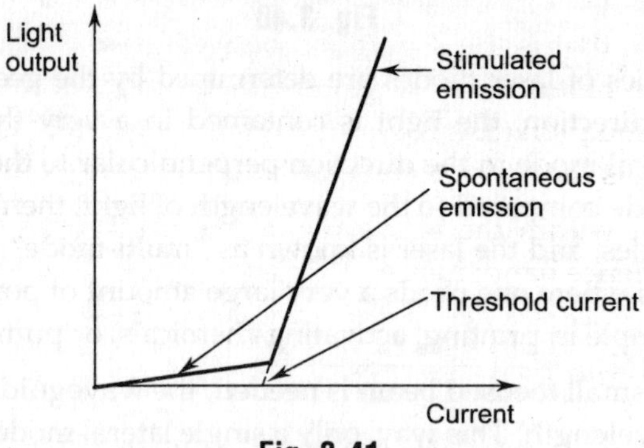

Fig. 3.41

The onset of laser action at the threshold current density is detected by an abrupt increase in the radiance of the emitting region, as shown in Fig. 3.41. Energy is channeled into one of the modes shown in Fig. 3.42 (*b*) so that a marked decrease in spectral width is noted above threshold, as illustrated in Fig. 3.42 (*c*). Fig. 3.42(*a*) shows radiant output as a function of frequency below threshold or spontaneous emission. Fig. 3.42(*b*) represents the laser modes at threshold and Fig. 3.42(*c*) represents the dominant laser mode above threshold.

An exact calculation of the threshold current for a semiconductor laser is complicated by the difficulty of defining by what is meant by a population inversion between two bands of energy levels. However, to simplify the problem and to gain some insight into the important factors, we use the idealized structure illustrated in Fig. 3.39. We let the active volume, where population inversion is maintained, have thickness t and the mode volume, where the generated electromagnetic mode is confined, be of thickness d ($d > t$). In other lasers, the mode volume is usually smaller than the volume within which the population inversion is maintained.

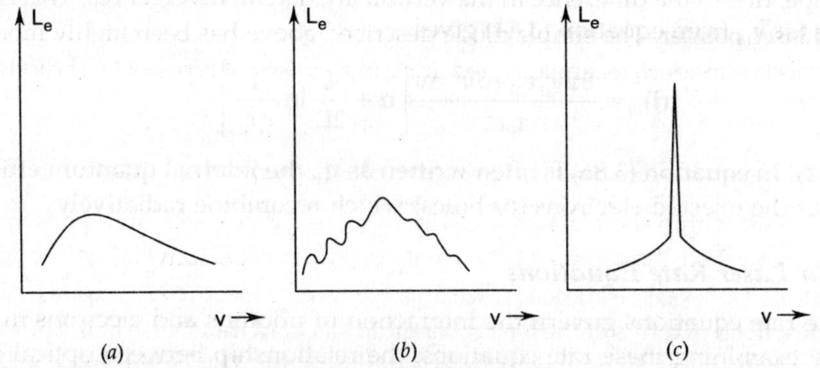

Fig. 3.42

As a result in semiconductor lasers, the portions of the mode propagating outside the active region may be absorbed. This offsets to some extent the gain resulting from those parts of the mode propagating within the active region. We allow for this by assuming that the effective population inversion within the mode volume ($d \times l \times \omega$) is given by reducing the actual population inversion in the active region by the factor t/d. The threshold condition is thus reached when

$$N_{th} = \left(N_2 - \frac{g_2}{g_1}N_1\right)_{th} = \frac{d}{t}\left(\frac{8\pi\nu_0^2\gamma_{th}\tau_{21}n^2\Delta\nu}{c^2}\right) \tag{3.83}$$

where $\Delta\nu$ is the linewidth, that is the separation between two points on the (frequency) curve where the function falls to half of its peak value which occurs at the frequency $\nu = \nu_0$, γ_{th} is the threshold gain (given by equation (3.44)), N is the refactive index of the medium, c is the velocity of light in vacuum and τ_{21} is the spontaneous radiation lifetime from level 2 to level 1.

Again, we assume that within the active region we can ignore N_1, that is, there is a large number of holes in the valence band; hence,

$$(N_2)_{th} = \frac{d}{t}\left(\frac{8\pi\nu_0^2\gamma_{th}\,\tau_{21}n^2\,\Delta\nu}{c^2}\right) \tag{3.84}$$

Now, if the current density flowing through the laser diode is J Am^{-2}, then the number of electrons per second being injected into the volume t, (which implies that the region is of thickness t and of unit cross-sectional area) of the active region is J/e. Thus the number density of electrons being injected per second is J/et electrons s^{-1} m^{-3}. The equilibrium number density of electrons in the conduction band required to give a recombination rate equal to this injection rate is N_2/τ_e, where τ_e is the electron lifetime.

The threshold current density is thus given by

$$\frac{(J)_{th}}{et} = \frac{(N_2)_{th}}{\tau_e}$$

and substituting from equation (3.84) we have

$$(J)_{th} = \frac{etd}{\tau_e t}\left(\frac{8\pi v_0^2 \gamma_{th}\, \tau_{21} n^2\, \Delta v}{c^2}\right) \tag{3.85}$$

Substituting for v_{th} from equation (3.44) gives

$$(J)_{th} = \frac{8\pi v_0^2 \tau_{21}\, edn^2\, \Delta v}{\tau_e c^2}\left[\alpha + \frac{1}{2L}\ln\frac{1}{r_1 r_2}\right] \tag{3.86}$$

The ratio τ_e/τ_{21} in equation (3.86) is often written as η_i, the internal quantum efficiency, which is the fraction of the injected electrons (or holes) which recombine radiatively.

Semiconductor Laser Rate Equations

The laser diode rate equations govern the interaction of photons and electrons in the active region. Hence by examining these rate equations, the relationship between optical output power and the diode drive current can be determined. The total carrier population is determined by carrier injection, stimulated emission, and spontaneous emission. For a *p-n* junction with a carrier-confinement region of depth *d*, the rate equations are given by

$$\frac{dp}{dt} = \chi np + R_{sp} - \frac{p}{\tau_{ph}} \tag{3.87}$$

= stimulated emission + spontaneous emission + photon loss

which governs the number of photons, and

$$\frac{dn}{dt} = \frac{J}{qd} - \frac{n}{\tau_{sp}} - \chi np \tag{3.88}$$

= injection + spontaneous recombination + stimulated emission

which governs the number of electrons n. Here, χ is the coefficient describing the strength of the optical absorption and emission interactions, τ_{ph} is the photon lifetime, τ_{sp} is the spontaneous recombination lifetime, R_{sp} is the rate of spontaneous emission into the lasing mode, and J is the injection-current density.

Equations (3.87) and (3.88) may be balanced by considering all the factors that affect the number of carriers in the laser cavity. The first term in equation (3.87) is a source of photons resulting from stimulated emission. The second term describes the number of photons produced by spontaneous emission. This term is relatively small compared with the first term. The third term represents the decay in the number of photons caused by loss mechanisms in the lasing cavity. The first term in equation (3.88) indicates the increase in electron concentration in the conduction band as current flows into the device. The second term gives the number of electrons lost from the conduction band owing to spontaneous transition, and the third term gives the same due to stimulated transition.

Solving these two equations for a steady-state condition, it will yield an expression for the output power. The steady-state is characterized by the left-hand sides of equations (3.87) and (3.88) being equal to zero. First from equation (3.87), assuming R_{sp} to be negligible and noting that dp/dt must be positive when p is small, we have

$$\chi n - \frac{1}{\tau_{ph}} \geq 0 \tag{3.89}$$

This shows that n must exceed a threshold value n_{th} in order for p to increase. Using equation (3.88), the threshold value can be expressed in terms of the threshold current J_{th} needed to maintain an inversion level $n = n_{th}$ in the steady state when the number of photons $p = 0$

$$\frac{n_{th}}{\tau_{sp}} = \frac{J_{th}}{qd} \tag{3.90}$$

The expression indicates the current required to sustain an excess electron density in the laser when spontaneous emission is the only decay mechanism.

Next, let us consider the photon and electron rate equations in the steady-state condition at the lasing threshold. Respectively, equations (3.88) and (3.89) become

$$0 = \chi n_{th} p_s + R_{sp} - \frac{p_s}{\tau_{ph}} \tag{3.91}$$

and

$$0 = \frac{J}{qd} - \frac{n_{th}}{\tau_{sp}} - \chi n_{th} p_s \tag{3.92}$$

where p_s is the steady-state photon density. The number of photons per unit volume may be obtained by adding equations (3.91) and (3.92), using equation (3.90) for the term n_{th}/τ_{sp}, and solving for p_s. Thus we get

$$p_s = \frac{\tau_{ph}}{qd}(J - J_{th}) + \tau_{ph} R_{sp} \tag{3.93}$$

The first term in equation (3.93) is the number of photons resulting from stimulated emission. The power from these photons is generally concentrated in one or a few modes. The second term indicates the photons generated by spontaneous emission. The power resulting from these photons is not mode-selective. They spread over all the possible modes of the volume, which are on the order of 10^8 modes.

Power Output of Semiconductor Lasers

As the injection current increases above the threshold, laser oscillations build up and the resulting stimulated emission reduces the population inversion until it is clamped at the threshold value. We can then express the power emitted be stimulated emission as

$$P = A\left[J - J_{th}\right]\frac{\eta_i h\nu}{e} \tag{3.94}$$

where A is the junction area, and ς_i is the internal quantum efficiency, which may be defined as:

$$\eta_i = \frac{\text{number of photons produced in the laser cavity}}{\text{number of injected electrons}} \tag{3.95}$$

Part of this power is dissipated inside the laser cavity and the rest is coupled out via the end crystal faces. These two components are proportional to α and $\frac{1}{2L}\ln\frac{1}{r_1 r_2}$ respectively. Hence, the output power may be written as

$$P = A\left[J - J_{th}\right]\frac{\eta_i h\nu}{e}\frac{\left[\frac{1}{2L}\ln\frac{1}{r_1 r_2}\right]}{\left[\alpha + \frac{1}{2L}\ln\frac{1}{r_1 r_2}\right]} \tag{3.96}$$

The external differential quantum efficiency η_{ex} is defined as the ratio of the increase in photon output rate resulting from an increase in the injection rate (*i.e.*, carriers per second), that is

$$\eta_{ex} = \frac{d\left(\frac{P_0}{h\nu}\right)}{d\left(\frac{A}{e}\left[J - J_{th}\right]\right)} \tag{3.97}$$

Assuming $r_1 = r_2$, from equation (3.96) we may write η_{ex} as

$$\eta_{ex} = \eta_i\left[\frac{\ln\frac{1}{r_1}}{\alpha L + \ln\frac{1}{r_1}}\right] \tag{3.98}$$

Equation (3.98) enables us to determine the internal quantum efficiency from the experimentally measured dependence of η_{ex} on L; η_i in GaAs is approximately equal to 0.7–1.0.

Now if the forward bias voltage applied to the laser is V_f, then the power input is $V_f AJ$ and the efficiency of the laser in converting electrical input to the laser output is

$$\eta = \frac{P_0}{V_f AJ} = \eta_i\left[\frac{J - J_{th}}{J}\right]\left(\frac{h\nu}{eV_f}\right)\frac{\ln\left(\frac{1}{r_1}\right)}{\alpha L + \ln\left(\frac{1}{r_1}\right)} \tag{3.99}$$

Normally, η_i is not a very well-defined quantity in laser diodes, but most measurements show that it is high (~0.6–0.7) at room temperature, and hence semiconductor lasers have a very high power efficiency. For standard semiconductor lasers, external differential quantum efficiencies of 15–20 percent per facet are typical. High-quality devices have that of 30–40 percent.

Resonant Frequencies

As the lasing threshold reaches, a steady-state oscillation takes place, and the magnitude and phase of the returned wave must be equal to those of the original wave. Hence for the amplitude and phase we get the following two conditions

$$I(2L) = I(0), \text{ for the amplitude,} \tag{3.100}$$

and

$$\exp(-j2\beta L) = 1, \text{ for the phase} \tag{3.101}$$

where I(0) is the initial optical field intensity and I(2L) is the same at a distance 2L. Equation (3.101) gives information about the resonant frequencies of the Fabry-Perot cavity. Now the condition in equation (3.101) holds when

$$2\beta L = 2m\pi \tag{3.102}$$

where m is an integer and β is the propagation constant given by $\beta = \frac{2\pi n}{\lambda}$. Hence, substituting the value of α in equation (3.102), we get

$$m = \frac{L}{\lambda / 2n} = \frac{2Ln}{c} \nu \tag{3.103}$$

where $c = \nu\lambda$. Equation (3.103) indicates that a standing wave pattern exists within the cavity (*i.e.*, the cavity resonates) when an integer number m of half-wavelengths spans the region between the mirrors.

Now as in all lasers, the gain is a function of frequency, there will a range of frequencies for which equation (3.103) holds. Each of these frequencies corresponds to a mode of oscillation of the laser. Depending on the laser structure, any number of frequencies can satisfy equations (3.100) and (3.101). Hence, some lasers are single-mode while the others are multimode. The relation between the gain and frequency can be assumed to have the Gaussian form

$$g(\lambda) = g(0) \exp\left[-\frac{(\lambda - \lambda_0)^2}{2\sigma^2} \right] \tag{3.104}$$

where λ_0 is the wavelength at the centre of the spectrum, σ is the spectral width of the gain, and the maximum gain $g(0)$ is proportional to the population inversion.

Now we find out the frequency, (or wavelength) spacing between the modes of a multimode laser. Hence, we consider only the longitudinal modes. However, for each longitudinal mode there may be several transverse modes that arise from one or more reflections of the propagating wave at the sides of the resonator cavity. We consider two successive modes of frequencies $\nu_m - 1$ and ν_m represented by the integers $(m - 1)$ and m respectively. From equation (3.103), we have

$$m - 1 = \frac{2Ln}{c} \nu_{m-1} \tag{3.105}$$

and

$$m = \frac{2Ln}{c} \nu_m \tag{3.106}$$

Subtracting these two equations,

$$1 = \frac{2Ln}{c} (\nu_m - \nu_{m-1}) = \frac{2Ln}{c} \Delta\nu \tag{3.107}$$

Thus we have the frequency spacing given by

$$\Delta\nu = \frac{c}{2Ln} \tag{3.108}$$

Again from the relation $\Delta\lambda/\lambda = \Delta\nu/\nu$, equation (3.108) may be written as

$$\Delta\lambda = \frac{\lambda^2}{2Ln} \tag{3.109}$$

$\Delta\lambda$ is the wavelength spacing as derived as equation (3.109).

Example 3.6. *A GaAs laser operating at 850 nm has a 550 μm length and a refractive index n = 3.8. What are the frequency and wavelength spacings? If, at the half power point, λ – λ₀ = 2 nm, what is the spectral width σ of the gain?*

Solution:

From equation (3.108), we have:

$$\Delta\nu = \frac{c}{2Ln} = \frac{3 \times 10^8}{2 \times 550 \times 10^{-6} \times 3.8} = 71.7\,\text{GHz}$$

Again, from equation (3.109), we have:

$$\Delta\lambda = \frac{\lambda^2}{2Ln} = \frac{\left(850\times10^{-9}\right)^2}{2\times550\times10^{-6}\times3.8} = 0.17 \text{ nm}$$

Using equation (3.104) with $g(\lambda) = 0.5g(0)$ yields:

$$g(\lambda) = g(0)\exp\left[-\frac{(\lambda-\lambda_0)^2}{2\sigma^2}\right]$$

or,
$$\sigma^2 = \frac{-(\lambda-\lambda_0)^2}{2\ln\dfrac{g(\lambda)}{g(0)}} = \frac{-(2)^2}{2\ln(0.5)} = 2.88$$

or,
$$\sigma = 1.69 \text{ nm}.$$

Laser Diode Structures

There is a great variety of different semiconductor lasers, spanning wide parameter regions and many different application areas. To name a few:

- Small *edge-emitting laser diodes*, generating a few milliwatts (or up to roughly half a watt) of output power in a beam with high beam quality, used *e.g.*, in laser pointers , CD players or for optical fiber communications.
- *External cavity diode lasers* contain a laser diode as the gain medium of a longer laser cavity. They are often wavelength-tunable and exhibit a small emission linewidth .
- *Broad-area laser diodes*, generating up to a few watts of output power, but with significantly poorer beam quality.
- High power *diode bars*, containing an array of broad-area emitters, generating tens of watts with poor beam quality.
- High power *stacked diode bars*, containing stacks of diode bars for the generation of extremely high powers of hundreds or thousands of watts.
- *Surface-emitting lasers (VCSELs)*, emitting the laser radiation in a direction perpendicular to the wafer, delivering a few milliwatts with good beam quality.
- Optically pumped *surface-emitting external cavity semiconductor lasers (VECSELs)*, being capable of multi-watt output powers with excellent beam quality, even in mode-locked operation.
- *Quantum cascade lasers*, operating on intraband transitions and emitting in the mid-infrared region, used *e.g.*, for trace gas analysis.

To discuss each of the above types in detail is out of scope of this book. We will try to discuss some of the most important types.

The threshold current density for the homojunction lasers is very large. This is due to poor optical and carrier confinement, which results in the parameters d and α in equation (3.86) to be large. The threshold current density can be reduced drastically, and higher efficiencies can be achieved by using lasers containing heterojunctions. A basic requirement of laser diodes is that, in addition to transverse optical and carrier confinement between heterojunction layers, the current flow must be restricted laterally to a narrow stripe along the length of the laser.

Stripe-Geometry Lasers

There are three basic optical-confinement methods used for bounding laser light in the lateral direction. They are illustrated in Fig. 3.43. In the first structure, a narrow electrode stripe (less than 8 μm wide) runs along the length of the diode. The injection of electrons and holes into the device alters the refractive index of the active layer directly below the stripe. These injected carriers create a weak, complex waveguide that confines the light laterally. These types of devices are often referred to as a *gain-guided* laser. Though these lasers emit high optical powers, but they have strong instabilities.

Rather, more stable structures use the configurations illustrated in Fig. 3.43 (*b*) and (*c*). Here, dielectric waveguide structures are fabricated in the lateral direction. The variations in the real refractive index of the various materials in these structures control the lateral modes in the laser. Hence, these devices are called *index-guided* lasers.

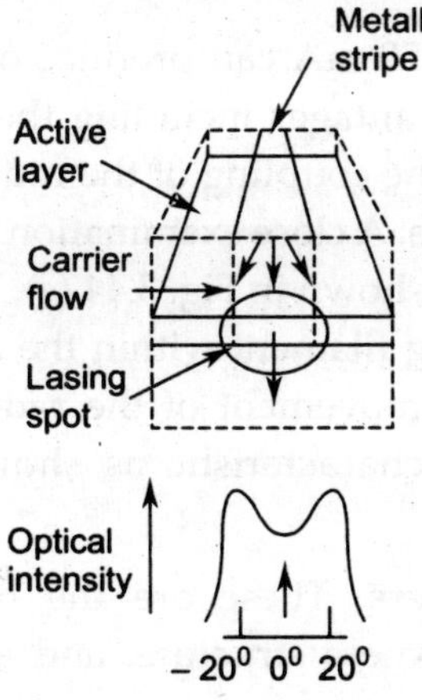

Fig. 3.43(*a*)

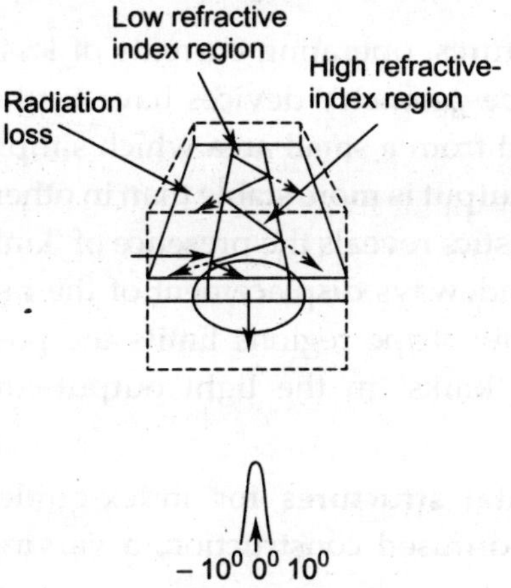

Fig. 3.43(*b*)

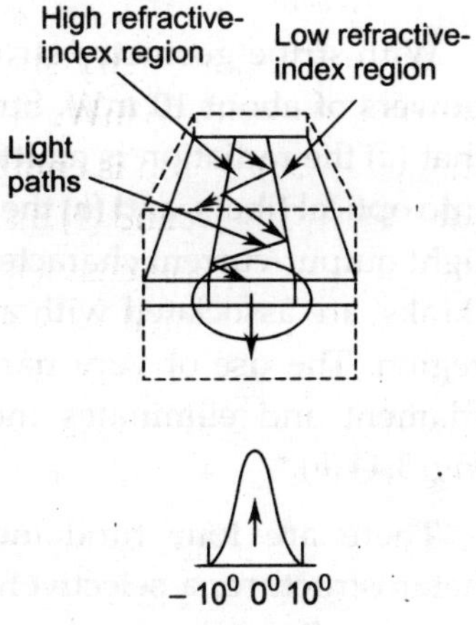

Fig. 3.43(*c*)

Index-guided lasers can be either negative-index or positive-index wave-guiding structures. Three fundamental structures for confining optical waves in the lateral direction are shown in Fig. 3.43. Fig. 3.43(*a*) illustrates in the gain-guided stripe; Fig. 3.43(*b*) shows the negative-index waveguide and has a lower refractive index the central portion of the active region and Fig. 3.43(*c*) shows the positive-index waveguide of higher refractive index in the central portion of the active region.In a negative-index waveguide, as the central portion of the active region has a lower refractive index than the outer regions, at the dielectric boundaries, part of the light is reflected and the rest is refracted into the surrounding materials and is thus lost. This radiation loss appears in the far-field radiation pattern as narrow side lobes to the main beam, as illustrated in Fig. 3.43(*b*). On the contrary, in a positive-index waveguide, the central region has a higher refractive index than the outer regions. Thus, all the guided light is reflected at the dielectric boundary, as it happens at the core-clad interface in an optical fiber. Of these two structures, the positive-index waveguide laser is more popular.

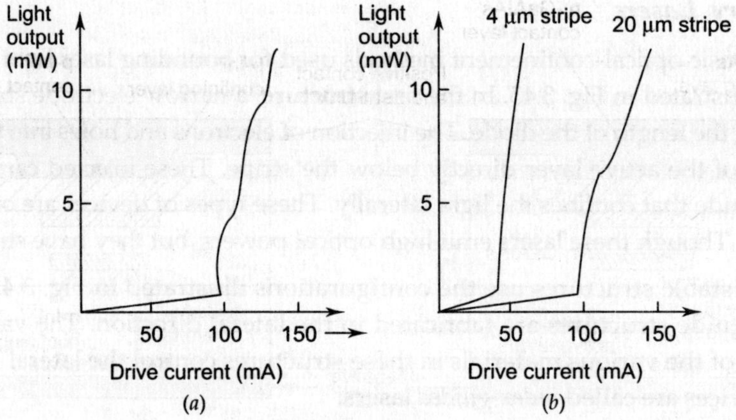

Fig. 3.44

With stripe-geometry structures, operating currents of less than 50 mA can produce output powers of about 10 mW. Stripe-geometry devices have further advantages including the facts that (*a*) the radiation is emitted from a small area which simplifies the coupling of the radiation into optical fibers, and (*b*) the output is more stable than in other lasers. A close examination of the light output-current characteristics reveals the presence of 'kinks' as shown in Fig. 3.44 (*a*). These 'kinks' are associated with a sideways displacement of the radiating filament within the active region. The use of very narrow stripe regions limits the possible movement of the radiating filament and eliminates the 'kinks' in the light output-current characteristic as shown in Fig. 3.44 (*b*).

There are four fundamental structures for index-guided lasers. These are the buried heterostructure, a selectively diffused construction, a varying-thickness structure, and a bent layer configuration.

Buried Heterostructure

By etching a narrow mesa stripe (1–2 μm wide) in a double-heterojunction material, a buried heterostructure laser can be constructed. This mesa is then embedded in high-resistivity lattice-matched *n*- type material with an appropriate bandgap and low refractive index. In 800-to-900 nm lasers, this material is GaAlAs with a GaAs active layer and in 1300-to-1600 nm lasers this is InP with an InGaAsP active layer. The generated light in a lateral waveguide is thus strongly trapped by this configuration. Buried heterostructure lasers for both the 800-to-900 nm range and 1300-to-1600 nm range are shown in Figs. 3.45(*a*) and 3.45(*b*) respectively.

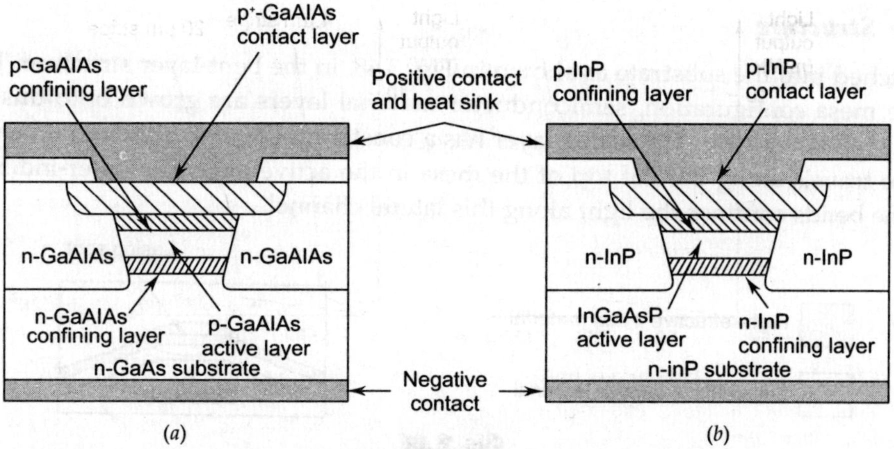

Fig. 3.45

Selectively Diffused Construction

In this structure a chemical dopant, *e.g.*, zinc for GaAlAs lasers and cadmium for InGaAsP lasers, is diffused into the active layer immediately below the metallic contact stripe. The dopant changes the refractive index of the active layer to form a lateral waveguide channel. Such a selectively diffused construction is illustrated in Fig. 3.46.

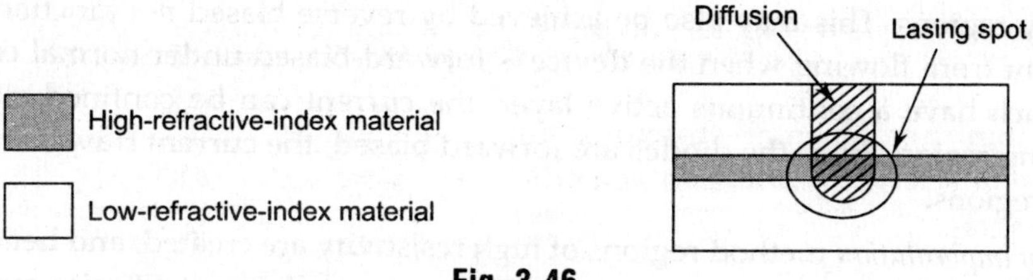

Fig. 3.46

Varying Thickness Structure

In the varying-thickness structure shown in Fig. 3.47, a channel (or a structure of mesa or terrace) is etched into the substrate. Then using the liquid-phase epitaxial technique, layers of cystals are regrown into the channel. This process fills in the depressions and partially dissolves the protrusions, thereby creating variations in the thicknesses of the active and confining layers. When an optical wave encounters a local increase in the thickness, as shown in Fig. 3.47, the thicker area acts as a positive-index waveguide of higher-index material.

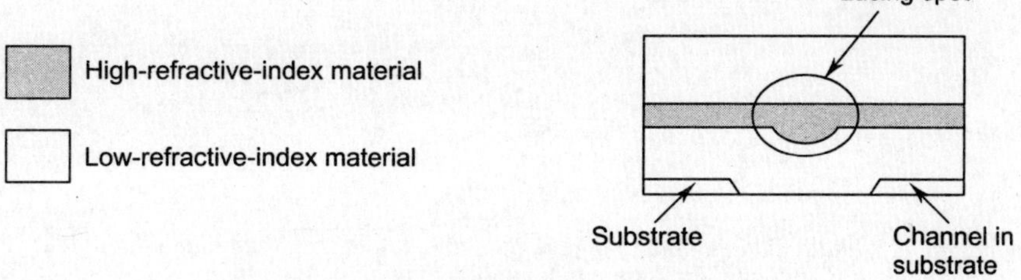

Fig. 3.47

Bent-Layer Structure

A mesa is etched into the substrate as shown in Fig. 3.48, in the bent-layer structure. To exactly replicate the mesa configuration, semiconductor material layers are grown onto this structure using vapour-phase epitaxy. The active layer has a constant thickness with lateral bends. As an optical wave travels along the flat top of the mesa in the active area, the lower-index material outside of the bends confines the light along this lateral channel.

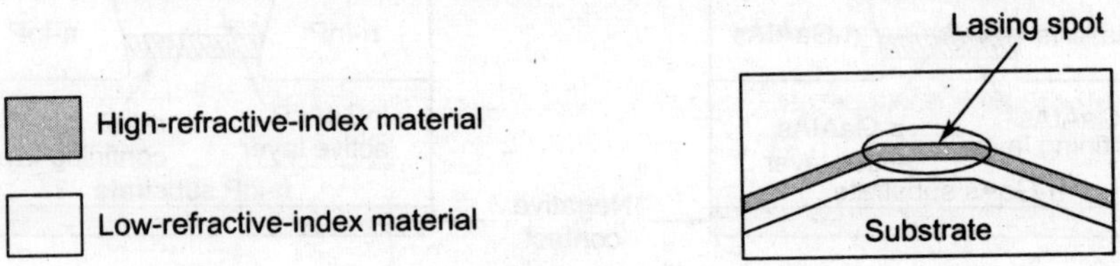

Fig. 3.48

Current Confinement by Stripe Geometry

In addition to optical confinement, one also needs to restrict the drive current tightly to the active layer. There are four basic current-confinement methods in laser diodes: (*a*) proton-implantation, (*b*) preferential-dopant diffusion, (*c*) inner-stripe confinement, and (*d*) regrowth of back-biased *p*-*n* junctions. They are illustrated in Fig. 3.49(*a*), (*b*), (*c*) and (*d*) respectively. In all the methods, the device architecture blocks current on both sides of the lasing region. This may be achieved by high-resistivity regions. This may also be achieved by reverse biased *p*-*n* junctions, which prevent the current from flowing when the device is forward-biased under normal conditions. For structures which have a continuous active layer, the current can be confined either below or above the lasing region. Since the diodes are forward biased, the current travels from the *p*-type to the *n*-type regions.

In the *proton implantation* method regions of high resistivity are created, and hence restricts the current to a narrow path between these regions. The *preferential-dopant diffusion* method partially diffuses a *p*-type dopant (*e.g.*, Cd or Zn) through an *n*-type capping layer, thus establishing a narrow path for the current. A lasing structure is grown above a channel etched into planar material in the *inner-stripe confinement* technique. *Back-biased p-n* junctions restrict the current on both sides of the channel. When the active layer is discontinuous, as in a buried heterostructure, current can be blocked on both sides of the mesa by growing *p*-*n* junctions that are reverse-biased when the device is operating. More than one current confining technique can be used for a laser diode.

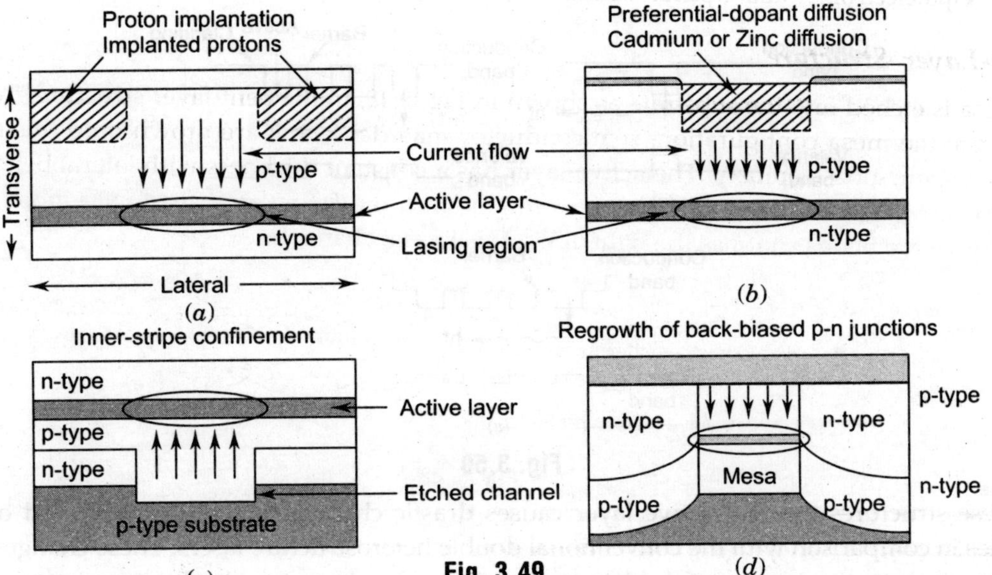

Fig. 3.49

In double-heterojunction laser, the waveguide thickness and the refractive-index differentials at the waveguide boundaries governs the highest-order transverse mode that can be excited. In fact, if the refractive-index differentials are kept at approximately 0.08, then only the fundamental transverse mode will propagate if the active area is thinner than 1 μm. While designing the width and thickness of the optical cavity, a tradeoff must be made between current density and output beam width. If either the width or the thickness of the active region is increased, the lateral or transverse beam widths become narrow respectively, but at the expense of an increase in the threshold current density.

Although the active layer is thin enough (1–3 μm) to confine the electrons and the optical field in a standard double-heterojunction laser, the electronic and optical properties remains the same as in the bulk material. This limits the attainable threshold current density, modulation speed and the linewidth of the device. Quantum-well lasers overcome these limitations by having an active layer thickness around 10 nm.

Quantum-Well Lasers

As already mentioned in the previous section, in recent times double heterostructure lasers have been fabricated with very thin active layer thickness of around 10 nm instead of the typical range for conventional double heterostructures of 0.1 to 0.3 μm. The carrier motion normal to the active layer in these devices is restricted, which results in a quantization of the kinetic energy into discrete energy levels for the carriers moving in that direction. This effect is similar to the well known quantum mechanical problem of a one dimensional potential well. Thus these devices are known as quantum-well lasers. Various quantum-well structures- single quantum-well, multiquantum-well and modified multiquantum-well structures are illustrated in Fig. 3.50(*a*), (*b*) and (*c*).

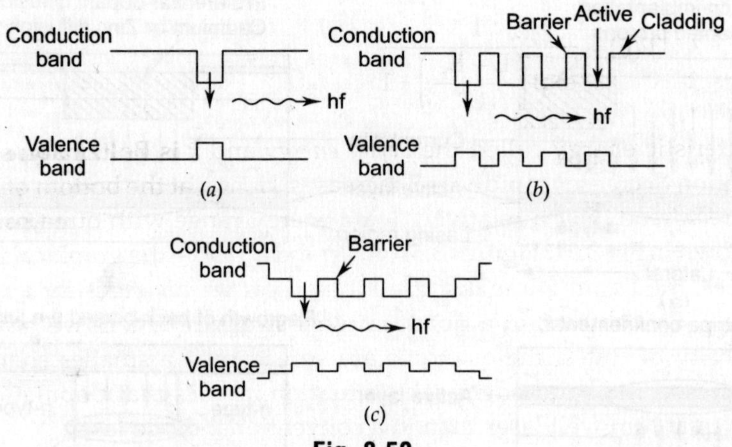

Fig. 3.50

In these structures the thin active layer causes drastic changes to the electronic and optical properties in comparison with the conventional double heterostructure lasers. These changes happen due to the quantized nature of the discrete energy levels with a step like density of states which differs from the continuum normally obtained. Thus, quantum-well lasers exhibit an inherent advantage over conventional double heterostructure devices in that they allow high gain at low carrier density, thus providing the possibility of significantly lower threshold currents.

In fact, interest in such devices is mainly due to the very low threshold currents that can result. To understand the underlying mechanism of this, we have to examine the electron energy levels. Let us suppose that the active layer has length and breadth L_x and L_y respectively and thickness L_z. Both L_x and L_y are much greater than L_z. Now, the total energy of an electron may be written in terms of three quantum numbers n_1, n_2 and n_3 as

$$E(n_1,n_2,n_3)=\frac{h^2}{8m_e}\left(\frac{n_1^2}{L_x^2}+\frac{n_2^2}{L_y^2}\right)+\frac{h^2}{8m_e}\left(\frac{n_3^2}{L_z^2}\right) \tag{3.110}$$

where m_e is the mass of an electron and h is the Planck's constant.

Because of the relative magnitudes of L_x, L_y and L_z, the lowest energy states will all have $n_3 = 1$ and will constitute effectively *a* 'two-dimensional' group of states with varying n_1 and n_2 values. Now, the number of states per unit energy interval is given by the density of states function Z(E). For a two-dimensional potential well Z(E) is given by $Z(E) = 4m_e/h^2$, which shows that the two-dimensional density of states is independent of E. The corresponding three-dimensional expression is given by

$$Z(E)=\frac{2\pi}{h^3}\left(2m_e^*\right)^{3/2}E^{1/2} \tag{3.111}$$

where E is measured relative to the bottom of the band, m_e^* is the effective mass of the electron. Thus we see that in this case the density of states varies as $E^{1/2}$. These expressions for the density of states apply to both for the electrons in the conduction band and to the holes in the valence band, with the appropriate modifications. This is true provided the hole energy is referenced to the top of the valence band and the hole mass replaced by an appropriate effective mass. Now, the probability of occupation of these levels is given by the Fermi-Dirac distribution function given as

$$F(E) = \frac{1}{\exp\left(\frac{E - E_F}{kT}\right) + 1} \tag{3.112}$$

where E_F is a characteristic energy called the *Fermi energy* and k is Boltzmann's constant. This probability of occupation decreases rapidly as E increases. Hence, at the bottom of the conduction band the electrons are spread over a relatively wide energy range with quite small densities at the band edges, as shown in Fig 3.51. Figures 3.51(*a*), (*b*) and (*c*) show the graphical representations of the density of states, the Fermi-Dirac distribution and carrier densities for an intrinsic semi-conductor respectively. By contrast, in a quantum well structure where we have effectively a two-dimensional density of states, the electrons are spread over a smaller energy range with relatively high densities at the band edges. This situation implies that population inversion is easier to achieve in a quantum well laser than in a conventional diode laser.

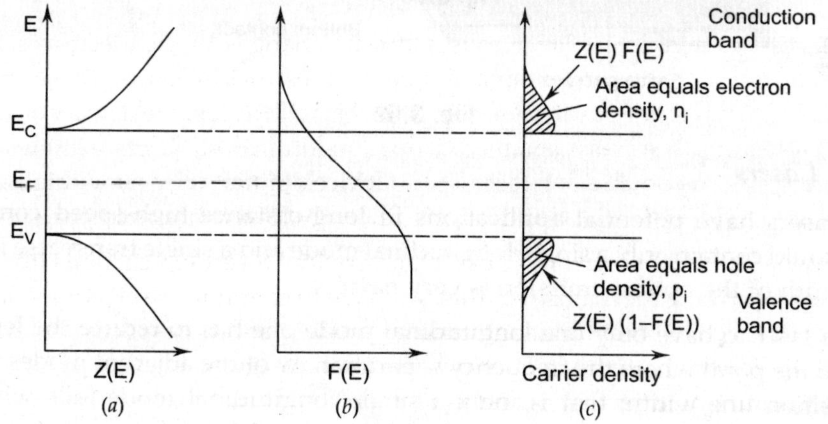

Fig. 3.51

Because of the extreme narrowness of the gain region in single quantum-well (SQW) lasers, optical confinement is very poor. Now this is a problem, which causes higher laser losses and hence tends to negate the potential advantage of a very low threshold current. To get rid of this difficulty multiple quantum-well (MQW) structures have to be fabricated. MQW structures allow a much higher optical confinement than SQW structures. In MQW structure, the layers separating the active layers are called barrier layers. Energy band diagrams for the active regions of these structures are illustrated in Fig. 3.50. In Fig. 3.50 (*c*), it may be observed that when the band-gap energy of the barrier layers differs from the cladding layer in a MQW device it is usually referred to as a modified multiquantum-well laser.

Edge-Emitting Lasers

Figure 3.52 shows the structure of a typical edge-emitting laser. The dimensions of the active region are 200 mm in length, 2–10 mm lateral width and 0.1 mm in transverse dimension. In reality there are many different designs of edge-emitting lasers. The current flows from the p- to the n-type semiconductor with electrons and holes being injected in the active region. A further advantage of the double heterostructure is that the large band-gap semiconductor has a lower refractive index than that in the active region giving index guiding in the transverse direction. In the plane of the active region the emission is confined by gain guiding, where the refractive index is modified by the carrier density. The formation of a stripe contact separated by semi-insulating regions of proton bombarded semiconductor allows the current to flow through a restricted portion of the active region. This also aids the optical confinement in the plane of the active region.

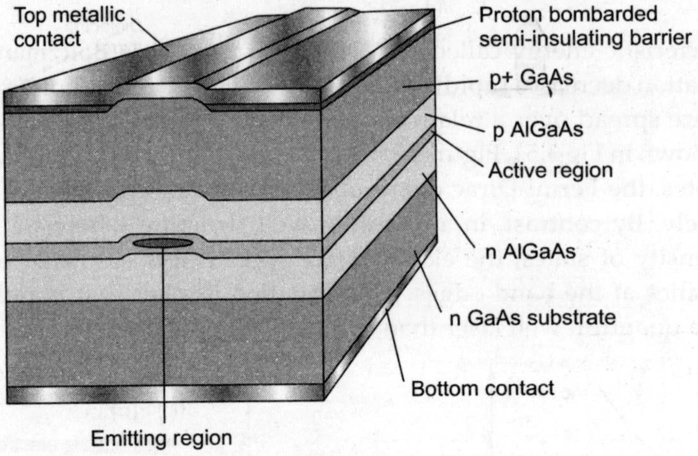

Fig. 3.52

Single-Mode Lasers

Single-mode lasers have potential applications in long-distance high-speed communication. These lasers should contain only a single longitudinal mode and a single transverse mode. Hence, the spectral width of the optical emission is very narrow.

To restrict a laser to have only one longitudinal mode one has to reduce the length L of the lasing cavity to the point where the frequency separation $\Delta\nu$ of the adjacent modes is larger than the laser transition line width; that is, only a single longitudinal mode falls within the gain bandwidth of the device. However reduction of the lengths of these devices makes them hard to handle and they are limited to optical output powers of only a few milliwatts.

Hence alternative devices are developed. The three mostly used structures are Vertical-Cavity Surface-Emitting Lasers (VCSEL), structures having built-in frequency-selective grating, and tunable lasers.

Vertical Cavity Surface Emitting Lasers (VCSEL)

The vertical cavity surface emitting laser has many potential advantages over the edge-emitting lasers. Its design allows the chips to be manufactured and tested on a single wafer. In the telecommunications industry, the VCSEL's uniform, single mode beam profile is desirable for coupling into optical fibers. However, along with these advantages come a number of problems particularly in the fabrication and operation at high powers. In this section, we look at the structure and operation of these devices and discuss the problems faced by the designer of such devices.

The earliest VCSEL was reported in 1965 by Melngailis. It consisted of a n^+pp^+ junction of InSb. When cooled to 10 K and subjected to a magnetic field to confine the carriers, the device emitted coherent radiation at a wavelength of around 5.2 µm. The early VCSEL devices had metallic mirrors with resulting high threshold current densities (44 kAcm^{-2}) and were cooled using liquid Nitrogen. Epitaxial mirrors for GaAs/AlGaAs VCSELs were pioneered in 1983. The pulsed room temperatures VCSELs were produced in the laboratory one year later. Reduction in the threshold current density was connected with reduction in the active volume of the cavity. Today, GaAs/AlGaAs VCSELs with oxide current confinement have threshold currents as low as 40μA.

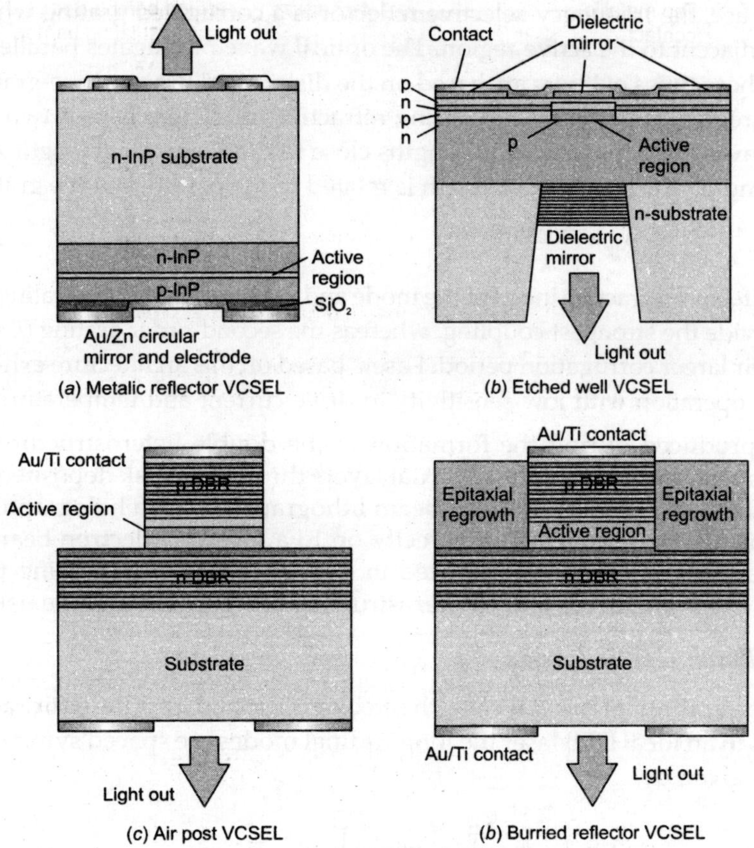

Fig. 3.53

Structure

There are many designs of VCSEL structures. However, they all have certain aspects in common. The special feature of a VCSEL is that the light emission is perpendicular to the semiconductor surface as shown in Fig. 3.53. The cavity length of VCSELs is very short typically 1–3 wavelengths of the emitted light. As a result, in a single pass of the cavity, a photon has a small chance of triggering a stimulated emission event at low carrier densities. Therefore, VCSELs require highly reflective mirrors to be efficient. In edge-emitting lasers, the reflectivity of the facets is about 30%. For VCSELs, the reflectivity required for low threshold currents is greater than 99.9%. Such a high reflectivity can not be achieved by the use of metallic mirrors. VCSELs make use Distributed Bragg Reflectors (DBRs). These are formed by laying down alternating layers of semiconductor or dielectric materials with a difference in refractive index. At the dispersion minima for optical fibers, semiconductor materials used for DBRs have a small difference in refractive index therefore many periods are required. Since the DBR layers also carry the current in the device, more layers increase the resistance of the device therefore dissipation of heat and growth may become a problem if the device is poorly designed. Some designs are shown in Fig. 3.53 showing the evolution VCSELs.

Built-in Frequency-Selective Reflector

In practice, three different types of laser configurations are used in a built-in frequency selective structure. In each case, the frequency-selective reflector is a corrugated grating which is a passive waveguide layer adjacent to the active region. The optical wave propagates parallel to the grating. The operations of these types of lasers are based on the distributed Bragg phase-grating reflector. A phase-grating is a region of periodically varying refractive index that causes two counter-propagating traveling waves to couple. For wavelengths close to the Bragg wavelength λ_B, the coupling becomes the maximum. The Bragg wavelength is related to the period Λ of the grating by

$$\lambda_B = \frac{2n_e\Lambda}{k} \tag{3.113}$$

where n_e is the effective refractive index of the mode and k is the order of the grating. The first-order gratings ($k = 1$) provide the strongest coupling, whereas the second-order grating ($k = 2$) are easier to fabricate due to their larger corrugation period. Lasers based on this architecture exhibit good single-mode longitudinal operation with low sensitivity to drive-current and temperature variations.

The grating is produced during the formation of the double heterostructure by chemically etching either the substrate or one of the epitaxial layers through a mask deposited on the surface. The mask may be formed either by electron-beam lithography or by a holographic technique. In the former case the electron beam writes directly on to a layer of 'electron-beam resist'. In the latter case, an interference pattern is produced in a layer of photoresist using two ultra-violet laser beams. The remaining layers of the heterostructure are deposited in the usual manner.

Distributed-Feedback (DFB) Lasers

In the DFB laser, the grating for the wavelength selector is formed over the entire active region. As shown in Fig. 3.54, in an ideal DFB laser the longitudinal modes are spaced symmetrically around λ_B at wavelengths given by

$$\lambda = \lambda_B \pm \frac{\lambda_B^2}{2n_e L_e}\left(m + \frac{1}{2}\right) \tag{3.114}$$

where $m = 0, 1, 2, \ldots$ is the mode order and L_e is the effective grating length. The amplitudes of successively higher-order lasing modes are greatly reduced from the zero-order amplitude.

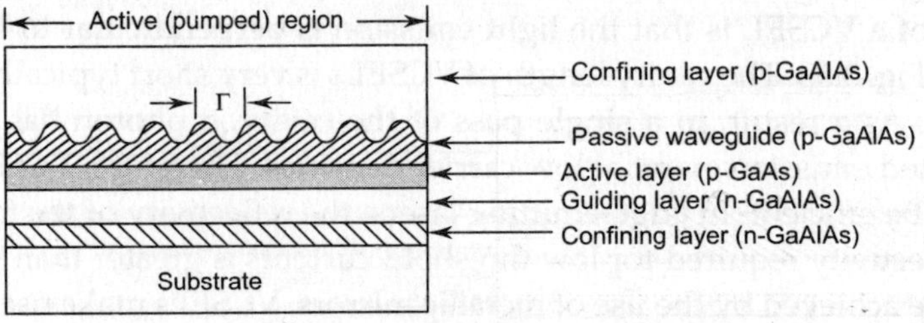

Fig. 3.54

Theoretically, in a DFB laser that has both ends coated with anti-reflection coating, the two zero-order modes on either side of the Bragg wavelength should experience the same lowest threshold gain and would lase simultaneously in an idealized symmetrical structure. But, in practice, the randomness of the cleaving process lifts the degeneracy in the modal gain and results in single-mode operation. The facet asymmetry can be further increased by putting a high-reflection coating on one end and a low-reflection coating on the other.

Because of the phase change that takes place on reflection, the Bragg condition given in equation (3.113) is not exactly correct. There are, in fact, two modes with propagation constants that require the same maximum gain to go into oscillation. If there is significant reflection from the end facets, one of the two modes is preferred. But the difference in the threshold gain for the two modes depends on the position of the facet relative to the grating. This cannot be controlled. An alternative way to select a single mode is to introduce an extra section of length Λ/2 in the centre of the grating, and coat with anti-reflection coating on both facets.

Distributed-Bragg-Reflector (DBR) Lasers

In the DBR lasers, the gratings are located at the ends of the normal active layer of the laser to replace the cleaved end mirrors used in the Fabry-Perot optical resonator. Thus the amplifying region is separated from the feedback region as shown in Fig. 3.55. The device comprises three parts: an amplifying region and two passive regions which control the phase and hence the wavelength of oscillation. Control of I_{DBR} allows step changes in the output wavelength, corresponding to different Bragg reflection peaks. Control of I_P gives a smooth variation of the path-length and hence of the wavelength linked to Bragg peak.

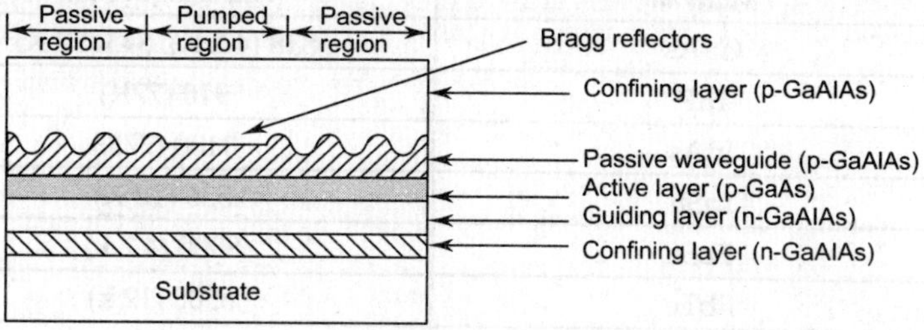

Fig. 3.55

Distributed-Reflector (DR) Lasers

The Distributed-Reflector (DR) laser consists of active and passive distributed reflectors as illustrated in Fig. 3.56. This structure improves the lasing properties of conventional DFB and DBR lasers, and has a high efficiency and high output capability.

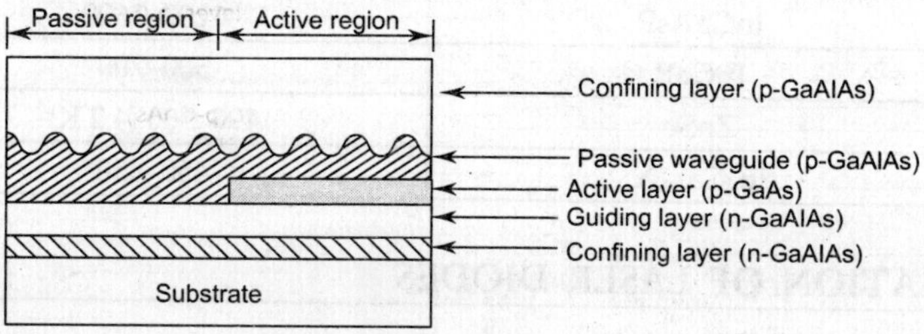

Fig. 3.56

Tunable Lasers

Tunable lasers are based on DFB or DBR structures. Since, the wavelength of a laser changes approximately 0.1 nm/°C, hence frequency tuning may be achieved by changing the temperature of the device. Tuning may also be done by altering the injection current into the active (gain) section or the passive section which yields a wavelength change of 0.8×10^{-2} to 4.0×10^{-2} nm/mA, or, equivalently, 1 to 5 GHz/mA. The latter method is generally used. A change in the effective refractive index takes place in this process and hence the peak output wavelength shifts. However, the maximum tuning range depends on the optical output power. A larger output level results in a narrower tuning range.

The tuning range can be estimated as

$$\frac{\Delta\lambda_{tune}}{\lambda} = \frac{\Delta n_{eff}}{n_{eff}} \tag{3.115}$$

where Δn_{eff} is the change in the effective refractive index. In practice, the maximum index change is around 1 percent, which results in a tuning range of 10–15 nm. Table 3.5 lists some representative laser materials and corresponding wavelengths.

Oscillation wavelengths and operating temperature of some semiconductor p-n junction lasers

Table 3.5

Material	*Wavelength (nm)*
GaAs	838 (4.2 K), 843 (77 K)
InP	910 (77K)
InAs	3100 (77K)
InSb	526 (10 K)
PbSe	8500 (4.2 k)
PbTe	6500 (12 k)
$Ga(As_xP_{1-x})$	650–840
$(Ga_xIn_{1-x})As$	840–3500
$In(As_xP_{1-x})$	910–3500
GaSb	1600 (77 K)
$Pb_{1-x}Sn_xTe$	(9.5–28 μm) (12 K)
$Ga_{1-x}Al_xAs$	690–850
InGaAsP	1200–1600
InGaP	500–700
ZnSe	490–500 (4.2 K)
$In_xGa_{1-x}N$	450–650

3.17 MODULATION OF LASER DIODES

The process of imposing information data on the light stream is called 'modulation'. The simplest and most widely used modulation scheme is called *On-Off Keying* (OOK), where the light stream is turned on or off, depending on whether the data bit is a 1 or 0. OOK modulated signals are usually realized in one of two ways: (*i*) by direct modulation of semiconductor laser or an LED, or (*ii*) by using an external modulator.

Though the direct modulation is simple and inexpensive technique, but it suffers from chirping effect. Chirp is a phenomenon wherein the carrier frequency of the transmitted pulse varies with time, and it causes a broadening of the transmitted spectrum. This enhanced pulse broadening of chirped pulses is significant enough to warrant the use of external modulators in high-speed (> 2.5 Gb/s), dispersion-limited communication systems.

A variety of external modulators are commercially available, either as a separate device or as an integral part of the laser transmitter package. An OOK external modulator is placed in front of a light source and turns the light signal on or off based on the data to be transmitted. The light source itself is continuously operated. This has the advantage of minimizing chirp.

The basic limitation on the direct modulation rate of the semiconductor laser diodes depends on the spontaneous (radiative) and stimulated carrier lifetimes and the photon lifetime. The spontaneous lifetime T_{sp} is a function of the semiconductor band structure and the carrier concentration. In GaAs-based materials for dopant concentrations on the order of 10^{19} cm^{-3}, at room temperature the value of T_{sp} is about 1 ns. The stimulated carrier lifetime T_{st} depends on the optical density in the lasing cavity and is on the order of 10 *ps*. The photon lifetime T_{ph} is the average time that the photon resides in the lasing cavity before being lost either by absorption or by emission through the facets. A typical value of T_{ph} is about 2 *ps*. This value sets the upper limit to the direct modulation capability of the laser diode.

Since the photon lifetime is much smaller than the carrier lifetime, a laser diode can readily be pulse modulated. If the laser is completely turned off after each pulse, the spontaneous carrier lifetime will limit the modulation rate. This is because, at the onset of a current pulse of amplitude I_p, a period of time t_d is required to achieve the population inversion necessary to produce a gain that is sufficient to overcome the optical losses in the lasing cavity. This delay time can be eliminated by dc-biasing the laser diode at the lasing threshold current. Pulse modulation is then carried out by modulating the laser only in the operating region above threshold. In this region, the carrier lifetime is now shortened to the stimulated emission lifetime, so that high modulation rates are possible. When using a directly modulated laser diode for high-speed transmission systems, the modulation frequency can be no larger than the frequency of the relaxation oscillations of the laser field. The relaxation oscillation depends on both the spontaneous lifetime and the photon lifetime. Since T_{sp} is about 1 ns and T_{ph} is on the order of 2 *ps* for a 300 μm-long laser, then when the injection current is about twice the threshold current, the maximum modulation frequency is a few GHz.

Analog modulation of laser diodes is carried out by making the drive current above threshold proportional to the baseband information signal. For this modulation scheme, a linear relation is required between the light output and the current input. However, signal degradation due to non-linearities that are a consequence of the transient response characteristics of laser diodes makes the implementation of analog intensity modulation susceptible to both intermodulation and cross-modulation effects.

3.18 DIODE LASER CHARACTERISTICS

To use the diode laser in optical communication, it is necessary to be aware of certain of its characteristics which may affect its efficient operation. In this section, we will discuss the major characteristics of the device, although there is substantial variation in behavior between devices of various materials and structures.

3.18.1 Temperature Effects

One of the most important factors to consider in the application of laser diodes is the temperature dependence of the threshold current I_{th}. In general terms the threshold current tends to increase with temperature in all types of semiconductor lasers. The temperature variation of I_{th} can be approximated by empirical expression

$$I_{th} \propto \exp\frac{T}{T_0}$$

or,
$$I_{th} = I_z \exp\left(\frac{T}{T_0}\right) \qquad (3.116)$$

where T is the device absolute temperature and T_0 is the threshold temperature coefficient which is a characteristic temperature describing the quality of the material, but which is also affected by the structure of the device and I_z is a constant. For a conventional stripe-geometry AlGaAs laser diode, T_0 is typically 120–165°C in the vicinity of room temperature, whereas for InGaAsP devices it is between 40 and 75°C. This shows the stronger temperature dependence of InGaAsP structures which is shown in Fig. 3.57(*a*). Smaller dependence of I_{th} on temperature has been demonstrated for AlGaAs quantum-well heterostructure lasers in Fig. 3.57(*b*). For these lasers T_0 may be as high as 437°C. The increase in threshold current with temperature for AlGaAs devices can be accounted for with reasonable accuracy by consideration of the increasing energy spread of electrons and holes injected into the conduction and valence bands. The intrinsic physical properties of the InGaAsP material system may cause its higher temperature sensitivity. It may be noted that it is important to obtain high values of T_0 for the devices in order to minimize temperature dependence.

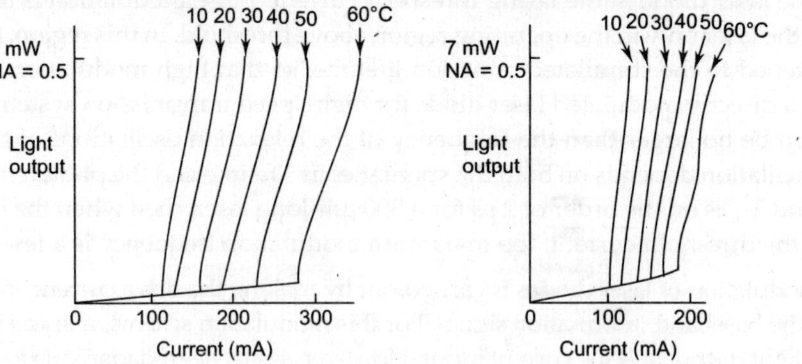

Fig. 3.57

Moreover, the lasing threshold can change with aging of laser diodes. Hence, if a constant optical power level is to be maintained as the temperature of the laser changes or as the laser ages, it is necessary to adjust the dc-bias current level. Some of the possible methods for achieving this automatically are optical feedback and feed-forward schemes, pre-distortion techniques, and temperature-matching transistors. Optical feedback can be incorporated by using a photodetector either to sense the variation in optical power emitted from the rear facet of the laser or to tap off and monitor a small portion of the fiber-coupled power emitted from the front facet. The photodetector compares the power output with a reference level and adjusts the dc-bias current level automatically to maintain a constant peak light output relative to the reference. The photodetector used should have a stable long-term responsivity which remains constant over a wide temperature range.

Figure 3.58 shows a feedback-stabilizing circuit, which may be used for a digital transmitter. Here, the light emitting from the rear facet of the laser diode is monitored by a *pin* photodiode. The electric input signal pattern is compared with the optical output level of the laser diode. This effectively prevents the feedback circuit from erroneously raising the bias current level during long sequences of digital zeros or during a period in which there is no input signal on the channel. Here in the circuit, the dc reference through resistor R_2 sets the bias current at the proper operating point during long sequences of zeros. When this bias current is added to the laser drive current, the desired peak output power from the laser is obtained. The resistor R_1 balances the signal reference current against the pin photocurrent from a 50-percent duty ratio at 25°C. As the lasing threshold changes because of aging or temperature variations, the bias current I_B is automatically adjusted to maintain a balance between the data reference and the *pin* photocurrent. However, a more sophisticated version of the circuit can control both the bias current and the modulation current simultaneously and independently.

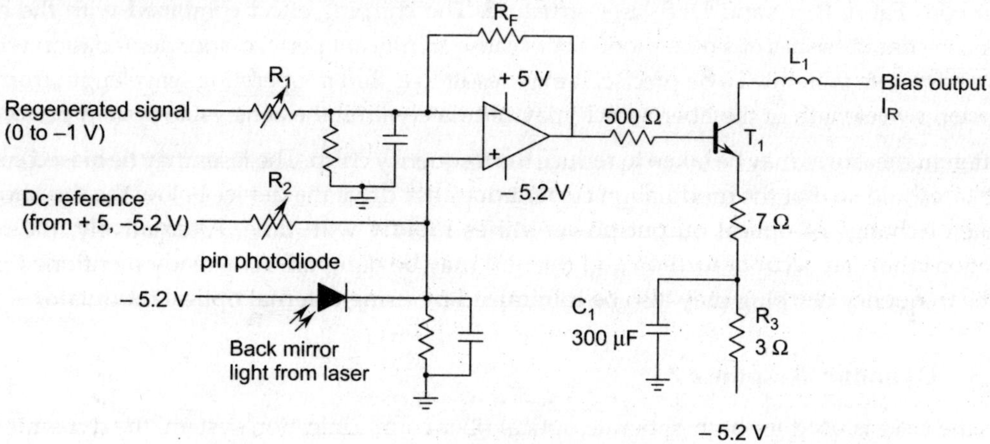

Fig. 3.58

Example 3.7. *Compare the ratio of the threshold current densities at 25°C and 75°C for a GaAs laser with T_0 = 165 K and the similar ratio for an InGaAsP device with T_0 = 50 K.*

Solution:

From equation (3.116) the threshold current density is given as:

$$I_{th} \propto \exp\frac{T}{T_0}$$

For the GaAs laser:

$$I_{th}\left(25°\text{ C}\right) \propto \exp\frac{297}{165} = 6.05$$

$$I_{th}\left(75°\text{ C}\right) \propto \exp\frac{348}{165} = 8.24$$

Thus the ratio of the current densities:

$$\frac{I_{th}\ (75°\text{C})}{I_{th}\ (25°\text{C})} = \frac{8.24}{6.05} = 1.36$$

For the InGaAsP laser:

$$I_{th}(25°\text{C}) \propto \exp\frac{297}{50} = 379.93$$

$$I_{th}(75°\,C) \propto \exp\frac{348}{50} = 1053.63$$

Thus the ratio of the current densities:

$$\frac{I_{th}\,(75°C)}{I_{th}\,(25°C)} = \frac{1053.63}{379.93} = 2.77$$

3.18.2 Frequency Chirp

As already discussed in Section 3.17, the direct current modulation of a single longitudinal mode semiconductor laser diode suffers from frequency chirping. It arises from gain-guided variations in the laser refractive index due to the strong coupling between the free carrier density and the index of refraction which is present in any semiconductor structure. Thus every small change in the carrier density will result in a phase shift of the optical field, resulting in an associated change in the resonance frequency within both Fabry-Perot and DFB laser structure. The chirping effect combined with the chromatic dispersion characteristics of single-mode fibers cause significant performance degradation within high transmission rate systems. To be precise, it may result in a shift in operating wavelength from the zero dispersion wavelength of the fiber, which may ultimately limit the achievable system performance.

Different measures may be taken to reduce the frequency chirp. The laser may be biased sufficiently above threshold so that the modulation current does not drive the device below the threshold where the rate of change of optical output power varies rapidly with time. Alternatively, the relaxation oscillations that can occur at turn-on and turn-off may be damped. As already mentioned in Section 3.17 the frequency chirping may also be minimized by using external optical modulator.

3.18.3 Dynamic Response

When the laser is used for high data rate optical fiber communication system, the dynamic behavior of the laser diode becomes critical. The application of a current step to the device results in a switch-on delay often followed by high frequency (~ 10 GHz) damped oscillations known as Relaxation Oscillations (RO) (mentioned in Section 3.18.2). These transient phenomena occur while the electron and photon populations within the structure come into equilibrium and are shown in Fig. 3.59.

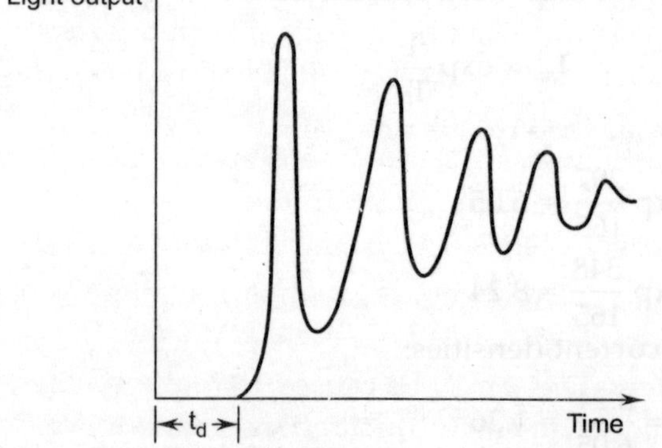

Fig. 3.59

Hence, the diode laser output can comprise several pulses as the electron density is repetitively built up and quickly reduced, thus causing RO. The switch-on delay t_d may last for 0.5 ns and the RO for perhaps twice that period. At data rates above 100 Mbit s^{-1} this behavior can produce a serious deterioration in the pulse shape. Hence, reducing t_d and damping the relaxation oscillations is highly desirable.

The switch-on delay is caused by the initial build-up of photon density resulting from stimulated emission. The switch-on delay may be reduced by biasing the laser near threshold. However, relaxation oscillations are basic laser phenomena, which vary with the device structures and the operating conditions. Narrow stripe geometry DH lasers and all the mode stabilized devices give RO damping, but it tends to coincide with a relatively slow increase in output power. However, RO damping and fast response may be obtained in BH structures with stripe widths less than the carrier diffusion length.

3.18.4 Noise

The noise behavior of the laser diode is another very important characteristic. This becomes more important when considering analog transmission. The different sources of noise are:

(*a*) phase or frequency noise;

(*b*) instabilities in operation such as kinks in the light output against current characteristic and self pulsation;

(*c*) reflection of light back into the device; and

(*d*) modal, mode partition and reflection noise.

However, by using mode stabilized devices and optical isolators, it is possible to reduce all types of noises, but phase or frequency noise. This phase noise is the intrinsic property of all laser types. It happens due to the discrete and random spontaneous or stimulated transitions which results in intensity fluctuations in the optical emission and is an unavoidable aspect of laser operation. It may be observed that the spectral density of this phase or frequency noise has a characteristic represented by 1/f to $1/f^2$ up to a frequency (*f*) of about 1 MHz, as shown in Fig. 3.60. At frequencies above 1 MHz the noise spectrum is flat or white and is associated with quantum fluctuations which are a principal cause of line broadening within semiconductor lasers.

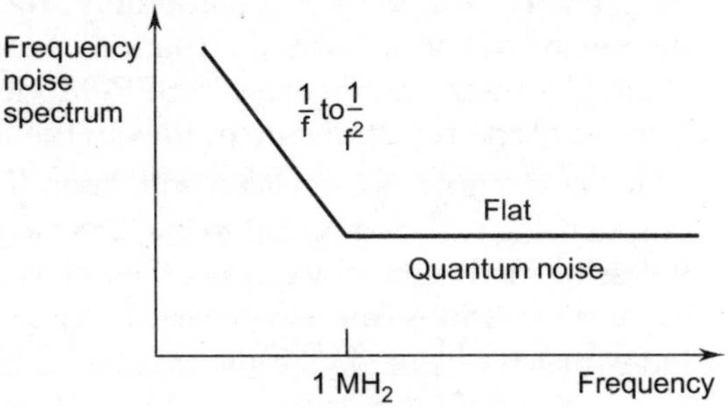

Fig. 3.60

Quantum noise levels are usually low for injection lasers operating at frequencies below 100 MHz, unless the device is biased within 10% of threshold. The noise spectrum is flat over this region. However, for wideband systems when the laser is operating above threshold, quantum noise becomes more pronounced. This happens mainly for multimode devices. The higher noise level would appear to result from a peak in the noise spectrum due to a relaxation resonance which typically occurs between 200 MHz and 1 GHz.

Optical intensity noise is also caused by the fluctuations in the amplitude or intensity of the output from semiconductor injection lasers. These fluctuations may be the result of the temperature variations or, alternatively may be caused from the spontaneous emission contained in the laser output. The random intensity fluctuations create a noise source referred to as Relative Intensity Noise (RIN), which may be defined in terms of the mean square power fluctuation $\overline{\delta P_e^2}$ and the mean optical power squared $(\bar{P}_e)^2$ which is emitted from the device following :

$$\mathrm{RIN} = \frac{\overline{\delta P_e^2}}{\left(\overline{P}_e\right)^2} \tag{3.117}$$

When light is launched into a multimode fiber from a coherent laser, a number of propagating modes of the fiber are normally excited. As long as these modes retain their relative phase coherence, the radiation pattern seen at the end of the fiber takes the form of a speckle pattern. This results due to the constructive and destructive interference between propagating modes at a given plane. The number of speckles in the pattern approximates the number of propagating modes. As the light travels along the fiber, a combination of mode dependent losses, changes in phases between modes, and fluctuations in the distribution of energy among the various fiber modes will change the modal interference and result in a different speckle pattern. Modal or speckle noise occurs when any losses that are speckle pattern dependent (viz. splices, connectors, microbends etc) are present in a link. Narrowband, high coherence sources, such as single-mode lasers result in more modal noise than broadband sources. Incoherent sources, such as LEDs, do not produce modal noise. The problem may be eliminated by using single-mode fibers.

Another noise phenomenon which occurs in multimode semiconductor lasers, where the modes are not stabilized is the mode partition noise. This is the dominant noise in single-mode fibers. Even when the total output power from a laser is maintained nearly constant, the temperature changes can cause the relative intensities of the various longitudinal modes in the laser's output spectrum to vary considerably from one pulse to the next. When combined with the fiber dispersion, these spectral fluctuations produce random distortion of received pulses on a digital channel, causing an increase in BER. Mode partition noise can also occur in single-mode lasers as a result of the residual side modes in the laser output spectrum.

Reflection noise is associated with laser diode output linearity distortion caused by some of the light output being reflected back into the laser cavity from the fiber joints. This reflected power couples with the lasing modes, thereby causing their phases to vary. Reflection noise problems can be greatly reduced by using optical isolators between the laser diode and the optical fiber transmission line or by using index-matching fluid in the gaps at fiber-to-fiber joints to eliminate reflections at the fiber-air interfaces.

3.18.5 Mode Hopping

Mode hopping occurs in all single-mode injection lasers due to increase in temperature of the device junction. The transition (hopping) from one mode to another is not a continuous function of the drive current but occurs suddenly over only 1 to 2 mA. The phenomenon alters the light output against current characteristics of the laser, and is responsible for the kinks already discussed. The single longitudinal mode output from a single-mode laser is shown in Fig. 3.61(*a*). The mode hopping to a longer wavelength as the current is increased above threshold is also shown in Fig. 3.61(*b*). During hops the mode tends to shift slightly with temperature in the range 0.05 to 0.08 nm K^{-1}. The mode hopping may be reduced with adequate heat sinking or thermoelectric cooling arrangement. However, at constant heat sink temperature, shifts due to thermal increase can only be fully controlled by the use of feedback from external or internal grating structures as previously discussed.

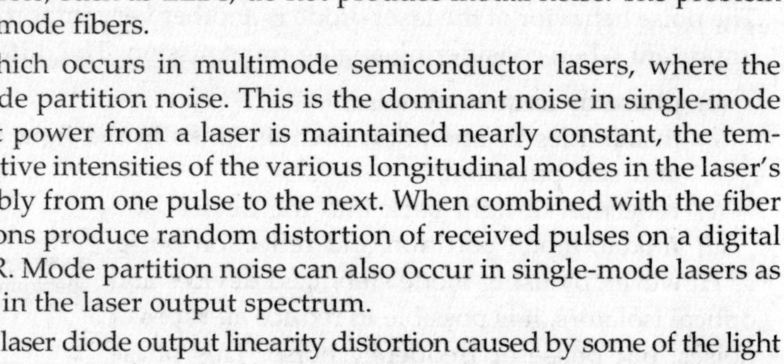

Fig. 3.61

3.18.6 Reliability Considerations

The reliability of semiconductor laser diodes are of major concern as they are one of the most important devices used in optical communication systems. Though the reliability conditions are extensively studied, but not all aspects of the failure mechanisms are fully understood. Lifetime tests of optical sources are carried out either at room temperature or at elevated temperatures (typically 70°C) to accelerate the degradation process. The two most commonly used techniques for determining the lifetime of the optical source is to either maintain a constant optical output by increasing the bias current automatically or to keep the current constant and monitor the optical output level. In the first case, the end of life of the device is assumed to be reached when the source can no longer provide a specific light output power at the maximum current value for CW operation. In the second case, the lifetime is determined by the time taken for the optical output power to decrease by 3 *d*B.

The degradation behaviour may be classified into two major processes known as 'catastrophic' and 'gradual' degradation. Mechanical damage of the mirror facets leads to catastrophic degradation, which ultimately damages the laser either partially or completely. This degradation actually reduces the laser mirror reflectivity and increase the non-radiative carrier recombination at the laser facets. This is caused by the average optical flux density within the structure at the facet and hence may be limited by using the device in a pulsed mode. This may severely restrict the operation to low optical power levels and the lifetime of the CW devices.

Gradual degradation mechanisms can be separated into two categories as: internal damage and ohmic contact degradation. These two mechanisms hold for both lasers as well as LEDs. These degradations are normally characterized by an increase in the threshold current for the laser which is often accompanied by a decrease in its internal quantum efficiency and increase in optical absorption.

Internal degradation arises from the migration of crystal defects into the active region of the light source. Defect formation in the active region can be promoted by the high density of recombining holes within the device. Internal damage may be caused due to the energy released, which may lead to possible presence of strain and thermal gradients by these nonradiative carrier recombination processes. Hence, the non-radiative electron-hole recombination at the damaged surface of a laser where it has been roughened accelerates the diffusion of the point defects into the active region of the device. These defect structures are generally named as 'Dark Spot Defects' (DSDs). Mobile impurities due to the precipitation process, such as oxygen, interstitial beryllium, copper or zinc atoms, may also be displaced into the active region of the laser. These atoms form cluster around existing dislocations increasing high local absorption of photons. Dark lines in the output spectrum of the device which are a major problem, is caused by this factor. This type of defects is normally named as 'Dark Line Defects' (DLDs). Both DSDs and DLDs are observed in ageing InGaAsP as well as AlGaAs lasers.

Degradation of the current confining junctions occurs in many index-guiding lasers structures which utilize current restriction layers to restrict most of the injected current to the active region. The current flowing outside the active region is called leakage current. Thus a type of degradation associated with this type of laser structures is an increase in leakage current which increases the device threshold with ageing.

Fabrication steps that can be taken to minimize internal degradation include the use of substrates with very low surface dislocation densities (less than 10^3 dislocations/cm^2), keeping work-damaged edges out of the diode current path, minimizing stresses in the active region to less than 10^8 dyn/cm^2, passivating the mirror facets to avoid surface related effects and mounting with soft solders to avoid external strain. For high-quality sources having lifetimes which follow a slow internal degradation mode, the optical power P decreases with time according to the relationship given below:

$$P = P_0 e^{-t/\tau_m} \tag{3.118}$$

where P_0 is the initial optical power at time $t = 0$, and τ_m is the time constant for the degradation process, which is approximately twice the 3 dB mean time to failure. Since the operating lifetime depends on both the current density J and the junction temperature T, internal degradation can be accelerated by increasing either one of these parameters.

The operating lifetime τ_s has been found experimentally to depend on the current density J through the relation

$$\tau_s \propto J^{-n} \tag{3.119}$$

where $1.5 \leq n \leq 2.0$. Since the degradation rate of optical sources increases with temperature, an Arrhenius relationship of the form is given as:

$$\tau_s = Ke^{E_A/k_B T} \tag{3.120}$$

Here, E_A is an activation energy characterizing the lifetime τ_s, k_B is Boltzmann's constant, T is the absolute at which τ_s was evaluated, and K is a constant. Activation energies for laser degradation reported in the literature have ranged from 0.3 to 1.0 eV. For most of the practical calculations the value of 0.7 eV is generally used. It may be noticed from equations (3.119) and (3.120) that, the lifetime of the light source may be increased by operating the devices at as low a temperature and current as is practicable. At higher temperatures the power output drops as a function of time.

As already mentioned a second fabrication-related degradation mechanism is ohmic contact degradation. In LEDs and laser diodes the thermal resistance of the contact between the light source chip and the device heat sink occasionally increases with time. This is a function of the solder used to bond the chip to the heat sink, the contact temperature and the current through the contact. The increase in thermal resistance results in rise of the junction temperature which in turn, leads to a decrease in the optical output power.

There are two types of facet damage: (*a*) *catastrophic facet degradation*, and (*b*) *facet erosion*. Catastrophic facet degradation had already been discussed. Facet erosion is a gradual degradation occurring over a longer period of time than catastrophic facet damage. Facet erosion decreases the internal quantum efficiency of the laser and increases the threshold current. Facet erosion occurs due to oxidation of mirror surface in AlGaAs lasers. Facet erosion may be minimized by depositing a half-wave-length-thick Al_2O_3 film on the facet. This type of coating acts as a moisture barrier and does not affect the mirror reflectivity or the lasing threshold current.

3.19 LASER DRIVE CIRCUITS (TRANSMITTER)

The normal transmitter module is made up of the mounted laser chip, its housing and the thermal management components, the bias and microwave drive circuitry, and the optical coupling of the output fiber pig-tail. In high performance systems, the laser is often mounted on a diamond header attached to a Peltier cooler. The diamond provides both good thermal conductivity and excellent insulation. Different configurations described for used as LED drive circuits for both analog and digital transmission may be adopted for semiconductor laser applications with only some minor changes. The laser, which is a threshold device however, have somewhat different drive current requirements from the LED.

For example, when digital transmission is considered, the laser is usually given a substantial applied bias, often referred to as prebias, in the off state. Although biasing near threshold causes spontaneous emission of light in the off state, but this does not create any problem for digital transmission, since the stimulated emission in the on state is much greater than that at the off state. Actually this prebias in the off state is provided to:

1. Reduce the switch-on delay and minimize any relaxation oscillations.
2. Reduce the junction heating caused by the digital drive current since the on and off currents are not widely different for most lasers.
3. Allow easy compensation for changes in ambient temperature and device ageing.

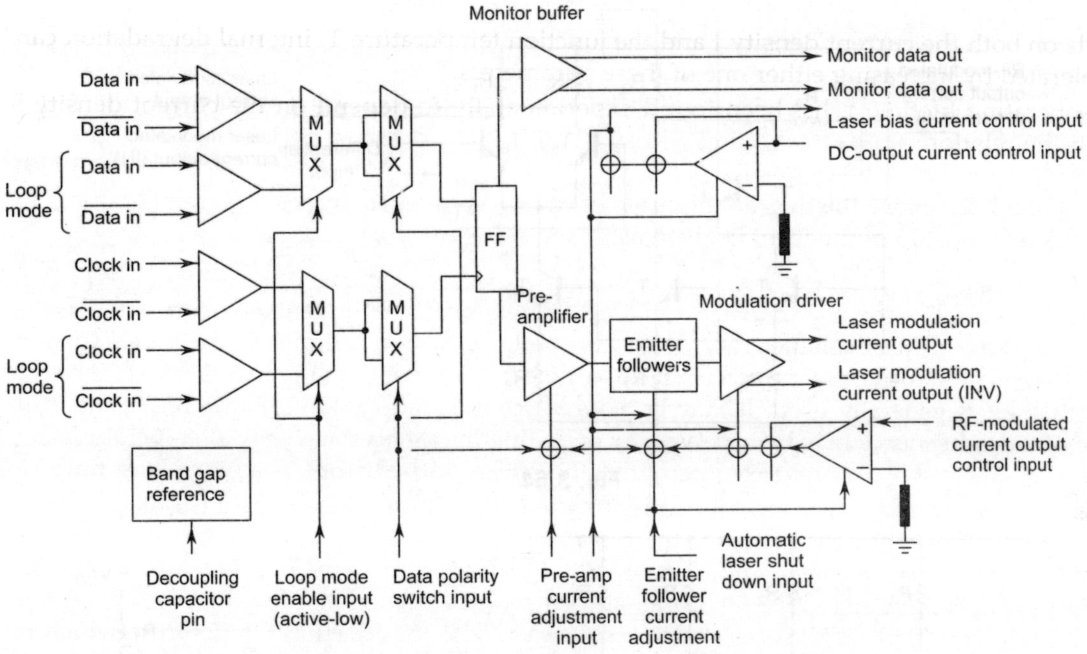

Fig. 3.62

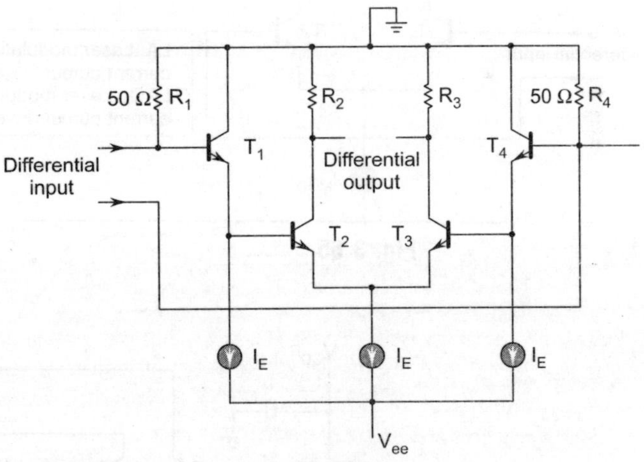

Fig. 3.63

The block diagram of a typical laser driver circuit is illustrated in Fig. 3.62. The main building blocks of the laser driver circuit in Fig. 3.62 are an input buffer, a current limiter, a laser control block, and a band gap reference. The multiplexer outputs vary in amplitude between 200 mV and 800 mV (the average is 400 mV) and are directly connected to the high impedance inputs of the laser driver circuit's buffer. In multiplexers, signals are to be accepted by the buffer inputs, and a 50 Ω pull up external resistor may be required to be connected between the inputs and the power supply. A typical schematic diagram of the buffer differential input stage is shown in Fig. 3.63. The output signal from the buffer circuit is used to drive the current of 60 mA to the laser modulation output and laser modulation output-inverted (illustrated in Fig. 3.64). In order to maintain the laser output optical power above the light emitting threshold level, a DC bias current of 90 mA (max.) may be required. The laser driver biasing circuit illustrated in Fig. 3.65 provides this current.

Although biasing near threshold causes spontaneous emission of light in the off state, this is not normally a problem for digital transmission because the stimulated emission in the on state is generally greater by, at least, a factor of 10.

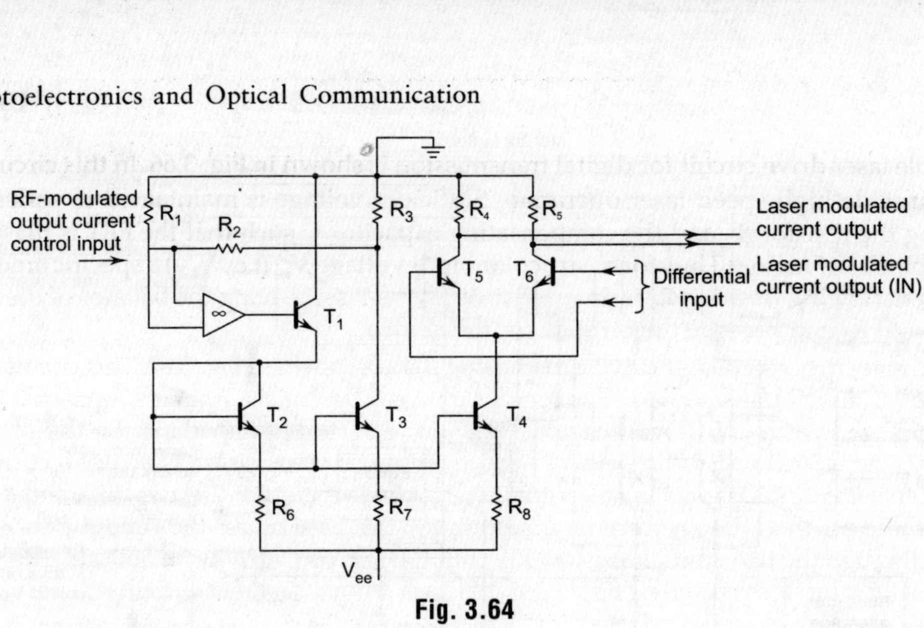

Fig. 3.64

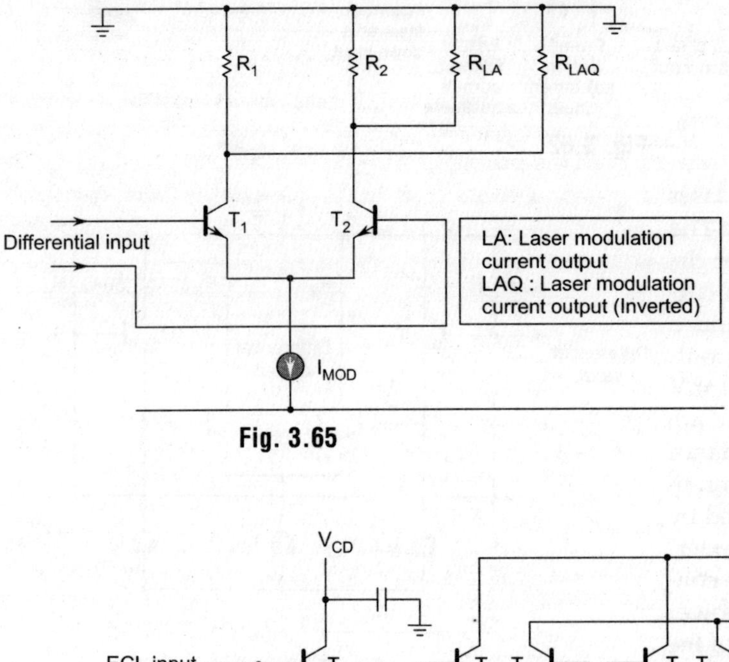

Fig. 3.65

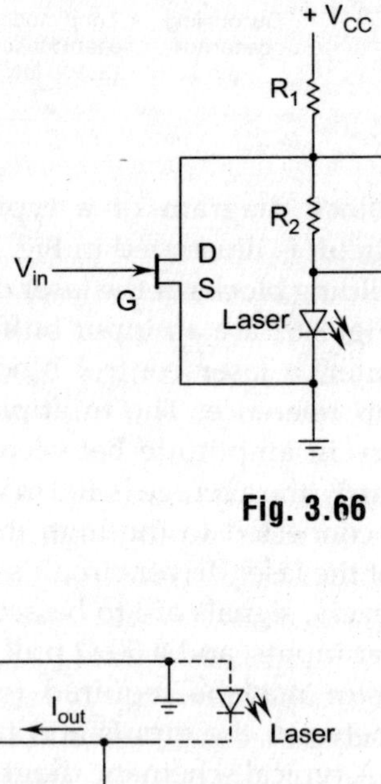

Fig. 3.66

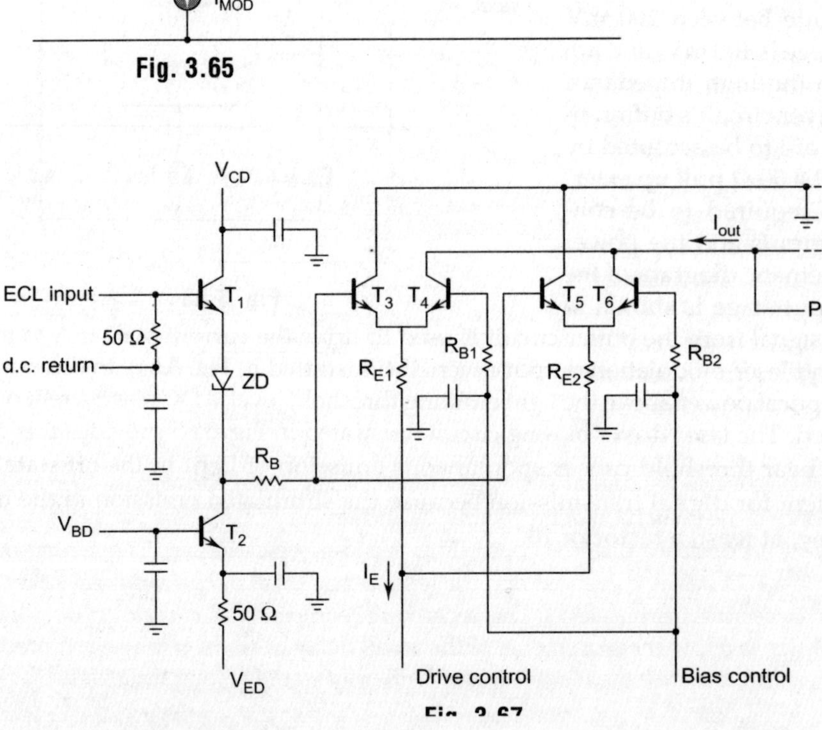

Fig. 3.67

A simple laser drive circuit for digital transmission is shown in Fig. 3.66. In this circuit a FET is used to provide high speed laser operation. Sufficient voltage is maintained in series with the laser using the resistor R_2 and the compensating capacitor C such that the FET is biased into its active or pinch-off region. Thus for a particular input voltage V_{in} (*i.e.*, V_{GS}) a specific amount of the total current flowing through R_1 is diverted around the laser leaving the balance of the current to flow through R_2 and provide the off state for the device.

An alternate high speed laser drive circuit using BJTs is shown in Fig. 3.67. This circuit configuration is also for digital transmission. This consists of two differential amplifiers connected in parallel. The input stage, which is ECL compatible, exhibits a 50 Ω input impedance by use of an emitter follower T_1 and a 50 Ω resistor in parallel with the input. The transistor T_2 acts as a current source with the Zener Diode (ZD) adjusting the signal level for ECL operation. The two differential amplifiers produce sufficient modulation current amplitude for the laser under the control of a d.c. control current I_E through the two emitter resistors R_{E1} and R_{E2}; I_E is provided by an optical feedback control circuit. Finally a pre-bias current is applied to the laser from a separate current source.

A major difference between the drive circuits shown in Figs. 3.66 and 3.67 is the absence and presence of feedback control for adjustment of the laser output level. So, the circuit illustrated in Fig. 3.66 is seldom used for a system application. Some form of feedback control is generally required for continuous laser operation because the device lasing threshold is a sensitive function of temperature. Again the threshold level tends to increase with ageing of the laser, leading to an increase in internal device losses. Though to compensate the temperature variation the laser may be cooled, but ageing cannot be so easily compensated. However, both the problems may be overcome by controlling the laser bias using a feedback technique. This may be achieved by using low speed feedback circuits.

There are different techniques of varying complexity to provide automatic output level control for the laser. One of the most common forms of laser drive circuit incorporating optical feedback is the mean power control circuit illustrated in Fig. 3.68. Often the monitor detector consists of a cheap, slow photodiode positioned next to the rear face of the laser package, as shown in Fig. 3.68. An alternate approach may be to use an optical coupler at the fiber input to direct some of the radiation emitted from the laser into the monitor photodiode. The detected signal is integrated and compared with a reference by an operational amplifier which is used to servo-control the d.c. bias applied to the laser. Thus by varying the threshold current level, the mean optical power may be maintained constant. This technique may be used for both analog and digital transmission.

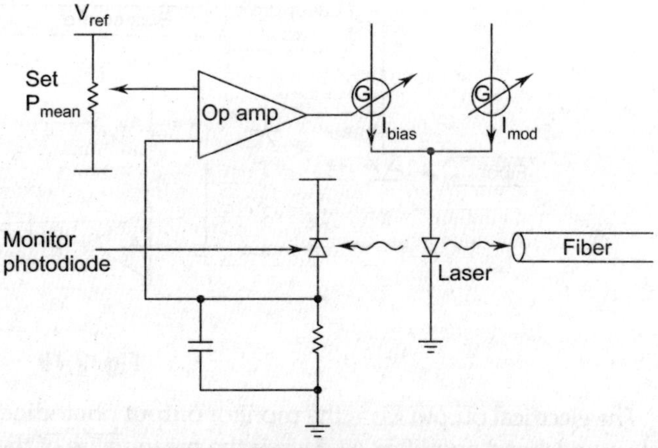

Fig. 3.68

A second method of control for digital systems which offers accurate threshold tracking and very little device dependence is the switch-on delay technique shown in Fig. 3.69. The circuit shown monitors the switch-on delay of an optical pulse for controlling the laser bias current. The switch-on delay is measured for a zero level set below threshold and the feedback is set to a constant fixed delay to control it. This signal is used to control the bias level. This technique requires a fast monitor photodiode as well as a wideband amplifier to allow measurements of the small delay periods. It is also required to set the zero level below threshold, because the feedback loop will only stabilize for a finite delay.

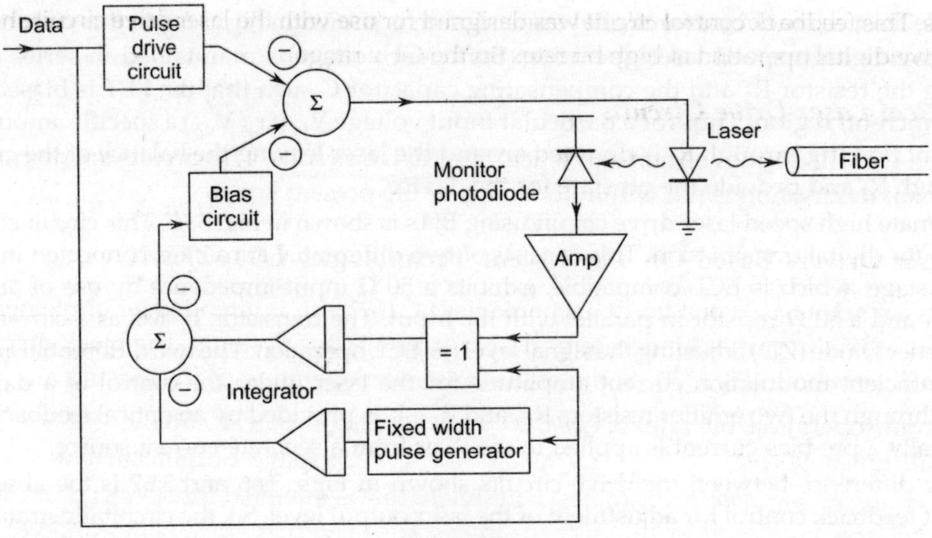

Fig. 3.69

When only the laser bias current is controlled, the variations in laser slope efficiency are not controlled. This is a major disadvantage. The modulation current for the device is preset and does not take into consideration any slope changes with ageing and temperature. To compensate for such changes, the a.c. and d.c. components of the monitored light output must be processed independently. This is especially important in the case of high bit rate digital systems. In Fig. 3.70, we illustrate a circuit which utilizes both a.c. and d.c. information in the laser output to control the device drive current and bias current independently.

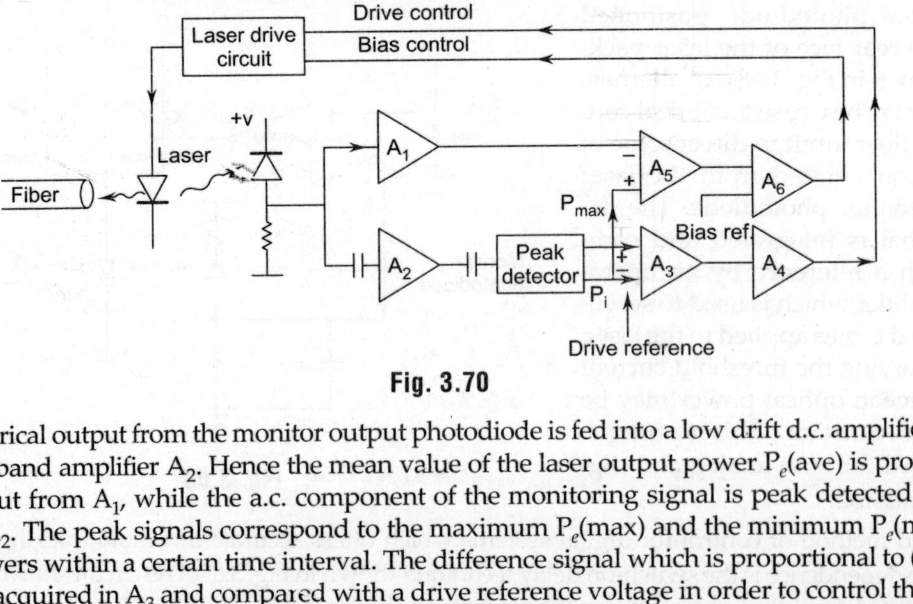

Fig. 3.70

The electrical output from the monitor output photodiode is fed into a low drift d.c. amplifier A_1 and into a wideband amplifier A_2. Hence the mean value of the laser output power P_e(ave) is proportional to the output from A_1, while the a.c. component of the monitoring signal is peak detected after the amplifier A_2. The peak signals correspond to the maximum P_e(max) and the minimum P_e(min) laser output powers within a certain time interval. The difference signal which is proportional to (P_e(max)-P_e(min)) is acquired in A_3 and compared with a drive reference voltage in order to control the current output from A_4 and, consequently the laser drive current. Hence, the modulation amplitude of the laser is controlled. Control of the laser bias current is achieved from the difference between the output signal of A_1 and P_e(min) which is acquired in A_5. The output voltage of A_5, which is proportional to P_e(min), is compared with a bias reference voltage in A_6 which supplies a current output to control the laser d.c. bias. This feedback control circuit was designed for use with the laser drive circuit shown in Fig. 3.67 to give digital operation at high bit rates (in the GHz range).

Some Practical Laser Drive Circuits

In this section we will discuss on some practical laser drive circuits. The circuits are discussed in brief, as a detail description is not within the scope of the present book.

3.19.1 Laser Drivers Based On The Lucent Technologies LG1627BXC

The laser driver circuit based on Lucent Technologies LG1627BXC is intended for implementation in high speed NRZ direct modulation optical fiber transmission systems. The block diagram of the device is illustrated in Fig. 3.71. Some of the important operating characteristics are:

1. Maximum operating bit rate: 3 Gb/s;
2. Single power supply: -5.2 V;
3. Rise/fall times: 90 ps;
4. Adjustable output current.

The device can also be used in optical test equipment that is intended for SONET/SDH applications. The circuit is composed of three stages in cascade. The first stage is designed to accept differential clock and data ECL inputs. The device is also capable of independently controlling the laser bias and modulation currents through external networks, while an internal 2.5 V band gap reference voltage compensates for supply and temperature variations. Data retiming and jitter minimization are accomplished through internal flip-flops.

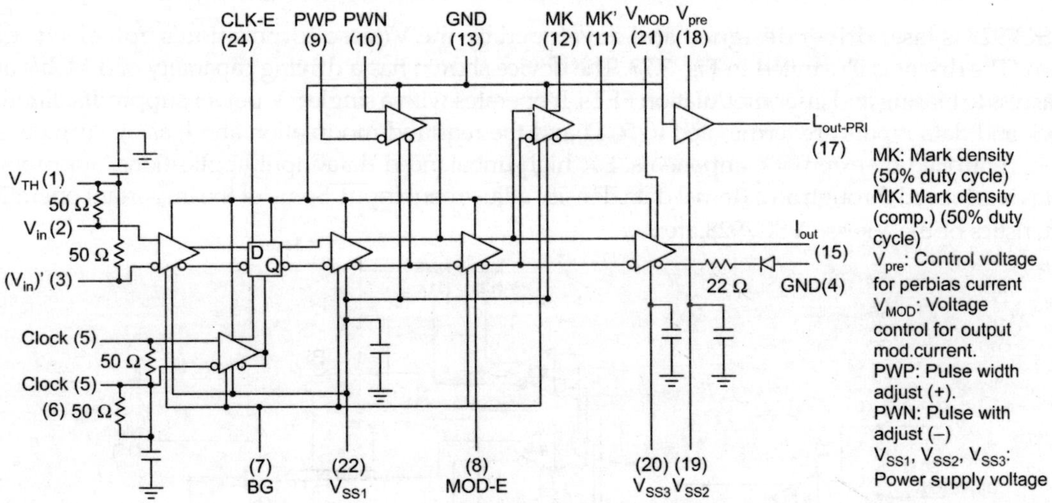

Fig. 3.71

3.19.2 Laser Drivers Based On The Nortel YA08

The following laser diode driver circuit is based on the NORTEL YA08 device. This device is able to handle data rates of up to 2.5 Gb/s for SONET-48/STM-16 applications. A block diagram of the device is illustrated in Fig. 3.72. The main building blocks of the above device are:

1. Input buffer;
2. Modulation switch;
3. Mean power control;
4. Bias current limiter;
5. Bias mirror output;
6. Laser shutdown input.

The device also provides a built-in reference voltage source for temperature and power supply stabilization.

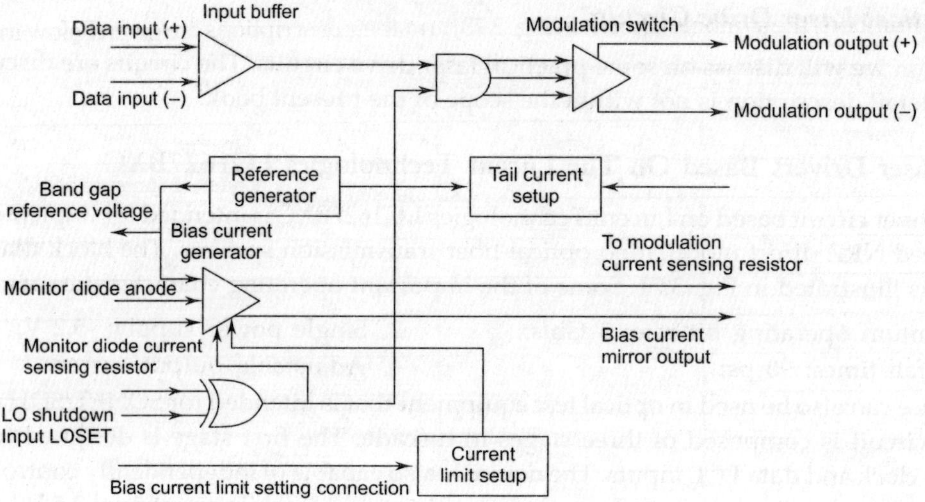

Fig. 3.72

3.19.3 Laser Drivers Based On The Vitesse VSC7928 3.2 Gb/s

The VSC7928 is laser driver designed and developed by the Vitesse corporation's optoelectronics division. The driver is illustrated in Fig. 3.73. The device shown has a driving capability of 3.2 Gb/s and direct access to biasing and laser modulation FETs. It operates with a single 5 V power supply. Its differential clock and data inputs are terminated to 50 Ω and the required modulation and biasing currents are precisely controlled by external components. For high unbalanced data input applications, appropriate laser bias is achieved through an external data density adjustment input. Some of the important operating characteristics of the Vitesse VSC7928 are:

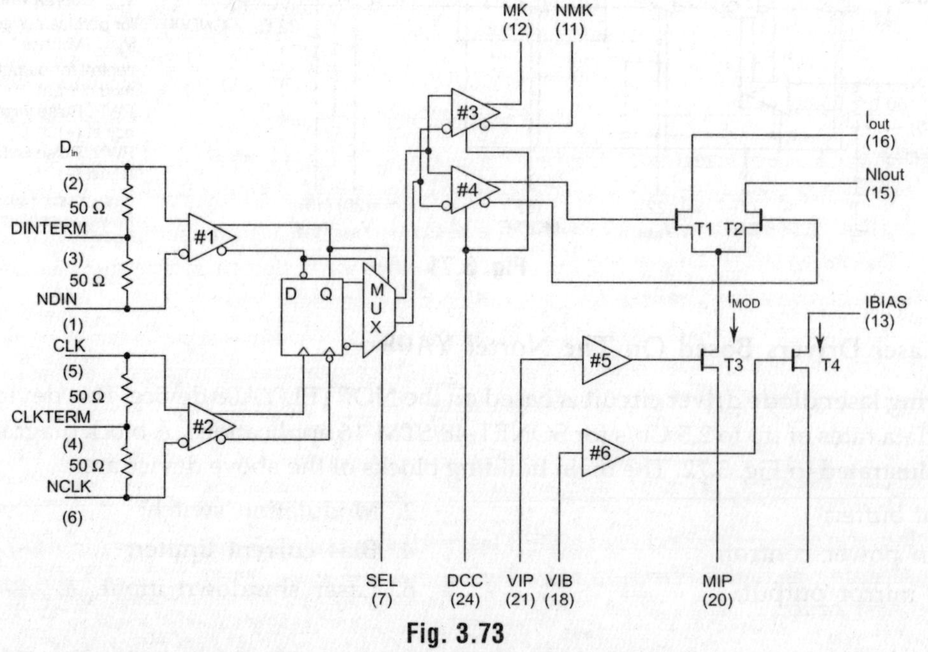

Fig. 3.73

1. Input data rate max: 3.2 Gb/s;
2. Rise time: < 100 ps;
3. Single or differential clock and data inputs.

The pin numbers, the symbols (shown in Fig. 3.73) and their descriptions are given below in Table 3.6.

Table 3.6

Pin Number	*Symbol*	*Description*
1	NDIN	Data reference (on chip)
2	DIN	Data input
3	DINTERM	Data reference
4	CLKTERM	Clock reference
5	CLK	Input clock
6	NCLK	Clock reference on chip
7	SEL	Clock-non-clock data select
8, 9, 10, 14, 17	GND	Positive voltage bus
11	NMK	Data density differential output (-)
12	MK	Data density differential output (+)
13	IBIAS	Laser bias output
15	NIOUT	Laser modulation current output (-)
16	IOUT	Laser modulation current output
18	VIB	Bias gate node
19	MIB	Bias source node
20	MIP	Modulation source node
21	VIP	Modulation gate node
22, 23	VSS	Negative voltage bus
24	DCC	Duty cycle control

3.20 CONCLUSION

In this chapter, we have started with the basic principles of lasers. The amplification of light, the requirements for such amplification have been dealt with in detail. Different types of lasers, viz., solid state, liquid, gas, and semiconductor have been discussed with practical applications. The comparative study has been done for the characteristics of different lasing systems. Basically lasers are one of the most fascinating devices, which have brought a revolution in optical communication. We have covered all those aspects in this chapter. We have specially put stress on the semiconductor lasers, as they are the backbone of optical communication. We have also discussed different practical circuits for the laser transmitters.

REVIEW QUESTIONS

1. Calculate the grating period Λ of a DFB laser to obtain single-mode operation at $\lambda_B = 1500$ nm, assuming that the effective refractive index for λ_B is $n_e = 3.1$, where the grating is of the first order.

(*a*) 123.3 nm (*b*) 201.2 nm (*c*) 241.9 nm (*d*) 300.5 nm

2. The ratio between the probabilities of spontaneous emission and stimulated emission is given by the relationship:

(*a*) $\frac{A_{ji}}{B_{ji}} = \frac{8\pi h v^3}{c^3}$ (*b*) $\frac{A_{ji}}{B_{ji}} = \frac{8\pi h c^3}{v^3}$

(*c*) $\frac{A_{ji}}{B_{ji}} = \frac{8\pi h v^3}{c^3}$ (*d*) $\frac{A_{ji}}{B_{ji}} = \frac{8\pi h v^2}{c^3}$

3. Determine the length of the optical cavity of an AlGaAs laser emitting at wavelength of 0.95 µm. The modes are separated by 180 GHz. The refractive index of AlGaAs is 3.32.

(*a*) 125 µm (*b*) 251 µm (*c*) 279 µm (*d*) 311 µm

4. The Hermite-Gaussian modes are also referred to as

(*a*) TE modes (*b*) TM modes (*c*) TEM modes (*d*) None of these

5. The first laser to be invented in 1960 by Maiman was

(*a*) He-Ne laser (*b*) Nd:YAG laser (*c*) Dye laser (*d*) Ruby laser

6. Carbon dioxide laser is a

(*a*) Molecular laser (*b*) Ionic laser (*c*) Atomic laser (*d*) None of these

7. A powerful laser which is basically always operated in the ultraviolet (UV) spectral region and generates nanosecond pulses is

(*a*) Solid state laser (*b*) Excimer laser

(*c*) Semiconductor laser (*d*) Gas laser

8. In a negative-index waveguide, the refractive index of the central portion of the active region compared to the outer regions is

(*a*) Lower (*b*) Same (*c*) Higher (*d*) Does not matter

9. Mode hopping occurs in all single-mode injection lasers when the temperature of the device junction

(*a*) Decreases (*b*) Remains unaltered

(*c*) Increases (*d*) None of these

10. ADH semiconductor laser has a cavity length of 0.6 mm, an effective loss coefficient of 1 mm^{-1}, a confinement factor of 0.9, and uncoated facet reflectivities of 0.33. Calculate the reduction in the threshold gain coefficient that occurs when the reflectivity of one of the facets is increased to 1

(*a*) 2.66 mm^{-1} (*b*) 2.03 mm^{-1} (*c*) 1.99 mm^{-1} (*d*) 1.03 mm^{-1}

11. What do you mean by *stimulated emission*? How does it differ from *spontaneous emission*? Estimate the ratio of the stimulated emission rate to the spontaneous emission rate for an incandescent lamp at a temperature of 850 K. It may be assumed that the average operating wavelength is 0.6 µm.

12. What is population inversion? How does it help in lasing action? Discuss the mechanism of optical feedback to provide oscillation and hence amplification within the laser.

13. Deduce the threshold condition for laser oscillation. For a GaAs laser, the reflectivities of the two facets are 0.25 for uncoated facets (*i.e.*, 25% of the radiation is reflected at the facets) and α = 10 cm^{-1}. If the laser diode is 750 µm long, estimate the threshold gain for the laser.

14. Discuss the principles of operation of the semiconductor lasers, with the help of proper diagram. What are the semiconductor materials used for emission over the wavelength range 800 to 1700 nm? Give reasons for their choice.

15. Discuss the techniques used to give both electrical and optical confinement in multimode injection lasers? Compare these techniques when used in gain-guided and index-guided lasers.

16. Briefly discuss on the following terms related to injection lasers:

(*a*) Temperature effects; (*b*) Frequency chirp;
(*c*) Dynamic response; (*d*) Noise;
(*e*) Mode hopping; (*f*) Relaxation oscillations.

17. What is lineshape function? Discuss briefly on *Doppler broadening*?

18. What is axial mode of laser? A ruby laser contains a crystal length 8 cm with a refractive index of 1.75. The peak emission wavelength from the laser is 0.55 μm. Determine the number of longitudinal modes and their frequency separation.

19. Discuss the principle of *mode locking*. What is active mode locking? Also discuss on passive mode locking.

20. What is Q-factor? What is the principle of Q-switching?

21. Describe the different methods of Q-switching.

22. Write short notes on:

(*a*) Solid state lasers; (*b*) Gas lasers;
(*c*) Excimer lasers; (*d*) Liquid lasers;
(*e*) Semiconductor lasers; (*f*) Tea co_2 laser.

23. Comment on the working principle of Nd:YAG laser. Briefly discuss the applications of Nd:YAG lasers.

24. What are fiber lasers? Discuss in brief.

25. Discuss the working principle of He-Ne lasers.

26. Describe the working principle of Ar lasers. Compare the working of CO_2 lasers with Ar lasers.

27. Comment on the power output of semiconductor lasers. A semiconductor laser has a GaAs active region with bandgap energy of 1.43 eV. Calculate the wavelength of optical emission from the device and determine its linewidth in Hertz when the measured spectral width is 0.1 nm.

28. Discuss the structure of stripe geometry lasers. The threshold current density for a stripe geometry laser is 3000 A cm^{-1} at a temperature of 15°C. Calculate the required threshold current at a temperature of 60°C when the threshold temperature coefficient T_0 for the device is 108 K, and the contact stripe is 20 × 100 μm.

29. Estimate the internal quantum efficiency of a GaAs semiconductor laser with a cavity of length 500 μm and a loss coefficient of 20 cm^{-1}. The measured differential quantum efficiency of the device is 45% and the refractive index of GaAs is 3.6.

30. What is a DFB laser? What is the difference of DFB lasers in comparison to DBR lasers? Write a short note on tunable lasers.

4

Photodetectors

4.1 INTRODUCTION

A photodetector is an optoelectronic device that receives optical energy and converts to electrical energy by producing the electrical current, normally termed as photocurrent. Generally three steps are involved in the photodetection process:

1. Absorption of optical energy and generation of carriers.
2. Transportation of photogenerated carriers across the absorption or transit region, with or without gain.
3. Carrier collection and generation of photocurrent, which flows through the external circuitry.

Sometimes the detection process is associated with demodulation, when a high frequency optical signal is converted into time-varying electrical signal and further processed and rectified. Optical signals ranging over a very wide range of the optical spectrum are detected by the photodetectors.

Photodetectors are widely used in optical communication systems. In this application, transmitted optical pulses are captured by the detectors and are converted to electrical pulses, with as little loss as possible, that can be used by telephone, computer or other output devices at the receiving side. The performance requirements of photodetectors are expected to be of high sensitivity, low noise, wide bandwidth, high reliability and low cost. Monitoring of laser transmitters is other common application of the photodetectors, where detectors of large area are placed near one facet of transmitting laser source to generate the photocurrent, which in turn is used in the circuit to maintain the laser output at near constant level in spite of temperature variation. Gain and bandwidth are two primary parameters to select the photodetectors for final applications. There are three main types of detectors, - (*i*) photoconductors, (*ii*) PIN diodes and (*iii*) avalanche photodiodes. Photoconductors and avalanche photodiodes inherit internal gain, whereas PIN diodes have no internal gain but can have large bandwidths. In this chapter, physical principles of these devices will be discussed and also some special types of photodetecctors like phototransistors modulated barrier photodiodes and Metal-Semiconductor- Metal (MSM) photodiodes will be described.

Photodetectors are also classified as *intrinsic* and *extrinsic* types. An *intrinsic* photodetector usually detects the light of wavelength close to the bandgap of the semiconductor. Electron-hole pairs are formed in the intrinsic photodetector by photoexcitation and produce photocurrent. On the other hand, an *extrinsic* photodetector detects light of energy smaller than the bandgap energy. In these devices, the transition corresponding to the absorption of photons involves deep impurity and defect levels within the bandgap. Absorption of a photon raises an electron from a deep level to the conduction band, or an electron from valence band to deep level, creating a hole in the valence band. Photocurrent is developed by the electron or hole in the respective band. There exists another type of *extrinsic* photodetector, where transitions are involved between the subband energies in a quantum well. The energy differences between the subbands is generally very small and of the order of 100meV, and therefore such devices are employed for the detection of far-infrared wavelengths. Figure 4.1(*a*) illustrates the intrinsic detection process, whereas Figs. 4.1(*b*) and 4.1(*c*) explain the extrinsic detection process. However, there are some drawbacks and limitation regarding the polarization of light that can be absorbed.

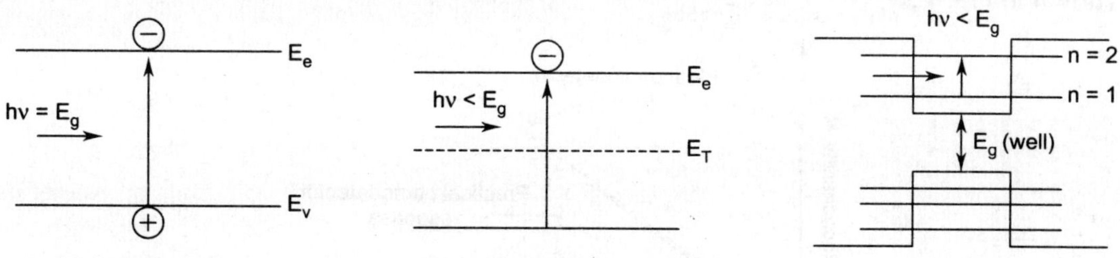

(*a*) Intrinsic detector

(*b*) Extrinsic detector using deep level

(*c*) Extrinsic detector using subbands

Fig. 4.1

4.2 QUANTUM EFFICIENCY

Two types of quantum efficiency may be defined during photodetection. The *internal quantum efficiency,* η_i is defined as the number of electron-hole pairs created divided by the number of photons absorbed. In pure defect-free materials, the internal quantum efficiency is observed to be very high and near to unity.

The *external quantum efficiency,* η_{ext} can be derived by the number of carriers (electro-hole pairs) collected to produce the photocurrent I_{ph} divided by the number of incident photons. If P_{inc} is the incident optical power, then

Number of carriers = I_{ph}/q and

Number of incident photons = $P_{inc}/h\nu$

Hence, the quantum efficiency η may be expressed as

$$\eta_{ext} = \frac{I_{ph}/q}{P_{inc}/h\nu} = \frac{I_{ph}}{q} \cdot \frac{h\nu}{P_{inc}} \tag{4.1}$$

The external quantum efficiency η_{ext} or η depends on the absorption coefficient of the material and the thickness of the absorbing region. It is related as per following expression.

$$\eta_{ext} \text{ or } \eta \propto (1 - e^{-\alpha a}) \tag{4.2}$$

where α is the absorption coefficient of the semiconductor and a is the thickness of the active region. From equation 4.2, it may be noted that the quantum efficiency is a function of photon wavelength. The *responsivity* of the detector R is defined as

$$R = \frac{I_{ph}}{P_{inc}} = \frac{\eta q}{h\nu}$$

or
$$R = \frac{\eta\lambda(\mu m)}{1.24} \text{ (A/W)} \tag{4.3}$$

ideally, when η has a fixed value, R should increase linearly in respect of λ as schematically shown in Fig. 4.2. However, in practical situation, η depends upon the absorption coefficient α, which in turn is dependent on the incident wavelength λ. Therefore, for practical detectors, the long-wavelength cut-off of the spectral response is determined by the absorption edge, or bandgap of the semiconductor. For short wavelength, the value of α is very large in most semiconductors and all the optical energy is absorbed near the surface. Due to this, short wavelength cut-off is also observed in the spectral response characteristics. The practical spectral response characteristic is also shown in Fig. 4.2.

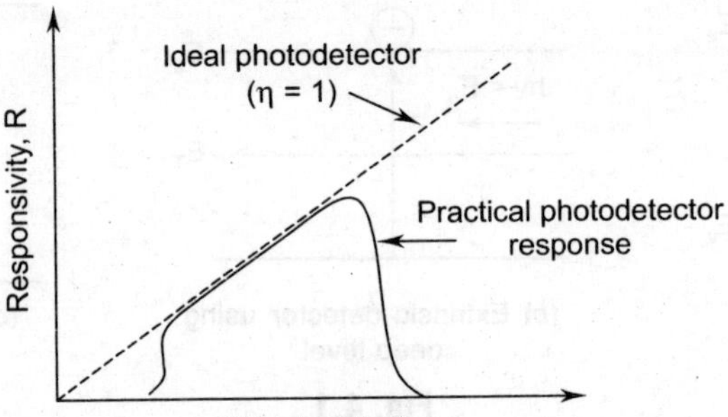

Fig. 4.2

Example 4.1. *Determine the long wavelength cut off of an InP photodetector.*

Solution: The photon energy must be greater the bandgap E_g to be absorbed by photodetector. *i.e.*, $h\nu \geq E_g$ where $\nu = C/\lambda$, C is the velocity of light. Therefore $hC/\lambda \geq E_g$. For InP photodetector $E_g = 1.35$ eV at room temperature. So long wavelength cut off $\lambda_c = hC/E_g = (6.626 \times 10^{-34} \times 3 \times 10^8) / (1.35 \times 1.6 \times 10^{-19})$ = 0.92 μm.

4.3 PHOTOCONDUCTORS

Photoconductor is the simplest type of optical detector, It exhibits an internal gain mechanism and also clearly demonstrates the gain-bandwidth limitation. Its operation based on the increase in conductivity of a specific region with photoexcitation. Photocurrent is developed when the photogenerated electrons and holes are collected by opposite contacts. The simplest form of a photoconductor is schematically shown at Fig. 4.3. On a high resistivity substrate, the active layer is formed epitaxially or by ion implantation and suitable contacts are fabricated on top by evaporation of metals and subsequent alloying. The thickness of the active layers optimized such that, it should be large enough so that it can absorb a significant amount of the incident light, but at the time small enough to minimize the noise current resulting from a low resistance of the semiconductor layer. The separation between the contact pads, either in linear or interdigitated form, plays an important role in the operation of the device. Photogenerated carriers produce electrical current when suitable bias is applied across the contacts. In InGaAs photoconductor, η_{ext} is observed typically in range of 80% to 90%, with an absorption region thickness of 2 μm. Quantum efficiency may be improved with the use of antireflective coating, or by forming a wide bandgap window layer on the surface of the absorbing region.

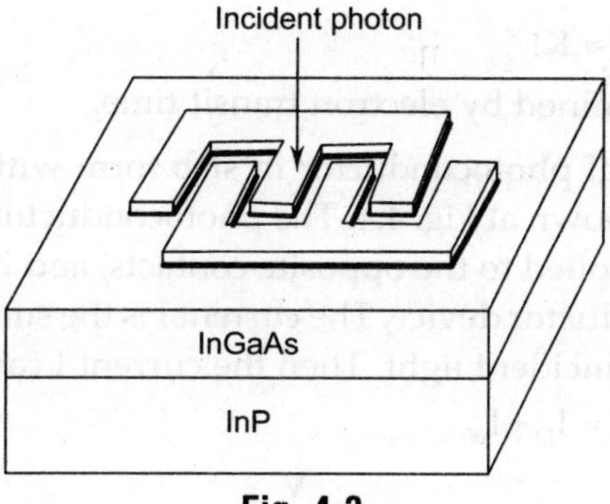

Fig. 4.3

Another important parameter that affects the performance of the device and needs to be considered is noise generation. *Johnson noise* or *thermal noise* is generated by the random motion of the carriers, which in turn produces large dark current in the device. The resulting noise current, i_j was formulated by Nyquist from thermodynamic arguments and is expressed as

$$i_j^2 = \frac{4k_B TB}{R_C} \tag{4.4}$$

where B is the bandwidth and R_C is the resistance of the photoconducting channel. Current is always present in the photoconductor and hence, the signal to noise ratio, S/N can be approximately expressed as

$$\frac{S}{N} = \frac{(\text{conductivity})_{\text{light}}}{(\text{conductivity})_{\text{dark}}} \tag{4.5}$$

So it is essential that dark current or dark conductivity be minimized to reduce noise.

The rate and density of data transmission depend on the response speed of the device. Under applied bias, the photogenerated electrons and holes move in opposite directions in the active region, resulting in photocurrent until the carriers are collected at the electrodes, or they are recombined before reaching at the respective contacts. The primary concern is the recombination of minority carriers and hence the minority carrier recombination time τ is the parameter of importance. The time for detection of the photogenerated current is governed by the transit time between the electrodes of the faster carrier—usually the electrons. Therefore, the shortest response time or maximum bandwidth can be obtained by minimizing the distance between the contacts. Applied bias is assumed to be such that the carriers travel at their saturation velocity. It may be noted that in a *n*-channel, photocurrent persists due to slower holes after the electrons are collected at the electrodes, resulting in an increase in response time. This means, the bandwidth is inversely proportional to the recombination time and expressed as

$$\text{Bandwidth} \propto \frac{1}{\tau} \tag{4.6}$$

where $\tau = \tau_h$. However, the continued persistence of hole in the channel will draw more electrons to maintain charge neutrality. This provides gain in photocurrent Γ_0 that can be defined as

$$\text{Gain } \Gamma_0 = \frac{\tau}{\tau_{tr}} \tag{4.7}$$

where τ_{tr} is the transit time for electrons. So, phenomenologically, it may be concluded that

$$\text{Gain } x \text{ Bandwidth} = K \tag{4.8}$$

where K is the constant determined by electron transit time.

Now let us consider an ideal photoconductor in slab form with contacts at the two opposite edges, schematic of which is shown at Fig. 4.4. The photoconductor has length L, thickness *a*, and width *b*. A bias voltage V is applied to the opposite contacts, and it is considered that current I is flowing through the photoconductor device. The current I is the summation of dark current I_D and photocurrent I_{ph} produced by incident light. Then the current I can be expressed as

$$I = I_D + I_{ph}$$

$$= abq(n\mu_e + p\mu_h)\frac{V}{L} \tag{4.9}$$

where *n* and *p* are the concentration of carriers, and μ_e and μ_h are the mobilities of the carriers. If P_{inc} is the optical power of the incident photons, then number of photons incident on the photoconductor device is $P_{inc}/h\nu$. If corresponding generation rate is considered to be G in the total volume V_v of the photoconductor, and all the incident photons are absorbed, then the internal quantum efficiency η_i may be related as

$$\eta_i = \frac{G}{R_{abs}} = \frac{GV_v}{P_{inc}/h\nu} \tag{4.10}$$

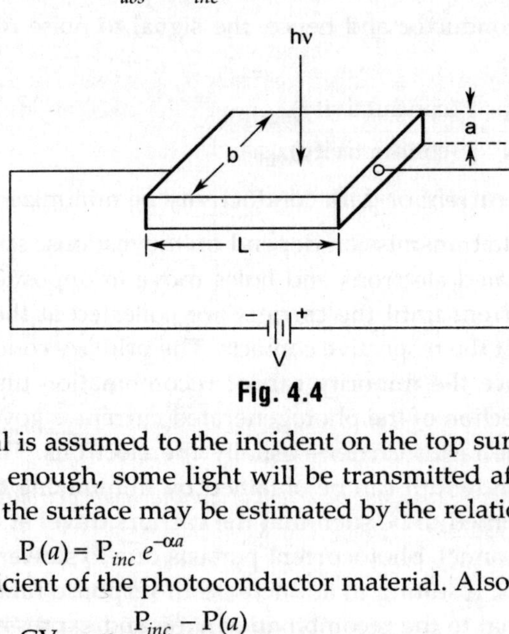

Fig. 4.4

Here the optical signal is assumed to the incident on the top surface. If the thickness a of the photoconductor is small enough, some light will be transmitted after absorption in the device. The light power leaving the surface may be estimated by the relation

$$P(a) = P_{inc}\, e^{-\alpha a} \tag{4.11}$$

α is the absorption coefficient of the photoconductor material. Also

$$GV_v = \eta_i \frac{P_{inc} - P(a)}{h\nu}$$

$$= \eta_i \frac{P_{inc}}{h\nu}\left(1 - e^{-\alpha a}\right) \tag{4.12}$$

and hence the quantum efficiency can be expressed as

$$\eta = \eta_i\,(1 - e^{-\alpha a}) \tag{4.13}$$

The spectral dependence of the absorption coefficient α of common semiconductors has been discussed in earlier chapters. The spectral dependence of the generation rate can be derived from the absorption coefficient. Figure 4.5 describes schematically the spectral dependence of absorption and the generation rate. These characteristics of the semiconductor ultimately determine the spectral response of photoconductor.

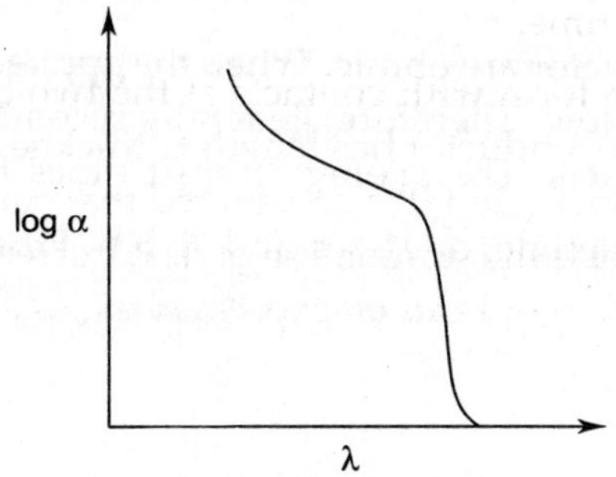

Fig. 4.5(*a*)

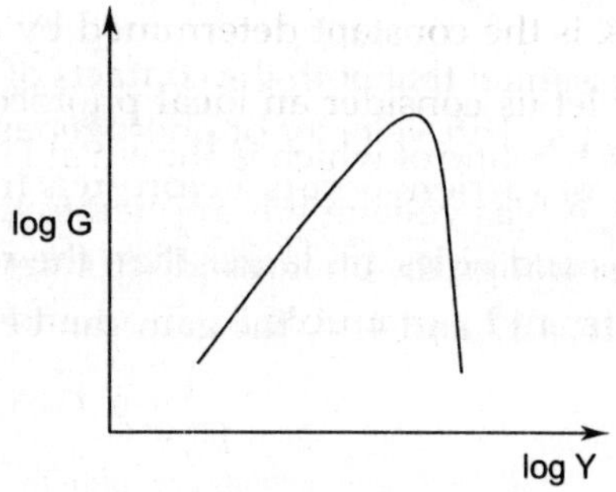

Fig. 4.5(*b*)

In the process of current gain in photoconductor, in order to calculate the increase of carrier concentration by the absorption of light, it is necessary to find out how long the carriers exist in the semiconductor before their recombination. Also injection of carriers due to applied bias is to be taken into consideration. Either of the two processes may be observed — reinjection of the faster carrier or loss by recombination. The photocurrent generated by the incident optical energy may be expressed as

$$I_{ph} = qGV_v\Gamma_G$$

$$= qP_{inc}\frac{\eta}{h\nu}\Gamma_G \tag{4.14}$$

If the carriers recombine before reaching the contacts, the value of gain factor Γ_G becomes less than 1, and Γ_G >1 when reinjection occur. Γ_G becomes equal to 1, if all the carriers are swept out before recombination and no reinjection takes place. In that case photocurrent may be reduced to

$$I_{ph} = qGV_v \tag{4.15}$$

Transit times for electrons t_{tr}^e and for holes t_{tr}^h of the device may be expressed as

$$t_{tr}^e = \frac{L}{V_e} = \frac{L^2}{\mu_e V} \tag{4.16}$$

and

$$t_{tr}^h = \frac{L}{V_h} = \frac{L^2}{\mu_h V} \tag{4.17}$$

V_e and V_h are the respective carrier velocities.

Example 4.2. *Calculate the rms thermal noise current of a photoreceiver circuit with a bandwidth of 1.5 GHz connected to a load resistance of 1 KΩ.*

Solution: Using the equation 4.4,

$$i_j^2 = \frac{4k_BTB}{R_C} = \frac{4\times 1.38\times 10^{-23}\times 300\times 1.5\times 10^9}{1\times 10^3} = 2.484\times 10^{-14}\ A^2$$

Therefore the rms noise current $i_J = 1.57\times 10^{-7}$ A = 157 nA

4.3.1 DC Photoconductor

In this section the operation and gain of a photoconductor will be examined when the photoexcitation or the incident light is unmodulated type. As discussed in the previous section, the gain Γ_G before recombination occurs, may be defined as

$$\Gamma_G = \frac{\text{path traveled by electron + path traveled by hole}}{L} \quad (4.18)$$

It is assumed that both the contacts of the photoconductor are ohmic. When the applied bias V is small, the drift velocity of photogenerated carriers is low. Therefore, generally recombination of most of carriers occurs before reaching at the contacts. The average transit times for both electrons and holes are larger than the recombination lifetime, or $t_{tr}^e > \tau$ and $t_{tr}^h > \tau$. From equations 4.16, 4.17 and 4.18, the gain can be written as

$$\Gamma_G = \frac{\mu_e (V/L)\tau + \mu_h (V/L)\tau}{L}$$

$$= \frac{\mu_e + \mu_h}{\mu_e}\frac{\tau}{t_{tr}^e} \quad (4.19)$$

$$= \frac{\tau}{t_{tr}^e} < 1 \quad \text{(as generally } \mu_h << \mu_e) \quad (4.20)$$

when bias voltage V is larger, as the mobility of electrons is higher than that of holes, the electrons reach the contact point faster, *i.e.*, $t_{tr}^e < \tau$ and $t_{tr}^h > \tau$. At this condition, there is the sweeping out and reinjection of electrons only. The gain OΓ_G of the equation 4.19 is modified to

$$\Gamma_G = \frac{\tau}{t_{tr}^e} > 1 \quad (4.21)$$

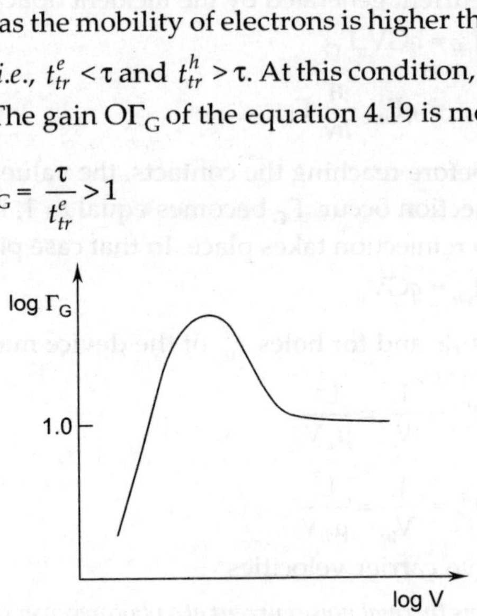

Fig. 4.6

finally, when very large bias is applied with the conditions $t_{tr}^e > \tau$ and $t_{tr}^h > \tau$, sweep-out as well as reinjection of both the carriers leads to space charge currents. The gain Γ_G also reduces to unity. The three conditions may be schematically described as in Fig. 4.6.

The above analysis is made assuming that both the two contacts are ohmic or resistive. Gain behavior is affected by the contact characteristics. Let us now consider that one contact is ohmic, while other is blocking or rectifying type, assuming that holes cannot be reinjected through this contact. Gain characteristics are similar to the previous for the first two cases, when applied bias is low to low-medium, due to higher mobility of electrons. However, when bias is made too large, as holes are not reinjected, gain Γ_G remains greater than unity. This is shown at Fig. 4.7.

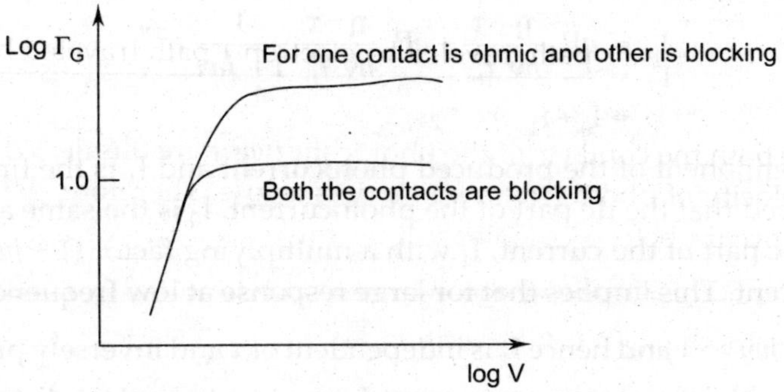

Fig. 4.7

Figure 4.7 also explains the gain behavior of the photoconductor with respect to applied bias when both the contacts are blocking type. There is no reinjection of either carrier through the contacts. For smaller bias, gain characteristics is similar to previous cases and $\Gamma_G < 1$, and the electron path is restricted to an average of L/2. The gain is expressed by

$$\Gamma_G = \frac{L/2 + \mu_h (V/L)\tau}{L} = (1/2 + \tau / t_{tr}^h) < 1 \tag{4.22}$$

for very large bias, $t_{tr}^e > \tau$ and $t_{tr}^h < \tau$, as there is no possibility of reinjection, the path lengths of both carriers are restricted and hence,

$$\Gamma_G = \frac{L/2 + L/2}{L} = 1 \tag{4.23}$$

This is similar to the normal operation of *p-i-n* diode.

4.3.2 AC Photoconductor

Let us now consider the case of the amplitude modulated light signal as the incident signal on the photoconductor. The power of the incident signal may be defined by

$$P_{inc} = P_0 + P_1 e^{jwt} \tag{4.24}$$

where P_0 is the incident power independent of time and P_1 in time-variant component of incident power.

Similarly the generation rate can be written as

$$G = G_0 + G_1 e^{jwt} \tag{4.25}$$

The electron concentration which is now time-dependent, may be expressed by the rate equation

$$\frac{dn}{dt} = G - \frac{n}{\tau} \tag{4.26}$$

The photogenerated electron concentration Δn is obtained from the equation (4.26) as

$$\Delta n = G_0\tau\left(1 - e^{-t/\tau}\right) + \frac{G_1\tau}{1 + jwt}\left(e^{jwt} - e^{-t/\tau}\right) \tag{4.27}$$

Under steady state condition, t >> τ. Then,

$$\Delta n = G_0\tau + \frac{G_1\tau}{1 + jwt} e^{jwt} \tag{4.28}$$

From equations (4.7) and (4.14), and behavior of electron concentration in respect of time as above due to modulating incident light, the produced photocurrent may be expressed as

$$I_{ph} = qP_0 \frac{\eta}{h\nu} \frac{\tau}{t_{tr}^e} + qP_1 \frac{\eta}{h\nu} \frac{\tau}{t_{tr}^h} \frac{1}{1+jwt} e^{jwt} \tag{4.29}$$

$$= I_0 + I_1 \tag{4.30}$$

where I_0 is the dc component of the produced photocurrent and I_1 is the time-variant ac component. It may be noticed that the dc part of the photocurrent, I_0 is the same as expressed earlier at equation 4.14. The ac part of the current, I_1 with a multiplying factor $(1 + jwt)$ has been superimposed on the dc current. This implies that for large response at low frequencies τ should be large. At high frequencies, $jwt >> 1$ and hence I_1 is independent of τ and inversely proportional to $t_{tr}^e = L^2/\mu_e V$. This means for obtaining higher ac current I_1, contact to contact distant L should be small and applied bias voltage V should be large. However, the space charge limited regime, as discussed earlier should be avoided. For this, it is necessary to ensure that the transit time t_{tr}^e is larger than the dielectric relaxation time $\tau_d = \rho\varepsilon_r\varepsilon_0$, where ρ is the resistivity and ε_r is the relative dielectric constant of the semiconductor.

4.3.3 Gain and Bandwidth

We have seen earlier that the photoconductive gain is defined by the equation (4.7). However it may be noted that the recombination process is probabilistic and is governed by the equation

$$\xi(t) = \frac{1}{\tau} e^{-t/\tau} \tag{4.31}$$

where $\xi(t)$ is the probability that a carrier will recombine after time t and τ is the average recombination time. The definition may be confirmed by the following derivation.

$$\tau = \int_0^\infty t\xi(t)\,dt = \int_0^\infty \frac{t}{\tau} e^{-t/\tau}\,dt = \tau$$

If an impulse of radiation is incident on the photoconductor, the resultant current will be of the form

$$i(t) = i_0 e^{-t/\tau} \tag{4.32}$$

In frequency response this can be expressed as

$$i(w) = \int_0^\infty i_0\, e^{-t/\tau} e^{-jwt}$$

$$= \frac{i_0\tau}{1 + jw\tau} \tag{4.33}$$

The power spectral density is given by

$$|i(w)|^2 = \frac{i_0^{\,2}\tau^2}{1 + w^2\tau^2} \tag{4.34}$$

The cut-off frequency f_c is defined as the frequency at which $w\tau = 1$. Therefore

$$f_c = \frac{1}{2\pi\tau} \tag{4.35}$$

The gain-bandwidth product of the photoconductor is

$$\Gamma_G f_c = \frac{1}{t_{tr}} \frac{1}{2\pi\tau} = \frac{1}{2\pi t_{tr}} \tag{4.36}$$

Therefore reduction in the spacing between the contacts or electrodes can increase the gain as well as the gain-bandwidth product.

Example 4.3. *A n-type GaAs photoconductor of length 25 μm is biased at 5 V. Assuming τ= 100 ns and μ_e and μ_h are 7500 cm²/Vsec and 400 cm²/Vsec respectively, calculate the gain. Compare the result using carrier velocities $v_e = 1.4 \times 10^7$ and $v_h = 2 \times 10^6$ at the applied field of 2 KV/cm.*

Solution: Using the equation 4.16, the transit time of electron

$$t_{tr}^e = \frac{L}{V_e} = \frac{L^2}{\mu_e V} \frac{\left(25\times10^{-6}\right)^2}{7500\times10^{-4}\times5} = 1.67\times10^{-10}\ \text{sec}$$

$$\text{Gain } \Gamma_G = \frac{\mu_e+\mu_h}{\mu_e}\frac{t}{t_{tr}^e} = \left(1+\frac{400}{7500}\right)\frac{100\times10^{-9}}{1.67\times10^{-10}} = 630.7$$

Again using the equation (4.16),

$$t_{tr}^e = \frac{L}{V_e} = \frac{25\times10^{-6}}{1.4\times10^7\times10^{-2}} = 1.8\times10^{-10}\ \text{sec}$$

and
$$\text{Gain } \Gamma_G = \frac{v_e+v_h}{v_e}\frac{t}{t_{tr}^e} = \left(1+\frac{2\times10^6}{1.4\times10^7}\right)\frac{100\times10^{-9}}{1.8\times10^{-10}} = 635$$

The discrepancy in the gain calculated using the carrier nobilities or velocities is almost negligible when applied field is low. However, at higher bias voltages the nearly constant carrier saturation velocities are reached and the discrepancy may be larger.

4.3.4 Noise in Photoconductors

Generation of noise associated with large dark current in photoconductive detector is one of the big drawbacks. Random motion of carriers contributes *resistance, thermal or Johnson noise* to the dark current with average energy of k_BT. The noise current due Johnson noise is already specified at equation (4.4). The corresponding random voltage produced across the open circuit terminal has a Gausssian distribution with a mean value of zero. Johnson noise is the dominant source of noise at high frequencies. The spectral density of Johnson noise is constant and is given by

$$\rho_J(f) = 2R_C k_B T\ (\text{V}^2/\text{Hz}) \tag{4.37}$$

fluctuations in generation and recombination processes within the photodetector result in fluctuation in the carrier concentration and hence fluctuations in the conductivity of the semiconductor. Thus undesired voltage fluctuation at the terminal is observed and termed as *generation noise* or *shot noise*. In small bandgap semiconductor the generation noise becomes more important, where unfortunately the dark current and Johnson noise are also high. The generation recombination noise current is expressed by

$$i_{GR}^{\ 2} = \frac{4q\Gamma_G I_0 B}{1+w^2\tau^2} \tag{4.38}$$

where I_0 is the steady state light induced output current. The generation recombination noise will be reduced by reduction of the operating temperature, as the probability of thermal generation of electron hole pairs is proportional to $\exp(-E_g/2k_BT)$. If the operating temperature T is less than $E_g/25k_B$, the generation recombination noise or shot noise becomes too low to consider. Also from the equation 4.38, it may be observed that generation recombination noise current becomes independent of frequency for $f << 1/2\pi\tau$. For $f > 1/2\pi\tau$, the noise current decreases with the increase in frequency with $1/f^2$ dependence. Thus at frequencies the Johnson noise becomes dominant.

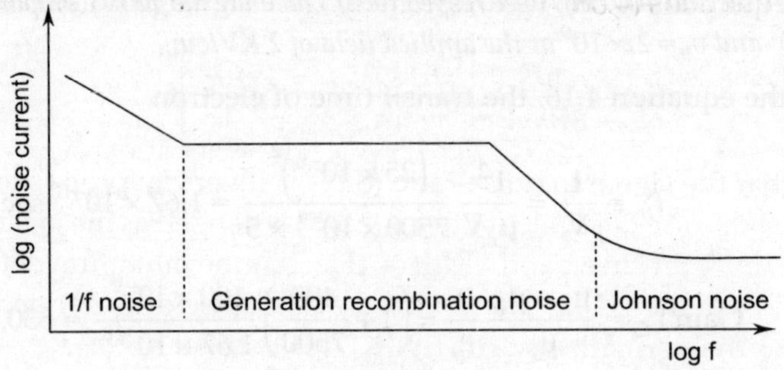

Fig. 4.8

At very low frequencies, less than 1 KHz, one type of noise is observed to be dominant, which arises from surface and interface defects and traps in the bulk of the semiconductor. This type of noise is called *flicker noise*. The spectral distribution of the flicker noise current is inversely proportional to frequency and hence is also known as $1/f$ noise. It is expressed as

$$i_f^2 \propto \frac{1}{f} \tag{4.39}$$

The spectral distribution of noise current in a photoconductor, as discussed above, may be summarized and graphically represented as in Fig. 4.8. At practical cases, thermal noise and shot noise determine the noise performance of a photodetector. The RF noise equivalent circuit has been illustrated in Fig. 4.9, assuming an ideal photoconductor with the simulation of different types of dominating noise currents. Here i is the total current available at the terminals and G_C represents the conductance resulting from dark current and photocurrent.

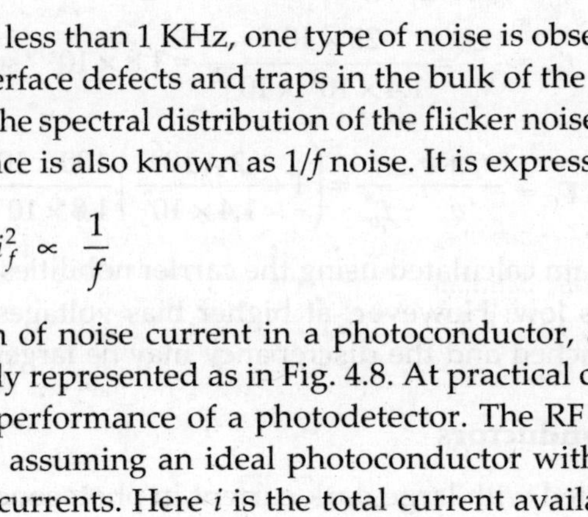

Fig. 4.9

Noise Equivalent Power (NEP) and *detectivity* (D*) are the two important parameters that describe the noise performance of a photoconductor. To estimate these, first the signal to noise ratio is considered, which may be expressed as

$$\frac{S}{N} = \frac{i_{ph}^{\ 2}}{i_j^2 + i_{GR}^{\ 2}} \tag{4.40}$$

Here the flicker noise contribution is neglected. If the incident modulated optical power on the photoconductor is $P_1 e^{j\omega t}$, the rms optical power is $P_1/\sqrt{2}$ and from equation (4.29) the rms photocurrent can be expressed as

$$I_{ph} = \frac{qP_1\eta}{\sqrt{2}h\nu} \frac{t}{t_{tr}} \frac{1}{\left(1 + w^2 t^2\right)} \tag{4.41}$$

Considering the equations (4.38), (4.40) and (4.41), the signal to noise ratio can be expressed as

$$\frac{S}{N} = \frac{\eta P_1^2}{8Bh\nu P_0}\left[1 + \frac{k_n T}{qA_G}\left(w^2 t^2\right)\frac{G_C}{I_0}\right]^{-1} \tag{4.42}$$

It is to be noted that the signal to noise ratio (S/N) is inversely proportional to the bandwidth of photoconductor. The Noise Equivalent Power (NEP) is defined as the incident rms optical power required to produce unity S/N in a bandwidth of 1Hz. It is the minimum detectable signal and can be deduced from the equation 4.42. The *detectivity* of a detector is defined by $D = (NEP)^{-1}$. The specific detectivity or the normalized detectivity D* is defined as

$$D^* = \frac{A^{1/2}\, B^{1/2}}{(NEP)}\left(cm.Hz^{-1/2}W^{-1}\right) \tag{4.43}$$

where A is the area of the photodetector on which the light is incident. The reference bandwidth is generally considered as 1Hz and area as $1cm^2$. D* is usually expressed as D*(λ, *f*, 1), where λ is the wavelength and *f* is the frequency of modulation.

Example 4.4. *A photoconductor of detector area of 20 mm^2 has noise equivalent power (NEP) of 15 pW at noise bandwidth of 2 KHz. If noise bandwidth is reduced to 200 Hz, what will be improvement at NEP?*

Solution: Using the equation 4.43,

$$D^* = \frac{A^{1/2}\, B^{1/2}}{(NEP)}\left(cm.Hz^{-1/2}W^{-1}\right)$$

$$= \frac{\left(\left(20 \times 10^{-2}\right) \times \left(2 \times 10^{3}\right)\right)^{1/2}}{15 \times 10^{-12}}\ cm.Hz^{-1/2}\,W^{-1}$$

$$= 1.33 \times 10^{12}\ cm.\ Hz^{-1/2}W^{-1}$$

At bandwidth of 200 Hz

$$NEP = \frac{A^{1/2}\ B^{1/2}}{D^*}\ W = \frac{\left(\left(20 \times 10^{-2}\right) \times (200)\right)^{1/2}}{1.33 \times 10^{12}}\ W = 4.75\ pW$$

Therefore, the improvement of NEP is over a factor of 3.

4.4 JUNCTION PHOTODIODES

The junction photodiodes are the most commonly used photodetection device for ordinary or specialized applications. Any semiconductor material may be employed to form *p-n* junction by any of the fabrication technique. Most photodiodes are intrinsic type where energy of photons absorbed is equal to the bandgap energy. Photodiodes can be made with elemental and compound semiconductors for a wide spectral range from visible to far infrared wavelengths.

The basic principle of operation of a junction photodiode is quite simple. It operates at reverse-biased condition. The depletion region is quite wide and a small dark current which is the reverse saturation current, flows through the device. When the *p-n* homojunction diode is excited by the incident optical signal, photons are absorbed mainly in the depletion region and also in the neutral regions, particularly at the region where the light is incident. Due to absorption of photons, electron-hole pairs are created. The generated carriers in depletion region are accelerated in opposite directions by reverse bias and produce photocurrent. The magnitude of the photocurrent depends on quantum efficiency. The quantum efficiency is said to be unity, if each photon generates one electron-hole pair. The carriers that are produced within a diffusion length of the depletion layer, on either side of it, also reach the depletion region by diffusion and are then accelerated by bias, thereby contributing to the photocurrent. Therefore an increase in reverse biased current of junction photodiode is the indication of photoexcitation. The process is schematically illustrated in Fig. 4.10(*a*). The photocurrent-voltage behavior is shown in Fig. 4.10(*b*). This type of device is commonly called a depletion layer photodiode.

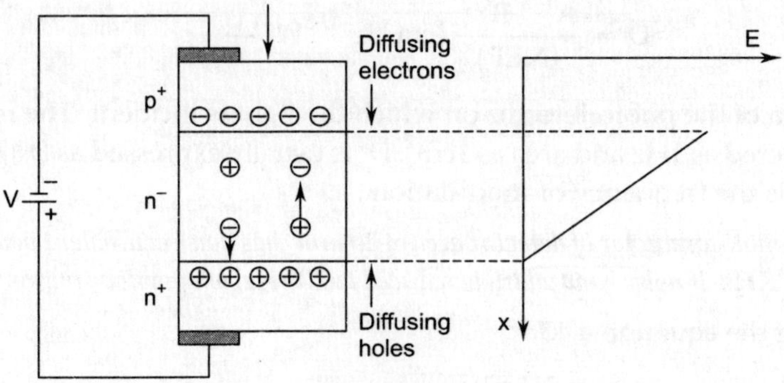

Fig. 4.10 (*a*)

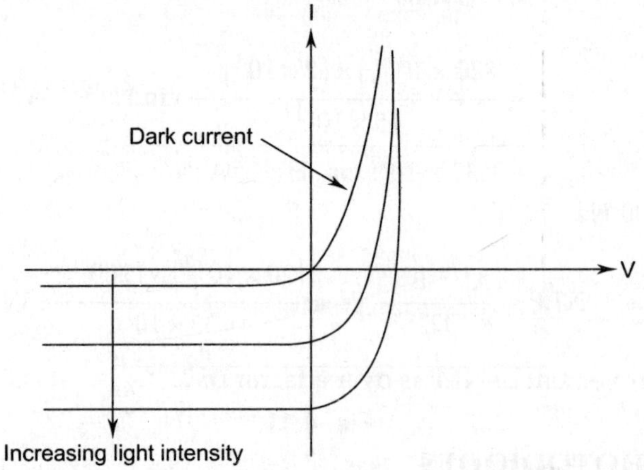

Fig. 4.10 (*b*)

It is obvious that to generate larger photocurrent, generation of larger number of electron-hole pairs is required, and hence it is desired to have a large depletion layer width, so that more photons can be absorbed. However, in this case the transit time of carriers across the depletion region will increase, resulting in increase in the response time of the diode. Thus high speed performance is degraded. In practical photodiodes, a compromise is made between the responsivity of the device, which is the measure of photocurrent produced, and the bandwidth, which is related to the transit time of carriers, junction capacitance, and the resistance of the diode and the circuit. It is also very common to have an asymmetric *p-n* junction where the depletion region is extended almost entirely in the lightly doped side. Its top layer is made with higher bandgap material, so that all the light is absorbed only in the depletion region of the lower bandgap material. Diffusion effect is eliminated in this type of heterojunction photodiodes and also response speed is improved. Photoresponse is also optimized, as all the photons are absorbed in the depletion layer. To achieve higher levels of photocurrent, or responsivity, or to detect very low level of light, avalanche photodiodes are used, in which the photocurrent is enhanced by avalanche multiplication. However, in avalanche process bandwidth is degraded, though gain increases, due to time consumed to build up avalanche process.

4.4.1 PIN Photodiodes

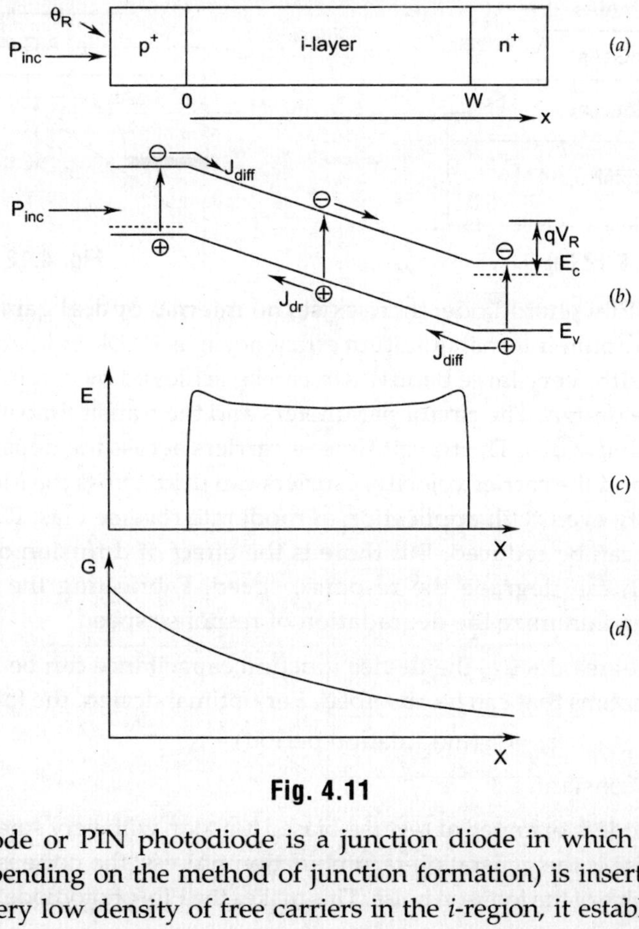

Fig. 4.11

The *p-i-n* photodiode or PIN photodiode is a junction diode in which a very lightly doped *i*-region (p^- or n^-, depending on the method of junction formation) is inserted between p^+ and n^+ regions. Because of very low density of free carriers in the *i*-region, it establishes high resistivity. The applied bias almost entirely drops across this region, which is fully depleted at zero or a very low value of reverse bias. The depletion layer width of the PIN diode should be such that all the photogenarated carriers are ensured to be absorbed as well as it can meet the requirements of photoresponse and bandwidth. For high response speed, the depletion width should be small, and for higher quantum efficiency or responsivity, the width is required to be large. Therefore the width of the depletion layer must be optimized.

The absorption and carrier generation processes in a PIN photodiode are explained schematically in Fig. 4.11. For practical applications, photoexcitation is provided through either an etched opening in the top contact, or etched hole in the substrate, as schematically shown in Figs. 4.12(*a*) (top illumination) and 4.12(*b*) (back illumination), thereby reducing the active area of the diode to the size of the incident light beam. Figure 4.13 illustrates the enhancement in excitonic absorption coefficient in a quantum well under slight tensile strain.

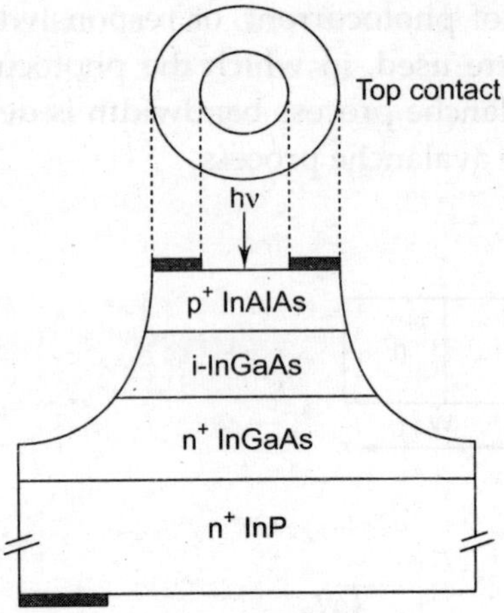

Fig. 4.12 (*a*)

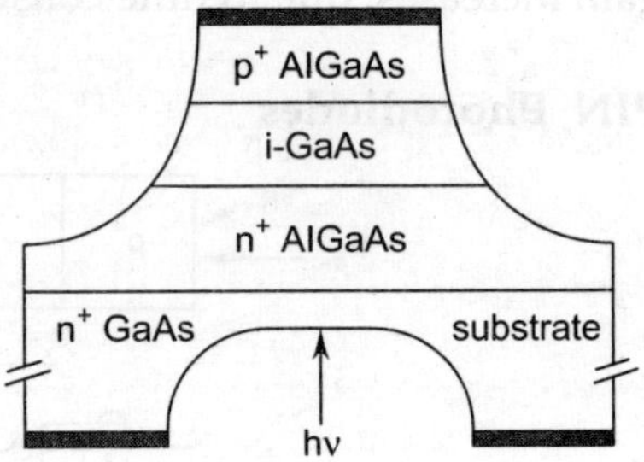

Fig. 4.12 (*b*)

In the operation of PIN photodiode, there exists no internal optical gain or it is an unity gain device. Hence the maximum internal quantum efficiency η_i is 100% and gain bandwidth product is equal to the bandwidth. Very large bandwidth can be achieved by careful selection of material parameters and device design. The circuit parameters and the transit time of the carriers limit the response speed and bandwidth. The transit time of carriers across the depletion region or *i*-layer depends on its width and the carrier velocity. Carriers can drift across the *i*-layer (of width ≈ 1μm) with saturation velocity even with application of moderate reverse bias. Reducing the thickness of i-layer transit time can be reduced. But there is the effect of diffusion of carriers outside the depletion layer, which can degrade the response speed. Fabricating the junction close to the illuminated surface can minimize the degradation of response speed.

Finally, using small area device, the device junction capacitance can be reduced. But this can limit the number of photons that can be absorbed. For optimal design, the following is considered.

$$\tau_{tr} = \tfrac{1}{2}\,(\text{modulation period}) = \tau_{RC} \tag{4.44}$$

where RC is the time constant.

Since a PIN photodiode is operated at reverse-biased junction with very small dark current, and its principle source of noise is the generation-recombination process, the noise is observed to be in the range of equal to or double of the Johnson noise. This makes the PIN photodiode a very sensitive device.

Quantum Efficiency

The operation of a PIN photodiode is shown schematically at Fig. 4.11 with its energy band diagram and optical power profile. Electron-hole pairs are formed by the absorption of light energy in the different regions of the device. Carriers produced within the depletion region or within the diffusion length will be separated by electric field and their movement leads to the current flow in external circuit. Photogenerated carriers make movement by drift and diffusion. The total current density through the reverse-biased depletion layer can be written as

$$J = J_{dr} + J_{diff} \tag{4.45}$$

For derivation of current density J, it is assumed that the thermal generation is within the depletion region and the absorption in the top p^+ layer is negligible. The absorption at top layer is made minimum by making the thickness of it much smaller than α^{-1}, where α is the absorption coefficient of the incident photoexcitation. The origin of the distant coordinate is considered at the $p^+ - i$ junction. It is also assumed that the internal quantum efficiency is unity (or $\eta_i = 1$), *i.e.*, each of the photons generates one electron-hole pair. Then the pair generation rate is expressed as

$$G(x) = \Phi_0 \alpha e^{-\alpha x}$$

$$= \frac{P_{inc}(1-\theta_R)}{Ah\nu} \alpha e^{-\alpha x} \tag{4.46}$$

where Φ_0 is the incident photon flux (photons/sec/cm^2), θ_R is the reflectivity of the surface and A is the surface area of the diode which is usually equal to junction area. In practice, anti-reflection coating is used at the top of the surface to minimize the value of θ_R to the possible extent. Now the drift current can be written as

$$J_{dr} = -q \int_0^W G(x)\,dx$$

$$= -q\Phi_0(1 - e^{-\alpha W}) \tag{4.47}$$

where W is the width of the *i*-layer. For calculation of diffusion current, only the diffusion of minority holes produced in the n^+ layer beyond the i-layer may be considered, as the production of carriers is assumed to be at the top p^+ layer with negligible absorption. The continuity equation for holes at n^+ layer at steady state may be used for derivation that is defined as

$$D_n \frac{\partial^2 p_N}{\partial x^2} - \frac{p_N - p_{N0}}{t_h} + G(x) = 0 \tag{4.48}$$

This equation can be solved with the boundary conditions

$$p_N = p_{N0} \text{ at } x = \infty \text{ and}$$

$$p_N = 0 \text{ at } x = W \tag{4.49}$$

The second boundary condition is established from the reason that the minority carrier density becomes zero at the edge of the depletion region in a reverse-biased diode. Then the solution of the equation 4.48 is of the form

$$p_N = p_{N0} - (p_{N0} + Ce^{-\alpha W})\, e^{(W-x)/Lh} + ce^{-\alpha x} \tag{4.50}$$

where $$L_h = \sqrt{D_h \tau_h} \tag{4.51}$$

and $$C = \frac{\Phi_0 \alpha L_h^2}{D_h\left(I - \alpha^2 L_h^2\right)} \tag{4.52}$$

The diffusion current is then derived as

$$J_{diff} = -qD_h \left(\frac{\partial p_N}{\partial x}\right) x = W$$

$$= -q\Phi_0 \frac{\alpha L_h}{1+\alpha L_h} e^{-\alpha W} - qp_{N0} \frac{D_h}{L_h} \tag{4.53}$$

Combining the equations (4.45), (4.47) and (4.53), the total current may be written as

$$J = J_{dr} + J_{diff}$$

$$= -q\Phi_0\left(1 - \frac{e^{-\alpha W}}{1 + \alpha L_h}\right) - qp_{N0}\frac{D_h}{L_h} \tag{4.54}$$

The value of the second term of the equation (4.54) is very small at normal operating conditions and this term may be neglected. In that case the total current is proportional to the incident flux, which is desired. The external quantum efficiency of the diode may be calculated as

$$\eta_{ext} = \frac{|J/q|}{P_{inc}/Ah\nu}$$

$$= (1 - \theta_R)\left(1 - \frac{e^{-\alpha W}}{1 + \alpha L_h}\right) \tag{4.55}$$

From the equation (4.55), it may be understood that a low reflection coefficient at top surface is desired to achieve high quantum efficiency. As already mentioned, this is realized by using special anti-reflection dielectric coatings, where $\theta_R \approx 0$. Also for high quantum efficiency, $\alpha W >> 1$, which implies that α or W should be large. This condition ensures that the entire photons incident on the device are absorbed. But for large W or the *i*-layer width, transit time of the carriers across the depletion region becomes large and device response speed is reduced. Therefore, the absorption coefficient α is desirable to be high. Unfortunately, the value of α for the semiconductors is determined by their bandstructure and cannot be changed by artificial way. Finally, considering the diffusion coefficient L_h very small, the equation (4.55) may be reduced to

$$\eta_{ext} = \eta_i\,(1 - \theta_R)\,(1 - e^{-\alpha W}) \tag{4.56}$$

However, there is a possible scheme for enhancement of bandedge coefficient that can be employed in strained quantum well materials. In *p-i-n* diode the *i*-region is realized as multiple quantum well in which well regions are under biaxial tensile strain. It has been observed that for a very small amount of tensile strain, the light and heavy hole resonances in excitonic spectra can be made to merge. This increases the effective mass and the density of states, thereby enhancing the oscillator strength of the transition from hole subband to electron subband as well as the absorption coefficient. The effect of quantum well under tensile strain is schematically shown at Fig. 4.13. It is possible to realize almost double the absorption coefficient at heavy-hole exciton wavelength. Thereby, PIN photodiodes with smaller areas and smaller thickness of *i*-region can be designed. Another technique of enhancing the absorption is to use microcavity photodiodes, which shall be discussed at later chapter.

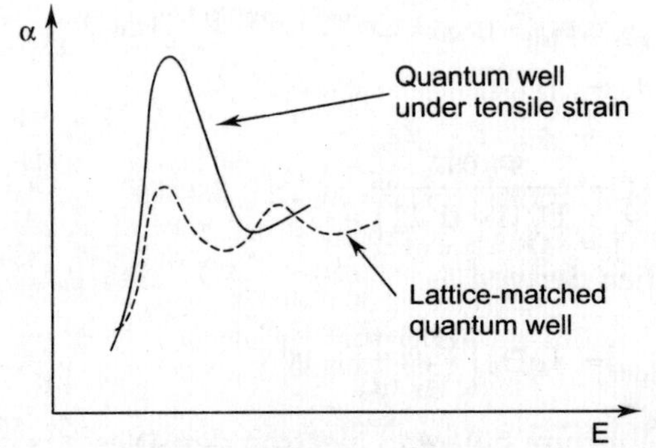

Fig. 4.13

Example 4.5. *A Si p-i-n photodiode with 8 μm receives the light of wavelength 0.87 μm and power of 0.5 W/cm². Calculate the photocurrent density. Assume that the top illuminated surface is coated with antireflection coating such that there is no reflection.*

Solution: Using the equation 4.54,

$$J = -q\Phi_0\left(1 - \frac{e^{-\alpha W}}{1+\alpha L_h}\right) - qp_{N0}\frac{D_h}{L_h}$$

Assuming that $\alpha L_h << 1$ and p_{N0} is very small above equation is reduced to

$$|J| = q\Phi_0\,(1 - e^{-\alpha W})$$

Now $$\Phi_0 = \frac{P_{inc}}{Ah\nu} = \frac{0.5 \times 0.87 \times 10^{-6}}{6.626 \times 10^{-34} \times 3 \times 10^{8}}\ s^{-t}\ cm^{-2} = 2.18 \times 10^{18}\ s^{-1}\ cm^{-2}$$

$$[h = 6.626 \times 10^{-34},\ \nu = C/\lambda,\ C = 3 \times 10^{8},\ \lambda = 0.87\ \mu m\ A = 1\ cm^2]$$

At 0.87 μm, $\alpha = 600\ cm^{-1}$. $W = 8\ \mu m = 0.0008$ cm. Therefore

$$|J| = 1.6 \times 10^{-19} \times 2.18 \times 10^{18}\ (1 - e^{-600 \times 0.0008})$$

$$= 1.33/cm^{-1}$$

Frequency Response

If the photodiode is designed to be sufficiently small with desired responsivity, then transit time effects of its carriers determine its bandwidth or frequency response. The photogenerated carriers require finite time to traverse the depletion layer which is referred as the transit time. Again, if the incident photoexcitation is modulated at high frequency, then there creates a phase difference between the photon flux and photocurrent. To analyze the frequency response of the photodiode, it is considered that the photogenerated carriers traverse the fully depleted *i*-region of width W at saturation velocity under the influence of the the applied reverse bias V_R. (Refer Fig. 4.11). The following continuity equations for both electrons and holes are considered.

$$\frac{\partial p}{\partial t} = \frac{-(p - p_0)}{\tau} + G - \frac{1}{q}\nabla . J_h \tag{4.57}$$

$$\frac{\partial n}{\partial t} = \frac{-(n - n_0)}{\tau} + G - \frac{1}{q}\nabla . J_e \tag{4.58}$$

And the current density equations are

$$J_h = q\mu_h pE - qD_h\frac{\partial p}{\partial x} \tag{4.59}$$

$$J_e = q\mu_h nE - qD_e\frac{\partial n}{\partial x} \tag{4.60}$$

$$J = J_h + J_e + \varepsilon_s\frac{\partial E}{\partial x} \tag{4.61}$$

where p_0 and n_0 are the equilibrium hole and electron densities, τ is the recombination time, E is the electric field, and μ_h and μ_e are the nobilities of holes and electrons respectively.

The analysis is based on the following assumptions.

(*i*) The transit time of carriers is much smaller than the bulk recombination lifetime, so that the recombination term can be neglected from the continuity equation.

(*ii*) The electric field is large enough, so that the diffusion current can be neglected.

(*iii*) Each incident photon produces an electron-hole pair. The incident photon flux is of the form of $\Phi_{inc} = \Phi_0 (1 + e^{jwt})$.

Then the generation rate and carrier densities can be expressed as

$$G(x,t) = G_0(x) + G_1(x)e^{jwt} \tag{4.62}$$

$$p(x,t) = p_{dc}(x) + p_1(x)e^{jwt} \tag{4.63}$$

$$n(x,t) = n_{dc}(x) + n_1(x)e^{jwt} \tag{4.64}$$

where the subscript '1' denotes the variable ac quantities and subscript 'dc' denotes the dc quantities. With these assumptions and from above equations the continuity equations can be rewritten as

$$\frac{\partial J_{e1}}{\partial x} - jw\frac{J_{e1}}{v_e} = -qG_1 \tag{4.65}$$

$$\frac{\partial J_{h1}}{\partial x} - jw\frac{J_{h1}}{v_h} = qG_1 \tag{4.66}$$

where v_e and v_h are the electron and hole velocities and $G_1 = \alpha\Phi_0 e^{-\alpha x}$. Using the boundary conditions $J_{e1}(0) = 0$ and $J_{h1}(W) = 0$, the following can be derived

$$J_{e1}(x) = -\alpha q\Phi_0\left[\frac{e^{jwx/v_e} - e^{-\alpha x}}{\alpha + jw/v_e}\right] \tag{4.67}$$

$$J_{h1}(x) = -\alpha q\Phi_0\left[\frac{e^{-\alpha x} - e^{-\alpha W + jw(W-x)/v_h}}{\alpha - jw/v_h}\right] \tag{4.68}$$

Maxwell's equation for Ampere's circuital law may be used which is of the form

$$\nabla \times H = J_e + \frac{\partial D}{\partial t} = J \tag{4.69}$$

where $\partial D/\partial t$ is the displacement current and $D = \varepsilon_s E$. Now $\nabla \cdot (\nabla \times H) = 0$. This means

$$\nabla \cdot J = 0 \tag{4.70}$$

and

$$J = \frac{1}{W}\int_0^W \left(J_e + \varepsilon_s\frac{\partial E}{\partial t}\right)dx \tag{4.71}$$

When applied bias is zero, *i.e.*, E is equal to zero, the short circuit current will be

$$J(w) = \frac{1}{W}\int_0^W \left[J_{e1}(x) + J_{h1}(x)\right]dx \tag{4.72}$$

By substitution of equations 4.67 and 4.68

$$J(w) = q\Phi_0\alpha W\left(\frac{1-e^{-\alpha w}}{\alpha W(\alpha W + jwt_{etr})} + \frac{1-e^{-jwt_{etr}}}{jwt_{etr}(\alpha W + jwt_{etr})}\right) \tag{4.73}$$

$$-q\Phi_0\alpha W\left(\frac{1-e^{-\alpha W}}{\alpha W(\alpha W + jwt_{hir})} + \frac{e^{-\alpha W}\left(1-e^{-jwt_{hir}}\right)}{jwt_{hir}(\alpha W + jwt_{hir})}\right)$$

where t_{etr} and t_{htr} are the electron and hole transit times respectively.

Now an equivalent circuit of the photodiode may be drawn as in Fig. 4.14, where the current source exhibits the current as per equation (4.73), with the inclusion of junction capacitance C_j, parasitic capacitance C_p, total series inductance L_s, shunt resistance R_D, series resistance R_S and load resistance R_L. The junction capacitance C_j originates from the depletion region. Parasitic capacitance C_p is external to wafer and is the sum of stray capacitance shunted across the wafer from leads, also called fringing capacitance and the package or encapsulation capacitance. The series inductance L_s originates from leads. R_D the shunt resistance is caused by carriers within the depletion region, which has a very small value at forward bias, but a very large value at zero and reverse bias. The series resistance R_S is the sum of contact resistance and resistance of the undepleted regions of the diode. Typical values of these elements at reverse biased diode are also given in the Fig. 4.14. The capacitance of the undepleted region under reverse bias is neglected in this equivalent circuit.

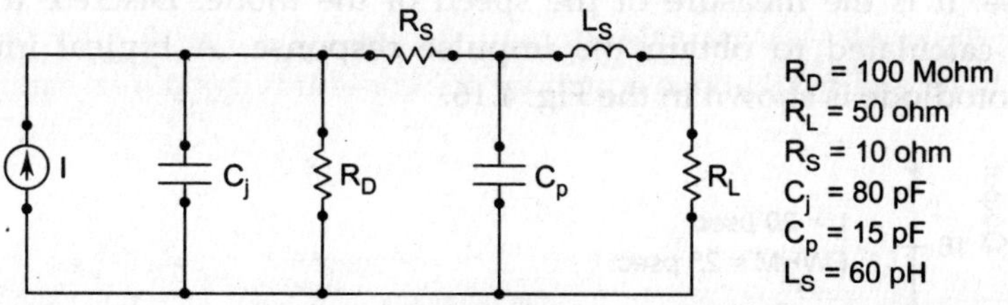

Fig. 4.14

With circuit elements as shown in Fig. 4.14 taken into account, the frequency response of these can be included in a transfer function is denoted as $H(w)$ and output current is $J_o(w)$ across the load may be expressed as

$$J_o(w) = J(w)\,H(w) \tag{4.74}$$

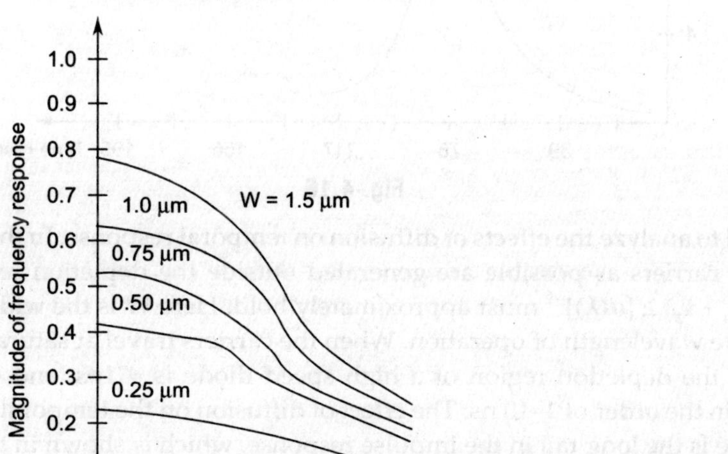

Fig. 4.15

The above expression of output photocurrent is in terms of intrinsic as well as extrinsic parameters of the diodes. The transfer function $H(w)$ of the equivalent circuit can be found by the circuit analysis as

$$H(w) = \frac{R_D}{A + jw\left(B - Cw^2\right) - Dw^2} \tag{4.75}$$

where
$$A = R_S + R_L + R_D$$
$$B = R_S R_L C_p + L_S + (R_S + R_L) R_D C_j + R_L R_D C_p$$
$$C = R_S L_S C_p R_D C_j$$
$$D = (R_S R_L C_p + L_S) R_D C_j + R_S L_S C_p + R_D C_p L_S$$

From the equations (4.73), (4.74) and (4.75), and the circuit parameters as discussed above the calculated output photocurrents are shown in Fig. 4.15 for different values of *i*-region thickness W.

Another important and very much relevant parameter for high-speed photodiodes is their impulse response. It is the measure of the speed of the diode. Discrete fourier transform of photocurrent is calculated to obtain the impulse response. A typical impulse response of $In_{0.53}Ga_{0.47}As$ photodiode is shown in the Fig. 4.16.

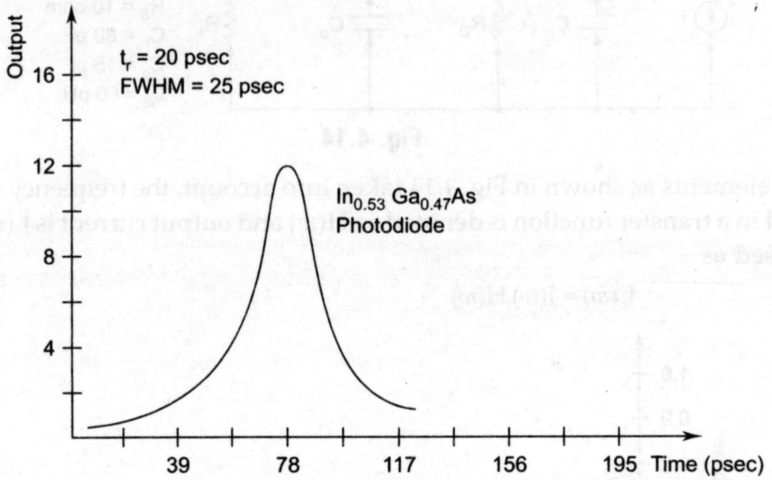

Fig. 4.16

It is also useful to analyze the effects of diffusion on temporal response. In this case, it is considered that as few carriers as possible are generated outside the depletion region. For this, the condition $W = (x_n + x_p) \geq [\alpha(\lambda)]^{-1}$ must approximately hold. Here W is the width of the depletion region and λ is the wavelength of operation. When the carriers travel at saturation velocities, the drift time across the depletion region of a high speed diode is $\approx$ 1ps, and on the other hand diffusion time is in the order of 1–10 ns. The effect of diffusion on the temporal response of a high speed photodiode is the long tail in the impulse response, which is shown in the Fig. 4.17.

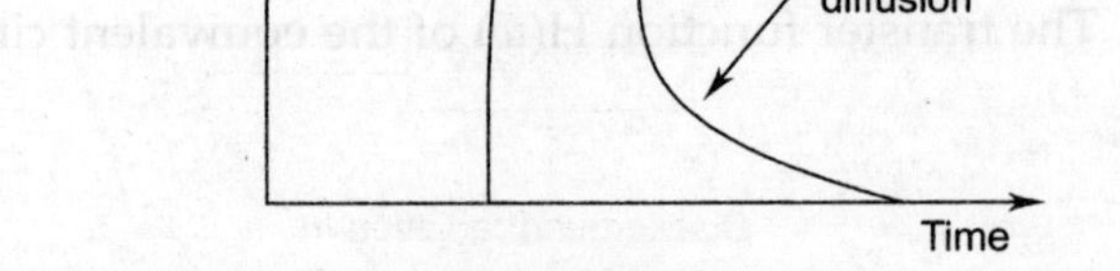

Fig. 4.17

The *p-i-n* photodiode has been largely discussed here, contents the i-layer of lightly doped *p*-type or *n*-type material. A related device is *p-n* photodiode, where it is configured as p^+-n-n^+ junction. Here the n-region, where the photons are absorbed, may not be fully depleted even at large bias. Therefore *p-n* photodiodes are observed to have the tendency of lower response speed than the *p-i-n* photodiodes, because of larger diffusion current in the former.

Noise Performance

In general, the PIN photodiode or any junction photodiode operates under reverse biased conditions, where the dark current is very small. Therefore the shot noise, which is developed only by generation within the depletion region, is very small. In these devices, the resistance or Johnson noise play the dominant role and even this can be minimized by optimizing the circuit parameters. To analyze the noise performance of PIN photodiode, let us consider the incident modulated optical power is $P(w) = P_{inc}(1 + e^{jwt})$, where the modulation index is unity. Then the rms optical power is $P_{inc}/\sqrt{2}$. or the rms photocurrent is given by

$$I_{ph} = \frac{q\eta P_{inc}}{\sqrt{2}hv} \tag{4.76}$$

The mean square Johnson noise current is given by

$$I_j^2 = \frac{4knTB}{R_{eq}} \tag{4.77}$$

where R_{eq} is the equivalent circuit resistance parameters comprising of shunt resistance R_D, series resistance R_S, load resistance R_L as shown at photodiode equivalent circuit in the Fig. 4.14 and also R_i, the input impedance of next circuit stage like preamplifier.

In addition to the resistance noise or the Johnson noise, other sources of noise current are the dark current I_D and noise current I_B due to background radiation in the ambience in which the detector is placed, though it is of very low intensity. The dark current is the reverse current, which flows due to thermally generated electron-hole pairs at the depletion region, and contributes to the total system noise and gives rise to random fluctuations of the output current. Because of the randomness of the generation of these currents, they give rise to shot noise at the output. The equivalent mean square shot noise current can be expressed as

$$I_s^2 = 2q(I_{ph} + I_B + I_D)B \tag{4.78}$$

where B is the bandwidth of the device. Note that the equation (4.80) is derived from the equation (4.38) with $w\tau = 1$. The corresponding frequency $f_c = 1/2\pi\tau$ is called the noise equivalent bandwidth. Figure 4.18(*a*) shows the different noise sources in a photodiode and Fig. 4.18(*b*) gives the equivalent circuit of the photodiode with the inclusion of all the noise current sources.

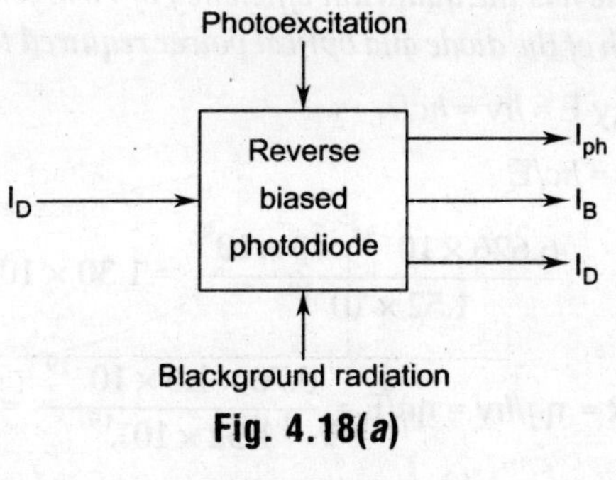

Fig. 4.18(*a*)

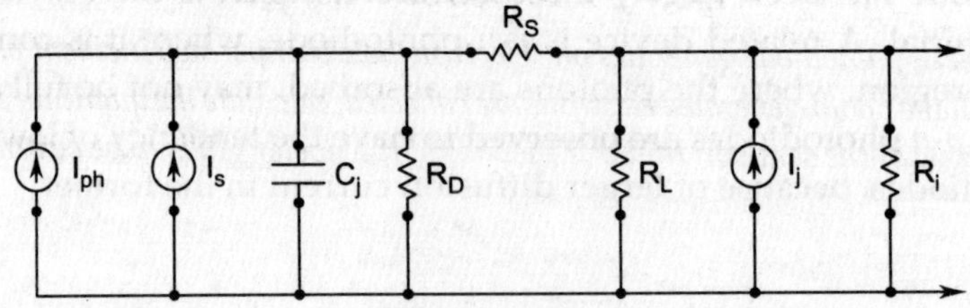

Fig. 4.18(*b*)

In the equivalent circuit as in Fig. 4.18(*b*), the series resistance R_S is very small to be neglected. Then the total equivalent resistance of the circuit R_{eq} may be calculated as

$$R_{eq} = (1/R_D + 1/R_L + 1/R_i)^{-1} \tag{4.79}$$

The signal power is given by $I_{ph}^2 R_{eq}$ and the total noise power is estimated as $\left(I_s^2 + I_j^2\right)R_{eq}$.

Therefore, from the equations (4.76), (4.77) and (4.78), the signal to noise ratio can be expressed as

$$\frac{S}{N} = \frac{\frac{1}{2}\left(\frac{q\eta P_{inc}}{h\nu}\right)^2}{2q\left(I_{ph} + I_B + I_D\right)B + \frac{4k_B TB}{R_{eq}}} \tag{4.80}$$

The sensitivity of a photodiode is an important characteristic and defined as the minimum optical input power required to achieve a signal-to-noise ratio greater than a given value. A convenient measurement of sensitivity is the measure of noise equivalent power or NEP. When the rms signal power is equal to NEP, the photocurrent is equal to the noise current and signal to noise ratio is unity. Thus $I_{ph} = I_N$, where $I_N^2 = I_s^2 + I_j^2$, I_N is the total noise current. Therefore,

$$\frac{q\eta P_{inc}}{\sqrt{2}\,h\nu} = I_N = \left(2q\left(I_{ph} + I_B + I_D\right)B + \frac{4k_B TB}{R_{eq}}\right)^{1/2} \tag{4.81}$$

From which NEP is derived for B = 1 Hz as

$$\text{NEP} = \left(\frac{h\nu}{q\eta}\left(2q\left(I_{ph} + I_B + I_D\right) + \frac{4k_B T}{R_{eq}}\right)^{1/2}\left(\text{W.Hz}^{-1/2}\right)\right) \tag{4.82}$$

It may be noticed from the equation 4.82 that to increase the sensitivity of the photodiode, η and R_{eq} should be as large as possible and the unwanted currents I_B and I_D should be as small as possible. Therefore the NEP is a measure of the minimum detectable power $(P_{inc})_{min}$.

Example 4.6. *A p-n photodiode has the quantum efficiency of 70% for photons of energy 1.52×10^{-19} J. Calculate the operating wavelength of the diode and optical power required to achieve the photocurrent of $3\mu A$.*

Solution: The photon energy $E = h\nu = hc/\lambda$

Therefore the wavelength $\lambda = hc/E$

$$= \frac{6.626 \times 10^{-34} \times 3 \times 10^8}{1.52 \times 10^{-19}} = 1.30 \times 10^{-6}\ \text{m} = 1.30\ \mu\text{m}$$

Responsivity $\quad R = \eta q/h\nu = \eta q/E = \dfrac{0.70 \times 1.6 \times 10^{-19}}{1.52 \times 10^{-19}} = 0.736$ A/W

Optical power $= I_p/R = 3 \times 10^{-6}/0.736 = 4.07\ \mu W$

Example 4.7. *A p-i-n photodiode generates one electron-hole pair per two incident photons at wavelength of 0.85 μm. Assuming all the photo-generated electrons are collected, calculate the quantum efficiency, the maximum possible bandgap energy of the semiconductor if the incident wavelength is to be cut-off, and the mean output photocurrent when incident optical power is 10 μW.*

Solution: Quantum efficiency $\eta = 1/2 = 0.5 = 50\%$

The bandgap energy $E_g = hc/\lambda = \dfrac{6.626 \times 10^{-34} \times 3 \times 10^8}{0.85 \times 10^{-6}}\ J = 2.33 \times 10^{-6}\ J = 1.46\ eV$

Mean output current $I_p = R\,P_{in} = \dfrac{\eta q}{h\nu} P_{in} = \dfrac{0.5 \times 1.6 \times 10^{-19}}{2.33 \times 10^{-19}} \times 10 \times 10^{-6} = 3.43 \times 10^{-6}$

$A = 3.43\ \mu A$

Example 4.8. *Photons of wavelength 0.9 μm are incident on a p-n photodiode at the rate of 5×10^{10}/s. The electrons are collected at the rate of 2×10^{10}/s at the terminal. Calculate the quantum efficiency and the responsivity of the diode at this wavelength.*

Solution: Quantum efficiency $\eta = 2 \times 10^{10}/5 \times 10^{10} = 0.4 = 40\%$

Responsivity $R = \eta q/h\nu = \eta q\lambda/hc = \dfrac{0.4 \times 1.6 \times 10^{-19}\ 0.9 \times 10^{-8}}{6.626 \times 10^{-34} \times 10^8}\ A/W = 0.29\ A/W$

4.4.2 Heterojunction Diodes

A junction photodiode is always operated under reverse biased condition and needless to say, that the reverse leakage current must be minimum. In photodiodes with large bandgap materials, it is easily achieved, as the reverse breakdown voltage is large enough and the diode can be operated at a bias far less than the breakdown voltage. For optical communication applications, photodiodes are usually made of InP based materials such as $In_{0.53}Ga_{0.47}As$ ($E_g = 0.74$ eV) or InGaAsP ($E_g = 0.74 - 1.1$ eV). Homojunctions made of these materials have a low breakdown voltage and a large reverse leakage current. Bandwidth requirement at high speed operation is another aspect of photodiode design. It is already established that for fast temporal response the area of the diode should be small as well as the thickness of the absorption layer should be small, while responsivity is to be maintained at the acceptable level. An optimal design of photodiode is to make the transit time of carriers in the absorption region or *i*-region approximately equal to the RC time constant of the diode and its equivalent circuit. The active area of the diode can be kept small by illuminating the diode with a well-collimated beam through a hole etched in the substrate. Using a heterojunction and placing the *p-n* junction either at heterojunction, or in the large bandgap material higher breakdown voltage can be achieved. These schemes are shown in Figs. 4.19(*a*) and 4.19(*b*).

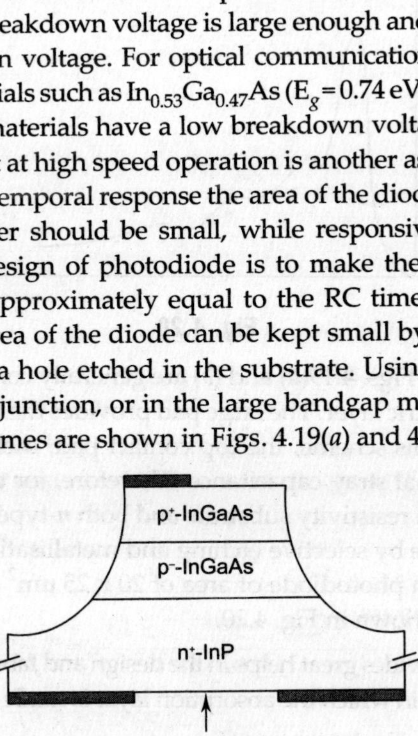

Fig. 4.19(*a*)

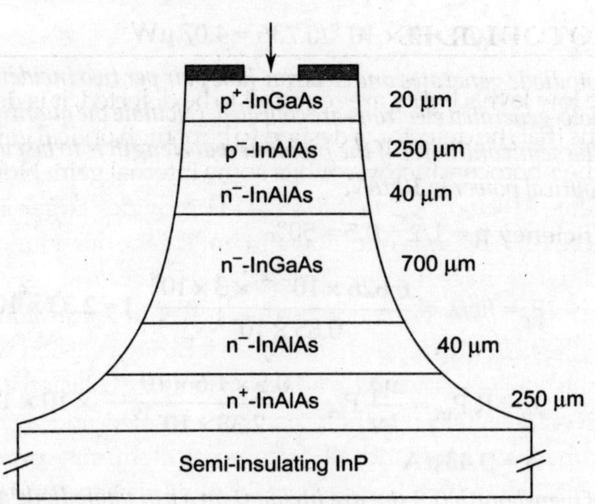

Fig. 4.19(*b*)

Excitation of the diode from substrate side ensures that the light is absorbed only in the n^--layer and the wider bandgap layers act as windows. In Fig. 4.19(*a*) the top p^+-layer is typically created by Zn-diffusion. In Fig. 4.19(*b*), the diode structure is completely realized by epitaxy. Mesa-shaped diodes are made by a combination of metal-deposition for contacts and subsequent alloying, etching and photolithography. A silicon dioxide or silicon nitride layer is usually deposited on the surface for passivation and reduction of surface leakage current. High speed diode has a typical diameter of 10 to 50 μm and the *i*-region (n^- or p^-) thickness is 0.5 to 1.0 μm.

Fig. 4.20

The top contacts as shown in Figs. 4.19(*a*) and (*b*) are generally connected monolithically to a large contact pad placed on the dielectric layer. The large pad provides the facility of bonding to the external circuit and power supply. In this scheme, the top contact pad, dielectric layer and the conducting substrate usually form additional stray capacitance. Therefore, for the high speed photodiodes, the structure is developed on a high resistivity substrate and both *n*-type and *p*-type contacts and contact pads are placed on the substrate by selective etching and metallisation. A typical temporal response performance of a heterojunction photodiode of area of 20 × 25 μm^2 and 0.75 μm InGaAs absorption region to a 1ps optical pulse is shown in Fig. 4.20.

The use of heterojunctions provides great helps in the design and fabrication of the photodiodes at near and mid-infrared spectral region in which the absorption layer is made of small bandgap semiconductor.

4.5 AVALANCHE PHOTODIODES

In many applications, where low levels lights are required to be detected, it is desirable to use a detector of high sensitivity. This means that the detector is desired to be of high optical gain. In the earlier sections, it has been mentioned that the photoconductor exhibits some internal gain. However the magnitude of this gain may not be sufficient. *Avalanche photodiode* or APD provides larger internal gain. Avalanche photodiodes also operates at reverse biased condition. But, unlike the other junction diodes that operate far below the reverse breakdown voltage, the avalanche diodes operate at very close to the breakdown voltage. Photogenerated carriers in the depletion region travel at saturation velocities, and if they acquire enough energy from the field during such transit, an ionizing collision with the lattice can occur. Depending on the material, the field necessary to produce an ionizing collision is in the range of 10^4 to 10^5 V/cm. During the process secondary electron-hole pairs are formed and again drift in the opposite direction along with the primary carriers, and all or some of them may produce new carriers. This process is called as *impact ionization*, which leads to carrier multiplication and gain. Depending on the semiconductor material and device design, very large avalanche gain (≈ 200 or more) can be achieved. Therefore the avalanche photodiode exhibits high sensitivity.

The overall noise performance of an avalanche photodiode is determined by the shot noise contributed by the unmultiplied dark current and the noise from the probabilistic avalanche process itself. There presents random fluctuation in actual distance between successive ionizing collisions. This fluctuation increases variations in the total number of secondary carriers generated for each primary carrier injected in the gain region. Therefore, noise in the total signal current becomes random in nature, the magnitude of which depends on the mean avalanche gain.

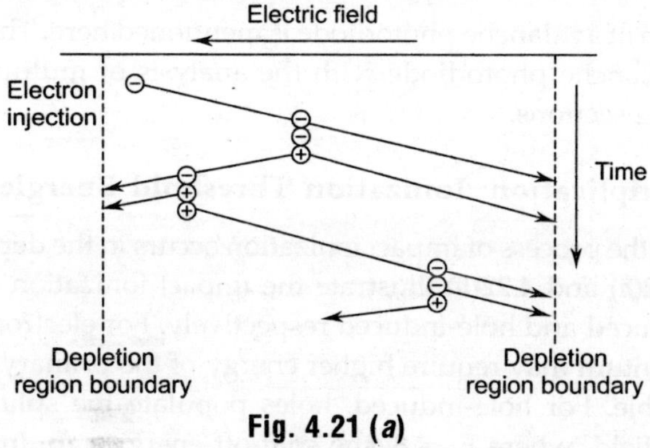

Fig. 4.21 (*a*)

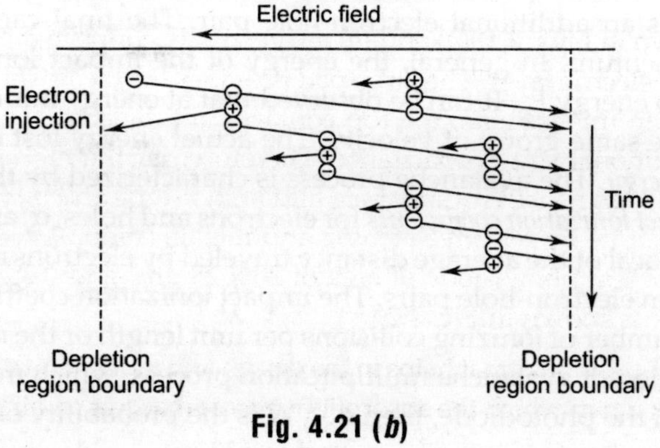

Fig. 4.21 (*b*)

The avalanche process is itself asymmetric. This means that the probability for initiating an avalanche is usually greater for one type of carrier than for the other. For example, in silicon, the ionizing collisions are 30-50 times more frequent with electrons than with holes. The asymmetry is characterized by the ratio α_e/α_h, where α_e and α_h are the impact ionization coefficient for electrons and holes respectively. Avalanche multiplication noise is observed to lowest in the devices where the carrier with higher ionization rate initiates the avalanche process. This leads to the fact that the value of α_e/α_h should be as large or as small as possible. In these cases, the recycling of carriers through the depletion region is minimized, and consequently the noise is reduced because of less fluctuation in the ionization process compared to the case when both the carriers participate. Also device response becomes faster when asymmetry is very large or very small.

In an avalanche photodiode, the largest bandwidth is obtained under unity gain condition and is limited the transit time. This is the condition similar to the *p-i-n* photodiode. In the avalanche process while ionization and gain increase, the bandwidth is reduced. This is due to that the avalanche process takes some time to build up, where a finite amount of time required for ionizing collision to occur. Also, if asymmetry ratio is approximately unity ($\alpha_e/\alpha_h \approx 1$), both types of carriers continuously recycle and persist in the depletion region, leading to long response time and consequentially a small bandwidth. For $\alpha_e/\alpha_h > 0$ or ∞, the response is limited only by duration of a single transit, and the best response is obtained. The time evolution of the avalanche photodiode is illustrated at Fig. 4.21(*a*) when $\alpha_e \approx \alpha_h$, and Fig. 4.21(*b*) for $\alpha_e >> \alpha_h$. In case of $\alpha_e \approx \alpha_h$, there is considerable recycling of both types of carriers, while in case of $\alpha_e >> \alpha_h$, only the injected and multiplied electrons produce impact ionizations. The holes, in the later case, simply drift in the opposite direction till they are collected (some scattering may be there).

The general description of avalanche photodiode is mentioned here. The detail of operation and characteristics of the avalanche photodiode with the analysis of multiplication process will be discussed in the following sections.

4.5.1 Avalanche Multiplication: Ionization Threshold Energies

In avalanche photodiode, the process of impact ionization occurs in the depletion region under high electric field. Figures 4.22(*a*) and 4.22(*b*) illustrate the impact ionization of three carrier collision process with electron-induced and hole-induced respectively. For electron-induced, the conservation of energy and momentum may require higher energy of the primary electron since state near $k = 0$ may not be available. For hole-induced, holes populate the split-off energy band under sufficiently high electric field, where $E_g = \Delta$, the split-off energy gap. In either case high energy primary carrier produces an additional electron-hole pair. The final carriers are left with finite kinetic energy and momentum. In general, the energy of the impact ionizing primary carrier is greater than the bandgap energy E_g. It can be observed that at energy minimization considerations the final carriers have the same group of velocity. The actual energy lost by the primary carrier is the *threshold ionization energy*. The avalanche process is characterized by the fundamental material parameters called as *impact ionization coefficients* for electrons and holes, α_e and α_h respectively. These coefficients are the reciprocal of the average distance traveled by electrons and holes in the direction of the electric field to form electron-hole pairs. The impact ionization coefficients, α_e and α_h are also defined as the average number of ionizing collisions per unit length or the number of secondaries. If W is the width of the region of avalanche multiplication process, which may be equal to the width of the depletion region of the photodiode, then $\alpha_{e(h)}$ W is the probability of ionizing collisions.

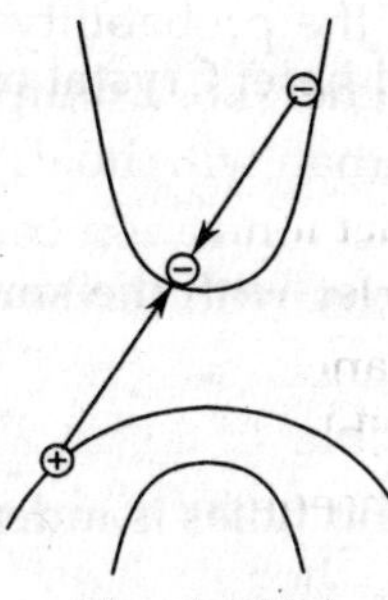

Fig. 4.22(*a*)

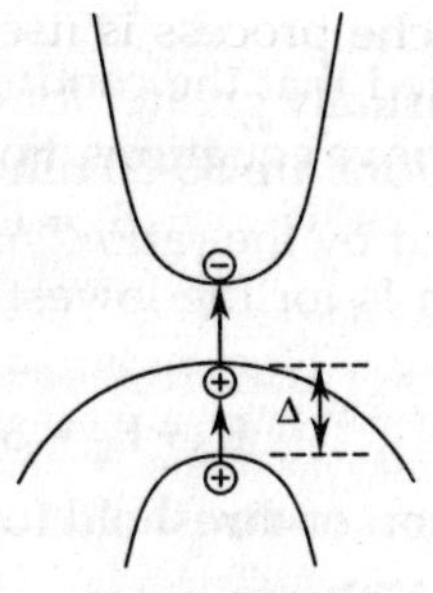

Fig. 4.22(*b*)

If an ensemble of carriers is injected into the region of avalanche multiplication process, $\alpha_{e(h)}$ is the probability of ionizing collisions for a carrier in the ensemble. Being a probability $\alpha_{e(h)}W$ must be less than unity, which physically signifies that carriers can be extracted from the region W without any carrier being injected. In Si and GaAs $\alpha_e > \alpha_h$, the behavior of α_e and α_h are shown in Fig. 4.23(*a*). However in most III-V compounds $\alpha_e \approx \alpha_h$, and due to the avalanche ionization coefficients the avalanche noise in the photodiodes made of these materials is increased. Figure 4.23(*b*) shows the ionizing impact coefficients in InP. Although there exists no analytical expression for dependence of these coefficients on electric field, the relation may be roughly established from measurement in units of inverse length as

$$\alpha_{e,h}(E) = \alpha_\infty \exp[-(b/E)^m] \tag{4.83}$$

where α_∞ is the value of $\alpha_{e,h}$ for $E \to \infty$, b and m are the arbitrary coefficients. As example, in GaAs, $\alpha_\infty = 1.3 \times 10^6\ cm^{-1}$, $b = 2 \times 10^6$ and $m = 2$. In GaAs, $\alpha_e \approx \alpha_h$, but in Si $\alpha_e = 30\alpha_h$ and in Ge $\alpha_e \approx 5\alpha_h$.

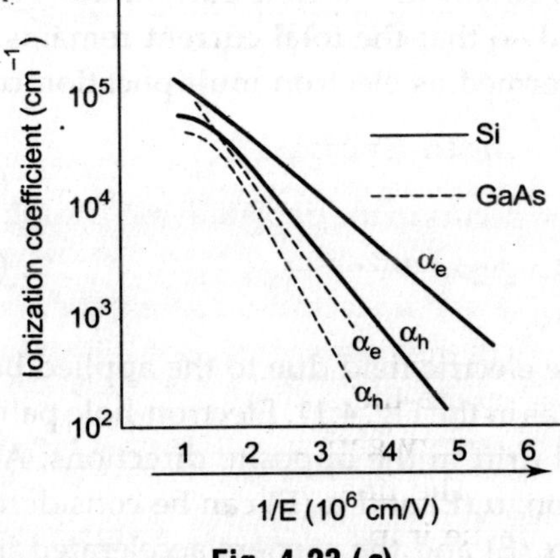

Fig. 4.23 (*a*)

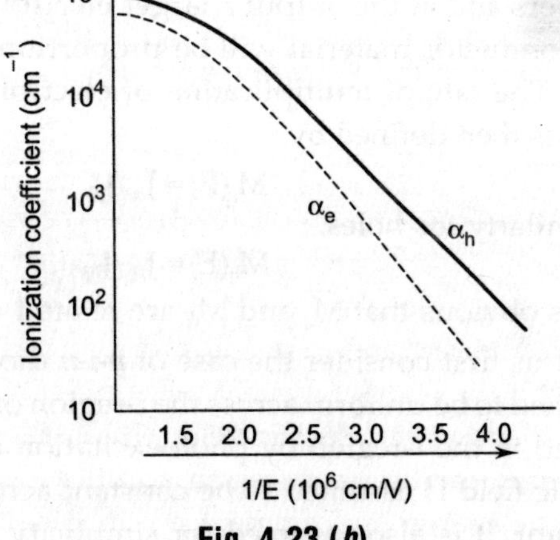

Fig. 4.23 (*b*)

The electric field required to impact ionization depends on the bandgap energy. At room temperature the required electric field ranges from 10^4 V/cm for InAs to more than 10^5 V/cm for GaP. The minimum energy needed for impact ionization is termed as threshold ionization energy E_i. The ionization rate of electrons and holes are affected by the value of E_i. For the simplest case of parabolic conduction and valence bands, described by $h^2k^2/2m^*$, the threshold ionization energy for electrons may be written as

$$E_{ie} = E_g(1 + m_e^*/2m_h^*) \tag{4.84}$$

and for holes

$$E_{ih} = E_g (1 + m_h^*/2m_e^*) \quad (4.85)$$

It may be noted that the conductivity effective masses are used here. Crystal orientation also influences the above equations. For $m_e^* = m_h^*$,

$$E_{ie} = E_{ih} = (3/2)\, E_g \quad (4.86)$$

This equation is for the lowest ionization threshold to the carrier with the smallest effective mass. Hence

$$E_{ie} + E_{ih} = 3E_g \quad (4.87)$$

The summation of threshold ion ization energies of electrons and holes is independent of the effective mass ratio.

For usual case of nonparabolic bands, the threshold ionization energies can no longer be defined by the simple equations as (4.84) and (4.85). For occurrence of impact ionization, the carrier energy must exceed the threshold ionization energy. As the impact ionization is probabilistic and can also be characterized by an impact ionization cross-section σ_i, the value of σ_i is zero when the carriers possess the threshold energy E_{th}. Above threshold, the value of σ_i is a strongly increasing function of electron energy. In other words, the probability of impact ionization per unit time for one type of carrier is proportional to the ionization cross-section for that carrier.

4.5.2 Multiplication and Ionization Coefficients in PIN Diodes and *pn*-Junction Diodes

The avalanche process is initiated by carrier multiplication in a high field region. Consider a piece of semiconductor material in which a number of electrons corresponding to an electron current J_{ei} is injected at one end. By avalanche multiplication process these electrons will be multiplied in numbers and at the output a larger electron current J_{eo} is obtained. The hole current through the semiconductor material will be proportionately adjusted so that the total current remains constant. The rate of multiplication of electrons, which is termed as electron multiplication coefficient, is then defined by

$$M_e(E) = J_{eo}/J_{ei} \quad (4.88)$$

Similarly for holes,

$$M_h(E) = J_{ho}/J_{hi} \quad (4.89)$$

It is obvious that M_e and M_h are related with α_e and α_h.

Let us first consider the case of *p-i-n* diode, where the electric field due to the applied bias is assumed to be uniform across the *i*-region of width of W, as in the Fig. 4.11. Electron-hole pairs are created in the *i*-region by photoexcitation and they will drift in the opposite directions. As the electric field is assumed to be constant across the *i*-region, $\alpha_e(E)$ and $\alpha_h(E)$ can be considered as constant. It is also assumed for simplicity that $\alpha_e(E) = \alpha_h(E)$ and the carriers accelerated in the opposite directions will suffer ionizing collisions creating more electron-hole pairs. The multiplication coefficient then can be expressed as

$$M = 1 + \alpha_e W + (\alpha_e W)^2 + (\alpha_e W)^3 + (\alpha_e W)^4 + \ldots + (\alpha_e W)^n + \ldots$$

$$= \frac{1}{1 - \alpha_e W} \ (\text{for } \alpha_e W < 1) \quad (4.90)$$

Avalanching occurs when current at breakdown goes to. This means $M \to \infty$. Hence the condition for avalanching is

$$\alpha_e W = 1 \quad (4.91)$$

In other words, breakdown occurs when every carrier suffers one ionizing collision. The lifetime of a particular generation of electrons is given by $\Delta t = (\alpha_e v_s)^{-1}$, where v_s is the saturation drift velocity. For example, if $v_s = 10^7$ cm/s and $\alpha_e = 2\text{x}10^4$ cm^{-1}, $\Delta t = 5\, ps$. The value of W can be calculated as $W = (\alpha_e)^{-1} = 0.5$ µm.

Next let us consider the case of p^+-n-n^+ diode, the electrical field distribution and various current components of which are shown in Fig. 4.24. The transport and collision of carriers are considered at n-region. The electrons drifting toward the n^+-layer will produce a current J_e and holes drifting towards the p^+-region will produce a current J_h. Then the total current J_T can be given by

$$J_T = J_e + J_h \tag{4.92}$$

And multiplication coefficient for electrons

$$M_e(E) = \frac{J_{e0}(W)}{J_{ei}} \tag{4.93}$$

W is the thickness of n-region. Now if an incremental slice in n-region of thickness Δx is considered, its current densities at input and output ends differ due to two components-electrons generated by electrons and electron generated by holes. In Fig. 4.24, $J_T = J_{e0}(W)$ neglecting the reverse saturation current. The increase in current at the slice Δx can be expressed as

$$\Delta J_e = J_e\alpha_e\Delta x + J_h\alpha_h\Delta x \tag{4.94}$$

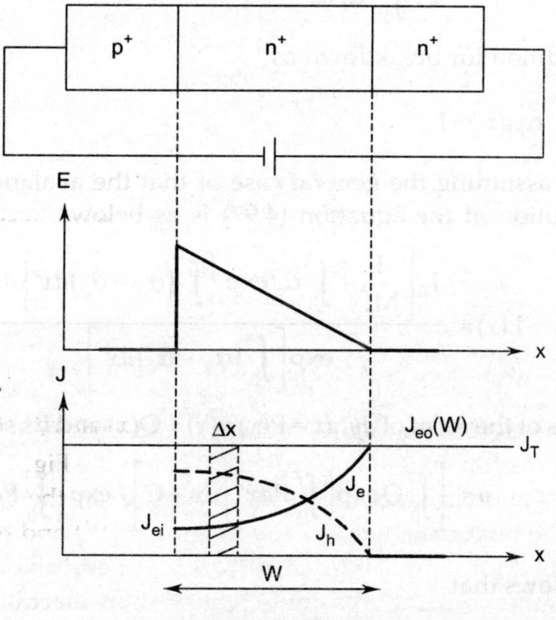

Fig. 4.24

Therefore,

$$\frac{dJ_e}{dx} = J_e\alpha_e + J_h\alpha_e \tag{4.95}$$

or,
$$\frac{dJ_e}{dx} = J_e\alpha_e + (J_T - J_e)\alpha_h \quad \text{[from equation 4.92]} \tag{4.96}$$

or,
$$\frac{dJ_e}{dx} + (\alpha_h - \alpha_e)\, J_e = \alpha_h J_T \tag{4.97}$$

The equation (4.97) is the general form of differential equation and it may be noted that α_e and α_h are the functions of electric field, which itself is not uniform across the incremental slice. However this equation is valid for the entire n-region or the region of multiplication.

(*i*) If $\alpha_e(E) = \alpha_h(E)$, the equation (4.97) reduces to

$$\frac{dJ_e}{dx} = \alpha_h J_T \tag{4.98}$$

The solution of the above equation is of the form

$$J_e(W) = J_T \int_0^W \alpha_e \, dx + J_{ei} \tag{4.99}$$

Since $\qquad J_{eo}(W) = J_e(W) = J_T,$

$$J_{eo}(W) = J_{e0} \int_0^W \alpha_e \, dx + J_{ei} \tag{4.100}$$

And from equation (4.93), the relation for electron multiplication coefficient becomes

$$M_e = M_e \int_0^W \alpha_e \, dx + 1 \tag{4.101}$$

or,

$$M_e = \frac{1}{1 - \int_0^W \alpha_e \, dx} \tag{4.102}$$

This leads to the condition for breakdown as

$$\int_0^W \alpha_e \, dx = 1 \tag{4.103}$$

(*ii*) For $\alpha_e(E) \neq \alpha_h(E)$, assuming the general case of that the avalanche process is initiated by electrons, the solution of the equation (4.97) is as below.

$$J_e(x) = \frac{J_T \left[\dfrac{1}{M_e} + \int_0^x \alpha_h \exp\left\{ \int_0^x (\alpha_h - \alpha_e) dx' \right\} dx \right]}{\exp\left[\int_0^x (\alpha_h - \alpha_e) dx' \right]} \tag{4.104}$$

[The equation (4.97) is of the form of $dy/dx = P(x)\, y(x) + Q(x)$ and its standard form of solution is

$$y = \left[\int_0^x Q \exp\left(\int_0^x P dx' \right) dx + C \right] / \exp\left(\int_0^x P dx' \right),$$ where C is the integration constant]

Since $J_e(W) = J_T$, it follows that

$$1 - \frac{1}{M_e} = \int_0^W \alpha_h \exp\left\{ \int_0^x (\alpha_h - \alpha_e) dx' \right\} dx \tag{4.105}$$

or,

$$M_e = \frac{1}{1 - \int_0^W \alpha_h \exp\left\{ \int_0^x (\alpha_h - \alpha_e) dx' \right\} dx} \tag{4.106}$$

And

$$M_h = M_e \exp\left\{ \int_0^x (\alpha_h - \alpha_e) dx' \right\} \tag{4.107}$$

Breakdown occurs when $M_e \rightarrow \infty$, which means

$$\int_0^W \alpha_h \exp\left\{\int_0^x (\alpha_h - \alpha_e)dx'\right\} dx = 1 \tag{4.108}$$

Similarly, if the avalanche process is initiated by holes, the breakdown condition is

$$\int_0^W \alpha_e \exp\left\{\int_0^x (\alpha_e - \alpha_h)dx'\right\} dx = 1 \tag{4.109}$$

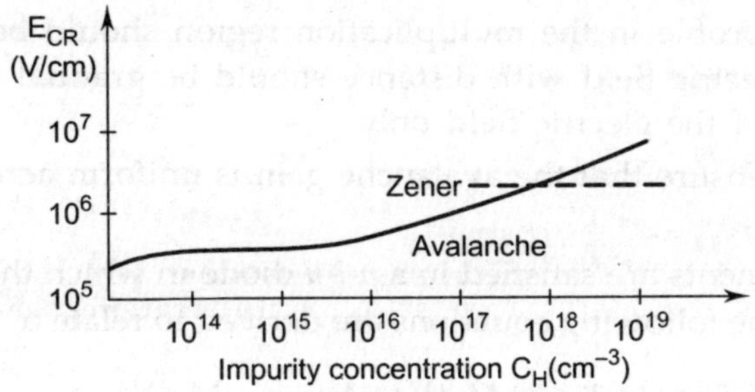

Fig. 4.25

It may be noticed that the equations (4.108) and (4.109) are equivalent. Therefore, the breakdown and avalanche process depends on the carrier population and collision rate in the depletion region, and not on the carrier type that initiates the avalanche process. When the avalanche breakdown condition is satisfied in a particular type of diode, *i.e.*, when the equations (4.91), (4.103), (4.108) and 4.109) are satisfied, the breakdown voltage can be calculated by Poisson's equation. Breakdown occurs whenever the maximum field in the depletion region E_M is equal to or larger than a critical field E_{CR}. For an one sided abrupt p^+-n junction. The reverse breakdown voltage V_{BR} can be expressed in terms of E_{CR} (=E_M) as

$$V_{BR} = \frac{\varepsilon_s \varepsilon_0 E^2_{CR}}{2qN_D} \tag{4.110}$$

where N_D is the donor doping density on the *n*-side and it is assumed that built-in voltage is much smaller than the breakdown voltage. The value of E_{CR} depends on the impurity concentration. Figure 4.25 shows the breakdown voltage with respect to the impurity concentration. It may be noted that, since the scattering rates are orientation dependent, the impact ionization process is also dependent on the crystal orientation.

The multiplication coefficient or factor M is therefore related to impact ionization coefficients and also termed as the *avalanche gain*. Very high values of avalanche gain can be obtained in the photodiodes at the range of 10^3 by biasing them near the breakdown voltage. However the avalanche process takes longer time to build up for a high avalanche gain and also persists longer. Thus it reduces the operating bandwidth and becomes unsuitable for detecting high frequency modulated light. This is the fundamental gain-bandwidth limitation of the avalanche photodiodes.

4.5.3 Measurement of Multiplication and Ionization Coefficients

The measurement of the impact ionization coefficients is not direct one and they are derived from the measurement of multiplication coefficients. To obtain reliable results certain experimental conditions as well as structural requirements regarding the diode are to be fulfilled which are as follows.

1. Pure electron and hole injection should be obtained in the same diode, rather than in complementary devices. The best technique of ensuring this is by photogeneration of the carriers to be injected.
2. Accurate measurement of photocurrent without avalanche gain is required. Then only the reliable values of M_e and M_h can be obtained.
3. The electric field profile in the multiplication region should be accurately known. The variation of the electric field with distance should be gradual to ensure that α_e and α_h are the functions of the electric field only.
4. It is important to ensure that the avalanche gain is uniform across the active area of the device.

All the above requirements are satisfied in a *p-i-n* diode in which the electric field is uniform across the i-region and the following equations are derived to relate α_e and α_h with M_e and M_h.

$$\alpha_e(E) = \frac{1}{W}\left[\frac{M_e(V)-1}{M_e(V)-M_h(V)}\right]\ln\frac{M_e(V)}{M_h(V)} \quad (4.111)$$

$$\alpha_h(E) = \frac{1}{W}\left[\frac{M_h(V)-1}{M_h(V)-M_e(V)}\right]\ln\frac{M_h(V)}{M_e(V)} \quad (4.112)$$

Where V = EW is the applied reverse bias and W is the thickness of the *i*-region. Similar equations are applicable for the abrupt *p-n* junctions.

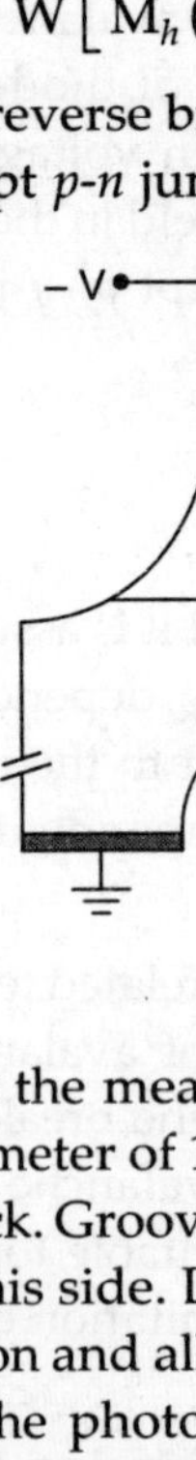

Fig. 4.26

A typical device structure for the measurement of impact ionization coefficients is shown in Fig. 4.26. Mesa diodes of the diameter of 100–300 μm are made by photolithography and etching. The *i*-region is typically 2 μm thick. Grooves are introduced on the backside by selective etching to facilitate photoexcitation from this side. Large area backside and annular top ohmic contacts are formed are formed by evaporation and alloying. The top p^+ layer and the n^+ layer on the substrate must be thick enough so that the photoexcitation incident from the either side is completely absorbed in these layers and does not reach the *i*-layer. Thus the pure carrier injection is ensured. This implies that when light is absorbed in the top p^+ layer, excess minority electrons are produced and injected into the avalanche *i*-region by the action of reverse bias. Similarly minority holes are produced and injected from the bottom n^+ layer.

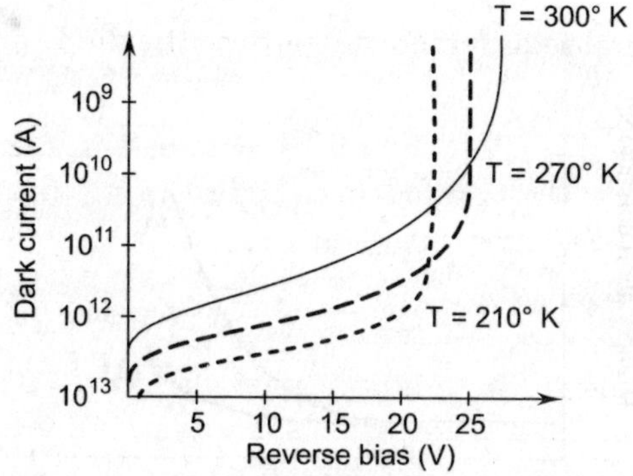

Fig. 4.27

For the measurement of multiplication coefficients, it is important to ensure that the avalanche multiplication is the dominant breakdown mechanism in the diode. Measurement of reverse current-voltage characteristics (I-V characteristics) of the diode with the function of temperature helps to check the dominant breakdown process. For avalanche multiplication at the dominant breakdown process, the reverse breakdown is higher at higher temperature. This occurs due to that the probability of non-ionizing collisions increases with the increase of temperature, and consequently the probability of ionizing collisions decreases. A competing process for breakdown is tunneling, either across the bandgap or via a trap level in the gap. But these processes are largely independent of temperature. Typical temperature dependent current-voltage characteristics for GaAs-AlGaAs *p-i-n* diodes in which I-region is a MQW are shown in Fig. 4.27, from which it may be observed that avalanching is the dominant breakdown mechanism.

A scheme for the measurement of the multiplication coefficients is shown in the Fig. 4.28. A beam splitter and chopper are incorporated between the excitation source and sample to allow phase-sensitive detection. The diode photocurrent is amplified by a current sensitive preamplifier and lock-in amplifier and is fed to the Y-axis of the X-Y recorder. The reverse bias is applied to the diode is fed to the X-axis of the recorder. The bias is varied from zero to near-breakdown and the corresponding photocurrent for both electron and hole injection is recorded. Appropriate stable unmultiplied values of the photocurrent at low bias are used as reference to obtain the electron and hole multiplication factors. Measured values of M_e and M_h in a *p-i-n* diode with a GaAs/AlGaAs with MQW *i*-region are shown in Fig. 4.29. From these values of M_e and M_h, the values of α_e and α_h are calculated using the equations (4.111) and (4.112).

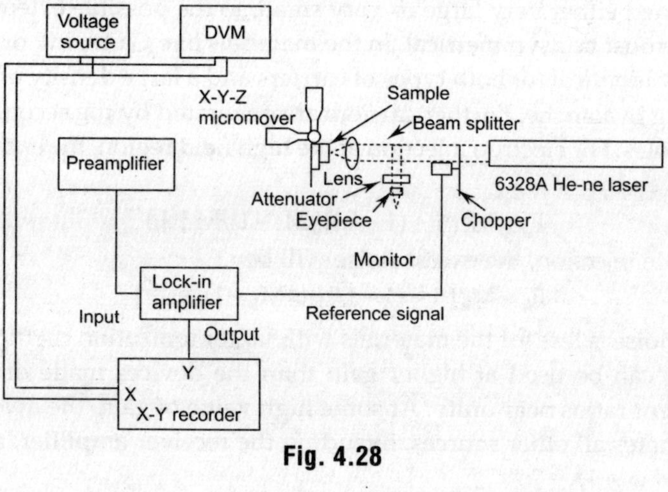

Fig. 4.28

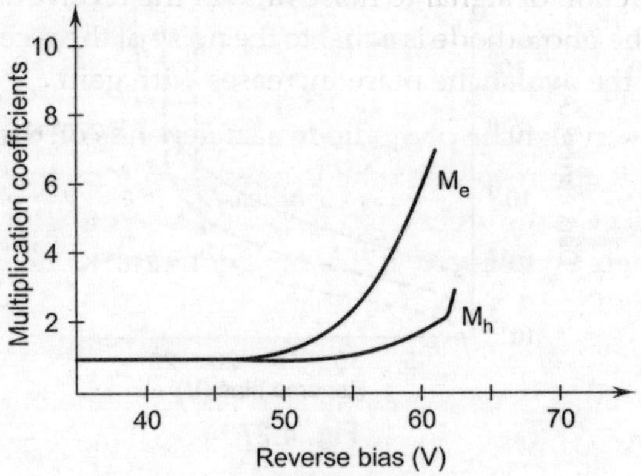

Fig. 4.29

The series resistance of the diode and the diode temperature are the limiting factors for the measurement of multiplication coefficients. An increase in temperature reduces α_e and α_h. It has been observed that experimentally obtained values of M obey the following relation.

$$M = \left\{1 - \left(\frac{V - IR}{V_{BR}}\right)\right\}^{-n} \tag{4.113}$$

Where V_{BR} is the breakdown voltage, V is the applied bias, R is an effective resistance which includes the series resistance of the diode and *n* is a parameter depending on the diode construction.

4.5.4 Noise Performance of Avalanche Photodiode

The principle source of noise in the avalanche photodiodes is the avalanche process itself that gives gain. However, these devices, like *p-i-n* diodes, are limited by shot noise associated with the primary, unmultiplied bulk signal, dark current and background current. As already discussed, the avalanche process is probabilistic and in any statistical process there are random fluctuations in the actual distance between the successive ionizing collisions. These fluctuations, in turn, cause variation in the total number of secondary carriers generated for each primary carrier injected into the gain region. This variation results in excess noise or avalanche noise in the total signal current. To minimize the avalanche noise, the electron to hole ionization coefficient ratio, $k_r = \alpha_h/\alpha_e$, should be maintained either very large or very small, to the possible extent. This means that the avalanche process must be asymmetrical. In the materials like GaAs, InP or Ge, the probability of ionization is nearly identical for both types of carriers and a large density of carriers of both types is generated during avalanche. Further carriers are generated by the secondary electrons as well as the secondary holes. For electron injection at the high field region, the excess noise generated by the electrons is given by

$$F_e = M_e[1 - (1 - k_r)\{(M_e - 1)/M_e\}^2] \tag{4.114}$$

Similarly for hole injection, the excess noise will be

$$F_h = M_h[1 - (1 - 1/k_r)\{(M_h - 1)/M_h\}^2] \tag{4.115}$$

The avalanche noise is less for the materials with larger ionization coefficient ratios, and therefore, such devices can be used at higher gain than the devices made of the materials having ionization coefficient ratios near unity. At some high value of gain, the avalanche noise from the photodiode dominates all other sources, including the receiver amplifier, and a further increase in gain results in the reduction of signal to noise ratio of the receiver. Optimization occurs when the avalanche gain from the photodiode is equal to the noise of the receiver, and the optimum gain depends on how rapidly the avalanche noise increases with gain.

The response time of an avalanche photodiode also depends on the ionization coefficient ratio k_r. For the best case $k_r \rightarrow 8$, the response time is limited only by the duration of a single transit across the depletion region. On the other hand the worst case situation is $k_r \rightarrow 1$, when continuous recycle of secondary carriers is observed at the depletion region and response time is long along with the high gain. The highest gain-bandwidth product of the avalanche photodiode is achieved for the devices and materials where the ionization coefficient ratio is maximized.

The expression for the signal to noise ratio of an avalanche photodiode is very much similar to that of the *p-i-n* diode as given by the equation (4.82). The avalanche gain of the diode enhances the dark current, the current produced by the optical signal and current produced by any background light. Similar to the equation (4.76), the photocurrent due to incident light is given by

$$I_{ph} = \frac{q\eta P_{inc}\Gamma_G}{\sqrt{2}h\nu} \tag{4.116}$$

The mean square shot noise is expressed as

$$I_s^2 = 2q\,(I_{ph} + I_B + I_D)\,\Gamma_G^2 FB \tag{4.117}$$

where F is the excess noise factor. The thermal noise is the same in the *p-i-n* diode. Therefore, for an input optical signal having unity modulation index, the signal to noise power ratio can be derived as

$$\frac{S}{N} = \frac{\dfrac{1}{2}\left(\dfrac{q\eta P_{inc}\,\Gamma_G}{h\nu}\right)^2}{2q\left(I_{ph} + I_B + I_D\right)\Gamma_G^2 FB + \dfrac{4k_B\,TB}{R_{eq}}} \tag{4.118}$$

The NEP for the avalanche photodiode is almost identical to that of *p-i-n* diode (equation 4.84), except that the term Γ_G appears at the denominator. Thus it is clear that the avalanche gain can reduce the NEP of the device and thereby enhance its sensitivity compared to the *p-i-n* photodiode. Among the photodiodes, the best sensitivity is obtained at avalanche photodiode. In a digital system the Bit Error Rate (BER) depends on the signal to noise ratio (S/N ratio). The BER is defined by the expression

$$\text{BER} = \frac{1}{2}\,\text{erfc}\left(\frac{S}{2\sqrt{2}N}\right) \tag{4.119}$$

4.5.5 Practical Avalanche Photodiodes

Reduction of dark current and enhancement of device speed are the two most important objectives in avalanche photodiode design. The best performance can be achieved from an avalanche photodetector, when several structural and materials requirements are fulfilled. First and foremost it is essential to ensure uniformity of carrier multiplication over the entire photosensitive area of the diode. The material in which the avalanching occurs must be defect free, and so great care must be taken during the device fabrication. Defects in the absorption and multiplication regions can give rise to *microplasmas*, which are the local areas in which the breakdown occurs at voltages lower than the rest of the junction. Such defects are caused by the bulk or edge dislocations propagating from the substrate, or initiated in the active area during growth and processing. Using small active region, typically between 10 to 100 μm in diameter, may reduce the defect density.

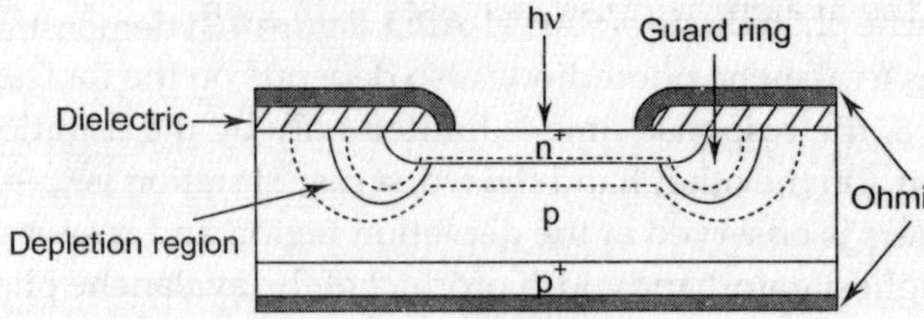

Fig. 4.30

The dark current in an avalanche photodiode is produced during generation-recombination at the depletion region and also due to leakage current and tunneling current. The dark current is also largely dependent on the material used. A common problem in avalanche photodiode is the excessive leakage current at the junction edges. A common technique used in the Si diodes to overcome this problem is to incorporate a guard ring, which is a *n-p* junction created by selective area diffusion around the periphery of the diode. A Si avalanche photodiode structure with guard ring is shown in Fig. 4.30. In addition, a SiO_2 film is deposited in mesa diodes to minimize the surface leakage current or edge leakage current.

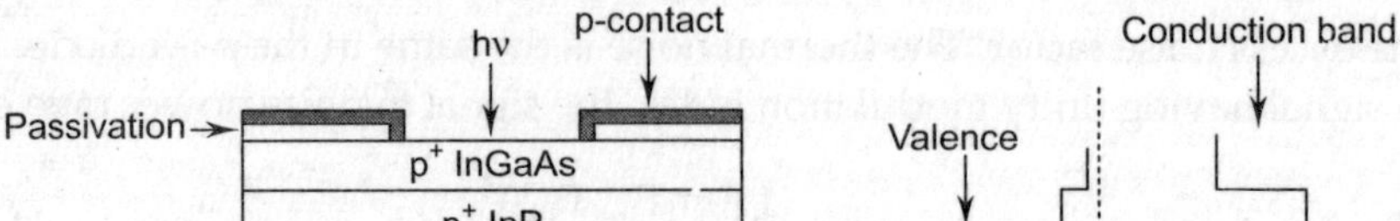

Fig. 4.31

The choice of material for an avalanche photodiode is dictated by the application. GaAs or AlGaAs diodes are used for the applications where the system is designed to operate at the near-infrared regions. For fiber-optic communication InGaAsP material is used, where bandgap E_g is between 0.74 and 1.35 eV. For example, $In_{0.53}Ga_{0.47}As$ has a bandgap of 0.74 eV' which is applicable for the wavelength of 155 μm. Homojunction diodes made of these low bandgap materials have high reverse leakage current at high voltages due to band to band tunneling. The tunneling probability of a semiconductor is exponentially proportional to $E_g^{-3/2}$. Therefore, incorporation of the high field avalanching region of the diode in the high-bandgap material can reduce the tunneling induced reverse leakage current. These considerations led to the design and realization of the separate *absorption and multiplication* avalanche photodiode or SAM-APD. These structures exhibit low leakage as the junction is placed in the high bandgap material (*e.g.* InP), with sensitivity at long wavelengths provided by the low bandgap absorption region (*e.g.* InGaAs). Dark currents in order of *p*A to *n*A can be obtained with gains of the order of 10. However, the operation of SAM-APD is associated with a potential problem. Holes are accumulated at the valence band discontinuity at the InGaAs /InP heterojunction and thereby increase the response time. To overcome this problem, a graded quaternary InGaAsP layer can be inserted between InP and InGaAs layer. This type of structure is referred as separate *absorption graded multiplication* (SAGM) avalanche photodiode or SAGM-APD. Figure 4.31 demonstrates the schematic of this type of structure.

To improve the performance of these devices, a thin delta-doped region can be incorporated between the absorption and avalanche regions. This region ensures the desired low-field in the absorption region. Gain-bandwidth product of over 100 MHz is achievable in the delta-doped SAGM-APDs. In the design of these devices, it is essential to ensure that all the absorption takes place in absorption region, and of the photogenerated electrons and holes, the carrier type of higher impact ionization coefficient in multiplication region is injected into it. This will ensure the low noise operation. In practical devices, absorption region of 1 to 2 μm thickness made undoped. The graded layer of 0.1 to 0.3 μm and avalanching layer of 1 to 2 μm are doped to the order of 10^{16} cm^{-3}. The p-layer can be thin and doped to the order of 10^{17} to 10^{18} cm^{-3}.

For high speed operation of these devices it is desirable to deplete all or most of the absorption region. Otherwise slow diffusion of holes may dominate the carrier transit, which eventually slower the process. For well-designed avalanche photodiodes, the temporal response exhibits a fast rise time in the order of 100 to 200 ps. This is determined by the transit and collection of the faster electrons and the time for the avalanche process to build up. The fall time, on the other hand, may be very large at the order of 10 to 100 ns, and this is assumed to be dominated by trapping and recombination of holes. SAGM-APD devices are suitable for use at optical communication systems.

The main drawback of the avalanche photodiodes is that they possess complicated structures compared to the photoconductors and PIN diodes, and therefore higher manufacturing cost is involved. The biasing voltage is generally quite high in the range of 20 to 50 V. Finally, the avalanche gain is extremely sensitive to temperature and applied bias. This makes their use and operation more expensive, as elaborate arrangements are necessary to compensate the temperature and to stabilize the supply voltage, for stable and reliable operation of the device.

Example 4.9. *10^{11} photons are incident per sec on an avalanche photodiode which detects 1.5 μm photoexcitation with a resposivity of 0.7 A/W and gain of 30. Calculate the quantum efficiency and the rms photocurrent.*

Solution: Using the equation (4.3),

$$R = \frac{\eta\lambda(\mu m)}{1.24}(A/W)$$

So,
$$\eta = \frac{1.24 \times R}{\lambda} = \frac{1.24 \times 0.7}{1.5} = 0.578$$

From equation (4.116),
$$I_{ph} = \frac{q\eta P_{inc}\Gamma_G}{\sqrt{2}\, h\nu}$$

where number of incident photons = $P_{inc}/h\nu = 10^{11}$ and $\Gamma_G = 30$

Therefore the rms photocurrent
$$I_{ph} = \frac{1.6 \times 10^{-19} \times 0.578 \times 10^{11} \times 30}{\sqrt{2}} = 0.196\ \mu A$$

Example 4.10. *An APD has the quantum efficiency of 40% at 1.3 μm. It produces an output photocurrent of 6 μA after avalanche gain when illuminated with optical power of 0.3μW. Calculate the multiplication factor.*

Solution: Multiplication factor

$$M = I/I_p = \frac{I}{P_{in} R} = \frac{I}{P_{in}\eta q\lambda / hc} = \frac{Ihc}{P_{in}\eta q\lambda} = \frac{6\times 10^{-6} \times 6.626\times 10^{-34} \times 3\times 10^{8}}{0.3\times 10^{-6} \times 0.4\times 1.6\times 10^{-19} \times 1.3\times 10^{-6}}$$
$$= 47.6$$

Example 4.11. *A silicon RAPD has the quantum efficiency of 90% at 0.8 μm with multiplication factor of 800 and dark current of 2 nA. Calculate the rate of photons to be incident on the device so that the output current after avalanche gain is greater than the dark current.*

Solution: Multiplication factor

$$M = I/I_p = \frac{Ihc}{P_{in}\eta q\lambda}$$

or,
$$\text{Photon rate} = \frac{P_{in}}{hc/\lambda} = \frac{I}{\eta q M} = \frac{2\times 10^{-9}}{0.9\times 1.6\times 10^{-19} \times 800} = 1.74\times 10^{7}\ /\ \text{sec}$$

4.5.6 Superlattice Avalanche Photodiodes

The best performance of an avalanche photodiode is obtained when $\alpha_e/\alpha_h \rightarrow 0$ or ∞. Unfortunately, this characteristic is available in none of the materials, where α_e/α_h is observed to be nearly equal to unity. Therefore different ways and means have been explored to enhance or reduce the value of α_e/α_h in artificially structured materials. Figure 4.32(*a*) demonstrates one of the viable multilayered structures, in which enhancement of α_e/α_h can be obtained.

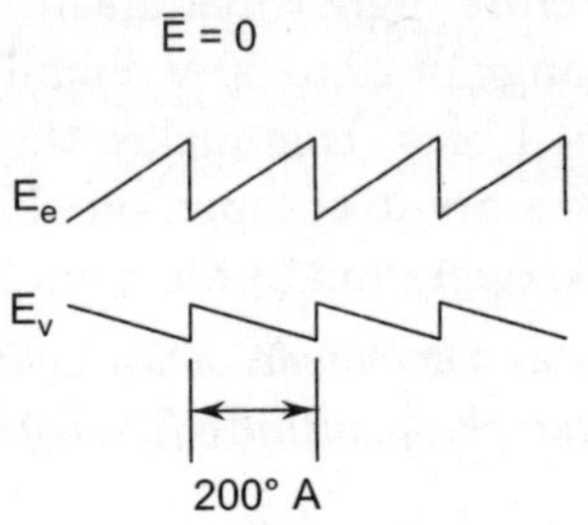

Fig. 4.32 (*a*)

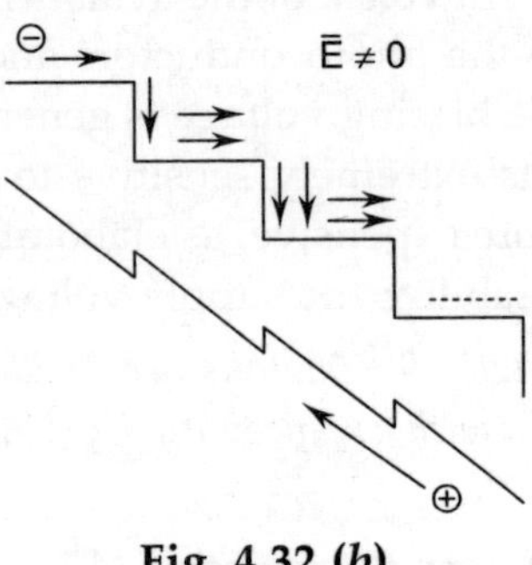

Fig. 4.32 (*b*)

This structure is called *staircase superlattice* and in a practical photodiode this material will form the avalanching region. The staircase superlattice structure is difficult to realize, as it involves the growth of precisely controlled quaternary graded layers in each period. To understand the process of impact ionization in a staircase superlattice, it is necessary to look at the reorganized band profile under applied bias, as shown in Fig. 4.32(*b*). The electrons gain energy at each step while drifting from left to right and undergo impact ionization. The process is repeated at each step and a large multiplication of the number of electrons is resulted. The holes, on the other hand, do not encounter such large potential steps, and their multiplication remains as in bulk semiconductors. Thus a large value of α_e/α_h can be achieved, and the structure behaves as a solid state photomultiplier of electrons.

A multiquantum well structure with large well and barrier thickness (200–500°A) and/or deep wells in the conduction band is another possibility for similar enhancement of α_e/α_h. Similarly α_h/α_e can be enhanced if band offset is higher in the valence band, resulting in deep quantum wells. The process of multiplication of one type of carrier is similar to that in a staircase superlattice. When well and barrier sizes become smaller (in the order of 100°A or less), enhancement in the multiplication factors as well as the value of α_e/α_h can still be achieved. However, if the conduction band offset is not very large, the mechanism for such enhancement is different. For example, in the structure of GaAs/$Al_{0.3}Ga_{0.7}As$ MQW, where E_c:E_v is 57:43, the electrons are not truly confined in the wells because of their small effective mass and electron behavior is similar to that of bulk semiconductor. On the other hand, the holes are confined and experience more scattering in the wells, and so their ionization rate is reduced. Thus an enhancement of α_e/α_h is obtained as overall effect.

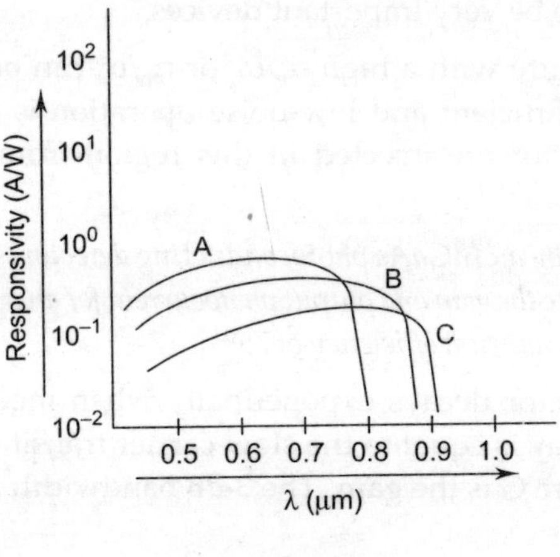

Fig. 4.33(*a*)

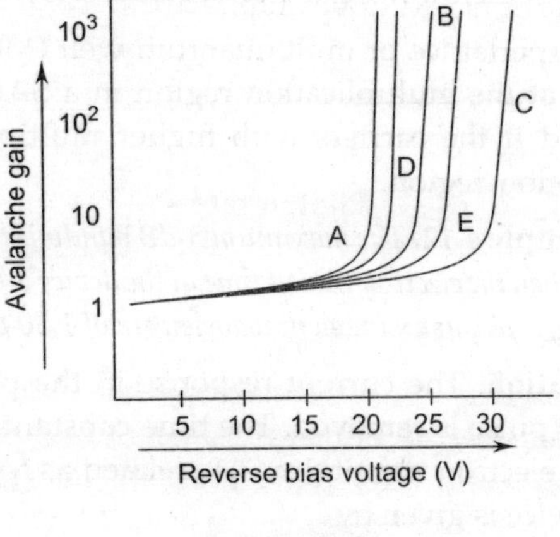

Fig. 4.33(*b*)

In addition to the low noise performance and gain of the superlattice avalanche photodiodes, they can also provide tunability of the spectral response. In these devices, multiquantum wells with varying sizes of wells and barriers are incorporated at the absorption and multiplication region. The measured spectral response and avalanche gain in respect to the applied reverse bias of GaAs/AlGaAs superlattice photodiodes, typically with 20–30 periods are shown in Figs 4.33(*a*) and 4.33(*b*) respectively. From Fig. 4.33(*b*), it may be observed that very large values of gain can be realized with avalanche photodiodes, when the stabilized bias supplies are used. The measured temporal response of a 80°A GaAs/90°A $Al_{0.1}Ga_{0.9}As$ superlattice avalanche photodiode of 30μm diameter to 10 ps excitation pulses is shown in Fig. 4.34.

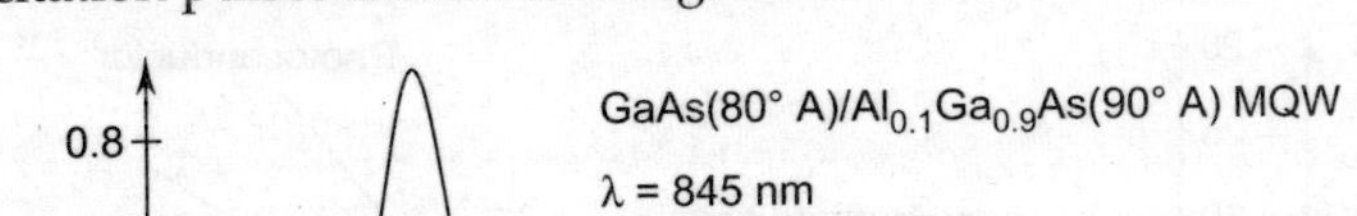

Fig. 4.34

The intrinsic speed of these devices is governed by the emission and collection of carriers created in well region, which is the sum of the emission or tunneling times from the wells and the carrier transit time across the entire absorption region of the diode. The carrier transit time at the absorption time is, however, is very small under large applied bias across the diode. Therefore, field-assisted thermionic emission and tunneling times set the intrinsic limit, which may be in the range of 60–100ps. The diode area, capacitance and resistance govern the extrinsic limit. Therefore, superlattice avalanche photodiodes with very large bandwidths can be obtained by proper design, and with their low noise performance they can prove to be very important devices.

A superlattice or multiquantum well (MQW) structure with a high α_e/α_h or α_h/α_e can be inserted at the multiplication region in a SAGM-APD. Efficient and low-noise operation is then ensured if the carriers with higher multiplication factor are injected in this region from the absorption region.

Example 4.12. *The maximum 3-dB bandwidth permitted by an InGaAs photoconducting detector is 450 MHz when the electron transit time in the device is 6 ps. Calculate the gain and output photocurrent for an optical power of 5 μW at an incident wavelength of 1.30 μm with the quantum efficiency of 75%.*

Solution: The current response in the photoconductor decays exponentially when incident optical pulse is removed. The time constant of this decay is equal to the slow carrier transit time t_s. The electron transit time t_f is related as $t_f = t_s/G$, where G is the gain. The 3-dB bandwidth Δf of the device is given by

$$\Delta f = 1/(2\pi t_s) = 1/(2\pi t_s G)$$

So, $$G = 1/(2\pi t_f \Delta f) = 1/(2\pi \times 6 \times 10^{-12} \times 450 \times 10^6) = 58.94$$

The output photocurrent $I = GI_p$

$$I = \frac{G\eta P_{in} e\lambda}{hc} = \frac{58.94 \times 0.75 \times 5 \times 10^{-6} \times 1.6 \times 10^{-19}\, 1.3 \times 10^{-6}}{6.626 \times 10^{-34} \times 3 \times 10^8}$$

$$= 232.1\mu A$$

4.6 COMPARISON OF DIFFERENT DETECTORS

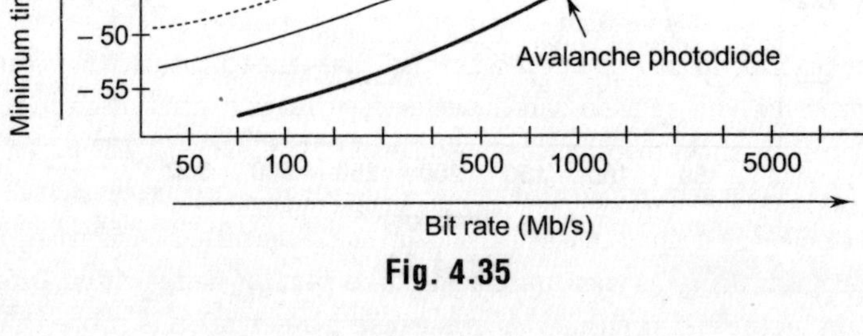

Fig. 4.35

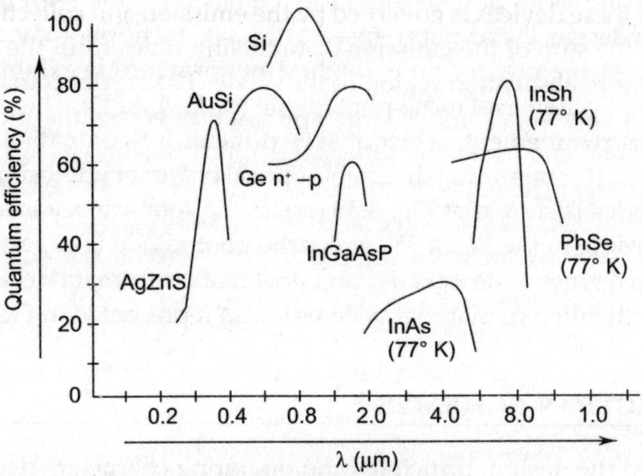

Fig. 4.36(a)

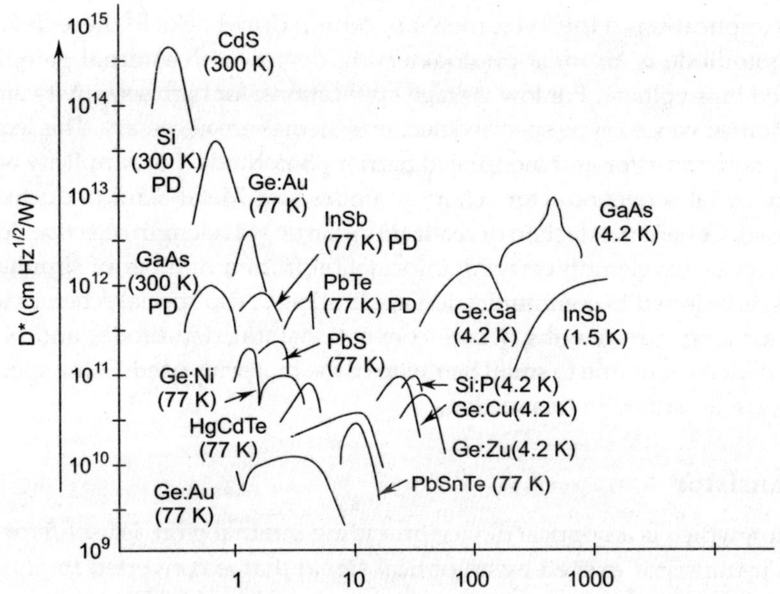

Fig. 4.36(b)

The choice of detectors depends on its application. An useful parameter to decide the selection of detectors is the minimum detectable time-averaged signal power. Considering the simplicity of fabrication and ease of operation, it has been seen that the photoconductor is an attractive device. However, its dark current as well as associated Johnson noise is high, which make it unsuitable for high-performance communication applications. For such applications, Bit Error Rate (BER) as defined by the equation (4.119), is the important parameter for performance index. In digital communication, bit error rate or BER is a measure of how often a transmitted '1' is mistakenly identified as '0' and vice versa. A BER of 10^{-9} (this means one mistake per 10^9 transmitted bits) which is considerably acceptable, can be achieved if signal to rms noise power ratio is more than six.

The combination of bandwidth and sensitivity is obtained by the *p-i-n* diode and avalanche photodiode. With proper design, the response speed of the avalanche photodiode nearly approaches to that of a *p-i-n* diode. At the present time, the best performance is exhibited by the separate-absorption-graded-multiplication avalanche photodiode (SAGM-APD). The main disadvantage of this device is higher bias requirement, which makes difficult for compatibility to the integrated circuit design. A comparison of minimum detectable optical power of photoconductors, *p-i-n* diodes and avalanche photodiodes is shown at Fig. 4.35 (assuming total capacitance of 0.5 pF and dark current of 50 nA). According to the Fig. 4.35, avalanche photodiode is expected to perform better compared to the other two types of devices. A comparison of performance characteristics of different devices fabricated with different material systems is also represented in Figs. 4.36(*a*) and 4.36(*b*).

4.7 SPECIAL DETECTION SCHEMES

So far we have discussed the design, principles and operating characteristics of three basic types of photodetectors, namely the photoconductor, the *p-i-n* photodiode and the avalanche photodiode (APD). These devices satisfy most of the detection needs in optical communication and other general applications. However, there are certain drawbacks in these devices. For example, avalanche photodiode is an ideal photodetecting device with internal gain, but it requires large and stabilized bias voltage. For low voltage applications, for high sensitivity and some special detection needs, more advanced type special detection schemes are necessary. This leads to the use of other devices like phototransistor and modulated barrier photodiode. For simplicity of operation and ease of fabrication, metal semiconductor schottky diodes and Metal-Semiconductor-Metal (MSM) diodes are being used. Coherent detection or realization of true wavelength selective detection (*i.e.*, the detection of preselected wavelength carrying information from a number of simultaneously transmitted waveforms) is required in communication applications. Also special schemes to design detectors are required for long wavelengths, where conventional junction diodes and photoconductors possess very high dark current due to small bandgap of the materials used. Some special devices and detection schemes are described in this section.

4.7.1 Phototransistor

The phototransistor, which is an optical device providing internal gain, is useful for many applications. It is a bipolar transistor excited by an optical signal that is converted to photocurrent at its collector. Usually there exists no base contact and therefore there is no base current. Actually, the optical signal acts as the base current in this device. This is called *floating-base* mode of operation. This device is analogous to a junction photodiode followed by an amplifier. In phototransistor, the light is absorbed at the reverse-biased base-collector junction that acts as the junction diode, and the generated photocurrent is amplified by the transistor action. Thus, in phototransistor, generation of photocurrent and its amplification are made integrated.

The band diagram of an *n-p-n* heterojunction phototransistor is shown in Fig. 4.37(*a*). The wide bandgap emitter serves two purposes. First, it acts as a window layer so that the optical signal is absorbed only in the base and base collector depletion region. Secondly, it enhances the current gain of the transistor. The current flow diagram is shown in Fig. 4.37(*b*). The base-collector junction acts as a reverse-biased photodiode and so the phototransistor as shown the diagram of Fig. 4.37(*b*) operates in a common emitter configuration.

In an *n-p-n* heterojunction bipolar transistor, the emitter injection efficiency is given as

$$\gamma_E = \left(1+\left(\frac{D_e L_h N_D}{D_e L_h N_A}\right)^{-1}\left(\frac{m_{eB}^* \, m_{hB}^*}{m_{eE}^* \, m_{hE}^*}\right)\exp\left(-\frac{\Delta E_g}{k_B T}\right)\tanh\left(\frac{W_b}{L_s}\right)\right)^{-1} \tag{4.120}$$

where D_e and D_h are the diffusion constants of electrons and holes respectively,

L_e and L_h are the diffusion lengths of electrons and holes respectively,

N_D and N_A are majority carriers doping concentrations,

m_{eB}^* and m_{hB}^* are the effective masses of electrons and holes respectively at base region,

m_{eE}^* and m_{hE}^* are the effective masses of electrons and holes respectively at emitter region,

W_b is the neutral base width.

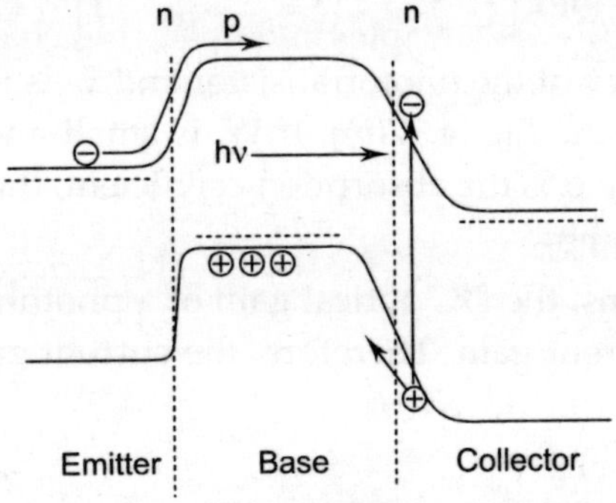

Fig. 4.37(*a*)

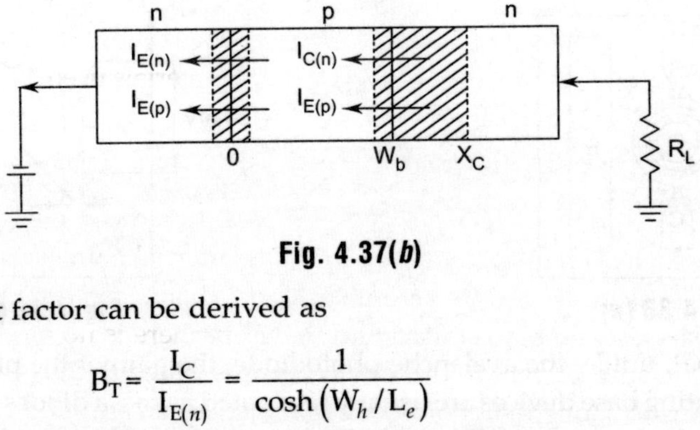

Fig. 4.37(*b*)

The base transport factor can be derived as

$$B_T = \frac{I_C}{I_{E(n)}} = \frac{1}{\cosh(W_h / L_e)} \tag{4.121}$$

The current transfer ratio may be derived from the relation $\alpha_T = \gamma_E B_T$ and common emitter current gain $\beta_T = \alpha_T/(1-\alpha_T)$.

The optical gain Γ_G of the phototransistor is the ratio of the number of photogenerated carriers creating the photocurrent to the number of incident photons. Therefore

$$\Gamma_G = \frac{h\nu}{q}\frac{I_{ph}}{P_{inc}} \tag{4.122}$$

where I_{ph} is the optically generated component of the collector current and P_{inc} is the incident optical power. I_{ph} is calculated with the assumptions that the base-emitter and base-collector depletion layer widths are much smaller than the thickness of these regions, and outside the depletion region the base remains neutral. The base is said to be neutral when the concentration of minority photogenerated electrons at the base region is much less than the majority holes. It is also assumed that steady sate conditions exist through the device. The optical excitation creates additional components of collector current due to absorption of photons in the base, the base-collector depletion region and the neutral collector region. In addition, some photogenerated holes move towards the emitter and modifies the emitter the potential as well as the emitter injection. Due to this, the optical excitation acts as a base current and can reduce the gain of the transistor. If ΔE_g is large enough, the γ_E is also large. If the base is very thin, then $W_b/L_e \ll 1$. Finally, it is also assumed that the absorption of photons at the collector region is only limited to depletion layer and not the neutral collector region. Under these conditions, the optical gain can be derived as

$$\Gamma_G = \eta\beta_T \left[1 + e^{-\alpha W_b}\left(1 - {}^{e - \alpha(x_c - W_b)}\right)\right] \tag{4.123}$$

where η is the quantum efficiency of the phototransistor and x_C is the edge of the depletion layer in the collector region [refer to the Fig. 4.37(*b*)]. If W_b is small and the depletion layer is large enough such that $x_C > 1/\alpha$, where α is the absorption coefficient, then equation (4.123) reduces to

$$\Gamma_G = \eta\beta_T \tag{4.124}$$

thus, with the above assumptions, the DC optical gain of a phototransistor is the product of the quantum efficiency and the current gain. Therefore, the current gain of a phototransistor is expressed as

$$\beta_T = \frac{h\nu}{q\eta}\frac{I_{ph}}{P_{inc}} \tag{4.125}$$

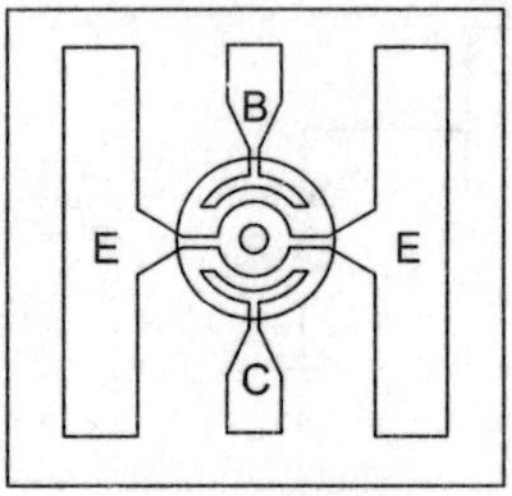

Fig. 4.38 (*a*)

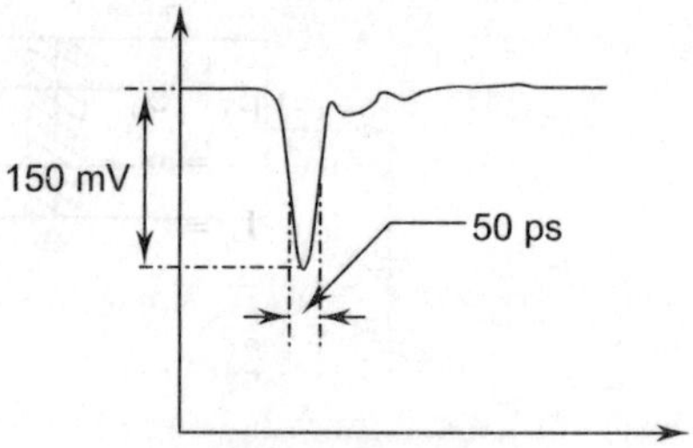

Fig. 4.38 (*b*)

It may be noticed that, unlike the avalanche photodiode, the gain of the phototransistor can be uniform with bias. Floating base devices are usually fabricated as mesa diodes with large areas. The optical gain Γ_G may be in the range of 100 to 500 in these devices. However, due to storage of the photogenerated minority carriers at the base, the response speed is poor. The modulation bandwidth of such phototransistors is generally less than 1 GHz. Better temporal response, or frequency bandwidth, can be achieved by a considerable reduction in the device dimensions and introduction of a base terminal to remove excess carriers. At the same time, the transit times and circuit parameters are required to be optimized as in the bipolar transistor. However, the gain of the device is greatly reduced. The layout of a high-speed GaAs/AlGaAs phototransistor is given in Fig. 4.38(*a*). A typical temporal response of the collector current of phototransistor is presented in Fig. 4.38(*b*).

The phototransistor is an attractive device of choice for optical communication applications because of its uniformity of the gain with bias voltage, in which InP/InGaAs heterojunction is used. The schematic diagram of a typical long-wave device is shown in Fig. 4.39. The dark current of the device may be quite large, which is a disadvantage in terms of noise performance.

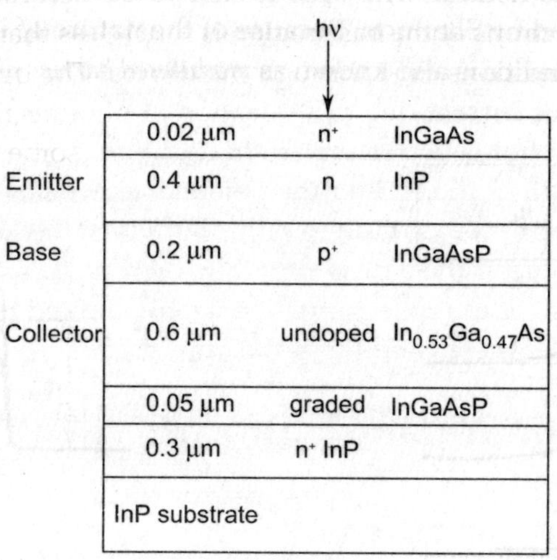

Fig. 4.39

Example 4.13. *A phototransistor with base width of 0.4 µm has the injection efficiency $\gamma_E = 0.998$ and $\eta = 0.7$. Calculate the reduction in optical gain if its the depletion length for electrons L_e changes from 8 µm to 0. 8 µm. γ_E and η are unchanged at both values of L_e.*

Solution: From the equation (4.121), for $L_e = 8$ µm, the base transport factor

$$B_T = \frac{I_C}{I_{E(n)}} = \frac{1}{\cosh(W_b/L_e)} = \frac{1}{\cosh(0.4/8)} = 0.9987$$

The current transfer ratio $\beta_T = \gamma_E\, B_T = 0.998 \times 0.9987 = 0.9967$

Current gain $\beta_T = \alpha_T/(1-\alpha_T) = 0.9967/(1-0.9967) = 302$

Optical gain $\Gamma_G = \eta_\beta = 0.7 \times 302 = 211.4$

For $L_e = 8$ µm, the base transport factor $B_T = 0.887$

The current transfer ratio $\alpha_T = 0.998 \times 0.887 = 0.885$

Current gain $\beta_T = 7.696$

Optical gain $\Gamma_G = 0.7 \times 7.696 = 5.38$

The optical gain drastically reduces if the depletion width reduces a little.

4.7.2 Modulated Barrier Photodiode

The Schottky diode is a majority carrier device that can be used as a high speed photodetector. The dominant current transport mechanism in this device is thermionic emission over the potential barrier formed at the metal-semiconductor interface. The barrier height cannot be changed by any means, but it is intrinsically determined by the metal-semiconductor work function difference. In practical device, it is pinned by interface states at the junction. A feasible variation of the Schottky diode is shown schematically in Fig. 4.40(*a*) alongwith the band diagram and charge distribution. The thickness and doping level of the p^+ layer is chosen such a way that it is fully depleted at zero bias. It may be noted that the barrier height can be altered by varying the doping level of p^+ layer. Figure 4.40(*b*) shows the variation of band diagram with bias. Under reverse bias condition there is significant lowering of the barrier. The dominant current transport mechanism of the device is still similar to the Schottky diode, *i.e.*, thermionic emission over the barrier. This type of device was first proposed by Shannon. Because of the nature of its band diagram, the device is called as *Camel diode*. The device is also known as *modulated barrier photodiode* due its barrier modulation possibilities.

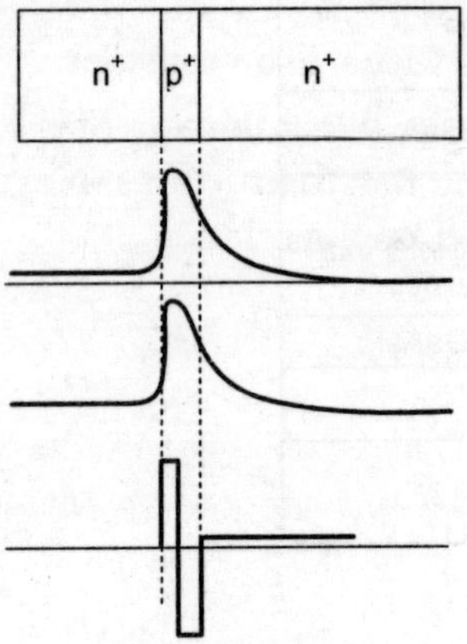

Fig. 4.40 (*a*)

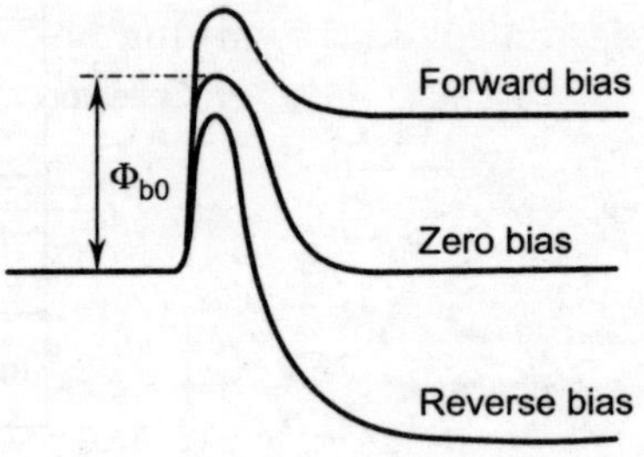

Fig. 4.40 (*b*)

Planner doped barrier or *triangular barrier diode* can be fabricated with similar configuration, in which the barrier is placed in the i-region of an n^+-i-n^+ structure with heavily doped p^+ layer. The schematic and band diagram of a triangular barrier diode are shown in Fig. 4.41. The p^+-region is fully depleted and the negative space charge of the ionized acceptors is balanced by the positive space charge of the *i*-regions. Using Poission equation with appropriate charge densities, it can be derived that the bias dependent barrier heights to electron flow from left to right and vice-versa are given respectively, by

$$\Phi_{b1} = \Phi_{b0} + \frac{d_1}{d_1 + d_2} V \qquad (4.126)$$

and

$$\Phi_{b2} = \Phi_{b0} + \frac{d_1}{d_1 + d_2} V \qquad (4.127)$$

where,

$$\Phi_{b0} = \frac{q N_A{}^S d_1 d_2}{\varepsilon_s \varepsilon_0 (d_1 + d_2)} \text{ volts} \qquad (4.128)$$

Φ_{b0} is the barrier height at zero bias, V is the applied bias, and $N_A{}^S$ is the sheet acceptor density in the p^+ layer in cm^{-2}.

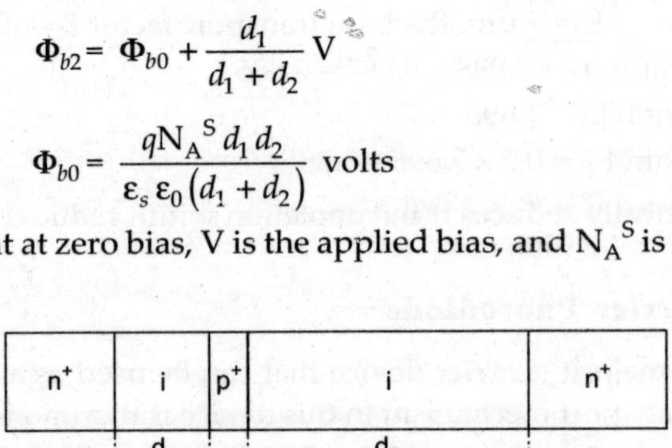

Fig. 4.41

If the barrier potential exceeds several times of $k_B T/q$, then the carrier transport over the resultant triangular potential barrier is by thermionic emission and the dark current density can be expressed as

$$J_{ed} = A^{**}T^2 \exp\left(-\frac{q\Phi_{b0}}{k_B T}\right)\left(\exp\left(\frac{q\alpha_2 V}{k_B T}\right) - \exp\left(\frac{q\alpha_1 V}{k_B T}\right)\right) \qquad (4.129)$$

where $\alpha_1 = d_1/(d_1 + d_2)$, $\alpha_2 = d_2/(d_1 + d_2)$ and A^{**} is the effective Richardson constant.

There are some unique features of the planner-doped barrier. First, the zero bias barrier height Φ_{b0} can be controlled by varying N_A^S, d_1 and d_2, or by using heterostructures. Next, the capacitance of the diode is independent of bias, which is estimated by the relation

$$C = \frac{\varepsilon_s \varepsilon_0 A}{d_1 + d_2} \qquad (4.130)$$

where A is the area of the diode. Finally, it may be noted that carrier injection in this device occurs in both the directions over the barrier. The parameters α_1 and α_2 are like diode ideality factors. It may also be noted that if $d_2 >> d_1$, the equation (4.129) reduces to that for a Schottky diode.

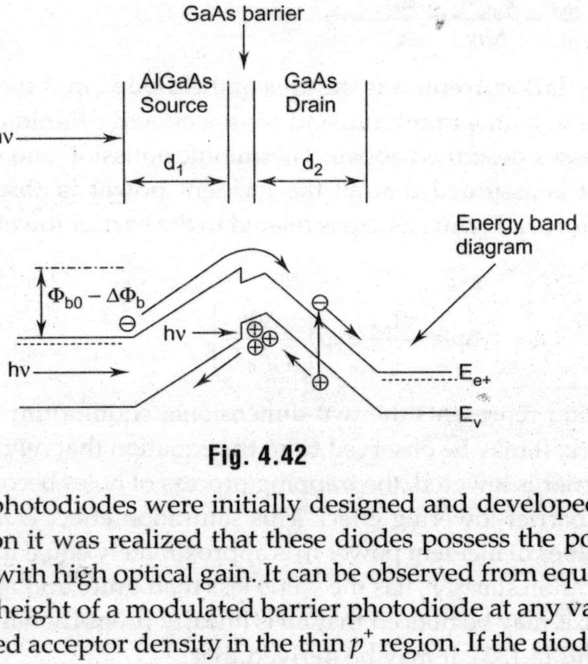

Fig. 4.42

The modulated barrier photodiodes were initially designed and developed for the rectification applications. However, soon it was realized that these diodes possess the potential to be used as very efficient photodiodes with high optical gain. It can be observed from equations (4.126), (4.127) and (4.128) that the barrier height of a modulated barrier photodiode at any value of applied bias is directly related to the ionized acceptor density in the thin p^+ region. If the diode is excited with light of appropriate wavelength, then it can be absorbed at the barrier region, producing electron-hole pairs. Referring to the Fig. 4.40, the electrons are swept away by the field, while the holes are drawn to the potential minimum at the p^+ region. The positively charged holes accumulate in the barrier region and neutralize some of the negative space charge, which results in the reduction of the barrier height. As a result, there is a large increase in current due to thermionic emission over the barrier, where the current is related exponentially to the barrier height. The increase in the current implies that there is the possibility of realization of an optical detector with a large internal gain.

Figure 4.42 represents the scheme of a device with its band diagram, where AlGaAs layer is used on the side of excitation by external illumination. This layer acts as a window layer for the photoexcitation, ensuring that most of the incident radiation is absorbed in the barrier region. The layer also serves the other purpose of formation of a more efficient well region for the holes to be trapped, thereby enhancing the barrier-lowering effect.

To analyze the photoresponse of the device, let us consider that the photoexcitation incident from left, of wavelength corresponding to bandgap of GaAs, will be mainly absorbed in the drain GaAs layer, since the thickness of the barrier GaAs layer is very small (=100°A). The absorbed photons will generate electron-hole pairs and the holes will also drift toward the potential minimum (for holes), beyond which they have to overcome a potential barrier of height ($\Phi_{b0} - \Delta\Phi_b$), where $\Delta\Phi_b$ is the barrier lowering due to the accumulation of photogenerated holes at the potential minimum. The trapped holes can be removed by a variety of processes, but each of the processes has a small probability. They can be removed by recombination with electrons, but the probability of this mechanism is very small, since electrons spend a very small amount of time in the barrier region. Holes can escape by overcoming the potential barrier by thermionic emission, or by diffusion across the source region. At the steady state condition, the rate of holes moving into the barrier region must be equal to the rate of holes moving out of the potential minimum. This can be expressed as

$$\frac{J_{hd}}{q} + \frac{P_{inc}\eta}{Ah\nu} = \frac{\Delta p}{\tau} \tag{4.131}$$

where J_{hd} is the hole dark current, A is the area of the diode, Δp is the two-dimensional density of holes trapped in the potential minimum and τ is the effective lifetime of the holes, determined by all the escape processes described above. Thermionic emission and diffusion are the most likely escape processes. It is assumed that all the incident power is absorbed in the device and the accumulated holes form a Fermi gas. Δp is related to the barrier lowering and can be expressed by the equation

$$\Delta p = \frac{\tau J_{hd}}{q} \exp\left(\frac{q\,\Delta\Phi_b}{n_f k_B T}\right) \tag{4.132}$$

The above equation represents the two-dimensional equilibrium hole density in the potential minimum in the dark. It may be observed from the equation that $\Delta\Phi_b$ will increase as Δp increases. However, when barrier is lowered, the trapping process of holes becomes less efficient and there is a saturation of the barrier-lowering effect. This saturation effect is accounted for by the ideality factor η_f. For low values of incident power η_f is approximately equal to unity. But at higher incident powers, when saturation starts, n_f has the value less than unity and approaches to zero. Also from the equation (4.132), it may be noticed that Δp is linearly proportional to $\Delta\Phi_b$, if $\Delta\Phi_b << k_B T/q$. From equations (4.131) and (4.132), it may be derived that

$$\Delta\Phi_b = \frac{n_f k_B T}{q} \ln\left(1 + \frac{P_{inc}\eta q}{AJ_{hd}h\nu}\right) \tag{4.133}$$

It may be noticed from the above expression of (4.133), that the amount of barrier-lowering can be expressed in terms of the incident power, and other parameters related to the device as well as the material. It may also be observed from the above equation, that barrier-lowering $\Delta\Phi_b$ is linearly proportional to incident power P_{inc}, for low values of P_{inc}, such that $P_{inc}\eta q/(AJ_{hd}h\nu) << 1$. On the other hand, when $P_{inc}\eta q/(AJ_{hd}h\nu) >> 1$, $\Delta\Phi_b$ is proportional to $\ln P_{inc}$. This implies the occurrence of saturation of barrier-lowering, which is due to the reduction in the trapping of holes at the barrier region.

Another important parameter of the device is its responsivity that is defined as

$$R = A(J_e - J_{ed})/P_{inc} \qquad (4.134)$$

where J_e and J_{ed} are the electron current densities under photoexcitation and in the dark, respectively. They are expressed as

$$J_e = A^{**}T^2 \exp\left(-\frac{q(\Phi_{b0} - \Delta\Phi_b)}{k_B T}\right)\exp\left(\frac{q\alpha_1 V}{k_B T}\right) \qquad (4.135)$$

and

$$J_{ed} = A^{**}T^2 \exp\left(-\frac{q\Delta\Phi_b}{k_B T}\right)\exp\left(\frac{q\alpha_1 V}{k_B T}\right) \qquad (4.136)$$

It is assumed that the currents are purely related to thermionic emission and the diodes in Figs. 4.41 and 4.42 are reverse biased (*i.e.*, the left hand side is biased negatively with respect to positive hand side). From equations (4.133), (4.134), (4.135) and (4.136), it may be deduced that

$$R = \frac{AJ_{ed}}{P_{inc}}\left(\left(1 + \frac{P_{inc}\eta q}{AJ_{hd} h\nu}\right)^{n_f} - 1\right) \qquad (4.137)$$

Earlier it has been mentioned that for low incident power level, $n_f = 1$, and $n_f < 1$ for higher incident power. From equation (4.137) above, it can be observed that responsivity R is nearly constant for very low incident power level. The maximum value of R at low power levels is

$$R_{max} = \frac{\eta q}{h\nu}\frac{J_{ed}}{J_{hd}} \qquad (4.138)$$

While at high power levels ($n_f < 1$), the responsivity decreases monotonically with incident power proportionately to $(P_{inc})^{n_f - 1}$.

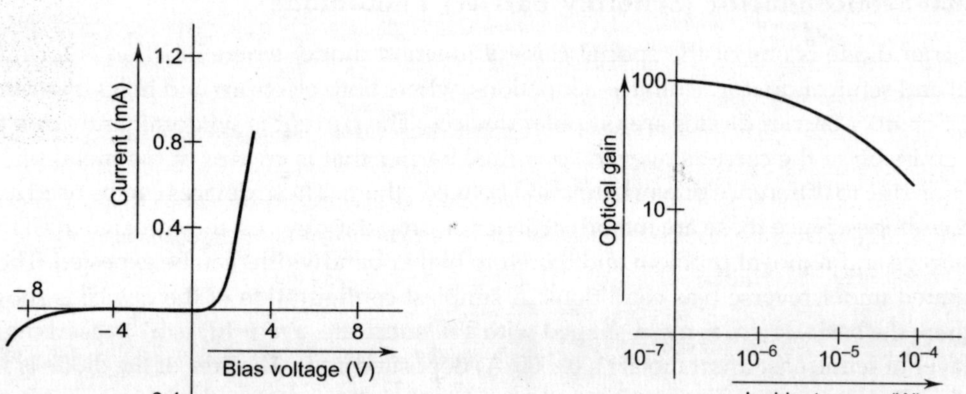

Fig. 4.43 (*a*) **Fig. 4.43 (*b*)**

The above analysis illustrates that the modulation barrier is an unique type of photodetection device, in which the responsivity or the optical gain increases with the decrease of incident optical power. This is contrary to the characteristics of a phototransistor, where responsivity decreases with the decrease of incident optical power. At low power levels, optical gain up to 100 can be achieved in the modulated barrier diodes. The current-voltage characteristics and the optical gain with respect to optical power is shown in Fig. 4.43, where decreasing optical gain with the increasing optical power may be observed.

It may be noted that a modulated barrier photodiode with majority carriers of holes and barrier of n^+ material, where electrons are in traps, possesses similar characteristics. The temporal response characteristics and bandwidth of the photodiode are determined by the intrinsic carrier transport mechanism and extrinsic circuit parameters. The temporal response of the device to a short pulse excitation in the order of few picoseconds is characterized by a very fast rise time excitation pulse and a slower fall time in the order of few hundreds of picoseconds. The slower fall time is due to the trapping of carriers in the barrier region. The device speed can be enhanced while operating at the lower gain.

The modulated barrier photodiode has several advantageous features. It has the potential of being operated at larger bandwidth systems compared to conventional phototransistor. This device does not have excess noise which is observed at avalanche process for avalanche photodiode. With its high gain at low incident power, this type of detector can play important role in low noise, low signal detection systems.

It may be relevant that the responsivity may be expressed in terms of barrier lowering $\Delta\Phi_b$. From equations (4.133) and (4.137)

$$R = \frac{AJ_{ed}}{P_{inc}}\left(\exp\left(\frac{q\Delta\Phi_b}{k_B T}\right) - 1\right) \tag{4.139}$$

Also the cross over power between low and high incident powers is given by

$$P^C_{inc} = \frac{AJ_{ed}h\nu}{nq} \tag{4.140}$$

4.7.3 Metal-Semiconductor (Schottky Barrier) Photodiode

Schottky barrier diode is one of the special class of junction diode, where junction is formed between metal and semiconductor. Unlike *p-n* junctions, where both electrons and holes contribute to the current, Schottky barrier diodes are unipolar devices. The current is predominantly due to the thermionic emission of the carriers over the potential barrier that is created at the metal-semiconductor junction due to difference of work function between them. These devices can be used as very efficient photodiodes. Since these are majority carrier or unipolar devices, they are free from minority carrier storage and removal problem and therefore higher bandwidths can be expected. They are usually operated under reverse bias conditions. A simplest configuration of the device is shown at Fig. 4.44, where the basic device is mesa-shaped with a n^+ substrate, a n^- (~10^{15} cm^{-3}) absorbing layer and a thin layer of semitransparent metal (100–300°A) deposited on it. The area of the diode is 10^{-3} to 10^{-5} cm^2.A thin dielectric antireflection coating is deposited on the metal, and large top contact pads are deposited through the opening in the same dielectric. Figure 4.44 shows a high-speed photodiode made on semi-insulating substrate. The equivalent circuit of the Schottky photodiode is similar to that of *p-i-n* diode as shown in Fig. 4.14, having the same circuit elements.

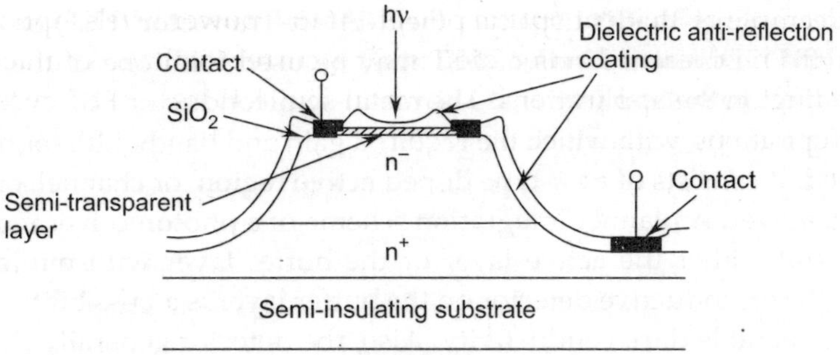

Fig. 4.44

The temporal response speed and bandwidth of the device are dependent upon the transit time of carriers through the absorption region and external circuit parameters. In high-speed diodes the absorption region is 0.2 to 0.5 μm thick, ensuring full depletion of this region even at low values of reverse bias and both electrons and holes travel at their respective saturation velocities. Approximately 1ps transit time is achieved under these conditions. Similarly, with optimized design of the diode and contact geometry, the time constant of the circuit can be made almost of the same order. The bandwidth of Schottky barrier diodes can be attained up to 100 GHz.

To analyze the photoresponse characteristics of the Schottky photodiode, let us consider a typical band diagram of InAs/GaAs heterojunction device as shown in the Fig. 4.45. The metal part of the Schottky diode is replaced by low bandgap semiconductor (InAs). At smaller values of photon energy, such that $q\Phi_b < h\nu < E_g$, electrons photoexcited in the low bandgap semiconductor (InAs) can overcome the barrier by thermionic emission, transit across the higher bandgap semiconductor (GaAs), and be collected at the ohmic contact. This process extends the spectral response of the diode to photon energies lower than the bandgap. However this is not desirable mode of operation from the point of view of device speed, as the thermionic process may be slow. But, with the use of semi-transparent low bandgap layer, photons are absorbed for $h\nu > E_g$. the electron-hole pair that are created by move in opposite directions with their respective saturation velocities and are collected. This is the most efficient mode of operation of Schottky diode and is similar to that of *p-i-n* diode. However, with comparison to the latter Schottky diode is simpler to operate. The entire InAs/GaAs heterojunction structure as described here can be realized by one-step epitaxy, which is a real advantage, and making contact to low bandgap *n*-type semiconductor is rather simple.

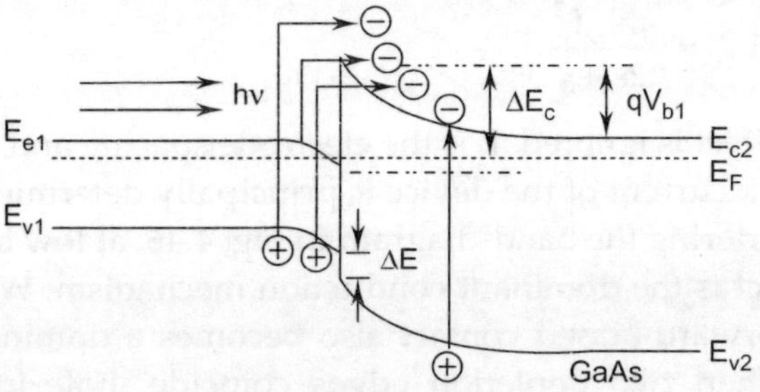

Fig. 4.45

4.7.4 Metal-Semiconductor-Metal (MSM) Photodiode

Optoelectronic integrated circuits are concerned to the integration of devices that involve the compatible fabrication and processing techniques. In photoreceivers, a photodetector is monolithically integrated with a preamplifier that is usually a Field-Effect-Transistor (FET) or a Heterojunction Bipolar Transistor (HBT). Different kinds of FET may be used with one of the several kinds of phodetectors depending on the applications. The metal-semiconductor FET or MESFET is one of the very simple configurations, with which the required gain and bandwidth for most applications can easily be obtained. It consists of an *n*-type doped active region, or channel, on an undoped or high-resistivity buffer layer. A planner integration scheme of a photoreceiver requires the design of a photodetector with either the active layer or the buffer layer with minimum number of processing steps. A photoconductive detector on the buffer layer is a possibility, but it is avoided due to higher than acceptable dark conductivity. Also, the detector generally shows a slow temporal response, when higher gain is required. A Metal-Semiconductor-Metal (MSM) photodiode is a suitable device for such applications.

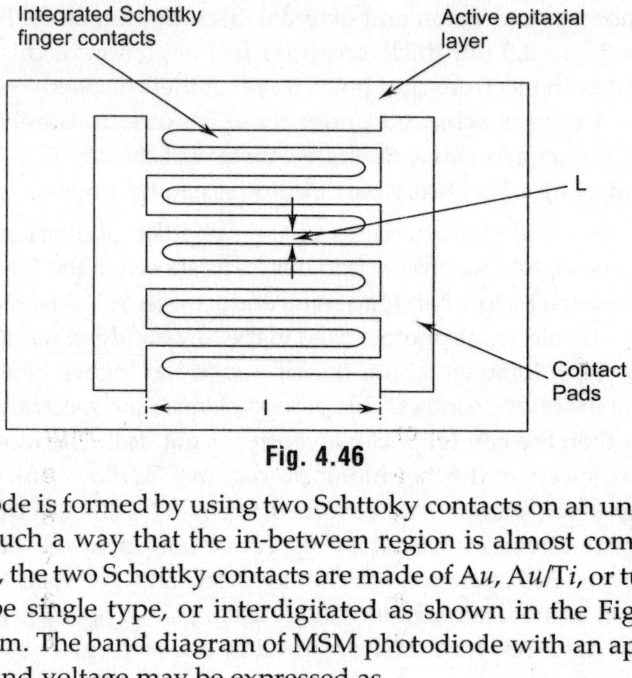

Fig. 4.46

The MSM photodiode is formed by using two Schttoky contacts on an undoped semiconductor layer. It is designed such a way that the in-between region is almost completely depleted. In a planner configuration, the two Schottky contacts are made of A*u*, A*u*/T*i*, or tungsten silicide (WSi). These contacts may be single type, or interdigitated as shown in the Fig. 4.46, with a contact spacing of 1 μm to 5 μm. The band diagram of MSM photodiode with an applied bias V is shown in Fig. 4.46. The flatband voltage may be expressed as

$$V_{FB} = \frac{qN_D L^2}{2\varepsilon_s \varepsilon_0} \tag{4.141}$$

Here two-dimensional effect is ignored. L is the electrode spacing and N_D is the donor density in the *n*-type layer. The dark current of the device is principally determined by thermionic emission over the barrier. Considering the band-diagram in Fig. 4.46, at low biases, electron injection at the reverse-biased contact is the dominant conduction mechanism. With the increase of bias, the hole injection at the forward-biased contact also becomes a dominant factor. After reach-through condition, *i.e.*, when two depletion edges coincide, hole injection becomes more dominant factor. The total current density under this condition may be estimated as

$$J = A_n^{**}T^2 \exp[-q(\Phi_{bn} - \Delta\Phi_{bn})/k_B T] + A_p^{**}T^2 \exp[-q(\Phi_{bp} - \Delta\Phi_{bp})/k_B T] \tag{4.142}$$

where A_n^{**} and A_p^{**} are the respective Richardson constants and $\Delta\Phi_{bn}$ and $\Delta\Phi_{bp}$ are the corresponding barrier height lowerings due to image force effect as shown in Fig. 4.47.

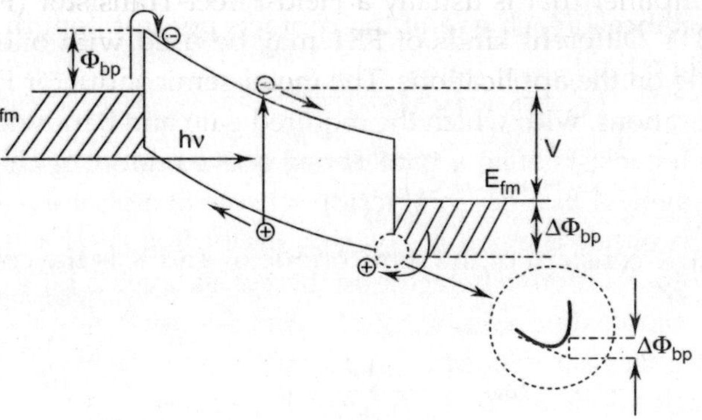

Fig. 4.47

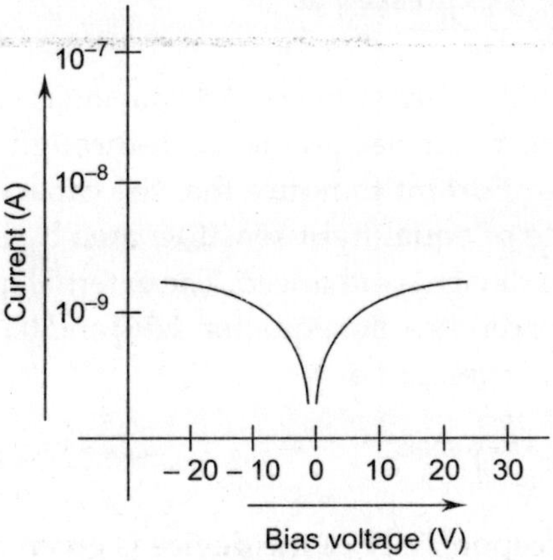

Fig. 4.48 (*a*)

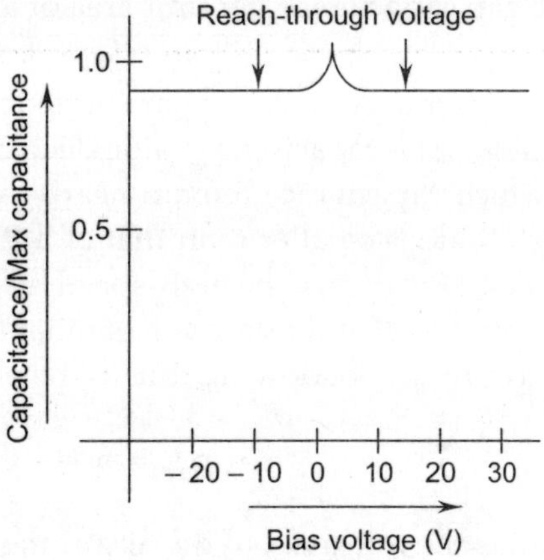

Fig. 4.48 (*b*)

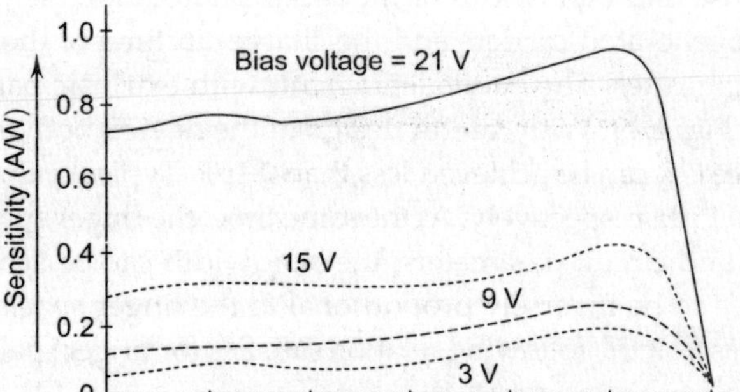

Fig. 4.48 (*c*)

The current-voltage characteristic of a GaAs metal-semiconductor-metal device is shown in Fig. 4.48(*a*). The dark current is in the order of 1nA or lower, which is comparable to *p-i-n* photodiode. The ideal capacitance for a source-drain configuration with contact or finger length l and finger spacing L as shown in Fig. 4.46 is calculated as

$$C = C_0 l \tag{4.143}$$

where C_0 is the capacitance at each gap of the contacts per unit length. Therefore, if there are N numbers of fingers, then total capacitance

$$C = (N-1)C_0 l \tag{4.144}$$

and

$$C_0 = \varepsilon_0 (1+\varepsilon_s) \frac{K(k)}{K(k')} \tag{4.145}$$

where ε_s is the dielectric constant of the semiconductor and K is the complete elliptic integral of the first kind with

$$k = \tan^{-1}\left(\frac{\pi W}{4(L+W)}\right) \tag{4.146}$$

and

$$k' = (1-k^2)^{1/2} \tag{4.147}$$

The gap capacitance per unit area of an MAM diode is expressed as

$$C_A = C_0 / (L+W) \tag{4.148}$$

The capacitance-voltage characteristic of a GaAs device with 3 μm contact spacing is shown in Fig. 4.48(*b*). The capacitance value decreases for both the polarities till the reach-through voltage, after which capacitance remain nearly constant. It is important to notice that the capacitance of a MSM diode is smaller than that of a PIN photodiode of equal light sensitive area by almost a factor of 4. Due to this, the high-speed response of MSM device is enhanced. The external quantum efficiency of a MSM diode is very similar to that of a *p-i-n* diode (refer equation 4.56) and has to take into account the shadowing due to the electrodes. It is expressed as

$$\eta_{ext} = \eta_i (1-\theta_R) \frac{L}{L+W} \left(1-e^{-\alpha d}\right) \tag{4.149}$$

where *d* is the thickness of the absorbing region. The responsivity of the device is given by

$$R = q\eta_{ext} / h\nu \tag{4.150}$$

The frequency response and bandwidth of the MSM photodiode are governed primarily by the transit time of the photogenerated carriers and the charge-up time of the diode. The latter is determined by the extrinsic parasites. The equivalent circuit with extrinsic parasitic elements of a MSM photodiode is shown in Fig. 4.49. With careful design of the device, both the parasitic capacitance C_p and the diode capacitance C_{pd} can be achieved less than 0.1pF. Typical values of R_S, R_D, R_L, and L_S are 50 Ω, 100 MΩ, 50 Ω and 0.1 nH respectively. At the same time, the finger spacing can be made less than 1μm. With these device and circuit parameters, the bandwidth can be achieved up to 20 to 50 GHz. Bandwidth Δf is expected to be inversely proportional to the finger spacing L and it is observed for GaAs and InGaAs devices that the following relation satisfies for finger spacing between 2 μm to 4 μm.

$$L(\Delta f) = \frac{14.17}{\Delta f - 0.314} \tag{4.151}$$

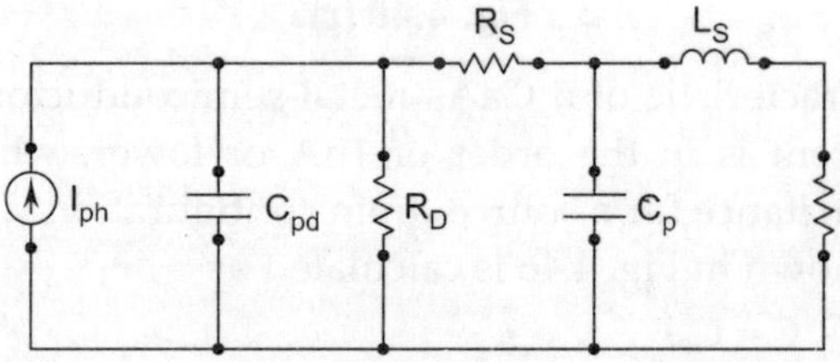

Fig. 4.49

It is also important to analyze the photoresponse characteristics of the device. The spectral response behavior is shown in Fig. 4.48(*c*). The increase in the responsivity with bias voltage indicates that the MSM diode may have internal gain, even at low bias values. There may be several reasons for gain at low bias. A photoconductive gain mechanism can be operative due to traps or surface defects having long lifetime or due to accumulation of electrons at the conduction minimum point lowering the barrier height for hole transit.

MSM photodiode is a versatile device that has all the desirable characteristics of a practical photodetector and can be fabricated by simple and planner integration techniques. Gain is obtained at a low dark current and it has a large bandwidth, suitable to be operated in the range of 1.3-1.6 μm. $In_{0.53}Ga_{0.47}As$ can also be used, in which a thin surface layer of $In_{0.52}Al_{0.48}As$ may be used to enhance the Schottky barrier height.

4.7.5 Detectors for Long Wavelength Applications

Detection capability is also desired at far-infrared (5–20 μm) operation, where a variety of applications are found. One successful way in such long-wave detection is to use a junction diode and band-to-band transitions in small bandgap materials. Examples are the antimony bearing III-V compounds, which have the lowest bandgaps and the II-VI alloys such as HgCdTe. Use of antimonides generally requires the use of strained semiconductors due to lack of suitable lattice-matched semiconductors. $InAs_{0.5}Sb_{0.5}$ has the lowest bandgap among III-V compounds. However, small bandgap materials with same properties produce a 'soft' material, which is very difficult to grow and process. The dark current in the device is very large due to small bandgap, which necessitates their operation at cryogenic temperatures. In spite of such difficulties, impressive progress in the design fabrication of long-wave detectors has been observed and detectivities D* is achieved in the range of 10^{10} $cmHz^{1/2}W^{-1}$ in these devices.

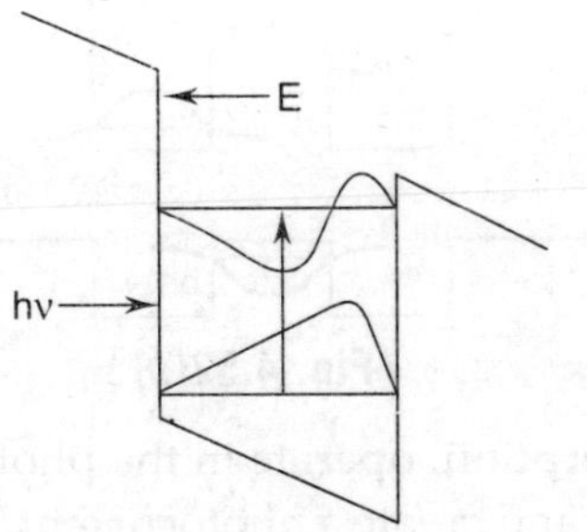

Fig. 4.50(*a*)

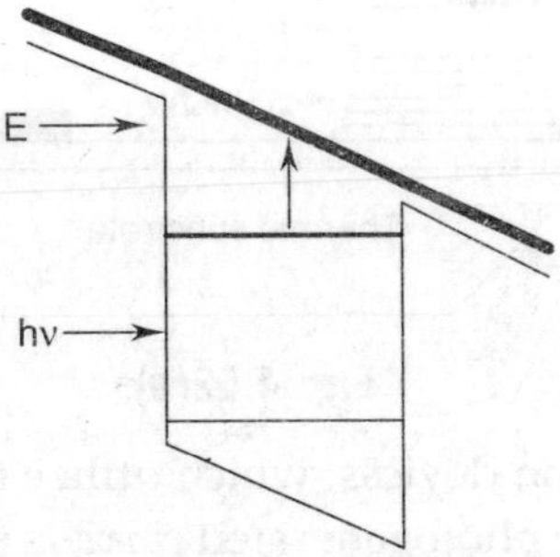

Fig. 4.50(*b*)

Another approach for designing detectors for long wave applications is schematically shown in Fig. 4.50(*a*). Transitions between the bound states in a quantum well are made use of. The absorption of photons having energy equal to the intersubband separation causes the transition of carriers from the ground state to the first excited state. For III–V compounds quantum wells of width of L_z = 100°A, the intersubband energy separation is in the order of 100 – 200 meV. The absorbance and intersubband transition energy measured in pseudomorphic $In_xGa_{1-x}As$/ $Al_{0.4}Ga_{0.6}As$ multiquantum wells are shown in Figs. 4.51(*a*) and 4.51(*b*). Two important points are to be noted. First, the oscillator strength of a transition from m^{th} to $(m+1)^{th}$ subband increases with the value of *m*, which is expressed as

$$f_{isb} = \frac{m_0}{m^*}\frac{64}{\pi^2}\frac{m^2(m+1)^2}{\left[(m+1)^2 - m^2\right]^3} \tag{4.152}$$

Secondly, the selection rules only allow absorption of electromagnetic radiations when the incident polarization is parallel to the growth (confinement) direction. This may cause difficulties in detecting a 2-D image, as radiation has to be absorbed in a waveguide geometry. However, illumination geometries as shown in the Fig. 4.52(*a*) can prevent his problem. Also, in the quasi-type II superlattice like InAs/ AlGaSb, whose equilibrium band diagram is shown in Fig. 4.52(*b*), the selection rules allow absorption of radiation with polarization parallel to the growth direction.

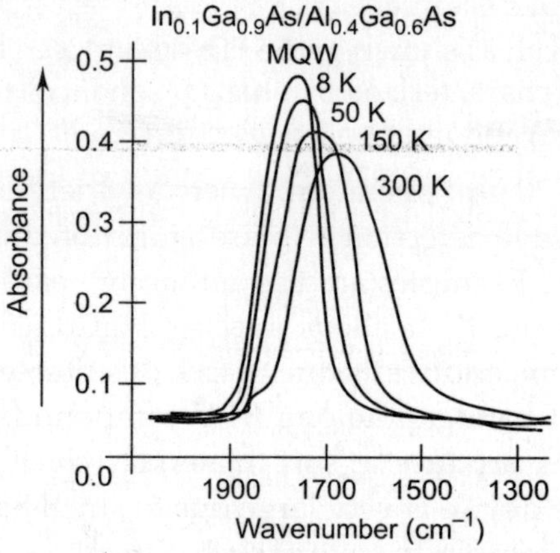

Fig. 4.51(*a*)

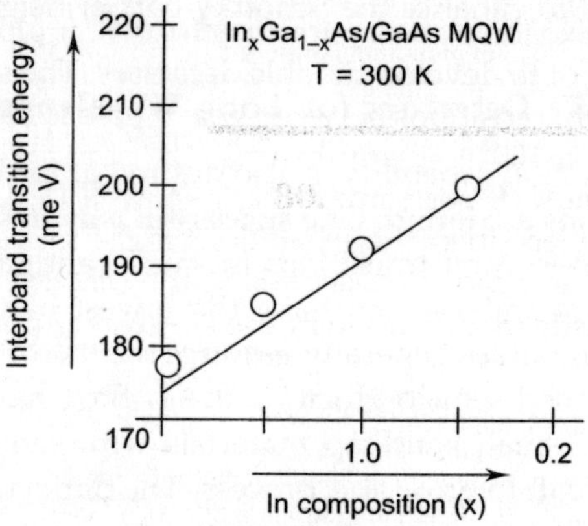

Fig. 4.51(*b*)

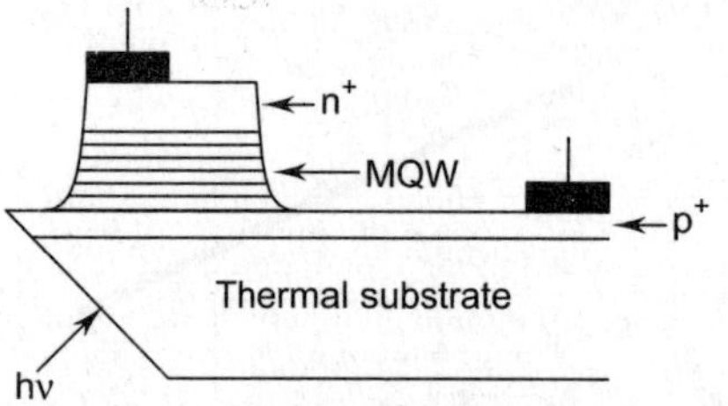

Fig. 4.52(*a*)

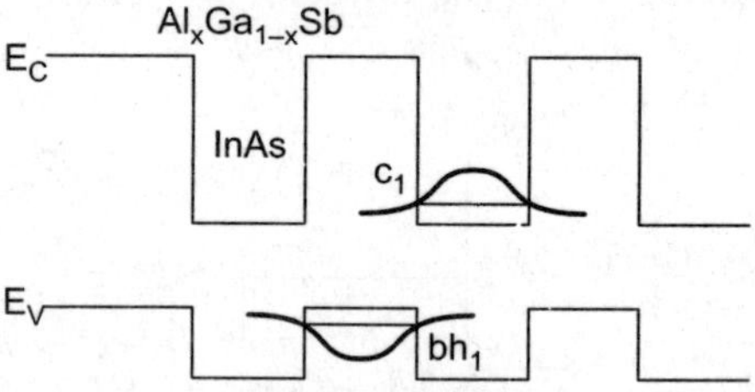

Fig. 4.52(*b*)

Detection devices, which utilize the intersubband absorption, operate in the photoconductive mode. The photogenerated carriers tunnel out of the wells and create a photocurrent. However, in the process of making this tunneling efficient, the tunneling of residual carriers is enhanced, which in turn increases the dark current. Such tunneling can be reduced by careful design of the barriers. With the utilization of photon absorption and carrier transitions between a bound state and the continuum outside the well, significant improvement has been observed in the performance of these devices. This scheme is illustrated in Fig. 4.50(*b*), where longer wavelength can be absorbed. With the use of small barrier heights, the absorption can be extended to even longer wavelengths. Additionally, as the photoexcited carriers are raised to the continuum, the bandwidth of the absorption as well as the detector responsivity is enhanced. The collection of the photocarriers in these detectors does not require any tunneling process, at the same time, the dark current can be reduced by using thick barriers.

4.7.6 Wavelength Selective Detection

Nowaday optical fibers are widely used for communication purpose, replacing conducting cables to utilize their enormous bandwidth capacity, which translate to an enhancement in the number of communication channels. Bandwidth of several hundreds of GHz can be handled by optical fibers in comparison to several hundreds of MHz of bandwidth in conducting cables. It is necessary to be able to selectively detect the information coming in at certain frequency or wavelength among the information of other frequencies or wavelengths coming in same communication channel. None of the detectors that are discussed so far has the ability of such selective detection of particular wavelength in their spectral response characteristics. Use of tuned filters and the use of coherent detection schemes overcome this difficulty. The use of the detectors with wavelength selectivity is important as well as feasible, and plays important role in system design.

Excitonic absorption spectrum of multiquantum well as shown in the Fig. 4.53(*a*), can be made use of to develop tunable detectors. The spectrum is characterized by sharp resonances resulting from transitions between heavy hole to electron ($e - hh$) and light hole to electron ($e - lh$) subbands. For a 100°A quantum well, the separation between such resonances is usually in the order of 10meV. If such multiquantum well (MQM) is incorporated in *p-i-n* diode, and reverse bias is applied, then the excitation peaks move to lower energies as shown in the Fig. 4.53(*b*). This phenomenon is described as the *Quantum Confined Stark Effect* (QCSE) and arises mainly due to a drastic modification of the electron and hole wavefunctions in the well region.

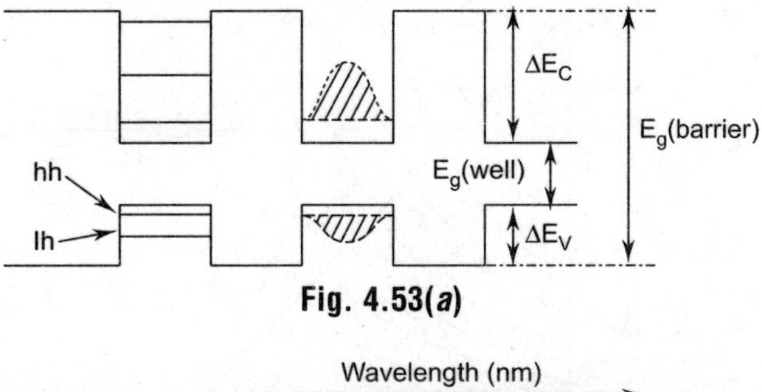

Fig. 4.53(*a*)

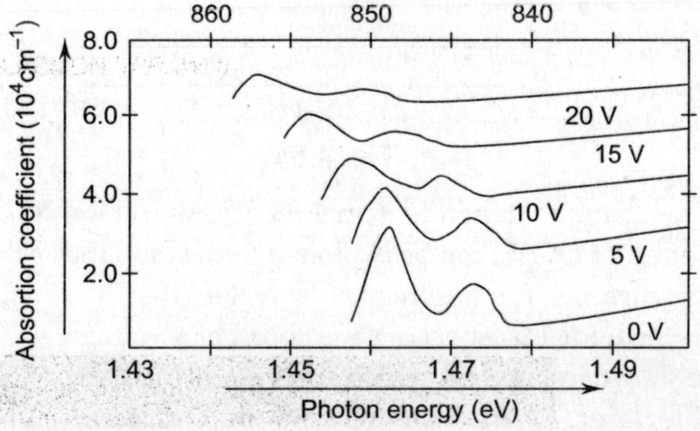

Fig. 4.53(*b*)

Figure 4.54, schematically describes the photocurrent-voltage characteristics of the diode. With illumination of energy slightly lower than that of HH peak in the absorption spectrum. At very low values of bias, the photocurrent is mainly due to recombination of photogenerated electrons and holes in the wells. With the increase in bias, the current is produced due to tunneling of photogenerated carriers out of the wells and their transit across the MQW region and eventual collection. Due to QCSE or the shift in absorption spectrum with bias, the photocurrent-voltage characteristics exhibit peaks similar to that of the absorption spectrum. The operation of the *p-i-n* diode with (MQW) is similar to that of a *p-i-n* diode, except that the photocurrent exhibits a Negative Differential Resistance (NDR). The QCSE and NDR effects can be exploited to develop efficient modulation and switching devices.

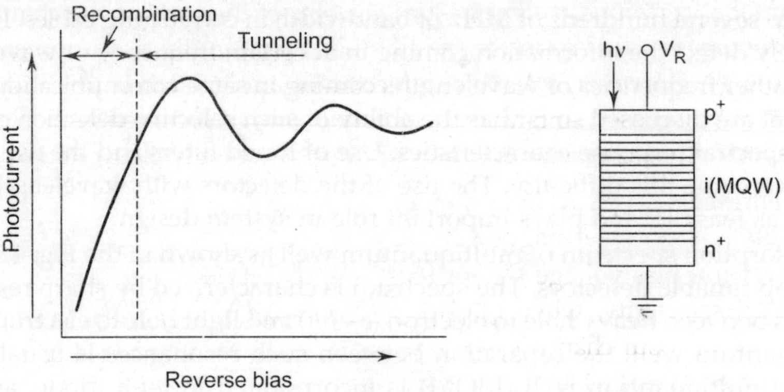

Fig. 4.54

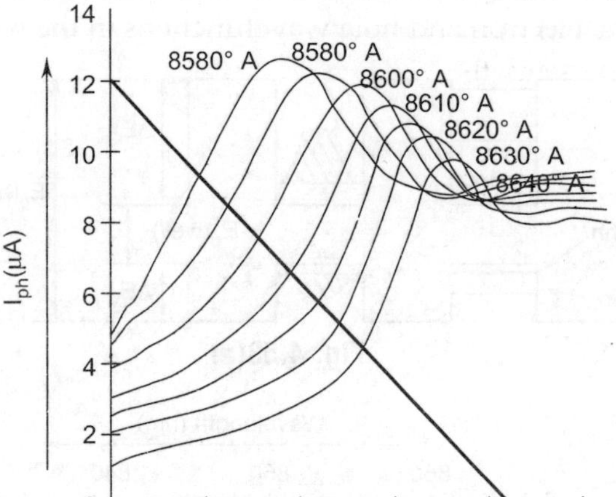

Fig. 4.55

The very strong wavelength dependent photocurrent-voltage characteristics, due the effect of QCSE, can be exploited to develop the tunable detectors. Photocurrents for a series of wavelengths of a reverse-biased *p-i* (MQW)-*n* diode in respect of variation of bias voltage at room temperature are shown in Fig. 4.56, which are of equal intensity generated by a tunable laser. It may be noticed that at room temperature, at a given bias condition the photocurrent is distinct for different wavelengths. With proper external load, the voltage across the *p-i-n* diode is available, which is strongly wavelength dependence (the load line is shown in the Fig. 4.56. It may be observed that a change in input wavelength of 100°A produces a change in output voltage of about 10V. For serial input information data, the output reference voltage in the circuit provides a very selective determination of certain

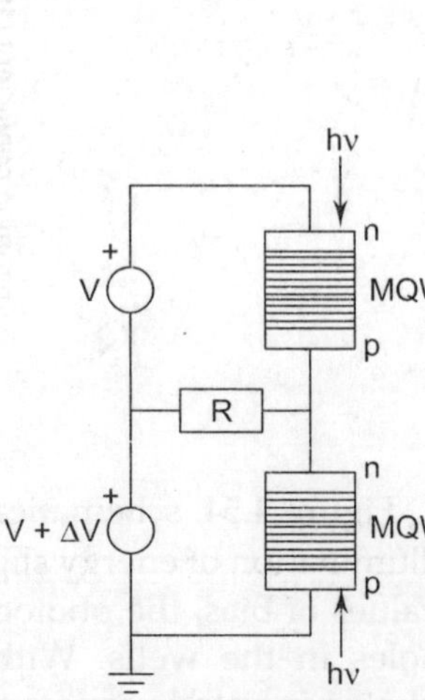

Fig. 4.56

wavelength. Thus the scheme enables the detection up to 50 wavelengths channels serially, with the variation of output reference voltage approximately by 0.2V.

In the scheme described as above, the serial detection of a given wavelength is possible, which is useful to many applications. But, this scheme is not suitable for detecting information selectively in a channel at certain wavelength λ_i regardless of the state of the channels at wavelength λ_j ($j \neq i$), as photocurrents corresponding to all other wavelengths are summed up at any particular reference voltage. However, selective detection of information is possible, if two *p-i-n* structures are biased at voltages V_i and $V_i + \Delta V$, where a simple subtraction of the currents provides the value of $\Delta I_{ph}/\Delta V$. The schematic diagram of this operation is illustrated in Fig. 4.56.

Figure 4.57 represents a typical plot of values of $\Delta I_{ph}/\Delta V$ at 220°K at the wavelength 8350°K with respect to reverse bias voltage V_R, for the scheme of parallel selectivity, when information is received through four channels at wavelengths 8350, 8370, 8410 and 8430°K. The light curves represent some of the results of $\Delta I_{ph}/\Delta V$ for a selected wavelength λ_i = 8350°K, depending on the states (on or off) of other channels, while the bold curve represents the value of $\Delta I_{ph}/\Delta V$, while only the channel λ_i = 8350°K is on. Therefore, if two MQWs are biased at the bias voltages at the region as indicated by the arrow mark (↑) in the curve of Fig. 4.57, with a slight difference between them, the information carried by the wavelength λ_i = 8350°A will be detected regardless of the states of the other channels.

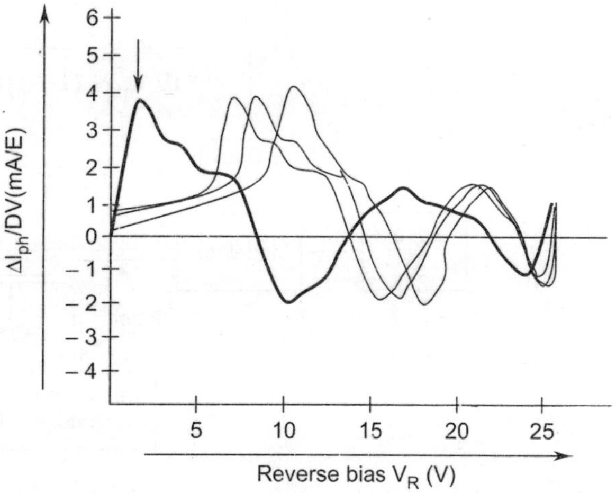

Fig. 4.57

Operation at low temperature improves the selectivity and hence increased number of channels can be simultaneously decoded. It may also be noted that the negative resistance region of the photocurrent-voltage characteristics causes the degree of nonselectivity in this scheme. If this region is not present, the selectivity can be increased to many more channels.

4.7.7 Coherent Detection

In earlier sections various types of detectors are described, where optical signal is converted to electrical signal in the device. This type of detection is known as direct detection. Although such detection is theoretically very sensitive, the actual sensitivity in the practical system is too far from the theoretically predicted values. In the radio communication system, coherent receiving technique of *heterodyning* type is commonly used. This technique recently has gained importance in the field of optical communication. In coherent detection, a weak received signal is mixed with a strong local oscillator wave of very close frequency, resulting in an effective signal gain through coherent phase interference. The gain is proportional to local oscillator amplitude. A schematic diagram of coherent detection technique is illustrated in Fig. 4.58. At the receiving side, the information modulated by the carrier signal is mixed with the local oscillator wave. A photodiode detects the composite wave and produces an electrical signal that is identical to the modulating information signal. Sharp electronic filters operating at beat frequency are used to selectively pick up electrical output of low frequency and amplified by high performance electronic circuitry. This gives wavelength selectivity, which enhances the channel density in the communication system.

However, there are some inherent problems in this scheme of detection. The optical gain obtained in a coherent system is very important, as it increases the sensitivity. Unfortunately, when the local oscillator signal power is increased, the shot noise in the detectors is also increased and this shot noise may be greater than the signal gain. Therefore, an optimum power of local oscillator must be used so that the highest signal to noise ratio as well as the maximum receiver sensitivity is achieved. A balanced detection configuration, as shown in the dashed box in Fig. 4.58, overcomes this problem. In balanced detector, two identical detectors are incorporated to enhance the beat signal, while suppressing the local oscillator noise. The signal to noise ratio of a balanced detection scheme can be expressed as

$$\frac{S}{N} = \frac{(1/h\omega)\tau(1-\tau)(\eta_1+\eta_2)^2 E_s^{\,2}}{B\left[\tau_{\eta 1} + (1-\tau)\eta_2 + (q\gamma_N/h\omega)E_L^{\,2}\left(\eta_1\tau - \eta_2(1-\tau)^2\right)\right]} \tag{4.153}$$

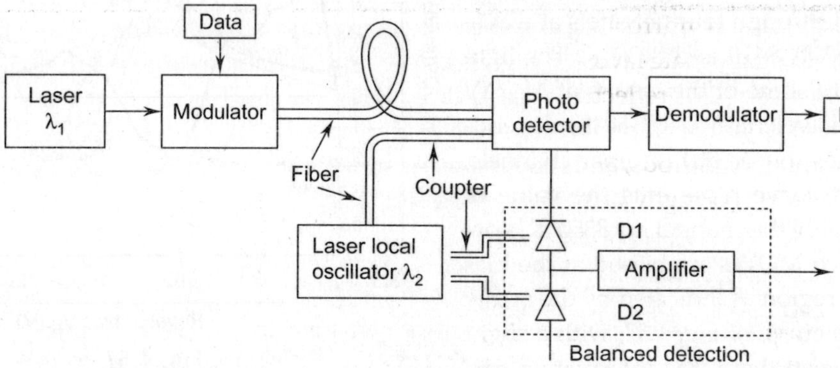

Fig. 4.58

where τ and $(1-\tau)$ respectively are the power and the reflection coefficients of the coupler, η_1 and η_2 are the quantum efficiencies of the two detectors, E_S and E_L are the signal amplitude and the local oscillator field amplitude, B is the bandwidth of the receiver, and γ_N is the parameter related to the intensity fluctuation of the local oscillator. γ_N has the unit of inverse of current and typical values of it ranges from 10^2 to 10^9 A^{-1}. For a special case, when quantum efficiencies of both detectors are equal, *i.e.*, $\eta_1 = \eta_2 = \eta$ and $T = 0.5$, then the expression for signal to noise ratio of the coherent heterodyne system reduces to

$$\frac{S}{N} = \frac{RE_S^{\,2}}{qB} \tag{4.154}$$

Where R is the responsivity of the detector.

The MSM or *p-i-n* diodes may be employed in the coherent detection system. The advantage of balanced detection in a coherent receiver system is its noise cancellation characteristics and it plays major role in the long distance optical communication due to its high sensitivity and selectivity. However, a stable local oscillator is required for heterodyne coherent receiver, and also required are the number of components, hybrid and monolithically integrated.

4.7.8 Microcavity Photodiodes

We have seen that the optical communication system is capable of high-speed data transmission. But this requires high-speed photodetectors. We have earlier also observed that there is an optimization between the quantum efficiency (responsivity) and bandwidth of a photodiode. Higher value of quantum efficiency requires the use of thick absorption layer (approximately 2 μm), whereas large bandwidth requires thin layer of absorption region (less than 0.5 μm). This contradictory situation can be avoided by the use of microcavity diodes as shown in the Fig. 4.59.

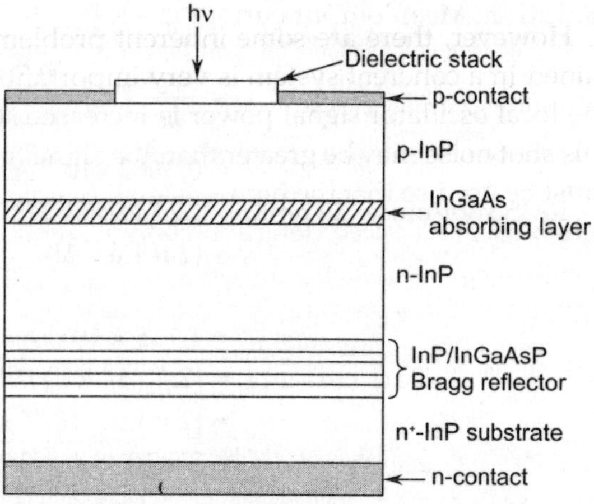

Fig. 4.59

In this type of diode configuration, a thin low bandgap absorption region is placed in the middle of a resonant cavity, which is formed by heavily doped wider bandgap regions and reflecting mirrors. The bottom-reflecting mirror is formed by a $\lambda/4$ — superlattice Distributed Bragg Reflector (DRR), while the upper mirror is formed by dielectric film or stack. The $\lambda/4$ — superlattice Reflector is made up of alternate layers of high and low refractive index, and the thickness of each layer is equal to $\lambda/4$. the light reflected at each layer interacts constructively at the surface. The reflective coefficient of entire superlattice reflector depends on the difference in the refractive indices and the number of periods, and it can be made quite large. The superlattice parameters are related to the design wavelength λ_0 by the expression

$$\lambda_0 = 2\,(n_{rH}\,d_H + n_{rL}\,d_L) \tag{4.155}$$

where d_H and d_L are the thickness of the layers of high and low refractive indices respectively.

The bandwidth and reflectance of the superlattice stack are respectively given by

$$\Delta\lambda = \frac{4}{\pi}\,\frac{n_{rH} - n_{rL}}{n_{rH} + n_{rL}}\,\lambda_0 \tag{4.156}$$

and

$$\theta_R = \left(\frac{1 - n_r^*}{1 + n_r^*}\right)^2 \tag{4.157}$$

where,

$$n_r^* = (n_{rL} / n_{rH})^{2p+1} \tag{4.158}$$

Here p is the number of periods, which is also related to the number of layers. The equation (4.158) holds good when the superlattice stack of the Fig. 4.59 is formed with $(p + 1)$ layers of InGaAs and p numbers of layers InP, each of which is $\lambda/4$ thick. These equations are valid for normal incidence. With GaAs/AlGaAs and InP/InGaAsP superlattices the reflectance can be achieved near to unity. In the InP cavity, as shown in the Fig. 4.59, a resonant is built up at those frequency components of the incoming light for which the reflected light have the phase shift of multiple of 2π. Therefore, at resonance frequency, the incoming light is reflected at two mirrors and the round-trip path is greatly enhanced, and so the absorption as well as the quantum efficiency. However, the transit length remains small as the photogenerated carriers only have to traverse the thickness of the absorption region to be collected. Hence it is possible to achieve large bandwidth as well as the high quantum efficiency in the microcavity photodiodes.

Example 4.14. *An InGaAs p-i-n photoconducting detector is operating at room temperature (300°K) at wavelength of 1.3 μm. Its quantum efficiency is 70% and incident optical power is 500 nW. Assuming the primary dark current as 5 nA, R_L as 1 KΩ and effective bandwidth of 25 MHz, calculate the rms value of shot noise current, dark current, thermal noise current and signal to noise ratio (S/N) at the input end of the amplifier of the receiver.*

Solution: Mean output current $I_p = R P_{in}$

$$= \frac{\eta q \lambda}{hc} P_{inc} = \frac{0.7 \times 1.6 \times 10^{-19} \times 1.3 \times 10^{-6}}{6.626 \times 10^{-34} \times 3 \times 10^{8}} \times 500 \times 10^{-9}$$

$$= 0.3663 \times 10^{-6}\ \text{A} = 0.3663\ \mu\text{A}$$

RMS shot noise current $i_s = [2qI_p(\Delta f)\,M^2F(M)]^{1/2}$

$$= [2 \times 1.6 \times 10^{-19} \times 0.3663 \times 10^{-6} \times 25 \times 10^{6} \times 1 \times 1]^{1/2}\ \text{A} \quad [M = 1 \text{ and } F(M)] = 1]$$

$$= 1.715 \times 10^{-9}\ \text{A} = 1.715\ n\text{A}$$

Dark current $i_d = [2qI_d(\Delta f)\,M^2F(M)]^{1/2}$

$$= [2 \times 1.6 \times 10^{-19} \times 5 \times 10^{-9} \times 25 \times 10^{6} \times 1 \times 1]^{1/2}\ \text{A}\ [M= 1 \text{ and } F(M)] = 1]$$

$$= 2 \times 10^{-10}\ \text{A} = 0.2\ n\text{A}$$

Thermal noise current $i_T = \left(\frac{4KT(\Delta f)}{R_L}\right)^{1/2} = \frac{4 \times 1.38 \times 10^{-23} \times 300 \times 25 \times 10^{6}}{1000}\ \text{A}$

$$= 20.34 \times 10^{-9}\ n\text{A} = 20.34\ n\text{A}$$

Sum of the mean square noise currents = $(1.715\ n\text{A})^2 + (0.2\ n\text{A})^2 + (20.34\ n\text{A})^2 = 416.7 \times 10^{-18}$

Signal to noise ratio (S/N) = Photocurrent$(I_p)^2 R_L$/Total noise current$^2 R_L$

$$= (0.3663 \times 10^{-6})^2 / 416.7 \times 10^{-180} = 322.$$

4.8 CONCLUSION

Different types of detectors and their constructions are discussed in this chapter. Minimum detectable time-averaged signal power is the primary factor for selection of detector. Photoconductors are simple to fabricate and easy to operate. But they are not suitable for high performance telecommunication applications for their higher dark current and Johnson noise. The *p-i-n* diode as well as the avalanche photodiode delivers good combination of bandwidth and sensitivity. With proper design, the response speed of the avalanche photodiode nearly approaches to that of a *p-i-n* diode. At the present time, the separate-absorption-graded-multiplication avalanche photodiode (SAGM-APD) exhibits the best performance. But it is difficult to make compatible to the integrated circuit design due to its higher bias requirement. Various detection schemes are also described in this chapter. Depending on the system applications they are chosen.

REVIEW QUESTIONS

1. Practically, what should be the energy of incident photon in order to create an electron-hole pair in a *p-n* diode?
   (*a*) Less than E_g (*b*) Equal to E_g
   (*c*) Greater than E_g (*d*) Much greater than E_g
2. What is the approximate bandgap of Si as it can absorb the highest wavelength of 1.12 µm?
   (*a*) 1.1 eV (*b*) 1.4 eV (*c*) 1.74 eV (*d*) 2.3 eV
3. What is the maximum wavelength that can be absorbed by Germenium (Ge) detector of bandgap of 0.67 eV?
   (*a*) 7080 nm (*b*) 4560 nm (*c*) 1850 nm (*d*) 1100 nm

4. If a *p-n* photodiode, on an average, generates one electron-hole pair per five incident photons at a wavelength of 900 nm, what is the quantum efficiency with the assumption that all the photo generated electrons are collected?
   (*a*) 20% (*b*) 30% (*c*) 40% (*d*) 50%
5. What is the responsivity of the *p-i-n* photodiode, when photons of wavelength of 0.85 μm are incident on it at the rate of 4×10^{10}/s and collection rate of electrons at the terminals of the diode is 2×10^{10}/s?
   (*a*) 0.15 A/W (*b*) 0.23 A/W (*c*) 0.34 A/W (*d*) 0.50 A/W
6. Which of the following detectors have amplified output?
   (*a*) *p-n* photodiode (*b*) *p-i-n* photodiode
   (*c*) Avalanche photodiode (*d*) Photovoltaic detector
7. What is photocurrent output at incident wavelength of 1 μm for a *p-i-n* photodiode of responsivity of 0.50 A/W with output power of 0.2 μW?
   (*a*) 0.1 μA (*b*) 1.0 μA (*c*) 10 μA (*d*) 1 A
8. Which of the following can be used for construction of a photoconducting detector?
   (*a*) An intrinsic semiconductor (*b*) An extrinsic semiconductor
   (*c*) Polycrystalline material (*d*) All of these
9. Which of the following is an inherent property of optical signal that cannot be eliminated even in principle?
   (*a*) Thermal noise (*b*) Shot noise
   (*c*) Environmental noise (*d*) Background noise
10. Which of the following material is most suitable for making a *p-n* diode?
   (*a*) A direct bandgap material (*b*) An indirect bandgap material
   (*c*) A metal (*d*) An insulator
11. Distinguish among a *p-n* photodiode, a *p-i-n* diode and an avalanche photodiode. Is it possible to use the same semiconductor to construct all these three types of detectors?
12. Define the quantum efficiency and responsivity of a *p-n* diode. How they are related to each other?
13. A *p-n* photodiode has the quantum efficiency of 50% at the wavelength of 900 nm. Calculate its responsivity, received optical power at the photocurrent of 10^{-6} A and number of received photons at this wavelength.
14. An APD of quantum efficiency of 50% at 1.3 μm produces an output photocurrent of 8 μA at the optical power of 0.4 μW after avalanche gain. What is the multiplication factor?
15. A typical photodiode with the active area of 2 mm^2 has the resposivity of 0.40 A/W for a He-Ne laser source (λ = 632.8 nm). What will be the output photocurrent if the incident flux is 100 $\mu W/mm^2$?
16. What are the factors responsible for deviation from ideal to practical Si diode in respect of responsivity versus wavelength curve? How the quantum efficiency may be improved for such a diode?
17. Compare the merits of APD and photoconducting detectors in optical communication and other applications.
18. What are the types of noises that are attributed during optical detection? How they are affecting the signal to noise ratio performance? How these noises can be minimized? Define the terms NEP and Detectivity.
19. How the choice of material and structure influence the performance of avalanche photodetector? What are the main disadvantages of avalanche photodetector?
20. Derive the expressions for NEP (Noise Equivalent Power) and Signal to Noise ratio for junction photodiode.

5

Optoelectronic Modulation Devices

5.1 INTRODUCTION

The advent of the laser and its increasing use in different applications has led to a demand for devices which can modulate a laser beam. A modulator is a device which changes the irradiance (or direction) of the light passing through it. There are several general types of modulator; viz. passive or dye modulators, mechanical choppers and shutters, electro-optic, acousto-optic and magneto-optic modulators. In this chapter we shall explain how the distortion produced in a crystal lattice by the application of an electric field or by the passage of a sound wave affects the propagation of light through the crystal. These effects, the electro-optic and acousto-optic effects are of considerable practical importance as they can be used to amplitude and phase modulate light beams, shift their frequencies and alter the direction in which they travel.

5.2 POLARIZATION

Light as we have already discussed is an electromagnetic wave, which has both electric field vector as well as magnetic field vector. Any real beam of light comprises many individual waves and in general the planes of vibration of their electric fields will be randomly oriented. In such a beam of light the resultant electric field vector changes its orientation randomly in time and is known as *unpolarized* light. However, it is possible to have light beams characterized by highly oriented electric fields and such light beam is referred to as *polarized* light beam. The simplest form of polarization is plane polarized light as shown in Fig. 5.1. The electric vector $(\overline{E})$ and the magnetic vector $(\overline{H})$ vibrate in orthogonal planes and perpendicularly to the direction of propagation.

If the electric filed vector of an electromagnetic wave propagating in free space vibrates in a specific plane, the wave is said to be plane polarized.

In many cases, an interesting phenomenon takes place. A beam of light may consist of two plane polarized wave trains with their planes of polarization at right angles to each other and which may also be out of phase.

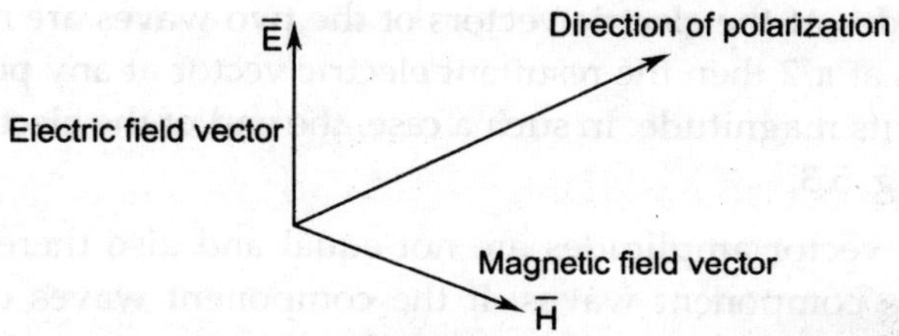

Fig. 5.1

Now, we consider a special case where the amplitudes of the two wave trains are equal but they have a phase difference of $\frac{\pi}{2}$. Hence, if the two wave trains are propagating in the z direction, the component electric fields may be written as

$$\overline{E_x} = \mathbf{i}E_0 \cos(kz - \omega t)$$

and

$$\overline{E_y} = \mathbf{j}E_0 \sin(kz - \omega t) \tag{5.1}$$

where $k = (2\pi/\lambda)$ is the propagation constant or wave number, ω is the angular frequency and **i** and **j** are unit vectors in the x and y direction respectively. The total electric field is the vector sum of the two components, and can be written as

$$\overline{E} = \overline{E_x} + \overline{E_y}$$

or

$$\overline{E} = E_0\left[\mathbf{i}\cos(kz - \omega t) + \mathbf{j}\sin(kz - \omega t)\right] \tag{5.2}$$

The resultant electric field vector expressed by equation (5.2) can be interpreted as a single wave in which the electric vector at a given point in space is constant in amplitude but rotates with angular frequency ω. These type of waves are said to be *circularly polarized*. Figure 5.2 shows the electric field vector (*a*) at a given point in space and (*b*) at a given instant of time.

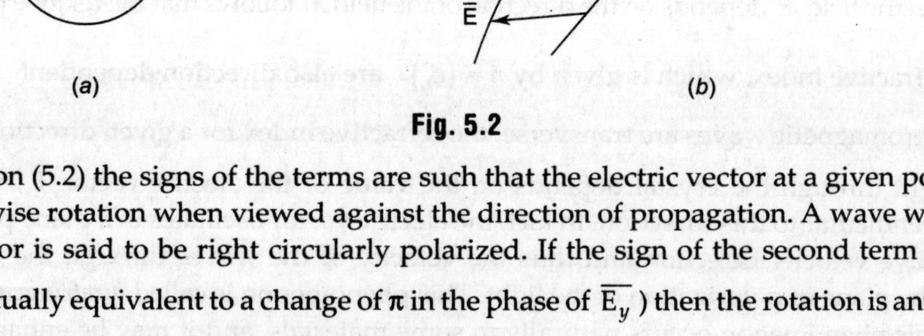

Fig. 5.2

In equation (5.2) the signs of the terms are such that the electric vector at a given point in time has a clockwise rotation when viewed against the direction of propagation. A wave with such an electric vector is said to be right circularly polarized. If the sign of the second term is changed (which is actually equivalent to a change of π in the phase of $\overline{E_y}$) then the rotation is anticlockwise and the wave is said to be left circularly polarized.

But when the amplitudes of the electric vectors of the two waves are not equal, whereas the phase difference remains at $\pi/2$ then the resultant electric vector at any point in space rotates at frequency ω but changes its magnitude. In such a case, the end of the electric vector describes an ellipse as illustrated in Fig. 5.3.

In general the electric vector amplitudes are not equal and also there is an arbitrary phase difference φ between the component waves. If the component waves can be represented by $\overline{E_x} = \mathbf{i} E_0 \cos(kz - \omega t)$ and $\overline{E_y} = \mathbf{i} E_0' \cos(kz - \omega t)$ with $E_0 \neq E_0'$, then the major and minor axes of the ellipse are parallel to the x and y axes. The resultant wave is said to be elliptically polarized. In fact the plane and circular polarization are special cases of elliptical polarization. We are not going in details of elliptically polarized light, since that is not possible within the scope of this chapter. The Nicol prism, Glan-Thomson prism, retardation plates, Soleil-Babinet compensator, Wollaston prism etc., are some examples of polarization devices.

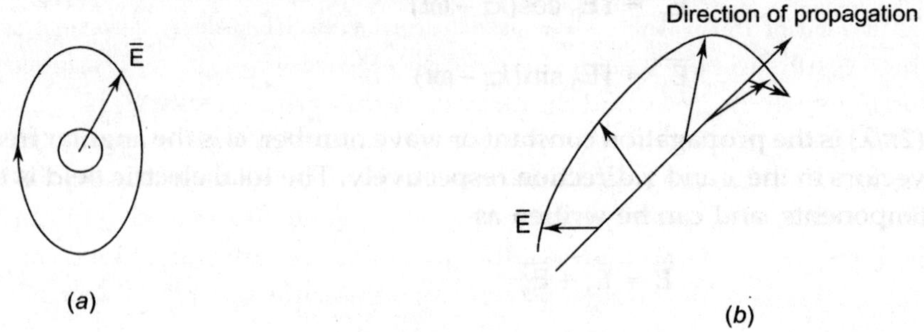

Fig. 5.3

5.3 BIREFRINGENCE

Wave propagation mechanism in anisotropic media differs from that in isotropic media. In isotropic media the velocity of propagation of an electromagnetic wave is independent of the direction of propagation, whereas that is not the case in anisotropic media. In anisotropic media, the velocity of propagation of the electromagnetic waves depends in general, on the direction of propagation as well as the state of polarization, and one can observe the phenomenon of *birefringence*.

Anisotropy in crystals is caused by the arrangement of atoms being different in different directions through the crystal. Consequently, the electric polarization **P** produced in the crystal by a given electric field $\overline{E}$ depends on the direction of the field. It follows that the relative permittivity ε_r and the refractive index, which is given by $n = (\varepsilon_r)^{\frac{1}{2}}$ are also direction dependent.

Since electromagnetic waves are transverse, the refractive index for a given direction of propagation of light through the crystal depends on the value of the electric vector $\overline{E}$, which is of course, perpendicular to this direction. In fact, the electric vector oscillates in a plane perpendicular to the wave velocity direction and thus the velocity of the waves through the crystal will depend on the plane of polarization of the light. This phenomenon is called *birefringence* or *double refraction*. This phenomenon occurs naturally in some materials, and it may be enhanced by the application of external forces such as those due to electric fields. It may even be induced in isotropic materials by the same methods.

Ordinary glass is isotropic in nature, but some important crystalline optical materials such as Quartz (SiO_2), calcite ($CaCO_3$) and KDP (KH_2PO_4) are anisotropic in nature. When a ray enters such a crystal, it splits up into two rays and they propagate in two different directions depending on the direction of polarization. These rays corresponding to these directions travel with different speeds and are mutually orthogonally polarized.

Between the two rays one of the rays follows Snell's laws of refraction and is called the ordinary ray (o-ray), whereas the other does not follow the Snell's laws of refraction and is called as the extraordinary ray (e-ray). If a point source is embedded in such a crystal, two different sets of wavefronts will be generated from the point source. One of them will be spherical corresponding to the o-ray, whereas the other will be an ellipsoid of revolution corresponding to the e-ray. The o-ray has same velocity for all directions of propagation and hence spherical wavefront is generated. On the other hand, the e-ray has different velocities along different directions of propagation which results in non-spherical wavefront. However, in these crystals there is one particular direction in which the o-ray and the e-ray have the same velocities, and this direction is called the *optic axis*. In these crystals there is only one optic axis and hence they are known as *uniaxial crystals*. In some crystals there are two directions along which the velocities may be equal. Such crystals are called *biaxial*, for example mica, sulphur, topaz etc. But we have to remember that in biaxial crystals the wave velocities (and not the ray velocities) are equal along the optic axes. However, in uniaxial crystals both the ray and the wave velocities are equal along the optic axis.

5.3.1 Positive and Negative Crystals

If the velocity of the o-ray is greater than the e-ray velocity along all directions then the crystal is called a positive uniaxial crystal. In such a case the spherical wavefront will be completely outside the ellipsoidal wavefront as shown in Fig. 5.4(*a*). For positive uniaxial crystal $n_o < n_e$, where n_o and n_e are the ordinary and extraordinary refractive indices respectively. On the other hand, if the velocity of the e-ray is greater than the o-ray velocity along all directions then the crystal is called a negative uniaxial crystal. In such a case the spherical wavefront lies completely inside the ellipsoidal wavefront as shown in Fig. 5.4(*b*). For negative uniaxial crystal, $n_o > n_e$. Quartz is an example of positive uniaxial crystal and calcite is a negative uniaxial crystal.

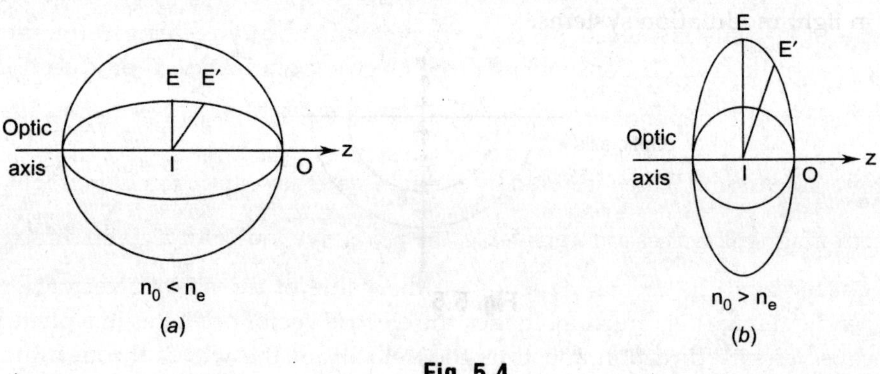

Fig. 5.4

The o-ray velocity can be written as

$$v_o = \frac{c}{n_o} \tag{5.3}$$

The e-ray velocity can be written as

$$\frac{1}{v_e^2(\theta)} = \frac{n_e^2(\theta)}{c^2} = \frac{\cos^2\theta}{c^2/n_o^2} + \frac{\sin^2\theta}{c^2/n_e^2} \tag{5.4}$$

or,
$$n_e^2(\theta) = n_o^2\cos^2(\theta) + n_e^2\sin^2(\theta) \tag{5.5}$$

where θ is the angle made by the e-ray with the optic axis.

The two directions of propagation of the incident light which gives only a single emergent beam are sometimes called the *fast* and *slow* axes or the *privileged directions* of the crystal. One direction corresponds to the propagation of the o-ray and will be the fast axis for positive crystal; the other direction corresponds to the direction of propagation of the e-ray and is the slow axis. The reverse is the case for negative crystals. For light incident along the direction of the optic axis the state of polarization of the emergent beam will be the same as that of the incident beam and there are no privileged directions.

5.3.2 Quarter-Wave Plate

A crystal plate cut as in Fig. 5.5 introduces a phase difference of $\pi/2$ between the o and e-rays. In this case there is no divergence of the o and e rays, but they become increasingly out of phase as they propagate through the crystal.This type of plate is called a *quarter-wave plate.* A phase difference of $\pi/2$ is equivalent to an optical path difference $|n_e d - n_o d| = \lambda/4$ where d is the plate thickness.

When a plane polarized light is incident on a quarter-wave plate, the emergent light is in general elliptically polarized. The axes of the ellipse are parallel to the privileged directions in the plate. If the plane of polarization of the incident beam, is inclined at 45° to the privileged directions then the emergent light is circularly polarized. If it is desirable to introduce a phase difference of π, *i.e.*, a path difference of $\lambda/2$, between two emerging beams, then it can be shown that the required thickness of the slab would be $d = \lambda/2(n_e - n_o)$. Such a device is known as half-wave plate and is useful for rotating the plane of vibration of plane-polarized beams. These types of plates are often used in light modulation systems.

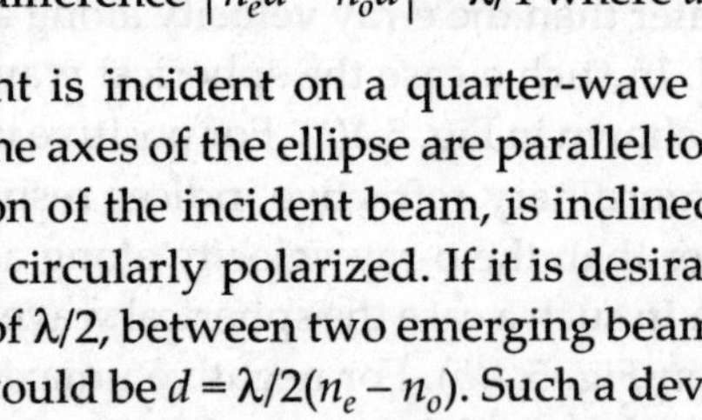

Fig. 5.5

Example 5.1. *Calculate the thickness of a quarter-wave plate made of quartz and to be used with light of wavelength $\lambda = 550$ nm. It is given that the principal refractive indices n_e and n_o for quartz are 1.553 and 1.544, respectively.*

Solution:

We have already discussed that quartz is a positive uniaxial crystal. Hence $n_e > n_o$.

$$d = \frac{\lambda}{4(n_e - n_o)} = \frac{550 \times 10^{-9}}{4(1.553 - 1.544)} = 0.0153 \text{ mm.}$$

Example 5.2. *A calcite half-wave plate is to used with light of wavelength $\lambda = 570$ nm. Calculate the thickness required. It is given that the principal refractive indices n_e and n_o for calcite are 1.486 and 1.658, respectively.*

Solution:

Calcite is a negative uniaxial crystal. Hence $n_e < n_o$.

$$d = \frac{\lambda}{2(n_o - n_e)} = \frac{570 \times 10^{-9}}{2(1.658 - 1.486)} = 0.0016 \text{ mm.}$$

5.4 ELECTRO-OPTIC EFFECT

There are certain materials which change their optical properties when subjected to an electric field. When an electric field is applied to a crystal, the ionic constituents move to new locations determined by the field strength, the charge on the ions and the restoring force. Unequal restoring forces along three mutually perpendicular axes in the crystal lead to anisotropy in the optical properties of the medium. When an electric field is applied to such a crystal, in general, it causes a change in the anisotropy.

This is caused by forces that distort the positions, shapes or orientations of the molecules constituting the material. Electro-optic effect is actually the change in refractive index resulting from the application of a dc or low-frequency electric field. A field applied to an anisotropic electro-optic material modifies its refractive indices and thereby its effect on the polarized light. This dependence of the refractive index on the applied electric field takes one of the two forms:

- The linear electro-optic effect or the *Pockels effect,* where the refractive index changes in proportion to the applied electric field.
- The quadratic electro-optic effect or the *Kerr effect,* where the change in refractive index is in proportion to the square of the applied electric field.

In this chapter we shall study in detail the Pockels effect and obtain expressions for the phase shift suffered by a beam propagating through a crystal which is being acted upon by an external electric field. We will show that under certain geometrical considerations, the applied electric field acts differently on two linearly polarized light waves passing through the crystal and thus an electric field dependent retardation may be introduced between the two polarizations. We shall also show that such an effect can be used to control the phase, intensity, etc., of the light beam in accordance with the applied electric field. Thus this gives us a method for imposing information on a light beam through the electro-optic effect. Such modulators are of great use in optical communications, where modulation of optical beams at very high frequencies (ranging from megahertz to gigahertz) is required.

Though the change in refractive index is typically very small, its effect on an optical wave travelling a distance much greater than a wavelength of light may be quite significant. Materials whose refractive index can be modified by means of an applied electric field are useful for producing electrically controllable optical devices. We can give some examples as:

- A prism whose beam bending ability is controllable can be used as an optical scanning device.
- A lens whose refractive index can be varied may be used to control focal length.
- Light transmitted through a transparent plate of controllable refractive index can have a controllable phase shift. Hence the plate can be used as an optical modulator.
- Similarly an anisotropic crystal whose refractive index can be varied serves as a wave retarder of controllable retardation.

There are several other examples also where this electro-optic property of those materials may be used such as directional couplers, optical switches, bistable devices, etc., in integrated optics and also in Q-switching and mode locking in lasers, etc.

5.4.1 Pockels and Kerr Effect

The refractive index $n(\mathrm{E})$ of an electro-optic medium is a function of the applied electric field E. Since this function varies only slightly with E, it may be expanded in a Taylor's series about E = 0. Hence we may write,

$$n(\mathrm{E}) = n + a_1\mathrm{E} + \frac{1}{2}a_2\mathrm{E}^2 + \Lambda \tag{5.6}$$

where the coefficients of expansion are $n = n(0)$, $a_1 = \left(\frac{dn}{de}\right)\Big|_{\mathrm{E}=0}$, and $a_2 = \left(\frac{d^2n}{de^2}\right)\Big|_{\mathrm{E}=0}$. It is convenient to write Equation (5.6) in terms of two new coefficients $r = -\frac{2a_1}{n^3}$ and $\xi = -\frac{a_1}{n^3}$. These two coefficients are known as the electro-optic coefficients, so that

$$n(\mathrm{E}) = n - \frac{1}{2}rn^3\mathrm{E} - \frac{1}{2}\xi n^3\mathrm{E}^2 + \Lambda \tag{5.7}$$

The second and higher-order terms of this series are typically many orders of magnitude smaller than n. Hence terms higher than the second order may be safely neglected. The values of the coefficients r and ξ depend on the direction of the applied electric field and the polarization of the light.

Pockels Effect

In many materials, for example naturally doubly refracting KDP crystal, or naturally isotropic GaAs crystals, the third term of equation (5.7) is negligible in comparison with the second. Hence we write:

$$n(\mathrm{E}) \approx n - \frac{1}{2}rn^3\mathrm{E} \tag{5.8}$$

The medium is then known as a Pockels medium (or a Pockels cell). We shall see that Pockels effect results only if the crystal lattice lacks a center of symmetry. So, some cubic crystals can exhibit the linear electro-optic effect.

The coefficient r is called the linear electro-optic coefficient or the Pockels coefficient. Typical values of γ lie in the range 10^{-12} to 10^{-10} m/V (1 to 100 pm/V). Changes induced by electric fields in these Pockels medium are indeed very small. For example for an applied voltage of 4000 V for a 1 cm wide crystal of KH_2PO_4 (or KDP), the change in refractive index is given below:

$$\Delta n = \frac{1}{2} \times 10.6 \times 10^{-12} \times 1.51^3 \times \frac{4000}{10^{-2}}$$

or

$$\Delta n = 7.29 \times 10^{-6}$$

Table 5.1

Material	*Pockels Coefficient, r (pm/V)*	*Refractive Index*		*Relative Permittivity*
		n_o	n_e	
KH_2PO_4 (KDP)	10.6	1.51	1.47	42
$LiNbO_3$	30.8	2.29	2.20	18
KD_2PO_4 (KD^*P)	26.4	1.51	1.47	50
$NH_4H_2PO_4$ (ADP)	8.5	1.52	1.48	12
$LiTaO_3$	30.3	2.175	2.18	43
GaAs	1.6	3.6	–	11.5
ZnS	2.1	2.32	–	16
CdTe	6.8	2.6	–	7.3

Some of the most commonly used crystals as Pockels cell include KH_2PO_4 (KDP), $LiNbO_3$ (Lithium niobate), KD_2PO_4 (KD^*P), $NH_4H_2PO_4$ (ADP), LiTaO3 (Lithium tantalate), GaAs (Gallium arsenide), ZnS (Zinc sulfide), and CdTe (Cadmium telluride). Some of the characteristics of the above mentioned materials are shown in Table 5.1 (values near a wavelength of 550 nm).

Kerr Effect

In this type of electro-optic effect the change in refractive index is proportional to the square of the applied electric field. This effect was discovered by J. Kerr in the year 1875. For the isotropic materials like gases, liquids and certain centro-symmetric crystals, $n(E)$ must be an even symmetric function, since it must be invariant to the reversal of E. The first derivative then vanishes, and hence the coefficient γ must be zero. Hence from Equation (5.7) we may write:

$$n(E) \approx n - \frac{1}{2}\xi n^3 E^2 \tag{5.9}$$

The medium is then known as *Kerr medium* (or a *Kerr cell*). The parameter ξ is called the Kerr coefficient or the quadratic electro-optic coefficient. Typical values of ξ lie in the range of 10^{-18} to 10^{-14} m^2/V^2 in crystals and 10^{-22} to 10^{-19} m^2/V^2 in liquids.

Liquid Kerr cells containing nitrobenzene have been used extensively for many years. But they require large power to operate. Mixed ferroelectric crystals operating at a temperature near to the Curie point may be used. Potassium tantalate niobate (KTN) is a mixture of two crystals with high and low Curie points giving a Curie point for the mixture near to the room temperature. KTN is used in Kerr-effect devices. Barium titanate ($BaTiO_3$) is another example of mixed crystal which is used as Kerr-effect device. However, in most of the practical electro-optic modulators the Pockels effect is used.

5.5 ELECTRO-OPTIC (POCKELS) MODULATORS AND SWITCHES

5.5.1 Phase Modulators

When a beam of light traverses a Pockels cell of length L to which an electric field E is applied, it undergoes a phase shift $\varphi = n(E)k_0L = 2\pi n(E)L/\lambda_0$, where λ_0 is the free-space wavelength. Using equation (5.8), we may write

$$\phi \approx \frac{2\pi L}{\lambda_0}\left(n - \frac{1}{2}rn^3 E\right)$$

$$= \frac{2\pi Ln}{\lambda_0} - \frac{\pi Lrn^3 E}{\lambda_0}$$

$$= \phi_0 - \pi\frac{rn^3 EL}{\lambda_0} \tag{5.10}$$

where $\phi_0 = \dfrac{2\pi Ln}{\lambda_0}$. If the electric field is obtained by applying a voltage V across two faces of the cell separated by distance d, then $E = \dfrac{V}{d}$, and from equation (5.10) we get

$$\phi = \phi_0 - \pi\frac{rn^3 LV}{\lambda_0 d}$$

$$= \phi_0 - \pi\frac{V}{V_\pi} \tag{5.11}$$

where, $$V_\pi = \frac{d\lambda_0}{Lrn^3} \tag{5.12}$$

The parameter V_π is known as the *half-wave voltage*. It is actually the applied voltage at which the phase shift changes by π. Form equation (5.11) we find a linear relationship between the optical phase shift and the voltage and can be graphically represented as shown in Fig. 5.6.

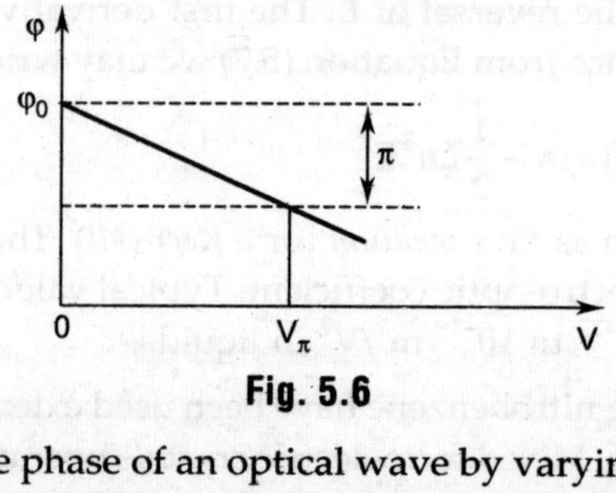

Fig. 5.6

Hence one can modulate the phase of an optical wave by varying the voltage V that is applied across a material through which the light passes. The parameter V_π is an important characteristic of the modulator. It depends on the material properties (n or r), on the wavelength λ_0, and on the aspect ratio d/L.

Now the electric field may be applied in a direction perpendicular to the direction of light propagation or parallel to that direction. In the former case it produces transverse modulators and in the later case it produces longitudinal modulators where d = L. The value of the electro-optic coefficient r depends on the directions of propagation and the applied electric field since the crystal is anisotropic. Typical values of the half-wave voltage range between 1 to a few kilovolts for longitudinal modulators, and hundreds of volts for transverse modulators.

Longitudinal Phase Modulators

To illustrate how an applied electric field affects an electro-optic crystal it is instructive to consider the specific important material KDP (potassium di-hydrogen phosphate, KH_2PO_4). The related material KD^*P (KD_2PO_4), ADP (ammonium di-hydrogen phosphate, $NH_4H_2PO_4$) and AD^*P ($NH_4D_2PO_4$) behave similarly.

We consider that light is propagating along the z-axis of a KDP crystal which is acted upon by an external electric field which is also along the z-direction, as shown in Fig. 5.7.

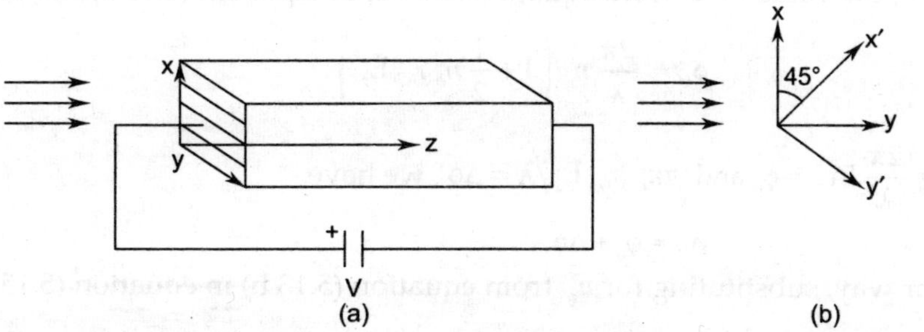

Fig. 5.7 *(a) Light propagating along the z-axis of a KDP crystal in which an external electric field is also applied along the same direction. (b) In the absence of the external filed light polarized normal to the z-axis propagates as a principal wave with refractive index n_o. In the presence of an electric field along the z-axis, the crystal becomes biaxial and the principal polarization directions are x′ and y′ which makes an angle of 45° with the old axis.*

This type of configuration is called *longitudinal* configuration. In the absence of the external electric field, since the crystal is uniaxial and its optic axis lies along the z-direction, a light wave polarized normal to the z-direction (*i.e.*, in the x-y plane) travels as a principal wave with the ordinary refractive index n_o. But on applying the external electric field E_z, along the z-direction, the crystal no longer remains uniaxial but becomes biaxial. The two linear polarization states which propagate through the crystal without changing the state of polarization lie along 45° to the x and y-axes. This means that they lie along the x' and y' axes as shown in Fig. 5.7. Now the refractive index $n_{x'}$ for a wave polarized along the x'-direction is given by:

$$n_{x'} = n_o + \frac{1}{2} n_o^3 r_{63} E_z \tag{5.13 a}$$

where r_{63} is the coefficient describing the electro-optic effect in this configuration. In the same manner the refractive index $n_{y'}$ for the wave propagating along the z-direction is given by:

$$n_{y'} = n_o - \frac{1}{2} n_o^3 r_{63} E_z \tag{5.13 b}$$

If the incident wave is represented by the equation $E = E_0 \cos(\omega t - kz)$, the components along the x' and the y' direction will be, respectively, given by

$$E_{x'} = \frac{E_0}{\sqrt{2}} \cos(\omega t - kz) \tag{5.14 a}$$

and
$$E_{y'} = \frac{E_0}{\sqrt{2}} \cos(\omega t - kz) \tag{5.14 b}$$

These components have refractive indices $n_{x'}$ and $n_{y'}$ respectively given by equations (5.13 a) and (5.13 b). Thus they will become increasingly out of phase as they propagate through the crystal. If the crystal thickness along the direction of propagation is l, the phase change experienced by these two components (at $z = l$) may be represented by the following relations:

$$\phi_{x'} = kn_{x'}l = \frac{2\pi}{\lambda} n_{x'}l \tag{5.15 a}$$

and
$$\phi_{y'} = kn_{y'}l = \frac{2\pi}{\lambda} n_{y'}l \tag{5.15 b}$$

Substituting the value of $n_{x'}$ from equation (5.13 a) in equation (5.15 a), we get

$$\phi_{x'} = \frac{2\pi}{\lambda} n_o l\left(1 + \frac{1}{2} n_o^2 r_{63} E_z\right) \tag{5.16 a}$$

Assuming $\frac{2\pi}{\lambda} \ln_o = \phi_o$ and $\pi n_o^3 l r_{63} E_z/\lambda = \Delta\phi$, we have

$$\phi_{x'} = \phi_o + \Delta\phi \tag{5.16 b}$$

In a similar way, substituting for $n_{y'}$ from equation (5.13 b) in equation (5.15 b), we get

$$\phi_{y'} = \frac{2\pi}{\lambda} n_o l\left(1 - \frac{1}{2} n_o^2 r_{63} E_z\right) \tag{5.17 a}$$

or,
$$\phi_{y'} = \phi_o - \Delta\phi \tag{5.17 b}$$

where
$$\Delta\phi = \frac{\pi}{\lambda} n_o^3 l r_{63} E_z = \frac{\pi}{\lambda} n_o^3 r_{63} V \tag{5.18}$$

and $V = E_z l$ is the applied voltage. We find that an extra phase shift $\Delta\phi$, due to the application of the electric field, for each component is directly proportional to the applied voltage V. Hence, if V oscillates with frequency ω_m, that is, if $V = V_0 \sin \omega_m t$, the phase shift will also vary sinusoidally and the peak value will be $\pi n_0^3 r_{63} V_0/\lambda$. Hence, the electro-optic effect may be used for phase modulation. The net phase shift or the total retardation (at $z = l$) between the two waves polarized in the x' - and y' - direction as a result of the application of voltage V to the crystal will be given by:

$$\Phi = \phi_{x'} - \phi_{y'} = 2\Delta\phi = \frac{2\pi}{\lambda} r_{63} n_0^3 V \tag{5.19}$$

In general, we know that two plane-polarized waves which are perpendicular to each other will produce elliptically polarized wave on superposition. Hence, in the case of equation (5.19) also, in general, elliptically polarized wave will emerge. However, if the superposition produces a phase difference of even multiple of $\pi/2$, the emergent beam will be plane-polarized, and if the phase difference is an odd multiple of $\pi/2$, the emergent beam will be circularly polarized.

The voltage $V = V_\pi$ required to introduce a phase difference of π between the two polarization components is called the half-wave voltage and may be obtained as follows:

$$\Phi = \pi = \frac{2\pi}{\lambda} r_{63} n_0^3 V_\pi$$

or,

$$V_\pi = \frac{\lambda}{2 r_{63} n_0^3} \tag{5.20}$$

Using equations (5.14 a), (5.14 b), (5.16 b) and (5.17 b), the equations for the components of the wave emerging from the crystal polarized in the x'- and y'- direction may be written as (omitting common phase factors):

$$E_{x'\,(z=l)} = \frac{E_0}{\sqrt{2}} \cos(\omega t + \Delta\phi) \tag{5.21 a}$$

and

$$E_{y'\,(z=l)} = \frac{E_0}{\sqrt{2}} \cos(\omega t - \Delta\phi) \tag{5.21 b}$$

Now let us now consider a linearly polarized plane wave propagating along the z-direction in a KDP crystal. A schematic diagram of a longitudinal electro-optic amplitude modulator using KDP is presented in Fig. 5.8. There is a plane polarizer at the output end of the KDP crystal oriented at right angles to the input polarizer. An external electric field is applied along the z-direction.

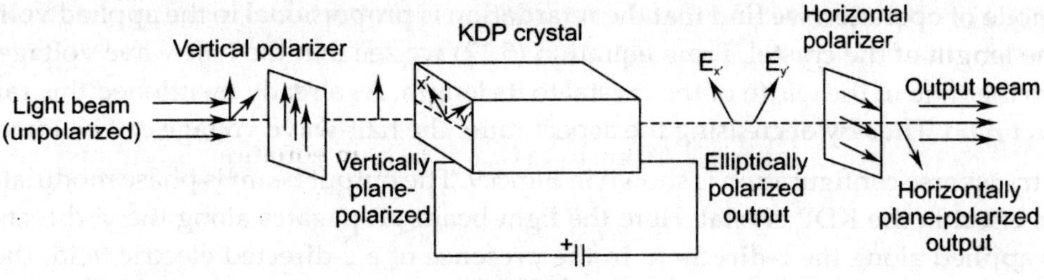

Fig. 5.8

From equation (5.19), we may write:

$$\Phi = \frac{2\pi}{\lambda} r_{63} n_0^3 V = 2\left(\frac{\omega}{2c}\right) n_o^3 r_{63} V_0 \sin\omega_m t = 2\zeta \sin\omega_m t \tag{5.22}$$

where,

$$\zeta = \left(\frac{\omega}{2c}\right) n_o^3 r_{63} V_0 \tag{5.23}$$

ζ is referred to as the phase modulation index. Thus, if a sinusoidal electric field is applied, it leads to a sinusoidal variation of the output light wave with a peak value ζ.

Example 5.3. *Estimate the change in refractive index due to longitudinal electro-optic effect for a 2-cm-wide KDP crystal for an applied voltage of 6 kV. If the wavelength of light being propagated through the crystal is 600 nm, calculate the net phase shift between the two polarization components after they emerge from the crystal. Also calculate V_π for the crystal. Assume $n_o = 1.51$ and $r_{63} = 10.5 \times 10^{-12}$ m/V for KDP.*

Solution:

With a voltage of 6 kV across 2 cm crystal, the electric field produced will be

$$E_z = \frac{V}{l} = \frac{6000}{2 \times 10^{-2}} = 3 \times 10^5 \text{ V/m}$$

The change in the refractive index will be:

$$\Delta n = \frac{1}{2} n_o^2 r_{63} E_z = \frac{1}{2} \times (1.51)^3 \times 10.5 \times 10^{-12} \times 3 \times 10^5$$

or $$\Delta n = 5.42 \times 10^{-6}$$

Therefore, $$\Delta\phi = \frac{2\pi}{\lambda} \Delta nl = \frac{2\pi}{600 \times 10^{-9}} \times 5.42 \times 10^{-6} \times 2 \times 10^{-2} = 0.36\pi$$

Thus, the net phase shift suffered by the two polarization components is given by equation (5.19) as:

$$\Phi = 2\Delta\phi = 0.72\pi$$

Now, the half-wave voltage V_π may be calculated from equation (5.20) as:

$$V_\pi = \frac{\lambda}{2 r_{63} n_0^3} = \frac{600 \times 10^{-9}}{2 \times 10.5 \times 10^{-12} \times (1.51)^3} \approx 8298 \text{ V} \approx 8.3 \text{ kV}$$

Transverse Phase Modulators

In this configuration the optical path is perpendicular to the direction of the applied electric field. In the previous configuration we have seen that the phase retardation is independent of the length of the crystal (equation 5.11) and depends only on the applied voltage. But in the transverse mode of operation we find that the retardation is proportional to the applied voltage as well as to the length of the crystal. From equation (5.12) we see that the half-wave voltage is proportional to the ratio of the width of the crystal to its length. As already mentioned this ratio is called the *aspect ratio*. Thus by decreasing the aspect ratio, the half-wave voltage can be lowered.

The transverse configuration is shown in Fig. 5.9. The output beam is phase modulated through Pockels effect in the KDP crystal. Here the light beam propagates along the y'-direction and the field is applied along the z-direction. In the presence of a z-directed electric field, the refractive indices for a wave propagating along the y'-direction and polarized along the x'- and z-directions are respectively given by equations (5.24) and (5.25).

$$n_{x'} = n_o + \frac{1}{2} n_o^3 r_{63} E_z \tag{5.24}$$

$$n_z = n_e \tag{5.25}$$

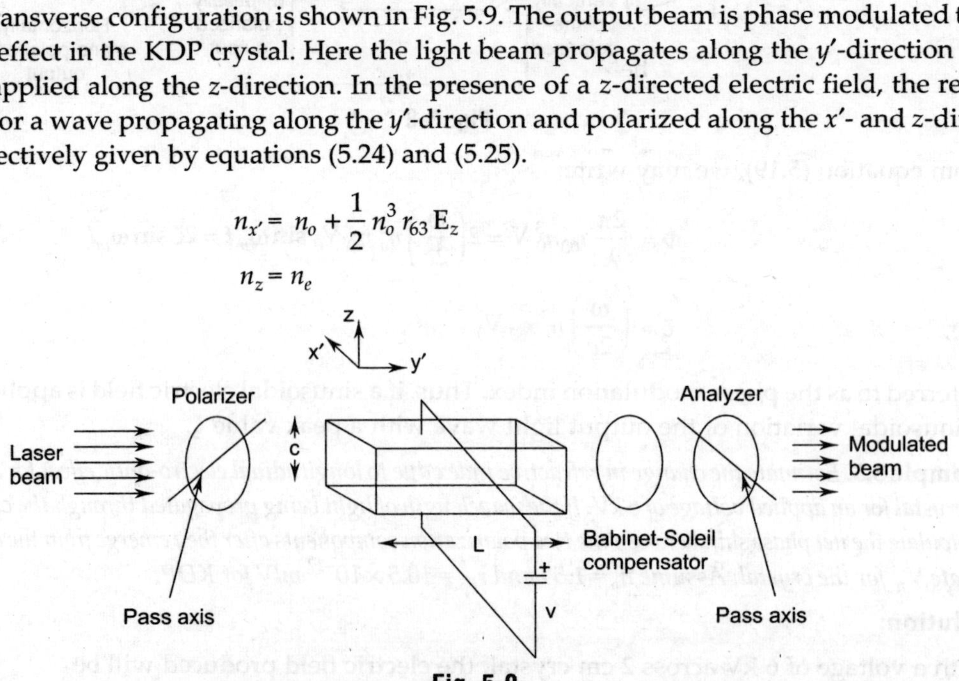

Fig. 5.9

Let us assume the light beam incident on the crystal is linearly polarized along 45° to the x'-direction (in the x'-z plane) and propagates along the y'-direction. The electric field is applied along the z-direction. The analyzer is placed such that it is normal to the direction of the polarizer. A Babinet-Soleil compensator is being introduced so that the modulator is biased in the linear region of the transmittance versus applied voltage curve.

The phase difference between the emergent beam components along the x'- and y'-direction after travelling a length l of the crystal is given by:

$$\Delta\phi = \phi_{x'} - \phi_z = \frac{2\pi}{\lambda} l(n_{x'} - n_z)$$

$$= \frac{2\pi}{\lambda}\left[n_o + \frac{1}{2} n_o^3 r_{63} E_z - n_e\right]$$

$$= \frac{2\pi}{\lambda} l(n_o - n_e) + \frac{\pi}{\lambda} n_o^3 r_{63}\left(\frac{V}{d}\right) l \qquad (5.26)$$

where V is the applied voltage across the width d of the crystal. It may be noted from equation (5.26), that even if the applied voltage V = 0, there is a finite retardation given by:

$$(\Delta\phi)_{V=0} = \frac{2\pi}{\lambda} l(n_o - n_e) \qquad (5.27\text{ a})$$

This is due to the intrinsic birefringence of the crystal. Thus the retardation induced by the external voltage is given by:

$$\Delta\phi = \frac{\pi}{\lambda} n_o^3 r_{63}\left(\frac{V}{d}\right) l \qquad (5.27\text{ b})$$

The half-wave voltage V_π for this configuration may be defined as the voltage required to produce a phase difference of π between the two polarization components (in excess to that produced by intrinsic birefringence). From equation (5.27 b), we may write:

$$\Delta\phi = \pi = \frac{\pi}{\lambda} n_o^3 r_{63}\left(\frac{V_\pi}{d}\right) l$$

or

$$V_\pi = \frac{\lambda}{n_o^3 r_{63}}\left(\frac{d}{l}\right) \qquad (5.28)$$

Transverse electro-optic modulators based on highly deuterated KDP crystals and ADP crystals are available commercially. They require low driving voltages.

Example 5.4. *A transverse electro-optic modulator with a ADP crystal is operating at a wavelength $\lambda = 550$ nm. The crystal has a length $l = 4$ cm and width $d = 0.4$ cm. The optical constants of the crystal may be obtained from Table 5.1. Estimate (a) the phase difference between the emergent field components with applied voltage V = 0, (b) the additional phase difference between the emergent field components with applied voltage V = 3 V, and (c) the half-wave voltage for the crystal.*

Solution:

(*a*) $(\Delta\phi)_{V=0}$ is given by equation (5.27 a) as:

$$(\Delta\phi)_{V=0} = \frac{2\pi}{\lambda} l(n_o - n_e) = \frac{2\times 4\times 10^{-2}}{550\times 10^{-9}}(1.52 - 1.48)\pi = 5.18\times 10^3\,\pi$$

(*b*) $\Delta\phi$ (due to external field) is given by equation (5.27 b) as:

$$\Delta\phi = \frac{\pi}{\lambda} n_o^3 r_{63}\left(\frac{V}{d}\right) l = \frac{(1.52)^3 \times 8.5 \times 10^{-12}}{550 \times 10^{-9}} \times \left(\frac{3 \times 4 \times 10^{-2}}{0.4 \times 10^{-2}}\right) = 1.628 \times 10^{-3}\,\pi$$

(*c*) From equation (5.28), we have:

$$V_\pi = \frac{\lambda}{n_o^3 r_{63}}\left(\frac{d}{l}\right) = \frac{550 \times 10^{-9}}{(1.52)^3 \times 8.5 \times 10^{-12}} \times \left(\frac{0.4 \times 10^{-2}}{4 \times 10^{-2}}\right) = 1842\ \text{V}$$

5.5.2 Intensity Modulators

If a beam of light is polarized along the y'-direction and traveling along the z-direction, then the output wave at $z = L$ will be given by

$$\varepsilon_{y'}(z = L) = \varepsilon_{y'}(0)\exp\left\{i\left[\omega t - n_{y'}\left(\frac{\omega}{c}\right)L\right]\right\}$$

$$= \varepsilon_{y'}(0)\exp\left\{i\left[\omega t - n_o\left(\frac{\omega}{c}\right)L - \left(\frac{\omega}{2c}\right)n_o^3 r_{63} E_z L\right]\right\} \tag{5.29}$$

Similarly, if a beam of light is polarized along the x'-direction and traveling along the z-direction, then the output wave at $z = L$ will be given by

$$\varepsilon_{x'}(L) = \varepsilon_{x'}(0)\exp\left\{i\left[\omega t - n_o\left(\frac{\omega}{c}\right)L - \left(\frac{\omega}{2c}\right)n_o^3 r_{63} E_z L\right]\right\} \tag{5.30}$$

If we now consider an incident wave is polarized along the y-direction, then this can be decomposed into two linearly polarized waves along the x' and the y' directions. These two components will be in phase at the input of the crystal, *i.e.*, at $z = 0$. They will also have equal amplitude. At any value $z = L$, the y' and x'-components will be given by equations (5.29) and (5.30). Hence, we see that the two components which were in phase at $z = 0$, will now develop a phase difference which is a function of the applied electric field. The retardation between the two components at $z = L$ will be

$$\gamma = \left(\frac{\omega}{c}\right) n_o^3 r_{63} E_z L = \left(\frac{\omega}{c}\right) n_o^3 r_{63} V \tag{5.31}$$

where $V = E_Z L$ is the voltage applied across the crystal. Hence, from the expression of retardation we find that it is independent of the length of the crystal and is dependent only on the externally applied voltage. And in the absence of any externally applied voltage, the phase retardation is zero.

We have already discussed in section 5.2, that if we superpose two linearly polarized waves, polarized along two perpendicular directions, then in general they will produce an elliptically polarized wave. Hence, in the present case also, the wave emerging at $z = L$ is, in general elliptical. Since the *half*-wave voltage V_π is defined as the voltage required to introduce a phase shift of π, hence, the half-wave voltage is given by

$$\gamma = \pi = \left(\frac{\omega}{c}\right) n_o^3 r_{63} V_\pi$$

or $$V_\pi = \frac{\lambda_0}{2n_o^3 r_{63}} \tag{5.32}$$

If now the elliptically polarized beam is passed through an analyzer, oriented perpendicular to the input polarization state (as shown in Fig. 5.10), then the intensity of the beam emerging from the analyzer will be modulated. Let us assume that the input polarization is along the y-direction, and the analyzer is placed along the x-direction. Hence the light wave emerging from the analyzer is given by

$$\varepsilon = \varepsilon_{x'} \cos\frac{1}{4}\pi - \varepsilon_{y'} \sin\frac{1}{4}\pi = \frac{1}{\sqrt{2}}\left(\varepsilon_{x'} - \varepsilon_{y'}\right) \tag{5.33}$$

The amplitudes of the waves polarized along the x' and y'-directions are equal, because the input beam is linearly polarized along the y-axis. Hence we have

$$\varepsilon_{x'}(0) = \varepsilon_{y'}(0) = \frac{A}{\sqrt{2}} \tag{5.34}$$

Substituting the values of $\varepsilon_{y'}$ and $\varepsilon_{x'}$ given by equations (5.29) and (5.30), we get

$$\varepsilon = \frac{1}{2} A\exp\left\{i\left[\omega t - \left(\frac{\omega}{c}\right)n_o L + \left(\frac{\omega}{2c}\right)n_o^3 r_{63} E_z L\right]\right\}\left[1 - \exp(-i\gamma)\right] \tag{5.35}$$

where, $$\gamma = \left(\frac{\omega}{c}\right)n_o^3 r_{63} E_z L = \pi\left(\frac{V}{V_\pi}\right) \tag{5.36}$$

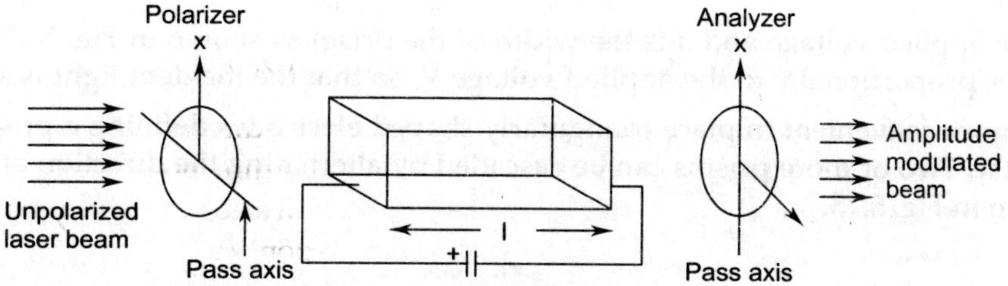

Fig. 5.10

Thus the intensity of the output beam is given by

$$I_o = \frac{1}{2}\text{Re}\left[\varepsilon\varepsilon^*\right] = \frac{1}{2}A^2 \sin^2\frac{1}{2}\gamma$$

$$= \frac{1}{2}A^2 \sin^2\left[\left(\frac{\omega}{2c}\right)n_o^3 r_{63} V\right] \tag{5.37}$$

where $V = E_z L$ is the applied voltage. The intensity of the input beam is given by

$$I_i = \frac{1}{2}A^2 \tag{5.38}$$

Thus

$$\frac{I_o}{I_i} = \sin^2\left[\left(\frac{\omega}{2c}\right)n_o^3 r_{63} V\right] \tag{5.39}$$

Using equation (5.32) we may write

$$\frac{I_o}{I_i} = \sin^2\left[\frac{\pi}{2}\left(\frac{V}{V_\pi}\right)\right] \tag{5.40}$$

Hence the output intensity varies sinusoidally with the applied voltage as shown in Fig. 5.11. Thus we can modulate the intensity of the emerging output beam by modifying the applied voltage. Both small and large aperture diameter electro-optic modulators are commercially available.

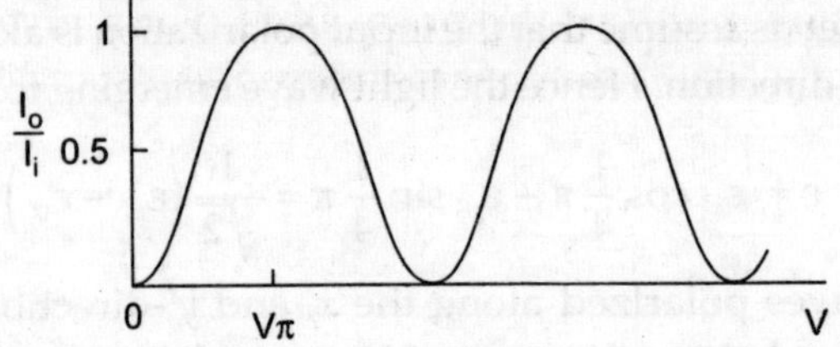

Fig. 5.11 *Variation of the transmitted intensity of the intensity modulator versus the externally applied voltage.*

5.5.3 Scanners

An optical beam can be deflected dynamically by using a prism with refractive index which can be controlled electrically. The angle of deflection introduced by a prism of small apex angle α and refractive index n is given by $\theta \approx (n-1)\alpha$. An incremental change of the refractive index Δn, caused by an applied electric field E corresponds to an incremental change of the deflection angle,

$$\Delta\theta = \alpha\Delta n = -\frac{1}{2}\alpha r n^3 E = -\frac{1}{2}\alpha r n^3 \frac{V}{d} \tag{5.41}$$

where V is the applied voltage and d is the width of the prism as shown in Fig. 5.12. Hence the angle $\Delta\theta$ varies proportionally to the applied voltage V, so that the incident light is scanned.

It is often more convenient to place triangularly shaped electrodes defining a prism on a rectangular crystal. Two or more prisms can be cascaded by alternating the direction of the electric field, as shown in Fig. 5.13.

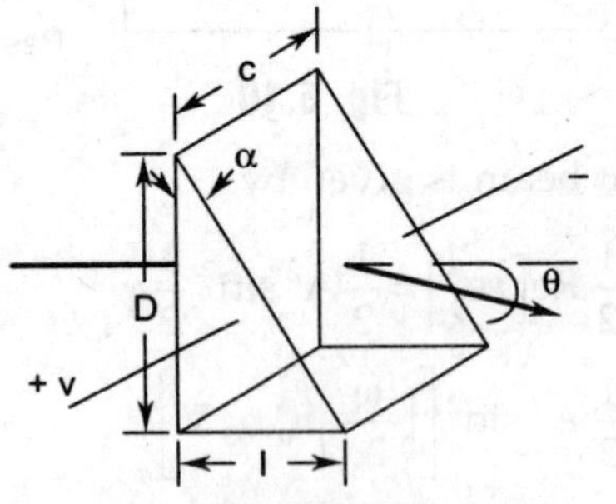

Fig. 5.12

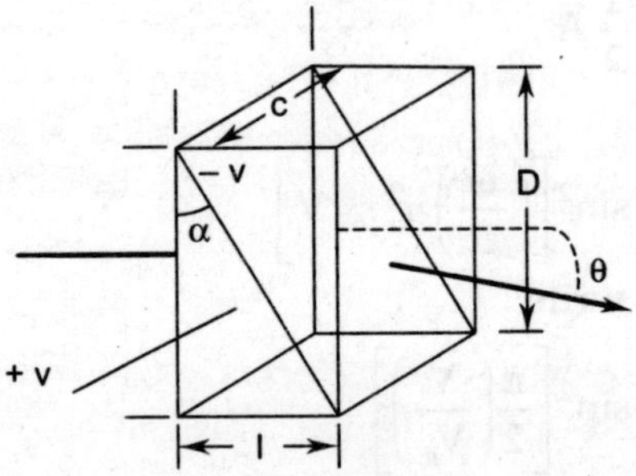

Fig. 5.13

One of the most important parameters of the scanner is its resolution. Resolution of a scanner is the number of independent spots it can scan. An optical beam of width D and wavelength λ_0 has an angular divergence $\delta\theta \approx \frac{\lambda_0}{D}$. To minimize that angle, the beam should be as wide as possible. Ideally it should cover the entire width of the prism itself. For a given maximum voltage V corresponding to a scanned angle $\Delta\theta$, the number of independent spots is given by

$$N \approx \frac{|\Delta\theta|}{\delta\theta} = \frac{\frac{1}{2}\alpha r n^3 \frac{V}{d}}{\frac{\lambda_0}{D}} \tag{5.42}$$

Substituting $\alpha \approx \frac{L}{D}$ and $V_\pi = \left(\frac{d}{L}\right)\left(\frac{\lambda_0}{rn^3}\right)$, we have

$$N \approx \frac{V}{2V_\pi} \tag{5.43}$$

From equation (5.43) we have $V \approx 2NV_\pi$. This is a disadvantage, since to scan *N* independent spots, a voltage 2N times greater than the half-wave voltage is necessary. Since the value of V_π is usually large, to make a useful scanner with N >> 1 requires unacceptably high voltages. Hence more popular scanners include mechanical and acousto-optic scanners.

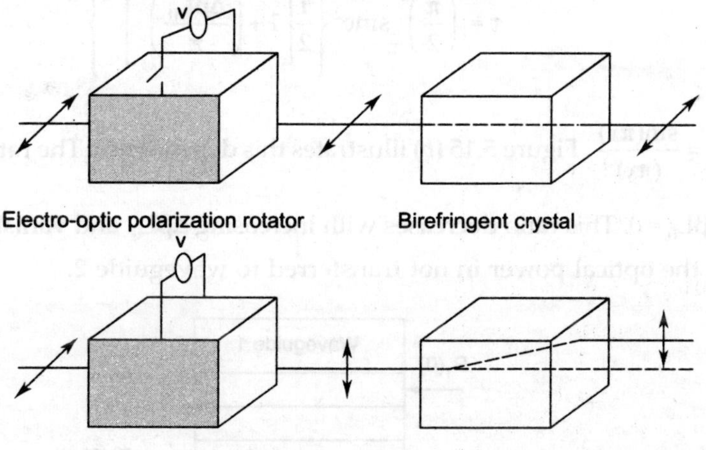

Fig. 5.14

The process of double refraction is anisotropic crystals introduces a lateral shift of an incident beam parallel to itself for one polarization and no shift for the other polarization. This effect can be used for switching a beam between two parallel positions by switching the polarizations. A linearly polarized optical beam is transmitted first through an electro-optic wave retarder acting as a polarization rotator and then through the crystal. The rotator controls the polarization electrically, which determines the beam is shifted laterally or not, as shown in Fig. 5.14.

5.5.4 Directional Couplers

A very important application of the electro-optic effect is in controlling the coupling between two parallel waveguides in an integrated-optical device. This can be used to transfer the light from one waveguide to the other, and the device serves as an electrically controlled directional coupler.

Let us consider the coupling of light between two parallel single-mode planar waveguides. The optical powers carried by the two waveguides, $P_1(z)$ and $P_2(z)$, are exchanged periodically along the direction of propagation z. Two parameters control the strength of this coupling process. They are the coupling coefficient ζ (which depends on the dimensions, refractive indices and wavelengths), and the mismatch of the propagation constants $\Delta\beta = \beta_1 - \beta_2 = \frac{2\pi\Delta n}{\lambda_0}$, where Δn is the difference between the refractive indices of the waveguides. If they are identical with $\Delta\beta = 0$ and $P_2(z) = 0$, then at a distance $z = L_0 = \frac{\pi}{2\zeta}$, called the transfer distance or coupling length, the power is transferred completely from the waveguide 1 into waveguide 2, *i.e.*, $P_1(L_0) = 0$ and $P_2(L_0) = P_1(0)$, as shown in Fig. 5.15 (*a*). Power is exchanged between two parallel identical weakly coupled waveguides with the same propagation constant β. AT $z = 0$ all of the power is in waveguide 1. At $z = L_0$ all of the power is transferred into waveguide 2.

For a waveguide of length L_0 and $\Delta\beta \neq 0$, the power transfer ratio $\tau = \frac{P_2(L_0)}{P_1(0)}$ is a function of the phase mismatch.

$$\tau = \left(\frac{\pi}{2}\right)^2 \text{sinc}^2\left\{\frac{1}{2}\left[1 + \left(\frac{\Delta\beta L_0}{\pi}\right)^2\right]^{1/2}\right\} \tag{5.44}$$

where $\text{sinc}^2 = \frac{\sin(\pi x)}{(\pi x)}$. Figure 5.15 (*b*) illustrates this dependence. The ratio has its maximum value of unity at $\Delta\beta L_0 = 0$. This ratio decreases with increasing $\Delta\beta L_0$, and vanishes when $\Delta\beta L_0 = \sqrt{3}\,\pi$, at which point the optical power in not transferred to waveguide 2.

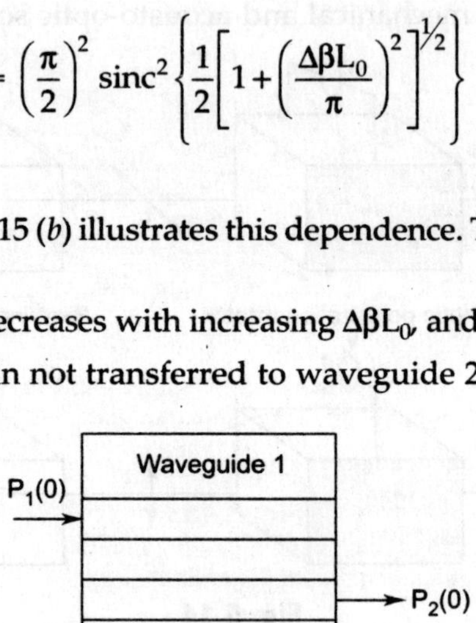

Fig. 5.15(*a*)

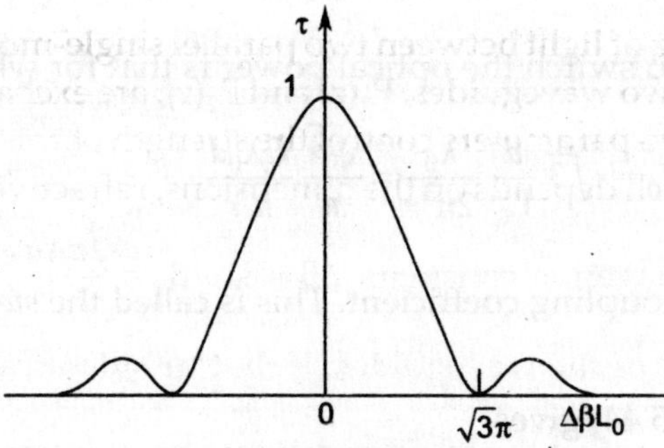

Fig. 5.15(*b*)

The key to make electrically controllable directional coupler is the dependence of the coupled power on the phase mismatch. If the phase mismatch $\Delta\beta L_0$ is switched from 0 to $\sqrt{3}\pi$, the light remains in waveguide 1. This electrical control of $\Delta\beta$ is achieved by using the electro-optic effect. The electrical field is applied in one of the two identical waveguides. This alters the refractive index by $\Delta n = -\frac{1}{2}n^3 rE$, where *r* is the Pockels coefficient. This results in a phase shift of

$$\Delta\beta L_0 = \Delta n \frac{2\pi L_0}{\lambda_0} = -\frac{\pi}{\lambda_0} n^3 r L_0 E \tag{5.45}$$

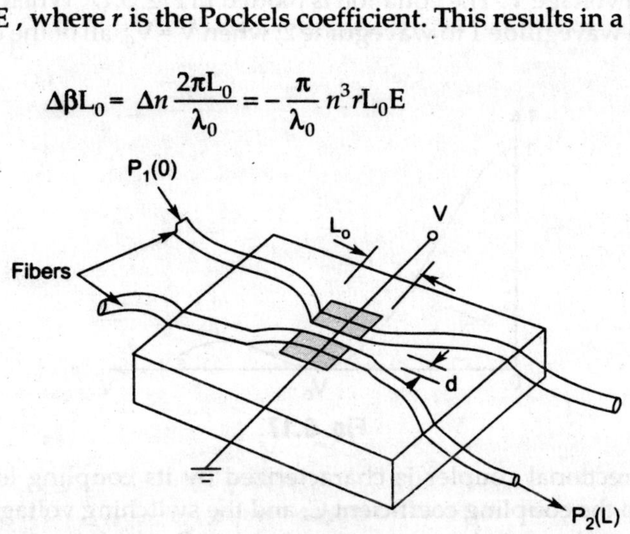

Fig. 5.16

A typical electro-optic directional coupler is shown in Fig. 5.16. The electrodes are laid over two waveguides separated by a distance *d*. If a voltage V is applied, then an electric field $E \approx \frac{V}{d}$ is created in one of the waveguides and $E \approx -\frac{V}{d}$ in the other. The refractive index is increased in one waveguide and decreased in the other. This results in a net refractive index difference of $2\Delta n = -n^3 r \frac{V}{d}$, corresponding to a phase mismatch factor $\Delta\beta L_0 = -\left(\frac{2\pi}{\lambda_0}\right) n^3 r \left(\frac{L_0}{d}\right) V$ which is proportional to the applied voltage V.

The voltage V_0 necessary to switch the optical power is that for which $|\Delta\beta L_0| = \sqrt{3}\pi$, *i.e.*,

$$V_0 = \sqrt{3}\frac{d}{L_0}\frac{\lambda_0}{2n^3 r} = \frac{\sqrt{3}}{\pi}\frac{C\lambda_0 d}{n^3 r} \tag{5.46}$$

where $L_0 = \dfrac{\pi}{2C}$ and C is the coupling coefficient. This is called the *switching voltage*. Substituting

$|\Delta\beta L_0| = \sqrt{3}\pi\dfrac{V}{V_0}$, equation (5.44) gives

$$\tau = \left(\frac{\pi}{2}\right)^2 \text{sinc}^2\left\{\frac{1}{2}\left[1 + 3\left(\frac{V}{V_0}\right)^2\right]^{1/2}\right\} \tag{5.47}$$

Equation (5.47) gives the coupling efficiency. This equation governs the coupling power as a function of the applied voltage V. The equation is plotted in Fig. 5.17. When V = 0, all of the optical power is coupled from waveguide 1 to waveguide 2; when $V = V_0$, all of the optical power remains in waveguide 1.

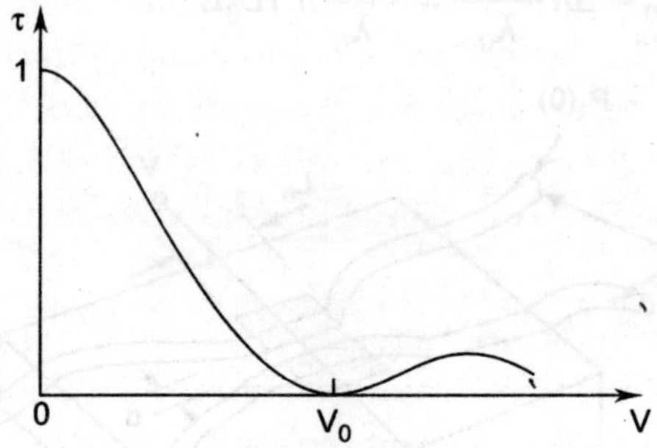

Fig. 5.17

An electro-optic directional coupler is characterized by its coupling length L_0, which is inversely proportional to the coupling coefficient C, and the switching voltage V_0, which is directly proportional to C. Hence the most important parameter is C, which is governed by the geometry and the refractive indices.

5.6 KERR MODULATORS

Many isotropic media, when placed in an electric field behave as uniaxial crystals with the optic axis parallel to the electric field. The difference in refractive indices for light polarized parallel to and perpendicular to the induced optic axis is given by:

$$\Delta n = n_p - n_s = \xi\lambda\varepsilon^2 \tag{5.48}$$

where ξ is the Kerr constant and λ is the vacuum wavelength.

The electric field induces an electric moment in nonpolar molecules and changes the moment of polar molecules. As a result there is then a reorientation of the molecules by the field which causes the medium as a whole to become anisotropic. This explains the delay that occurs between the application of the field and the appearance of the maximum effect. For nonpolar liquids the delay is very small and probably less than 10^{-11} seconds. However the delay can be several seconds. A Kerr cell filled with one of such liquids and placed between crossed polaroids can be used as a modulator or an optical switch. Modulation at frequencies up to 10^{10} Hz has been obtained.

In the optical frequency Kerr effect, the refractive index change is brought about by an applied optical frequency field. An interesting possibility is there of using one beam of light to switch another, if the refractive index change can be used to eliminate the second beam. Using a Fabry-Perot interferometer with an electro-optic material as the spacer between the reflecting plates such optical beam switching can be achieved. Intense maxima in the interference pattern formed by light passing through the interferometer occurs when $p\pi = \left(\frac{2\pi}{\lambda}\right) nd \cos\theta$, where p is the order of interference, d is the plate separation and θ is the angle between an internally reflected ray and the surface normal. A small change in the refractive index n of the spacer material induced by an optical beam would detune the interferometer and switch off the transmitted beam.

5.7 ACOUSTO-OPTIC EFFECT

The acousto-optic effect is the change in the refractive index of a medium caused by the mechanical strains produced by the passage of an acoustic wave through the medium. As the strain in the acoustic wave varies periodically with a wavelength Λ, the refractive index of the medium also does the same. This leads to the creation of a dynamic refractive index grating. The refractive index changes are caused by the photoelastic effect which occurs in all materials due to the application of a mechanical stress. The change in refractive index is proportional to the square root of the total acoustic power.

When a light beam is incident on such a refractive index grating, diffraction occurs. This will produce either single order diffraction or multiple order diffraction. The former is referred to as Bragg diffraction, which usually takes place at high acoustic frequencies. This is analogous to the Bragg diffraction of X-rays in crystals. The latter is usually observed at low acoustic frequencies and is referred to as Raman-Nath diffraction. Many useful devices make use of the acousto-optic effect; these include acousto-optic modulators, switches, deflectors, filters, isolators, frequency shifters for heterodyning, spectrum analyzers, Q-switching and mode-locking in lasers.

Fig. 5.18

We can consider a situation where a monochromatic light wave of wavelength λ, is incident upon a medium in which an acoustic wave has produced sinusoidal variations of wavelength Λ in the refractive index. A schematic illustration of acousto-optic modulation is shown in Fig. 5.18, where the solid horizontal lines represent acoustic peaks (pressure maxima) and the dashed horizontal lines represent acoustic wave troughs (pressure minima). As the light enters the medium, the portions of the wavefront near to the pressure peak will encounter a higher refractive index and hence advance with a lower velocity than those portions of the wavefront which experiences pressure minima. Hence the wavefront in the medium soon acquires the wavy nature as shown by the dashed curve in Fig. 5.18.

Since optical frequencies are much higher compared to the acoustic frequencies, the variations of the refractive index in a medium perturbed by sound are usually very slow in comparison with an optical period. There are therefore two significantly different time scales for light and sound. As a consequence, it is possible to use an adiabatic approach in which the optical propagation problem is solved separately at every instant of time during the relatively slow course of the acoustic cycle, always treating the material as if it were a static (frozen) inhomogeneous medium.

As elements of light wave propagate in a direction normal to the local wavefront, almost all the wave elements will suffer a change in direction leading to a re-distribution of the light flux, which tends to concentrate near regions of compression. In effect, the acoustic wave sets up a diffraction grating within the medium so that optical energy is diffracted out of the incident beam into various orders. This periodic refractive index grating has the same period as the acoustic wave and is also propagating at the same velocity as the acoustic wave. Typically the refractive index variations are about 10^{-4} around the mean refractive index value. As already mentioned there are two main cases of interest, viz. the *Raman-Nath regime* and *the Bragg regime*.

If the length of interaction L between the optical and acoustic waves satisfies

$$L << \frac{k}{K^2} = \frac{\Lambda^2 n_0}{2\pi\lambda_0}$$

$$K = \frac{2\pi}{\Lambda}, \quad k = \frac{2\pi n_0}{\lambda_0} \tag{5.49}$$

where Λ is the acoustic wavelength, λ_0 is the free space optical wavelength and n_0 is the refractive index of the medium, then the incident light waves diffract into multiple orders. This is referred to as *Raman-Nath* diffraction.

On the other hand, only one diffraction order is produced if

$$L >> \frac{k}{K^2} \tag{5.50}$$

This too will occur only when the Bragg condition is satisfied.

The corresponding angle of incidence θ is close to

$$\theta_B = \sin^{-1}\left(\frac{\lambda_0}{2n_0\Lambda}\right) \tag{5.51}$$

This is referred to as *Bragg* diffraction.

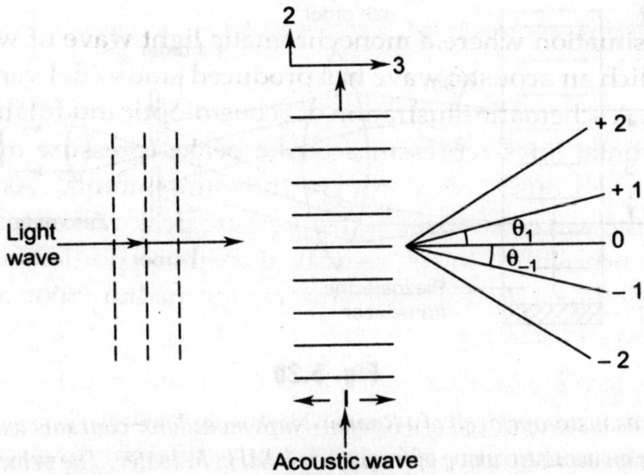

Fig. 5.19

The geometry of Raman-Nath diffraction is shown in Fig. 5.19 that leads to multiple diffraction. In the Raman-Nath regime, the acoustically perturbed medium acts as a thin phase diffraction grating. In fact, the grating produced is so 'thin' that before leaving the modulator, the diffracted light suffers no further redistribution. The light is diffracted as from a simple plane grating such that

$$\sin \theta_m = \frac{m\lambda_0}{n_0 \Lambda} \tag{5.52}$$

where n is the refractive index of the medium in the absence of the acoustic wave, $m = 0, \pm 1, \pm 2, \ldots$, is the order number and θ_m is the corresponding angle of diffraction, as shown in Fig. 5.20.

In these orders, the irradiance I of the light depends on the 'ruling depth' of the acoustic grating. This, in turn, is related to the amplitude of the acoustic grating, which is again related to the amplitude of the acoustic modulating wave. If I_0 is the transmitted irradiance in the absence of the acoustic wave, then the faction of the light removed from the zero order beam is given by $\eta = \frac{(I_0 - I)}{I_0}$.

Hence amplitude variations of the acoustic wave are transformed into irradiance variations of the optical beam. For small acoustic powers, the relative intensity in the first order is given by:

$$\eta \approx \frac{\pi^2 (\Delta n)^2 L^2}{\lambda^2} \tag{5.53}$$

where Δn is the peak change in refractive index of the medium due to the acoustic wave and L is the width of the acoustic beam, normally equal to the length of the medium. It may be shown that $(\Delta n)^2$ is proportional to the acoustic power. Hence, if the acoustic wave is amplitude-modulated, the first-order diffracted beam (corresponding to $m = \pm 1$) will be intensity modulated.

If the light diffracted from the incident beam is extensively re-diffracted before leaving the acoustic field, it forms the physical basis of the Bragg regime. Under these conditions, the acoustic field acts like a 'thick' diffraction grating, made up of planes rather than lines. The situation is then very similar to that of Bragg diffraction (or 'reflection') of X-rays from planes of atoms in a crystal.

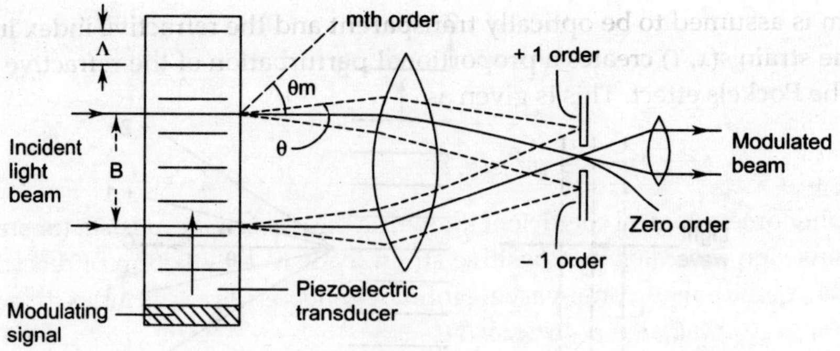

Fig. 5.20

Example 5.5. *A typical acousto-optic cell of a Raman-Nath modulator contains water. A piezoelectric crystal bonded to the cell generates an acoustic wave of frequency 6 MHz in water. The velocity of the acoustic wave in water is 1500 ms^{-1} and the thickness of the cell is 2 cm. If a He-Ne laser beam of wavelength 633 nm is incident on the cell, calculate the (a) angle between the first-order diffracted beam and the direct beam and (b) relative intensity of the diffracted beam of the first order if $\Delta n = 10^{-5}$. Take $n_0 = 1.33$ for water.*

Solution:

(*a*) From equation (5.52), we have $\sin\theta_m = \dfrac{m\lambda_0}{n_0\Lambda}$

Here, $m = 1$, $\lambda_0 = 633$ nm, $n_0 = 1.33$ (for water).

Now, $$\Lambda = \frac{\text{Velocity of acoustic wave in water}}{\text{Frequency of acoustic wave in water (Hz)}} = \frac{1500\ \text{ms}^{-1}}{6\times10^6\ \text{s}^{-1}} = 250\times10^{-6}\ \text{m}$$

$= 250$ μm

Hence, $$\theta_1 = \sin^{-1}\left[\frac{1\times633\times10^{-9}}{1.33\times250\times10^{-6}}\right] = 0.11°$$

(*b*) From equation (5.53),

$$\eta \approx \frac{\pi^2(\Delta n)^2 L^2}{\lambda^2} = \frac{\pi^2\left(10^{-5}\right)^2\times\left(2\times10^{-2}\right)^2}{\left(633\times10^{-9}\right)^2} = 0.985$$

5.7.1 Bragg Diffraction

Consider an acoustic plane wave traveling in the x-direction in a medium with velocity v_s, frequency f, and wavelength $\Lambda = \dfrac{v_s}{f}$. The strain (relative displacement) at position x and time t is

$$s(x, t) = S_0\cos(\Omega t - qx) \tag{5.54}$$

where S_0 is the amplitude, $\Omega = 2\pi f$ is the angular frequency, and $q = \dfrac{2\pi}{\Lambda}$ is the wave number. The acoustic intensity (W/m^2) is

$$I_s = \frac{1}{2}\rho v_s^3 S_0^2 \tag{5.55}$$

where ρ is the mass density of the medium.

The medium is assumed to be optically transparent and the refractive index in the absence of sound in n. The strain $s(x, t)$ creates a proportional perturbation of the refractive index, which is analogous to the Pockels effect. This is given as

$$\Delta n(x,t) = -\frac{1}{2} p n^3 s(x,t) \tag{5.56}$$

where p is the phenomenological coefficient known as the *photoelastic constant* (or strain-optic coefficient). The minus sign indicates that positive strain leads to a reduction of refractive index. As a consequence, the medium has a time-varying inhomogeneous refractive index in the form of a wave

$$n(x,t) = n - \Delta n_0 \cos(\Omega t - qx) \tag{5.57}$$

with amplitude

$$\Delta n_0 = \frac{1}{2} p n^3 S_0 \tag{5.58}$$

Now, substituting from equation (5.55) into (5.58), we find that the change of the refractive index is proportional to the square root of the acoustic intensity,

$$\Delta n_0 = \left(\frac{1}{2} \mathrm{M} I_s\right)^{\frac{1}{2}} \tag{5.59}$$

where

$$\mathrm{M} = \frac{p^2 n^6}{\rho v_s^3} \tag{5.60}$$

is a material parameter representing the effectiveness of sound in altering the refractive index. M is a figure of merit for the strength of the acousto-optic effect in the material.

Now we consider an optical plane wave traveling in this medium with frequency ν, angular frequency $\omega = 2\pi\nu$, free-space wavelength $\lambda_0 = \frac{c_0}{\nu}$, wavelength in the unperturbed medium $\lambda = \frac{\lambda_0}{n}$ corresponding to a wave number $k = \frac{n\omega}{c_0}$, and wave vector k lying in the x-z plane and making an angle θ with the z-axis, as shown in Fig. 5.21. Bragg diffraction due to acoustic plane wave acts as a partial reflector of light when the angle of incidence θ satisfies the Bragg condition.

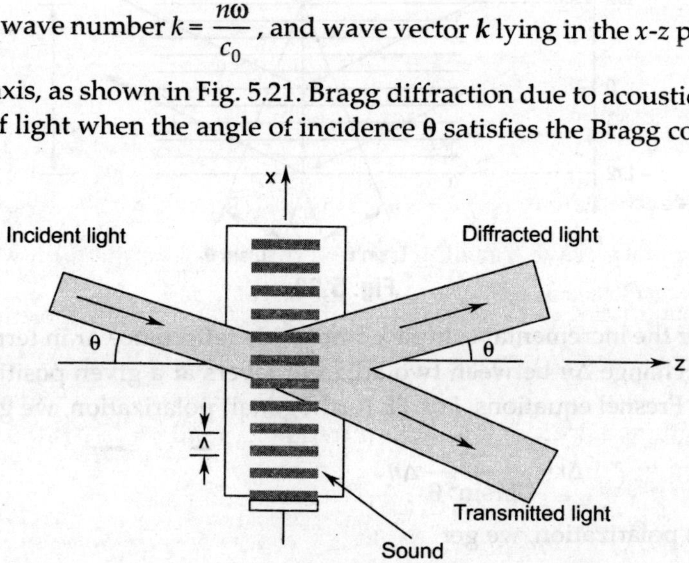

Fig. 5.21

Since the acoustic frequency f is much smaller than the optical frequency ν (by at least five orders of magnitude), an adiabatic approach for studying light-sound interaction may be adopted. We regard the refractive index as a static "frozen" sinusoidal function

$$n(x) = n - \Delta n_0 \cos(qx - \phi) \tag{5.61}$$

where φ is affixed phase. We determine reflected light from this inhomogeneous (graded-index) medium and track its slow variation with time by taking $\varphi = \Omega t$.

To determine the amplitude of the reflected wave we divide the medium into two incremental planar layers orthogonal to the x-axis. The incident optical plane wave is partially reflected at each layer because of the refractive index change. We assume that the reflectance is sufficiently small so that the transmitted light from one layer approximately maintains its original magnitude (*i.e.*, is not depleted) as it penetrates through the following layers of the medium.

If $\Delta r = \left(\dfrac{dr}{dx}\right)\Delta x$ is the incremental complex amplitude reflectance of the layer at position x, the total complex amplitude reflectance for an overall length L (see Fig. 5.22) is the sum of all incremental reflectances,

$$r = \int_{-L/2}^{L/2} e^{j2kx\sin\theta} \frac{dr}{dx}\, dx \tag{5.62}$$

The phase factor $e^{j2kx \sin\theta}$ is included since the reflected wave at a position x is advance by a distance $2x \sin\theta$ (corresponding to a phase shift $2kx \sin\theta$) relative to the reflected wave at $x = 0$ (see Fig. 5.22). The wavenumbers for the incident and the reflected waves are taken to be the same.

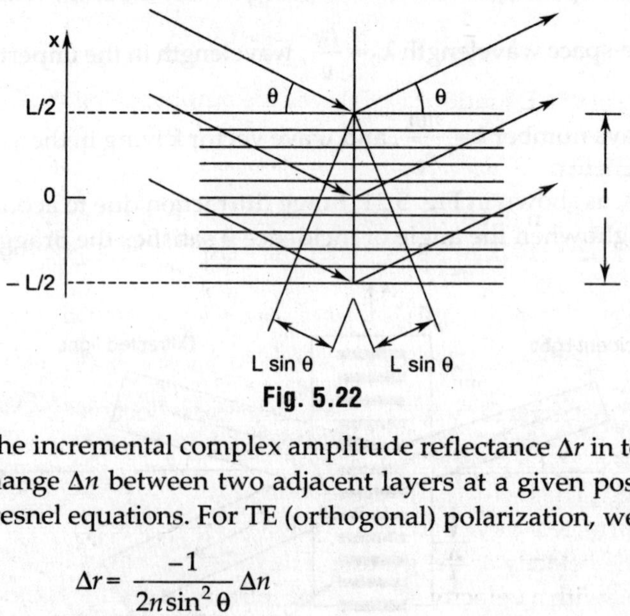

Fig. 5.22

An expression for the incremental complex amplitude reflectance Δr in terms of the incremental refractive-index change Δn between two adjacent layers at a given position x may be determined by using the Fresnel equations. For TE (orthogonal) polarization, we get

$$\Delta r = \frac{-1}{2n\sin^2\theta}\Delta n \tag{5.63}$$

For TM (parallel) polarization, we get

$$\Delta r = \frac{-\cos 2\theta}{2n\sin^2\theta}\Delta n \tag{5.64}$$

In most acousto-optic devices value of θ is very small, so that $\cos 2\theta \approx 1$, making equation (5.63) applicable to both polarizations.

Using equations (5.61) and (5.63), we obtain

$$\frac{dr}{dx} = \frac{dr}{dn}\frac{dn}{dx} = \frac{-1}{2n\sin^2\theta}\left[q\,\Delta n_0 \sin(qx-\varphi)\right] = r' \sin(qx-\varphi) \tag{5.65}$$

where
$$r' = \frac{-q}{2n\sin^2\theta}\Delta n_0 \tag{5.66}$$

Finally we substitute equation (5.65) into equation (5.62), and use complex notation to write as $\sin(qx-\varphi) = \dfrac{e^{j(qx-\varphi)} - e^{-j(qx-\varphi)}}{2j}$, thereby obtaining

$$r = \frac{1}{2}jr'e^{j\varphi}\int_{-L/2}^{L/2} e^{j(2k\sin\theta - q)x}\,dx - \frac{1}{2}jr'e^{-j\varphi}\int_{-L/2}^{L/2} e^{j(2k\sin\theta + q)x}\,dx \tag{5.67}$$

The first term in equation (5.67) has its maximum value when $2k\sin\theta = q$, whereas the second term is maximum when $2k\sin\theta = -q$. If L is sufficiently large, these maxima are sharp, so that any slight deviation from the angles $\theta = \pm\sin^{-1}(q/2k)$ makes the corresponding term negligible. Thus only one of these two terms may be significant at a time, depending on the angle θ. The conditions $2k\sin\theta \approx q$ and $2k\sin\theta \approx -q$ are called the upshifted and downshifted reflections respectively. We first consider the upshifted condition, $2k\sin\theta \approx q$, for which the second term is negligible. Performing the integral in the first term of equation (5.67) and substituting $\varphi = \Omega t$, we obtain

$$r = \frac{1}{2}jr'\mathrm{L}\,\mathrm{sinc}\left[(q - 2k\sin\theta)\frac{\mathrm{L}}{2\pi}\right]e^{j\Omega t} \tag{5.68}$$

where $\mathrm{sinc}(x) = \sin(\pi x)/(\pi x)$. Equation (5.68) gives the amplitude reflectance for upshifted case.

5.7.2 Bragg Condition

The sinc function in equation (5.68) has its maximum value of 1.0 when its arguments is zero, *i.e.*, when $q = 2k\sin\theta$. This occurs when $\theta = \theta_B$, where $\theta_B = \sin^{-1}(q/2k)$ is the Bragg angle. Since $q = 2\pi/\Lambda$ and $k = 2\pi/\lambda$,

$$\sin\theta_B = \frac{\lambda}{2\Lambda} \tag{5.69}$$

The Bragg angle is the angle for which the incremental reflections from the planes separated by an acoustic wavelength Λ have a phase shift of 2π so that they interfere constructively.

The acoustic waves, which create the diffraction grating, are moving through the medium. As a consequence the diffracted wave behaves as if it had been reflected from a 'mirror' which is also moving with the same velocity as the grating. The reflected waves therefore appears to originate from a source moving with a velocity at twice the mirror (or grating) velocity. Thus the frequency of the reflected beam is changed by the Doppler effect and is given by

$$\nu' = \nu_0\left(1 \pm 2V_a\left(\frac{c}{n}\right)\right) \tag{5.70}$$

where $\pm V_a$ is the component of velocity of the acoustic wave along (or away from) the original; beam direction and n is the refractive index of the medium. The frequency shift is thus given as

$$\Delta\nu = \nu' - \nu_0 = \frac{\pm 2\nu_0 V_a n}{c} \tag{5.71}$$

If the light is incident at an angle $\theta = \theta_B$ to the acoustic wave, then $V_x = V_a \sin\theta_B$, where V_a is the acoustic wave velocity. The frequency shift is then given as

$$\Delta\nu = \frac{\pm 2\nu_0 V_a \sin\theta_B n}{c} \tag{5.72}$$

Combining equations (5.72) and (5.69) gives a frequency shift of $\pm V_a/\Lambda$ or $\pm f_0$, where f_0 is the acoustic wave frequency. This change in frequency can be used as the basis of a frequency modulator.

5.8 ACOUSTO-OPTIC DEVICES

We have already discussed the basic principle behind acousto-optic interaction. In this section we will discuss some acousto-optic devices. The intensity of the diffracted light depends on the acoustic power. Hence by changing the acoustic power the intensity of the diffracted light can be modulated correspondingly. This is the basic principle of the acousto-optic modulator. Again the angle of diffraction depends on the acoustic frequency. Hence if we have an acoustic beam whose frequency is changed, the corresponding diffracted light will appear along different directions. This forms the basis of acousto-optic deflector. Similarly, if an acousto-optic transducer is fed simultaneously with a signal containing different frequencies then corresponding to each frequency, the diffracted light appears along different directions. This principle is used in acousto-optic spectrum analyzer.

5.8.1 Modulators

Raman-Nath Acouto-Optic Modulator

Figure 5.23 shows an acousto-optic modulator based on Raman-Nath diffraction. The signal carrying the information modulates the amplitude of the acoustic wave propagating through the medium. The light beam incident on the cell is diffracted. At the output the zero order beam is blocked using a stop as shown in Fig. 5.23. We are not going in the detail calculation as that is beyond the scope of our current book. We find

$$\eta = \frac{\pi^2 M_2}{2\lambda_0^2}\left(\frac{L}{H}\right)P_a \tag{5.73}$$

where η is the diffraction efficiency, L/H is the aspect ratio, P_a is the acoustic power, and M_2 is the figure of merit given by

$$M_2 = \frac{n^6 \bar{p}^2}{\rho v_a^3} \tag{5.74}$$

where n represents the effective refractive index, $\bar{p}$ is the effective acousto-optic coefficient, ρ is the density of the medium and v_a is the acoustic velocity in the medium. From equation (5.73), for small acoustic powers η is proportional to P_a. Thus the diffracted power will be modulated correspondingly, as the acoustic beam is modulated. To obtain low acoustic powers M_2 should be large and the aspect ratio L/H should also be large.

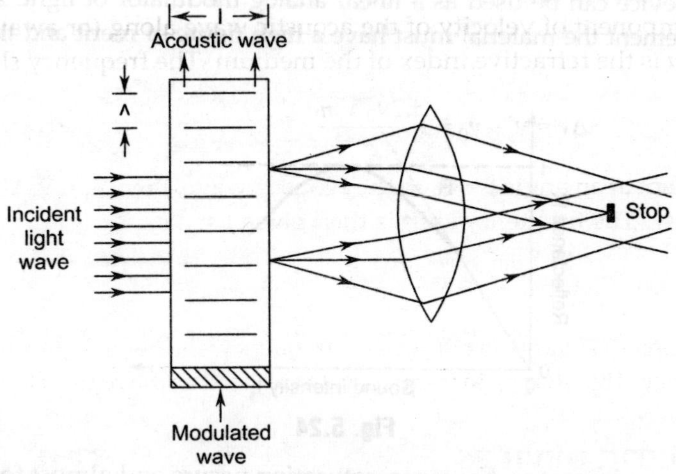

Fig. 5.23

In Raman-Nath interaction, one of the main drawbacks is the restriction on the length of interaction. For Raman-Nath diffraction, we must have

$$L \ll \frac{n\Lambda^2}{2\pi\lambda_0} = \frac{nv_a^2}{2\pi\lambda_0 f^2}, \tag{5.75}$$

where $\Lambda(= v_a/f)$ is the acoustic wavelength and f and v_a are the acoustic frequency and acoustic velocity respectively. From equation (5.75), we find that for low frequencies, the restriction imposed is not too severe, but for higher frequencies the maximum allowable value of L becomes too small.

Hence Raman-Nath modulators can be used at relatively low acoustic frequencies and also with limited modulation bandwidths. For higher carrier frequencies and higher modulation bandwidths, Bragg modulators are advantageous over Raman-Nath modulators.

Bragg Acouto-Optic Modulator

If the interaction length satisfies the condition

$$L \gg \frac{nv_a^2}{2\pi\lambda_0 f^2} \tag{5.76}$$

then the acousto-optic interaction falls in the Bragg regime. Again in this regime the light wave must be incident at the Bragg angle θ_B given be equation (5.69) so that it may be coupled to the diffracted wave. The diffraction efficiency for incidence at the Bragg angle is given by

$$\eta = \sin^2\left[\frac{\pi}{\lambda_0 \cos\theta_B}\left(\frac{M_2}{2}\frac{L}{H}P_a\right)^{\frac{1}{2}}\right] \tag{5.77}$$

For small values of P_a, the above equation reduces to

$$\eta \approx \frac{\pi^2 M_2}{2\lambda_0^2 \cos^2\theta_B}\left(\frac{L}{H}\right)P_a \tag{5.78}$$

Equations (5.78) and (5.73) are almost same. Hence, the intensity of the diffracted light is proportional to the acoustic power, if the acoustic power is sufficiently weak. Using an electrically controlled acoustic transducer, as shown in Fig. 5.25(*a*), the intensity of the diffracted beam can be varied proportionally. The device can be used as a linear analog modulator of light. As before, for a low acoustic power requirement the material must have a high figure of merit and the aspect ratio (L/H) should be large.

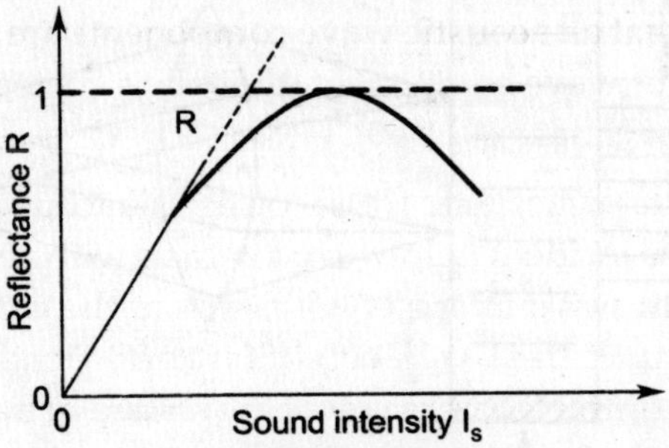

Fig. 5.24

As the acoustic power increases, however, saturation occurs and almost total reflection can be achieved (see Fig. 5.24). The figure shows the dependence of the reflectance R_e of the Bragg reflector on the intensity of sound I_s. When I_s is small $R_e = R$, which is a linear function of I_s. The modulator then serves as an optical switch, which by switching the sound off and on, turns the reflected light off and on, and the transmitted light on and off respectively, as shown in Fig. 5.25(*b*).

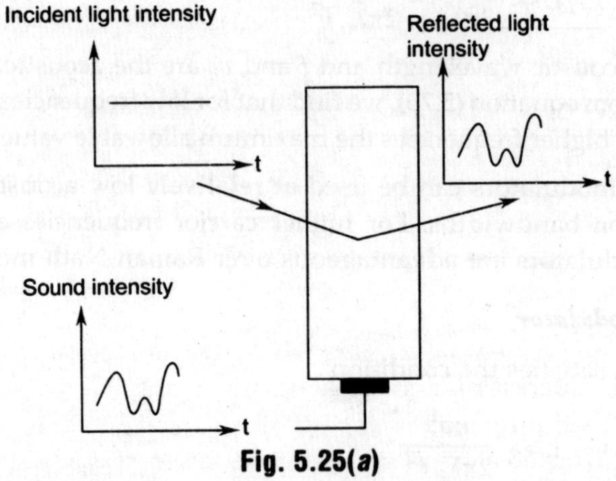

Fig. 5.25(*a*)

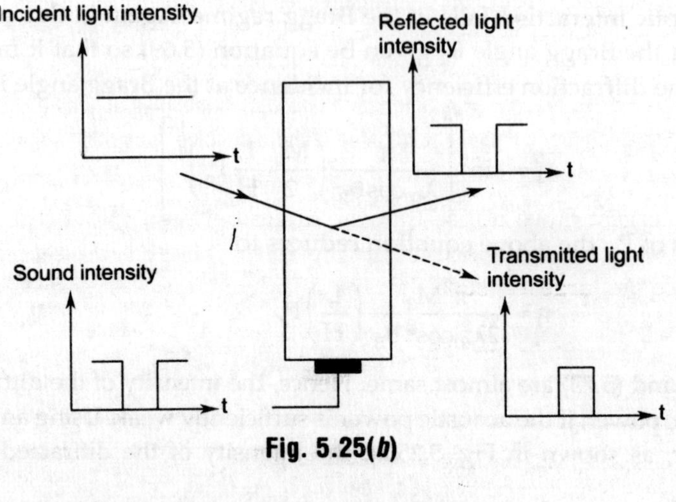

Fig. 5.25(*b*)

Modulation Bandwidth

The bandwidth of a modulator is one of the most important characteristics. We consider a case when both the incident light wave and the acoustic wave have finite angular divergences of $\delta\theta_o$ and $\delta\theta_a$ respectively. In order that all acoustic wave components are used in Bragg diffraction and also that for every incident light wave component, there be a corresponding acoustic wave component at Bragg angle, we must choose $\delta\theta_a \geq \delta\theta_o$.

Now, we know that a modulated acoustic wave consists of acoustic wave components at different frequencies. Hence when the incident light wave interacts with an amplitude modulated acoustic wave, then the incident light wave interacts with each of the component acoustic frequencies and Bragg diffraction will occur if the Bragg condition corresponding to the various frequency components is satisfied. As the diffracted light waves are shifted in frequency by the corresponding acoustic wave frequencies, when the diffracted light waves interfere, there will be a beating between the waves which ultimately results in the intensity modulation of the diffracted light.

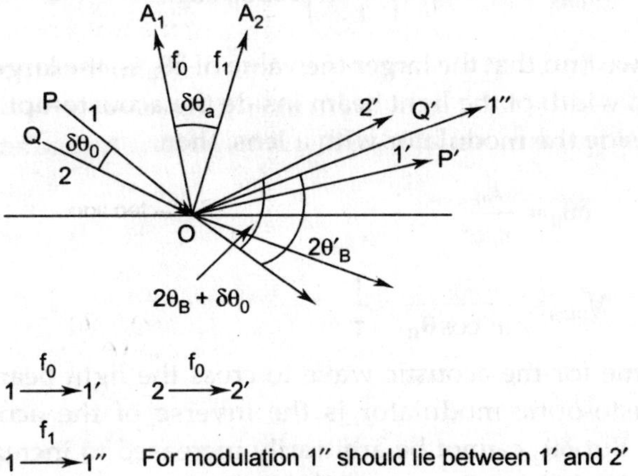

Fig. 5.26

In Fig. 5.26, we show an incident light wave which has an angular divergence $\delta\theta_0$ interacting with acoustic waves at frequencies f_0 and f_1 (each with an angular divergence $\delta\theta_a$). The incident light waves which have their propagation vectors along 1 and 2 interact with the acoustic waves at frequency f_0 and are diffracted along 1′ and 2′ directions. If we consider $f_1 > f_0$ then as the Bragg angle increases with acoustic frequency, the light wave incident along direction 1 may pick an acoustic wave at f_1 which satisfies the Bragg condition and be diffracted along a different direction (say 1″ as shown in Fig 5.26). Some of the incident light waves between directions 1 and 2 will pick corresponding acoustic waves at f_1 and give the diffracted light wave. From Fig. 5.26 it can be seen that the largest f_1 such that the light waves diffracted by acoustic waves at f_0 and f_1 overlap angularly is such that the diffracted light wave corresponding to $\nu + f'$ lies along direction 2′.

To calculate the largest f_1 we notice that the angle between the undiffracted and diffracted wave is twice the Bragg angle. Hence according to the above argument overlap will occur if $2\theta'_B < 2\theta_B + \delta\theta_0$. Hence the largest f_1 such that the diffracted wave is modulated corresponds to

$$\theta'_B = \sin^{-1}\left(\lambda_0 f_1 / 2n_0 v_a\right) = \theta_B + \frac{1}{2}\delta\theta_0$$

or
$$f_1 = \left(\frac{2n_0 v_a}{\lambda_0}\right) \sin\left(\theta_B + \frac{1}{2}\delta\theta_0\right)$$

$$\approx f_0 + \left(\frac{n_0 v_a}{\lambda_0}\right) \cos\theta_B \delta\theta_0 \tag{5.79}$$

Here we have assumed $\delta\theta_0 << 1$ and also used the relation

$$f_0 = \left(\frac{2n_0 v_a}{\lambda_0}\right) \sin\theta_B \tag{5.80}$$

Since an acoustic wave containing both frequencies f_0 and f_1 simultaneously corresponds to amplitude modulation at frequency $f_1 - f_0$, from equation (5.79) we get for the maximum modulation frequency

$$\Delta f_{max} = f_1 - f_0 = \left(\frac{n_0 v_a}{\lambda_0}\right) \cos\theta_B \, \delta\theta_0 \tag{5.81}$$

From equation (5.81) we find that the larger the value of $\delta\theta_0$ is, the larger Δf_{max} will be. Now if w represents the minimum width of the light beam inside the acousto-optic cell, when the incident light beam is focused inside the modulator with a lens, then

$$\delta\theta_0 \approx \frac{\lambda_0}{n_0 w} \tag{5.82}$$

Thus,
$$\Delta f_{max} \approx \frac{v_a}{w/\cos\theta_B} = \frac{1}{\tau} \tag{5.83}$$

where τ is the transit time for the acoustic wave to cross the light beam. Thus the modulation bandwidth of the acousto-optic modulator is the inverse of the acoustic propagation time through the light beam. But $\delta\theta_0$ cannot be arbitrarily increased to increase the bandwidth since if $\delta\theta_0$ becomes too large, the diffracted and undiffracted light waves may start overlapping. Hence the maximum modulation bandwidth will be such that there is no such overlap. It can be shown that the largest modulation bandwidth is half the centre frequency. Hence, for large bandwidths, f_0 must be large implying Bragg regime of operation.

5.8.2 Acousto-Optic Deflector

If a light wave is incident on an acoustic wave at the Bragg angle θ_B then the angle of deflection θ_D (*i.e.*, the angle between the diffracted and the incident wave) is $2\theta_D$. Hence assuming $\theta_B << 1$, we have

$$\theta_D = 2\theta_B \approx \left(\frac{\lambda_0}{n_0 v_a}\right) f \tag{5.84}$$

where f is the acoustic frequency. Thus angle of deflection θ_D is directly proportional to the acoustic frequency f (see Fig. 5.27). Thus if the Bragg condition can be satisfied over a frequency range Δf, then the corresponding range over which the light beam may be deflected will be

$$\Delta\theta_D = \left(\frac{\lambda_0}{n_0 v_a}\right) \Delta f \tag{5.85}$$

Since the diffracted light wave will have a finite angular spread, the most interesting quantity of a deflector is the number of resolvable spots N, *i.e.*, the number of independently addressable directions. If $\delta\theta_0$ is the angular divergence of the optical beam then we have

$$N \approx \frac{\Delta\theta_d}{\delta\theta_0} \approx \frac{\lambda_0}{n_0 v_a} \Delta f \frac{n_0 w_0}{\lambda_0} = \tau\Delta f \tag{5.86}$$

where we have used equation (5.78) for $\delta\theta_0$ and $\tau(\approx w_0/v_a)$ represents the acoustic propagation time through the light beam which is also the access time of the beam deflector, *i.e.*, it represents the minimum time required to change the position of the deflected spot randomly. For a good resolution $\delta\theta_0$ must be small, hence the light wave must have a large width w_0. Thus in order to satisfy the Bragg condition over a range of frequencies Δf, the acoustic beam is chosen to have an angular divergence

$$\delta\theta \approx \Lambda/L \tag{5.87}$$

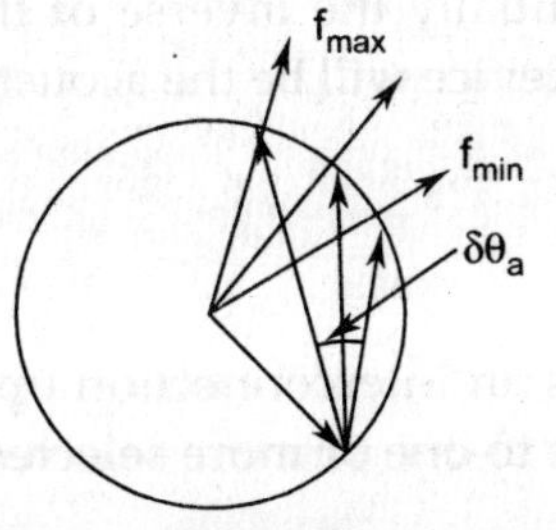

Fig. 5.27

A large Δf requires a small acoustic transducer length. On the other hand a reduction of length also leads to a reduction in diffraction efficiency (see equation (5.78)). In order to overcome this, various beam steering techniques have been devised in which the sound beam changes the ángle of propagation as the frequency changes so as to satisfy the corresponding Bragg angle.

5.8.3 Acousto-Optic Spectrum Analyzer

From equation (5.84), we see that the angle at which the diffracted beam appears is proportional to the acoustic frequency. When the acoustic transducer is fed with different frequencies f_1, f_2, f_3, ..., etc., simultaneously, then the Bragg diffracted light waves will propagate along different angles. If we arrange the configuration shown in Fig. 5.28, the different diffracted light beams will focus at different points on the focal plane of the lens. In addition, since the diffraction efficiency is proportional approximately to the power in the corresponding acoustic wave component, the diffracted light intensity pattern on the output plane will give the Fourier spectrum of the signal feeding the acoustic transducer. Hence the configuration shown in Fig. 5.28 leads to a real time spectral analysis of wide band signals and is referred to as a spectrum analyzer.

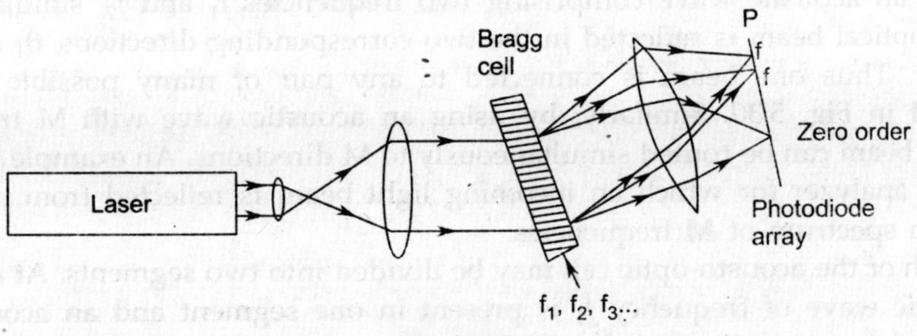

Fig. 5.28

Spectrum analyzer is essentially a deflector. Hence we can use the results obtained for deflector in the previous section. We have seen that the change in the angle of deflection $\Delta\theta_D$ for a change in acoustic frequency Δf is given by Equation (5.85). If $\delta\theta_0$ as given by equation (5.82) represents the angular divergence of the optical beam then the minimum resolvable frequency difference will be the value of Δf when $\Delta\theta_D = \delta\theta_0$. Naming this frequency resolution δf, we obtain from equations (5.82) and (5.85),

$$\left(\frac{\lambda_0}{n_0 v_a}\right)\delta f = \frac{\lambda_0}{n_0 w}$$

or

$$\delta f = \frac{v_a}{w} = \frac{1}{\tau} \tag{5.88}$$

i.e., the frequency resolution is essentially the inverse of the acoustic transit time through the optical beam. The bandwidth of the device will be the acoustic frequency range Δf over which the Bragg condition could be satisfied.

5.8.3 Interconnections

An acousto-optic cell can be used as an interconnection optical switch that routes information carried by one or more optical beams to one or more selected directions. Several interconnection schemes are possible:

- An acousto-optic cell in which the frequency of the acoustic wave is one of N possible values f_1, f_2, ..., or f_N, deflects an incident optical beam to one of N corresponding directions, θ_1, θ_2, ..., or θ_N. This is illustrated in Fig. 5.29. The device routes one beam to any of N directions. For example, the optical beam is defined by an angle θ_4 and routed to point 4 in the figure.

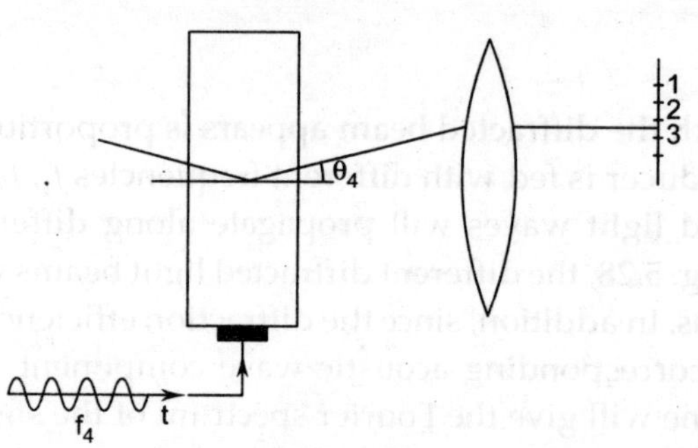

Fig. 5.29

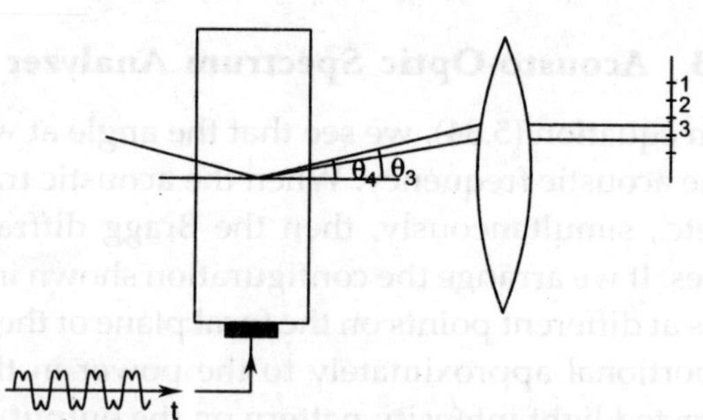

Fig. 5.30

- By using an acoustic wave comprising two frequencies f_1 and f_2, simultaneously, the incident optical beam is reflected in the two corresponding directions, θ_1 and θ_2, simultaneously. Thus one beam is connected to any pair of many possible directions as illustrated in Fig. 5.30. Similarly, by using an acoustic wave with M frequencies the incoming beam can be routed simultaneously to M directions. An example is the acoustic spectrum analyzer for which an incoming light beam is reflected from a sound wave carrying a spectrum of M frequencies.
- The length of the acousto-optic cell may be divided into two segments. At a certain time, an acoustic wave of frequency f_1 is present in one segment and an acoustic wave of frequency f_2 is present in the other. This can be accomplished by generating the acoustic wave from a frequency-shift-keyed electric signal in the form of two pulses: a pulse of

frequency f_1 followed by another of frequency f_2, each lasting duration $\frac{T}{2}$, where $T = \frac{W}{v_s}$ is the transit time of sound through the cell length W (see Fig. 5.31). When the leading edge of the acoustic wave reaches the end of the cell, the cell processes two incoming optical beams by deflecting the top beam to the direction θ_1 corresponding to f_1, and the bottom beam to the direction θ_2 corresponding to f_2. This is a switch that connects each of the two beams to any of many possible directions. By placing more than one frequency component in each segment, each of the two beams can itself be routed simultaneously to several directions.

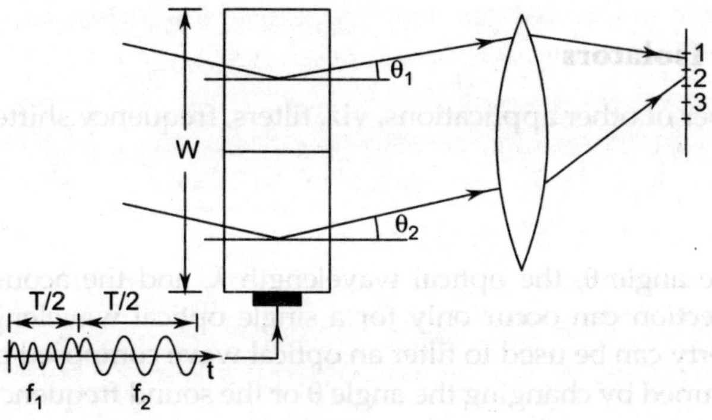

Fig. 5.31

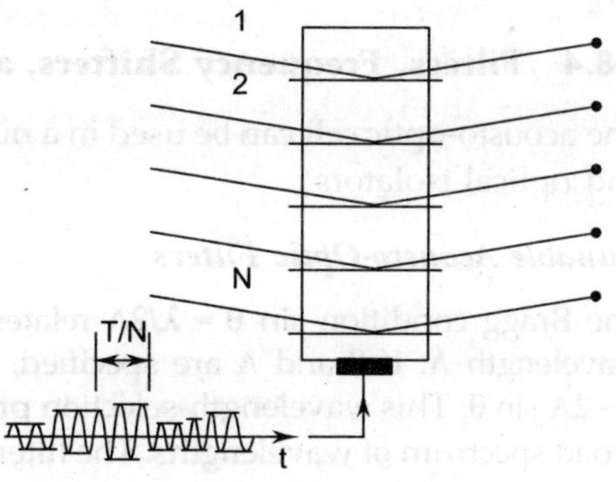

Fig. 5.32

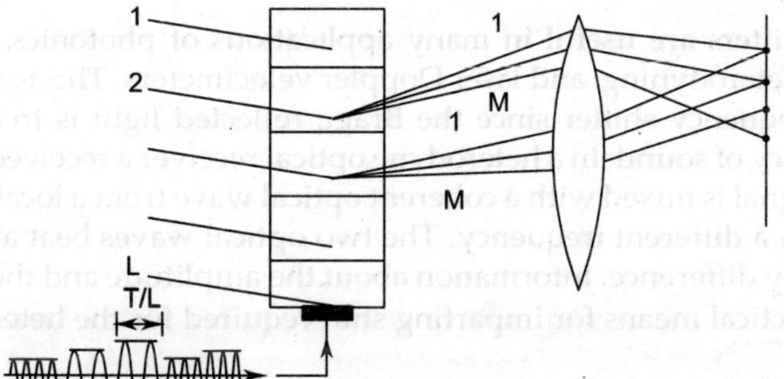

Fig. 5.33

- The cell may also be divided into N segments, each carrying a harmonic acoustic wave of the same frequency f but with a different amplitude. The result is a spatial light modulator that modulates the intensities of N input beams (see Fig. 5.32). Spatial Light Modulators (SLM) are useful in optical signal processing.
- The most general interconnection architecture is one for which the cell is divided into L segments, each of which carries an acoustic wave with M frequencies. The device acts as a random access switch that routes each of L incoming beams to M directions simultaneously. This is shown in Fig. 5.33.

Interconnection Capacity

There is an upper limit to the number of interconnections that may be established by the acousto-optic device. If an acousto-optic cell is used to route each of L incoming light beams to a maximum of M directions simultaneously, then the product ML cannot exceed the time-bandwidth product N = TB, where T is the transit time through the cell and B is the bandwidth of the acoustic wave.

$$ML \leq N \tag{5.89}$$

This upper bound on the number of interconnections is called the interconnection capacity of the device.

As acousto-optic cell with L segments uses an acoustic wave composed of L segments each of time duration T/L, for each segment to address M independent points the acoustic wave must carry M independent frequency components per segment. For a signal of duration T/L there is an inherent frequency uncertainty of L/T hertz. The M frequency components must therefore be separated by at least that uncertainty. For the M components to be placed within the available bandwidth B, we must have $M(L/T) \leq B$, from which $ML \leq TB$, and hence equation (5.89) follows.

5.8.4 Filters, Frequency Shifters, and Isolators

The acousto-optic cell can be used in a number of other applications, viz. filters, frequency shifters, and optical isolators.

Tunable Acousto-Optic Filters

The Bragg condition $\sin\theta = \lambda/2\Lambda$ relates the angle θ, the optical wavelength λ, and the acoustic wavelength Λ. If θ and Λ are specified, reflection can occur only for a single optical wavelength $\lambda = 2\Lambda \sin\theta$. This wavelength-selection property can be used to filter an optical wave composed of a broad spectrum of wavelengths. The filter is tuned by changing the angle θ or the sound frequency f.

Frequency Shifters

Optical frequency shifters are useful in many applications of photonics, including optical FM modulators, optical heterodyning, and laser Doppler velocimeters. The acousto-optic cell may be used as a tunable frequency shifter since the Bragg reflected light is frequency shifted (up or down) by the frequency of sound. In a heterodyne optical receiver a received amplitude- or phase-modulated optical signal is mixed with a coherent optical wave from a local light source, acting as a local oscillator with a different frequency. The two optical waves beat and the detected signal varies at the frequency difference. Information about the amplitude and the phase of the acousto-optic cell offers a practical means for imparting shift required for the heterodyning process.

Optical Isolators

An optical isolator is one-way optical valve, which is often used to prevent reflected light from retracing its path back to the original light source. Optical isolators are often used with semiconductor lasers since the reflected light can interact with the lasing process and create noise effects. The acousto-optic cell can serve as an isolator. If part of the frequency-upshifted Bragg-diffracted light is reflected onto itself by a mirror and traces its path back into the cell, as shown in Fig. 5.34, it undergoes a second Bragg diffraction accompanied by a second frequency upshift. Since the frequency of the returning light differs from that of the original light by twice the sound frequency, a filter may be used to block it. Even without a filter, the lasing process may be insensitive to the frequency-shifted light.

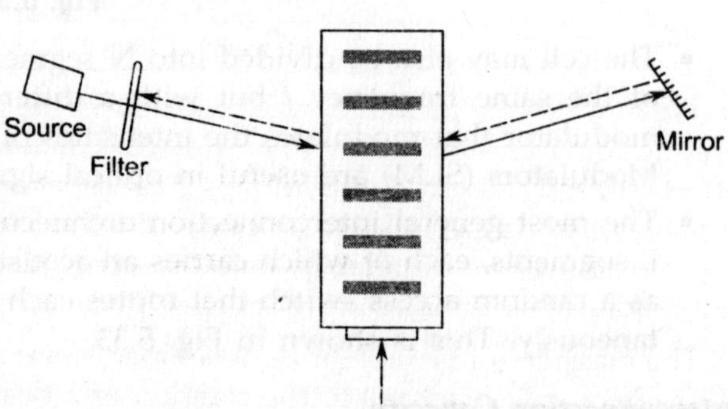

Fig. 5.34

5.9 MAGNETO-OPTIC DEVICES

The presence of magnetic fields may also affect the optical properties of some substances thereby giving rise to a number of useful devices. In general, as electric fields are easier to generate than magnetic fields, electro-optic devices are usually preferred to magneto-optic devices.

5.9.1 Faraday Effect

This is the simplest magneto-optic effect. This magneto-optic effect is used for optical modulators. This deals with the change in refractive index of a material subjected to a steady magnetic field. Faraday found that when a beam of polarized light passes through a substance subjected to a magnetic field, its plane of polarization is observed to rotate by an amount proportional to the magnetic field component parallel to the direction of propagation. The sense of rotation of the plane of polarization in the Faraday effect is independent of the direction of propagation. Hence here the rotation can be doubled by reflecting the light back through the Faraday Effect device.

The rotation of the plane of polarization is given as

$$\theta = VBL \tag{5.90}$$

where V is the Verdet constant, B is the magnetic flux parallel to the direction of propagation and L is the path length in the material. Some of the Verdet constants of commonly used materials are given in Table 5.2. The Faraday effect is small and wavelength dependent. We can also express θ in terms of the refractive indices n_r and n_l, where n_r and n_l are the different refractive indices for certain materials for right and left circularly polarized lights respectively. Thus θ may be expressed as

$$\theta = \frac{2\pi}{\lambda}(n_r - n_l)L \tag{5.91}$$

Table 5.2

Material	*V (rd m⁻¹ T⁻¹)*
Zinc sulfide (ZnS)	82
Quartz (SiO_2)	4.0
Flint glass	23
Sodium chloride (NaCl)	9.6
Crown glass	6.4

We now discuss certain devices which uses the Faraday effect. Optical isolators, circulators etc are such devices, which we will now discuss in detail. One potential application of magneto-optic effect currently receiving attention is large capacity computer memories. Such memories must be capable of storing very large amounts of information in a relatively small area and permit very rapid readout and, preferably, random access. The usual magnetic memories have a number of limitations of size and reading speed. Optical techniques can overcome both constraints.

The magneto-optic memories developed so far are read via the Faraday effect or the magnetic Kerr effect, which relates to the rotation of a beam of plane polarized light reflected from the surface of a material subjected to a magnetic filed. In either case a magnetized ferro- or ferromagnetic material rotates the plane of polarization of laser light incident on it.

By heating the memory elements on the storage medium to a temperature above the Curie point using a laser beam, writing may be achieved. The element is then allowed to cool down in the presence of an external magnetic field thereby acquiring a magnetization in a given direction. Magnetizations of the elements in one direction may represent 'zeros', in the opposite direction 'ones'. To read the information the irradiance of the laser beam is reduced and then directed to the memory elements. The direction of the change in the polarization of the laser beam on passing through or being reflected from the memory elements depends on the directions of magnetization; therefore we can decide if a given element is storing 'zero' or 'one'.

Prototype systems, incorporating, for example, a 50 mW He-Ne laser with a Pockels modulator and manganese bismuth (MnBi) thin film storage elements have enabled information to be stored, read and erased at rates in excess of 1 Mb s^{-1}

5.9.2 Isolators

Most passive optical devices (*e.g.*, couplers) are reciprocal devices, in that the devices work exactly the same way if their inputs and outputs are reversed. However, in many systems there is a need for a passive nonreciprocal device. An isolator is an example of such a device. The main function of an isolator is to allow transmission in one direction through it but block all transmissions in other direction. Isolators are used in systems at the output of optical amplifiers and lasers primarily to prevent reflections from entering these devices, which would degrade their performance. The two key parameters of an isolator are its insertion loss, which is the loss in the forward direction, and which should be as small as possible (typically 0.15 dB), and its isolation, which is the loss in the reverse direction, and which should be as large as possible. Typical isolation loss is around 40 dB–70 dB.

In operation an isolator uses the properties of polarization of light. In the forward direction, an input polarizer converts the unpolarized light into vertically polarized light. This polarized light passes through a Faraday rotator, which rotates the plane of polarization of the light by 45°. An output polarizer (or analyzer) allows 45° polarized light to pass and the light passes through the unit with minimum loss. In the reverse direction, an output polarizer converts unpolarized back reflected light into 45° polarized light. The Faraday rotator turns the plane of polarization another 45°, so that the back reflected light becomes horizontally polarized light. But the input polarizer only allows the vertically polarized light to pass through; therefore, horizontally polarized back reflected light will be rejected. This is how an isolator protects the optical path from back reflected light. The principle of operation of an isolator is shown in Fig. 5.35. The polarizer can be realized using crystals, called dichroics, which have the property of selectively absorbing light with one State of Polarization (SOP).

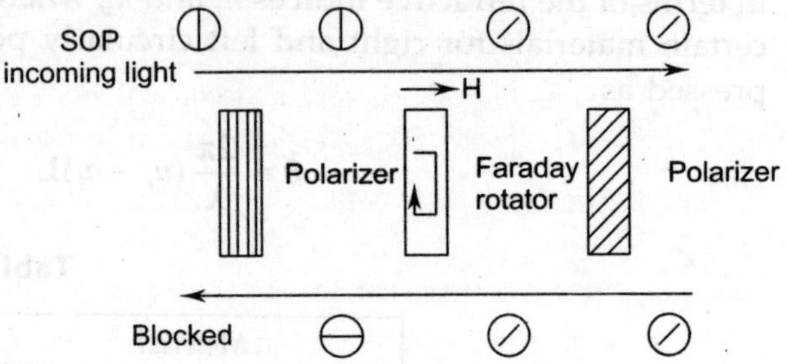

Fig. 5.35

To what extent an input polarizer rejects horizontally polarized back-reflected light depends on the extinction ratio of the device. This is the ratio of the intensities of two orthogonally polarized beams and can be more than 100000, resulting in an isolation > 50 dB.

The core of the device is a Faraday rotator, which consists of a piece of optically transparent material and a magnet surrounding this material. The magnet applies a magnetic field parallel to the direction of light propagation. Polarized light passing through a Faraday rotator undergoes the rotation of its plane of polarization. The angle of rotation (Φ) is given by

$$\Phi = \rho H L \tag{5.92}$$

where H is the strength of the magnetic field along the direction of the light in A/m, L(m) is the length of interaction between the light and the magnetic field, and ρ is the Verdet's constant of the material. We need to obtain $\Phi = 45°$, which is accomplished by choosing the proper materials and applying a magnetic field with the appropriate strength (H).

We have so far assumed a particular State of Polarization (SOP) for the input light signal. In practice we cannot control the SOP of the input, and so the isolator must work regardless of the input SOP. One such design for a miniature polarization-independent isolator is shown in Fig. 5.36(*a*) for propagation from left to right and in Fig. 5.36(*a*) for propagation from right to left.

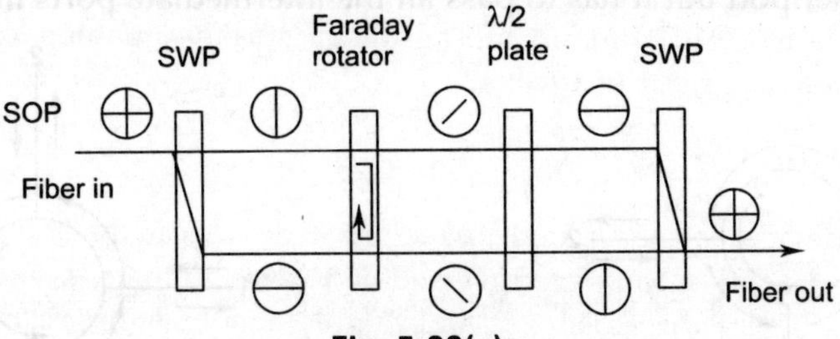

Fig. 5.36(*a*)

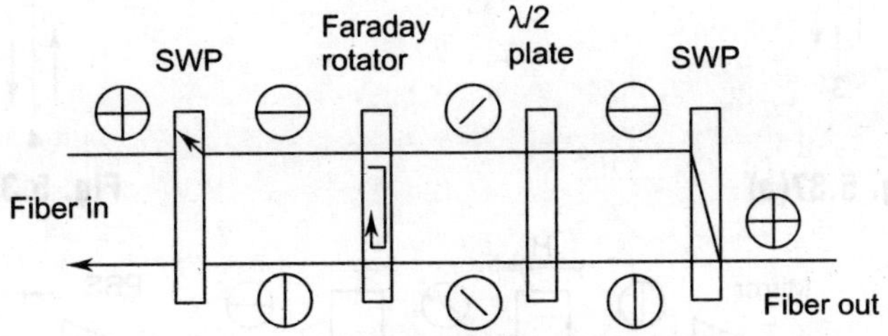

Fig. 5.36(*b*)

The main feature of a Faraday rotator is that it is nonreciprocal, which means that the direction of rotation of the plane of light polarization is determined by the direction of the magnetic field and not by the direction of the light polarization.

The input signal in Fig. 5.36, with an arbitrary SOP is first sent through a spatial walk-off polarizer (SWP). The SWP splits the signal into two orthogonally polarized components. Such an SWP can be realized using birefringent crystals whose refractive index is different for the two components. When light with an arbitrary SOP is incident on such a crystal, the two orthogonally polarized components are refracted at different angles. Each component goes through a Faraday rotator, which rotates the SOPs by 45°. The Faraday rotator is followed by a half-wave plate. The half-wave plate is a reciprocal device and it rotates the SOPs by 45° in the clockwise direction for signals propagating from left to right, and by 45° in the anticlockwise direction for signals propagating from right to left. Hence, the combination of the Faraday rotator and the half-wave plate converts the horizontal polarization into a vertical polarization and vice-versa, and the two signals are combined by another SWP at the output. For reflected signals in the reverse direction, the half-wave plate and the Faraday rotator cancel each other's effects, and the SOPs remain unchanged as they pass through these two devices and are thus not recombined by the SWP at the input.

A problem is the dependence on wavelength of the characteristics of the isolator. The range within which the characteristics of an isolator hold their values is often called the bandwidth of the isolator.

5.9.3 Circulators

A circulator is similar to an isolator, except that it has multiple ports. Circulators are the nonreciprocal devices that direct a light signal from one port to another sequentially in only one direction, as shown in Fig. 5.37. Three, four, and six-port circulators are commercially available for use in recent optical networks. Circulators are used in dispersion-compensating units and in add/drop passive multiplexers/demultiplexers. Circulators are necessary components in the construction of functional network modules.

Circulators can redirect a light signal from port 1 to port 2 and from port 2 to port 3, as illustrated in Fig. 5.37 (*a*). Fig. 5.37 (*b*) shows a four-port configuration. Light from any port can be redirected to any other port but it has to pass all the intermediate ports in a sequential order.

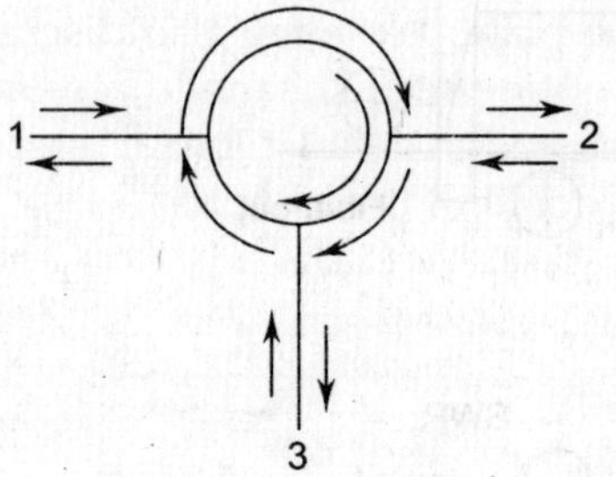

Fig. 5.37(*a*)

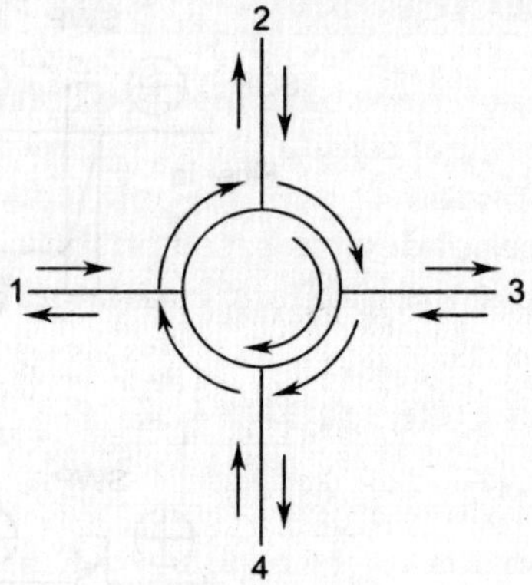

Fig. 5.37(*b*)

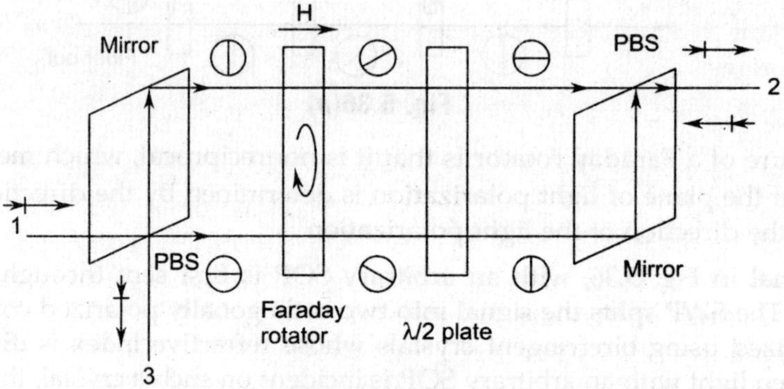

Fig. 5.37(*c*)

The principle of operation of a circulator is described in Fig. 5.37 (c). Observe that unpolarized light entering port 1 is split by a Polarization Beam Splitter (PBS) into horizontally and vertically polarized beams. Fig. 5.37 (*c*) shows that horizontally polarized light travels along the bottom of the diagram and that vertically polarized light travels along the top. A Faraday rotator turns the SOP of both beams to a 45° angle. A half-wave plate turns the SOP of both beams another 45° so that the vertically polarized light becomes horizontally polarized and vice-versa. These two beams are recombined by another PBS and leave the device at port 2. Light entering port 2 experiences the same turn at the quartz polarizer ($\lambda/2$), but an opposite turn at the Faraday rotator, so that beams keep their original state of polarization. The operation with this light is not shown in Fig. 5.37 (*c*). The PBS recombines the two beams and directs them to port 3. The beams cannot go to port 1 because the PBS does not recombine them to head in that direction. Thus we see, circulators have much in common with isolators.

To characterize an optical circulator, the following main parameters need to be considered. First an optical circulator produces isolation in an undesired direction — from port 2 to port 1, for example (Fig. 5.37 (*a*)). The value of this parameter can be as high as 70 dB. Second, this device, like any device inserted in a light path, causes insertion loss, which can be as low as 0.6 dB. Third, circulators with Polarization Dependent Loss (PDL) less than 0.05 dB are available. This is obviously very important because circulators must be polarization independent. Return loss is also an important factor because a circulator has reflecting surfaces. Circulators function over a range of operating wavelengths (bandwidths). Circulators are usually designed to work around 1550 nm or 1310 nm with a typical bandwidth of ±20 nm.

5.10 CONCLUSION

In this chapter, we have discussed on the basic properties of light, viz. polarization, and properties like birefringence, the phase retarders like quarter wave plate. We have also discussed on the electro-optic effect, acousto-optic effect and the magneto-optic effect. The modulators and various other types of devices based on these effects are also discussed in detail. Optical inter-connective devices like the isolators, circulators are also covered in detail in this chapter.

REVIEW QUESTIONS

1. Consider an electromagnetic wave is travelling in the *z*-direction. The orthogonal components of its resultant E-vector are along the *x*-direction and *y*-direction and of same amplitude but of a phase difference of $\pi/2$. What will be the resultant E-vector at any point in space?
   (*a*) Constant in amplitude rotating with an angular frequency ω
   (*b*) Stationary
   (*c*) Variation in amplitude rotating with an angular frequency ω
   (*d*) Random
2. What is the behaviour of the electromagnetic wave of question 5.1?
   (*a*) Unpolarized (*b*) Plane-polarized
   (*c*) Circularly polarized (*d*) Elliptically polarized
3. What are the behaviours of *e*-ray and *o*-ray in a birefringent crystal?
   (*a*) *o*-ray follows Snell's law but *e*-ray does not
   (*b*) *e*-ray follows Snell's law but *o*-ray does not
   (*c*) Both follow Snell's law
   (*d*) Both does not follow Snell's law
4. An uniaxial crystal has
   (*a*) One principal refractive index and no optic axis
   (*b*) One principal refractive index and one optic axis
   (*c*) Two principal refractive indices and one optic axis
   (*d*) Three principal refractive indices and two optic axes
5. In a doubly refracting crystal, what is the direction of optic axis?
   (*a*) v_o is greater than v_e (*b*) v_o is equal to v_e
   (*c*) v_o is less than v_e (*d*) v_o and v_e are random
6. How much phase difference will be introduced by a quarter wave plate between two emerging beams?
   (*a*) $\pi/4$ (*b*) $\pi/2$ (*c*) π (*d*) $3\pi/2$
7. How much path difference will be introduced by a half wave plate between two emerging beams?
   (*a*) $\lambda/4$ (*b*) $\lambda/2$ (*c*) $3\lambda/2$ (*d*) λ

8. In a longitudinal electro-optic modulator, how much phase shift will be introduced between two polarization components by half wave components?
   (*a*) $\pi/4$ (*b*) $\pi/2$ (*c*) π (*d*) 2π
9. In a transverse electro-optic modulator, which are the depending factors of modulator crystal for V_π?
   (*a*) Length *l* and width *d* (*b*) Only length *l*
   (*c*) Only width *d* (*d*) Ratio *d*/*l*
10. In a Raman-Nath modulator, the acousto-optic grating is
   (*a*) So thin that it behaves almost like a plane transmission grating
   (*b*) So thick that it behaves almost like a Bragg's crystal grating
   (*c*) Analogous to a concave Rowland's grating
   (*d*) Quite complicated
11. What do you mean by polarization of light? Distinguish between plane polarized, circularly polarized and elliptically polarized light. How circularly polarized light can be produced?
12. What is electro-optic effect? How can this effect be used for phase modulation of an optical signal? How can the amplitude of the optical signal be modulated?
13. Calculate the thickness of a half wave plate made of quartz used with sodium light of λ = 589.3 nm, if the principal indices, n_e and n_o of quartz are 1.533 and 1.544 respectively. Calculate the same when λ = 1500 nm.
14. Calculate the thickness of a quarter wave plate made of calcite used with sodium light of λ = 589.3 nm, if the principal indices, n_e and n_o of quartz are 1.658 and 1.486 respectively.
15. Calculate the change in refractive index due to longitudinal electro-optic effect for a lithium niobate crystal of 5 mm long with an applied voltage of 100 V. Also calculate the net phase polarization components emerging from crystal if propagating wavelength through crystal is 550 nm.
16. (*a*) What is the acousto-optic effect? Distinguish between the modulators with acousto-optic effect and the modulators operating in Raman-Nath and Bragg regimes.
   (*b*) A typical acousto-optic cell of a Raman-Nath modulator contains water. A piezoelectric crystal bonded to the cell generates an acoustic wave of frequency 5 MHz in water. The velocity of the acoustic wave in water is 1500 ms^{-1} and the thickness of the cell is 3 cm. If a laser beam of wavelength 700 nm is incident on the cell, calculate the (*i*) angle between the first-order diffracted beam and the direct beam and (*ii*) relative intensity of the diffracted beam of the first order if $\Delta n = 10^{-5}$. Take n_0 = 1.33 for water.
17. A typical electro-optic modulator with lithium niobate crystal operates at 550 nm. Its refractive indices are 2.29 (n_o) and 2.20 (n_e) and relevant electro-optic coefficient is 30.8×10^{-12} (r_{33}).
   (*a*) What should be the length of the crystal to create a phase difference of $\pi/2$ between the emergent field components at the applied voltage of 0 V?
   (*b*) What will be the width of the crystal to produce additional phase difference of $\pi/2$ for an applied voltage of 20 V?
   (*c*) Calculate the half wave voltage V_π for the crystal.
18. A Raman-Nath modulator uses a cell with the thickness of 1.2 cm containing water (n_0 =1.33). An acoustic transducer is bonded to the cell generates a wave of frequency of 4 MHz and velocity of 1500 m/s in water. Considering a He-Ne laser beam of wavelength of 633 nm is incident on the cell, find the angle of first order diffracted beam.
19. What is Faraday effect? How an optical circulator can be constructed using Faraday effect?
20. Discuss the operation of an optical isolator using magneto-optic effect.
21. What is birefringence? What is the difference between a positive and a negative crystal?

6

HOLOGRAPHY

6.1 INTRODUCTION

The word *'holo'* means entire, full or complete; *'graphy'* means recording. Hence holography means recording of the complete information of an object. A simple photograph carries information of light intensity distribution on the object only. The phase which carries the information of the morphology (*i.e.*, relative depth of different parts) of the object is not being recorded in the photograph. Basically a photograph represents a two-dimensional recording of a three-dimensional scene. Recording of the complete information of an object; *i.e.*, its amplitude and phase is called holography.

Although *Holography* was developed prior to laser, the requirement of holography is light with very high degree of spatial and temporal coherence. This has closely linked the development of holography to that of lasers. *Holography* is a method of recording information from a three-dimensional object in such a way that a three-dimensional image may be subsequently reconstructed. The phenomenon is often known as *wavefront reconstruction*.

6.2 BASIC PRINCIPLE OF HOLOGRAM

In the recording of a hologram, one superimposes on the object wave another wave called the reference wave (usually a plane wave). In fact a portion of a coherent beam is allowed to be scattered by the object and the other portion is reflected by a mirror. The former corresponds to the object wave and the latter to the reference wave. The object and the reference wave are made to interfere and the photographic plate is made to record the resulting interference pattern. This recorded interference pattern forms the hologram and contains information about the amplitude as well as the phase of the object wave. However, the hologram bears little resemblance to the object. Actually the information about the object is coded into the hologram.

To view the image we have to illuminate the hologram with another wave, called the reconstruction wave. In most of the cases this reconstruction wave is identical to the reference wave used during the formation of the hologram. The process of viewing the hologram is called reconstruction. This reconstruction process leads, in general, to a real and a virtual image of the object scene. The real image can be photographed just by placing a light sensitive medium at the position where the real image is formed, without the aid of lenses. The virtual image bears all the characteristics of the object. We will discuss different types of holograms, their recording and reconstruction techniques in detail in due course of time.

6.3 COHERENCE REQUIREMENTS

Since holography actually involves an interference phenomenon, certain coherence requirements are to be met. Thus the object beam and the reference beam must be spatially coherent over the whole region. In addition, if stable interference fringes are to be formed the maximum path difference between the object wave and the reference wave must be less than the coherence length.

During reconstruction if the original wave is to be reconstructed with minimum aberration, the wave used for reconstruction should be identical to the recording wave. For maximum resolution, the reconstructing light source must be coherent to the recording source.

6.4 THE IN-LINE (GABOR) HOLOGRAM

The first hologram was recorded by Gabor in 1948. The setup used by Gabor is shown in Fig. 6.1. In the setup, the coherent light source as well as the object, which is a transparency containing small opaque details on a clear background, are located along the axis normal to the photographic plate. When the object is illuminated with a uniform parallel beam of light, the light which is transmitted by the transparency consists of two parts.

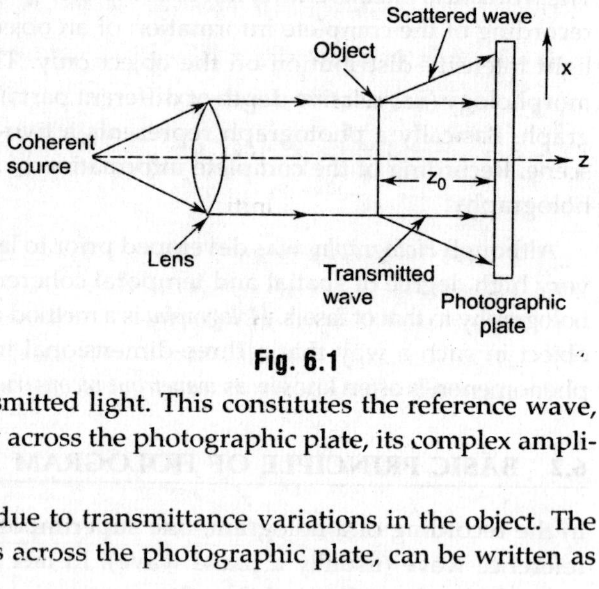

Fig. 6.1

The first part is relatively strong uniform plane wave corresponding to the directly transmitted light. This constitutes the reference wave, and since, its amplitude and phase do not vary across the photographic plate, its complex amplitude can be written as a real constant r.

The second part is a weak scattered wave due to transmittance variations in the object. The complex amplitude of this wave, which varies across the photographic plate, can be written as $o(x, y)$ where $|o(x, y)| \ll r$.

Hence, the resultant complex amplitude at any point on the photographic plate is the sum of these two parts, so that the intensity at this point is

$$I(x, y) = |r + o(x, y)|^2$$

$$= r^2 + |o(x, y)|^2 + ro(x, y) + ro^*(x, y) \tag{6.1}$$

where $o^*(x, y)$ is the complex conjugate of $o(x, y)$.

A positive transparency is made from this recording. Let us assume, for simplicity, that the amplitude transmittance t (the ratio of the transmitted amplitude to that incident on it) is a linear function of the intensity and can be written as

$$t = t_0 + \beta TI \tag{6.2}$$

where t_0 is a constant background transmittance, T is the exposure time and β is a parameter determined by the photographic material used and the processing conditions. The amplitude transmittance of the transparency is

$$t(x, y) = t_0 + \beta T\left[r^2 + |o(x, y)|^2 + ro(x, y) + ro^*(x, y)\right] \tag{6.3}$$

6.4.1 Reconstruction of the In-Line (Gabor) Hologram

To view the reconstructed image, this transparency is replaced in the same position as the original photographic plate and illuminated once again with the same parallel beam of monochromatic light used to record the hologram as shown in Fig. 6.2. Since the complex amplitude at any point in this beam is, apart from a constant factor, the same as that in the reference beam, the complex amplitude transmitted by the hologram can be written as

$$u(x, y) = rt(x, y)$$

$$= r\left(t_0 + \beta Tr^2\right) + \beta Tr\,|o(x, y)|^2 + \beta Tr^2 o(x, y) + \beta Tr^2 o^*(x, y) \tag{6.4}$$

The complex amplitude of the transmitted wave consists, therefore of four distinct components.

(*i*) The first of these $r(t_0 + \beta Tr^2)$, is a uniformly attenuated plane wave. It is the directly transmitted beam.

(*ii*) The second term $\beta Tr|o(x, y)|^2$, is extremely small in comparison to the other terms, since it has been assumed initially that $|o(x, y)| << r$. Under these conditions, it can be neglected.

(*iii*) Except for a constant factor, the third term $\beta\ Tr^2 o(x, y)$, is the same as the original wavefront from the object incident on the photographic plate. Hence, it gives rise to a reconstructed image of the object in its original position. Since this is located behind the transparency at a distance z_0 from it, the reconstructed wave appears to diverge from it. It is actually a virtual image.

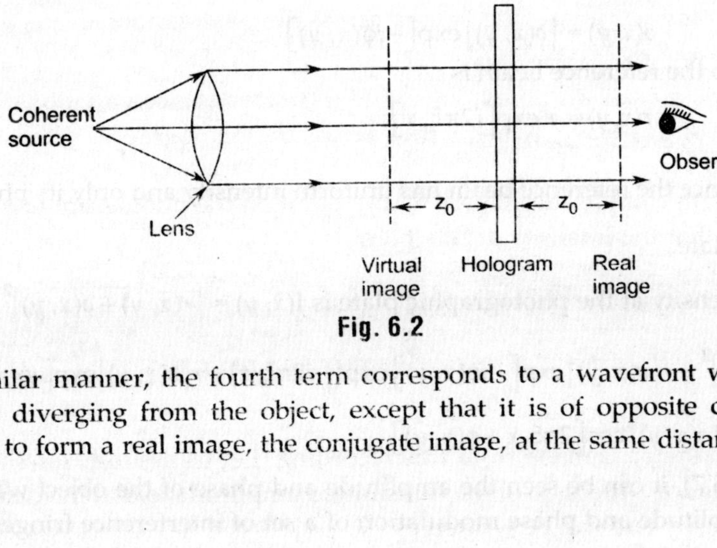

Fig. 6.2

(*iv*) In the similar manner, the fourth term corresponds to a wavefront which resembles the wavefront diverging from the object, except that it is of opposite curvature. Hence, it converges to form a real image, the conjugate image, at the same distance z_0 in front of the hologram.

6.4.2 Disadvantages of the In-Line (Gabor) Hologram

There are certain limitations of the Gabor hologram. They are discussed below:

(*i*) In such a system, an observer focusing on one image finds it superposed on the out-of-focus twin image as well as a strong coherent background. This forms the most serious limitation.

(*ii*) The object should have a high average transmittance, if the second term on the right hand side of equation (6.4), which has been taken as negligible, is not to interfere with the reconstructed image. Typically, it is possible to form images of fine opaque lines on a transparent background, but not vice versa.

(*iii*) The hologram used to reconstruct the image must be a positive transparency. Since the image forming waves interfere with the background in the process of reconstruction, if the plate which has been exposed in the recording step is used directly (in which case β in equation (6.2) is negative), the reconstructed image will also be a negative.

6.5 OFF-AXIS (LEITH-UPATNIEKS) HOLOGRAM

The successful technique for separating the twin images was developed by Leith and Upatnieks. The setup for such recording is shown in Fig. 6.3. Here a separate reference beam derived from the same coherent source is allowed to fall on the photographic plate, during the recording process, at an offset angle θ to the beam from the object. The reference beam is assumed to be a collimated beam of uniform intensity.

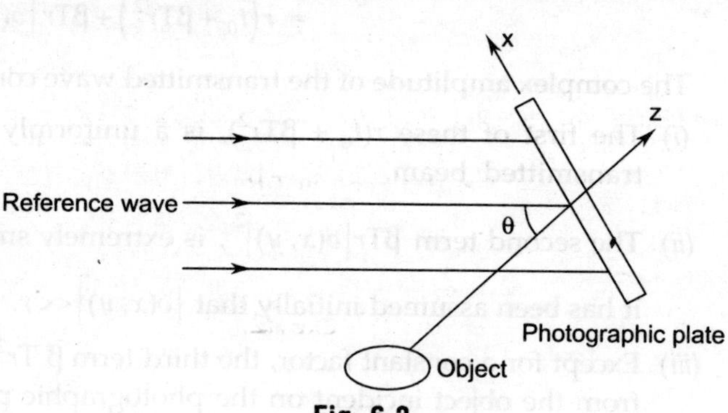

Fig. 6.3

At any point (x, y) on the photographic plate, the complex amplitude due to the object beam is given by

$$o(x,y) = |o(x,y)| \exp[-i\phi(x,y)] \tag{6.5}$$

While that due to the reference beam is

$$r(x,y) = r \exp[i2\pi\xi_r x] \tag{6.6}$$

where $\xi_r = \dfrac{\sin\theta}{\lambda}$, since the reference beam has uniform intensity and only its phase varies across the photographic plate.

The resultant intensity at the photographic plate is $I(x, y) = |r(x,y) + o(x,y)|^2$

$$= |r(x,y)|^2 + |o(x,y)|^2 + r|o(x,y)| \exp[-i\phi(x,y)] \exp(-i2\pi\xi_r x) + r|o(x,y)| \exp[i\phi(x,y)] \exp(i2\pi\xi_r x)$$

$$= r^2 + |o(x,y)|^2 + 2r|o(x,y)| \cos[2\pi\xi_r x + \phi(x,y)] \tag{6.7}$$

From equation (6.7), it can be seen the amplitude and phase of the object wave are encoded, respectively, as amplitude and phase modulation of a set of interference fringes equivalent to a spatial frequency equal to ξ_r.

If, as in equation (6.2), it is assumed that the amplitude transmittance of the photographic plate after processing is linearly related to the intensity in the interference pattern, the amplitude transmittance of the hologram can be written as:

$$t(x,y) = t_0 + \beta T\left\{r^2 + |o(x,y)|^2 + r|o(x,y)|\exp[-i\phi(x,y)]\exp(-i2\pi\xi_r x) + r|o(x,y)|\exp[i\phi(x,y)]\exp(i2\pi\xi_r x)\right\} \quad (6.8)$$

where T is the exposure time, t_0 is the constant background transmittance and β is the slope of the amplitude transmittance versus exposure characteristic of the photographic material.

6.5.1 Reconstruction of the Off-Axis (Leith-Upatnieks) Hologram

To reconstruct the image, the hologram is to be illuminated again as shown in Fig. 6.4 with the same reference beam as was used to record it. In this case also, the complex amplitude $u(x, y)$ of the transmitted wave is the sum of four terms. Each of these four terms corresponds to one of the terms of equation (6.8) and can be written as

$$\begin{aligned} u(x,y) &= r(x,y)\,t(x,y) \\ &= u_1(x,y) + u_2(x,y) + u_3(x,y) + u_4(x,y) \end{aligned} \quad (6.9)$$

Hence,

$$u(x,y) = r\exp(i2\pi\xi_r x)\Big[t_0 + \beta T r^2 + \beta T|o(x,y)|^2 + \beta T r|o(x,y)|\exp[-i\phi(x,y)]\exp(-i2\pi\xi_r x) + \beta T r|o(x,y)|\exp[i\phi(x,y)]\exp(i2\pi\xi_r x)\Big]$$

$$\therefore \quad u(x,y) = \left\{(t_0 + \beta T r^2)r\exp(i2\pi\xi_r x)\right\} + \left\{\beta T r|o(x,y)|^2\exp(i2\pi\xi_r x)\right\} + \left\{\beta T r^2 o(x,y)\right\} + \left\{\beta T r^2 o^*(x,y)\exp(i4\pi\xi_r x)\right\} \quad (6.10)$$

Therefore,

$$u_1(x,y) = (t_0 + \beta T r^2)r\exp(i2\pi\xi_r x) \quad (6.11)$$

$$u_2(x,y) = \beta T r|o(x,y)|^2\exp(i2\pi\xi_r x) \quad (6.12)$$

$$u_3(x,y) = \beta T r^2 o(x,y) \quad (6.13)$$

$$u_4(x,y) = \beta T r^2 o^*(x,y)\exp(i4\pi\xi_r x) \quad (6.14)$$

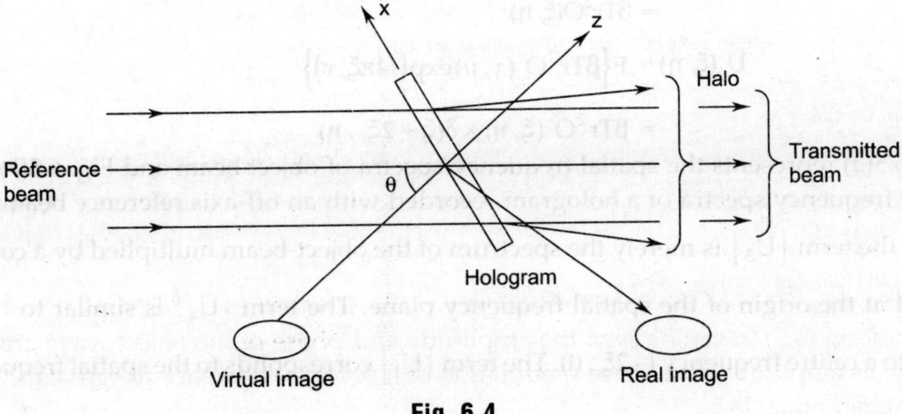

Fig. 6.4

The first term $u_1(x, y)$ is only the attenuated reference beam, which is a plane wave directly transmitted through the hologram. This $u_1(x, y)$ is surrounded by a halo caused by the second term $u_2(x, y)$, which is spatially varying. The angular spread of this halo is determined by the angular extent of the object. The third term $u_3(x, y)$ is identical with the original object wave, except for a constant factor. This generates a virtual image of the object in its original position. This wave makes an angle θ with the directly transmitted wave. The fourth term $u_4(x, y)$ gives rise to the conjugate real image. However, the fourth term includes an exponential factor, exp $(i4\pi\xi_r x)$. This indicates that the conjugate wave is directed off the axis at an angle twice that which the reference wave makes with it.

6.5.2 Advantages of Leith-Upatnieks Technique over Gabor Hologram

Even though two images are still reconstructed in this setup, they are separated angularly from each other, as well as the directly transmitted beam. If the offset angle θ of the reference beam is made large enough, it is possible to ensure that there is no overlap. This method therefore overcomes the drawbacks of Gabor in-line arrangement.

The minimum value of θ required to ensure that each of the images can be observed distinctly without any interference from its twin image or from the directly transmitted image and the halo of the scattered beam is determined by the minimum spatial carrier frequency (ξ_r) for which there is no overlap between the angular spectra of the third and fourth terms and those of the first and second terms.

These angular spectra are actually the Fourier transforms of these terms and can be written as follows.

$$U_1(\xi, \eta) = F\left\{\left(t_0 + \beta T r^2\right) r \exp\left(i2\pi\xi_r x\right)\right\}$$

$$= \left(t_0 + \beta T r^2\right) r\,\delta(\xi + \xi_r, \eta) \tag{6.15}$$

$$U_2(\xi, \eta) = F\left\{\beta T r \left|O(x, y)\right|^2 \exp\left(i2\pi\xi_r x\right)\right\}$$

$$= \beta T r\left[O(\xi, \eta) * O(\xi, \eta) \times \delta(\xi + \xi_r, \eta)\right] \tag{6.16}$$

where $O(\xi, \eta) = F\{o(x, y)\}$ is the spatial frequency spectrum of the object beam, and the symbols $*$ and $\times$ denote, the operations of correlation respectively.

$$U_3(\xi, \eta) = F\{\beta T r^2 O(x, y)\}$$

$$= \beta T r^2 O(\xi, \eta) \tag{6.17}$$

$$U_4(\xi, \eta) = F\left\{\beta T r^2 O^*(x, y) \exp\left(i4\pi\xi_r x\right)\right\}$$

$$= \beta T r^2 O^*(\xi, \eta) \times \delta(\xi + 2\xi_r, \eta) \tag{6.18}$$

Figure 6.5(a) represents the spatial frequency spectra of object beam and Fig. 6.5(b) illustrates the spatial frequency spectra of a hologram recorded with an off-axis reference beam. It may be noted that the term $|U_3|$ is merely the spectrum of the object-beam multiplied by a constant and is centered at the origin of the spatial frequency plane. The term $|U_4|$ is similar to $|U_3|$ but is displaced to a centre frequency $(-2\xi_r, 0)$. The term $|U_1|$ corresponds to the spatial frequency of the carrier fringes and is a δ function located at $(-\xi_r, 0)$. The term $|U_2|$ is centered on this δ function. Being proportional to the auto-correlation function of $O(\xi, \eta)$, it has twice the extent of the object-beam spectrum.

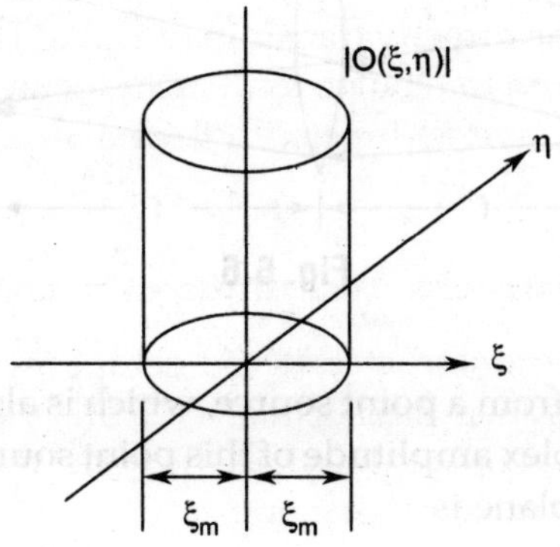

Fig. 6.5 (*a*)

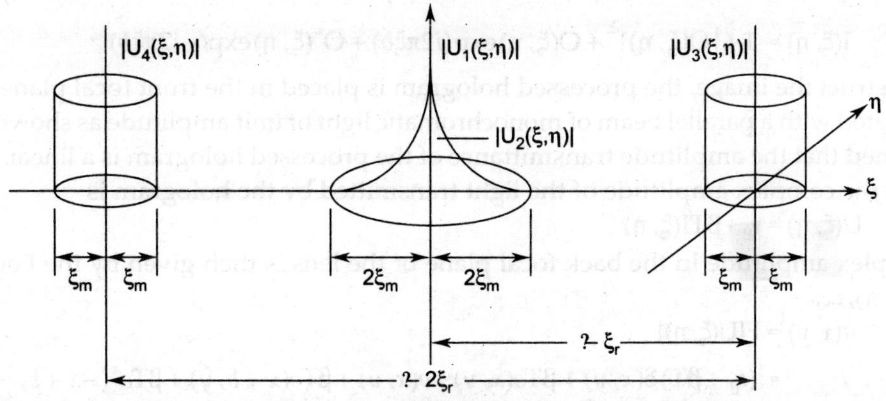

Fig. 6.5 (*b*)

Hence, $|U_3|$ and $|U_4|$ will not overlap with $|U_1|$ and $|U_2|$ if the spatial carrier frequency ξ is chosen so that

$$\xi_r \geq 3\xi_m, \tag{6.19}$$

where ξ_m is the highest frequency in the spatial frequency spectrum of the object beam.

6.6 FOURIER TRANSFORM HOLOGRAM

We can have another very interesting category of holograms, where the waves that interfere at the hologram planes are actually the Fourier transforms of the object wave and the reference wave. This implies that the object in a single plane, or of limited thickness. The optical arrangement for recording such a hologram is shown in Fig. 6.6.

The object is located in the front focal plane of a lens and is illuminated by a parallel beam of monochromatic light. If the complex amplitude leaving the object plane is $o(x, y)$, the complex amplitude of the object wave at the photographic plate located in the back focal plane of the lens is

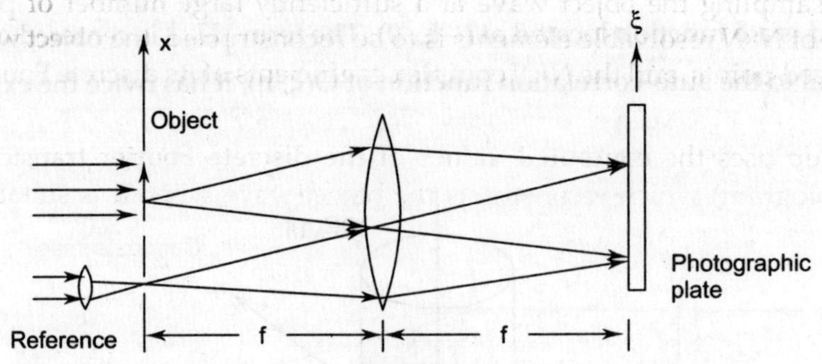

Fig. 6.6

$$O(\xi, \eta) = F\{o(x, y)\} \tag{6.20}$$

The reference beam is derived from a point source, which is also located in the front focal plane of the lens. If $\delta(x + b, y)$ is the complex amplitude of this point source, the complex amplitude of the reference wave at the hologram plane is

$$R(\xi, \eta) = \exp(-i2\pi\xi b) \tag{6.21}$$

The intensity in the interference pattern formed by these two waves is therefore given by

$$I(\xi, \eta) = 1 + |O(\xi, \eta)|^2 + O(\xi, \eta)\exp(i2\pi\xi b) + O^*(\xi, \eta)\exp(-i2\pi\xi b) \tag{6.22}$$

To reconstruct the image, the processed hologram is placed in the front focal plane of the lens and illuminated with a parallel beam of monochromatic light of unit amplitude as shown in Fig. 6.7. If it is assumed that the amplitude transmittance of the processed hologram is a linear function of $I(\xi, \eta)$, then the complex amplitude of the light transmitted by the hologram is

$$U(\xi, \eta) = t_0 + \beta T I(\xi, \eta) \tag{6.23}$$

The complex amplitude in the back focal plane of the lens is then given by the Fourier transform of $U(\xi, \eta)$, *i.e.*

$$\begin{aligned} u(x, y) &= F\{U(\xi, \eta)\} \\ &= (t_0 + \beta T)\delta(x, y) + \beta T o(x, y) * o(x, y) + \beta T o(x - b, y) + \beta T o^*(-x + b, -y) \end{aligned} \tag{6.24}$$

The first term on the right hand side of equation (6.24) is focused on the axis, while the second term forms a halo around it. The third term corresponds to the original object wave shifted down wards by a distance b. The fourth term is the conjugate of the original object wave inverted and shifted upwards by the same amount b.

These types of holograms have a very useful property that the reconstructed image is stationary even when the hologram is translated in its own plane.

6.7 COMPUTER-GENERATED HOLOGRAM

Holograms can be produced by means of digital computer. The technique has been discussed in detail by Lee [1978], Yaroslavskii and Merzlyakov [1980] and Dallas [1980]. In this method the generation of wavefronts with any prescribed amplitude and phase distribution is permitted. These are therefore extremely useful for testing optical surfaces as well as in applications such as laser-beam scanning and optical spatial-filtering. Basically this generation of hologram using computers involves two major steps.

The first step is to calculate the complex amplitude of the object wave at the hologram plane. Now this is usually taken as the Fourier transform of the complex amplitude in the object plane.

This is done by sampling the object wave at a sufficiently large number of points. Thus, if an image consisting of $N \times N$ resolvable elements is to be reconstructed, the object wave is sampled at $N \times N$ equally spaced points, and the $N \times N$ complex coefficients of its discrete Fourier transform are evaluated.

The second step uses the computed values of the discrete Fourier transform to produce a transparency (hologram) which reconstructs the object wave when it is suitably illuminated.

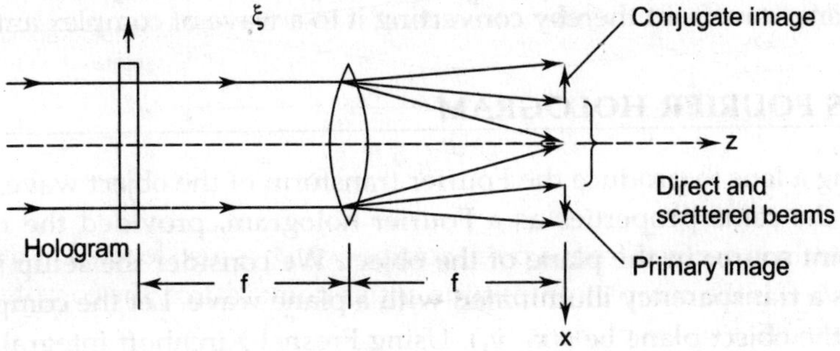

Fig. 6.7

Two approaches can be followed for this purpose. The first is like the off-axis optical holography, which is discussed in the previous section. The complex amplitudes of a plane reference wave and the object wave in the hologram plane are added and the squared modulus of their sum is evaluated. Hence, we get a transparency which has real and positive amplitude transparencies everywhere. Since, the first two terms on the right hand side of equation (6.7) have no contribution to the reconstructed image, hence it is only necessary to compute the last term at regular intervals and add a constant bias to all the samples to make them positive.

There is an alternative approach which is possible with only computer generated hologram. Here, a transparency is produced which records both the phase and amplitude of the object wave in the hologram plane. This can be considered as the superposition of two transparencies, one of constant thickness having a transmittance proportional to the amplitude of the object wave, and the other having thickness variations corresponding to the phase of the object wave, but no transmittance variations. This type of hologram has the advantage that it can form a single on-axis image.

In either case, the computer is used to control a plotter which produces a large-scale version of the hologram. This is then photographically reduced to produce the required transparency. To reconstruct the object waveform an optical system as shown in Fig. 6.7 may be used.

6.8 THE KINOFORM

In the case where the object is diffusely illuminated, the amplitudes of the Fourier coefficients are relatively unimportant, and the object may be reconstructed using only the values of their phase information. This led to the concept of a completely different type of computer generated hologram. This type of hologram is called *kinoform*, after Lesem, Hirsch and Jordan.

Like normal hologram it can display three-dimensional images. It however, differs from normal hologram in that, it can diffract all the illumination it receives into a single diffraction order. A basic assumption is that an object wave complex amplitude $o(x, y)$ can be satisfactorily recorded by considering the wave to have constant amplitude. Its complex amplitude is then given by

$$o(x, y) = \text{constant} \times \exp[i\varphi(x, y)] \tag{6.25}$$

The computed phase $\varphi(x, y)$, has to be recorded. However, before this is done, the computed phase is adjusted so that it varies only between 0 and 2π rad from a phase origin, over the entire xy plane. Integral multiples of 2π rad is subtracted from the relative phase computed for any point (x, y) on the plane. The phase record, the kinoform, is then a thin, transparent sheet whose optical thickness is altered according to the computed phase $\varphi(x, y)$ of the object wave where $0 \le \varphi(x, y) \le 2\pi$. When illuminated with a plane wave, the kinoform impresses the phase function $\varphi(x, y)$ onto the plane wavefront thereby converting it to a wave of complex amplitude $o(x, y)$.

6.9 LENS-LESS FOURIER HOLOGRAM

Even without using a lens to produce the Fourier transform of the object wave, a hologram may be obtained with the same properties as a Fourier hologram, provided the reference wave is produced by a point source in the plane of the object. We consider the setup shown in Fig. 6.8 where the object is a transparency illuminated with a plane wave. Let the complex amplitude of the wave leaving the object plane be $o(x_1, y_1)$. Using Fresnel-Kirchhoff integral, it can be shown that the complex amplitude at the photographic plate due to the wave can be written as

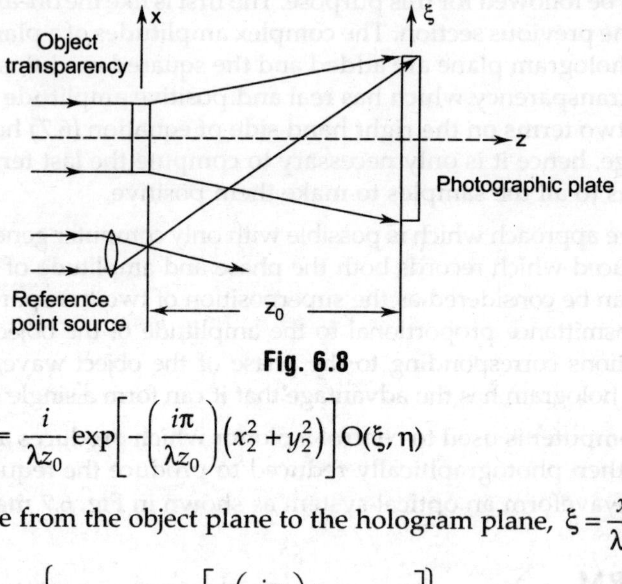

Fig. 6.8

$$O(x_2, y_2) = \frac{i}{\lambda z_0} \exp\left[-\left(\frac{i\pi}{\lambda z_0}\right)(x_2^2 + y_2^2)\right] O(\xi, \eta) \tag{6.26}$$

where z_0 is the distance from the object plane to the hologram plane, $\xi = \dfrac{x_2}{\lambda z_0}$, $\eta = \dfrac{y_2}{\lambda z_0}$, and

$$O(\xi, \eta) = F\left\{O(x_1, y_1) \exp\left[-\left(\frac{i\pi}{\lambda z_0}\right)(x_1^2 + y_1^2)\right]\right\} \tag{6.27}$$

This is actually the Fourier transform of the object wave, modified by a spherical phase factor which depends on the distance from the object to the hologram. In a similar manner, it can be shown that the complex amplitude at the photographic plate due to the reference wave is

$$r(x_2, y_2) = r \exp\left[-\left(\frac{i\pi}{\lambda z_0}\right)(x_2^2 + y_2^2)\right] \exp(-i2\pi\xi b) \tag{6.28}$$

where b is the distance of the reference point source from the z axis.

Hence the intensity of the resulting pattern is given by

$$I(x_2, y_2) = r^2 + |O(x_2, y_2)|^2 + \left(\frac{i}{\lambda z_0}\right) O(\xi, \eta) \exp(i2\pi\xi b) + \left(\frac{i}{\lambda z_0}\right) O^*(\xi, \eta) \exp(-i2\pi\xi b) \tag{6.29}$$

A comparison with equation (6.22) shows that this is very similar to that obtained with the conventional setup used to produce a Fourier hologram, and the resulting hologram has essentially the same properties.

In the recording configuration, the effect of the spherical phase factor associated with the near-field or Fresnel diffraction pattern of the object transparency is eliminated by using a spherical reference wave with the same curvature.

6.10 FRAUNHOFER HOLOGRAM

When the object is small enough as shown in Fig. 6.9, for its far-field diffraction pattern to be formed at the photographic plate, a Fraunhofer hologram is formed. In fact this is a situation in which in-line holography can be used without the usual problems associated with the presence of the conjugate image. For this hologram, the distance z_0, of the plate from the object, must satisfy the far-field condition $z_0 = >> \frac{(x_0^2 + y_0^2)}{\lambda}$, where x_0 and y_0 define the maximum dimensions of the object.

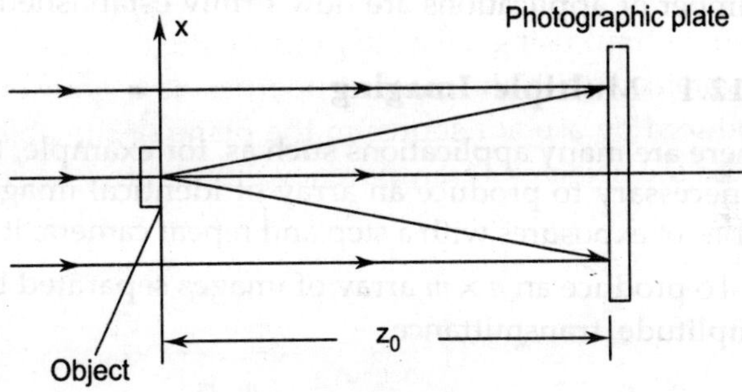

Fig. 6.9

A typical example is a hologram of a group of point scatters such as aerosol particles. The particles are so small that even at distances of α few millimeters the hologram can be considered to lie in far field of any individual particle. In such a case, the diffracted light due to the conjugate image is spread over such a large area in the primary image plane that it only contributes a weak, uniform background. As result the image is free from the interference of its conjugate.

6.11 IMAGE HOLOGRAM

When a hologram of the real image of an object is recorded instead of the object itself, some significant advantages are achieved. The optical setup for the formation of an image hologram is shown in Fig. 6.10. While recording the hologram with such a setup shown in Fig. 6.10, it is possible to position the projected image of the object on the photographic plate. The reconstructed primary image is then formed in the same position, so that one half of it appears to be behind the plate while the other half is in front of it.

Fig. 6.10

Such an image hologram has very useful property, that it can be illuminated with an incoherent source of appreciable size and spectral bandwidth and will still produce an appreciable sharp image. It has another advantage, that it has increased image luminance. However this is partially offset by the fact that the angle over which the image can be viewed is limited by the angular aperture of the imaging lens.

6.12 APPLICATIONS OF HOLOGRAPHY

Undoubtedly, the full potential of the holographic technique is still to be realized although a number of applications are now firmly established.

6.12.1 Multiple Imaging

There are many applications such as, for example, the production of integrated circuits, where it is necessary to produce an array of identical images. While this is normally done by making a series of exposures with a step and repeat camera, it is also possible to use holographic techniques.

To produce an $n \times m$ array of images separated by intervals x_0, y_0, a hologram is used with an amplitude transmittance

$$H(\xi, \eta) = \sum^{n}\sum^{m} \exp\left[-i2\pi\left(nx_0\xi + my_0\eta\right)\right] \tag{6.30}$$

This hologram represents a set of plane waves travelling in directions corresponding to the centers of the images in the array.

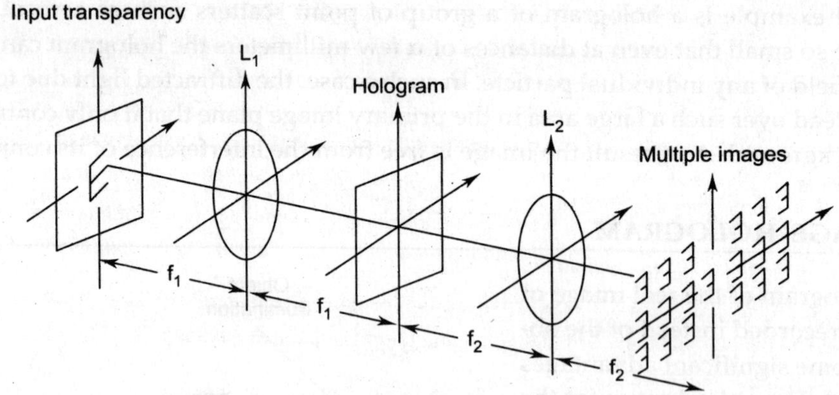

Fig. 6.11

This hologram is placed in the back focal plane of a lens L_1 in the optical setup shown in Fig. 6.11. If a transparency with an amplitude transmittance $f(x, y)$ located in the front focal points of L_1 and illuminated with a collimated beam is used as an input, its Fourier transform $F(\xi, \eta)$ is displayed in the back focal plane of L_1.

The waveform emerging from the hologram can be written as

$$G(\xi, \eta) = F(\xi, \eta)\, H(\xi, \eta) \tag{6.31}$$

A second Fourier transform operation by the lens L_2 then produces a set of multiple images

$$g(x, y) = f(x, y) \times \sum^{n}\sum^{m} \delta\left(x - nx_0, y - my_0\right)$$

$$= \sum^{n} \sum^{m} f(x - nx_0, y - my_0) \tag{6.32}$$

6.12.2 Character Recognition

The property of associative storage makes it possible to use the hologram for character recognition. This is essentially a spatial filtering operation in which the hologram acts as a matched filter. To produce the matched filter, a transparency of the set of characters to be identified is placed in the input plane and a hologram of this transparency is recorded in the Fourier transform plane using a point reference source. We assume that the complex amplitude in the input plane due to the transparency is a one-dimensional distribution given by:

$$f(y) = \sum_{i=1}^{N} f_i(y - c_i) \tag{6.33}$$

where $f_i(y - c_i)$ is the complex amplitude due to a typical character centered at c_i, while that due to the reference source is $\delta(y + b)$. If linear recording takes place, the transmittance of the hologram can be written as

$$t(\eta) = t_0 + \beta T\left[1 + |F(\eta)|^2 + F^*(\eta)\exp(-i2\pi\eta b) + F(\eta)\exp(i2\pi\eta b)\right] \tag{6.34}$$

where, $F(\eta) \leftrightarrow f(y)$

The hologram is replaced in exactly the same position in which it was recorded after being processed. Then it is illuminated by a single character of the set centered on the axis. If the amplitude due to this character in the input plane is $f_l(y)$, the transmitted amplitude of the hologram is given by:

$$O(\eta) = F_l(\eta)t(\eta)$$

$$= (t_0 + \beta T)F_l(\eta) + \beta T F_l(\eta)|F(\eta)|^2 + \beta T F_l(\eta)F^*(\eta)\exp(-i2\pi\eta b)$$
$$+ \beta T F_l(\eta)F(\eta)\exp(i2\pi\eta b)] \tag{6.35}$$

Hence the complex amplitude in the output plane is the Fourier transform of equation (6.32), and is given by:

$$o(y) = (t_0 + \beta T)\, f_l(y) + \beta T f_l(y) \times [f(y) * f(y)] + \beta T f_l(y) * f(y) \times \delta(y + b)$$
$$+ \beta T f_l(y) \times f(y) \times \delta(y - b) \tag{6.36}$$

The only term of interest in equation (6.34) is the third one on the right hand side which corresponds to the correlation of $f_l(y)$ with all the characters of the set. If we ignore the constant factor βT, this can be expanded as

$$f_l(y) * f(y) \times \delta(y + b) = \left[f_l(y) * \sum_{i=1}^{N} f_i(y - c_i) \right] \times \delta(y + b)$$

$$= [f_l(y) * f_l(y)] \times \delta(y + c_l + b) + \left[f_l(y) * \sum_{i=1}^{N} f_i(y - c_i) \right] \times \delta(y + b) \tag{6.37}$$

If the autocorrelation of the character presented is sharply peaked, the first term on the right hand side of Equation (6.34) represents a bright spot of light. This is the reconstructed image of the reference source, located at $y = -c_l - b$. The presence of the bright spot in the output plane corresponds to the recognition of the character.

6.12.3 Holographic Interferometry

Holographic interferometry is an extension of interferometric measurement techniques in which at least one of the waves which interfere is reconstructed by the hologram. The wavefronts which were actually separated in time or space can be compared by holographic interferometry, since holography permits storing a wavefront for reconstruction at a later time. Hence, changes in shape of the objects with quite rough surfaces can be studied with interferometric precision. There are a number of recognizable different types of holographic interferometry which we will now describe in brief.

Real-Time Holographic Interferometry

In this technique, after processing the hologram it is placed in exactly the same position in which it was recorded. Now if it is illuminated with the original reference beam, the virtual image will coincide with the object. If, however, the shape of the object has changed very slightly, two sets of light waves will reach the observer. One of them is the directly transmitted wave from the object in its present state, and the other being the reconstructed wave corresponding to the object before the change. The two wave amplitudes will add at the points where the optical path difference is zero or a whole number of wavelengths, and will cancel at some other points in between. As a result, a pattern of interference fringes occur, which is actually the contour map of the changes in the shape of the object.

Although real time holographic interferometry provides a sensitive tool for measuring the strains of an object as they actually deform, it suffers from a number of problems. The biggest problem is the difficulty of replacing the plate exactly in its original place. The distortion of the photographic emulsion during processing also is a problem.

Double-Exposure Holographic Interferometry

Double-exposure holographic interferometry is often used in industries to measure very small displacements or distortions of an object by counting interference fringes. In double-exposure holographic interferometry, interference takes place between the wavefronts reconstructed by two holograms of the object recorded on the same photographic plate. The first exposure is made with the object in its initial, unstressed condition, and the second exposure is made with a stress or displacement applied to the object. When the processed hologram is illuminated with the same reference beam, it reconstructs two images. One of them corresponds to the object in its unstressed condition, and the other corresponds to the stressed state, of the object. The resultant interference pattern reveals the change of the object between the two exposures.

Double-exposure holographic interferometry is much easier than real-time holographic interferometry, because the two interfering waves are always reconstructed in exact register. The distortion of the photographic emulsion is equal in both the images and hence no special care is to be taken. In addition, since the two diffracted wavefronts are similarly polarized and have almost the same amplitude, the fringes have very good visibility.

Time-Average Holographic Interferometry

This technique is very useful in examining the spatial characteristics of low amplitude variations of an object. Normally, during exposure in holography, the object should remain stationary, whereas that is not the case in this technique. The object moves continuously during exposure. Hence, the resultant hologram may be considered as the limiting case of a large number of exposures for many different positions of the object. The fringes which are produced represent contour lines of equal amplitude of variation of the surface. The brightest fringes occur at the nodes where the surface remains stationary. Elsewhere, providing the period of exposure covers many variations of the surface, there is a variation in irradiance due to the surface motion, with almost zero irradiance at the antinodes.

6.12.4 Holographic Computer Memories

Potentially, holographic computer memories have a very high storage capacity. Theoretically, it can go up to 10^{10} bits mm^{-3}. Moreover, holographic memory records and reads out a large number of bits in parallel. The schematic diagram of a holographic memory system is shown in Fig. 6.12.

A page composer is a very important device which may be thought of as an array of light valves. These light valves may be open or closed corresponding to 'ones' and 'zeros' respectively. The page composer transforms the information to be stored in a two-dimensional array of bits. This array of up to perhaps 10^4 bits is then stored at one time in a particular location of the holographic memory. During recording, the light modulators allow maximum irradiance in both the signal and reference beams. To store a different array as another hologram on the storage medium the beam deflector moves the beams to the appropriate position.

By addressing the hologram only with the reference beam the read-out of the data takes place. Modulator 1 is partly closed to reduce the irradiance of the light reaching the holographic storage medium. Modulator 2 is closed to cut off the signal beam. The deflector directs the beam to the hologram to be read via the tracking mirror and an image representing the arrays of 'ones' and 'zeros' is produced. The image is focused by a lens and projected onto the detector array. Each bit is originally stored in the page composer is incident on one photodetector in the array. The stored information is hence converted into electrical signals.

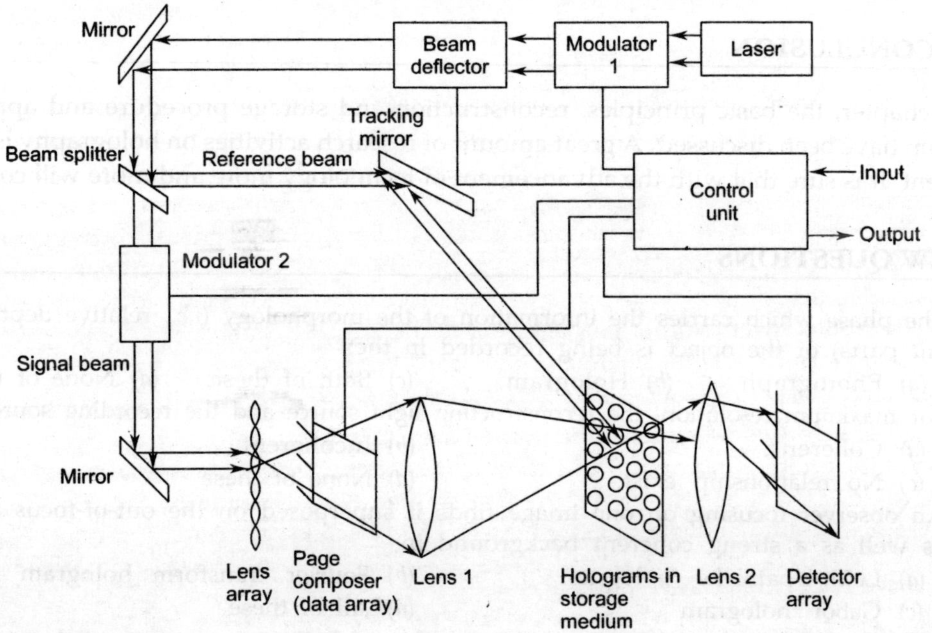

Fig. 6.12

Advantages of Holographic Computer Memories

There are a number of advantages in this storage system compared to a bit-oriented memory. Firstly the information about a particular array is distributed in a holographic fringe pattern across the entire hologram. Hence, the hologram is not sensitive to small scratches or particles of dust which might otherwise cause the loss of bits of information in a bit-oriented memory.

Secondly, the holographic recording and construction is insensitive to the exact position of the reference beam on the hologram. The hologram may be moved, but the focused spots remain stationary. Hence holographic memory systems are easier to align and less subject to problems of vibration than other optical memories.

A third advantage is that the information can be recovered in parallel. Hence read-out data rate is very high. Each position to which the beam is deflected represents 10^4 bits of data. Hence a 10^8 bit memory requires only 10^4 separate locations. Presently available deflectors can satisfy this criterion. Electro-optical deflectors which have access times in the region of microseconds can be used for addressing.

Finally it is possible to record several holograms in different planes of one thick sensitive material provided that different reference beam directions are used to record each hologram. The appropriate hologram is then read with a readout beam aligned at precisely the same angle as the original reference beam in the recording process.

Though there are several advantages of the holographic memory system, but the availability of suitable storage medium is limited to date. Although photographic emulsion is satisfactory, it provides a permanent storage only. Hence this cannot be easily updated. But ideally a storage medium should be easily alterable as well as stable. At the same time the write energy required should be low and readout efficiency should be high. Some examples of read/write/erase optically sensitive materials are magneto-optic MnBi, thermoplastic-photo-conducting layers and electro-photochromatic KCl.

6.13 CONCLUSION

In this chapter, the basic principles, reconstruction and storage procedure and application of hologram have been discussed. A great amount of research activities on holography is in process at present. It is sure that with the advancement of technology more and more will come.

REVIEW QUESTIONS

1. The phase which carries the information of the morphology (*i.e.*, relative depth of different parts) of the object is being recorded in the:
   (*a*) Photograph (*b*) Hologram (*c*) Both of these (*d*) None of these
2. For maximum resolution, the reconstructing light source and the recording source must be:
   (*a*) Coherent (*b*) Incoherent
   (*c*) No relationship exist (*d*) None of these
3. An observer focusing on one image finds it superposed on the out-of-focus twin image as well as a strong coherent background in
   (*a*) Leith-Upatnieks hologram (*b*) Fourier transform hologram
   (*c*) Gabor hologram (*d*) All of these
4. The two images formed during reconstruction of hologram are separated angularly from each other in
   (*a*) Fourier transform hologram (*b*) Gabor hologram
   (*c*) Both (*a*) and (*b*) (*d*) Leith-Upatnieks hologram
5. The form of hologram which can diffract all the illumination it receives into a single diffraction order is

(a) Kinoform (b) Fourier transform hologram
(c) Gabor hologram (d) Leith-Upatnieks hologram

6. When the object is small enough, for its far-field diffraction pattern to be formed at the photographic plate, the hologram formed is
   (a) Fourier hologram (b) Fraunhofer hologram
   (c) Image hologram (d) None of these
7. If a transparency with an amplitude transmittance *f(x, y)* located in the front focal points of lens L_1 and illuminated with a collimated beam is used as an input, its Fourier transform $F(\xi, \eta)$ is displayed at a distance equal to the
   (a) Half the back focal plane of L_1 (b) Back focal plane of L_1
   (c) Twice the back focal plane of L_1 (d) None of these
8. If stable interference fringes are to be formed in holography the maximum path difference between the object wave and the reference wave must be
   (a) Less than the coherence length
   (b) Equal to the coherence length
   (c) Greater than the coherence length
   (d) No such requirement is there
9. In the reconstruction of in-line Gabor hologram, the third term $\beta T r^2 o(x, y)$ actually gives rise to a
   (a) Uniformly attenuated plane wave
   (b) Real image at the same distance in front of the hologram
   (c) Reconstructed image of the object in its original position
   (d) None of these
10. The first term in the reconstruction of Leith-Upatnieks hologram gives rise to
   (a) A halo
   (b) Virtual image of the object in its original position
   (c) Conjugate real image
   (d) Attenuated reference beam
11. What is the basic difference between a normal photograph of an object and the hologram of the same object? Discuss in brief the basic principle of hologram.
12. With the aid of suitable diagram describe how in-line Gabor hologram is recorded. How is the image reconstructed in Gabor hologram? Illustrate with proper diagram.
13. With the aid of suitable diagram illustrate the mechanism of recording of the off-axis hologram. How is the image reconstructed in the off-axis hologram? Discuss with proper diagram.
14. Compare between Gabor and Leith-Upatnieks hologram. Which is advantageous and why?
15. Discuss the technique of recording and reconstruction of Fourier transform hologram.
16. Write a short note on:
   (a) Computer generated hologram; (b) Kinoform;
   (c) Lens-less fourier hologram; (d) Fraunhofer hologram;
   (e) Image hologram.
17. How can multiple imaging be implemented using hologram technique? Discuss with proper diagram.
18. Discuss on the application of holography for character recognition.
19. What do you mean by holographic interferometry? Discuss in brief different types of holographic interferometry.
20. How can holography be used in computer memories? Discuss with proper diagram. What are the advantages of holographic computer memories?

7

Optoelectronic Display Devices

7.1 INTRODUCTION

Visual display is one of the basic requirements of the recent systems. As the technology advances and most of the systems are based on the electronic devices, development of display systems is not lagging behind. Depending on the applications, various display devices have been developed to meet the requirements. Display devices can be divided into two broad categories: (*a*) those that emit radiation their own (active devices) and (*b*) those that in some way modulate incident radiation to provide display information (passive devices). However all the display devices are characterized by the general term as *Luminescence,* which is used to describe the emission of radiation from a solid when it is supplied with some form of energy. The type of luminescence may be distinguished by its method of excitation, as described below:

- **Photoluminescence:** Excitation arises from the absorption of photons.
- **Cathodoluminescence:** Excitation by the bombardment with a beam of electrons.
- **Electroluminescence:** Excitation from the application of an electric field.

Laser devices are the examples of the *photoluminescence* as their principle of operation is based on the absorption of photons. In previous chapters, the construction, operation and applications of various types of Lasers have been discussed. Light Emitting Diodes or LEDs that are excited by the application of electric field, are of *electroluminescence* type and have been discussed in previous chapters in detail. They are available with various colors are intensities. According to the application they are widely being used in single or in clusters. Clusters of LED are formed to give certain shapes in visual display, like seven-segment display to represent the numerical, or alphanumeric display or in the application of traffic signals. But one of the disadvantages of using LEDs is relatively higher power consumption. For greater display size, number LEDs are more and so the power consumption as well as the heat generation cannot be unnoticed. Therefore the technology has made advancement to find alternate and compatible types of display devices. Liquid Crystal Display or LCD, Plasma etc., are such display devices that agree widely used nowadays. In this chapter, some display devices and their applications will be described.

7.2 CATHODE RAY TUBE

Cathode Ray Tube is based on *cathodoluminescence* — excitation by the bombardment with a beam of electrons. In *cathodoluminescence*, the emission process is similar to the process of photoluminescence, but the excitation mechanism is somewhat involved. When a beam of energetic electrons (say with energy greater than 1KeV) hits a solid, a fraction (about 10%) is backscattered. The remainder penetrate into the solid where they rapidly loss energy, mainly by causing bound electrons to be ejected from their present ions. These secondary electrons in turn may generate further secondary electrons, provided they have sufficient energy. The final stages in the secondary generation process consist of the excitation of electrons from states at the top of the valence band with energy E_v to those at the bottom of the conduction band of energy E_c. According to the energy conservation considerations, the exciting electrons must have the energy of at least $E_c - E_v = E_g$ above E_c, if they are to create electron-hole pairs and still remain in the conduction band. Additionally, momentum conservation requires a somewhat higher minimum energy of approximately $E_c + 3E_g/2$. When electrons have energies between this value and E_{c+}, they can only loose energy by exciting lattice vibration (phonons). Energy may still be wasted in phonon generation even when electrons have the higher energies than $E_c + 3E_g/2$. It has been observed for a range of semiconductors that the total number of electron-hole pairs generated may be related as $E_B/\beta E_g$, where E_B is the total electron beam energy and $\beta \approx 3$. Such inefficiency in the generation of electron-hole pairs makes the cathodoluminescence process less efficient than the photoluminescence.

The primary electrons rapidly loose energy and they penetrate only a little into the solid they are exciting. It has been found experimentally that the penetration depth, or range, R_c of an electron beam of energy is given by

$$R_c = KE_B{}^b$$

The parameters K and b depend on the material. As example, for the material ZnS, $K = 1.2 \times 10^{-4}$ and $b = 1.75$, the electron beam of 10KeV has a range of penetration is equal to $1.3 \times 10^{-4} \times 10^{1.75}$ or 0.7 μm.

It is often found that the cathodoluminescence efficiency increases with the increase in beam voltage. This may be attributed mainly to the fact that at low beam voltages most of the electron-hole pairs are generated close to the surface of the luminescent material, where there is often relatively high concentration of nonluminescent recombination centers.

The relative ease with which a beam of electrons can be directed and focused led to the early development and retention of the *Cathode Ray Tube* (CRT). This is an important tool for the analysis of rapidly varying electrical signals as well as providing versatile optical display device. The principle of operation of a CRT is described with the basic construction as shown in Fig. 7.1. Electrons are generated at *electron gun* by thermionic emission by heating a specially impregnated cathode surface, which is generally based on oxides of barium and strontium, and then focused onto the viewing screen by a series of metal electrodes held at various potentials. The electron gun assembly usually also includes a grid for the control of the electron flow. The electron beam is scanned across the viewing screen in a series of lines; when one line scan is completed the beam is rapidly switched to the start of the line below. The beam deflection is controlled by electrostatic or electromagnetic fields acting at right angles to the beam direction. The electrostatic deflection enables to achieve the highest beam deflection, whereas the electromagnetic deflection enables to employ higher beam accelerating potentials, resulting in a smaller spot size and higher screen brightness. When the beam strikes the viewing screen, radiation is generated by cathodoluminescence. The screen consists of a thin layer of phosphor granules of the dimensions of approximately 5 μm with the layer of aluminum of the thickness of approximately 0.1 μm, evaporated at the gun side. This layer serves two purposes; firstly it prevents charge build-up on the phosphor granules (which generally have low conductivities), and secondly it helps to reflect light emitted in a direction away from the observer back towards him.

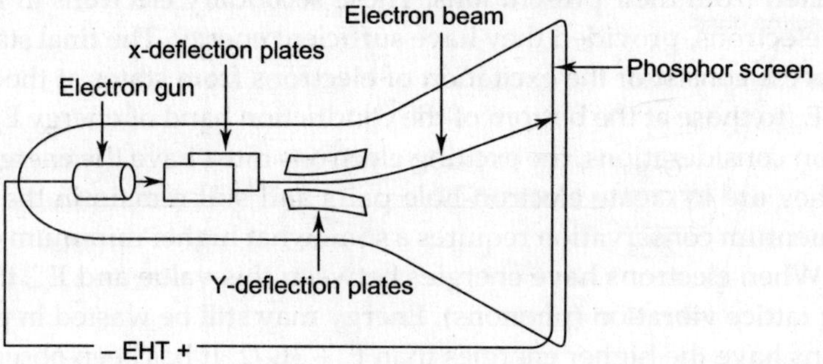

Fig. 7.1

The aluminum and phosphor granule layers are elaborated in Fig. 7.2. The thickness of these layers are fairly critical. If the thickness of aluminum is too thick, an appreciable fraction of the electron beam energy will be absorbed within it, while reflectivity becomes poor if it is too thin. On the other hand, higher thickness of phosphor layer reduces the light output due to scattering and absorption, and a too thin layer of it results in incomplete coverage of the screen area.

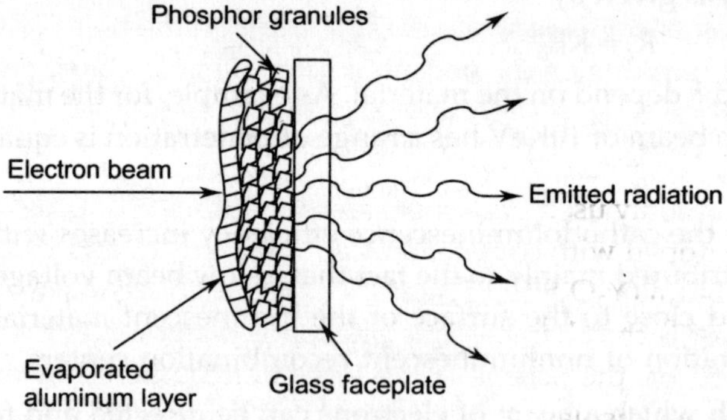

Fig. 7.2

For normal display operations as in television, the beam is scanned line by line over the viewing area. In video applications the display consists of 625 lines according to European standard and 525 lines for American standard. To avoid the flicker effect of the image, the picture must be renewed at a rate greater than about 45 Hz. However it is possible to reduce the scan rate by splitting the scanning of picture into two interlaced halves. In this arrangement, the lines 1, 3, 5, 7 etc., are scanned in first $t_s/2$ seconds and the lines 2, 4, 6, 8 etc., are scanned at next $t_s/2$ seconds. Thus the complete picture may be scanned at t_s seconds. As two images are effectively superimposed, the eye treats the repetition rate of the scanning as $2/t_s$ Hz instead of $1/t_s$ Hz. The reduction in the picture repetition rate of picture information for trouble-free or flicker-free viewing is very much useful because it halves the transmission frequency bandwidth. Varying light irradiences are obtained by varying the beam current. Ideally, the phosphor used should have a luminescent decay time shorter than the picture cycle time, otherwise streaking effects due to image persistence are obtained. CRT displays can be made sufficiently bright for them to be visible under nearly all ambient lighting conditions. The brightness limits are usually reached when the phosphor screen rapidly deteriorates under high beam currents.

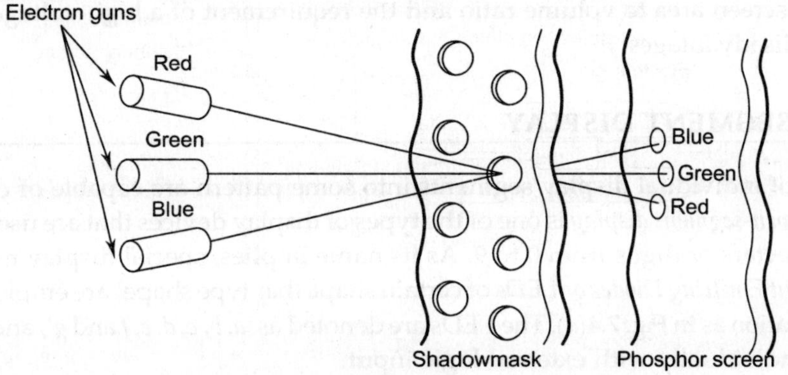

Fig. 7.3

Color displays for home video viewing are obtained using the 'shadowmask' principle, in which three electron guns are used that are slightly inclined to each other so that their beams coincide at the plane of the shadowmask. The shadowmask is a metal screen, with holes in it, placed in front of the phosphor screen as shown in the Fig. 7.3. Three beams of primary colors *i.e.*, Red, Green and Blue colors coincide while passing through the holes of the shadowmask screen and again diverge while striking on the phosphor screen. Thus each of three phosphor dots emits one of the primary colors and so, desired color can be generated by varying the relative excitation intensities.

Some of the most commonly used phosphors are zinc sulphide doped with silver (ZnS:Ag-blue), zinc cadmium silphide doped with copper ($Zn_sCd_{1-s}S$:Cu-green) and yttrium oxysulphide doped with europium and terbium (Y_2O_2S:EU,Tb-red). The first two are the noncharacteristic phosphor materials while the latter is the characteristic phosphor material. As the controlled colored dots are physically displayed on the screen, some loss is encountered in the resolution capability compared to monochrome display. Furthermore, the alignment of the shadowmask with the guns and the phosphor screen is critical, and can be spoilt by harsh environmental conditions such as stray magnetic fields. Several modifications have been made to the basic shadowmask principle to overcome these disadvantages.

Another type of color CRT display has been developed based on the principle of 'penetration phosphors'. Here electron penetration depth is controlled by varying the beam voltage. For example, a two-color display may be obtained by using a mixture of two different kinds of phosphor particle. Let us consider an ordinary phosphor granule of red emitting type and another green emitting phosphor with a non-luminescent coating. At low beam voltage the display will show only red, as low beam energy is insufficient to penetrate the non-luminescent coating of green phosphor. At higher voltages, electron penetration into green phosphor will enhance both red and green emissions. So at high voltages green will be predominated if the red phosphor is less efficient. However, there are considerable difficulties in obtaining any desired intermediate color at any desired level. Furthermore, rapid variation as necessary and desired for video applications cannot be achieved. For these disadvantages, this type of display is not useful for video or multicolor applications. However this technique finds its applications for relatively static display of few fixed colors. This technique offers good resolution capability and relative freedom from magnetic interference.

CRT provides a very versatile display that can readily cope with complex displays. Its comparatively low screen area to volume ratio and the requirement of a high voltage power supply are the major disadvantages.

7.3 SEVEN SEGMENT DISPLAY

Combinations of individual display segments into some pattern are capable of conveying more information. *Seven-segment display* is one of the types of display devices that are used to display the numerical characters or digits from 0 to 9. As its name implies, special display modules consisting of seven *Light Emitting Diodes or LEDs* of certain shape (bar type shape) are employed and placed in certain orientation as in Fig. 7.4(*a*). The LEDs are denoted as '*a, b, c, d, e, f* and *g*', and each LEDs can be controlled individually with external logic input.

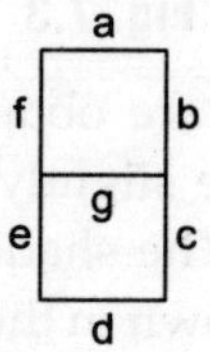

Fig. 7.4 *(a) (Orientation of seven LEDs in a seven segment LED display)*

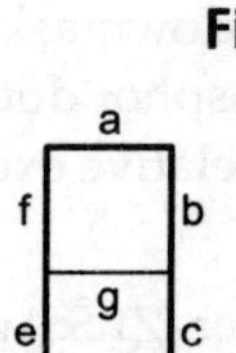

Fig. 7.4 *(b) (Digit 0)*

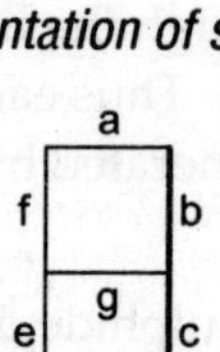

Fig. 7.4 *(c) (Digit 1)*

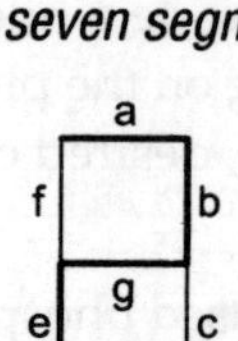

Fig. 7.4 *(d) (Digit 2)*

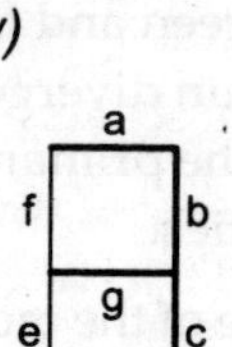

Fig. 7.4 *(e) (Digit 3)*

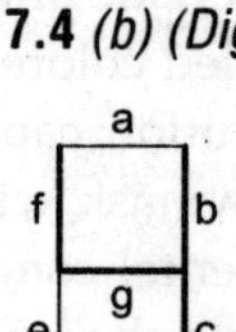

Fig. 7.4 *(f) (Digit 4)*

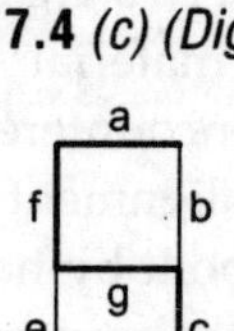

Fig. 7.4 *(g) (Digit 5)*

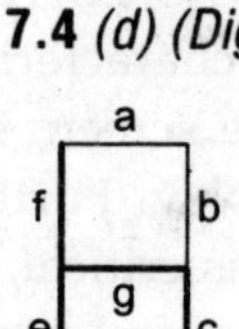

Fig. 7.4 *(h) (Digit 6)*

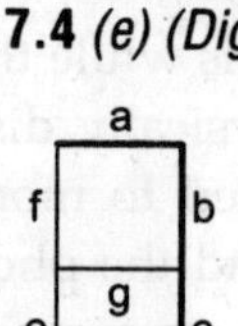

Fig. 7.4 *(i) (Digit 7)*

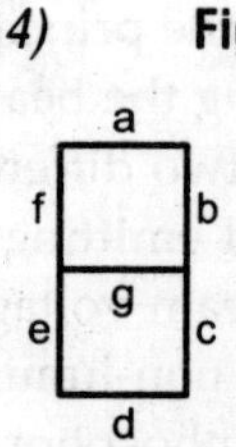

Fig. 7.4 *(j) (Digit 8)*

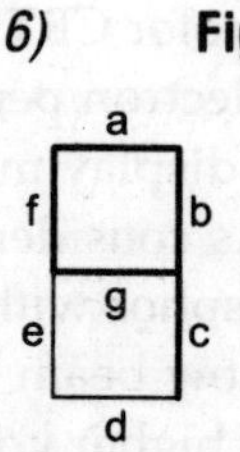

Fig. 7.4 *(k) (Digit 9)*

Decimal digits 0 to 9 can be displayed by glowing some particular LED segments. As example, digit '0' may be represented by glowing the segments *a, b, c, d, e* and *f* as in Fig. 7.4(*b*). Digit '1' may be represented by glowing *b* and *c* as in Fig. 7.4(*c*). Other digits can also be displayed by glowing certain segments as illustrated in Figs. 7.4(*d*) to 7.4(*k*). In the figures thick segments represent the glowing LEDs.

Two types of seven segment display modules are available — *Common Cathode* type and *Common Anode* type; the equivalent electronic circuits are shown at Figs. 7.5(*a*) and 7.5(*b*). From the equivalent circuit, it is clear that to glow a particular LED of common cathode type, logic 1 or high level of voltage is to be applied at the anode of that LED as all the cathodes are grounded. Alternatively, logic 0 low level of voltage is to be applied to glow certain LED of common anode type, as all the anodes are connected to high voltage Vcc.

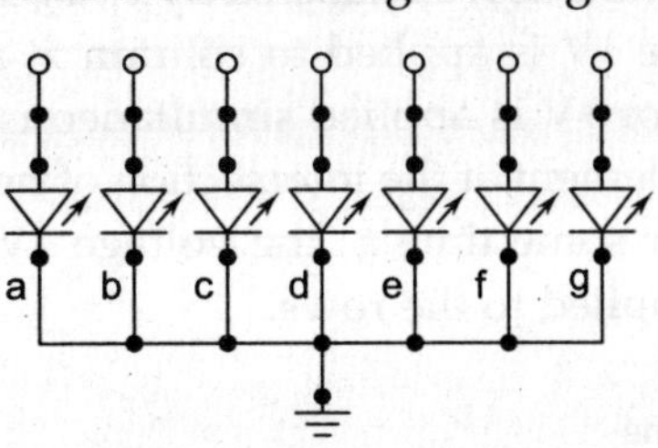

Fig. 7.5 *(a): Common cathode LED*

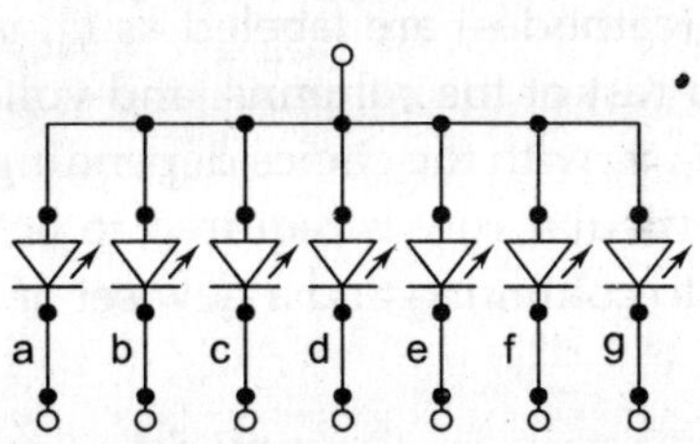

Fig. 7.5 *(b): Common anode LED*

It may be noted that each segment bar might consist of several numbers of discrete LED elements depending upon the size of the display device. Also these display devices are commercially available that are directly compatible with TTL circuits or CMOS logic circuits.

7.4 ALPHANUMERIC DISPLAY

By increasing the number of display elements, more complex characters are possible to be displayed. One of simplest example is to arrange 35 LEDs in 7 × 5 matrix form, *i.e.*, LEDS are arranged in seven rows and five columns as shown in the Fig. 7.6(*a*). With proper selection of LEDs for glowing all the alphabets A to Z in capital and small letters as well as the numerals from 0 to 9 are possible to display. Figures 7.6(*b*) to 7.6(*e*) demonstrates the examples to display the characters 'A', 'B', '*a*' and '*b*' respectively, by bold indication for LEDs that are glowing and blanking those LEDs that are not glowing. Numerals 0 to 9 can also be displayed by similar technique.

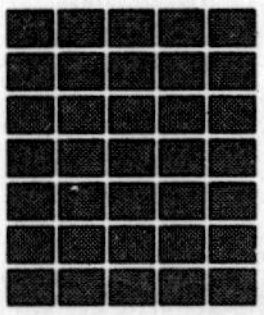

Fig. 7.6(*a*)

Fig. 7.6(*b*)

Fig. 7.6(*c*)

Fig. 7.6(*d*)

Fig. 7.6(*e*)

There are two basic methods of wiring up a matrix display. The simpler is known as *common anode* or *common cathode* and interconnection of it is shown in the Fig. 7.7(*a*). The anodes (or cathodes) are all wired together, while each cathode (or anode) has a separate connection to make available externally. Thus for N elements, the number of external connections is N + 1. A technique requiring fewer connections is the coordinate connected (or matrix connected) method that is shown at Fig. 7.7(*b*). In this all the anodes of the elements in each column are connected together as are all the cathodes of the elements in each row. Thus for a square array of N elements, $N^{1/2}$ external connections are required. If N is large, number of connections is considerably less compared to common anode or common cathode connections. However it is obvious that a general display pattern cannot be maintained merely by applying static voltages to the leads. A scanning mechanism must be employed. Consider that the columns (anodes) are labeled by x_1, x_2, x_3 etc and the rows (cathodes) are labeled as y_1, y_2, y_3 etc., a voltage +V is applied to column x_1 and zero voltage to rest of the columns, and voltages of either –V or +V is applied simultaneously to the rows y_1, y_2, y_3 with the choice depending on whether the element at the intersection of he column x_1 and particular row is required to be illuminated. After some time t_c, the voltage +V is to be switched to column x_2 and a new set of voltage is to be applied to the rows.

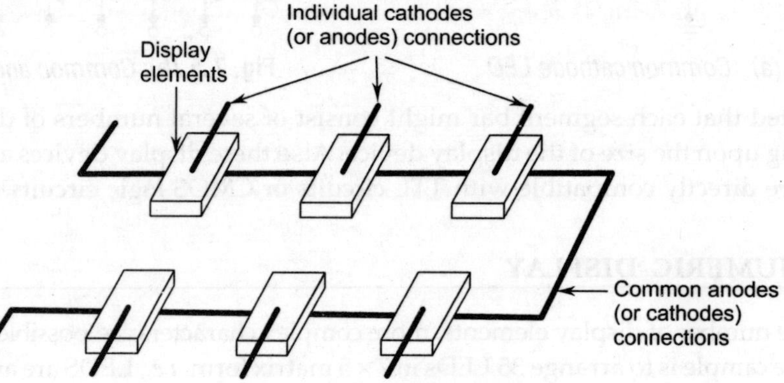

Fig. 7.7(*a*)

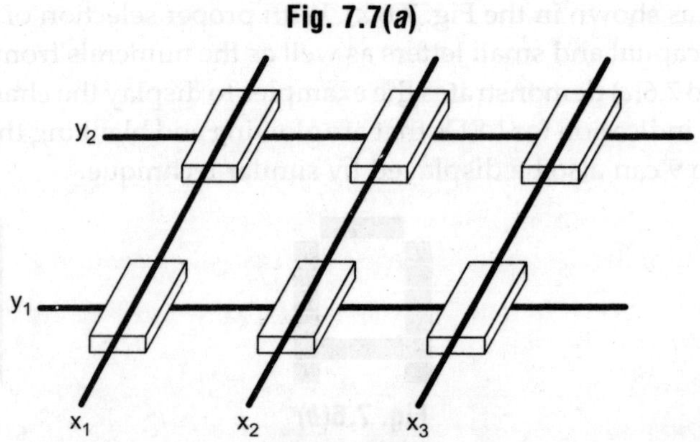

Fig. 7.7(*b*)

If the number of columns is N_c, then the all the LED elements can be scanned in a time $N_c t_c$. According to the persistence of vision of human eyes, $N_c t_c$ must be less than or equal to 1/45 sec, and hence $t_c = 1/(45N_c)$ sec. If N_c is very large, t_c may approach to the response time of the device resulting in a reduced device output. Thus the size of N_c is limited, unless a much-reduced brightness is acceptable.

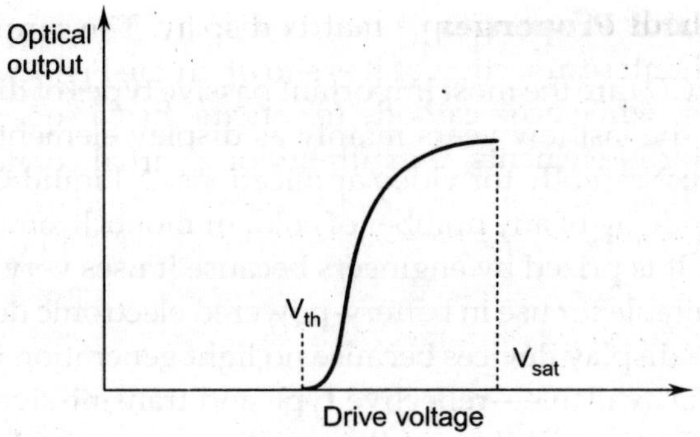

Fig. 7.8

It may be noted that even when an element is OFF, it still may have a voltage V across it. An ideal element should show no output up to a threshold voltage V_{th} and then give its maximum saturated output at a voltage V_{sat}, where $V_{sat} = 2 \times V_{th}$ as shown in the Fig. 7.8. However, V_{sat} is often greater than $2V_{th}$ and so it is not possible to drive the device to saturation, resulting in the reduction of the maximum brightness available. In addition V_{th} may vary with the operating temperature and the safe value is somewhat below the actual value may have to be adopted, again leading to the reduced output.

7.5 LIQUID CRYSTAL DISPLAY (LCD)

Liquid crystal materials were first discovered in 1888 by an Austrian botanist, F. Renitzer. However those liquid crystals were not suitable for any commercial usage, and it is only 25 years ago since the first material suitable for electronically driven displays was developed. The first room-temperature nematic liquid crystal was observed in the late 1960s. Unfortunately this crystal had quite a short temperature range as it was affected by impurities. Occasionally in homologous series the temperature range could reach from –40 to +100 degrees Celsius. Unfortunately these mixtures were very unstable and they possessed a negative dielectric anisotropy not useful in the twist cell.

The major breakthrough came when cyanobiphenyl materials were discovered a few years later. The more stable phase had a large positive dielectric anisotropy as well as a strong birefringence nearly ideal for the twist cell. During the 1970s and 1980s several liquid crystal compounds and phases were discovered, primarily by the industry, but also in several research programs on liquid crystal materials in colleges and universities around the world.

The Ferroelectric Chiral Smecic (FLC) phase was discovered in 1975 and proved to have a unique form of ferroelectricity. The first display based on the FLC phase was actually patented in 1980. Another example of new liquid crystal phases also discovered during this intense research period, are the forms of polymer dispersions. Also a new effect, the electroclinic effect, was discovered during this period and is now being carefully studied for possible future display applications.

Lately several new materials has been discovered such as the retardation film which is extremely important for the supertwisted nematic (STN) and Twisted Nematic (TN) displays. Other types of rapidly growing display types are the Electrically Controlled Birefringence (ECB) and Polymer-Dispersed Liquid Crystals (PDLCs). Due to the relatively recent technology discovered in producing these cells, they have not yet reached a fully commercial usage.

7.5.1 LCD Structures and Properties

Liquid Crystal Displays or *LCDs* are the most important passive types of display devices. These have come into prominence in the last few years mainly as display elements for digital watches and pocket calculators, and most recently for video applications. A Liquid Crystal Display (LCD) is a thin, flat display device made up of any number of color or monochrome pixels arrayed in front of a light source or reflector. It is prized by engineers because it uses very small amounts of electric power, and is therefore suitable for use in battery-powered electronic devices. LCDs consume the least power of all common display devices because no light generation is required in them. There are two basic types of LCD available—reflective type and transmissive type. The reflective type LCD requires front illumination, while rear illumination is required for the tranmissive type. Most reflective types utilize ambient light for illumination with provision for secondary illumination by a small incandescent lamp or LED, if ambient levels become too low. The basic principle of all LCDs is the formation of a cell between two glass plates each with a conductive coating. The cell has the thickness of about 10 μm or less and filled with a liquid crystal material.

Liquid Crystal Displays (LCDs) offer several advantages over traditional cathode-ray tube isplays. LCDs are flat, and they use only a fraction of the power required by CRT's. They are easier to read and more pleasant to work with for long periods of time than most ordinary video monitors. One should also know that there are several tradeoffs, such as limited view angel, brightness, and contrast, not to mention high manufacturing costs. As research continues, this limitations are slowly becoming less significant.

What is liquid crystal? There are three common states of matter that we know about: solid, liquid and gas. Liquid crystal is a fourth "state" that certain kinds of matter can enter into under the right conditions. The molecules in solids exhibit both positional and orientational order, in other words the molecules are constrained to point only certain directions and to be only in certain positions with respect to each other. In liquids the molecules do not have any positional or orientational order, the direction the molecules point and positions are random.

The liquid crystal "phase" exists between the solid and liquid phase, the molecules in liquid crystal do not exhibit any positional order, but they do possess a certain degree of orientational order. A large number of organic materials exhibit the liquid crystal state over a restricted temperature range. At lower end of temperature range the material becomes a crystalline solid, while at upper temperature range the material changes into a clear solid. Within this temperature range, it has a milky yellowish appearance and exhibits some optical properties of solids in combination with the fluidity of liquids. All the liquid crystal compounds have the major characteristics of the rod-like shape of their molecules. At the liquid crystal phase, these molecules can take up certain orientations relative to each other as well as to the liquid crystal surface. The molecules do not all point the same direction all the time. They tend to point more in one direction over time than other directions. This orientation is generally described by the term—*director*, an unit vector pointing along the time-averaged preferred orientation of the molecules in any small volume. The amount of order is measured by the order parameter of the liquid crystal. This order parameter is highly dependent on the temperature of the sample. Figure 7.9 may be referred for a typical order vs. temperature relationship. T_c is the temperature of transition between the liquid crystal and the liquid states.

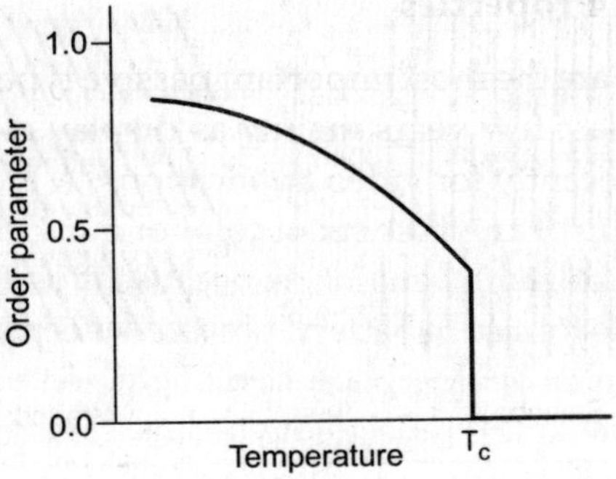

Fig. 7.9

Not all substances can have a liquid crystal phase. Molecules that tend to be candidates for having the phase, are long and have a rigid central region and ends that are slightly flexible. Liquid crystals are essentially more like liquids than they are like solids. This is evident from the latent heat of transition (the amount of energy needed for a phase transition to occur).

There are three basic types of ordering in liquid crystals, — *nematic, cholesteric* and *smectic*, depending on the arrangement of the molecules. They are also birefringent, meaning that it possesses two different indices of refraction. One index of refraction correspondents to light polarized along the director of the liquid crystal, and the other is for light polarized perpendicular to the direction. Nematic liquid crystal is a common type and has a thread-like formation. A *nematic* liquid crystal molecule usually consists of two benzene rings linked with a central group. A typical example is 4-methoxybenzylidene-4-butylanaline (MBBA), which has chemical formula as

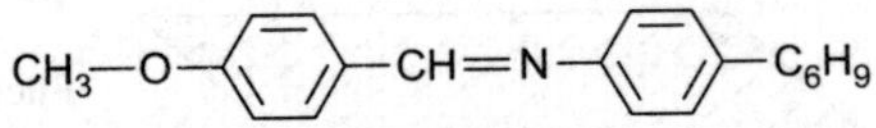

Within the temperature range of 20°C to 47°C, liquid crystal behavior is observed in MBBA. The molecular orientation (and hence the material's optical properties) can be controlled with applied electric fields. The molecular orientation at liquid crystal phase is shown in Figs. 7.10 and 7.11.

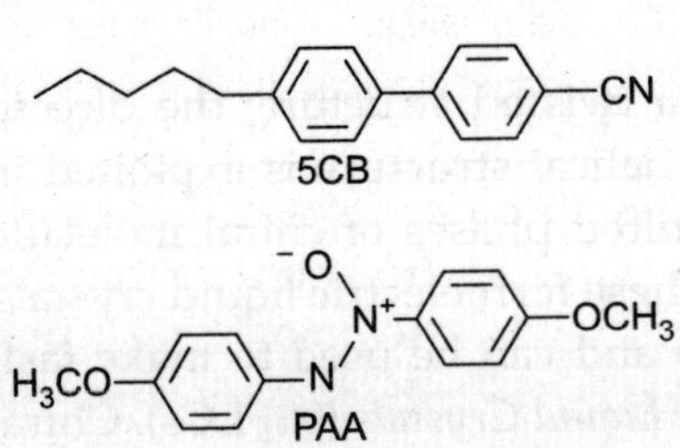

Fig. 7.10

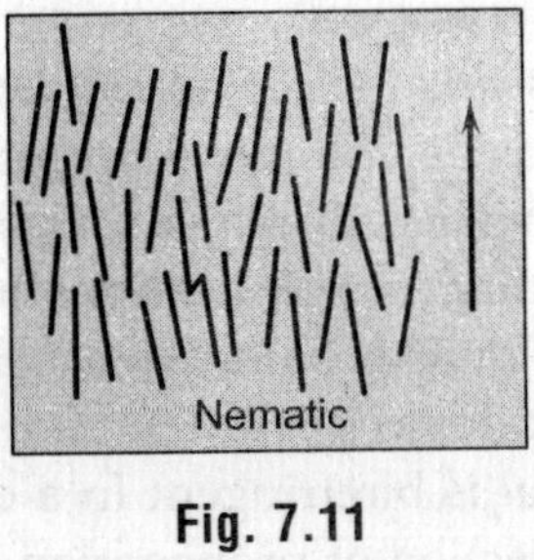

Fig. 7.11

Smectic liquid crystal has a soapy texture and is found at lower temperatures than the nematic. In the Smectic A phase, the molecules are orientated along the layer normal, while in the Smectic C phase they are tilted away from the layer normal as illustrated in Fig. 7.12.

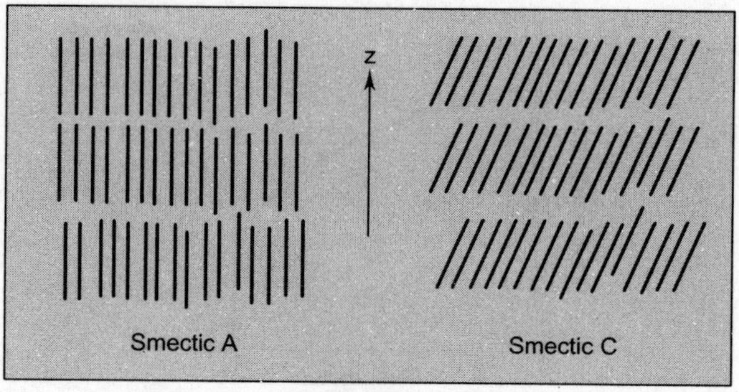

Fig. 7.12

For *cholesteric* or *Chiral nematic* type ordering, the material is considered to be made of large number of planes each having a nematic like structure, but each plane showing a progressive change in the director direction from the one below. This is illustrated in Fig. 7.13. The director directions thus display a helical twist through the material. The distance between the planes of similar director direction is called the *pitch, p*. Cholesteric liquid crystal exhibit some interesting color effects. If light of wavelength λ is incident on the director planes, then strong Bragg reflection will occur when $p = n\lambda$, where n is an integer, but not otherwise. Therefore, if white light is incident on a cholesteric liquid crystal, it can appear strongly colored. In addition to this, the pitch is usually temperature dependent, and so the color of the reflected light will also be temperature dependent. Obviously this phenomenon can form the basis of a thermometer. Most liquid crystal displays are of so-called twisted nematic type.

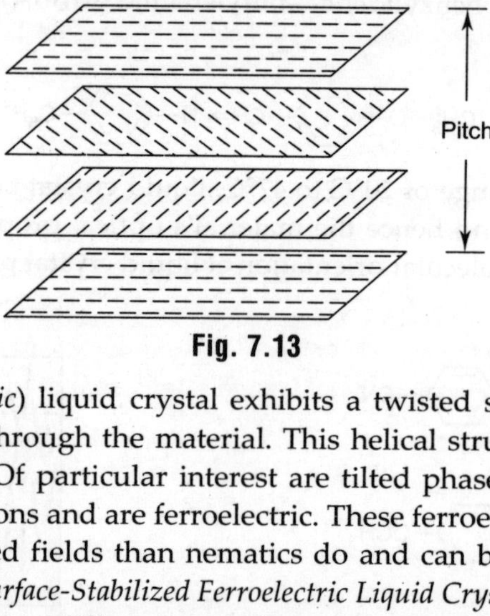

Fig. 7.13

Chiral nematic (or *cholesteric*) liquid crystal exhibits a twisted structure; the director rotates about an axis as you move through the material. This helical structure is exploited in making modern flat-panel displays. Of particular interest are tilted phases of chiral molecules, which possess permanent polarizations and are ferroelectric. These ferroelectric liquid crystals respond much more quickly to applied fields than nematics do and can be used to make fast, bistable electro-optic devices called *Surface-Stabilized Ferroelectric Liquid Crystals* (SSFLCs). Chiral nematic liquid crystal is birefringent in a different way. Supposing that the helical structure is aligned with the direction of propagation of the light, circularly polarized light will travel through the crystal at different speeds depending on whether it is right-circularly polarized or left-circularly polarized (referring to the direction the polarization rotates around the axis of propagation). This is called circular birefringence, and it is exhibited in chiral nematic liquid crystal because of the helical structure the molecules form.

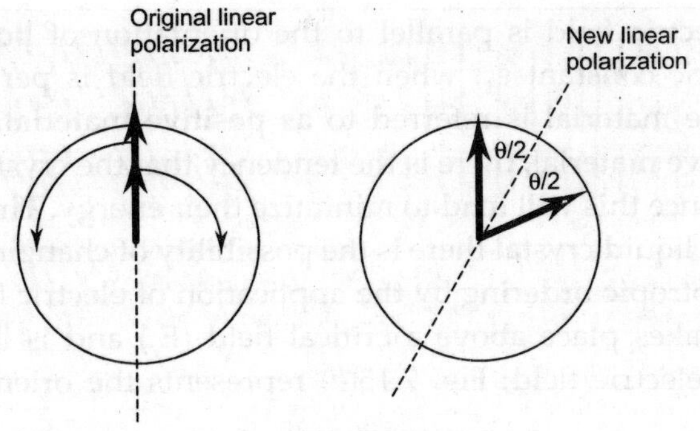

Fig. 7.14

The circular birefringence is highly wavelength dependent, so light of different colors gets modified in different amounts. Circularly polarized light can be thought of containing tow components, one of left- and one the right-circularly polarized light. If the light is linearly polarized, these components are equal. If we send linearly light through a chiral nematic liquid crystal, the component of circularly polarized light that matches the chirality of the crystal structure will travel faster than the other component, having the ultimate effect that its polarization will rotate faster with respect to the other component's polarization. When the two components emerge, their polarization will again rotate at the same rate and the light will again be linearly polarized, but since one of the rotations got ahead of the other, the light will now be linearly polarized along a different angle. The new angle will depend on how far out of "circular phase" the two components were, which is directly dependent on thickness of the liquid crystal. This effect is called the optical activity. It is a measure of the change in polarization angle per unit thickness, and it is heavily wavelength dependent. Figure 7.14 describes the visualization of optical activity. The two arrows represent the directions of the polarization of the components of left and right circularly polarized light at a particular point in time. After passing through the liquid crystal (right), the two components sum to create linear polarization in a different direction.

When a nematic liquid crystal material comes into contact with a solid surface, the directors often become aligned either perpendicular to the surface (*homeotropic* ordering) or parallel to the surface (*homogeneous* ordering). By suitable treatment of the surface these two forms of alignment can be produced. In case of homogeneous ordering, this treatment typically consists of a thin polymer layer that is unidirectionally rubbed using a soft fabric like cotton before it comes to contact with the liquid crystal material. The liquid crystal molecules then align parallel to the direction of rubbing.

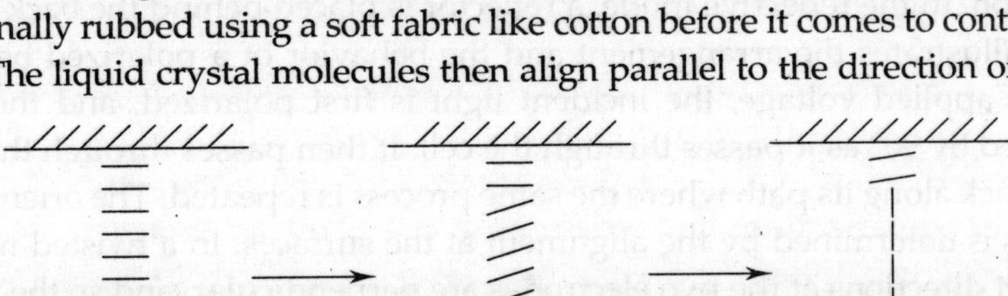

Fig. 7.15

One of the most important characteristics of liquid crystal materials is that they exhibit different dielectric constants depending on the externally applied electric field. If the dielectric constant

$\varepsilon_{\parallel}$, when the applied electric field is parallel to the orientation of liquid crystal molecules, is greater than the dielectric constant $\varepsilon_{\perp}$, when the electric field is perpendicular to the crystal molecule orientation, the material is referred to as positive material. With the application of electric field to the positive material, there is the tendency that the crystal molecules are oriented along the electric field, since this will tend to minimize their energy. Therefore, it may be noticed that for positive material liquid crystal there is the possibility of changing the homogeneous type of ordering into a homeotropic ordering by the application of electric field perpendicular to the surface. The transition takes place above a critical field (E_c) and is illustrated at in Fig. 7.15. Figure 7.15(*a*) is for no electric field; Fig. 7.15(*b*) represents the orientation when $E > E_c$; and Fig. 7.15(*c*) is for $E >> E_c$.

In the twisted nematic cell of liquid crystal, the composite walls of molecular alignment directions at right angles to each other are treated to produce homogeneous arrangement. Thus the molecules undergo a 90° rotation across the cell as shown in Fig. 7.16(*a*). When a beam of polarized light is incident on the cell, the strong optical anisotropy of liquid causes the polarization to undergo 90° rotations. With strong enough electric field ($E >> E_c$) across the cell, the molecular arrangement becomes as shown in Fig. 7.16(*b*) and in this state the molecular arrangement will have no effect on an incident polarized light beam.

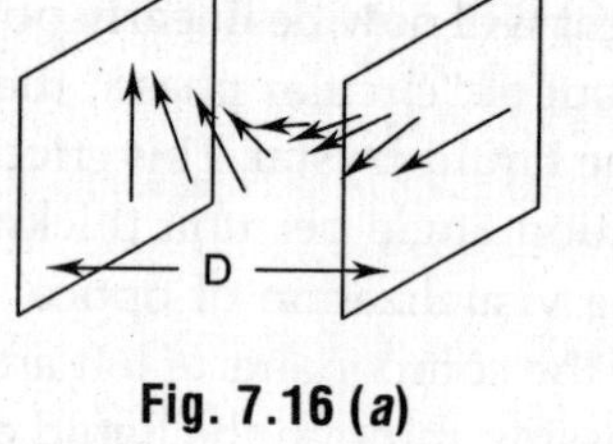

Fig. 7.16 (*a*)

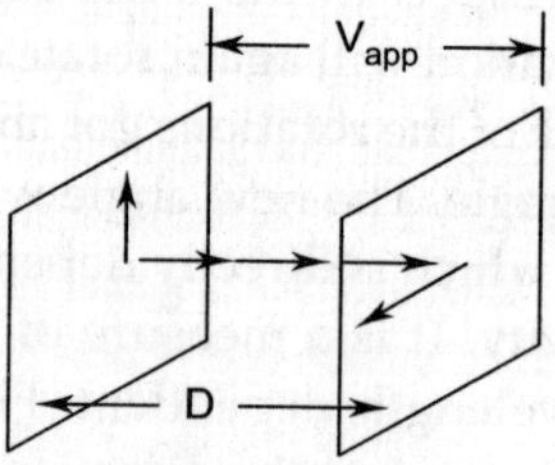

Fig. 7.16(*b*)

At practical situation, each cell of the LCD consists of a layer of perpendicular molecules aligned between two transparent electrodes, and two polarizing filters or polaroids, the axes of polarity of which are perpendicular to each other. With no liquid crystal between the polarizing filters, light passing through one filter would be blocked by the electrodes. The surfaces of the electrodes that are in contact with the liquid crystal material are treated so as to align the liquid crystal molecules in a particular direction. In the reflective mode, a reflector is placed behind the back sheet of the polaroid. Figure 7.17 illustrates the arrangement and the behavior of a polarized beam traversing the system. With no applied voltage, the incident light is first polarized, and then its polarization direction is rotated by 90° as it passes through the cell. It then passes through the second polarizer and is reflected back along its path where the same process is repeated. The orientation of the liquid crystal molecules is determined by the alignment at the surfaces. In a twisted nematic device, the surface alignment directions at the two electrodes are perpendicular, and so the molecules arrange themselves in a helical structure, or twist. Because the liquid crystal material is birefringent, light passing through one polarizing filter is rotated by the liquid crystal helix as it passes through the liquid crystal layer, allowing it to pass through the second polarized filter. Half of the light is absorbed by the first polarizing filter, but otherwise the entire assembly is transparent.

When a voltage is applied across the electrodes, a torque acts to align the liquid crystal molecules parallel to the electric field distorting the helical structure (this is resisted by elastic forces since the molecules are constrained at the surfaces). This reduces the rotation of the polarization of the incident light, and the device appears gray. If the applied voltage is large enough, the liquid crystal molecules are completely untwisted and the polarization of the incident light is not rotated at all as it passes through the liquid crystal layer. This light will then be polarized perpendicular to the second filter, and thus be completely blocked and the pixel will appear black. By controlling the voltage applied across the liquid crystal layer in each pixel, light can be allowed to pass through in varying amounts, correspondingly illuminating the pixel. Therefore, with no field applied, the device reflects the incident radiation and appears bright. When an electric field is applied, the direction of polarization of light traversing the cell is not rotated and hence cannot pass through the second polarizer and it will appear dark.

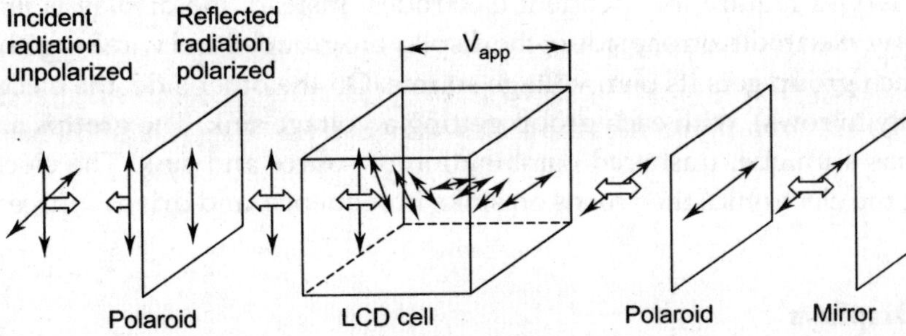

Fig. 7.17

With a twisted nematic liquid crystal device it is usual to operate the device between crossed polarizers, such that it appears bright with no applied voltage. With this setup, the dark voltage-on state is uniform. The device can be operated between parallel polarizers, in which case the bright and dark states are reversed (in this configuration, the dark state appears blotchy).

The amount of the reflected light is the function of the applied voltage and it is schematically shown in Fig. 7.18. The reflectance that is initially constant falls rapidly beyond a critical voltage V_c, where $V_c = E_cD$, D is the thickness of the cell. The reflectance again becomes constant beyond a voltage V_{sat}. A typical value of V_c is 3V. Both the liquid crystal material and the alignment layer material contain ionic compounds. If an electric field of one particular polarity is applied for a long period of time, this ionic material is attracted to the surfaces and degrades the device performance as well as the operating lifetime. This is avoided by applying either an alternating current, or by reversing the polarity of the electric field as the device is addressed (the response of the liquid crystal layer is identical, regardless of the polarity of the applied field). Generally a square waveform of the frequency range of 25 Hz to 1 KHz is used.

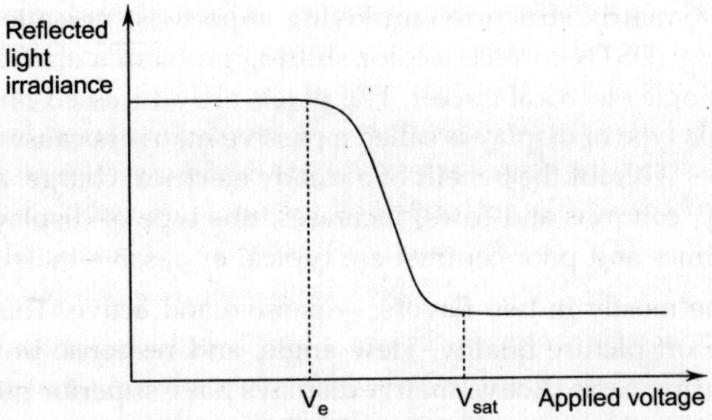

Fig. 7.18

The operation principle of transmission type LCD is similar to that of reflective type, but they do not have the reflector and therefore they require rear illumination. The use of polarizers in the twisted nematic cell substantially reduces the maximum amount of light that can be reflected from it. In addition, the viewing angle is restricted to about ± 45°. In the recent development of 'supertwist' liquid crystal displays, the viewing angle is observed to be wider. The 'supertwist' LCD is basically nematic displays, where the directors undergo a rotation of 270° across the cell, instead of 90°. This is achieved by some special treatment of the surfaces as well as the addition of some cholesteric material to the nematic material.

When a large number of pixels is required in a display, it is not feasible to drive each directly since each pixel would require independent electrodes. Instead, the display is *multiplexed.* In a multiplexed display, electrodes on one side of the display are grouped and wired together (typically in columns), and each group gets its own voltage source. On the other side, the electrodes are also grouped (typically in rows), with each group getting a voltage sink. The groups are designed so that each pixel has a unique, unshared combination of source and sink. The electronics, or the software driving the electronics then turns on sinks in sequence, and drives sources for the pixels of each sink.

7.5.2 Color Displays

In color LCDs each individual pixel is divided into three cells, or subpixels, which are colored red, green and blue respectively, by additional filters (pigment filters, dye filters and metal oxide filters). Each subpixel can be controlled independently to yield thousands or millions of possible colors for each pixel. Older CRT monitors employ a similar method. Color components may be arrayed in various pixel geometries, depending on the monitor's usage. If software knows which type of geometry is being used in a given LCD, this can be used to increase the apparent resolution of the monitor through subpixel rendering. This technique is especially useful for text anti-aliasing.

7.5.3 Passive Matrix and Active Matrix

LCDs with a small number of segments, such as those used in digital watches and pocket calculators or general purpose alphanumeric LCD with two lines of 16 characters have a single electrical contact for each segment. An external dedicated circuit supplies an electric charge to control each segment. This display structure is unwieldy for more than a few display elements.

Small monochrome displays such as those found in personal organizers, or older laptop screens have a passive-matrix structure employing supertwist nematic (STN) or double-layer STN (DSTN) technology (DSTN corrects a color-shifting problem with STN). Each row or column of the display has a single electrical circuit. The pixels are addressed one at a time by row and column addresses. This type of display is called a passive matrix because the pixel must retain its state between refreshes without the benefit of a steady electrical charge. As the number of pixels (and, correspondingly, columns and rows) increases, this type of display becomes less feasible. Very slow response times and poor contrast are typical of passive-matrix LCDs.

Today's LCDs come mostly in two flavors — passive and active. The less expensive passive matrix displays trade off picture quality, view angle, and response time with power requirements and manufacturing costs. Active matrix displays have superior picture quality and viewing characteristics, but need more power to run and are much more expensive to fabricate.

Passive Matrix Displays

The most common kind of display is the twisted nematic display. In this display, the liquid crystal molecules lie parallel to the glass plates, and the glass is specially treated so that the director of the crystal is forced to point a particular direction near one of the plates and perpendicular to that direction near the other plate. This forces the director to twist by 90° from the back to the front of the display, forming a helical structure similar to chiral nematic liquid crystals. Some chiral nematic crystal is added to make sure all of the twists go the same direction.

This thin film of twisted nematic liquid crystal is circularly birefringent. When linearly polarized light passes through, the optical activity of the material causes the polarization of the light to rotate by a certain angle. The thickness of the film, typically around 6 or 8 micrometers, can be controlled to produce a rotation of the polarization of exactly 90° for visible light. Therefore, when the film is placed between crossed polarizers, this arrangement allows light to pass through. However, when an electric field is applied across the film, the director will want to align with the field. The crystal will lose its twisted structure and, consequently, its circular birefringence. Therefore, linearly-polarized light entering the crystal will not have its polarization rotated (it is only rotated very slightly), so light will not be able to penetrate through the other polarizer. When the field is turned off, the crystal will relax back into its twisted structure and light will again be able to pass through. In some displays, the polarizers are parallel to each other, thus reversing the on and off states. If red, green, and blue colored filters are used on groups of 3 pixels, color displays can be created.

Figure 7.19 shows the basic layers of STN LCD. Each color pigment is turned on or off by supplying current through a timed operation method to electrodes oriented in length and width. TFT LCDs may be noted for their outstanding screen image, but STN LCDs certainly offer superior economy. STN is also known as passive matrix LCD technology.

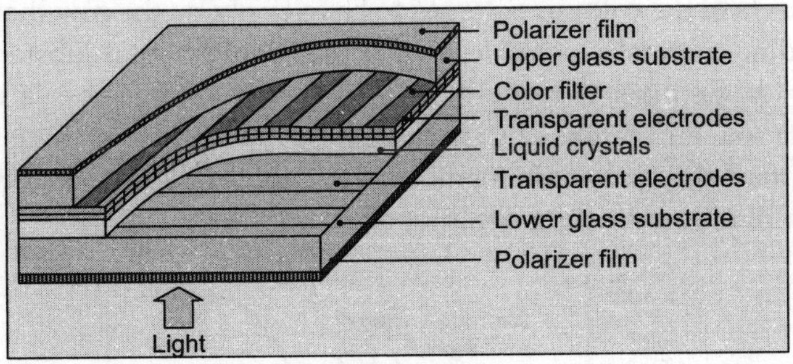

Fig. 7.19

The twisted nematic system coupled with multiplexed addressing is used in many of today's so-called passive matrix LCDs. Even though it is the most popular kind, it does have a number of disadvantages as well. First of all, the use of polarizers reduces the potential brightness since they allow less than half of the light incident on the display to pass through. The effective viewing angle of the display can be very small because the optical activity and the polarizers are tuned to work best only on light that is propagating perpendicular to the display. The voltage-brightness response curve is often not very sharp, leading to reduced contrast. The display is also affected by crosstalk where voltage meant for a certain pixel can leak through "sneak paths" to nearby pixels, causing a ghosting effect. And finally, switching speed of the liquid crystal is often not as high as might be desired—typically around 150 milliseconds. Lower switching speeds are necessary when doing multiplexing since one want the crystal to respond to voltages over the whole scanning cycle to reduce flicker, but such low speeds make passive matrix displays unusable for many applications (such as full motion video).

The degree of multiplexing in twisted nematic displays has a huge influence on the contrast of the display. The liquid crystal will respond to the average voltage applied to it over a certain period of time, depending on its viscosity. Assuming the liquid crystal responds to voltages over one frame period, we can calculate the average voltage felt by a pixel that is on and a pixel that is off. A row gets a voltage of 1 if it is being addressed and 0 if it is not. During each row pulse, the column voltages are set according to which pixels in that row are on or off. If a certain pixel is on, it receives a column voltage of –1, otherwise it is 0.

It is now possible to calculate the average voltage felt by a specific pixel assuming that other pixels in the same column have equal probability of being on or off. The calculation will show that the average voltage is of a 1/2 over a frame period. Since the voltage-brightness response curve of the crystal is not very steep these voltages will not lead to as high of a difference in brightness as would be possible if each pixel received either a full 2 volts or 0 volts continuously over the entire frame period. Refer Fig. 7.20 for the response curve for light transmission vs. voltage.

Ideally, one would want the response curve to have near infinite slope between the average on and off voltages, but this is very difficult to achieve. A certain minimum ratio is needed to get good contrast. Since the ratio is a function of the number of rows, the more rows, the less contrast the display will have. Active matrix displays solves this problem by putting a full switch at every pixel in the entire Display (as described above). This special element "actively" drives the voltage to the pixel continuously over the frame period. Essentially, a thin-film transistor at each pixel allows the effects of the column voltages to be felt only by the row that is being addressed. Therefore, outstanding contrast is possible, and a faster liquid crystal substance can be used since it is not necessary for the pixel to respond to the average voltage over a whole frame period. Crosstalk is also cut to a minimum. With increased contrast and switching speed, active matrix displays are more than able to handle many kinds of information, including full-motion video which requires a minimum switching speed of 50 milliseconds.

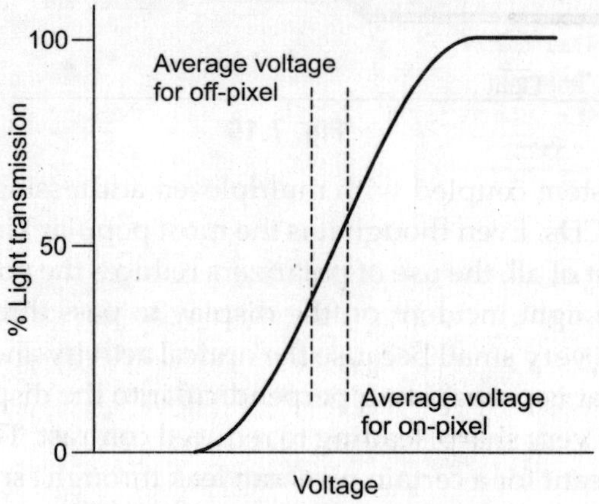

Fig. 7.20

Active Matrix Display

The limitations of passive matrix displays are described here and active matrix displays are able to over come those problems. However active matrix displays are extremely expensive to manufacture and the process of creating large sheets of thin-film transistors is far from perfect. Therefore, much research has been done to try to improve the characteristics of the more affordable passive matrix displays. In some very recent cases, the quality of passive matrix displays has gotten very close to that of active matrix displays.

Active matrix liquid crystal displays are standard on most new laptop computers. Two properties of liquid crystal are used as tiny switches to turn picture elements (pixels) off and on. First the crystals are transparent but can alter the orientation of polarized light passing through them. Second, the alignment of their molecules (and their polarization properties) is changed by applying an electric field.

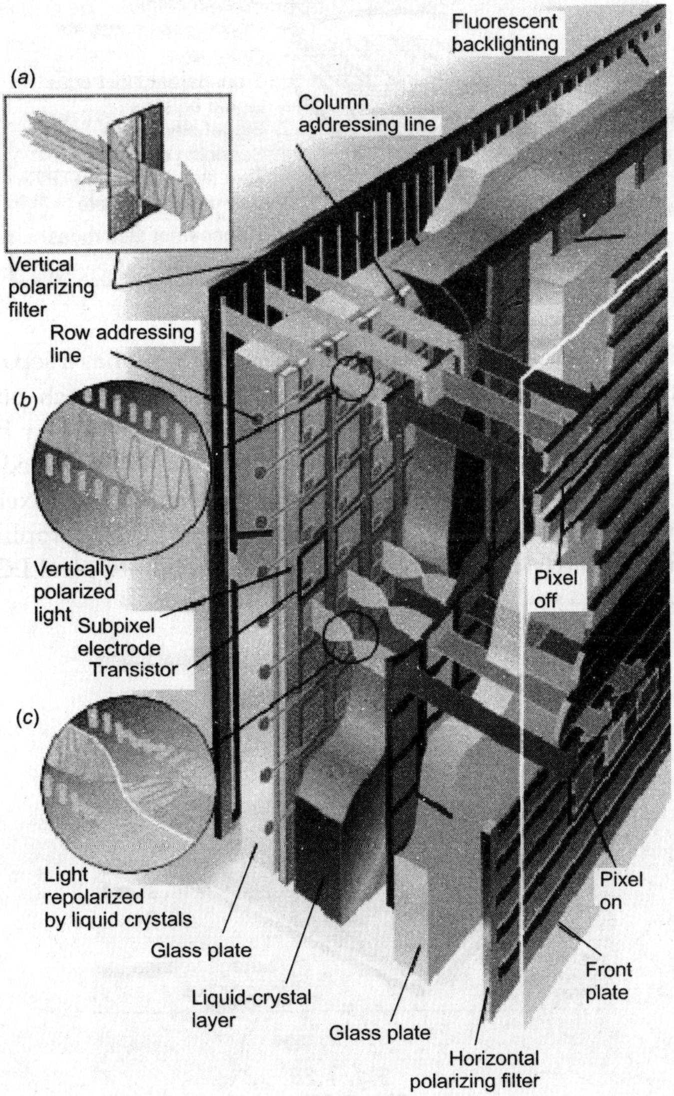

Fig. 7.21

In a color display the liquid crystals are held between two glass plates or transparent plastics. These plates are usually manufactured with transparent electrodes, typically made of indium tin oxide that makes it possible to apply an electric field across small areas of the film of liquid crystal. The outsides are coated with polarizing filters. Only light with a perpendicular polarization can pass through these filters. See (*a*) of Fig. 7.21

Inside the plates are transparent electrodes and color filters, which form very small picture element regions called subpixels. A grouping of a red, a green and a blue subpixel defines the color that the pixel transmits. Fluorescent backlighting illuminates a display from the rear. In pixels that are off, light passes through the rear-polarizing filter, the crystals (*b*) at Fig. 7.21 and the color filters, only to be blocked (absorbed) by the front polarizing filter. To the eye, these pixels appear dark. When a pixel is turned on, the liquid crystals reorient their position, and they in turn repolarize the light so that it can pass through the front polarizing filter (*c*) in Fig. 7.21.

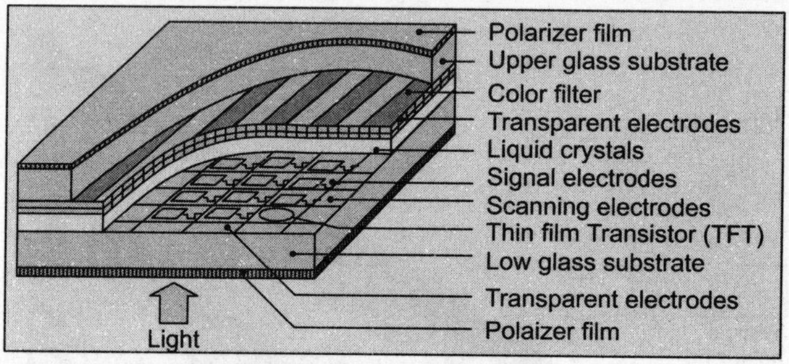

Fig. 7.22

A display made with active matrix TFT technology is a LCD that has a separate and independent transistor for each pixel on the display. Having a transistor at each pixel means that the current that triggers pixel illumination can be smaller and therefore it may be switched on and off more quickly. The pixel emit colored light which passes through a RGB (Red, Green, Blue) filter. The result of this is a screen with a very small response time and direct pixel addressing is also possible which makes the TFT LCD suitable for video and fast graphic applications. Basic layer construction for an active matrix LCD is shown in Fig. 7.22 and a typical LCD screen layout is shown in Fig. 7.23.

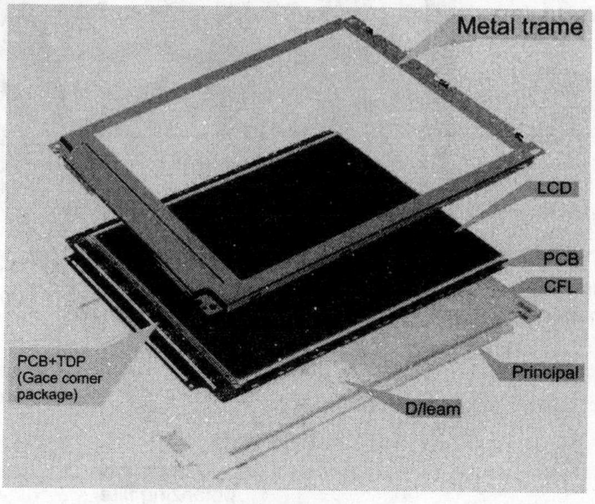

Fig. 7.23

The active matrix provides a superior method of electronically addressing (turning on) an array of pixels. For an image to appear on screen, one row of pixels receives the appropriate voltage. At the same time, software in the computer dictates that voltage be applied to those columns holding active subpixels. Where an activated row and column intersect, a transistor turns on a subpixel electrode, generating an electrical field that controls the orientation of the liquid crystal. This process repeats sequentially for each of the rows in an advanced display, which can take 16 to 33 milliseconds.

Direct-Active Addressing

In addition to the issues of display design concerning direct addressing there is a direct connection to every element in the display, which is good since we would have direct control over the pixels, but it is also bad because in large displays there could be thousands or even millions of pixels that would require separate connections. The method used in the vast majority of large modern displays is multiplexing. In this method, all the pixels across each row are connected together on the plate on one side of the liquid crystal film, and all the pixels in each column are connected on the opposite side. The rows are then "addressed" serially by setting all of the column voltages separately for each row and turning on the row voltages in sequence. This scheme leads to very low voltage selection ratios and, low contrast.

A new method called *active addressing* has been developed to tackle this problem. In this scheme, several rows are addressed simultaneously in different patterns corresponding to orthogonal functions.

High-resolution color displays such as modern LCD computer monitors and televisions use an active matrix structure. A matrix of thin-film transistors (TFTs) is added to the polarizing and color filters. Each pixel has its own dedicated transistor, allowing each column line to access one pixel. When a row line is activated, all of the column lines are connected to a row of pixels and the correct voltage is driven onto all of the column lines. The row line is then deactivated and the next row line is activated. All of the row lines are activated in sequence during a refresh operation. Active-matrix displays are much brighter and sharper than passive-matrix displays of the same size, and generally have quicker response times, producing much better images.

Comparison of Various Passive Matrix Lcd Properties

The main thing that limits the widespread of active matrix LCDs today is the high cost and complex manufacturability. In all other areas except size, active matrix LCDs wins the comparison between the two LCD types. However it is believed that in the not so distant future, passive matrix LCDs might achieve the same results as active matrix LCDs when it comes to brightness and contrasts, which will make them take the lead in commercial products. A comparison has been drawn in Table 7.1.

Table 7.1

Type	*Passive*	*Active*
Contrast	10–20	100+
Viewing angle	limited	wide
Gray scale	16	256
Response time	100-200 ms	<50 ms
Multiplex ratio	~480	>1000
Size	up to 19″	<14″
Manufacturability	simple	complex
Cost	moderate	high

7.5.4 Some Liquid Crystal Materials

Twisted Nematic (TN)

Twisted nematic displays contain liquid crystal elements, which twist and untwist at varying degrees to allow light to pass through. When no voltage is applied to a TN liquid crystal cell, the light is polarized to pass through the cell. In proportion to the voltage applied, the LC cells twist up to 90 degrees changing the polarization and blocking the light's path. By properly adjusting the level of the voltage almost any gray level or transmission can be achieved.

Super Twisted Nematic (STN)

Twisted nematic displays rotate the director of the liquid crystal by 90° , but super-twisted nematic displays employ up to a 270° rotation. This extra rotation gives the crystal a much steeper voltage-brightness response curve and also widens the angle at which the display can be viewed before losing much contrast. With the sharper response, it is possible to achieve higher contrast with the same voltage selection ratio. Therefore, the degree to which multiplexing is possible is greatly increased. The largest common super-twist displays have up to 500 rows. Figure 7.24 shows the voltage vs. light transmission charateristics in super-twisted nematic LCDs.

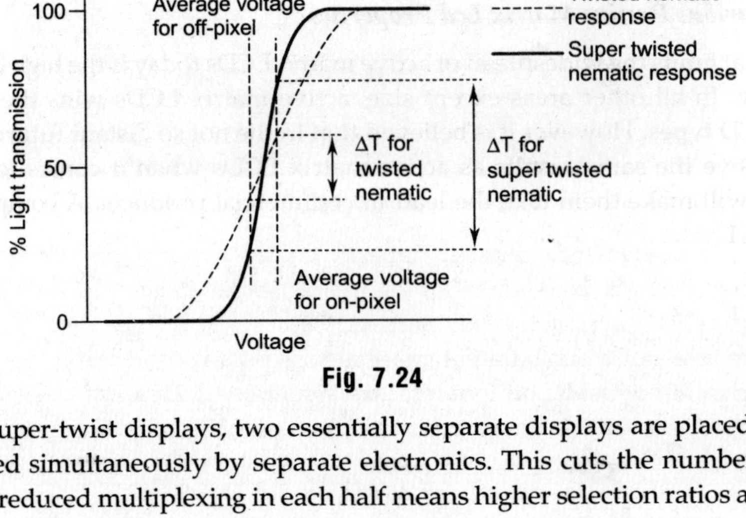

Fig. 7.24

In dual-scan super-twist displays, two essentially separate displays are placed adjacent to each other and scanned simultaneously by separate electronics. This cuts the number of rows in each part by half. The reduced multiplexing in each half means higher selection ratios and consequently, even better contrast. In a double-cell super-twist display, there are two cells on top of each other with helical directions opposite each other. Since the circular birefringence is wavelength dependent as mentioned earlier, this provides a way to undo the wavelength disturbance created when passing through only one layer, thereby providing slightly improved contrast and a true black and white display. Also mentioned earlier if red, green, and blue colored filters are used color displays can be created.

Ferroelectric Liquid Crystal (FLC)

The Nematic and Smectic A liquid crystal phases are too symmetric to allow any vector order, such as ferroelectricity. The tilted smectics, however, do allow ferroelectricity if they are composed of chiral molecules. Figure 7.25 shows the original ferroelectric liquid crystal, DOBAMBC, while Fig. 7.26 shows the modern compound W 314.

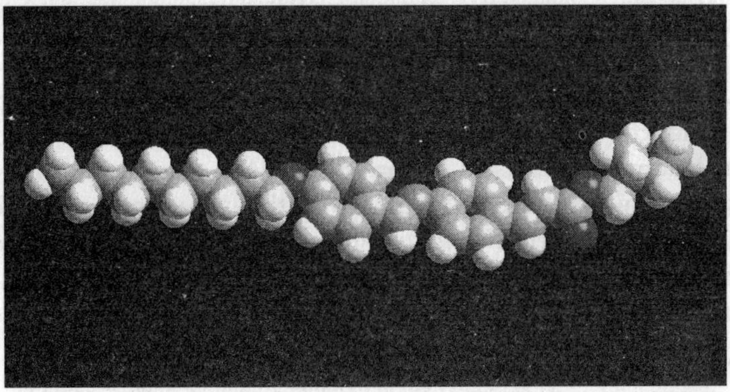

Fig. 7.25

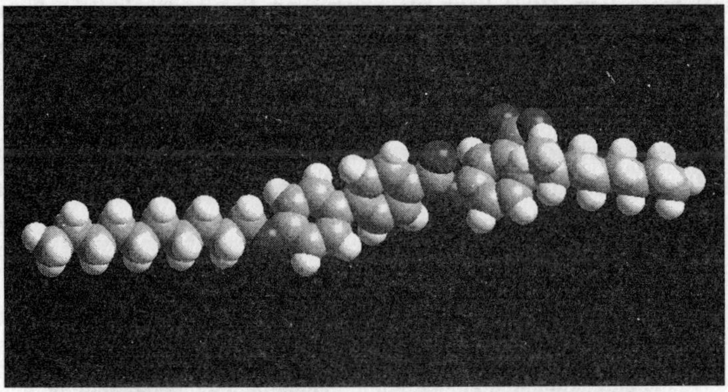

Fig. 7.26

In the simplest case, the Chiral Smectic (Smectic C), the average long molecular axis is tilted from the layer normal z by a fixed angle but the molecules are free to rotate on the so-defined tilt cone. The phase has a C2 symmetry axis perpendicular to both the molecular director and the layer normal. The molecules exhibit a net spontaneous polarization along this axis. The magnitude of the polarization depends on temperature, generally decreasing as the tilt angle goes to zero at the Smectic C-Smectic A phase transition. Figure 7.27 shows the geometry of the chiral Smectic C phase.

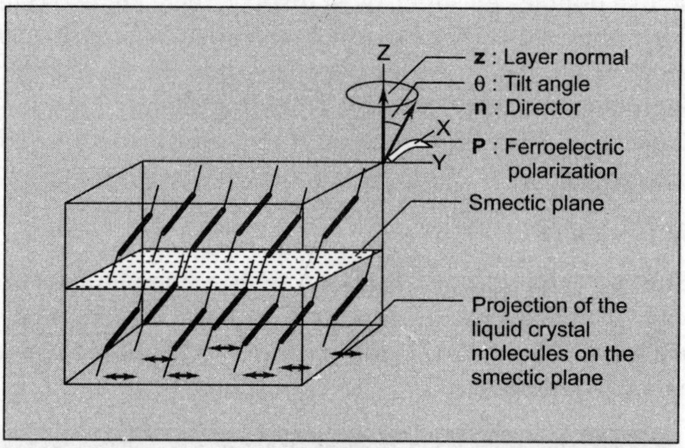

Fig. 7.27

Ferroelectric liquid crystals also exhibit a sponteneous helixing of the polarization, so that over macroscopic distances (a few microns) the polarization averages to zero. Since the coupling of the polarization to applied fields is linear in the field, this means that Ferroelectric liquid crystals can be made to switch quickly (typically within a few microseconds) and in a bipolar manner. This makes Ferroelectric liquid crystals ideally suited to electrooptic applications. Ferroelectric liquid crystals are now included in several display technologies, the most popular of which use the surface-stabilized (SSFLC) geometry.

Surface-Stabilized Ferroelectric Liquid Crystal (SSFLC)

The molecular director in bulk ferroelectric liquid crystals adopts a helical structure, Noel Clark and Sven Lagerwall found in 1980 that by confining the ferroelectric liquid crystals material between closely-spaced glass plates (spaced closer than the ferroelectric helix pitch), the natural helix could be suppressed. This principle is illustrated in the polarized micrograph Fig. 7.28, where helix lines are largely absent in the thinner (upper right) part of the cell. Clark and Lagerwall found that the smectic layers were oriented approximately perpendicular to the glass. Furthermore, they discovered that such cells could be switched rapidly between two optically distinct, stable states simply by alternating the sign of an applied electric field.

Fig. 7.28

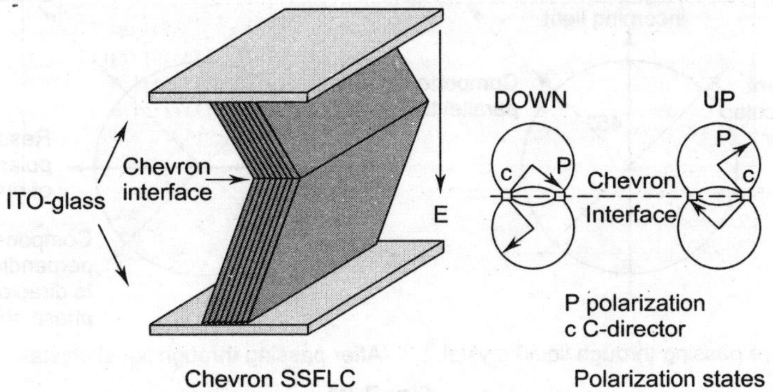

Fig. 7.29

The surface-stabilized ferroelectric liquid crystal display shows great promise. It uses chiral smectic liquid crystal, a type in which the director rotates around a cone of directions 22.5° from the perpendicular of the layers. The hope is to make the film so thin that the director will not rotate. The planar layers are perpendicular to the glass plates of the display. The liquid crystal molecules have a permanent electric dipole perpendicular to the long axis of the molecule (the director). When a positive electric field is applied across the film, the dipole tends to align toward one of the glass plates. This makes the director point a particular direction that is parallel to the glass plates. When the opposite electric field is applied, the dipole will rotate 180° at the other direction. The director will then be 45° away from where it used to be. Another polarized micrograph of ferroelectric liquid crystal is given in Fig. 7.30.

Fig. 7.30

If the entering light is polarized along the director in one of these states, then no light will be able to penetrate a crossed polarizer on the other side. When the molecule is rotated, the polarization of the light will be at a 45 degree angle to the director. This means that the two components of the light, one parallel and one perpendicular to the director, will travel at different speeds through the liquid crystal. If the thickness can be controlled so that the component perpendicular to the director will be 180° out of phase from where it started, then the resulting light's polarization will be 90° rotated and will be able to penetrate through the crossed polarizer. Refer Fig. 7.31.

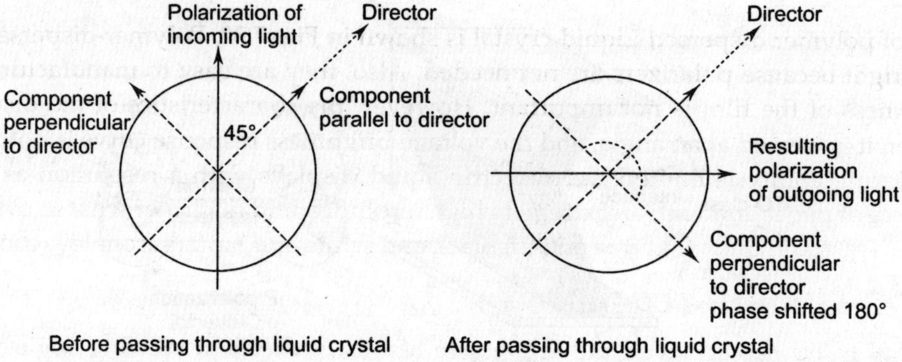

Fig. 7.31

Before passing through the liquid crystal, the components of light polarized parallel and perpendicular to the director can be separated. After passing through the liquid crystal film, the perpendicular component is shifted in phase by 180°. The resulting linear polarization is rotated 90° from its original direction. Surface-stabilized ferroelectric displays have the distinct advantage that switching speeds are very high since electric field is necessary for both turning a pixel on and for turning it off. However, problems in development of this kind of display have persisted. The most difficult issue to resolve has been the orientation of the director throughout all the layers of the film.

Polymer-Dispersed LCD (PDLC)

Also under active development is the polymer-dispersed liquid crystal display. In this kind of display, very small bubbles of liquid crystal are produced inside a transparent polymer. If no field is applied, these bubbles of liquid crystal will take on many different orientations. Since at least one of the two indices of refraction of the liquid crystal and the index of refraction of the polymer must differ, there will be some reflection of incoming light. Now, if a field is applied, the director of the liquid crystal will align with the field. If the materials are carefully controlled so that the index of refraction of the liquid crystal for light polarized perpendicular to the director matches the index of refraction of the polymer, then light will propagate all the way through the material without being reflected at the bubbles.

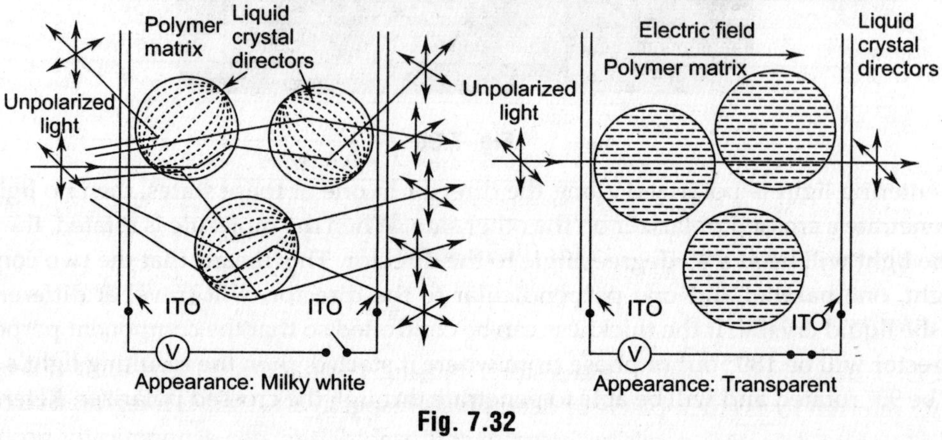

Fig. 7.32

Action of polymer dispersed Liquid crystal is shown in Fig. 7.32. Polymer-dispersed displays are very bright because polarizers are not needed. Also, they are easy to manufacture since the exact thickness of the film is not important. However, the characteristics of the display break down when it is viewed at an angle, and the voltage-brightness response curve is not very steep, so multiplexing ability is limited. Ferroelectric liquid displays with a resolution as good as a printed piece of paper and with an extremely high resolution and no flicker what so ever has been developed. The trade off is that it is only in black and white and suffers from low contrast.

Vertically Aligned Nematic LCD (VAN LCD)

Lately, VAN LCDs has achieved the impressive results of excellent viewing angles, up to 120 degrees, in full color. These are developed mainly by Toshiba and Stanley and are available up to a 14-inch diagonal size. Since the cell gap isn't critical it is also easy to manufacture. Major disadvantages are the limited temperature range, low response time (~250 msec) and low transmission (~1.5-20% for color VAN). Due to the low transmission efficiency, and slow response time use will be limited to AC-powered monitors with non-video applications. The future for these LCDs is unsure and they may only be used in special niche markets, if at all. Most probably they will be left behind as faster LCDs showing the same results in large viewing angles catch up with them.

Comparison Table of LCD Technologies

In table 7.2, several comparisons are made between the faster, direct addressed active matrix TFT LCD and the passive matrix LCD. In order to compare them correctly, both active and passive matrix LCDs are run with multiplex addressing, this due to the fact that passive matrix LCDs aren't able to use direct pixel addressing.

Table 7.2

Type	*TN LCD*	*STN LCD*	*VAN LCD*	*FLCD*	*PDLCD*
Diagonal	Up to 17″	Up to 17″	Up to 17″	Up to 19″	Unlimited
Resolution (lines)	64 lines	400 lines	400 lines	1000 lines	8 lines
Contrast ratio (Direct drive)	> 100:1	> 100:1	> 50:1	> 50:1	> 10:1
Multiplexed	> 3:1	> 20:1	> 20:1	> 50:1	> 5:1
Response time (ON/OFF)	60 ms	200 ms	400 ms	0.1 ms	100 ms
Suitable application	Video	mouse speed	quasi static	line address	mouse speed
Color*	Poor	average	good	good	average
Viewing angle	Poor	average	good	good	–
Cost	Low	moderate	moderate	high	low
Status	mass prod.	mass prod.	development	development	R & D stage

*Color achieved with absorptive color filters within the LCD.

7.5.5 Quality Control

Some LCD panels have defective transistors, causing permanently lit or unlit pixels that are commonly referred to as stuck pixels or dead pixels respectively. Unlike integrated circuits, LCD panels with a few defective pixels are usually still usable. It is also economically prohibitive to discard a panel with just a few defective pixels because LCD panels are much larger than ICs. Manufacturers have different standards for determining a maximum acceptable number of defective pixels. The maximum acceptable number of defective pixels for LCD varies a lot (such as zero-tolerance policy dead-pixel policy) from one brand to another, often a hot debate between manufacturers and customers. To regulate the acceptability of defects and to protect the end user, ISO released the ISO 13406-2 standard [2]. However, not every LCD manufacturer conforms to the ISO standard and the ISO standard is quite often interpreted in different ways.

Examples of Defects in LCD Displays

LCD panels are more likely to have defects than most ICs due to their larger size. As example, a 12" SVGA LCD has 8 defects and a 6" wafer has only 3 defects. However, 134 of the 137 dies on the wafer will be acceptable, whereas rejection of the LCD panel would be a 0% yield. The standard is much higher now due to fierce competition between manufacturers and improved quality control. An SVGA LCD panel with 4 defective pixels is usually considered defective and customers can request an exchange for a new one. Some manufacturers, notably in South Korea where some of the largest LCD panel manufacturers, such as LG, are located, now have "zero defective pixel guarantee" and would replace a product even with one defective pixel. Even where such guarantees do not exist, the location of defective pixels is important. A display with only a few defective pixels may be unacceptable if the defective pixels are near each other. Manufacturers may also relax their replacement criteria when defective pixels are in the center of the viewing area.

7.5.6 Zero-Power Displays

The Zenithal Bistable Device (ZBD), developed by QinetiQ (formerly DERA), can retain an image without power. The crystals may exist in one of two stable orientations (Black and White) and power is only required to change the image. ZBD Displays is a spin-off company from QinetiQ who manufacture both grayscale and color ZBD devices.

A French company, Nemoptic, has developed another zero-power, paper-like LCD technology, which has been mass-produced since July 2003. This technology is intended for use in applications such as Electronic Shelf Labels, E-books, E-documents, E-newspapers, E-dictionaries, Industrial sensors, Ultra-Mobile PCs, etc. Zero-power LCDs are a category of electronic paper.

Kent Displays has also developed a "no power" display that uses Polymer Stabilized Cholesteric Liquid Crystals (ChLCD). The major drawback to the ChLCD display is slow refresh rate, especially with low temperatures.

7.5.7 Drawbacks

LCD technology still has a few drawbacks in comparison to some other display technologies:

- While CRTs are capable of displaying multiple video resolutions without introducing artifacts, LCD displays produce crisp images only in their "native resolution" and, sometimes, fractions of that native resolution. Attempting to run LCD display panels at non-native resolutions usually results in the panel scaling the image, which introduces blurriness or blockiness.
- Although LCD displays typically have more vibrant images, and better "real-world" contrast ratios, (being able to maintain contrast and variation of color in bright environments)

than CRTs, they do have a lower contrast ratio than that on a plasma display for example. CRTs can also have higher contrast ratios if it is compared how deep their blacks are — since a contrast ratio is the difference between a completely on (white) and off (black) pixel and LCDs can at times have "backlight bleed". This is due to their "light valve" nature: some light (usually seen most around corners of the display) leaks out and turns black into gray. In brightly lit rooms the contrast of LCD monitors can, usually, exceed CRT displays due to higher maximum brightness.

- Some less expensive LCDs can not 'truly' display as many colors as CRTs and Plasmas can, typically ones that have lower-end panel types such as Twisted Nematic panels (TN), that were mainly used to reduce ghosting before Overdrive technology started becoming used often. The circuitry of these types of panels feed the full-color image to the panel, but it cannot really display it as CRTs, like higher-end LCDs and Plasmas do. Skin tones look a lot less natural because of this on these monitors. Subtle details also suffer because of this. The primary colors (red, green blue) especially suffer. A bright, red car looks a lot more 'smooth' on a TFT panel with low-end panel types, or than a CRT that will show subtle nuances in color a lot better. Note that many recent LCDs can display subtle details in colors such as skin tones but not with the accuracy of CRTs.
- LCDs have longer response time than their plasma and CRT counterparts, especially older displays creating visible ghosting when images rapidly change; this drawback, however, is continually improving as the technology progresses and is hardly noticeable in current LCD displays with "overdrive" technology. Most newer LCDs have response times less than 8 ms.
- In addition to the response times, some LCD panels have significant input lag, and if the lag delay is large enough, makes them unsuitable for fast and time-precise mouse operations (CAD design, FPS gaming) as compared to CRTs or smaller LCD panels with a negligible amount of input lag; although note that input lag effects a somewhat small percent of LCDs-usually 20" or larger. Some speculate that Overdrive technology, to reduce ghost trails, contributes to input lag.
- Overdrive technology on some panels can produce artifacts across regions of rapidly transitioning pixels (eg. video images) that looks like increased image noise or halos. This is a side effect of the pixels being driven past their intended brightness value (or rather the intended voltage necessary to produce this necessary brightness/color) and then allowed to fall back to the target brightness in order to enhance response times.
- Historically LCD display panels had a limited viewing angle, thus reducing the number of people who could conveniently view the same image, laptop screens being a prime example. Most modern LCD panels, however, especially those that have been developed for televisions, typically offer viewing angles of >170 degrees horizontal and vertical. As the viewer moves closer to the limit of the viewing angle, the colors, contrast, and brightness deteriorate. However, this negative has actually been capitalized upon in two ways. Some vendors offer screens with intentionally reduced viewing angle, to provide additional privacy, such as when someone is using a laptop in a public place. Such a set can also show two different images to one viewer, providing a three-dimensional effect.
- Some users of older (around pre-2000) LCD monitors complain of migraines and eyestrain problems due to flicker from fluorescent backlights fed at 50 or 60 Hz. This does not happen with most modern displays that feed backlights with high-frequency current.

- a small percentage of LCD screens occasionally suffer from image persistence, which is similar to screen burn on CRT and plasma displays. This is becoming less of a problem as technology advances, with newer LCD panels using various methods to reduce the problem. Sometimes the panel can be restored to normal by displaying an all-white pattern for extended periods of time.
- Some light guns do not work with this type of display since they do not have flexible lighting dynamics that CRTs have. However, the field emission display will be a potential replacement for LCD flat-panel displays since they emulate CRTs in some technological ways.
- Some panels are incapable of displaying low-resolution screen modes (such as 320 × 200). However, this is due to the circuitry that drives the LCD rather than the LCD itself.

Consumer LCD monitors are more fragile than their CRT counterparts, with the screen especially vulnerable. However, lighter weight makes falling less dangerous, and some displays may be protected with glass shields.

7.6 PLASMA DISPLAYS

The plasma display panel was invented at the University of Illinois at Urbana-Champaign by Donald L. Bitzer, H. Gene Slottow, and graduate student Robert Willson in 1964 for the PLATO Computer System. The original monochrome (usually orange or green, sometimes yellow) panels enjoyed a surge of popularity in the early 1970s because the displays were rugged and needed neither memory nor circuitry to refresh the images. A long period of sales decline followed in the late 1980s as semiconductor memory made CRT displays cheaper than plasma displays. Nonetheless, plasma's relatively large screen size and thin profile made the displays attractive for high-profile placement such as lobbies and stock exchanges.

In 1983, IBM introduced a 19-inch orange-on-black monochrome display (model 3290 'information panel') which was able to show four simultaneous IBM 3270 Virtual Machine (VM) terminal sessions. That factory was transferred in 1987 to startup company Plasmaco, which Dr. Larry F. Weber, one of Dr. Bitzer's students, founded with Stephen Globus, and James Kehoe, who was the IBM plant manager. In 1992, Fujitsu introduced the world's first 21-inch full-color display. It was a hybrid, based upon the plasma display created at the University of Illinois at Urbana-Champaign and NHK STRL, achieving superior brightness. In 1996, Matsushita Electrical Industries (Panasonic) purchased Plasmaco, its color AC technology, and its American factory. In 1997, Pioneer started selling the first plasma television to the public. In popular culture plasma televisions are often seen around the home and are being introduced thinner and in greater sizes, in order to try and compete with projector screens.

Screen sizes have increased since the 21-inch display in 1992. The largest plasma video display in the world was shown at the 2006 Consumer Electronics Show in Las Vegas, Nevada, U.S.A, a 103 inch unit manufactured by Matsushita Electrical Industries (Panasonic).

Until quite recently, the superior brightness, faster response time, greater color spectrum, and wider viewing angle of color plasma video displays, when compared with LCD televisions, made them one of the most popular forms of display for HDTV Flat panel displays. For a long time it was widely believed that LCD technology was suited only to smaller sized televisions, and could not compete with plasma technology at larger sizes, particularly 40 inches and above.

However, since then, improvements in LCD technology have narrowed the technological gap. The lower weight, falling prices, higher available resolution, which is important for HDTV, and often lower electrical power consumption of LCDs make them competitive against plasma displays in the television set market. As of late 2006, analysts note that LCDs are overtaking plasmas, particularly in the important 40" and above segment where plasma had previously enjoyed strong dominance a couple of years before. Till now the number of manufacturers of plasma displays is still limited.

7.6.1 Basic Working Principle of Plasma Display

Plasma Displays rely on the glow produced when an electrical current passed through a gas (usually Neon). Free electrons and ionized gas atoms are present during the discharge. The central element in a fluorescent light is plasma, a gas made up of free-flowing ions (electrically charged atoms) and electrons (negatively charged particles). Under normal conditions, a gas is mainly made up of uncharged particles. That is, the individual gas atoms include equal numbers of protons (positively charged particles in the atom's nucleus) and electrons. The negatively charged electrons perfectly balance the positively charged protons, so the atom has a net charge of zero. With the application of external electrical field, the electrons acquire a high kinetic energy and when they collide with gas atoms (or ions) they transfer energy to the atoms, thereby exciting them into levels above ground state. The atoms then lose energy radiatively and return to the ground state, causing them to release photons of energy.

AC excitation is commonly used, although sometimes DC excitation is used. The basic construction of AC plasma display element is shown at Fig. 7.33, in which the electrodes are external to the gas cavity. (When DC excitation is used, the electrodes must be inside the gas cavity). Cavity width is typically 100 μm and gas pressure is 400 torr. Application of firing voltage of about 150 V initiates the discharge activity. However, once the discharge process is commenced, it may be sustained at lower voltage V_s of about 90V. A suitable voltage waveform that may be applied is shown in Fig. 7.34.

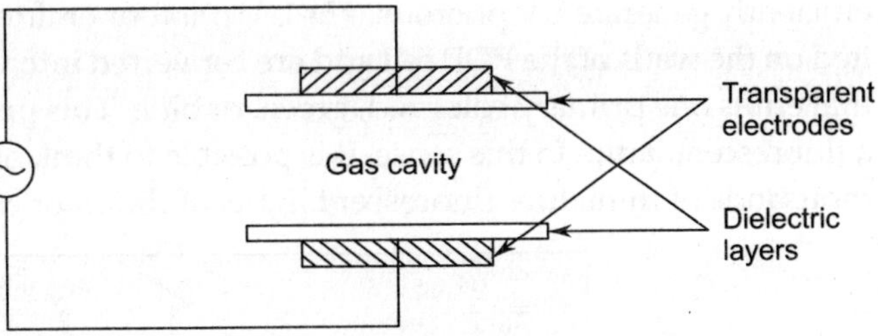

Fig. 7.33

If the initial voltage pulse is relatively wide then an appreciable amount of charge accumulates on the interior of the dielectric layer. This charge build-up results is an internal electric field that is opposite to the external field, this may be strong enough to extinguish the discharge if the pulse width is too long. Such accumulated charge takes longer time to dissipate and this effect can be used to enable the device for memory action. Thus, after applying the first pulse of initiating voltage (V_f), may be termed as writing pulse, charge accumulates and at subsequent time, when external voltage of opposite polarity is applied, the field from stored charge will add to the external field and discharge will again commence, but now at lower voltage than V_f. This is illustrated in Fig. 7.34.

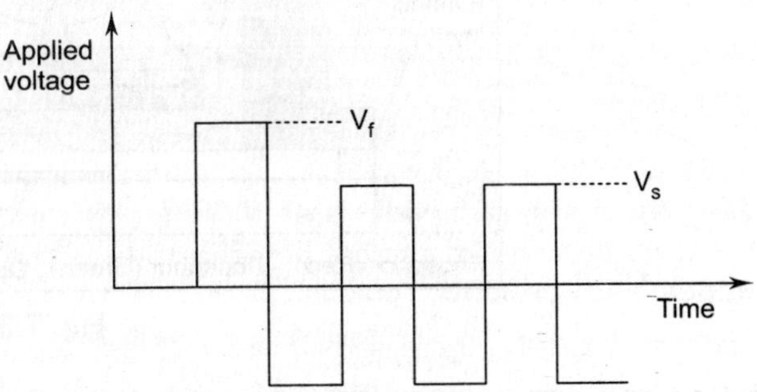

Fig. 7.34

7.6.2 Plasma Cell Structure

Typical plasma displays consist of two glass plates, each with parallel electrodes deposited on their surfaces. The electrodes are covered with a dielectric film. The plates are sealed together with their electrodes at right angles, and the gap between the plates is filled with an inert gas mixture composed of helium, neon or xenon. Dike-like barriers, or ribs, keep the glass plates parallel and separate. Groups of electrodes are placed at right angles between the panes, forming rectangular compartments, or cells, between the glass sheets. The plasma cell structure is schematically in Fig. 7.35.

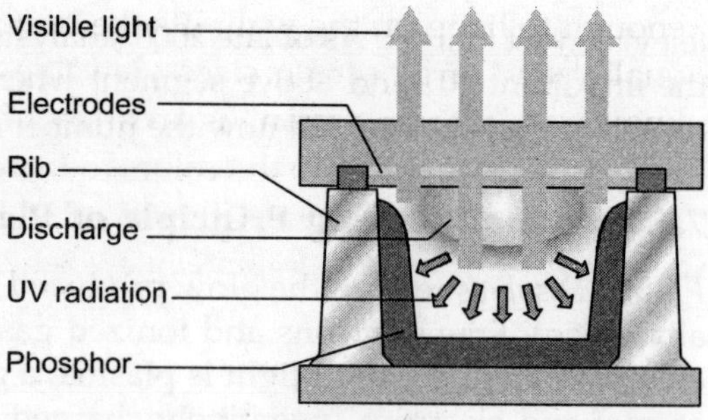

Fig. 7.35

A protective MgO layer is deposited above the dielectric film. The role of this layer is to decrease the breakdown voltage caused by the high secondary electron emission coefficient of MgO. Phosphors embedded within each cell individually emit red, green or blue light and collectively create a single color pixel when excited, which is shown in Fig. 7.36. The discharge is initiated by applying a voltage pulse to the electrodes. Selectively applying voltages to the electrodes causes them to generate a discharge in the panel's dielectric layer and on its protective surface. This generates ultraviolet light that excites the phosphors, stimulating them to emit light. Inert gas mixtures are used to efficiently generate UV photons. The UV photons emitted by the discharge hit the phosphors deposited on the walls of the PDP cell and are converted into visible photons. Each cell contains phosphor that emits one primary color red, green, or blue. This principle of operation is very similar to that of a fluorescent lamp. In this sense, it is possible to think of a plasma display as a screen incorporating thousands of miniature fluorescent lights of different color.

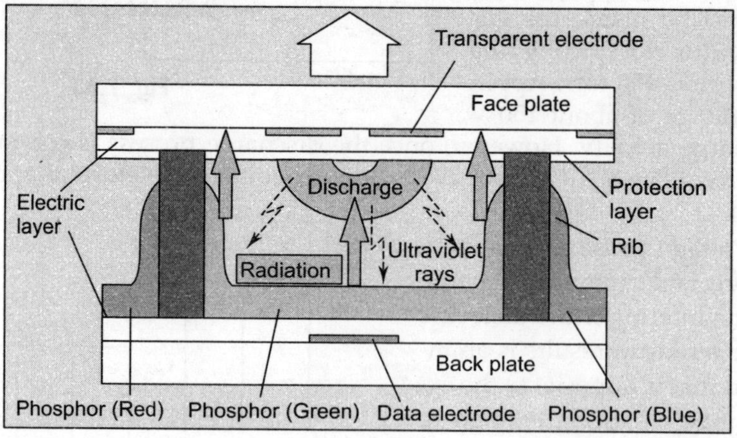

Fig. 7.36

7.6.3 How Plasma Display Works?

The plasma display principle leads to the development of *Plasma Display Panel* (PDP), which is one of the leading candidates in the competition for large-size, high-bright-ness flat panel displays, suitable for high-definition television (HDTV) monitors. Their advantages are high resolution, fast response, wide viewing angle, low weight, and simple manufacturing process for fabrication. The fact that they are expected to be the next generation of TV displays is evident in the remarkable recent progress of PDP technology development and manufacturing. Plasma display technology is a new "emmissive" flat panel display technology that gives the rich, accurate color fidelity of conventional Cathode Ray Tube (CRT) monitors in a large plasma display that is thin enough to hang on the wall. It's the best way to achieve flat panel display with excellent image quality and large screen sizes viewable in any environment. This plasma display technology known as "plasma vision" is an array of cells, known as pixels, which are composed of 3 sub-pixels, corresponding to the colors red, green and blue. The basic construction, plasma action and creation of visible light are illustrated in Figs. 7.37(*a*) to 7.37(*j*) in step by step.

Fig. 7.37(*a*) : A plasma panel is two sheets of 1/16 inch glass. Edge view of the glass is shown.

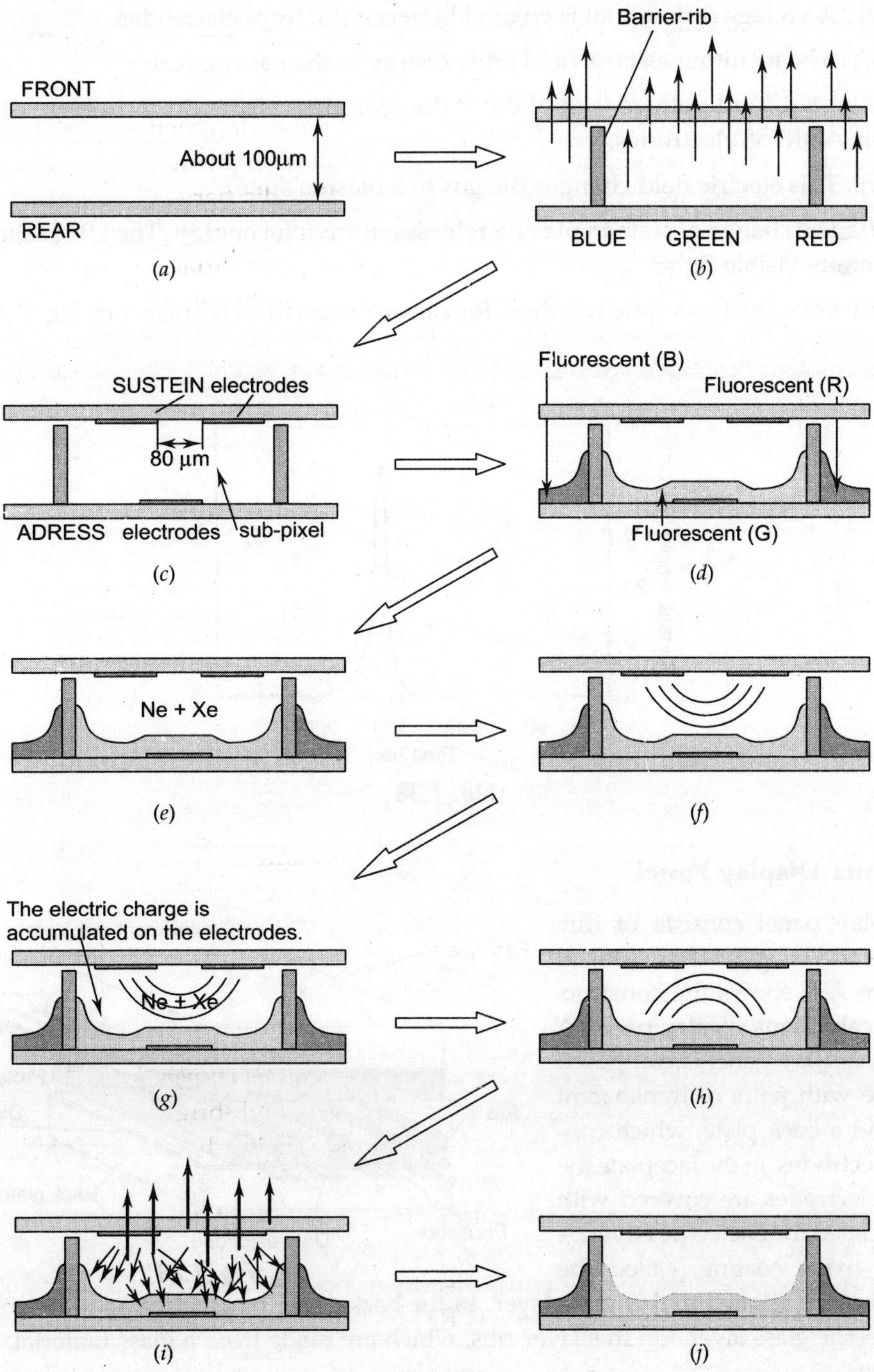

Fig. 7.37

Fig. 7.37(*b*) : uniform spacers that also create R-G-B sub-pixels separate the sheets of glass.

Fig. 7.37(*c*) : Transparent electrodes are coated inside the glass, two on the front and one on the rear glass.

Fig. 7.37(*d*) : Inside of the rear glass is coated with phosphor that resembles CRT phosphor.

Fig. 7.37(*e*) : The sub-pixel is filled with a mixture of neon and xenon.

Fig. 7.37(*f*) : A voltage differential is created between the front electrodes.

Fig. 7.37(*g*) : The resulting electric field adds energy to the gas mixture.

Fig. 7.37(*h*) : A second voltage is applied to the rear glass electrode. A voltage differential is created on the ADRESS electrodes.

Fig. 7.37(*i*) : This electric field changes the gas to a plasma state.

Fig. 7.37(*j*) : The change of state to plasma releases ultraviolet energy. The UV excites the phosphor and it emits visible light.

Typical current density vs time required for plasma formation is shown in Fig. 7.38.

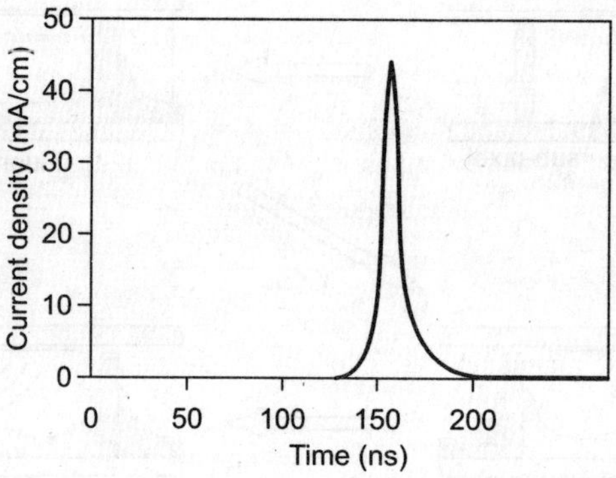

Fig. 7.38

7.6.4 Plasma Display Panel

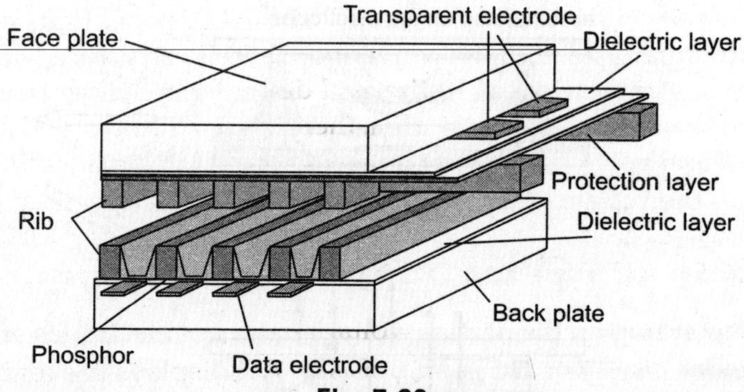

Fig. 7.39

Plasma display panel consists of tiny pixels or cells and subpixels as described above. Figure 7.39 shows the construction of a typical plasma display panel. A color plasma display panel is composed of a faceplate with pairs of transparent electrodes and a back plate, which contains metal electrodes. In the faceplate the transparent electrodes are covered with a glass layer called the Dielectric Layer. A magmesium oxide coating, called the Protective Layer, is applied to this glass layer. In the back plate, metal electrodes are covered with another Dielectric glass layer. On this layer ribs, which are made from a glass material, are placed.

These barrier ribs serve to separate the plasma panel into separate discharge cells. Red, green and blue phosphors are alternately deposited between the ribs.

Each individual plasma cell is switched on and off by its own electrode. An HDTV (high definition television) with plasma technology can have up to 3 million of these cells. As a plasma panel is illuminated at the sub-pixel level, images are extremely accurate, and the panel's light output is both high and consistent across the entire screen area. Display width is also very much thinner compared to traditional CRT display. That is why the plasma display finds its wide application in image reproduction, television and advertising system. Plasma TVs also provide very wide horizontal and vertical viewing angles, especially when compared to conventional rear-projection TVs. Picture quality looks sharp and bright from virtually anywhere in the room. Because plasma TV screens do use a phosphor coating (like direct-view and projection CRT TVs), the potential for image burn-in exists, so it's important to follow the manufacturer's recommen-dations on day-to-day use. Until recently, all flat-panel TVs were progressive displays - at any given moment all of the pixels are illuminated. But plasma TVs based on innovative *AliS technology (Alternate Lighting of Surfaces)* are proving that a non-progressive picture can look outstanding. Typical plasma panels have a strip of electrodes for each horizontal row of plasma cells, while ALiS panels share an electrode strip between two rows of cells. At any given instant only half the panel's pixels are turned on. It's somewhat similar to interlaced-scanning on a CRT-based TV. In fact, ALiS technology was developed as a way to make a simpler, lower-cost plasma panel capable of displaying interlaced HDTV signals.

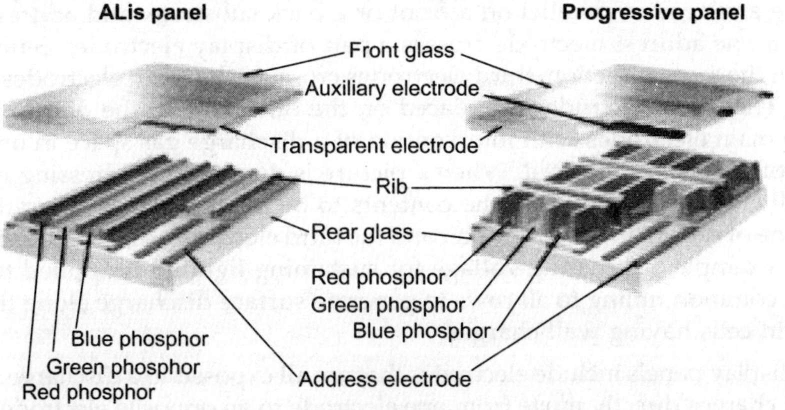

Fig. 7.40

Figure 7.40 illustrates the structure of plasma display panel of ALiS technology with the comparison with progressive panel. The ALiS panel (left) is a simpler design that uses a "trough" structure instead of the conventional "cell" structure, while also streamlining the drive electronics (fewer electrodes). Both panel designs can deliver outstanding picture quality. Since electrode strips take up less screen area, there is more illuminated area, and as a result, ALiS panels provide a seamlessly clear, bright image. Other ALiS benefits include high-efficiency - they require only half the voltage of conventional plasma drive systems as well as extended panel life.

7.6.5 AC Type and DC Type Plasma Displays

Depending on the driving voltage nature of electrodes of plasma displays, they are termed as *AC plasma display* or *DC plasma display*. AC displays use an internal dielectric glass layer to limit the current. This dielectric glass forms a small capacitor that is in series with every gas discharge in an AC Plasma display. Because the dielectric glass is an excellent insulator, no dc current can flow, so that an AC voltage must be applied to maintain a discharge. The major difference between AC and DC is that the DC electrodes are in intimate contact with the gas and the AC electrodes are insulated from it by the dielectric glass.

The AC drive type plasma display is generally composed of a front plate, a back plate, and a discharge space region that is formed between the front and back plates and that has a great number of cells partitioned by walls. The front plate has formed therein a plurality of pairs of display electrodes. The back plate has formed therein a plurality of address electrodes that are substantially perpendicular to these display electrodes. AC-driven plasma display panels are classified into *two-electrode type* plasma display panels, which perform selective discharge (address discharge) and sustain discharge using two electrodes, and *three-electrode* type plasma display panels that perform address discharge using a third electrode. The three-electrode types plasma display panels are further classified into a type with the third electrode formed on a substrate on which the first and second electrodes for performing sustain discharge are laid out and a type with the third electrode formed on another substrate opposite to the substrate of the first and second electrodes. AC type PDP utilizes a memory function of a dielectric layer that covers display electrodes. Addressing is performed for controlling charge quantity of a cell in accordance with display data before applying a sustaining voltage V_s having alternating polarity to a pair of display electrodes. The sustaining voltage V_s satisfies the following inequality. A surface discharge AC type PDP has a three-electrode structure in which first and second display electrodes to be anodes and cathodes in display discharge for determining light emission quantity in a cell are arranged in parallel on a front or a back substrate, and address electrodes are arranged so that one address electrode crosses a pair of display electrodes. Since the main electrodes extend in the same direction, third electrodes crossing the main electrodes are required for selecting cells. The third electrodes are placed on the other one of the pair of substrates to be opposed to the main electrodes with intervention of a discharge gas space in order that electrostatic capacity of the cells is reduced. When a picture is displayed, addressing is carried out for controlling wall charges according to the contents to be displayed, by generating address discharge across one of each main electrode pair and the third electrode. After addressing is performed line by line, for example, alternating voltage for sustaining lighting is applied to the main electrode pairs at a common timing to all rows to generate surface discharge along the surface of the substrate only in cells having wall charges.

DC plasma display panels include electrodes that are all exposed to a discharge space. In the DC PDPs, electrical charges directly move from one electrode to an opposite electrode. In the DC-type plasma display device, discharge occurs in all periods during which a voltage is being applied to an electrode. The most significant difference between the DC PDP and the AC PDP is that, in the DC PDP, the current directly flows in the discharge area while applying voltages, because electrodes are exposed to the discharge spaces. Since the DC PDPs have electrodes exposed in the discharge space, they allow the current to flow in the discharge space while the voltage is supplied. Therefore, a resistor that restricts the current must be used outside of the DC PDP. On the other hand, in the AC PDP, the current is restricted because capacitance is natually formed by the dielectric layer covering the electrodes. The AC PDP has a longer life than the DC PDP because the electrodes are protected against the ion shocks generated by the discharge. Since the AC PDPs have electrodes covered by a dielectric layer, capacitances are naturally formed to restrict the current, and the electrodes are protected from ion shocks in the case of discharging. Accordingly, they have a longer lifespan than the DC PDPs. In terms of high definition, large screen size and facilitation of production, the surface discharge AC type PDP has become mainstream under present conditions. The AC type plasma display device is suitable for high definition because barrier ribs for partitioning individual discharge cells from each other within the display screen are sufficient to be formed into, for example, stripe shapes, and is also advantageous in that since surfaces of electrodes for discharge are covered with a dielectric layer, the electrodes are less worn, to thereby prolong the service life.

AC and DC type Plasma technology each have inherent advantages and limitations. DC Plasma provides higher contrast and the ability to reproduce a greater number colors but is inherently complex in structure and requires a coarse pixel array (similar to coarse dot pitch specification in a CRT-based monitor). AC Plasma offers a simpler structure and finer (more detailed) display but does so at the expense of lower contrast and more limited ability to display shades of color.

Another difference between the two has to do with the difference between resistive and capacitive charging methods. The capacitive method of AC Plasma requires a longer time for cells to charge and discharge while the resistive method of DC allows a faster drive capability that is better suited to moving images in high resolution and HDTV. This faster response time comes at the expense of higher power consumption. Development efforts for both AC plasma panels are leading toward delivering better contrast and bigger displayable color palettes. In the case of DC plasma, efforts are concentrated on lowering its power usage and simplifying its manufacturing process.

7.6.6 General Characteristics

Plasma displays are bright (1000 lux or higher for the module), have a wide color gamut, and can be produced in fairly large sizes, up to 262 cm (103 inches) diagonally. They have a very high "dark-room" black level, creating the "perfect black" desirable for watching movies. The display panel is only about 6 cm (2½ inches) thick, while the total thickness, including electronics, is less than 10 cm (4 inches). Plasma displays use as much power per square meter as a CRT or an AMLCD television. Power consumption will vary greatly depending on what is watched on it. Bright scenes (say a football game) will draw significantly more power than darker scenes (say a movie scene at night). Nominal measurements indicate 400 watts for a 50" screen.

Plasma display panel, like familiar phosphor-based TV sets, gently lose some brightness over the years. The lifetime of the latest generation of plasma displays is estimated at 60,000 hours (or 27 years at 6 hours of use per day) of actual display time. More precisely, this is the estimated half life of the display, the point where the picture has degraded to half of its original brightness. It is watchable after this point, but is generally considered the end of the functional life of the display. In real life, six-hour-a-day use, Plasmavision sets will be half as bright after about 14 years. They'll lose a bit more brightness in the 14 years after that. Typical lifetime of a plasma display panel is shown in Fig. 7.41.

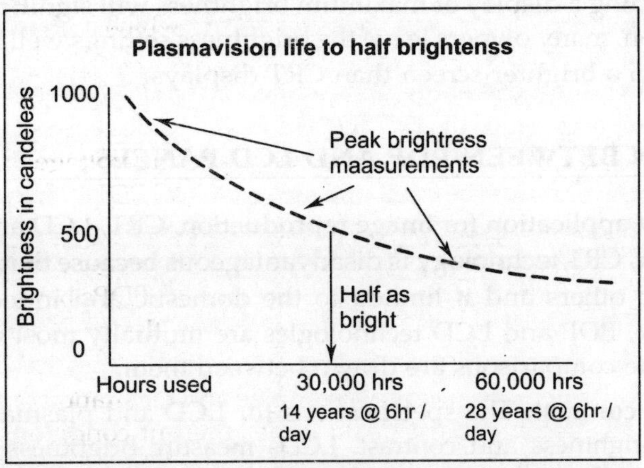

Plasma displays deliver very bright pictures for a long long time.

Fig. 7.41

Another concern is the pixel life. While a pixel or two on a plasma monitor might not light at the time of installation, additional pixel failure after installation is virtually unheard of. The reason is the inert nature of the plasma pixel. There are no switches, gates or circuits associated with individual PDP pixels. All of the electronic operations take place in the video boards, display drivers and other components — *not* within the plasma display panel itself. These electronics are all either at the edges of the panel or behind the sealed glass. Qualified service experts can replace these components if they fail. LCD pixels, by contrast, are prone to failure over the lifespan of the screen. Dot electrodes, located at each of the one-million-plus sub-pixels, control them. Any of these is prone to failure.

Competing displays include the CRT, OLED, AMLCD, DLP, SED-tv and field emission flat panel displays. The main advantage of plasma display technology is that a very wide screen can be produced using extremely thin materials. Since each pixel is lit individually, the image is very bright and has a wide viewing angle.

Contrast Ratio

Contrast ratio is the difference between the brightest and darkest parts of an image, measured in discrete steps, at any given moment. Generally, the higher the contrast ratio, the more realistic the image is. Contrast ratios for plasma displays are often advertised as high as 10,000:1. On the surface, this is a significant advantage of plasma over other display technologies. Although there are no industry-wide guidelines for reporting contrast ratio, most manufacturers follow either the ANSI standard or perform a full-on-full-off test. The ANSI standard uses a checkered test pattern whereby the darkest blacks and the lightest whites are simultaneously measured, yielding the most accurate "real-world" ratings. In contrast, a full-on-full-off test measures the ratio using a pure black screen and a pure white screen, which gives higher values but does not represent a typical viewing scenario. Manufacturers can further artificially improve the reported contrast ratio by increasing the contrast and brightness settings to achieve the highest test values. However, a contrast ratio generated by this method is misleading, as content would be essentially unwatchable at such settings.

The underlying technology of plasma displays allows for a high contrast ratio. Similar to CRTs, plasmas can achieve nearly true black because a cell/pixel can be switched off completely. This contrasts with LCD technology, where black pixels generated by a light polarization method are unable to completely block the underlying backlight. However, one shortcoming with plasma technology is that running a display at maximum brightness will significantly reduce the panel's lifespan. For this reason, many owners leave the brightness settings well below maximum, which typically still results in a brighter screen than CRT displays.

7.7 COMPARISON BETWEEN PDP AND LCD PANELS

In large screen display application for image reproduction, CRT, LCD and PDP technologies are widely used. However, CRT technology is disadvantageous because of its bulkness, thogh it is of low cost compared to others and it limited to the domestic and indoor use. For commercial application now-a-day, PDP and LCD technologies are mutually most competitive one against the other. Hereby some comparisons are drawn between them.

- **Brightness and contrast:** The specifications for LCD and plasma displays are measured differently for brightness and contrast. LCDs measure brightness according to the Video Electronics Standards Association (VESA) Flat Panel Display Measurements (FPDM) Standard Version 2.0 (June 1, 2001), by using a full screen white pattern. Contrast ratio according to the VESA standard is measured as the difference between full screen white and full screen black in a dark room. Plasma display specifications are measured in the same way a Cathode Ray Tube (CRT) would be measured. The brightness is specified by using a peak value rather than a typical value. This is done by generating a small white square on the screen, concentrating all of the display's energy in this small area. The contrast ratio is calculated as the difference between the small white area and the black area surrounding it. The VESA FPDM standard is more of a "real world" test since there aren't many applications that use a small portion of the screen while leaving the rest blank. In a true comparison using the VESA FPDM standard method the results in dark room conditions would be:

Brightness (full screen white) 450 cd/m² and 100 cd/m² for typical 40" LCD and typical 42" plasma respectively. Contrast ratio (full screen white to full screen black) 600:1 and 200:1 respectively.

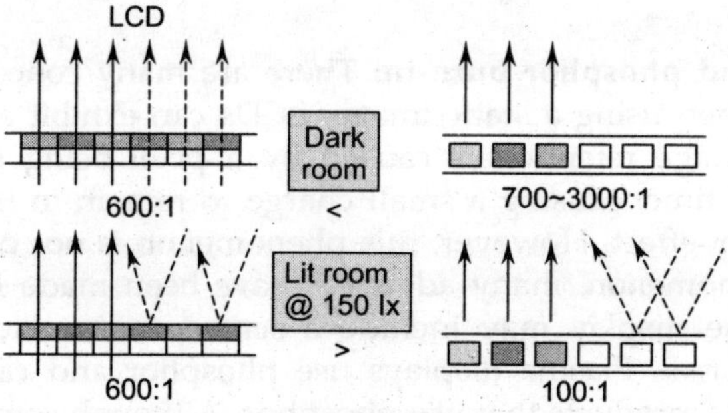

Fig. 7.42

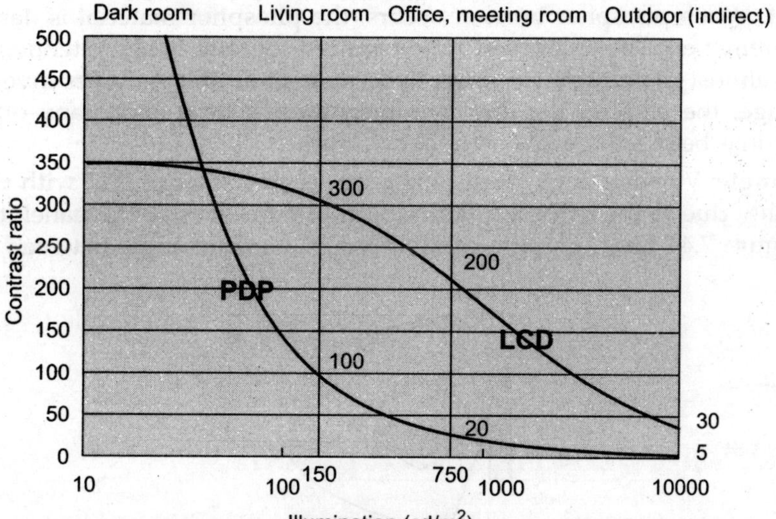

Fig. 7.43

LCD displays also perform better than plasma displays under higher ambient-lit conditions (Fig. 7.42). An LCD display absorbs ambient light while a plasma display reflects it. The contrast of both displays is reduced under these conditions, but it is reduced to a much higher degree for the plasma. Plasma display performs better in the dark room (refer Fig. 7.43).

- **Color reproduction:** The color spectrum of a display affects its ability to reproduce colors accurately. The color spectrum of displays varies between LCD and plasma manufacturers and is also affected by other devices (DVD player, computer video card, etc.) in the graphics system. Based on all of these factors, as well as the perception of the human eye and the application being used, it is usually left to the user to determine their preference and calibration. The plasma display has better green representation, while the LCD has better red and blue.
- **Brightness and color uniformity:** Since LCD displays use backlights that consist of many fluorescent bulbs; many precautions have to be designed into the diffusion layer to distribute light evenly. To prevent "shading" or "hot spots," many bulbs and more than one diffusion layer are used. The result is a uniform image in both brightness and color, which is different than the LCDs of the past and comparable to the uniformity of a plasma display.
- **Image retention and phosphor burn-in:** There are many concerns with both LCD and plasma displays when using a static image. LCDs can exhibit a phenomenon known as image retention. Image retention is caused by a pixel being in the "on" state for an extended period of time, causing a small charge to remain in the cell in the "off" state, similar to a memory effect. However, this phenomenon is not permanent. To reduce the chance of this phenomenon, many advances have been made in the liquid crystal material itself, and the display may include a screen saver pixel-shifting technology to further reduce the risk. Plasma displays use phosphor and can suffer from phosphor burn-in just like CRT products that use phosphor. Although some manufacturers include pixel-shifting technologies to reduce the chance of this happening, the end result is permanent. When phosphor burn-in occurs, the phosphor material is damaged and this cannot be removed. The amount of time required for both image retention and phosphor burn-in is almost impossible to calculate because of all the variables involved. The color of the image, the pattern, the display's brightness setting and many other factors can affect the time before these phenomenons appear.
- **Viewing angle:** Viewing angle of plasma panel is almost a full 180° with virtually no loss of readability due to the emissive nature of it. In other words, the panel itself is the light source. Figure 7.44 gives graphic comparison of viewing angle between PDP and LCD display.

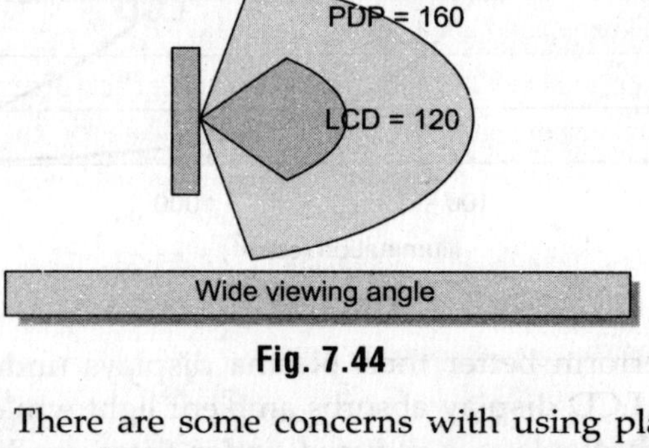

Fig. 7.44

- **Altitude limitations:** There are some concerns with using plasma displays at high altitudes. Most plasma manufacturers have an operational altitude specification between 6500 and 9500 feet, while an LCD's operating altitude specification is more than 13,000 feet. This allows LCD technology to be used and transported in some areas where plasma displays cannot be used because of this limitation to the technology.

- **Power consumption:** The energy cost to operate an LCD or plasma display can be a determining factor in large installations. The power consumption of desktop displays has always been an advantage for LCDs when compared to CRT technology. Large-screen LCDs continue to be the more cost-effective solution in regards to power consumption when compared to plasma as well. Power consumption of 40″ LCD is 235 watts, while that of 42″ XGA plasma is 370 watts. The LCD monitor provides a 36% decrease in wattage and costs.
- **Resolution:** Although the viewable image size of an LCD or plasma may be identical, the native resolution can be quite different. Many of the 40″ to 43″ plasma displays have a native resolution of 853 × 480, and others have a native resolution of 1024 × 768. LCDs are capable of higher resolutions within the same size parameters and have a native resolution of 1280 × 768. The native resolution affects how much information can be displayed on the screen at one time.

7.8 CONCLUSION

One of the most critical issues in PDP research and technology development is the improvement of luminance and luminous efficiency, which is dependent on the gas mixture composition, phosphor efficiency, driving voltage characteristics, and cell geometry. PDP cells can operate only if the applied voltage is held within certain limits. The minimum and maximum values of the applied voltage define the margin of the panel. These limits are determined by the breakdown voltage. In some PDP designs, reducing the breakdown voltage may be of higher priority than is increasing the efficiency, because of the high cost of high-voltage driving circuits. Research in PDP technology is still in progress.

REVIEW QUESTIONS

1. What is the method of excitation for Cathodoluminescence?
   (*a*) Excitation arises from the absorption of photons
   (*b*) Excitation from the application of an electric field
   (*c*) Excitation by the bombardment with a beam of electrons
   (*d*) None of these
2. What is the excitation process of Cathode Ray Tube (CRT)?
   (*a*) Photoluminescence (*b*) Cathodoluminescence
   (*c*) Electroluminescence (*d*) None of these
3. In video application, what is the minimum rate of renewal of picture to avoid flickering effect?
   (*a*) 15 Hz (*b*) 25 Hz (*c*) 45 Hz (*d*) 75 Hz
4. How many numbers of connections are required for N number of elements of alphanumeric display with matrix oriented format?
   (*a*) N^2 (*b*) N + 1 (*c*) N (*d*) $N^{1/2}$
5. What type of illumination is required for transmissive type LCD?
   (*a*) Front illumination (*b*) Rear Illumination
   (*c*) Both of these (*d*) None of these
6. Which of the following types of LCD exhibits the maximum contrast ratio?
   (*a*) STN LCD (*b*) VAN LCD (*c*) FLCD (*d*) PDLCD

7. Which of the following gas is not used in plasma display cell?
   (*a*) Neon (*b*) Argon (*c*) Xenon (*d*) Nitrogen
8. What is the purpose of MgO protective layer on dielectric film of plasma display?
   (*a*) For enhancement charge accumulation
   (*b*) To increase the charge accumulation time
   (*c*) To decrease break-down voltage
   (*d*) To reduce the discharge time of charged particle
9. What is the purpose of third electrode at plasma display cells?
   (*a*) To decrease break-down voltage
   (*b*) For enhancement charge accumulation
   (*c*) To reduce the discharge time of charged particle
   (*d*) To reduce electrostatic capacity of the cells
10. Which of the display exhibits highest viewing angle?
    (*a*) CRT (*b*) LED (*c*) LCD (*d*) PDP
11. According to the method of excitation, how many classification of luminescence are defined? Describe their methods of excitation with examples.
12. Describe the working principle of CRT with diagram. What is the importance of structure and thickness of its viewing screen? How color display is achieved in CRT?
13. Explain with circuit schematic how digits can be displayed by seven segment LED. How number of connections is reduced for display of matrix configuration.
14. What are advantages and limitations of LCD? Explain the working principle of LCD. Mention the materials which suitable for LCDs. Make a comparisons chart for the characteristics in respect of materials
15. What are the basic types of liquid crystal? Describe their structures and principles of operation. What is supertwist LCD?
16. What are the differences between passive and active matrix displays? Describe their structural and connection schematics. Discuss their advantages and disadvantages. How zero power display is achieved?
17. Describe the plasma cell structure and operating procedure of plasma display.
18. Discuss general characteristics of plasma display? Explain the working principle of AC and DC types of plasma displays
19. How PDPs are constructed? Elaborate the ALiS technology.
20. Make a comparison of plasma display panel and LCD panel.

8

Introduction to Optical Fiber Communication

8.1 INTRODUCTION

The transfer of information from one point to another is the basic concept of communication. There are different modes of communication. Perhaps the optical mode of communication is the most ancient form of communication. Simple systems such as signal fires, reflecting mirror, signaling lamps, etc. proved to be quite well to convey information form one end to the other within short distance. But when the information is to be sent to a greater distance a proper communication system is required. Within a communication system the information to be transferred is superimposed on a carrier which is an electromagnetic wave of higher frequency. The modulated carrier is then transmitted through the information channel to the receiver, where it is demodulated and the original signal is retrieved. If the carrier is chosen from optical frequency range, the communication is termed to be optical communication. The electromagnetic spectrum is shown in Fig. 8.1.

The term 'fiber optics' was first used by N. S. Kapany in 1956, when he along with his colleagues developed an image-transmitting device called the 'flexible fiberscope'. Then there came one of the most important developments in optical communication, the Laser. The first pulsed ruby laser was demonstrated by T. Maiman in 1960. Then the first semiconductor laser had been realized in 1962 by researchers working almost independently at various research laboratories. However, another eight years went before laser diodes could be produced for application in optical communication. Then in 1966 Charles Kao and Charles Hockham proposed for optical communication via dielectric waveguides or optical fibers fabricated from glass to avoid degradation of the optical signal by the atmosphere. Initially the optical fibers showed very high attenuation of about 1000 $dB\ km^{-1}$. Hence, in this form the optical fibers were not sufficient to replace then existing coaxial cables which exhibited attenuation of about 5 to 10 $dB\ km^{-1}$. Moreover at that time jointing of optical fiber cables was also a serious problem. However, within a decade optical fiber losses could be reduced to below 5 $dB\ km^{-1}$ and suitable low loss jointing techniques were invented.

However, with the development of optical fibers, focus was there also to develop proper optical source as well as optical detector which will work together to complete an optical communication system. As optical frequencies accompany extremely small wavelengths the development of all these optical components required a new technology. Thus the optical sources like the semiconductor laser diodes and light emitting diodes and also the detectors like photodiodes and in some cases phototransistors comparable in size with optical fibers were designed for the optical fiber communication system. Initially the semiconductor lasers worked at 0.8 and 0.9 µm. But subsequently it was found that the optical fibers exhibited enhanced performance at higher wavelengths like 1.1 to 1.6 µm. To take the advantage of the fact, the laser diodes were then fabricated with different alloys to emit light at these higher wavelengths.

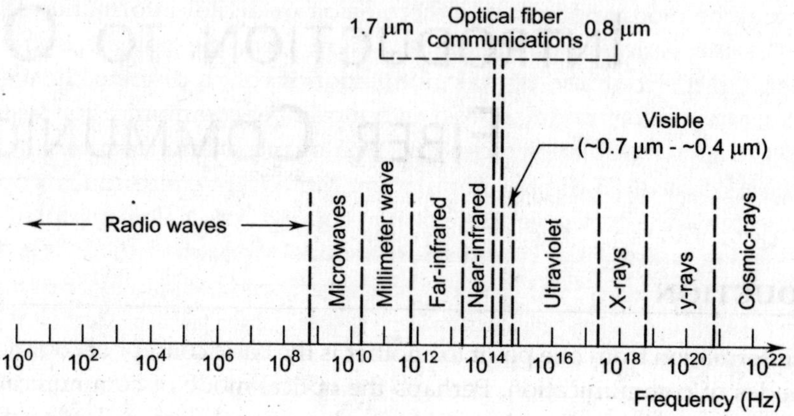

Fig. 8.1

In order that the optical fiber communication system becomes commercially viable, the optical fibers should exhibit much lower losses. The major breakthrough came when Dr Robert Maurer, Dr Donald Keck and Dr Peter Schultz of Corning Glass Corporation of USA succeeded in producing pure glass fibers in 1970 which exhibited an attenuation less than 20 *d*B/km. A lot of development took place since then and now a days optical fiber showing attenuation less than 0.2 *d*B/km is obtained in the longer wavelength window at 1.55 µm. Unfortunately intramodal dispersion is greater at this wavelength, thereby limiting the maximum bandwidth attainable. However to obtain both low loss and low dispersion at the same operating wavelength, different new advanced fiber structures are being implemented. In fact high performance, reliable optical fiber communication systems are now widely deployed.

8.2 GENERAL OPTICAL FIBER COMMUNICATION SYSTEM

Fiber-optic communication is similar to any other type of communication system form the basic concept. A schematic block diagram of fiber-optic communication system is shown in Fig. 8.2.

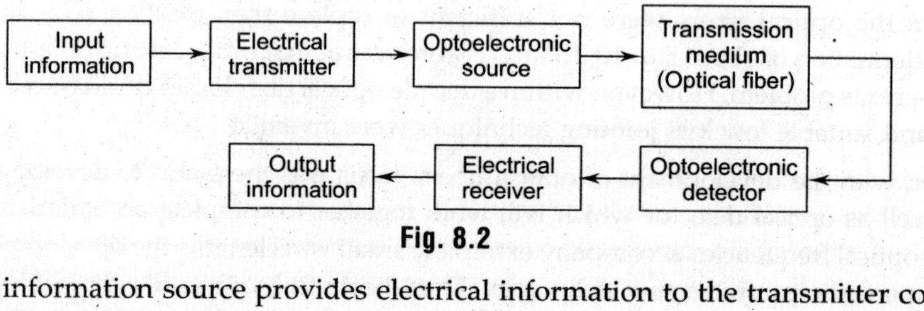

Fig. 8.2

The input information source provides electrical information to the transmitter consisting of an electrical stage. This transmitter in turn drives an optoelectronics source like LED (light emitting diode) or semiconductor Laser. This optoelectronic source as previously discussed in

Chapter 2 and Chapter 3 provides the electrical-optical conversion. The transmission medium generally consists of optical fiber. For free-space optical communication, this transmission medium is free-space. The output from the optical fibers falls on optoelectronic detector which further drives an electrical receiver. Generally *p-n*, *p-i-n* or avalanche photodiodes, and in some cases, photodiodes and photoconductors as already discussed in Chapter 4, are used as optoelectronic detectors. These detectors provide the optical-electrical conversion.

Optical carrier may be modulated using either analog or digital information signal. Analog modulation involves the variation of the emitted light from the optoelectronic source in a continuous manner. On the contrary, digital modulation involves discrete changes (*i.e.*, on-off pulses) in light intensity. Analog modulation in an optical fiber communication system is far less efficient compared to digital modulation technique. Analog modulation requires far higher signal to noise ratio at the receiver than digital modulation. Moreover semiconductor optical sources cannot always provide the high degree of linearity required for analog modulation. For these reasons generally analog optical fiber communication links are limited to shorter distances and lower bandwidths compared to digital links. The block diagram shown in Fig. 8.2 may be used for analog optical fiber communication. For digital links, there should be a provision for the input digital signal coming from the information source to be properly encoded. Similarly at the detection end there should be the proper decoder to decode the output digital signal.

8.3 ADVANTAGES OF FIBER-OPTIC COMMUNICATION

Due to several advantages of optical fiber communication system, it was that the idea of optical fibers as a means of communication was originally conceived. Subsequently with the advancement of the technology, it was found to be one of the most promising communication techniques. Thus it is useful to consider the advantages and special features of optical fiber communication over the prevalent electrical communication. We now discuss some of the advantages.

1. **Bandwidth:** The optical carrier frequency is in the range 10^{13} to 10^{16} Hz, which indicates a very high potential bandwidth of the system. Though the bandwidth available to the fiber systems cannot be fully utilized at present, but the transmission bandwidth and hence information-carrying capacity of optical fiber systems is much greater than the best copper cable systems.
2. **Isolation in electrical environment:** Optical fibers are fabricated from glass or plastic polymer. These materials are all electrical insulators. So they do not create arcing or spark hazard at short circuits. Thus the optical fibers are ideal for communication purpose in electrically hazardous environments
3. **Small size**: Optical fibers have very small diameters compared to their copper cable counterparts. Thus they are smaller and lighter than the copper cables.
4. **Immunity to interference:** Since optical fibers are made from dielectrics, they are free from electromagnetic interference (EMI), radiofrequency interference (RFI), etc. Thus optical fibers do not need any shielding from EMI or RFI. Radiofrequency interference is basically caused by television or radio broadcasting stations, radars or other signals originating in electronic equipment. Electromagnetic interference may be caused by the above mentioned sources of radiation as well as from industry machinery, or even from lightning sparking.
5. **Security in signal:** Since light do not radiate significantly from the optical fibers, they provide high signal security. Moreover a transmitted optical signal cannot be obtained

without drawing optical power from the fiber. Hence, any attempt to acquire a message from an optical fiber may be detected.

6. **Less transmission loss:** Development in fiber optic technology over the last four decades has produced optical fibers which exhibit even less than 0.2 *d*B/km transmission losses. This property of optical fibers helps in implementation of communication links with fewer intermediate repeaters or line amplifiers.
7. **Reliability:** due to usage of fewer repeaters, system reliability is high. Moreover, since the predicted lifetimes of optical components are very high, hence the reliability of these components is also not a problem.
8. **Ruggedness:** Optical fibers are manufactured with high tensile strength. Even the fibers, though primarily made of glass, may be bent or twisted to small radii without any damage. Moreover, cable structures have been developed which are flexible, compact and extremely rugged.
9. **Low cost:** For a system to be commercially successful and to be used on large scale, the two most important considerations are availability of raw materials and the overall cost of the system. The basic raw material for the fabrication of glass optical fibers is silica, which is quite easily available. Regarding the overall cost of the system, it is seen that for long-haul communication, optical fiber communication is cheaper than its copper cables counterpart.

8.4 CONCLUSION

Optical fiber communication has been developed over the past four decades in a series of generations mainly identified by the operating wavelengths they employed. The first generation optical-fiber systems used the wavelength of 0.85 μm. This wavelength corresponds to the 'first low-loss window' in the silica glass based optical fibers. This window was widely used at that time, since the LED as well as the photo detector technologies at that wavelength was available then. However this window gradually became less attractive as it exhibited relatively high loss of the order of 3 *d*B/km.

Moreover for long-distance communication along with the attenuation consideration, one of the other most important considerations is the dispersion of the pulses by the fiber. It was found that there is a 'second low-loss window' at 1.3 μm which theoretically exhibited zero dispersion for silica fibers and also less attenuation of about 0.5 *d*B/km.

Again in 1977, a 'third low-loss' window was found at 1.55 μm, where the attenuation was further lowered to 0.2 *d*B/km. However this window was not zero-dispersion window. Originally, short-distance optical fiber links were designed around multimode graded-index fibers; but in the late 1980s single-mode fibers operating at 1.31 μm were used. Later on use of dispersion-shifted fibers with low dispersion and low attenuation at 1.55 μm became the popular choice for long-distance communication. This will be discussed in detail in Chapter 10. As an optical fiber has tremendous potential bandwidth, there are possibilities for future optical fiber systems and their applications.

9

OPTICAL WAVEGUIDES AND FIBERS

9.1 INTRODUCTION

Light is the medium of communication in fiber optic technology. As we know, light has a fascinating property: light is treated as electromagnetic waves in classical wave theory, whereas the quantum theory treats light as photons, *i.e.*, quanta of electromagnetic energy. This is known as the wave-particle dual nature of light. The transmission of light through a dielectric waveguide was first proposed and investigated at the beginning of the twentieth century. Several theoretical and experimental works had been done thereafter. In the initial years the structure of the waveguides proposed resulted in excessive losses and were of not much use. However, in mid-1950s proposals for a clad dielectric rod came up to overcome the previous problems. The structure is illustrated in Fig. 9.1. The waveguide consists of a transparent core of refractive index n_1 surrounded by a transparent cladding of slightly lower refractive index n_2.

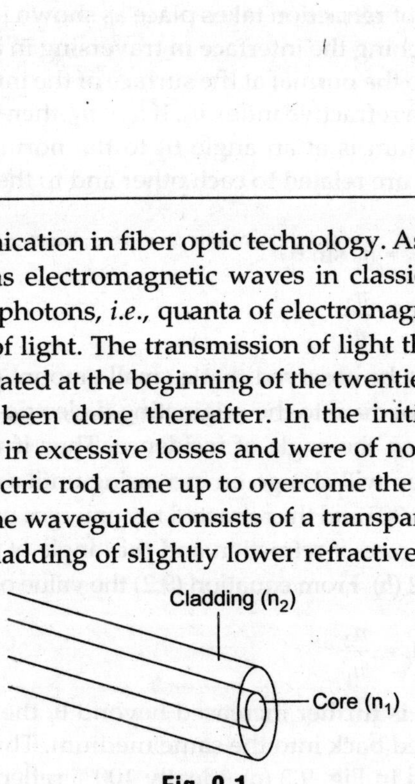

Fig. 9.1

To appreciate the transmission mechanism of light through optical fibers, it is necessary to consider the optical wave-guiding of a cylindrical fiber. Such a fiber may be analyzed utilizing ray theory of light. Ray optics uses the geometry of a straight line to explain the phenomenon of reflection, refraction, etc. Thus, it is also called *geometrical optics*. However, the concepts of geometrical optics are not sufficient when considering all types of optical fiber, and electromagnetic mode theory is to be used to give a complete picture. Let us now review, in brief, certain laws of geometrical optics, which helps in understanding of ray propagation in optical fibers.

9.2 REVIEW OF BASIC OPTICAL LAWS AND DEFINITIONS

To understand the propagation of light within a dielectric medium the most important parameter that has to be taken into account is the refractive index of the medium. It is defined as the ratio of the velocity of light in vacuum (c) to the velocity of light in the medium (v). We may write:

$$n = \frac{c}{v} \tag{9.1}$$

As c is always greater than v, n is always greater than unity. A ray of light travels more slowly in an optically dense medium than in one which is less dense. The refractive index gives a measure of this effect.

Whenever a ray of light is incident on the interface of two dielectric media differing in refractive indices, the phenomenon of refraction takes place as shown in Fig. 9.2 (*a*). It may be seen from the figure that the ray approaching the interface in traversing in a dielectric medium of refractive index n_1 and is at an angle θ_1 to the normal at the surface of the interface. Now the dielectric on the other side of the interface has a refractive index n_2. If $n_2 < n_1$, then the refraction is such that the ray path in the lower index medium is at an angle θ_2 to the normal, where $\theta_2 > \theta_1$. The angles of incidence θ_1 and refraction θ_2, are related to each other and to the refractive indices by Snell's law of refraction, given by:

$$n_1 \sin \theta_1 = n_2 \sin \theta_2$$

or
$$\frac{\sin \theta_1}{\sin \theta_2} = \frac{n_2}{n_1} \tag{9.2}$$

From Fig. 9.2 (*a*), it may also be observed that a small amount of light is reflected back from the interface of the two dielectric media into the originating dielectric medium. As $n_1 > n_2$, the angle of refraction is always greater than the angle of incidence. Thus if we go on increasing the angle of incidence, the angle of refraction will also go on increasing until a critical situation is reached, when for a certain value of $\theta_1 = \theta_c$, θ_2 is 90° and the refracted ray emerges parallel to the interface between the dielectrics. This is the limiting case of refraction and this angle of incidence is known as the critical angle θ_c as illustrated in Fig. 9.2 (*b*). From equation (9.2) the value of the critical angle is given as:

$$\sin \theta_c = \frac{n_2}{n_1} \tag{9.3}$$

If the angle of incidence θ_1 is further increased beyond θ_c the ray is no longer refracted in the second medium but is reflected back into the same medium. This phenomenon is called the *total internal reflection* as illustrated in Fig. 9.2 (*c*). Ideally 100% reflection is expected, but in practice, however there is always some light energy which passes through the interface. The wave carrying this tunneled energy is called the *evanescent wave*. The phenomenon of total internal reflection is responsible for the propagation of light through optical fibers.

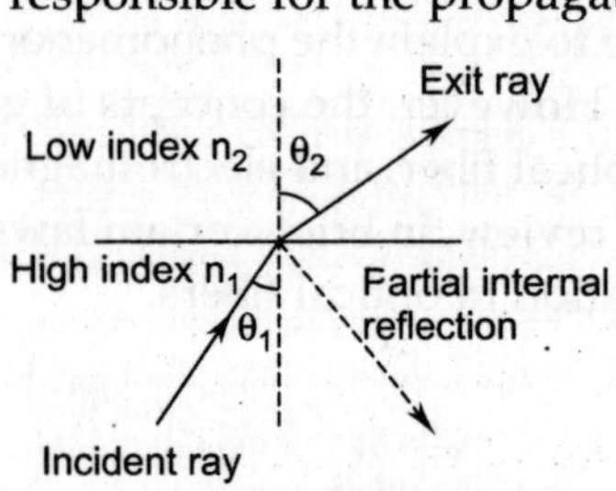

Fig. 9.2 (*a*)

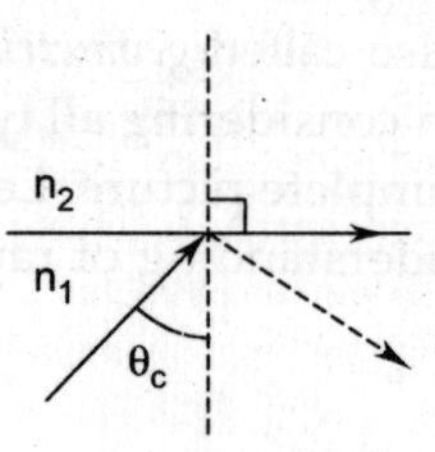

Fig. 9.2(*b*)

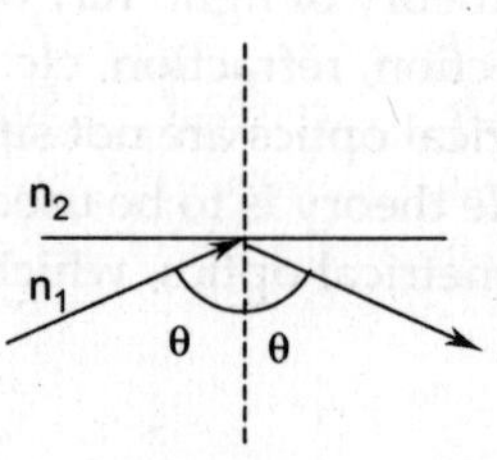

Fig. 9.2(*c*)

9.3 RAY THEORY

When light ray propagates through an optical fiber, two different categories of ray may pass through, viz. *meridional rays* and *skew rays*. If the light ray passes through the axis of the fiber core, it is called *meridional rays*. This type of ray is the simplest to describe and is generally used when discussing the fundamental transmission properties of optical fibers. However the ray which is transmitted without passing through the fiber axis is termed as *skew rays*. These rays greatly outnumber the meridional rays and follow a helical path through the fiber. The passage of both the rays is illustrated in Fig. 9.3 (*a*) and (*b*).

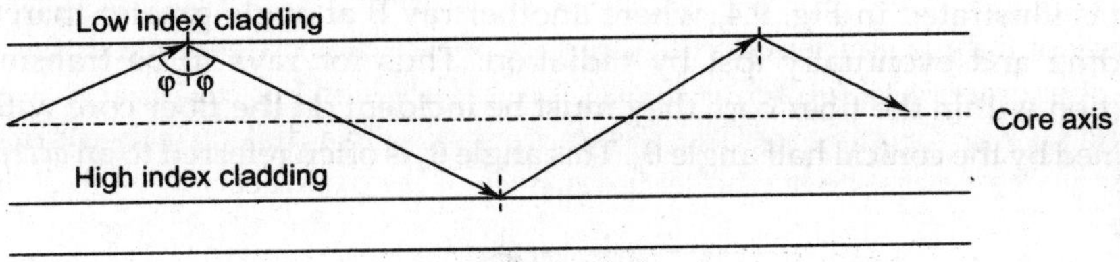

Fig. 9.3(*a*)

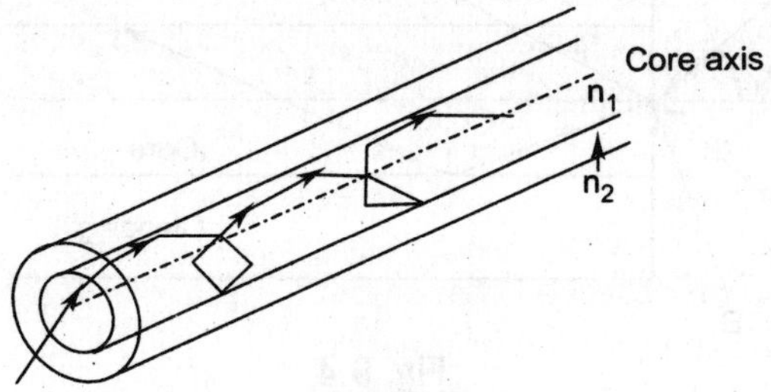

Fig. 9.3(*b*)

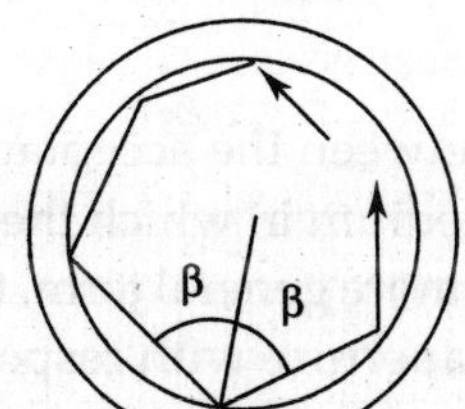

Fig. 9.3(*c*)

It is difficult to visualize the path of the skew rays in two dimensions. We try to understand the skew rays from Fig. 9.3 (*c*), where it is shown that the helical path traced through the fiber gives a change in direction of 2β at each reflection where β is the angle between the projection of the ray in two dimensions and the radius of the fiber core at the point of reflection. Thus the point of emergence of skew rays from the fiber in air will depend upon the number of reflections they undergo rather than the input conditions to the fiber. Therefore if the light input to the fiber is nonuniform, skew rays will tend to have a smoothening effect on the distribution of the light as it is transmitted, producing a more uniform output. This effect of smoothening depends on the number of reflections encountered by the skew rays.

9.3.1 Acceptance Angle

We now want to introduce a very important measurement for optical fiber communication, *i.e., acceptance angle*. To discuss about acceptance angle we want to consider an example of launching a light ray into an optical fiber as shown in Fig. 9.4. A meridional ray *A* is launched into the fiber in such a way so that it falls at the critical angle θ_c within the fiber at the core-cladding interface. This ray enters the fiber core at an angle θ_a to the fiber axis and is refracted at the air-core interface before transmission to the core-cladding interface at the critical angle. Thus all the rays which are incident at the air-core interface at an angle greater than θ_a will be transmitted to the core-cladding interface at an angle less than θ_c. Thus those rays will not be totally internally reflected. This situation is illustrated in Fig. 9.4, where another ray B at angle greater than θ_a is refracted into the cladding and eventually lost by radiation. Thus for rays to be transmitted by total internal reflection within the fiber core they must be incident on the fiber core within an acceptance cone defined by the conical half angle θ_a. This angle θ_a is often referred to an *acceptance angle* for the fiber.

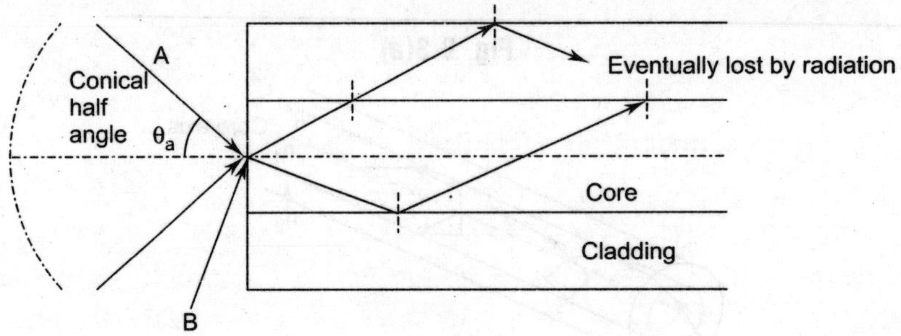

Fig. 9.4

9.3.2 Numerical Aperture

It is possible to obtain a relationship between the acceptance angle of the fiber and the refractive indices of the core, cladding and the medium in which the fiber is lying (*e.g.,* air) by using the ray theory analysis. This actually leads to a more general term, the *Numerical Aperture* (NA) of the fiber. We will now discuss about numerical aperture with respect to the meridional ray analysis of the fiber.

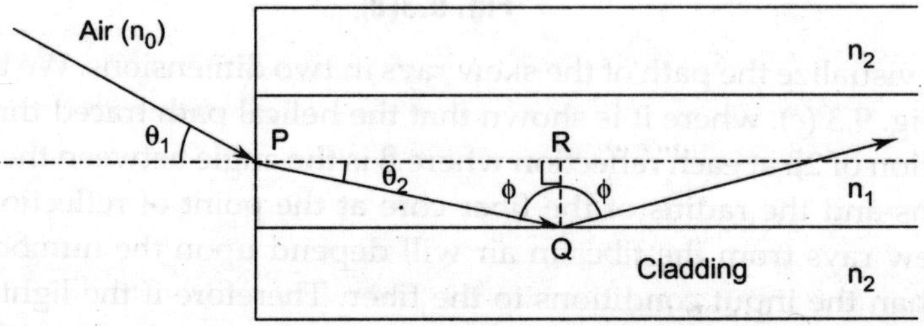

Fig. 9.5

In Fig. 9.5, a light ray is shown to be incident on the fiber core at an angle θ_1 to the fiber axis. This angle is less than the fiber acceptance angle θ_a. Thus the ray will propagate along the fiber. The ray enters the fiber from a medium (air) of refractive index n_0. The fiber core has the refractive index n_1, which is slightly greater than the refractive index n_2 of the cladding. Assuming the entrance face at the fiber core to be normal to the axis, and using Snell's law of refraction at the air-core interface we have

$$n_0 \sin\theta_1 = n_1 \sin\theta_2 \tag{9.4}$$

Considering the right-angled triangle PQR indicated in Fig. 9.5, we may write

$$\phi = \frac{\pi}{2} - \theta_2 \tag{9.5}$$

where ϕ is greater than the critical angle at the core-cladding interface. Thus equation (9.4) becomes

$$n_0 \sin\theta_1 = n_1 \cos\phi \tag{9.6}$$

or,

$$n_0 \sin\theta_1 = n_1 \sqrt{(1 - \sin^2\phi)} \tag{9.7}$$

For the limiting case of total internal reflection, ϕ becomes equal to θ_c *i.e.*, the critical angle for the core-cladding interface. Also in the limiting case θ_1 becomes the acceptance angle for the fiber θ_a. Considering these limiting cases into equation (9.7) we have

$$n_0 \sin\theta_a = \sqrt{(n_1^2 - n_2^2)} \tag{9.8}$$

From equation (9.8), we define the numerical aperture (NA) as:

$$\text{NA} = n_0 \sin\theta_a = \sqrt{(n_1^2 - n_2^2)} \tag{9.9}$$

Numerical aperture actually determines the light-gathering capacity of the fiber. However, the numerical aperture may also be defined in terms of the relative refractive index difference Δ between the core and the cladding given by:

$$\Delta = \frac{n_1^2 - n_2^2}{2n_1^2} \tag{9.10}$$

Hence,

$$\text{NA} = n_1\sqrt{2\Delta} \tag{9.11}$$

Both the definitions of numerical aperture given by equations (9.9) and (9.11) are independent of the fiber core diameter and holds for diameters as small as 8 μm. But for still smaller diameter geometric optics approach becomes invalid. In such cases electromagnetic mode theory must be applied.

Example 9.1. *An optical fiber with a core diameter 30 μm has a core refractive index of 1.50 and a cladding refractive index of 1.48. Assume the fiber is kept in air. Calculate (i) the NA for the fiber; (ii) the critical angle at the core-cladding interface; (iii) the acceptance angle in air for the fiber.*

Solution:

Since the core diameter of the fiber is greater than 8 μm, hence the geometric optics approach is valid for the analysis of the given fiber.

(*i*) The numerical aperture is given by:

$$\text{NA} = \sqrt{(n_1^2 - n_2^2)} = \sqrt{(1.50^2 - 1.48^2)} = \sqrt{(2.25 - 2.19)}$$

$$= 0.24$$

(*ii*) The critical angle at the core-cladding interface is given by:

$$\theta_c = \sin^{-1}\frac{n_2}{n_1} = \sin^{-1}\frac{1.48}{1.50}$$

$$= 80.63°$$

(*iii*) The acceptance angle in air for the fiber is given by equation (9.9) as:

$$\theta_a = \sin^{-1} \text{NA} = \sin^{-1} 0.24$$

$$= 13.88°.$$

Example 9.2. *For an optical fiber designed for long distance transmission, the typical relative refractive index difference is 1%. Estimate the NA and the solid acceptance angle in air for the fiber when the core refractive index is 1.50. Also calculate the critical angle at the core-cladding interface within the fiber. It may be assumed that the concepts of geometric optics hold for the fiber.*

Solution:

The numerical aperture is given by:

$$\text{NA} = n_1\sqrt{2\Delta} = 1.50\sqrt{0.02}$$

$$= 0.21$$

For small angles the solid acceptance angle in air ζ is given by:

$$\zeta \approx \pi\theta_a^2 = \pi \sin^2\theta_a = \pi(\text{NA})^2$$

$$= 0.13 \text{ rad}$$

The relative refractive index gives:

$$\Delta \approx \frac{n_1 - n_2}{n_1} = 1 - \frac{n_2}{n_1}$$

Thus,
$$\frac{n_2}{n_1} = 1 - \Delta = 1 - 0.01$$

$$= 0.99$$

Thus the critical angle at the core-cladding interface is:

$$\theta_c = \sin^{-1}\frac{n_2}{n_1} = \sin^{-1} 0.99$$

$$= 81.9°.$$

9.4 ELECTROMAGNETIC MODE THEORY

So far we have studied the propagation of light signals through optical fibers using the ray model. But as we have already mentioned that for smaller diameter ray model approach becomes invalid. In such cases in order to obtain an improved model for the propagation of light in an optical fiber, electromagnetic mode theory must be applied. The basis for such discussions is provided by Maxwell's equations. So let us begin with a brief discussion on the matter.

9.4.1 Maxwell's Equations

There are a set of four equations. For a medium with zero conductivity these equations may be written as:

$$\nabla \times \mathbf{E} = -\frac{\partial \mathbf{B}}{\partial t} \tag{9.12}$$

$$\nabla \times \mathbf{H} = \frac{\partial \mathbf{D}}{\partial t} \tag{9.13}$$

$$\nabla \cdot \mathbf{D} = 0 \text{ (no free charges)} \tag{9.14}$$

$$\nabla \cdot \mathbf{B} = 0 \text{ (no free poles)} \tag{9.15}$$

where **E** is the electric field vector, **H** is the magnetic field vector, **D** is the electric flux density and **B** is the magnetic flux density. ∇ is the nabla operator defined as

$$\nabla = \hat{i}\frac{\partial}{\partial x} + \hat{j}\frac{\partial}{\partial y} + \hat{k}\frac{\partial}{\partial z} \tag{9.16}$$

where $\hat{i}$, $\hat{j}$, and $\hat{k}$ are the unit vectors along the x, y, and z-axis, respectively. The four field vectors are related by the relations:

$$\mathbf{D} = \varepsilon \mathbf{E} \tag{9.17 a}$$

$$\mathbf{B} = \mu \mathbf{H} \tag{9.17 b}$$

where ε is the dielectric permittivity and μ is the magnetic permeability of the medium. The first two relations of Maxwell are curl conditions, whereas the last two are divergence relations. Now taking curl of equation (9.12), we get

$$\nabla \times \nabla \times \mathbf{E} = -\nabla \times \frac{\partial \mathbf{B}}{\partial t} = -\frac{\partial (\nabla \times \mathbf{B})}{\partial t}$$

$$= -\mu \frac{\partial (\nabla \times \mathbf{H})}{\partial t} = -\mu \frac{\partial}{\partial t}\left(\frac{\partial \mathbf{D}}{\partial t}\right)$$

$$= -\mu\varepsilon \frac{\partial^2 \mathbf{E}}{\partial t^2} \tag{9.18}$$

But $$\nabla \times \nabla \times \mathbf{E} = \nabla(\nabla \cdot \mathbf{E}) - \nabla^2 \mathbf{E} \tag{9.19}$$

Where ∇^2 is a Laplacian operator given by:

$$\nabla^2 \equiv \frac{\partial^2}{\partial x^2} + \frac{\partial^2}{\partial y^2} + \frac{\partial^2}{\partial z^2}$$

and $$\nabla \cdot \mathbf{E} = \nabla \cdot (\mathbf{D}/\varepsilon) = 0 \quad \text{[using equation 9.14]} \tag{9.20}$$

Hence, from equations (9.18)–(9.20), we have

$$\nabla^2 \mathbf{E} = \mu\varepsilon \frac{\partial^2 \mathbf{E}}{\partial t^2} \tag{9.21}$$

In a similar way it can be shown that

$$\nabla^2 \mathbf{H} = \mu\varepsilon \frac{\partial^2 \mathbf{H}}{\partial t^2} \tag{9.22}$$

For rectangular Cartesian and cylindrical polar coordinates the above wave equations hold for each component of the field vectors **E** (*i.e.*, E_x, E_y, E_z) or **H** (*i.e.*, H_x, H_y, H_z). If we write ψ to represent any component of **E** or **H** field, then the vector wave equations (9.21) and (9.22) become a set of scalar wave equations

$$\nabla^2 \psi = \mu\varepsilon \frac{\partial^2 \psi}{\partial t^2} \tag{9.23}$$

These are linear wave equations, which, in the absence of boundary conditions, have solutions that take the form of plane waves that propagate with a phase velocity given by:

$$v_p = \frac{1}{\sqrt{\mu\varepsilon}} = \frac{1}{\sqrt{\mu_r\mu_0\varepsilon_r\varepsilon_0}} \tag{9.24}$$

where μ_r and ε_r are the relative permeability and permittivity for the dielectric medium and μ_0 and ε_0 are the permeability and permittivity of free space. In vacuum or free space, $\mu_r = \varepsilon_r = 1$, so that the velocity of light in free space c is therefore

$$c = \frac{1}{\sqrt{\mu_0\varepsilon_0}} \tag{9.25}$$

From equations (9.23) and (9.24), it may be written:

$$\nabla^2\psi = \frac{1}{v_p^2}\frac{\partial^2\psi}{\partial t^2} \tag{9.26}$$

The basic solution of the wave equation is a sinusoidal wave, the most important form of which is a uniform plane wave given by:

$$\psi = \psi_0 \exp\{j(\omega t - \beta z)\} \tag{9.27}$$

This represents a plane wave propagating in the positive z-direction with a phase velocity $v_p = \omega/\beta$. Here ω is the angular frequency of the field; β is the propagation phase constant and $j = \sqrt{-1}$.

9.4.2 Planar Optical Waveguide

Planar optical waveguides are the simplest type of optical waveguides. The geometric configuration of such waveguide is shown in Fig. 9.6. It consists of a thin slab of dielectric of refractive index n_1 and thickness $2a$ sandwiched between two dielectric slabs of lower refractive index n_2 and infinite thickness. We assume that the waveguide is so oriented that the wave propagates along the z-direction. The y and z dimensions of the guide are assumed to be extended to infinity.

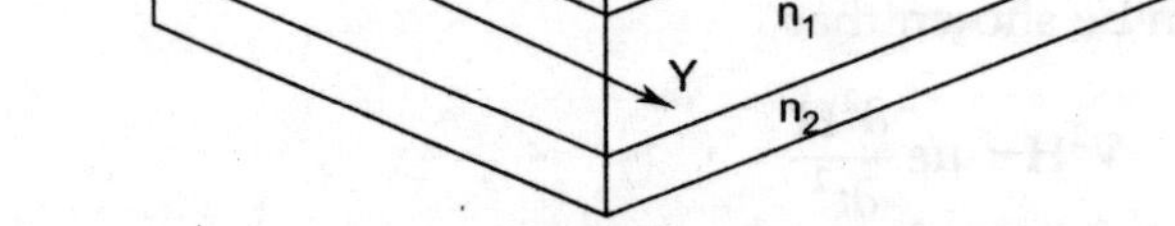

Fig. 9.6

A ray of light launched into the optical waveguide would progress by multiple reflections as illustrated in Fig. 9.7. The conceptual transition from ray to wave theory is aided by consideration of a plane monochromatic wave propagating in the direction of the ray path within the guide. As the refractive index within the guide layer is n_1, the optical wavelength in this layer is reduced to $\lambda_g = \lambda/n_1$, where λ is the wavelength of light in vacuum. The propagation constant is thus increased to $\beta_1 = kn_1$, where $k = 2\pi/\lambda$ is the vacuum propagation constant or the propagation vector. When the propagation vector or the equivalent ray makes an angle θ with the guide axis (which is actually the z-axis), the plane wave can be resolved into two component plane waves propagating in the z and x directions as shown in Fig. 9.7.

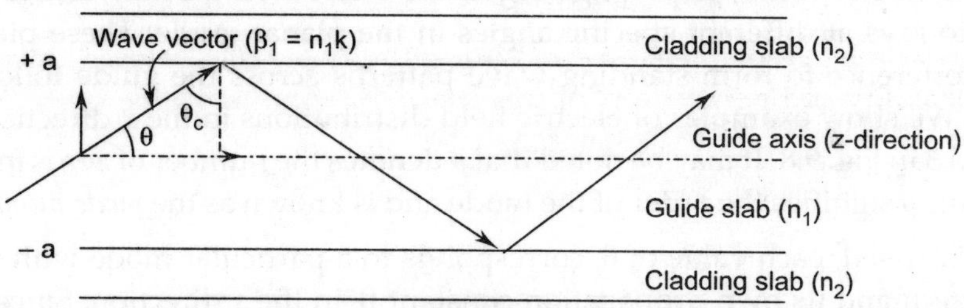

Fig. 9.7

The component of the plane wave propagating in the z-direction is given by:

$$\beta_z = n_1 k \cos\theta = \beta_1 \cos\theta \tag{9.28}$$

The limiting value of θ, *i.e.*, θ_m is related to the critical angle θ_c at the interface of the guide layer and the cladding slabs by the relation:

$$\sin\theta_c = \cos\theta_m = n_2/n_1 \tag{9.29}$$

Hence the minimum value of β in the z-direction, *i.e.*, β_{min}, will be determined by the maximum value of θ. So we get

$$\beta_{min} = \beta_1 \cos\theta_m = \beta_1 \frac{n_2}{n_1} = \beta_2 \tag{9.30}$$

The maximum value of β may be β_1, which happens when $\theta = 0$. Hence, $\beta_{max} = \beta_1$. Thus, it is expected that β lies between β_1 and β_2, or $\beta_2 < \beta < \beta_1$. Now, the component of the plane wave propagating in the x-direction is given by:

$$\beta_x = n_1 k \sin\theta = \beta_1 \sin\theta \tag{9.31a}$$

or
$$\beta_x = \frac{2\pi}{\lambda_g} \sin\theta \tag{9.31b}$$

This component of the plane wave in the x-direction is reflected at the interface between the guide layer of higher refractive index and the cladding slabs of lower refractive index. If the total phase change after two successive reflections at the upper and lower interfaces is equal to $2p\pi$ radians, where p is an integer, then constructive interference takes place and a standing wave pattern is generated in the x-direction. This stable field distribution in the x-direction with only periodic z dependence is known as a *mode*. Thus only a finite number of discrete modes will propagate through the guide which satisfies the above mentioned condition of constructive interference. Thus we may write

$$4a\beta_x = 2p\pi$$

or
$$4a \sin\theta_i = p\lambda_g \tag{9.32}$$

Hence, the light propagating within the guide is formed into discrete modes, each typified by a distinct value of θ_i. These modes have a periodic z dependence of the form exp $(-j\beta_z)$ where β_z becomes the propagation constant for the mode as the modal field pattern is invariant except for a periodic z dependence. It is a common accepted practice to denote the mode propagation constant by β, where $\beta = \beta_z$. Now if we assume time dependence for the monochromatic electromagnetic light field with angular frequency ω of the form $\exp(j\omega t)$, then the combined factor $\exp j(\omega t - \beta_z)$ denotes a mode propagating in the z direction.

To understand different modes propagating in the z direction we may consider plane waves corresponding to rays at different specific angles in the planar guide. These plane waves give constructive interference to form standing wave patterns across the guide following a sine or cosine formula. We show examples of electric field distributions in the x direction for such rays for $p = 0, 1, 2$, and 3 in Fig. 9.8. It may be noted that p denotes the number of zeros in this transverse field pattern. Thus p signifies the order of the mode and is known as the *mode number*.

As already discussed, each value of θ_i corresponds to a particular mode with its own characteristic field pattern and its own propagation constant β_i in the z direction. Since the maximum value that θ_i can take is θ_m, the number of guided modes is limited to

$$M = p_{\max} = \frac{4a \sin\theta_m}{\lambda_g} = \frac{4an_1 \sin\theta_m}{\lambda} = \frac{4a}{\lambda}\sqrt{\left(n_1^2 - n_2^2\right)} \tag{9.33}$$

It is to be noted that for the pth mode to be propagated, it is required that $p \le (4a/\lambda)\sqrt{\left(n_1^2 - n_2^2\right)}$.

The mode corresponding to the highest value of p, *i.e.*, $p_{\max}$, does not meet the condition for total internal reflection, as the value of θ_m corresponds exactly to the critical angle θ_c and is refracted at the interface. However, it can propagate freely in the cladding layers and is said to be a radiation mode.

Light may be described as an electromagnetic wave which consists of a periodically varying electric field **E** and magnetic field **H** which are oriented at right angles to each other. The transverse modes shown in Fig. 9.8 illustrate the case when electric field is perpendicular to the direction of propagation and hence $E_z = 0$, but a corresponding component of the magnetic field **H** is in the direction of propagation. These modes are said to be Transverse Electric (TE). On the contrary, if a component of the **E** field exists in the direction of propagation, but $H_z = 0$, the modes formed are called Transverse Magnetic (TM). The mode numbers are incorporated into the nomenclature by referring to the TE_m and TM_m modes, as shown for the transverse electric modes in Fig. 9.8. If the total field lies in the transverse plane, transverse electromagnetic (TEM) waves exist where both E_z and H_z are zero. However, TEM waves are very uncommon in optical waveguides, and generally occur in metallic conductors.

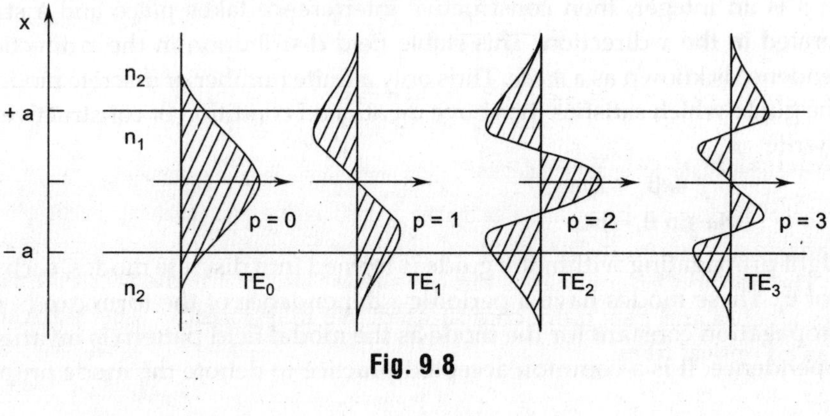

Fig. 9.8

9.4.3 Phase Velocity And Group Velocity

For all electromagnetic waves there are points of constant phase which form a surface referred to as a wavefront. If a monochromatic light wave propagates along a waveguide in the z-direction these points of constant phase travel at a phase velocity v_p given by:

$$v_p = \frac{\omega}{\beta} \tag{9.34}$$

where $d\omega$ is the angular frequency of the wave. But in practice, it is not possible to produce perfectly monochromatic light waves. Hence, light energy is generally made up of plane wave components of different frequencies. In such cases when a group of closely similar frequencies propagate their resultant forms a packet of waves. Such wave packet does not travel at the phase velocity of the individual waves but moves at a group velocity v_g given by:

$$v_g = \frac{d\omega}{d\beta} \tag{9.35}$$

where $d\omega$ is the rate of change of angular frequency, $d\beta$ is the rate of change of propagation constant. When a packet wave is propagated through an optical waveguide of infinite length that has a refractive index n_1 surrounded by cladding of refractive index n_2 ($n_1 > n_2$), then the propagation constant may be written as:

$$\beta = n_1 \frac{2\pi}{\lambda} = \frac{n_1 \omega}{c} \tag{9.36}$$

where c is the velocity of light in free space. Using equation (9.34) we obtain the following relationship for the phase velocity

$$v_p = \frac{c}{n_1} \tag{9.37}$$

Similarly, for a packet wave, group velocity is expressed as:

$$\begin{aligned} v_g &= \frac{d\lambda}{d\beta}\frac{d\omega}{d\lambda} = \frac{d}{d\lambda}\left(n_1 \frac{2\pi}{\lambda}\right)^{-1}\left(-\frac{\omega}{\lambda}\right) \\ &= -\frac{\omega}{2\pi\lambda}\left(\frac{1}{\lambda}\frac{dn_1}{d\lambda} - \frac{n_1}{\lambda^2}\right)^{-1} \\ &= \frac{c}{\left(n_1 - \lambda \dfrac{dn_1}{d\lambda}\right)} = \frac{c}{N_g} \end{aligned} \tag{9.38}$$

The parameter N_g is known as the guide group index of the guide.

9.4.4 Cylindrical Optical Waveguide

We have already discussed how an electromagnetic wave propagates through an ideal Step-Index (SI) planar waveguide. In this section, we want to study wave propagation in cylindrical waveguides. The method of analysis is almost similar except for the fact that the mathematics involves more algebra and gets more complicated. We may consider a cylindrical waveguide to be the same as an optical fiber.

We consider an ideal step-index fiber consisting of a uniform cylindrical dielectric core of radius a and refractive index n_1, surrounded by an infinitely thick uniform dielectric cladding of refractive index n_2. Now we attempt a solution of the wave equation for magnetic and electric fields for the modes propagated by such a fiber. It may be shown that the equation for an electric field in an isotropic, linear, non-conducting, and non-magnetic, but inhomogeneous medium is given by:

$$\nabla^2 \mathbf{E} + \nabla\left\{\frac{1}{\varepsilon_r}\nabla(\varepsilon_r)\cdot \mathbf{E}\right\} - \mu_0\varepsilon_0\varepsilon_r \frac{\partial^2 \mathbf{E}}{\partial t^2} = 0 \tag{9.39}$$

In a similar manner, the equation for a magnetic field may be written as:

$$\nabla^2 \mathbf{H} + \frac{1}{\varepsilon_r}\{\nabla(\varepsilon_r)\times(\nabla\times\mathbf{H})\} - \mu_0\varepsilon_0\varepsilon_r \frac{\partial^2 \mathbf{H}}{\partial t^2} = 0 \tag{9.40}$$

From equations (9.39) and (9.40) it is evident that the different components of **E** and **H** are coupled. In this present case, the refractive index $n = \sqrt{\varepsilon_r}$ is constant ($= n_1$) up to $r \le a$ and equal to n_2 for points $r > a$, but there is a discontinuity at $r = a$. If this discontinuity is small (*i.e.*, $n_1 \approx n_2$), then this is called a weakly guiding approximation. In this approximation, the second term on the left-hand side of equations (9.39) and (9.40) may be neglected, and each Cartesian component of **E** and **H** satisfies the scalar wave equation. Putting $\varepsilon_r = n^2$,

$$\nabla^2\Psi = \varepsilon_0\mu_0 n^2 \frac{\partial^2\Psi}{\partial t^2} \tag{9.41}$$

where Ψ represents the scalar E or H field. The fiber boundary conditions have cylindrical symmetry and we assume that the direction of propagation of the electromagnetic waves is along the axis of the fiber, which is taken to be the z-axis. In the scalar wave approximation, the modes may be assumed to be nearly transverse and they may possess an arbitrary polarization state. These linearly polarized modes are referred to as LP modes. The propagation constants of the TE and TM modes are almost equal. In cylindrical coordinates, equation (9.41) may be written as

$$\nabla^2\Psi - \varepsilon_0\mu_0 n^2 \frac{\partial^2\Psi}{\partial t^2} = \frac{\partial^2\Psi}{\partial r^2} + \frac{1}{r}\frac{\partial\Psi}{\partial r} + \frac{1}{r^2}\frac{\partial^2\Psi}{\partial\phi^2} + \frac{\partial^2\Psi}{\partial z^2} - \varepsilon_0\mu_0 n^2 \frac{\partial^2\Psi}{\partial t^2} = 0 \tag{9.42}$$

The solution of equation (9.42) may be written as:

$$\Psi(r,\phi,z,t) = \Psi(r,\phi)\exp j(\omega t - \beta z) \tag{9.43}$$

Substituting the value of Ψ from equation (9.43) into equation (9.42), we get

$$\frac{\partial^2\Psi}{\partial r^2} + \frac{1}{r}\frac{\partial\Psi}{\partial r} + \frac{1}{r^2}\frac{\partial^2\Psi}{\partial\phi^2} + \left[\varepsilon_0\mu_0\omega^2 n^2 - \beta^2\right]\Psi = 0 \tag{9.44}$$

Putting $\varepsilon_0\mu_0 = 1/c^2$ and $\omega/c = k$, in equation (9.44), we get

$$\frac{\partial^2\Psi}{\partial r^2} + \frac{1}{r}\frac{\partial\Psi}{\partial r} + \frac{1}{r^2}\frac{\partial^2\Psi}{\partial\phi^2} + \left[n^2k^2 - \beta^2\right]\Psi = 0 \tag{9.45}$$

As the fiber under consideration has cylindrical symmetry, the variables may be separated and written as:

$$\Psi(r,\phi) = R(r)\Phi(\phi) \tag{9.46}$$

where we consider that R is a function of only r and Φ is a function of only ϕ. Now by substituting Ψ from equation (9.46) in equation (9.45), we get

$$\Phi \frac{\partial^2 R}{\partial r^2} + \frac{1}{r}\Phi\frac{\partial R}{\partial r} + \frac{R}{r^2}\frac{\partial^2 \Phi}{\partial \phi^2} + \left[n^2k^2 - \beta^2\right] R\Phi = 0 \tag{9.47}$$

As the derivatives involved are dependent either on r or ϕ only, the partial derivatives may be replaced by full derivatives. Again, dividing the entire LHS by RΦ, we have

$$\frac{1}{R}\frac{d^2R}{dr^2} + \frac{1}{Rr}\frac{dR}{dr} + \frac{1}{r^2}\frac{1}{\Phi}\frac{d^2\Phi}{d\phi^2} + \left[n^2k^2 - \beta^2\right] = 0$$

or

$$\frac{r^2}{R}\left(\frac{d^2R}{dr^2} + \frac{1}{r}\frac{dR}{dr}\right) + r^2\left[n^2k^2 - \beta^2\right] = -\frac{1}{\Phi}\frac{d^2\Phi}{d\phi^2} = l^2 \text{ (say)} \tag{9.48}$$

Here, l is a constant and known as an azimuthal eigenvalue.

The dependence of Φ on ϕ is of the form exp $(jl\phi)$. In order that the function is single-valued, *i.e.*, $\Phi(\phi + 2\pi) = \Phi(\phi)$, the constant l is required to be an integer, *i.e.*, $l = 0, 1, 2, 3, \ldots$ Thus the complete transverse field is given by:

$$\Psi(r, \phi, z, t) = R(r)\, e^{jl\phi} e^{j(\omega t - \beta z)} \tag{9.49}$$

The radial part of equation (9.48) may be written as

$$\frac{r^2}{R}\left(\frac{d^2R}{dr^2} + \frac{1}{r}\frac{dR}{dr}\right) + r^2\left[n^2k^2 - \beta^2\right] = l^2$$

This may be rearranged as:

$$r^2\frac{d^2R}{dr^2} + r\frac{dR}{dr} + \left[r^2\left(n^2k^2 - \beta^2\right) - l^2\right]R = 0 \tag{9.50}$$

Now, we know that $n = n_1$ for $r \le a$ (*i.e.*, the core) and $n = n_2$ for $r > a$ (*i.e.*, the cladding). Hence, substituting the value of n in equation (9.50), for the case of a step-index fiber, we obtain:

$$r^2\frac{d^2R}{dr^2} + r\frac{dR}{dr} + \left[r^2\left(n_1^2k^2 - \beta^2\right) - l^2\right]R = 0, \quad \text{for } r \le a \tag{9.51}$$

and

$$r^2\frac{d^2R}{dr^2} + r\frac{dR}{dr} + \left[r^2\left(n_2^2k^2 - \beta^2\right) - l^2\right]R = 0, \quad \text{for } r > a \tag{9.52}$$

To simplify the above equations, we put

$$U \equiv a\left(n_1^2k^2 - \beta^2\right)^{\frac{1}{2}} \tag{9.53}$$

and

$$W \equiv a\left(\beta^2 - n_2^2k^2\right)^{\frac{1}{2}} \tag{9.54}$$

U is also referred to as the radial phase parameter or the radial propagation constant, whereas W is known as the cladding decay parameter. The sum of the squares of U and W defines a very useful quantity, which is referred to as the normalized frequency V where

$$V = \left(U^2 + W^2\right)^{\frac{1}{2}} = ka\left(n_1^2 - n_2^2\right)^{\frac{1}{2}} = \frac{2\pi}{\lambda}a(\text{NA}) \tag{9.55}$$

The normalized frequency may also be expressed in terms of relative index difference Δ as:

$$V = \frac{2\pi}{\lambda} a n_1 (2\Delta)^{\frac{1}{2}} \tag{9.56}$$

This normalized frequency is a dimensionless parameter and is also sometimes called the *V* number of the fiber. Substituting the values of U and W in equations (9.51) and (9.52), we get

$$r^2 \frac{d^2R}{dr^2} + r\frac{dR}{dr} + \left(\frac{U^2 r^2}{a^2} - l^2\right) R = 0 \quad \text{for } r \le a \tag{9.57}$$

and

$$r^2 \frac{d^2R}{dr^2} + r\frac{dR}{dr} - \left(\frac{W^2 r^2}{a^2} + l^2\right) R = 0 \quad \text{for } r > a \tag{9.58}$$

Equations (9.57) and (9.58) are second-order equations and hence should possess two independent solutions. But considering all the facts, a valid solution of equation (9.57) would be given by first kind of Bessel function of order *l* and that of equation (9.58) would be given by the second kind of modified Bessel function of order *l*. Thus we may write:

$$R(r) = AJ_1\left(\frac{Ur}{a}\right), \quad \text{for } r < a \tag{9.59}$$

and

$$R(r) = BK_1\left(\frac{Wr}{a}\right), \quad \text{for } r > a \tag{9.60}$$

Since, Ψ is continuous at $r = a$, R(r) must be continuous at $r = a$. Imposing this condition, we can get the values of constants A and B. Hence, from equations (9.59) and (9.60) we may write:

$$A = \frac{R(a)}{J_1(U)} \tag{9.61}$$

and

$$B = \frac{R(a)}{K_1(W)} \tag{9.62}$$

Substituting the values of R and Φ in equation (9.46), we get the transverse dependence of the modal fields as below:

$$\Psi(r, \phi) = AJ_1\left(\frac{Ur}{a}\right)\begin{bmatrix}\cos(l\phi)\\ \sin(l\phi)\end{bmatrix}; \quad r < a \tag{9.63}$$

and

$$\Psi(r, \phi) = BK_1\left(\frac{Wr}{a}\right)\begin{bmatrix}\cos(l\phi)\\ \sin(l\phi)\end{bmatrix}; \quad r > a \tag{9.64}$$

Again, at $r = a$, $\partial\Psi/\partial r$ is continuous. Hence,

$$\left.\frac{\partial\Psi}{\partial r}\right|_{r=a} = A\frac{U}{a} J_1'\left(\frac{Ur}{a}\right)\cos l\phi \,\Big|_{r=a} = A\frac{U}{a} J_1'(U)\cos l\phi$$

and

$$\left.\frac{\partial\Psi}{\partial r}\right|_{r=a} = B\frac{W}{a} K_1'\left(\frac{Wr}{a}\right)\cos l\phi \,\Big|_{r=a} = B\frac{W}{a} K_1'(W)\cos l\phi$$

Thus, substituting the values of A and B from equations (9.61) and (9.62), we have:

$$\frac{R(a)}{J_1(U)}\frac{U}{a} J_1'(U)\cos l\phi = \frac{R(a)}{K_1(W)}\frac{W}{a} K_1'(W)\cos l\phi$$

or
$$\frac{UJ_1'(U)}{J_1(U)} = \frac{WK_1'(W)}{K_1(W)} \tag{9.65}$$

Hence, the continuity of Ψ and $\partial\Psi/\partial r$ at the core-cladding interface leads to an eigenvalue equation for the LP modes of the form shown in equation (9.65). Hence, the equation may be written in either of the following two forms:

$$\frac{UJ_{l\pm1}(U)}{J_l(U)} = \pm\frac{WK_{l\pm1}(W)}{K_l(W)} \tag{9.66}$$

The values of U and W for various values of l and the corresponding values of the propagation constant β may be obtained from equation (9.66). For values of β lying within the range $\beta_2^2\left(=n_2^2k^2\right) < \beta^2 < \beta_1^2\left(=n_1^2k^2\right)$, the radial part of the field, $R(r)$, in the core of the fiber is given by the Bessel function $J_l(x)$, which is oscillatory in nature. Thus there exist m allowed solutions of β for each value of l. Hence, each allowed value of β is characterized by two integers l and m. The first integer l is associated with two circular functions cos $l\phi$ and sin $l\phi$ corresponding to the azimuthal part of the solution. The second integer m is associated with the mth root of the eigenvalue equation corresponding to the radial part of the solution. These are known as guided modes.

From equation (9.55), we know that $V^2 = U^2 + W^2$. Hence, the solution of the transcendental equations (for given values of l and V) will give universal curves describing the dependence of U or W on V. The value of β may be calculated by substituting the values of U (or W) and V in the defining equations. Alternatively, it is also possible to define the normalized propagation constant b for a fiber as:

$$b = \frac{\beta^2 - \beta_2^2}{\beta_1^2 - \beta_2^2} = \frac{\beta^2 - n_2^2k^2}{n_1^2k^2 - n_2^2k^2} = \frac{W^2}{V^2} = 1 - \frac{U^2}{V^2} \tag{9.67}$$

As β lies between $\beta_1(=n_1k)$ and $\beta_2(=n_2k)$ for the guided modes, the value of b will lie between 0 (for $\beta = \beta_2$) and 1(for $\beta = \beta_1$). Considering the limit of mode propagation when $\beta = n_2k$, then the mode phase velocity is equal to the velocity of light in the cladding and the mode is no longer properly guided. In this case the mode is said to be cut off and the eigenvalue W = 0. A mode ceases to be guided when $\beta < n_2k$, and hence W is imaginary. Such modes are called *radiation modes* or *unguided modes*. In terms of the ray theory, these modes correspond to rays that undergo refraction, rather than total internal reflection, at the core-cladding interface. Nevertheless, wave propagation does not cease abruptly below cutoff. Modes exist where $\beta < n_2k$ but the difference is very small, such that some of the energy loss due to radiation is prevented by an angular momentum barrier formed near the core-cladding interface. Solutions of the wave equation giving these states are called *leaky modes*. Alternatively, as β is increased above n_2k, less power is propagated in the cladding until at $\beta = n_1k$ all the power is confined to the fiber core. This range of values for β signifies the guided modes of the fiber.

The lower order modes obtained in a cylindrical homogeneous core waveguide are shown in Fig. 9.9. Both the LP notation and the traditional HE, EH, TE and TM mode notations are indicated. Additionally the Bessel functions J_0 and J_1 are plotted and where they cross the zero gives the cutoff point for the various modes.

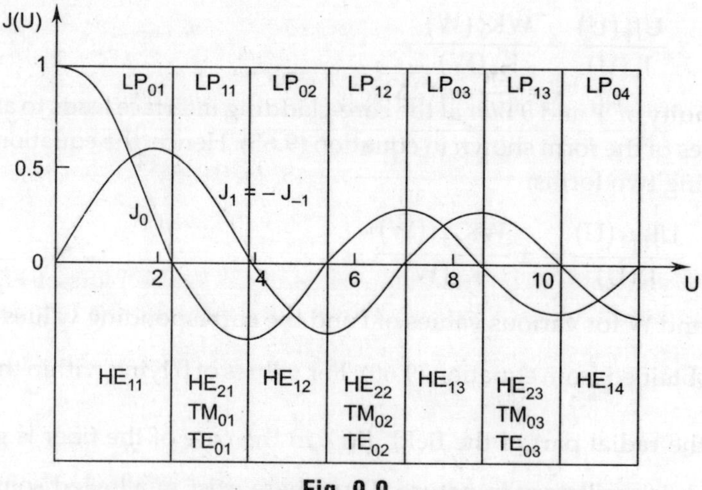

Fig. 9.9

The cutoff point for a particular mode corresponds to a distinctive value of the normalized frequency (*i.e.*, $V = V_c$) for the fiber. Figure 9.9 shows allowed regions for the LP modes of order l = 0, 1 for a circular optical waveguide with a constant refractive index core (step index fiber) (Gloge 1971). It may be seen from the figure that the value of V_c is different for different modes. For example, the first zero crossing J_1 occurs when the normalized frequency is 0 and this corresponds to the cutoff for the LP_{01} mode. On the other hand, the first zero crossing for J_0 is when the normalized frequency is 2.405, giving a cutoff value V_c of 2.405 for the LP_{11} mode. Similarly, the second zero of J_1 corresponds to a normalized frequency of 3.83, giving a cutoff value V_c for the LP_{02} mode of 3.83. Hence, fibers may be produced with particular values of normalized frequency which allow only certain modes to propagate. This is further illustrated in Fig. 9.10 which shows the normalized propagation constant b for a number of LP modes as a function of V. It is observed that the cutoff value of normalized frequency V_c which occurs when $\beta = n_2k$ corresponds to $b = 0$.

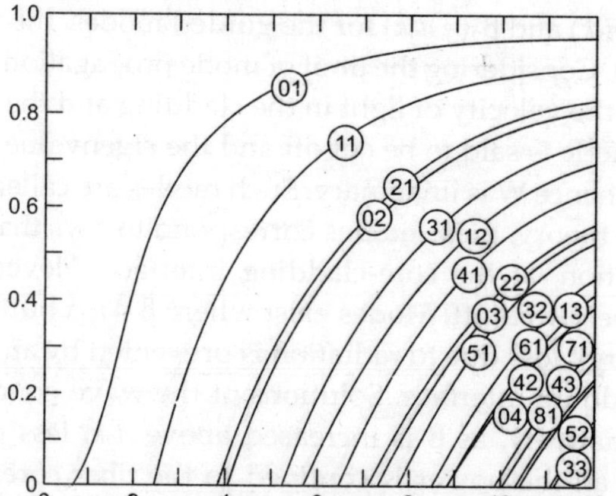

Fig. 9.10

Example 9.3. *A step-index fiber has a core refractive index of 1.45, a relative refractive index difference of 1.5% and a core diameter of 8.1 μm. A light of wavelength 1.55 μm is used to excite the modes in the fiber. Find (i) normalized frequency parameter V, (ii) the propagation constants β for these modes, and (c) the phase velocity of each mode.*

Solution:

(*i*) We know, $$V = \frac{2\pi}{\lambda} a n_1 (2\Delta)^{\frac{1}{2}}$$

$$= \frac{\pi \times 8.1 \times 10^{-6}}{1.55 \times 10^{-6}} \times 1.45(2 \times 0.015)^{1/2}$$

$$= 4.12$$

(*ii*) If we draw a vertical line on the graph of b versus V (the plots of Fig. 9.10) at $V \approx 3$, the line will intersect the curves for only two modes, namely, the LP_{01} and LP_{11} modes. This means that the fiber supports only these two modes.

(*iii*) We have $$b_{lm} = \frac{\beta_{lm}^2 - \beta_2^2}{\beta_1^2 - \beta_2^2}$$

Again $$n_2 = n_1(1-\Delta) = 1.45(1-0.015) = 1.4281$$

Now, $$\beta_1 = kn_1 = \frac{2\pi}{\lambda} n_1 = \frac{2\pi}{1.55 \times 10^{-6}} \times 1.45 = 5.878 \times 10^6 \text{ m}^{-1}$$

$$\beta_2 = kn_2 = \frac{2\pi}{1.55 \times 10^{-6}} \times 1.4281 = 5.789 \times 10^6 \text{ m}^{-1}$$

Now, $$\beta_{lm} = \left[\beta_2^2 + b_{lm}\left(\beta_1^2 - \beta_2^2\right)\right]^{\frac{1}{2}}$$

$$= \left[33.512 \times 10^{12} + b_{lm}\left(1.0384 \times 10^{12}\right)\right]^{\frac{1}{2}} \text{ m}^{-1}$$

From Fig. 9.10, we get approximately $b_{01} = 0.62$ and $b_{11} = 0.18$. Hence,

$$\beta_{01} = \left[33.512 + 0.62(1.0384)\right]^{\frac{1}{2}} \times 10^6 \text{ m}^{-1}$$

$$= 5.8443 \times 10^6 \text{ m}^{-1}$$

and $$\beta_{11} = \left[33.512 + 0.18(1.0384)\right]^{\frac{1}{2}} \times 10^6 \text{ m}^{-1}$$

$$= 5.8051 \times 10^6 \text{ m}^{-1}$$

(*iv*) Phase velocity $$v_p = \frac{\omega}{\beta} = \frac{2\pi c}{\lambda\beta}$$

Thus, $$(v_p)_{01} = \frac{2\pi \times 3 \times 10^8}{1.55 \times 10^{-6} \times 5.8443 \times 10^6} = 2.08 \times 10^8 \text{ ms}^{-1}$$

and $$(v_p)_{02} = \frac{2\pi \times 3 \times 10^8}{1.55 \times 10^{-6} \times 5.8051 \times 10^6} = 2.09 \times 10^8 \text{ ms}^{-1}$$

9.4.5 Mode Coupling

Though all the discussions so far are based on perfect dielectric waveguides, but in practice, waveguides are never perfect. There may be various perturbations like deviations of the fiber axis from straightness, irregularities at the core-cladding interface, variations in the core diameter and refractive index variations. These may change the propagation characteristics of the fiber. These in effect cause the energy of one traveling mode to be coupled to another mode depending on the specific perturbation. Hence, individual modes do not normally propagate throughout the length of the fiber without large energy transfers to adjacent modes, even if the fiber is exceptionally good quality and is not strained or bent by its surroundings. This mode conversion is known as mode coupling or mixing. But this treatment is out of the scope of this book. Interested readers may go through the book titled *Theory of Dielectric Optical Waveguides* by D.Marcuse.

9.5 PROPAGATION OF RAYS IN STEP-INDEX FIBERS

Optical fibers considered in the preceding sections with a core of constant refractive index n_1 and a cladding of slightly lower refractive index n_2 is known as step index fiber. This is because the refractive index profile for this fiber forms a step at the core-cladding interface as shown in the Fig. 9.11. The refractive index profile may be defined as:

$$\begin{aligned} n(r) &= n_1 \quad r < a \text{ (core)} \\ &= n_2 \quad r \geq a \text{ (cladding)} \end{aligned} \tag{9.68}$$

Figure 9.11(*a*) shows a single-mode step index fiber which allows the propagation of only one transverse electromagnetic mode (typically HE_{11}). The core diameter is of the order of 2 to 10 μm. The propagation of the single mode is illustrated in the figure as the axial ray. Figure 9.11(*b*) shows a multimode step index fiber with a core diameter of around 50μm or greater. This is large enough to allow the propagation of many modes within the core.

As single-mode fibers support only one mode, these type of fibers do not exhibit Intermodal dispersion, whereas with multimode step index fibers considerable dispersion may take place due to differing group velocities of the propagating modes. This will be discussed in detail in the next Chapter 10. Though multimode step index fibers suffer from dispersion, for lower bandwidth applications these fibers have definite advantages over single-mode step index fibers. To name a few:

(*a*) Larger numerical apertures, as well as core diameters facilitate easier coupling to optical sources;

(*b*) Spatially incoherent optical sources like LEDs may be efficiently coupled with multimode step index fibers, etc.

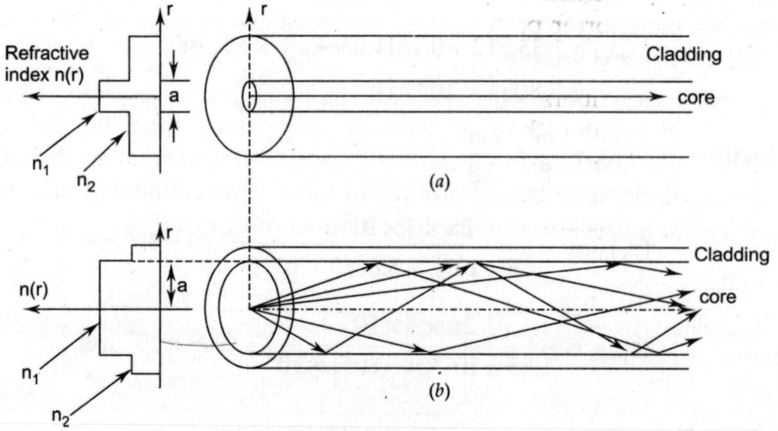

Fig. 9.11

The number of guided modes in multimode step index fibers depends upon the relative refractive index difference, core radius of the fiber and the wavelength of the transmitted light. These are all included in the normalized frequency V for the fiber.

9.6 PROPAGATION OF RAYS IN GRADED-INDEX FIBERS

The refractive index of the core varies constantly for a graded index fiber. The refractive index has a maximum value of n_1 on the fiber core axis and gradually decreases with radial distance towards the cladding and attains a constant value of n_2 in the cladding. This index variation is represented as:

$$n(r) = n_1\left[1 - 2\Delta\left(r/a\right)^{\alpha}\right]^{\frac{1}{2}} \qquad r < a \quad \text{(core)}$$

$$= n_1\left[1 - 2\Delta\right]^{\frac{1}{2}} = n_2 \qquad r \geq a \quad \text{(cladding)} \tag{9.69}$$

Here $\Delta = \left(n_1^2 - n_2^2\right)/2n_1^2 \approx \left(n_1 - n_2\right)/n_1$ is the relative refractive index difference and α is the profile parameter. The value of $\alpha = 2$ denotes a parabolic profile, while a triangular profile is being described by $\alpha = 1$. As the value of α increases, the profile tends towards the step-index profile. In fact a step-index profile is the result when $\alpha = \infty$. It has been seen that the best results for multimode optical propagation are obtained for a near parabolic refractive index profile core with $\alpha = 2$. In our discussions throughout this text, we consider the parabolic refractive index profile for graded index fibers, unless otherwise specified.

A multimode graded index fiber with a parabolic index profile core is shown in Fig. 9.12. It is seen that the meridional rays shown appear to follow curved paths through the fiber core. This may be described using the concepts of geometrical optics. The gradual decrease in refractive index from the centre of the core creates a large number of refractions of the rays as they are effectively incident on many high to low index interfaces.

Multimode graded index fibers exhibit much less intermodal dispersion than multimode step index fibers. Though many different modes are excited in the graded index fibers, the different group velocities of the modes tend to be normalized by the index grading. Considering ray theory, the rays traveling close to the fiber axis have shorter paths compared with the rays which travel in the outer region of the core of the fiber. However, the near axial rays are transmitted through a region of higher refractive index and hence travel with lower velocity than the extreme rays. This compensates for the shorter path lengths and reduces dispersion in the fiber. This will be discussed in detail in the next chapter.

Multimode graded index fibers with parabolic index profile cores exhibit transmission bandwidths which is much greater than multimode step index fiber bandwidths. Though the multimode graded index fibers cannot attain the bandwidths similar to the single-mode fibers, but these multimode fibers have the advantage of large core diameters of the order of 30μm or above along with bandwidths suitable for long haul communication.

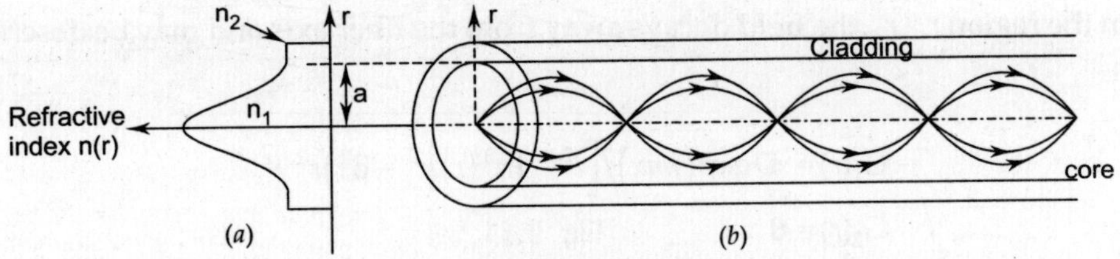

Fig. 9.12

The parameters like *numerical aperture, relative index difference, V number, etc,* defined for step index fibers may also be applied to graded index fibers. However, it is to be noted that for graded index fibers the situation is much more complicated as because the numerical aperture in this case is a function of the radial distance from the fiber axis. Hence, with the same relative index difference, graded index fibers accept less light than corresponding step index fibers.

In general, for a cylindrical optical fiber, the electric and magnetic fields are governed by equations (9.39) and (9.40). The exact solutions of these equations for graded-index fibers are difficult to obtain. However, using the WKB (Wentzel-Kramers-Brillouin) approximation method from quantum mechanics, it is possible to study the propagation characteristics of these fibers. Employing the WKB method modal solutions of the guided wave are achieved by expressing the field in the form:

$$E_x = \frac{1}{2}\left[G_1(r)\exp\left[jS(r)\right] + G_2(r)\exp\left[-jS(r)\right]\right]\begin{pmatrix}\cos l\phi \\ \sin l\phi\end{pmatrix}\exp(j\beta z) \quad (9.70)$$

Here G and S are assumed to be real functions of the radial distance r. It may be observed from the ray diagram shown in Fig. 9.12 that a light ray propagating in a graded index fiber does not necessarily reach every point within the fiber core. The ray is contained within two cylindrical caustic surfaces and for most of the rays a caustic does not coincide with the core-cladding interface. Thus the caustics define the turning points of the light ray within the graded fiber core. The turning points may be designed as occurring at $r = r_1$ and $r = r_2$.

In the region $r_1 < r < r_2$, the result of the WKB approximation yields an oscillatory field, where:

$$G_1(r) = G_2(r) = d\Big/\left[\left(n^2(r)k^2 - \beta^2\right)r^2 - l^2\right]^{\frac{1}{4}} \quad (9.71)$$

where D is an amplitude coefficient and

$$S(r) = \int_{r_1}^{r_2}\left[\left(n^2(r)k^2 - \beta^2\right)r^2 - l^2\right]^{\frac{1}{2}}\frac{dr}{r} - \frac{\pi}{4} \quad (9.72)$$

Outside the interval $r_1 < r < r_2$ the field solution have an evanescent form. In the region inside the inner caustic defined by $r < r_1$ and assuming r_1 is not too close to $r = 0$, the field decays towards the fiber axis giving:

$$G_1(r) = D\exp(jmx)\Big/\left[l^2 - \left(n^2(r)k^2 - \beta^2\right)r^2\right]^{\frac{1}{4}} \quad (9.73)$$

$$G_2(r) = 0 \quad (9.74)$$

Here the integer m is the radial mode number and

$$S(r) = j\int_{r}^{r_1}\left[l^2 - \left(n^2(r)k^2 - \beta^2\right)r^2\right]^{\frac{1}{2}}\frac{dr}{r} \quad (9.75)$$

Again in the region $r > r_2$, the field decays away from the fiber axis and may be described by the equations:

$$G_1(r) = D\exp(jmx)\Big/\left[l^2 - \left(n^2(r)k^2 - \beta^2\right)r^2\right]^{\frac{1}{4}} \quad (9.76)$$

$$G_2(r) = 0 \quad (9.77)$$

$$S(r) = j\int_{r_2}^{r}\left[l^2 - \left(n^2(r)k^2 - \beta^2\right)r^2\right]^{\frac{1}{2}}\frac{dr}{r} \quad (9.78)$$

The WKB method does not initially provide valid solutions of the wave equation in the vicinity of the turning points. However, D.Marcuse in his book titled *Light Transmission Optics* used Hankel functions of the first and second kind of order $\frac{1}{3}$ to obtain the solutions at the turning points. This helps in joining together of the two separate solutions described previously for inside and outside the interval $r_1 < r < r_2$. Hence, an approximate eigenvalue equation for the propagation constant β of the guided modes may be obtained using WKB method. The WKB eigenvalue equation of which β is a solution is given by:

$$\int_{r_1}^{r_2}\left[\left(n^2(r)k^2-\beta^2\right)r^2-l^2\right]^{\frac{1}{2}}\frac{dr}{r}=(2m-1)\frac{\pi}{2} \tag{9.79}$$

Here the radial mode number $m = 1, 2, 3, 4 \ldots$ and determines the number of maxima of the oscillatory field in the radial direction. In most of the cases the eigenvalue equation is solved either approximately or by using numerical techniques.

The amplitude coefficient D may be expressed in terms of the total optical power P_G within the guided mode. After D.Marcuse, D.Gloge and E.A.J.Marcatili the D may be written as:

$$D=\frac{4(\mu_0/\varepsilon_0)^{\frac{1}{2}}P_G^{\frac{1}{2}}}{n_1\pi a^2 I} \tag{9.80}$$

where,

$$I=\int_{r_1/a}^{r_2/a}\frac{x\,dx}{\left[\left(n^2(ax)k^2-\beta^2\right)a^2x^2-l^2\right]^{\frac{1}{2}}} \tag{9.81}$$

The WKB method may be used to calculate the propagation constants for the modes in a parabolic refractive index profile core where, from equation (9.69) we have:

$$n^2(r)=n_1^2\left(1-2\left(\frac{r}{a}\right)^2\Delta\right) \quad \text{for } r<a \tag{9.82}$$

Substituting equation (9.82) into equation (9.79), we get:

$$\int_{r_1}^{r_2}\left[n_1^2k^2-\beta^2-2n_1^2k^2\left(\frac{r}{a}\right)^2\Delta-\frac{l^2}{r^2}\right]^{\frac{1}{2}}\frac{dr}{r}=\left(m+\frac{1}{2}\right)\pi \tag{9.83}$$

The integral shown in equation (9.83) may be evaluated using a change of variable from r to $u=r^2$. As the square root term in the resulting expression goes to zero at $r=r_1$ and $r=r_2$, then we may write:

$$\left[\frac{a\left(n_1k^2-\beta^2\right)}{4n_1k\sqrt{2\Delta}}-\frac{l}{2}\right]\pi=\left(m+\frac{1}{2}\right)\pi \tag{9.84}$$

Solving equation (9.84) for β^2 gives:

$$\beta^2=n_1^2k^2\left[1-\frac{2\sqrt{2\Delta}}{n_1ka}(2m+l+1)\right] \tag{9.85}$$

Equation (9.85) allows us to consider the mode number plane spanned by the radial and azimuthal mode numbers m and l. Such a plane is illustrated in Fig. 9.13, where each mode of the fiber described by a pair of mode numbers is represented as a point in the plane. The mode number plane contains all types of modes, *e.g.*, the guided, leaky and radiation modes. The mode boundary separates the guided modes from the leaky and radiation modes. All the points in the mode number plane which are lying below the mode boundary line represent the guided modes, whereas the region above the line is occupied by leaky and radiation modes. Hence, we may count the total number of guided modes within the fiber.

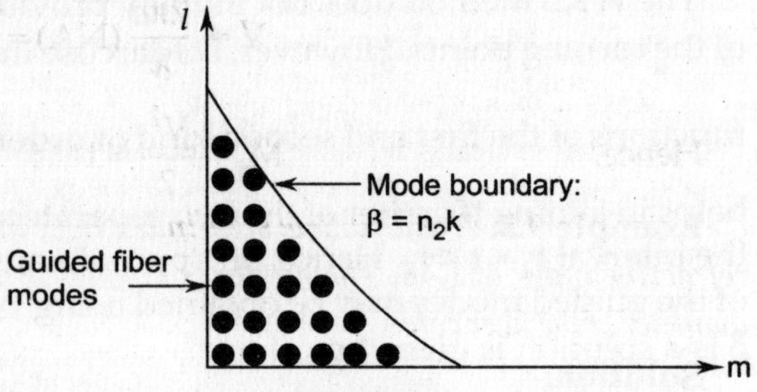

Fig. 9.13

For every pair of mode numbers m and l the corresponding mode field can have azimuthal mode dependence cos $l\phi$ or sin $l\phi$ and can exist in two possible polarizations. Thus the modes may be said to be fourfold degenerate. If we define the mode boundary as the function $m = f(l)$, then the total number of guided modes M will be given by

$$M = 4\int_0^{l_{max}} f(l)\,dl \tag{9.86}$$

Equation (9.86) helps to calculate the total number of guided modes M_g supported by the graded index fiber. This may be given as:

$$M_g = k^2 n_1^2 a^2 \Delta\left(\frac{\alpha}{\alpha+2}\right) \tag{9.87}$$

Again normalized frequency V for the fiber is given by equation (9.56) as:

$$V = kn_1 a(2\Delta)^{1/2} \tag{9.88}$$

Hence, substituting equation (9.88) into equation (9.87), we have:

$$M_g \approx \left(\frac{\alpha}{\alpha+2}\right)\left(\frac{V^2}{2}\right) \tag{9.89}$$

Equation (9.90) gives the approximate modal volume or the number of modes guided by the α-profile graded index fiber. For a step index fiber, where $\alpha = \infty$, $n = n_1$ in the core and $n = n_2$ in the cladding. Hence, for a step index fiber, the modal volume M_s is given by:

$$M_s \approx \frac{V^2}{2} \tag{9.90}$$

For a parabolic refractive index profile core fiber ($\alpha = 2$), we have the mode volume as:

$$M_s \approx \frac{V^2}{4} \tag{9.91}$$

Example 9.4. *A multimode step-index fiber has a numerical aperture of 0.18 and a core diameter of 80μm. Estimate the normalized frequency parameter of the fiber when light of wavelength 0.85μm is transmitted through it. Also calculate the number of guided modes propagating in the fiber.*

Solution:

Normalized frequency is given by:

$$V = \frac{2\pi a}{\lambda}(\text{NA}) = \frac{\pi \times 80 \times 0.18}{0.85} = 53.22$$

Hence, $$M_s \approx \frac{V^2}{2} = 1416$$

Example 9.5. *A multimode step-index fiber has a relative refractive index difference of 1.5% and a core refractive index of 1.48. The number of modes propagating at a wavelength of 1.5 μm is 1000. Find out the diameter of the fiber core.*

Solution:

We know $$M_s \approx \frac{V^2}{2} \approx \frac{1}{2}\left[\frac{2\pi a}{\lambda} n_1 \sqrt{2\Delta}\right]^2$$

Thus, $$2a = \frac{\lambda}{\pi n_1}\left(\frac{M_s}{\Delta}\right)^{1/2} = \frac{1.5}{\pi \times 1.48}\left(\frac{1000}{0.015}\right)^{1/2} \approx 82.3\ \mu\text{m}$$

Example 9.6. *A multimode graded index fiber has a core with a parabolic refractive index profile with a diameter of 60 μm. The fiber has a numerical aperture of 0.25. Estimate the total number of guided modes propagating in the fiber when it is operating at a wavelength of 1 μm.*

Solution:

We have $$V = \frac{2\pi a}{\lambda}(\text{NA}) = \frac{\pi \times 60 \times 0.25}{1} = 47.1$$

Hence, the mode volume is given by:

$$M_g \approx \frac{V^2}{4} = 555$$

Hence, the fiber supports approximately 555 guided modes.

Example 9.7. *A multimode graded index fiber with a parabolic profile supports the propagation of 800 guided modes. The fiber has a relative refractive index difference of 1.5%, a core refractive index of 1.48 and a core diameter of 80μm. Calculate the wavelength of light propagating in the fiber.*

Solution:

We know $$M_g \approx \frac{V^2}{4}$$

Hence, $$V \approx 2\sqrt{M_g} = 2\sqrt{800} = 56.57$$

Thus, $$\lambda = \frac{2\pi a}{V} n_1 \sqrt{2\Delta} = \frac{\pi \times 80 \times 1.48}{56.57}\sqrt{2 \times 0.015} \approx 1.1\ \mu\text{m}$$

9.7 SINGLE-MODE FIBERS

For long haul optical fiber communication, the single mode fibers have emerged as the mostly preferred medium. The most important cause for this is the very large bandwidth exhibited by this type of fibers. Moreover, the advantage of using a single mode optical fiber is that the signal dispersion caused by the delay differences, between different modes in a multimode fiber may be avoided. Due to mode coupling between different modes at both input mismatches and fiber imperfections within a multimode fiber, it is not possible to use a multimode fiber as a single mode fiber. Thus for the transmission of a single mode the fiber must be designed to allow propagation of only one mode, while all other modes will be attenuated by leakage or absorption.

In fact an optical fiber that supports only the fundamental LP_{01} or HE_{11} mode is called a single-mode fiber. This type of fiber is designed such that all higher order modes are cut off at the operating wavelength. From the preceding discussions, we know that the number of modes supported by a fiber is governed by the *V* parameter and the cut off condition of various modes is also determined by V value. The cut-off normalized frequency for the LP_{11} mode in a step-index fiber is 2.405. Hence, single-mode operation in a step-index fiber is possible over the normalized frequency range $0 < V < 2.405$, as there is no cut-off for the fundamental LP_{01} mode. An important fact to mention here is that there are in fact two modes with orthogonal polarization over this range, and the term single-mode applies to propagation of light of a particular polarization. It is noted that the normalized frequency for the fiber may be adjusted to within the range by reduction of the core radius, and the relative refractive index difference following equation (9.56) which, for single-mode fibers is usually less than 1%. However, both these factors create practical difficulties with single-mode fibers. With small core diameters, it becomes difficult to launch light into the fiber and also to join the fibers. Moreover, it is very difficult to fabricate a single-mode fiber with reduced relative refractive index difference.

Alternatively higher values of cut-off normalized frequency and hence core diameter may be obtained by using graded index fibers. It may be shown that the cut-off value of normalized frequency V_c to support a single mode in a graded index fiber is given by:

$$V_c = 2.405\left(1+\frac{2}{\alpha}\right)^{1/2} \tag{9.92}$$

Hence, for a parabolic core refractive index profile (*i.e.*, $\alpha = 2$), V_c will increase by a factor of $\sqrt{2}$, whereas for a triangular core refractive index profile with $\alpha = 1$, the value of V_c will increase by a factor of $\sqrt{3}$.

Moreover, there is another problem associated with single-mode fibers with low relative index differences and low V values is that the modal field of the LP_{01} mode extends appreciably into the cladding. It is seen that for V-values less than 1.4, more than half the modal power propagates through the cladding. Hence, the evanescent field, which should decay exponentially, may extend considerably in the cladding. In order that the power propagating in the cladding should not be lost, the cladding should have a suitable thickness (of the order of 50μm and above). Further, the absorption and scattering losses in the cladding should also be less.

Alternatively V value for a single-mode fiber may also be increased above 2.405 in **W** fiber. The refractive index profile for such a fiber is shown in Fig. 9.14. Here two cladding regions exist. By using of such two step cladding, the loss threshold between the desirable and undesirable modes may be increased substantially. The fundamental mode will be fully supported with small cladding loss when its propagation constant lies in the range $kn_3 < \beta < kn_1$.

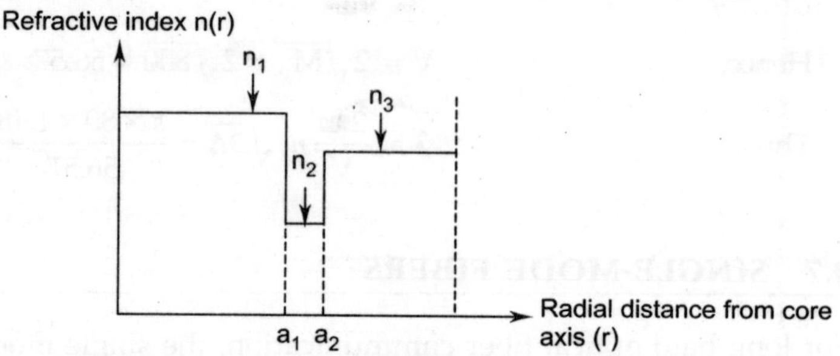

Fig. 9.14

If the undesirable higher order modes have values of propagation constant $\beta < kn_3$, they will leak through the barrier layer between a_1 and a_2 (Fig. 9.14) into the outer cladding region n_3. These modes will lose power by radiating into the lossy surroundings. **W** fibers can provide larger core diameters compared to conventional single cladding approach. Thus these fibers ease out the jointing difficulties. Figure 9.14 shows the refractive index profile for a single-mode **W** fiber.

Most commonly used single-mode fibers employ a step index profile design and are dispersion optimized for operating in the 1.3 μm wavelength region. These fibers are either a Matched Cladding (MC) or a Depressed-Cladding (DC) design as shown in Fig. 9.15. In the conventional MC fibers, the region external to the core has a constant uniform refractive index slightly lower than the core region, typically consisting of pure silica. Alternatively, if the core region is made up of pure silica then the lower index cladding may be obtained by fluorine doping. Another MC fiber design employs a segmented core as illustrated in Fig. 9.15(*b*). Such a structure provides conventional single-mode dispersion optimized performance at wavelengths around 1.3 μm, but exhibits multimode operation with two to three modes in the shorter wavelength region around 0.8 μm. In the DC fibers shown in Fig. 9.15 the cladding region immediately adjacent to the core is of lower refractive index than that of an outer cladding region. Figures 9.15(*c*) and 9.15(*d*) show depressed-cladding design and profile specifications of a depressed-cladding fiber respectively.

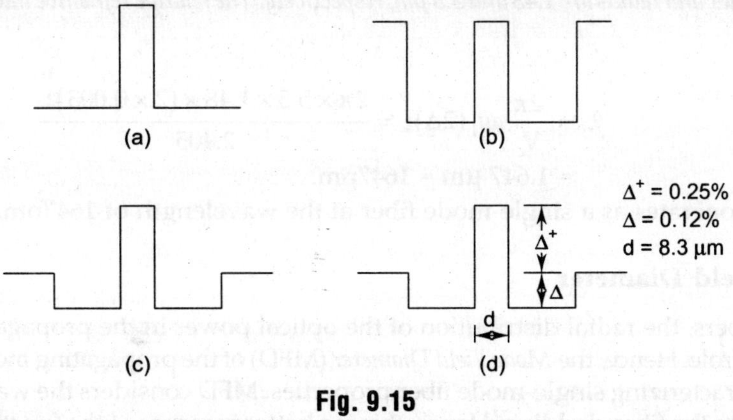

Fig. 9.15

Example 9.8. *Consider a fiber with core refractive index 1.50 and relative refractive index difference of 1.48%. Estimate the maximum core radius for the optical fiber in order that it may be suitable for single-mode operation. The fiber is operating at a wavelength of 1.5 μm. Further, calculate the new maximum core radius for single-mode operation when the relative refractive index difference is reduced by a factor of 10.*

Solution:

The maximum V value for a fiber which gives single-mode operation is 2.495. Thus the core radius a is given as:

$$a = \frac{V\lambda}{2\pi n_1 (2\Delta)^{1/2}} = \frac{2.405 \times 1.5}{2\pi \times 1.5 \times (2 \times 0.0148)^{1/2}} = 2.2 \text{ μm}.$$

Reducing the relative refractive index difference by a factor of 10 we have:

$$a = \frac{2.405 \times 1.5}{2\pi \times 1.5 \times (2 \times 0.00148)^{1/2}} = 7.0 \text{ μm}.$$

9.8 SINGLE-MODE FIBERS CHARACTERISTIC PARAMETERS

9.8.1 Cutoff Wavelength

By rearrangement of equation (9.56), it is noted that single-mode operation may be obtained above a theoretical cutoff wavelength λ_c given by:

$$\lambda_c = \frac{2\pi}{V_c} a n_1 (2\Delta)^{\frac{1}{2}} \tag{9.93}$$

where V_c is the cutoff normalized frequency. Thus a particular fiber becomes single-mode fiber above the wavelength given by λ_c. Dividing equation (9.93) by equation (9.56) for the same fiber we obtain the inverse relationship:

$$\frac{\lambda_c}{\lambda} = \frac{V}{V_c} \tag{9.94}$$

For step index fiber the value of $V_c = 2.405$. Hence, the cutoff wavelength is given by:

$$\lambda_c = \frac{V\lambda}{2.405} \tag{9.95}$$

Example 9.9. *Calculate the cutoff wavelength for a step index fiber to exhibit single-mode operation when the core refractive index and radius are 1.48 and 5.5 μm, respectively. The relative refractive index difference is 0.3%.*

Solution:

We know:
$$\lambda_c = \frac{2\pi}{V_c} a n_1 (2\Delta)^{\frac{1}{2}} = \frac{2\pi \times 5.5 \times 1.48 \times (2 \times 0.003)^{\frac{1}{2}}}{2.405}$$
$$= 1.647\ \mu\text{m} = 1647\text{nm}$$

Thus the fiber operates as a single-mode fiber at the wavelength of 1647nm.

9.8.2 Mode Field Diameter

In single-mode fibers, the radial distribution of the optical power in the propagating fundamental mode plays a vital role. Hence, the *Mode Field Diameter* (MFD) of the propagating mode is an important parameter for characterizing single-mode fiber properties. MFD considers the wavelength dependent penetration into the fiber cladding. Hence, this is a better measure of the functional properties of the single-mode fibers than the core diameter. This is also known as *mode spot size*.

For step-index and parabolic profile graded index single-mode fibers operating near the cut-off wavelength λ_c, the field distribution of the fundamental mode may be approximated by a Gaussian distribution as:

$$\psi(r) = \psi_0 \exp(-r^2/w^2) \tag{9.96}$$

where $\psi(r)$ is the electric or magnetic field at the radius r, ψ_0 is the axial field (at $r = 0$), and w is the mode field radius, which is the radial distance from the axis where ψ_0 drops to ψ_0/e. Since the power is proportional to ψ^2, the MFD may also be defined as the radial distance between the $1/e^2 = 0.135$ power points (in the power versus radius graph shown in Fig. 9.16). Another very important parameter which is directly related to the MFD of a single-mode fiber is the spot size (or mode field radius) w. Hence, the MFD is $2w$, where w is the normal half width of the input excitation as illustrated in Fig. 9.16.

If the power per unit area at a radial distance r in the fundamental mode is $P(r)$, the power in the annular ring between r and $(r + dr)$ will be $2\pi r P(r) dr$. Hence the MFD may be defined as:

$$\text{MFD} = 2w = 2\left[\frac{\int_0^\infty 2\pi r^2 P(r)\,dr}{\int_0^\infty 2\pi r P(r)\,dr}\right]^{\frac{1}{2}} = 2\left[\frac{\int_0^\infty r^2 \psi^2(r)\,dr}{\int_0^\infty r \psi^2(r)\,dr}\right]^{\frac{1}{2}} \tag{9.97}$$

Here, $\psi(r)$ is the field distribution in the fundamental mode.

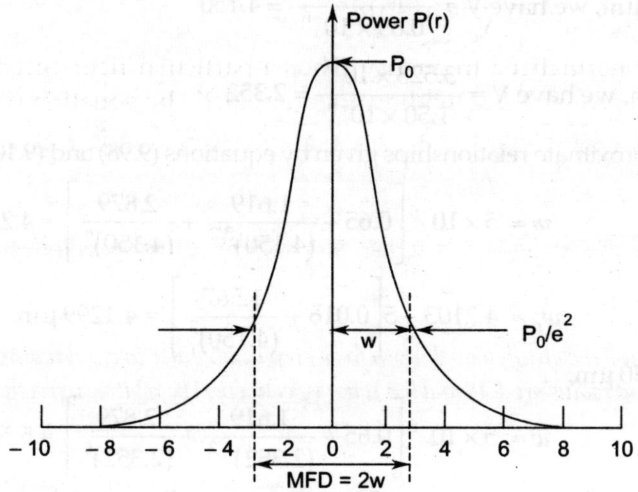

Fig. 9.16

For a step-index fiber, the MFD w may be approximately given after Marcuse as:

$$w \approx a\left[0.65 + \frac{1.619}{V^{3/2}} + \frac{2.879}{V^6}\right] \tag{9.98}$$

where a is the core radius. For V-values in the range 0.8–2.5, the above formula gives within about 1% of w value. Often, for practical fibers and those with arbitrary refractive index profiles the radial field distribution is not strictly Gaussian. Hence, alternative techniques have been proposed for such purpose. Though different definitions of MFD for non-Gaussian field distributions have been proposed, the general definition recommended by CCITT is called Petermann II definition. This definition is based on the second moment of the far field and is defined by Petermann in 1983 as:

$$(\text{MFD})_p = 2w_p = 2\left[\frac{2\int_0^\infty r^2\psi^2(r)\,dr}{\int_0^\infty r\left(d\psi/dr\right)^2 dr}\right]^{\frac{1}{2}} \tag{9.99}$$

Here w_p is called the Petermann II mode radius. The empirical relation for w_p which is accurate to within 1% for a step-index fiber with V-values between 1.5 and 2.5 is given as:

$$w_p \approx w - a\left[0.016 + \frac{1.567}{V^7}\right] \tag{9.100}$$

Here w is the mode field radius.

Example 9.10. *A step-index fiber has a core refractive index of 1.45, a core diameter of 10 μm, and a relative refractive index difference of 0.003. Calculate w and w_p at $\lambda = 0.85\ \mu m$ and $1.50\ \mu m$.*

Solution:

We know:

$$V = \frac{2\pi a}{\lambda} n_1\sqrt{2\Delta} = \frac{2\pi \times 5\times 10^{-6} \times 1.45\sqrt{2\times 0.003}}{\lambda(m)} = \frac{3.528\times 10^{-6}}{\lambda(m)}$$

Hence, for $\lambda = 0.85\ \mu m$, we have $V = \dfrac{3.528 \times 10^{-6}}{0.85 \times 10^{-6}} = 4.150$

And for $\lambda = 1.50\ \mu m$, we have $V = \dfrac{3.528 \times 10^{-6}}{1.50 \times 10^{-6}} = 2.352$

Now we use the approximate relationships given by equations (9.98) and (9.100) at $\lambda = 0.85\ \mu m$, to get

$$w \approx 5 \times 10^{-6} \left[0.65 + \frac{1.619}{(4.150)^{3/2}} + \frac{2.879}{(4.150)^{6}} \right] = 4.2103\ \mu m$$

and

$$w_p \approx 4.2103 - 5 \left[0.016 + \frac{1.567}{(4.150)^{7}} \right] = 4.1299\ \mu m$$

Similarly, at $\lambda = 1.50\ \mu m$,

$$w \approx 5 \times 10^{-6} \left[0.65 + \frac{1.619}{(2.352)^{3/2}} + \frac{2.879}{(2.352)^{6}} \right] = 5.5792\ \mu m$$

and

$$w_p \approx 5.5792 - 5 \left[0.016 + \frac{1.567}{(2.352)^{7}} \right] = 5.4795\ \mu m.$$

9.8.3 Birefringence in Fiber

We have already discussed in Section 9.6 that a single-mode fiber actually supports two orthogonal linearly polarized modes that are degenerate. For a single-mode fiber of a perfectly cylindrical core of uniform diameter, these modes have the same mode refractive index n and also the same propagation constant β. Thus they have the same traveling velocity. But practical fibers do not have perfectly cylindrical core. Due to non-uniformity along the length of the fiber, they exhibit variation in the diameter. In effect, the propagation constants of the two polarization components become different and hence the fiber exhibit birefringence. Assuming arbitrarily that two modes are polarized along the x- and y-directions, modal birefringence may be defined as:

$$\delta n = |n_x - n_y| \tag{9.101}$$

Here, n_x and n_y are the mode refractive indices for the two polarization components. The corresponding change in the propagation constants may be given by:

$$\delta\beta = \frac{2\pi}{\lambda}\,\delta n \tag{9.102}$$

Here λ is the wavelength of light in vacuum.

The length, L_b of the fiber over which the polarization rotates through an angle of 2π radians is called the *beat length* of the fiber. It is given as:

$$L_b = \frac{2\pi}{\delta\beta} \tag{9.103}$$

In conventional single-mode fibers, birefringence changes over the length of the fiber randomly in both magnitude as well as direction. Hence, linearly polarized light launched in such a fiber quickly attains an arbitrary state of polarization.

Example 9.11. *A beat length of 12 cm is observed in a single-mode fiber when light of wavelength 1.3 μm is launched into it. Calculate (i) the difference between the propagation constants of the two orthogonally polarized modes, and (ii) the modal birefringence.*

Solution:

(*i*) We know:
$$L_b = \frac{2\pi}{\delta\beta}$$

or,
$$\delta\beta = \frac{2\pi}{L_b} = \frac{2\pi}{12 \times 10^{-2}} = 52.36\ \text{m}^{-1}$$

(*ii*) Modal birefringence is:
$$\delta\beta = \frac{2\pi}{\lambda}\delta n$$

or,
$$\delta n = \frac{\lambda}{2\pi}\delta\beta = 52.36 \times \frac{1.3 \times 10^{-6}}{2\pi} = 13.333 \times 10^{-6}$$

9.8.4 Effective Refractive Index

The effective refractive index, n_{eff} is defined by the ratio of the propagation constant of the fundamental mode to that of the vacuum propagation constant as:

$$n_{eff} = \frac{\beta}{k} \tag{9.104}$$

This is sometimes referred to as a phase index or normalized phase change or normalized phase change coefficient. The rate of change of phase of the fundamental LP_{01} mode propagating along a straight fiber is determined by the phase propagation constant β. This is directly related to the wavelength of the fundamental mode λ_{01} by the factor 2π. We have:

$$\beta\lambda_{01} = 2\pi, \text{ or } \lambda_{01} = \frac{2\pi}{\beta} \tag{9.105}$$

Thus, the wavelength of the fundamental mode is smaller than the vacuum wavelength λ by the factor $1/n_{eff}$ where:

$$\lambda_{01} = \frac{\lambda}{n_{eff}} \tag{9.106}$$

Since the fundamental mode propagates in a medium with a refractive index $n(r)$ dependent on the distance r from the fiber axis, the effective refractive index may be considered as an average over the refractive index of the medium.

At long wavelengths for normally clad fibers, the mode-field diameter is large compared to the core diameter and thus the electric field extends far into the cladding region. In this case the propagation constant β will be approximately equal to n_2k and the effective index will be similar to the refractive index of the cladding n_2. On the contrary, at short wavelengths, the field is concentrated in the core region and the propagation constant β is approximately equal to the maximum wave-number n_1k. Hence, the propagation constant in a single-mode fiber varies over the interval $n_2k < \beta < n_1k$. Thus, the effective refractive index will vary over the range $n_2 < n_{eff} < n_1$.

A relationship between the effective refractive index and the normalized propagation constant b defined in equation (9.67) may be written as:

$$b = \frac{(\beta/k)^2 - n_2^2}{n_1^2 - n_2^2} = \frac{\beta^2 - n_2^2k^2}{n_1^2k^2 - n_2^2k^2} \tag{9.107}$$

Since, $\beta \approx n_1k$, we may write from equation (9.107)

$$b \approx \frac{\beta/k - n_2}{n_1 - n_2} \tag{9.108}$$

Substituting the value of n_{eff} from equation (9.104) in equation (9.108), we have:

$$b \approx \frac{n_{eff} - n_2}{n_1 - n_2} \tag{9.109}$$

9.9 CONCLUSION

In this chapter, we have discussed on the basic laws of optics like, Snell's laws of refraction, etc. We have also discussed in detail about the ray theory approach of propagation of light and defined the meridional and skew rays. The different parameters like numerical aperture, acceptance angle of fibers have also been discussed. We have also studied the electromagnetic mode theory of optical wave propagation. The Maxwell's equations have been discussed and the equation of the optical wave has been derived. Wave propagation in planar optical waveguide and cylindrical waveguide has been discussed. The phase velocity and group velocity has also been dealt with. Propagation of optical waves in step-index and graded index fibers has been shown. Details about single-mode fibers and their characteristics have also been discussed. In effect we have covered almost all the aspects of optical waveguides and fibers required for the students to make their base in optical communication. In the next chapter we will discuss in detail about the characteristics of optical fibers.

REVIEW QUESTIONS

1. A ray of light is passing from a material of refractive index 1.48 to another material of refractive index 1.46. What is the range of angles (measured with respect to the normal to the interface) for which the ray will suffer total internal reflection?
   (*a*) 180° – 360° (*b*) 0° – 70° (*c*) 81° – 90° (*d*) 110° – 115°
2. A step-index fiber has a core with a refractive index of 1.50 and the refractive index of the cladding is 1.48. The value of the numerical aperture is
   (*a*) 0.244 (*b*) 0.346 (*c*) 0.415 (*d*) 0.178
3. Within the core of a step-index fiber light is guided by
   (*a*) Change in the speed of light within the core
   (*b*) Total internal reflection at the outer surface of the cladding
   (*c*) Refraction at the core-cladding interface
   (*d*) Total internal reflection at the core-cladding interface
4. A step-index fiber has a core refractive index 1.50 and a cladding refractive index of 1.48. The core diameter is 90 μm. How many guided modes are supported by the fiber if the wavelength of light is 0.85 μm?
   (*a*) 2040 (*b*) 3298 (*c*) 4105 (*d*) 6596
5. A graded-index fiber has a triangular profile with $n_0 = 1.48$ and $\Delta = 0.02$. If it is excited by a source of $\lambda = 1.0$ μm, what is the range of phase propagation constants for the modes constants for the modes supported by the fiber?
   (*a*) 8.620–9.299 μm^{-1} (*b*) 2.160 – 3.982 μm^{-1}
   (*c*) 6.308–8.918 μm^{-1} (*d*) 4.166 – 9.982 μm^{-1}
6. An unclad fiber with a core refractive index of 1.46 and core diameter of 60 μm is placed in air. What is the normalized frequency for the fiber when light of wavelength 0.85 μm is transmitted through it?
   (*a*) 54.77 (*b*) 221.65 (*c*) 178.55 (*d*) 787.12
7. For symmetric TE modes supported by a symmetrical step index planar waveguide, the stable electric-field distribution inside the guide distribution inside the guide layer takes
   (*a*) An exponential form (*b*) A tangential form
   (*c*) A sinusoidal form (*d*) A cosinusoidal form

8. The maximum number of TE modes supported by a symmetrical SI planar waveguide is an integer greater than
   (*a*) V (*b*) $V/2\pi$ (*c*) $2V/\pi$ (*d*) None of these.
9. In a multimode SI fiber, the higher order modes propagate within the fiber with
   (*a*) Same phase velocity as that of lower order modes
   (*b*) Lower group velocity than the lower order modes
   (*c*) Higher group velocity than the lower order modes
   (*d*) None of the above
10. For any multimode optical fiber, what is the range of the normalized propagation parameter?
    (*a*) 1–5
    (*b*) 0–1
    (*c*) Cannot be calculated unless the profile parameter and wavelength of light is known
    (*d*) None of these.
11. What is relative refractive index difference for an optical fiber? Show how it may be related to the numerical aperture. Define the normalized frequency for an optical fiber and explain its use in the determination of the number of guided modes propagating within a step index fiber.
12. With the help of a suitable diagram briefly discuss what is meant by the acceptance angle for an optical fiber. Show how this is related to the fiber numerical aperture and the refractive indices for the fiber core and cladding.
13. Using simple ray theory, describe the mechanism for the transmission of light within an optical fiber. Briefly indicate with the aid of suitable diagrams the difference between meridional and skew ray paths in step index fibers.
14. Giving an expression for the possible refractive index profile, explain what is meant by a graded index optical fiber. Using simple ray theory concepts, discuss the transmission of light through the fiber. What are the major advantages of this type of fiber with regard to multimode propagation?
15. In relation to a planar optical waveguide, explain the concept of electromagnetic modes in relation to a planar optical waveguide. Discuss the modifications that may be made to electromagnetic mode theory in a planar waveguide in order to describe optical propagation in a cylindrical fiber.
16. An optical fiber has a numerical aperture of 0.20 and a cladding refractive index of 1.59. Estimate:
    (*a*) The acceptance angle for the fiber in a liquid which has a refractive index of 1.33;
    (*b*) The critical angle at the core-cladding interface.
17. A step index fiber with a large core diameter compared with the wavelength of the transmitted light has an acceptance angle in air of 22° and a relative refractive index difference of 3%. Calculate: (*a*) the numerical aperture; and (*b*) the critical angle at the core-cladding interface for the fiber.
18. A step index fiber in air has a numerical aperture of 0.16, a core refractive index of 1.45 and a core diameter of 60 μm. (*a*) Determine the normalized frequency for the fiber when light at a wavelength of 900 nm is transmitted; (*b*) Further, calculate the number of guided modes propagating in the fiber.
19. A graded index fiber with a parabolic index profile supports the propagation of 742 guided modes. The fiber has a numerical aperture of 0.3 in air and a core radius of 35 μm. (*a*) What is the wavelength of the light propagating in the fiber? (*b*) Also calculate the maximum radius of the fiber which gives single-mode operation at the same wavelength.
20. A single-mode step index fiber has a core diameter of 7 μm and a core refractive index of 1.49. Estimate the shortest wavelength of light which allows single-mode operation when the relative refractive index difference for the fiber is 1%.

10

Characteristics of Optical Fibers

10.1 INTRODUCTION

We have already discussed about the basic transmission mechanisms of various types of optical waveguides and fibers in Chapter 9. But we have not yet dealt with the factors influencing the performance of optical fibers in Chapter 9. However, without the knowledge of these characteristics in detail, it is almost impossible to design an optical fiber communication system. Two most important transmission characteristics are that of attenuation or loss and the dispersion in optical fibers, which is in fact related to the bandwidth of the optical fibers.

The concept of using optical waveguides as transmission medium was contemplated in the mid-1960s. Until that time, the idea of transmission through optical waveguides seemed somewhat ludicrous because of their very large optical attenuation. It was reported that such waveguides, which were operating at the 800 to 900 nm wavelength window, exhibited an average attenuation on the order of 1000 dB/km. However, careful investigation of the attenuation showed that it was largely due to absorption in the glass, caused by impurities such as iron, manganese, copper, and other transition metals which are generally found in the third row of the periodic table. Thus researches were carried out towards a new generation of *pure* glasses to be used in the optical communication.

The first major breakthrough came in 1970, when attenuation below 20 dB/km was reported for a fiber. Further improvements both in material and fabrication techniques, significantly improved fiber performance to attenuation levels of 0.2 dB/km. Thus, comparatively low loss fibers were incorporated into optical communication systems throughout the world. As the fundamental lower limits for attenuation in silicate glass fibers have virtually been achieved, to achieve lower attenuation, optical fibers other than those that have a silicon base were proposed and experimentally tested. Results showed that the fluoride glass fibers that operate at the 2500 nm (infrared) wavelength window, exhibited attenuation levels on the order of 0.01 dB/km. Such mid-infrared (and possibly far-infrared) transmitting fibers could eventually provide for extremely long-haul repeater-less communication, provided the optical source and detector requirements are also satisfactorily met.

The other very important characteristic of optical fiber is the dispersion, which is in turn related to the bandwidth of the fiber. Dispersion within the fiber determines the bit rate in the fiber. Thus, once the attenuation was reduced to an acceptable level, attention was devoted to the dispersion properties of the fiber. In this chapter, we discuss the attenuation and dispersion mechanisms within the fiber in detail. We will also discuss about the modal noise, polarization and nonlinear phenomena.

10.2 ATTENUATION

One of the most critical factors that determine optical fiber applicability in communication systems is signal attenuation, as it plays a major role in determining the maximum transmission distance between a transmitter and a receiver or an in-line amplifier. Though the fiber attenuation started at a very high value, it was reduced progressively to 5 dB/km. At this level, fiber attenuation was in parity with copper wire lines. However, today attenuation is further decreased to the order of 0.2 dB/km.

The basic attenuation mechanisms in a fiber are material absorption, material scattering (linear and nonlinear), and radiative losses (curve and microbending loss, mode coupling radiation loss and losses due to leaky modes) of the optical energy. These mechanisms are influenced by the material composition, the preparation and purification technique, and the waveguide structure. Also there are losses at connectors and splices. Optical attenuation is referred to as the progressive amplitude reduction of the light ray traveling through the fiber. This amplitude reduction or signal attenuation is expressed in terms of power loss in the logarithmic unit of the decibel. As the light travels along a fiber, its power decreases exponentially with distance. If P_i is the optical power at the input to the fiber, then the optical power, P_o at the output of the fiber of length L is given by

$$P_o = P_i \exp(-\alpha_p L) \tag{10.1}$$

where

$$\alpha_p = \frac{1}{L} \ln\left[\frac{P_i}{P_o}\right] \tag{10.2}$$

is the fiber attenuation coefficient given in units of, for example, km^{-1}.

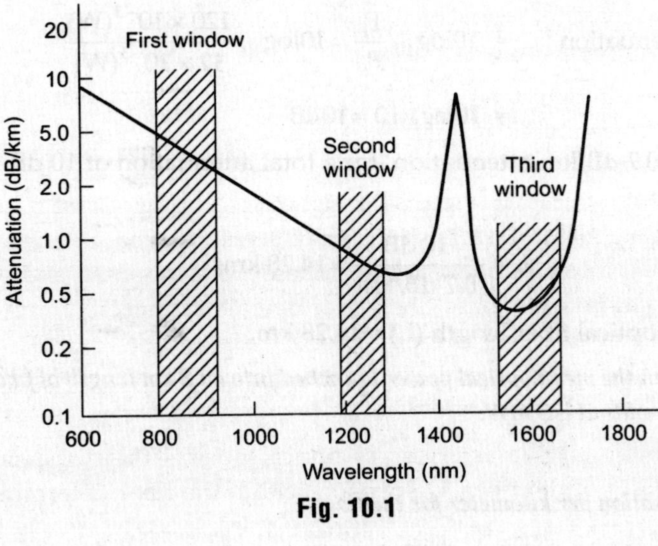

Fig. 10.1

For simplicity in calculating optical signal attenuation in a fiber, generally the attenuation coefficient is expressed in units of dB/km. Designating this parameter by α, we may write

$$\alpha(\text{dB/km}) = \frac{10}{L}\log_{10}\left[\frac{P_i}{P_o}\right] \tag{10.3}$$

This parameter is normally referred to as the *fiber attenuation* or the *fiber loss*. The parameter is a function of wavelength, which is shown in Fig. 10.1. The logarithmic unit has the advantage that the operations of multiplication and division reduce to addition and subtraction, while powers and roots reduce to multiplication and division. However, addition and subtraction require a conversion to numerical values which may be obtained by the relationship:

$$\frac{P_i}{P_o} = 10^{(\text{dB}/10)} \tag{10.4}$$

Example 10.1. *The optical power launched at the input of a 20 km fiber is 70 mW. If the optical power measured at the output of the fiber is 10 mW, find out the (i) total fiber attenuation; (ii) attenuation per kilometer of the fiber.*

Solution:

(*i*) The total fiber attenuation $= 10\log_{10}\frac{P_i}{P_o} = 10\log_{10}\frac{70\times10^{-3}(\text{W})}{10\times10^{-3}(\text{W})}$

$= 10\log_{10}7 = 8.45\,\text{dB}$

(*ii*) The total fiber length is 20 km, and the total attenuation has been found out to be 8.45 *d*B.

Hence, from equation 10.3,

$$\alpha(\text{dB/km}) = \frac{8.45}{20} = 0.42\ \text{dB/km}.$$

Example 10.2. *Estimate the maximum length of an optical fiber that exhibits 0.7 dB/km attenuation if the output power is 12 mW and the power launched at the input is 120 mW.*

Solution:

The total fiber attenuation $= 10\log_{10}\frac{P_i}{P_o} = 10\log_{10}\frac{120\times10^{-3}(\text{W})}{12\times10^{-3}(\text{W})}$

$= 10\log_{10}10 = 10\,\text{dB}$

The fiber exhibits 0.7 dB/km attenuation, for a total attenuation of 10 dB, the fiber length is as follows:

$$L = \frac{10\ \text{dB}}{0.7\ \text{dB/km}} = 14.28\ \text{km}$$

Hence, maximum optical fiber length (L) = 14.28 km.

Example 10.3. *When the mean optical power launched into an 8 km length of fiber is 150 mW, the mean optical power at the fiber output is 3 mW.*

Determine:

(*i*) *the signal attenuation per kilometer for the fiber;*

(ii) *the overall signal attenuation or loss in dB through 10 km of fiber assuming there are no connectors or splices;*

(iii) *the overall signal attenuation for a 10 km optical link using the same fiber with splices at 1 km intervals, each having an attenuation of 1 dB;*

(iv) *the numerical input/output power ratio in (iii).*

Solution:

(i) The overall signal attenuation in decibels through the 8 km fiber is:

$$\text{Signal attenuation} = 10\log_{10}\frac{P_i}{P_o} = 10\log_{10}\frac{150\times10^{-3}(\text{W})}{3\times10^{-3}(\text{W})}$$

$$= 10\log_{10} 50 = 16.99 \text{ dB}$$

Hence, using equation 10.3,

$$\alpha(\text{dB/km}) = \frac{16.99}{8} = 2.12 \text{ dB/km}.$$

(ii) As α(dB/km) = 2.12 dB/km, the loss incurred along 10 km of the fiber is given by

$$\alpha(\text{dB/km})\,L = 2.12\times10 = 21.2 \text{ dB}.$$

(iii) The total loss incurred along 10 km of the fiber is 21.2 dB.

According to the problem, the link has nine splices (at 1 km intervals), each with an attenuation of 1 dB. Hence, the loss due to the splices is 9 dB.

Thus the overall signal attenuation for the link is:

Signal attenuation = 21.2 + 9 = 30.2 dB.

(iv) Equation 10.4 may be used to obtain a numerical value for the input/output power ratio as follows:

$$\frac{P_i}{P_o} = 10^{(30.2/10)} = 1047.1.$$

Example 10.4. *A 40 km long optical fiber has an attenuation of 0.4 dB/km at 1550 nm. Calculate the optical output power P_{out} if the input power launched into the fiber is 240 μW.*

Solution:

Since, no other loss mechanism has been specified; we may consider the loss due to only attenuation of the fiber. The total loss for 40 km of the fiber is:

$$\text{Total loss} = 40 \text{ km}\times0.4 \text{ dB/km} = 16 \text{ dB}.$$

Hence, using equation 10.4, we may write:

$$P_o = P_i\times10^{-(16/10)} = 240\times10^{-6}\times10^{-1.6}$$

$$= 6.02\ \mu\text{W}.$$

10.3 ABSORPTION

Absorption is a loss mechanism related to the material composition and the fabrication process for the fiber. Absorption can be divided into three categories:

1. Absorption by atomic defects in the glass composition.
2. Intrinsic absorption by the basic constituent atoms of the fiber material.
3. Extrinsic absorption by impurity atoms in the fiber material.

10.3.1 Absorption by Atomic Defects

There lie imperfections in the atomic structure of the fiber material which create the atomic defects. Usually the absorption losses due to the atomic effects are negligible compared to the other two absorption effects - intrinsic and extrinsic absorption. However, if the fiber is exposed to ionizing radiation, this absorption might become significant. Radiation damages a material by changing its internal structure. The damage effects depend on the energy of the ionizing rays or particles, the fluence (*i.e.*, particles per square centimeter), and the radiation flux. The total dose a material receives is expressed in units of *rad(Si)*, which is a measure of the radiation absorbed in bulk silicon. Basically, fibers when exposed to radiation shows increase in absorption due to the creation of atomic defects, or attenuation centers that absorb the optical energy. However, these attenuation centers will relax or anneal out with time.

10.3.2 Intrinsic Absorption

Intrinsic absorption is the result of the interaction of free electrons and the operating wavelength within the basic fiber material (*e.g.*, pure SiO_2). This is the principal physical factor which defines the transparency window of a material over a specified spectral region. Basically intrinsic absorption occurs when the material is in a perfect state with no impurities, density variations and so on. Hence, intrinsic absorption sets the lower limit on absorption for any particular material. The propagating light wave covers a wide wavelength spectrum staring from the ultraviolet to the infrared region. Thus, any particular wavelength within the optical spectrum interacts differently with the atoms of the fiber material.

For example, the ultraviolet region interacts with the outer shell electrons in the atoms, while the infrared region interacts with the lattice structure in the atoms. Absorption occurs when a photon interacts with an electron in the valence band and excites it to a higher energy level. The absorption observed for both amorphous and crystalline materials at the ultraviolet region may be expressed by the empirical relationship

$$\alpha_{uv} = Ce^{\frac{\lambda_{uv}}{\lambda}} \tag{10.5}$$

where α_{uv} is the absorption at the ultraviolet region, C is the constant (1.108×10^{-3} dB/km), λ_{uv} is 4.582 μm, and λ is the wavelength in the *uv* region (160 nm – 400 nm). The magnitude and characteristic exponential decay of the ultraviolet absorption are illustrated in Fig. 10.2. The ultraviolet loss is small compared with scattering loss in the near-infrared region. The absorption of the operating wavelength, due to its interaction with the material lattice structure is given by another empirical relationship given by:

$$\alpha_{(lt)} = \alpha_{(lo)}e^{\frac{\lambda_{(lt)}}{\lambda}} \tag{10.6}$$

where, $\alpha_{(lo)}$ is the constant (4×10^{-11} dB/km), $\alpha_{(lt)}$ is the absorption due to lattice structure (dB/km), $\lambda_{(lt)}$ is the constant reference wavelength (48 μm) and λ is the operating wavelength.

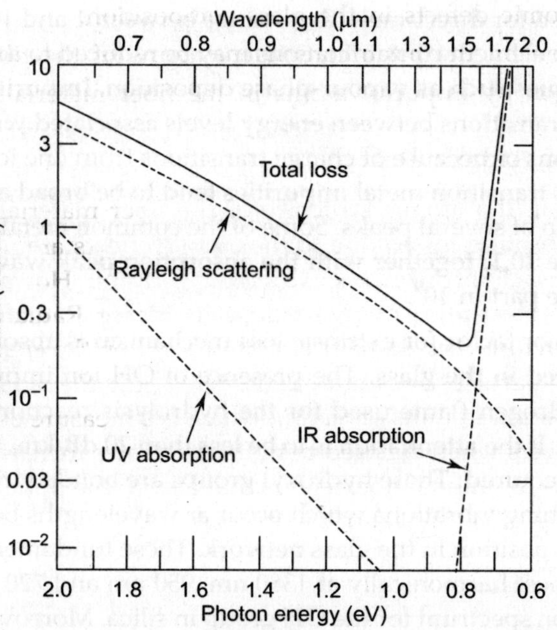

Fig. 10.2

In the near-infrared region above 1.2 μm, the optical waveguide loss is mainly determined by the presence of OH ions and the inherent infrared absorption of the constituent material. This inherent infrared absorption is again in turn associated with the characteristic vibration frequency of the particular chemical bond between the atoms of which the fiber is composed. This absorption is quite strong due to many bonds present in the fiber. An absolutely pure silicate glass has little intrinsic absorption loss in the near-infrared region due to its basic material structure. But, it has two major intrinsic absorption windows over the 0.8 to 1.7 μm wavelength range as optical fiber attenuation spectra in pure GeO_2-SiO_2 glass for the intrinsic loss mechanism is shown in Fig. 10.2. It may also be observed that there is a fundamental absorption edge, whose peaks are centered in the ultraviolet wavelength region. The tail of this peak may extend into the window region at the shorter wavelengths as shown in Fig. 10.2. The absorption bands occur in the infrared and far-infrared region due to the interactions of photons with molecular vibrations within the glass. The strong absorption bands occur due to the oscillations of structural units like P-O (8.1 μm), B-O (7.2 μm), Si-O (9.2 μm) and Ge-O (11 μm) within the glass. Thus, above 1.5 μm the tails of these largely far-infrared absorption peaks tend to cause most of the pure glass loss. However, the effects of both the processes may be minimized by suitable choice of both core and cladding compositions.

10.3.3 Extrinsic Absorption

Extrinsic absorption is attributed to impurities which are unintentionally injected into the optical fiber mix during the direct-melt method. Impurity absorption results mainly from transition metal ions, like iron, cobalt, chromium, and copper. The presence of metal ions in the optical fiber alters the characteristic transmission properties of the fiber, resulting in unacceptable optical power loss. The metal ion concentration in the fiber mix is so critical that it should not exceed the level of one part per billion (*ppb*). However, the transition metal impurities which are present in the starting materials used for direct-melt fibers range between 1 and 10 *ppb*, causing losses from 1 to 10 dB/km. Transition element contaminations may be reduced to acceptable limits (*i.e.*, 1 *ppb*) by glass refining techniques such as vapour-phase deposition. Impurity absorption losses occur either due to electronic transitions between energy levels associated with the incompletely filled inner subshell of these ions or because of charge transitions from one ion to another. The absorption peaks of the various transition metal impurities tend to be broad and this situation worsens further due to the overlap of several peaks. Some of the common metallic impurity ions found in glasses are listed in table 10.1, together with the absorption peak wavelengths and the absorption losses caused by one part in 10^9.

However, another major factor for extrinsic loss mechanism is absorption due to hydroxyl or OH ions (water) dissolved in the glass. The presence of OH ion impurities in the fiber results mainly from the oxyhydrogen flame used for the hydrolysis reaction of the $GeCl_4$, $SiCl_4$, and $POCl_3$ starting materials. If the attenuation is to be less than 20 dB/km, then OH ion impurities of less than a few ppb are required. These hydroxyl groups are bonded into the glass structure and have fundamental stretching vibrations which occur at wavelengths between 2700 nm and 4200 nm depending on group position in the glass network. These fundamental vibrations give rise to overtones appearing almost harmonically at 1380 nm, 950 nm and 720 nm as shown in Fig. 10.3. This shows the absorption spectrum for the OH group in silica. Moreover, combinations between the overtones and the fundamental SiO_2 vibration occur at 1240 nm, 1130 nm and 880 nm, completing the absorption spectrum illustrated in Fig. 10.3.

Between the absorption peaks there are regions of low attenuation. The peaks and valleys in the attenuation curve resulted in the designation of various "transmission windows" to optical fibers. By reducing the residual OH content of fibers to around 1 *ppb*, single-mode fibers have nominal attenuations of 0.5 dB/km in the 1300 nm window and 0.3 dB/km in the 1550 nm window as shown by the curve in Fig. 10.1.

Table 10.1

Metallic Ion	*Peak Wavelength (nm)*	*Parts Per Billion (dB/km)*
Fe^{3+}	400	0.15
Fe^{2+}	1100	0.68
Cr^{3+}	625	1.60
Ni^{2+}	650	0.10
C^{2+}	685	0.10
Mn^{3+}	460	0.20
V^{4+}	725	2.70
Cu^{2+}	850	1.10

From Fig. 10.3, it may also be observed that the only significant absorption band in the region below wavelength of 1000 nm is the second overtone at 950 nm which causes attenuation of about 1 dB/km for one ppm (parts per million) of hydroxyl. At longer wavelengths the first overtone at 1380 nm and its sideband at 1240 nm are strong absorbers which gives attenuation of about 2 dB/km *ppm* and 4 dB/km *ppm* respectively.

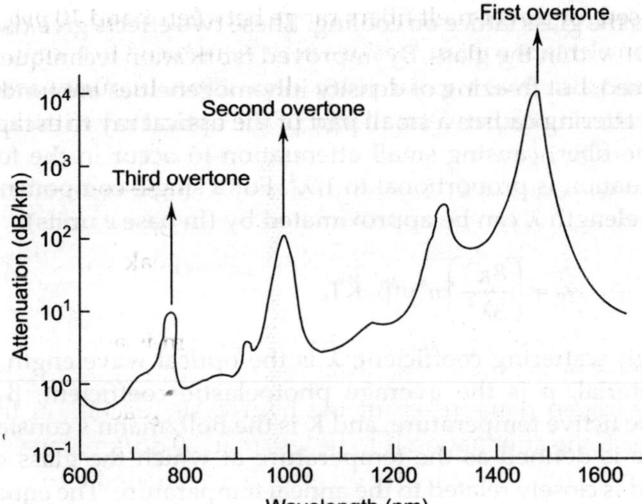

Fig. 10.3

10.4 SCATTERING

Another very important phenomenon which affects the system performance, and that occurs during the transmission of optical waves through the optical fibers, is that of scattering. The scattering losses in glass arise primarily due to the microscopic variations in the material density, the compositional fluctuations, and the structural inhomogeneities or defects occurring during fiber manufacture. Scattering may be divided into two broad categories: linear scattering and nonlinear scattering.

10.4.1 Linear Scattering

Linear scattering mechanisms occurs when some or all of the optical power is transferred linearly (*i.e.*, proportionally to the input optical power injected into the dominant mode) from the dominant propagating mode to adjacent modes. This process results in attenuation of the transmitted light as the transfer may be to a leaky or radiation mode which may not continue to propagate within the fiber core, and is radiated from the fiber. But being a linear process, no change of frequency on scattering takes place. This optical energy transfer promotes intersymbol interference (ISI), which contributes to an increase in the BER. Linear scattering may be divided into two major categories: *Rayleigh scattering* and *Mie scattering*. Both Rayleigh and Mie scattering are phenomena based on physical anomalies embedded into the optical fiber material structure as a result of imperfect fabrication techniques.

Rayleigh Scattering

Rayleigh scattering is the dominant intrinsic loss mechanism in the low absorption window, which is in between the UV and IR absorption tails. Rayleigh scattering is caused when material anomalies within the fiber material are 1/10 of the diameter (or less) of the operating wavelength. Glass is composed of randomly connected network of molecules. In such a structure naturally there are regions where the molecular density is either higher or lower than the average density in the glass. Additionally, since glass is made up of several oxides, compositional fluctuations can occur and be frozen into the glass lattice on cooling. These two effects give rise to refractive-index fluctuations which occur within the glass. By improved fabrication techniques, the compositional variations may be reduced, but freezing of density inhomogeneities are fundamental and cannot be avoided. Rayleigh scattering causes a small part of the optical ray to escape from its predetermined path through the fiber, causing small attenuation to occur in the forward propagating optical fiber. This attenuation is proportional to $1/\lambda^4$. For a single component glass the Rayleigh scattering loss at a wavelength λ can be approximated by (in base *e* units):

$$\gamma_R = \left(\frac{8\pi^3}{3\lambda^4}\right) n^8 p^2 \beta_T K T_F \tag{10.7}$$

where γ_R is the Rayleigh scattering coefficient, λ is the optical wavelength, *n* is the refractive-index of the fiber material, *p* is the average photoelastic coefficient, β_T is the isothermal compressibility, T_F is the fictive temperature, and K is the Boltzmann's constant (1.38×10^{-23} J/K). The fictive temperature is defined as the temperature at which the glass can reach a state of thermal equilibrium and is closely related to the anneal temperature. The equation 10.7 is given in units of *nepers* (*i.e.*, base *e* units). To change this to decibel for optical power attenuation calculations, multiply this equation by 10 log *e* = 4.343. For multi-component glasses the Rayleigh scattering is given by

$$\gamma = \left(\frac{8\pi^3}{3\lambda^4}\right) (\delta n^2)^2 \, \delta V \tag{10.8}$$

where the square of the mean-square refractive-index fluctuations $(\delta n^2)^2$ over a volume of δV is

$$(\delta n^2)^2 = \left(\frac{\partial n}{\partial \rho}\right)^2 (\delta\rho)^2 + \sum_{i=1}^{m} \left(\frac{\partial n^2}{\partial C_i}\right)^2 (\delta C_i)^2 \tag{10.9}$$

Here, δp is the density fluctuation and δC_i is the concentration fluctuation of the *i*th glass component. Since Rayleigh scattering follows characteristic $1/\lambda^4$ dependence, it decreases drastically with increasing wavelength as illustrated in Fig. 10.2. This Rayleigh scattering is the dominant loss mechanism in a fiber for wavelength below about 1000 nm. At wavelengths longer than 1000 nm, infrared absorption effects tend to dominate optical signal attenuation.

Furthermore, the Rayleigh scattering coefficient is used to establish the transmission loss factor (transmissivity) of the fiber (Γ), expressed by the equation

$$\Gamma = \exp(-\gamma_R L) \tag{10.10}$$

where L is the length of the fiber.

Example 10.5. *Determine the theoretical attenuation in dB/km of an optical fiber fabricated with silicon core, due to the Rayleigh scattering in silica at optical wavelengths of 650 nm, 1300 nm and 1550 nm. Silica has fictive temperature of 1450 K with an isothermal compressibility of 9×10^{-11} m²/N. The photoelastic coefficient and the refractive-index of silica are 0.288 and 1.46 respectively. The length of the fiber is 1 km and the Boltzmann's constant is 1.38×10^{-23}.*

Solution: The Rayleigh scattering coefficient may be obtained from equation (10.7) for each wavelength. Since, the only variable in each case is the wavelength, hence, the constant of proportionality of equation (10.7) applies in all cases. Thus:

$$\gamma_R = \left(\frac{8\pi^3}{3\lambda^4}\right) n^8 p^2 \beta_T K T_F$$

$$= \frac{8 \times 3.14^3 \times 1.46^8 \times 0.288^2 \times 9 \times 10^{-11} \times 1.38 \times 10^{-23} \times 1450}{3 \times \lambda^4}$$

$$= \frac{2.546 \times 10^{-28}}{\lambda^4} \text{ m}^{-1}.$$

At a wavelength of 650 nm:

$$\gamma_R = \frac{2.546 \times 10^{-28}}{650^4 \times 10^{-36}} = 1.426 \times 10^{-3} \text{ m}^{-1}$$

The transmission loss factor for one kilometer of the fiber may be obtained using equation (10.10),

$$\Gamma_{km} = \exp(-\gamma_R L) = \exp\left(-1.426 \times 10^{-3} \times 10^3\right) = 0.240$$

The attenuation due to Rayleigh scattering in dB/km may be obtained as:

$$\text{Attenuation} = 10\log_{10}(1/\Gamma_{km}) = 10\log_{10} 4.166 = 6.2\ dB/\text{km}$$

At a wavelength of 1300 nm:

$$\gamma_R = \frac{2.546 \times 10^{-28}}{1300^4 \times 10^{-36}} = 8.914 \times 10^{-5} \text{ m}^{-1}$$

Using equation (10.10),

$$\Gamma_{km} = \exp(-\gamma_R L) = \exp\left(-8.914 \times 10^{-5} \times 10^3\right) = 0.915$$

The attenuation due to Rayleigh scattering in dB/km may be obtained as:

$$\text{Attenuation} = 10\log_{10}(1/\Gamma_{km}) = 10\log_{10} 1.093 = 0.39\ dB/\text{km}$$

At a wavelength of 1550 nm:

$$\gamma_R = \frac{2.546 \times 10^{-28}}{1550^4 \times 10^{-36}} = 4.411 \times 10^{-5} \text{ m}^{-1}$$

Using equation (10.10),

$$\Gamma_{km} = \exp(-\gamma_R L) = \exp\left(-4.411 \times 10^{-5} \times 10^3\right) = 0.957$$

The attenuation due to Rayleigh scattering in *d*B/km may be obtained as:

$$\text{Attenuation} = 10\log_{10}(1/\Gamma_{km}) = 10\log_{10} 1.045 = 0.19\ \text{dB/km}$$

Mie Scattering

Mie scattering occurs when the size of the inhomogeneities within the optical fiber are comparable to the size of the guided wavelength. When this scattering inhomogeneity size is larger than 1/10 of the operating wavelength, the scattered intensity which has an angular dependence can be very large. These imperfections result from the imperfect cylindrical structure of the waveguide and may be caused by fiber imperfections such as irregularities in the core-cladding interface, the differences in the core-cladding refractive indices along the fiber length, strains and bubbles, diameter fluctuations, etc. Mie scattering occurs mainly in the forward direction.

Recent advancements in optical fiber manufacturing techniques have significantly contributed to the improvements in areas such as:

(*a*) wave guidance by increasing the relative refractive index difference and precise control of coating;

(*b*) reduction of all relevant glass manufacturing imperfections.

Today, fiber manufacturing techniques have advanced so much, that Mie scattering is almost completely eliminated.

10.4.2 Nonlinear Scattering

Optical fibers do not always exhibit the liner phenomena, where the output optical power is directly proportional to the input optical power. Rather, usually at high optical power levels, several nonlinear phenomena occur, and the optical power at the output of a fiber does not change in proportion to the input optical power. Such behaviour leads to nonlinear scattering losses. This nonlinear scattering causes the optical power from one mode to be transferred in either the forward or reverse direction to the same, or other modes, at a different frequency. The level of loss is a function of critical optical power level launched into the input of the fiber. This power is referred to as threshold optical power.

Two basic types of nonlinear scattering exist within the optical fiber. These are Stimulated Raman Scattering (SRS) and Stimulated Brillouin Scattering (SBS). Both of them are usually only observed at high optical power densities in long single mode fibers. These scattering mechanisms in fact give optical amplification, but with a shift in frequency, thus contributing to attenuation for light transmission at a specific wavelength.

Stimulated Raman Scattering

Stimulated Raman Scattering (SRS) in the result of the interaction between light waves and vibrational modes of silica molecules. When a photon with energy $h\nu_i$ is incident on a molecule having vibrations frequency ν_v, the molecule can absorb some energy from the photon. In the process, the photon is scattered and attains a lower frequency ν_s and a corresponding lower energy $h\nu_s$. The modified photon is named Stokes *photon*. Since the optical signal injected into the fiber is the source of the interacting photons, it is often called the *pump wave*.

SRS contributes to signal attenuation at the operating wavelength. However, if another signal is present at this shifted wavelength, the SRS light will amplify it. Hence, the negative effect of SRS must be eliminated during the propagation of the optical wave through the fiber, while the amplification phenomena at the shifted wavelengths must be fully exploited in the design and fabrication of all optical amplifiers. SRS can occur in both the forward and backward directions in an optical fiber, and may have a high optical power threshold. The threshold optical power ($P_{R(Th)}$) for SRS in a long single-mode fiber is given by:

$$P_{R(Th)} = 5.9 \times 10^{-2}\, \lambda d^2\, \alpha \text{ watts} \qquad (10.11)$$

where λ is the operating wavelength in μm, d is the core diameter of the fiber in μm, and α is the fiber attenuation in dB/km. From equation (10.11), we see that if the fiber core diameter is increased while the wavelength is kept constant, the threshold power will also increase. This can also be verified from Example 10.6. In fact a proper combination of selected operating wavelengths and core diameter can practically eliminate SRS.

Example 10.6. *Determine the stimulated Raman scattering threshold power in a long single mode optical fiber with core diameter 6 μm, and attenuation 0.4 dB/km when operating at a wavelength of 0.85 nm. Compute the SRS for the same parameters, but with an operating wavelength of 1.3 μm and 1.55 μm.*

Solution: The threshold optical power for SRS may be obtained from equation (10.11), where:

$$P_{R(Th)} = 5.9 \times 10^{-2}\, \lambda d^2\, \alpha$$

$$= 5.9 \times 10^{-2} (0.85) (6)^2 (0.4)$$
$$= 722.16 \text{ mW}$$

At operating wavelength 1.3 μm

$$P_{R(Th)} = 5.9 \times 10^{-2} (1.3) (6)^2 (0.4) = 1.1 \text{ W}$$

At operating wavelength 1.55 μm

$$P_{R(Th)} = 5.9 \times 10^{-2} (1.55) (6)^2 (0.4) = 1.31 \text{ W}$$

Stimulated Brillouin Scattering

Stimulated Brillouin Scattering (SBS) occurs when the incident optical power at the input of a single mode optical fiber is modulated along the traveling path by thermal molecular vibrations within the fiber. The scattered light appears as upper and lower sidebands which are separated from the incident light by the modulation frequency. The resultant scattered wave propagates principally in the backward direction in single mode fibers making SBS a mainly backward process. The incident photon in this scattering process generates another scattering photon along with a phonon of acoustic frequency. The backward scattered light experiences gain from the forward propagating signals, which lead to the depletion of the signal power. The frequency of the scattered light experiences a Doppler shift given by

$$\nu_B = 2nV_s/\lambda \qquad (10.12)$$

where V_s is the velocity of sound in the material and n is the refractive index of the material. The optical frequency shift varies with the scattering angle because the frequency of the sound wave varies with acoustic wavelength.

System impairment starts when the amplitude of the scattered wave is comparable to the signal power. The degradation due to this scattering may be limited by using optical isolators to prevent backward scattered signal from reentering the system. One criterion for determining at what point SBS becomes a problem is to consider the SBS threshold power $P_{B(Th)}$, given by:

$$P_{B(Th)} = 4.4 \times 10^{-3} \lambda^2 d^2 \alpha B_w \qquad (10.13)$$

where λ is the operating wavelength in μm, d is the core diameter of the fiber in μm, α is the fiber attenuation in dB/km, and B_w is the bandwidth of the source in GHz. The threshold power is defined as the signal power at which the backward scattered light equals the fiber input power. From equation (10.13), we find that either increasing the operating wavelength, the fiber core diameter, or both can increase the SBS threshold. In this context, operation at the longest possible wavelength is advantageous although this may be offset by the reduced fiber attenuation (from Rayleigh scattering and material absorption) normally obtained. This can be verified from Examples 10.7 and 10.8. SRS and SBS are usually not observed in multimode fibers because their relatively large core diameters make the threshold optical power levels extremely high.

Example 10.7. *Determine the stimulated Brillouin scattering threshold power in a long single mode optical fiber with core diameter 6 μm, attenuation 0.5 dB/km, and laser diode bandwidth of 1 GHz when operating at a wavelength of 0.85 nm. Compute the SBS for the same parameters, but with an operating wavelength of 1.3 μm and 1.55 μm.*

Solution: The threshold optical power for SBS may be obtained from equation (10.13), where:

$$P_{B(Th)} = 4.4 \times 10^{-3} \lambda^2 d^2 \alpha B_w$$
$$= 4.4 \times 10^{-3} (0.85)^2 (6)^2 (0.5) (1) = 57.22 \text{ mW}$$

At operating wavelength 1.3 μm

$$P_{B(Th)} = 4.4 \times 10^{-3}\ (1.3)^2\ (6)^2\ (0.5)\ (1) = 133.85 \text{ mW}$$

At operating wavelength 1.55 μm

$$P_{B(Th)} = 4.4 \times 10^{-3}\ (1.55)^2\ (6)^2\ (0.5)\ (1) = 190.28 \text{ mW.}$$

Example 10.8. *Use the same parameters as Example 10.7, but use a core diameter of 5 μm and 8 μm respectively. The operating wavelength is 1.3 μm.*

Solution: For core diameter 5 μm

$$P_{B(Th)} = 4.4 \times 10^{-3}\ (1.3)^2\ (5)^2\ (0.5)\ (1)$$
$$= 92.95 \text{ mW.}$$

For core diameter 8 μm

$$P_{B(Th)} = 4.4 \times 10^{-3}\ (1.3)^2\ (8)^2\ (0.5)\ (1)$$
$$= 237.95 \text{ mW.}$$

10.5 BENDING LOSS

Fiber bending is another major factor which contributes significantly to optical fiber power loss. In fact optical fibers suffer radiation losses at bends or curves on their paths. When optical fiber is bent, a portion of the optical energy escapes from the core to the cladding. If the optical signal attenuation due to bending is to be avoided, both the waves traveling through the core and the cladding must propagate with the same velocity and must form a common wavefront at the output of the fiber. This requires the velocity of the wave traveling through the cladding to be greater than the velocity of the light in that medium in order to cover the longer distance. This is impossible. Hence, the wave traveling through the cladding will be lost in the form of radiation. This radiation loss is an optical power loss. The situation is illustrated in Fig. 10.4.

There are two categories of bending loss: (*a*) macroscopic bending loss, arising due to bends having radii that are large compared with the fiber diameter, *e.g.*, the bends that occur when a fiber cable turns a corner, and (*b*) microscopic bending loss, arising from the random microscopic bends of the fiber axis that can arise when the fibers are incorporated into cables.

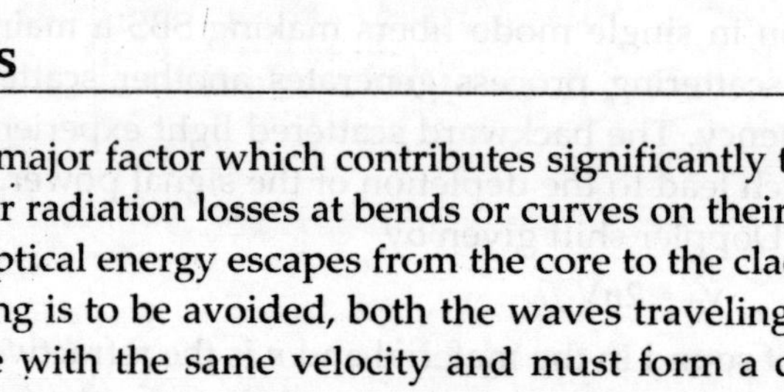

Fig. 10.4

10.5.1 Macrobending Loss

Let us first examine large curvature losses, which are known as macrobending losses or simply bending losses. Macrobending losses are subject to damages to the fiber during the packaging, transportation, and field installation processes. Fiber losses caused by macroscopic bending can be expressed in terms of an attenuation coefficient given by:

$$\alpha_B = K_1 e^{-K_2 R} \tag{10.14}$$

where α_B is the attenuation coefficient due to macroscopic bending, K_1, K_2 are constants independent of R, and R is the radius of curvature of the fiber bend.

Comparing the effect of macrobending for both single mode and multimode fibers shows that multimode fibers are more susceptible to macrobending than single mode fibers. The amount of optical radiation from a bent fiber depends on the field strength x_b at that bend point and on the radius of curvature R. In multimode fibers since, higher order modes are bound less tightly to the fiber core than lower order modes, the higher order modes radiate out of the fiber first. Hence, the total number of modes supported by a bend fiber is less than a straight fiber. We can get the following expression for the effective number of modes N_{eff} that are guided by a bend multimode fiber of radius r after Gloge as:

$$N_{eff} = N_\infty \left\{ 1 - \frac{\alpha + 2}{2\alpha\Delta} \left[\frac{2r}{R} + \left(\frac{3}{2n_2 kR} \right)^{2/3} \right] \right\} \tag{10.15}$$

where α is the graded index profile, n_2 is the refractive index of the cladding, Δ is the core-cladding index difference, $k = 2\pi/\lambda$ is the wave propagation constant, and N_∞ is the total number of modes in a straight fiber given by:

$$N_\infty = \frac{\alpha}{\alpha + 2} (n_1 kr)^2 \Delta \tag{10.16}$$

The attenuation coefficient in equation (10.14) becomes significant beyond a specific value of the radius of curvature (R), referred to as the *critical radius*. The critical radius (R_{CMM}) of a multimode fiber is expressed as:

$$R_{CMM} \approx \frac{3\lambda n_1^2}{4\pi \left(n_1^2 - n_2^2\right)^{\frac{3}{2}}} \tag{10.17}$$

where λ is the operating wavelength, n_1 is the core refractive index, and n_2 is the cladding refractive index. From equation (10.17), it may be observed that macrobending is proportional to the operating wavelength and core refractive index, and also inversely proportional to the refractive index difference between core and cladding. Hence, an increase of the operating wavelength together with an increase of the refractive index difference between core and cladding will reduce macrobending loss to an appreciable amount.

The above criteria for the reduction of macrobend losses also apply to single mode fibers. For single mode fibers, the critical bending radius (R_{CSM}) is expressed as:

$$R_{CSM} \approx \frac{20\lambda}{(n_1 - n_2)^{\frac{3}{2}}} \left(2.748 - 0.996 \frac{\lambda}{\lambda_c} \right)^{-3} \tag{10.18}$$

where λ_c is the cut-off wavelength for the single mode fiber, λ is the operating wavelength, n_1 is the core refractive index, and n_2 is the refractive index of the cladding.

Example 10.9. *Two step index fibers exhibit the following parameters:*

(a) A multimode fiber with a core refractive index of 1.550, a relative refractive index difference of 2.5 % and an operating wavelength of 0.85 μm.

(b) A 6 μm core diameter single-mode fiber with a core refractive index the same as (a), a relative refractive index difference of 0.25 % and an operating wavelength of 1.3 μm.

Compute the critical radius of curvature at which large bending losses occur in both cases.

Solution:

(*a*) The relative refractive index difference is given by:

$$\Delta = \frac{n_1^2 - n_2^2}{2n_1^2}$$

Hence $$n_2^2 = n_1^2 - 2\Delta n_1^2 = 2.4025 - (0.05 \times 2.4025) = 2.282$$

Using equation (10.17) for the multimode fiber critical radius of curvature:

$$R_{CMM} \approx \frac{3\lambda n_1^2}{4\pi\left(n_1^2 - n_2^2\right)^{\frac{3}{2}}} = \frac{3 \times 0.85 \times 10^{-6} \times 2.4025}{4\pi(0.1205)^{\frac{3}{2}}}$$

$$= 11.65\ \mu\text{m}.$$

(*b*) Again, we have:

$$n_2^2 = n_1^2 - 2\Delta n_1^2 = 2.4025 - (0.005 \times 2.4025)$$

$$= 2.390$$

The cutoff wavelength for the single mode fiber is given by:

$$\lambda_c = \frac{2\pi r n_1 (2\Delta)^{\frac{1}{2}}}{2.405}$$

$$= \frac{2\pi \times 3 \times 10^{-6} \times 1.550 \times (0.005)^{\frac{1}{2}}}{2.405} = 0.859\ \mu\text{m}.$$

Substituting into equation (10.18) for the critical radius of curvature for the single mode fiber gives:

$$R_{CSM} \approx \frac{20\lambda}{(n_1 - n_2)^{\frac{3}{2}}}\left(2.748 - 0.996\frac{\lambda}{\lambda_c}\right)^{-3}$$

$$\approx \frac{20 \times 1.3 \times 10^{-6}}{(0.004)^{\frac{3}{2}}}\left(2.748 - \frac{0.996 \times 1.3 \times 10^{-6}}{0.859 \times 10^{-6}}\right)^{-3}$$

$$= 53.8\ \text{mm}$$

10.5.2 Microbending Loss

Another form of radiation loss in optical fibers results from mode coupling caused by random microbends of the optical fiber. Microbends are repetitive small-scale fluctuations occurring in the radius of curvature of the fiber axis, as is shown in Fig. 10.5. Microbending can be induced into the fiber during the fabrication process or by the nonuniform lateral pressures created during the installation process or maintenance of the fiber cables. The major factor that causes microbending is temperature. Temperature variations can cause local expansions and contractions of the core material, which may slightly alter the physical characteristics of the core, and ultimately affecting the optical fiber performance. The physical variations of the core can cause modal coupling which may significantly contribute to modal dispersion in multimode optical fiber.

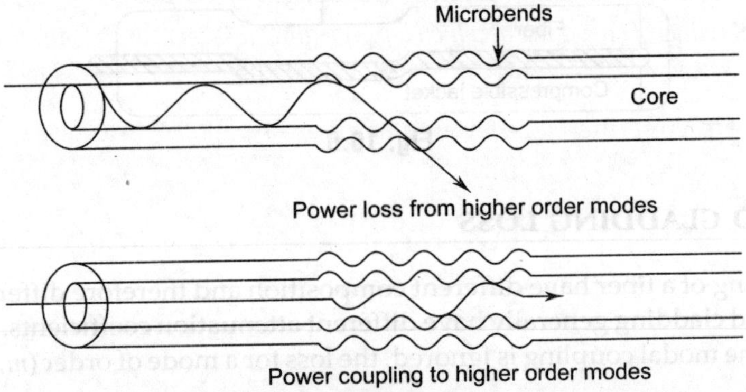

Fig. 10.5

Modal coupling is actually a process where optical energy from one mode of transmission is transferred to other modes. Signal attenuation increases due to modal coupling. In fact microbending is susceptible to the optical power modal distribution, which is categorized as either uniform or equilibrium mode distribution. Uniform mode distribution is basically a process where all the available modes in the fiber are launched simultaneously with maximum sustainable optical power. On the other hand, equilibrium mode distribution is the process where the optical power carried by the individual modes in the fiber remains constant beyond one kilometer of the fiber length and remains independent of the input optical power. This means that any increase in the input optical power launched into a particular mode will be coupled to other modes. Thus a constant modal optical power is maintained.

Thus, we find that microbending is modal dependent and hence wavelength dependent. Recent researches show that optical fiber attenuation due to microbending decreases with an increase in the operating wavelength. For single mode fibers, increase in attenuation due to microbending follows almost exponential relationship with the wavelength beyond 1500 nm wavelength window, while minimum attenuation is encountered at wavelengths below the 1300 nm range. To minimize the microbending effect in optical fibers, certain precautions must be taken to eliminate externally imposed permanent pressures due to the cabling process. At the same time, optical fibers must be fully protected from environmental changes, especially temperature variations. One method of minimizing microbending losses is by including a compressible jacket over the fiber. In such a situation, when external pressure is applied, the jacket will be deformed but the fiber will tend to stay almost straight as illustrated in Fig. 10.6. After Gloge, we may write that for a multimode graded index fiber having a core radius a, outer radius b (excluding the jacket), and refractive index difference Δ, the microbending loss α_{MB} of a jacketed fiber is reduced from that of an unjacketed fiber by a factor

$$F(\alpha_{MB}) = \left[1 + \pi\Delta^2 \left(\frac{b}{a}\right)^4 \frac{E_f}{E_j}\right]^{-2} \tag{10.19}$$

where E_f and E_j are the Young's moduli of the fiber and jacket respectively.

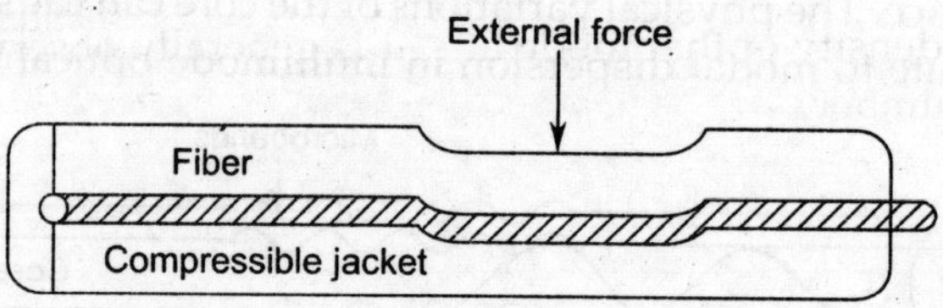

Fig. 10.6

10.6 CORE AND CLADDING LOSS

The core and cladding of a fiber have different composition and therefore differ in their refractive indices. The core and cladding generally have different attenuation coefficients, denoted by α_1 and α_2, respectively. If the modal coupling is ignored, the loss for a mode of order (m, n) for a step index waveguide is

$$\alpha_{mn} = \alpha_1 \frac{P_{core}}{P} + \alpha_2 \frac{P_{clad}}{P} \qquad (10.20)$$

Gloge has shown that, based on the weakly guided mode approximation, which has accuracy on the order of the index difference Δ between the core and cladding, the relative core and cladding powers for a particular mode m is given by

$$\frac{P_{clad}}{P} = 1 - \frac{P_{core}}{P} \qquad (10.21)$$

where P is the total power in the mode m. Using equation (10.21), equation (10.20) may be written as

$$\alpha_{mn} = \alpha_1 + (\alpha_2 - \alpha_1)\frac{P_{clad}}{P} \qquad (10.22)$$

The total loss of the waveguide may be found by summing over all the modes weighted by the fractional power in that mode.

The situation becomes much complicated for graded index fiber. In this case, both the attenuation coefficients and the modal power tend to be functions of the radial coordinate. At a distance r from the core axis the loss is

$$\alpha(r) = \alpha_1 + (\alpha_2 - \alpha_1)\frac{n^2(0) - n^2(r)}{n^2(0) - n_2^2} \qquad (10.23)$$

where α_1 and α_2 are the axial and cladding attenuation coefficients, respectively, and $n(r)$ is given by:

$$n(r) = n_1\left[1 - 2\Delta\left(\frac{r}{a}\right)^{\alpha}\right]^{1/2} \quad \text{for} \quad 0 \le r \le a$$

$$= n_1(1 - 2\Delta)^{1/2} \approx n_1(1 - \Delta) = n^2 \quad \text{for} \quad r \ge a \qquad (10.24)$$

Here, r is the radial distance from the fiber axis, a is the core radius, n_1 is the refractive index of the core axis, n_2 is the refractive index of the cladding, and α is the shape of the index profile, which is a dimensionless quantity. The loss encountered by a given mode is given by

$$\alpha_{gi} = \frac{\int_0^\infty \alpha(r)p(r)r\,dr}{\int_0^\infty p(r)r\,dr} \qquad (10.25)$$

where $p(r)$ is the power density of that mode at r. It is generally observed that the loss increases with increasing mode number.

10.7 DISPERSION

Dispersion is the phenomenon whereby an optical signal, both analog and digital, becomes broadened while traveling through the core of the optical fiber. An illustration using the digital pattern 10011 of the broadening of light pulses as they are transmitted along a fiber is given in Fig. 10.7: (*a*) fiber input; (*b*) fiber output at a distance L_1; (*c*) fiber output as a distance $L_2 > L_1$. The causes of the pulse dispersion are several and can be classified into two major categories as *intramodal* and *intermodal*. These dispersion effects can be explained by examining the behaviour of the group velocities of the guided modes, where the *group velocity* is the speed at which energy in a particular mode travels along the fiber. From Fig. 10.7, it may be observed that each pulse broadens and overlaps with its neighbours, eventually becoming indistinguishable at the receiver input. This effect is known as *intersymbol interference* (ISI). Thus an increase in the system BER may occur on the digital optical channel as the ISI becomes more and more prominent. The error rate is also a function of the signal attenuation on the link and the subsequent signal to noise ratio (SNR) at the receiver. However, signal dispersion alone limits the maximum possible bandwidth attainable with a particular optical fiber to the point where individual symbols can no longer be distinguished. The relationship between the dispersion and attainable bandwidth is given by equation 10.26 as:

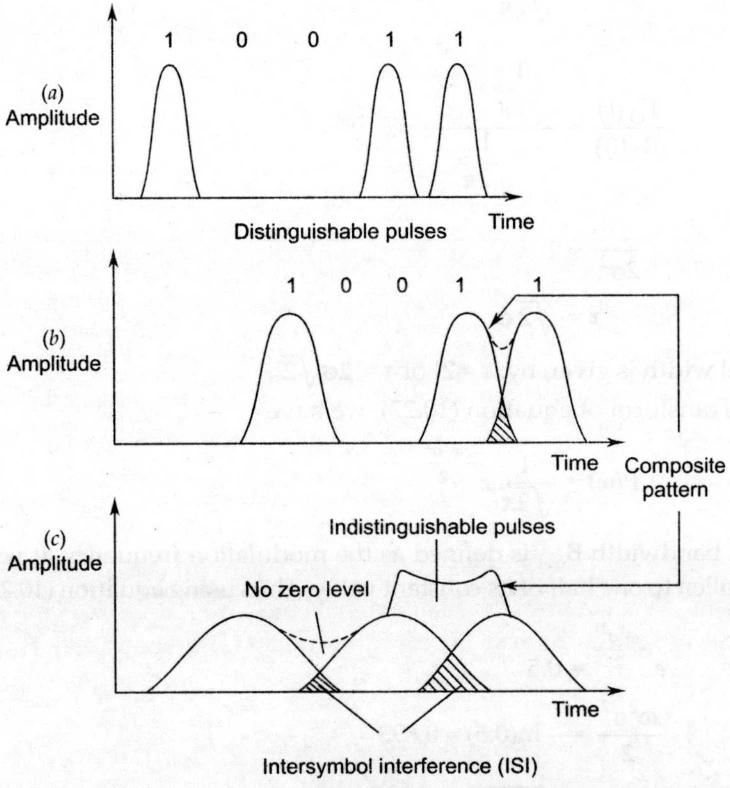

Fig. 10.7

$$B_T \leq \frac{1}{2\tau} \tag{10.26}$$

where B_T is the channel bit rate and τ is the pulse duration. Thus, equation (10.26) gives a conservative estimate of the maximum bit rate that may be obtained on an optical fiber link as $1/2\tau$.

A more accurate determination of the maximum channel bit rate based on dispersion is given by assuming the optical pulse at the output of the fiber to be Gaussian with standard deviation (σ). The rms pulse width is equal to the standard deviation. Unlike the relationship given in equation (10.26), this analysis allows for the existence of a certain amount of signal overlap on the channel, while avoiding any SNR penalty which occurs when ISI becomes pronounced. The optical output power variation with reference to time (t) at the output of the fiber is expressed by equation (10.27).

$$P_O(t) = \frac{1}{\sqrt{2\pi}} e^{-\frac{t^2}{2\sigma^2}} \tag{10.27}$$

where $P_O(t)$ = optical output power, t = time, σ = standard deviation, and σ^2 = distribution variance.

Let $\frac{P_O(t)}{P_O(0)} = \tau$

where τ is the pulse duration.

Hence, $P_O(t) = \frac{1}{\sqrt{2\pi}}$

or $$\frac{P_O(t)}{P_O(0)} = \frac{\frac{1}{\sqrt{2\pi}} e^{-\frac{t^2}{2\sigma^2}}}{\frac{1}{\sqrt{2\pi}}} = e^{-\frac{t^2}{2\sigma^2}}$$

or $$\frac{t^2}{2\sigma^2} = 1$$

or $$\tau = \sqrt{2}\,\sigma$$

Hence, pulse full width is given by: $\tau = 2t$ or $\tau = 2\sigma\sqrt{2}$.

Taking Fourier Transform of equation (10.27), we have:

$$P(\omega) = \frac{1}{\sqrt{2\pi}} e^{-\frac{\omega^2\sigma^2}{2}} \tag{10.28}$$

The 3 dB optical bandwidth B_{opt} is defined as the modulation frequency at which the received optical power has fallen to one half of its constant value. Thus using equation (10.28), we may write:

$$e^{-\frac{\omega^2\sigma^2}{2}} = 0.5$$

or $$\frac{\omega^2\sigma^2}{2} = -\ln(0.5) = 0.693$$

or $$\omega = \sqrt{\frac{1.386}{\sigma^2}} = \frac{1.177}{\sigma}$$

Since, we know $\omega = 2\pi B_{opt}$, we have $2\pi B_{opt} = \dfrac{1.177}{\sigma}$.

Thus, $$B_{opt} = \frac{0.187}{\sigma} \text{ Hz.} \tag{10.29}$$

When employing RZ pulse where the maximum bit rate $B_T(max) = B_{opt}$, then:

$$B_{T(max)} = \frac{0.187}{\sigma} \text{ bit/sec} \tag{10.30}$$

With reference to pulse width, $\tau = 2\sigma\sqrt{2}$

or $$\sigma = \frac{\tau}{2\sqrt{2}} \tag{10.31}$$

Substituting equation (10.31) in (10.30) yields

$$B_{T(max)} = \frac{0.187}{\dfrac{\tau}{2\sqrt{2}}} = \frac{0.187 \times 2\sqrt{2}}{\tau} = \frac{0.529}{\tau} \tag{10.32}$$

Hence, maximum bit rate is also expressed by equation (10.32). Equation (10.30) gives a reasonably good approximation for other pulse shapes which may occur on the channel resulting from the various dispersive mechanisms within the fiber.

In fact the conversion of bit rate to bandwidth in hertz or vice versa, depends on the digital coding format used. For metallic conductors when a NRZ code is used the maximum bandwidth B is one half the maximum data rate or

$$B_{T(max)} = 2B \tag{10.33}$$

As already discussed, when a RZ code is employed, the data rate is equal to the bandwidth in hertz, and hence $B_{T(max)} = B$.

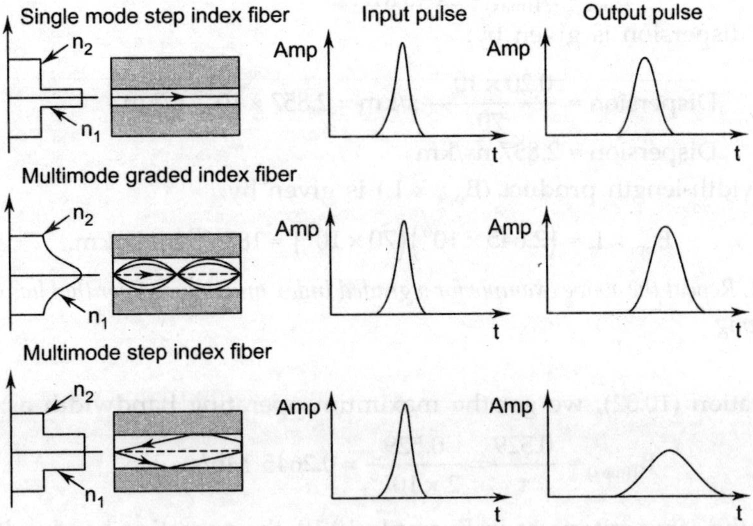

Fig. 10.8

Figure 10.8 illustrates the respective pulse broadening associated with the three most common optical fiber structures, viz., single-mode step index, multimode graded index, and multimode step index. It is observed that the single mode step index fiber exhibits the best performance resulting in the minimum pulse broadening, and is thus capable of the greatest transmission bandwidths which are currently in the GHz range. It is also observed that the multimode step index fiber exhibits the greatest dispersion of a transmitted light pulse, whereas the multimode graded index fiber shows a considerably improved performance. Transmission through multimode step index fiber is usually limited to bandwidths of a few tens of MHz.

However, amount of pulse broadening through an optical fiber is also subject to its operating length, *i.e.*, the distance the pulse travels within the fiber. Hence, the measurement of the dispersive properties of a particular fiber is usually stated as the pulse broadening in time over a unit length of the fiber (*i.e.*, ns/km). Thus, the information carrying capacity of the fiber is restricted by the amount of pulse dispersion per unit length. If filtering or mode coupling is absent, the broadening of pulse increases linearly with fiber length and thus the bandwidth is inversely proportional to the distance. This leads to a more useful parameter called the bandwidth-length product (*i.e.* $B_{opt} \times L$), for the information carrying capacity of the optical fiber. Figure 10.8 indicates that the step index single mode fibers have the best bandwidth length product, estimated at 100 GHz-km, followed by the graded index multimode fiber at 1 GHz-km, and the step index multimode fibers, which have the worst bandwidth length performance of 20 MHz-km.

Example 10.10. *Compute: (a) the maximum operating bandwidth (B_{opt}), (b) the pulse dispersion of a step index single-mode fiber that exhibits a dispersion of 0.20 μs and has a total fiber length of 70 km, and (c) the bandwidth-length product ($B_{opt} \times L$).*

Solution:

(*a*) Using equation (10.32), we get the maximum operating bandwidth as:

$$B_{T(max)} = \frac{0.529}{\tau} = \frac{0.529}{0.20 \times 10^{-6}} = 2.645 \text{ Mb/s}$$

Assuming a zero ISI and an RZ input data signal, the operating bandwidth is equal to the system bit rate. Hence,

$$B_{opt} = B_{T(max)} = 2.645 \text{ MHz}.$$

(*b*) The pulse dispersion is given by:

$$\text{Dispersion} = \frac{0.20 \times 10^{-6}}{70} \text{ s/km} = 2.857 \times 10^{-9} \text{ s/km}$$

$$\therefore \quad \text{Dispersion} = 2.857 \text{ ns/km}$$

(*c*) The bandwidth-length product ($B_{opt} \times L$) is given by:

$$B_{opt} \times L = \left(2.645 \times 10^6\right)\left(70 \times 10^3\right) = 185.15 \text{ MHz/km}.$$

Example 10.11. *Repeat the above example for a graded index multimode fiber that has a total dispersion of 2 μs and is 70 km long.*

Solution:

(*a*) Using equation (10.32), we get the maximum operating bandwidth as:

$$B_{T(max)} = \frac{0.529}{\tau} = \frac{0.529}{2 \times 10^{-6}} = 0.2645 \text{ Mb/s}$$

Assuming the same criteria as in Example 10.10, the operating bandwidth is equal to the system bit rate. Hence,

$$B_{opt} = B_{T(max)} = 0.2645 \text{ MHz}.$$

(*b*) The pulse dispersion is given by:

$$\text{Dispersion} = \frac{2 \times 10^{-6}}{70} \text{ s/km} = 28.57 \times 10^{-9} \text{ s/km}$$

∴ Dispersion = 2.857 ns/km

(c) The bandwidth-length product ($B_{opt} \times L$) is given by:

$$B_{opt} \times L = \left(0.2645 \times 10^{6}\right)\left(70 \times 10^{3}\right) = 18.515 \text{ MHz/km}.$$

10.7.1 Intramodal Dispersion

Intramodal dispersion or *chromatic dispersion* is a phenomenon of pulse spreading that occurs within a single mode. Since this dispersion is a result of the group velocity, which is actually a function of the wavelength, this phenomenon is also known as *Group Velocity Dispersion* (GVD). Since, intramodal dispersion depends on the wavelength, its effect on signal distortion increases with the spectral linewidth of the optical source. Optical sources, in fact, do not emit just a single frequency but a band of frequencies. For LEDs the rms spectral width is about 5 percent of the central wavelength. However injection laser diodes have much narrower spectral widths, corresponding to only a fraction of a percent of the central wavelength. There may be propagation delay differences between the different spectral components of the transmitted signal. This causes broadening of each transmitted mode which in turn causes intramodal dispersion. The two main causes of intramodal dispersion are: *material dispersion* which may be caused by the dispersive properties of the waveguide material, and *waveguide dispersion* which is due to the guidance effects within the fiber structure.

Material Dispersion

Pulse broadening is the result of a nonlinear change of the phase velocity in reference to wavelength of a plane wave that is composed of different frequency components and is propagating through a dielectric medium. The different spectral components launched into the fiber from the optical source reflect different group velocities which directly contribute to pulse broadening at the output of the fiber. A material is said to exhibit material dispersion when the second differential of the refractive index with respect to wavelength is not zero. Hence, the pulse width broadening may be expressed by the total group delay τ_g at the output of the fiber. The group delay is the reciprocal of the group velocity and is given by:

$$\tau_g = \frac{1}{c}\left(n_1 - \lambda \frac{dn_1}{d\lambda}\right) \tag{10.34}$$

where n_1 is the refractive index of the core material, λ is the operating wavelength, and c is the velocity of light in free space. Hence, the pulse delay τ_m at the end of an optical fiber of length L is given by:

$$\tau_m = \frac{L}{c}\left(n_1 - \lambda \frac{dn_1}{d\lambda}\right) \tag{10.35}$$

Equation (10.35) indicates that pulse broadening due to material dispersion is proportional to the length of the fiber. For a more detailed evaluation of the pulse broadening due to material dispersion, the source rms spectral line width σ_λ and the mean wavelength λ are taken into account to form an approximate expression given by:

$$\sigma_m \approx \sigma_\lambda \frac{d\tau_m}{d\lambda} \tag{10.36}$$

The pulse spread may be evaluated by considering the dependence of the pulse delay τ_m on λ. From equation (10.35), we may write:

$$\frac{d\tau_m}{d\lambda} = \frac{L\lambda}{c}\left(\frac{dn_1}{d\lambda} - \frac{d^2 n_1}{d\lambda^2} - \frac{dn_1}{d\lambda}\right) = -\frac{L\lambda}{c}\frac{d^2 n_1}{d\lambda^2} \tag{10.37}$$

Hence, substituting equation (10.37) in (10.36), the pulse broadening due to material dispersion is obtained as:

$$\sigma_m \approx \frac{\sigma_\lambda L}{c}\left|\lambda \frac{d^2 n_1}{d\lambda^2}\right| \tag{10.38}$$

The material dispersion is sometimes quoted as a value for $\left|d^2 n_1/d\lambda^2\right|$ or $\left|\lambda^2\left(d^2 n_1/d\lambda^2\right)\right|$. The material dispersion parameter is defined as:

$$M_{dp} = \frac{1}{\lambda c}\left|\lambda^2 \frac{d^2 n_1}{d\lambda^2}\right| \tag{10.39}$$

M is expressed in units of ps/nm/km.

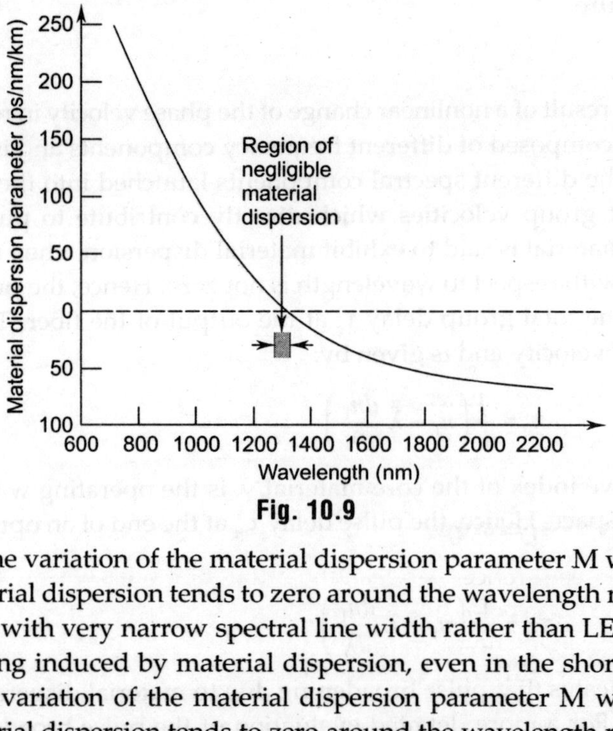

Fig. 10.9

Figure 10.9 shows the variation of the material dispersion parameter M with wavelength λ. It is observed that the material dispersion tends to zero around the wavelength region of 1300 nm. Also the use of laser diodes with very narrow spectral line width rather than LED as the optical source reduce pulse broadening induced by material dispersion, even in the shorter wavelength region.
Figure 10.9 shows the variation of the material dispersion parameter M with wavelength λ. It is observed that the material dispersion tends to zero around the wavelength region of 1300 nm. Also the use of laser diodes with very narrow spectral line width rather than LED as the optical source reduce pulse broadening induced by material dispersion, even in the shorter wavelength region.

Example 10.12. *An LED source with an rms spectral linewidth of 30 nm is coupled into an optical fiber that exhibits material dispersion of 0.025. Compute (i) the material dispersion parameter at the wavelength of 870 nm, and (ii) the rms pulse broadening per kilometre for the 870 nm operating wavelength.*

Solution:

(*i*) The material dispersion parameter (M_{dp}) is given by:

$$M_{dp} = \frac{1}{\lambda c}\left|\lambda^2 \frac{d^2 n_1}{d\lambda^2}\right| = \frac{0.025}{870 \times 3 \times 10^5} \text{ s nm}^{-1}\text{km}^{-1} = 95.8 \text{ ps nm}^{-1} \text{ km}^{-1}.$$

(*ii*) From equations (10.38) and (10.39), the rms pulse broadening is given by:

$$\sigma_m \approx \sigma_\lambda L M_{dp}$$

Hence, the rms pulse broadening per kilometre due to material dispersion is:

$$\sigma_m(1\text{km}) = 30 \times 1 \times 95.8 \times 10^{-12} = 2.87 \text{ns km}^{-1}.$$

Example 10.13. *Estimate the rms pulse broadening per kilometre for the fiber in Example 10.12 when the optical source used is an injection laser with a relative spectral width σ_λ/λ of 0.0014 at a wavelength of 870 nm. Rest of the data of Example 10.12 remains the same. Compare the result obtained to the result of the previous example.*

Solution: The rms spectral width is obtained from the relative spectral width as:

$$\sigma_\lambda = 0.0014\lambda = 0.0014 \times 870 \times 10^{-9}$$
$$= 1.22 \text{ nm}$$

The rms pulse broadening in terms of material dispersion parameter following Example 10.12 is given by:

$$\sigma_m \approx \sigma_\lambda L M_{dp}$$

Hence, the rms pulse broadening per kilometer due to material dispersion is:

$$\sigma_m(1\text{km}) = 1.22 \times 1 \times 95.8 \times 10^{-12} = 0.16 \text{ ns km}^{-1}$$

Hence, in this example the rms pulse broadening is reduced by a factor of around 18 compared to that obtained with the LED source of example 10.12.

Waveguide Dispersion

The waveguide dispersion results from the variation in group velocity with wavelength for a particular mode. If we consider the ray theory approach it is equivalent to the angle between the incident ray and the fiber core axis varying with wavelength which leads to a variation in the transmission times for the rays. Hence, dispersion occurs. Waveguide dispersion is mostly noticeable in single-mode fibers. Multimode fibers are almost free of wavelength dispersion since, majority of the modes propagate far below the cut off wavelength.

10.7.2 Intermodal Dispersion

The propagation delay differences between the different modes within a multimode fiber results in pulse broadening. This type of dispersion is called *Intermodal dispersion*. The different modes which constitute a pulse in the multimode fiber travel along the fiber with different group velocities. Since, these modes follow different transmission paths; they arrive at the output of the fiber at different times. Hence, the pulse width at the output is dependent upon the transmission times of the slowest and fastest modes.

Multimode step index fibers are mush more susceptible to intermodal dispersion than multimode graded index or the single mode fibers. This is because of the different modes that arrive at the output at different times, which gives the greatest pulse broadening. However, intermodal dispersion in multimode fibers may be reduced by using graded index profile, since the varying index profile of the fiber core compensates for the different group velocities generated by the propagating modes. Hence, the pulse broadening is significantly reduced than that obtained in multimode step index fibers (typically by a factor of 100). Under purely single mode operation there is no intermodal dispersion because only one mode of transmission is used. In single mode fibers pulse broadening is solely due to the intramodal dispersion mechanisms. Since such fibers exhibit the least pulse broadening, they have the maximum possible bandwidths.

10.8 PULSE BROADENING IN SINGLE-MODE FIBERS

Since only a single mode is allowed to propagate through single-mode fibers, the pulse broadening in this case is almost entirely due to intramodal or chromatic dispersion. Thus the bandwidth is limited by the finite spectral width of the source. The specific group delay τ_g for a light pulse propagating along a unit length of a single-mode fiber is given by:

$$\tau_g = \frac{1}{c}\frac{d\beta}{dk} \tag{10.40}$$

where β is the propagation constant for a mode within the fiber core of refractive index n_1, c is the velocity of light in vacuum and k is the propagation constant. The total first order dispersion parameter of a single-mode fiber, D_T, is given by the derivative of the specific group delay with respect to the vacuum wavelength λ as:

$$D_T = \frac{d\tau_g}{d\lambda} \tag{10.41}$$

D_T is often expressed in units of ps nm^{-1} km^{-1}. If the variable λ is replaced by ω, then the total dispersion parameter becomes:

$$D_T = -\frac{\omega}{\lambda}\frac{d\tau_g}{d\omega} = -\frac{\omega}{\lambda}\frac{d^2\beta}{d\omega^2} \tag{10.42}$$

Now, β may be expressed in terms of the relative refractive index difference Δ and the normalized propagation constant b as:

$$\beta = n_1 k\left[1 - 2\Delta(1-b)\right]^{1/2} \tag{10.43}$$

In a fiber of length L, the rms pulse broadening caused by intramodal dispersion is given by the derivative of the group delay with respect to wavelength as:

$$\text{Total rms pulse broadening} = \sigma_\lambda L\left|\frac{d\tau_g}{d\lambda}\right|$$

$$= \frac{\sigma_\lambda L 2\pi}{c\lambda^2}\frac{d^2\beta}{dk^2} \tag{10.44}$$

where σ_λ is the source rms spectral linewidth centered at a wavelength λ.

The detailed calculations of the derivatives give the final expression for the total dispersion parameter, which may be separated into three composite dispersion components. The effects are as follows:

(*i*) The waveguide dispersion parameter D_W, which is given by:

$$D_W = -\left(\frac{n_1 - n_2}{\lambda c}\right)V\frac{d^2(Vb)}{dV^2} \tag{10.45}$$

where V is the normalized frequency for the fiber. The normalized waveguide dispersion coefficient $Vd^2(Vb)/dV^2$ depends on V.

(*ii*) The material dispersion parameter D_M, which is defined by:

$$D_M = \frac{\lambda}{c}\left|\frac{d^2 n}{d\lambda^2}\right| \tag{10.46}$$

where $n = n_1$ or n_2 for the core or cladding respectively.

(*iii*) A profile dispersion parameter D_P proportional to $d\Delta/d\lambda$. However, the profile dispersion parameter can be quite small, especially at long wavelengths and hence is often neglected in rough estimates of total dispersion within single-mode fibers.

Thus the total first order dispersion D_T in a practical single-mode fiber may be written as:

$$D_T = D_W + D_M + D_P \ (\text{ps nm}^{-1}\ \text{km}^{-1}) \tag{10.47}$$

However, in standard single-mode fibers the total dispersion tends to be dominated by the material dispersion of fused silica. This parameter is shown plotted against wavelength in Fig. 10.9. It may be seen that the characteristic goes through zero at a wavelength of about 1270 nm. This Zero Material Dispersion (ZMD) point can be shifted anywhere in the wavelength range 1200 to 1400 nm by the addition of suitable dopants. With zero material dispersion the pulse broadening is dominated by the waveguide dispersion coefficient $Vd^2\ (Vb)/dV^2$.

The total fiber dispersion, which depends on both the fiber material composition and dimensions, may be minimized by trading off material and waveguide dispersion while limiting the profile dispersion. For wavelengths longer than the ZMD point, D_M is positive whereas D_W is negative as illustrated in Fig. 10.10. However, the total dispersion D_T is approximately equal to the sum of D_W and D_M. Hence, for a particular wavelength, λ_0 which is slightly larger than the ZMD point wavelength, the waveguide dispersion compensates the material dispersion and the total first order dispersion parameter becomes zero. By careful control of the fiber profile and core diameter the wavelength λ_0 may be chosen in the range 1300 to 2000 nm.

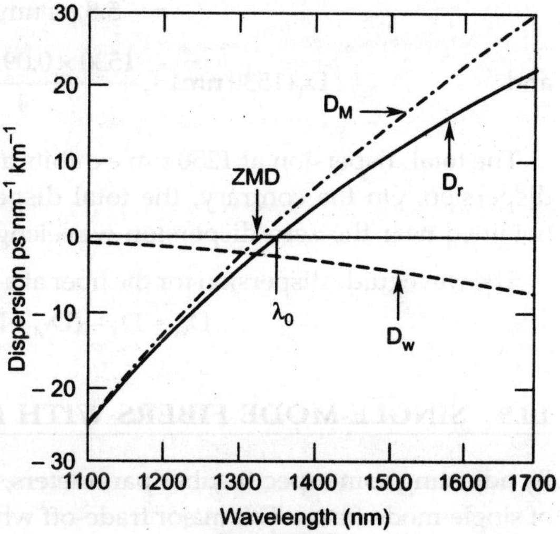

Fig. 10.10

By the combination of three techniques, the wavelength λ_0 at which the first order dispersion is zero may be extended to 1550 nm. The techniques are:

(*i*) The relative refractive index difference Δ for the fiber is increased;

(*ii*) The silica is suitably doped with germanium;

(*iii*) The normalized frequency (V value) for the fiber is decreased.

By these techniques the bandwidth-length product for such single-mode fibers increases to be in excess of 100 GHz km^{-1}.

The intramodal dispersion variation with wavelength is usually characterized by the second order dispersion parameter or dispersion slope S which is written as:

$$S = \frac{dD_T}{d\lambda} = \frac{d^2\tau_g}{d\lambda^2} \tag{10.48}$$

An important value of the dispersion slope $S(\lambda)$ is obtained at the wavelength of minimum intramodal dispersion λ_0 as:

$$S_0 = S(\lambda_0) \tag{10.49}$$

where S_0 is called the zero-dispersion slope. When the two parameters λ_0 and S_0 are specified, the total chromatic dispersion at an arbitrary wavelength can be written as:

$$D_T(\lambda) = \frac{\lambda S_0}{4}\left[1 - \left(\frac{\lambda_0}{\lambda}\right)^4\right] \tag{10.50}$$

Example 10.14. *A single-mode fiber has a zero-dispersion wavelength (λ_0) of 1310 nm with a dispersion slope of 0.09 ps nm^{-2} km^{-1}. Compare the total first order dispersion for the fiber at the wavelengths of 1250 nm and 1530 nm. When the material dispersion and profile dispersion at the latter wavelength are 13.0 ps nm^{-1} km^{-1} and 0.5 ps nm^{-1} km^{-1}, respectively, determine the wavelength dispersion at this wavelength.*

Solution: The total first order dispersion for the fiber at the two wavelengths may be obtained from equation (10.50) as:

$$D_T(1250\text{ nm}) = \frac{\lambda S_0}{4}\left[1 - \left(\frac{\lambda_0}{\lambda}\right)^4\right] = \frac{1250 \times 0.09 \times 10^{-12}}{4}\left[1 - \left(\frac{1310}{1250}\right)^4\right]$$

$$= -5.8 \text{ ps nm}^{-1}\text{ km}^{-1}$$

and

$$D_T(1530\text{ nm}) = \frac{1530 \times 0.09 \times 10^{-12}}{4}\left[1 - \left(\frac{1310}{1530}\right)^4\right] = 15.9 \text{ ps nm}^{-1}\text{ km}^{-1}$$

The total dispersion at 1250 nm exhibits a negative sign due to the influence of the waveguide dispersion. On the contrary, the total dispersion at 1530 nm is considerably greater than that obtained near the zero-dispersion wavelength.

The waveguide dispersion for the fiber at a wavelength of 1530 nm is given by equation (10.47) as:

$$D_W = D_T - (D_M + D_P) = 15.9 - (13.0 + 0.5) = 2.4 \text{ ps nm}^{-1}\text{ km}^{-1}$$

10.9 SINGLE-MODE FIBERS WITH MODIFIED DISPERSION

By adjusting some specific fiber parameters, it is possible to modify the dispersion characteristics of single-mode fibers. The major trade-off which will occur in this process between the waveguide dispersion and material dispersion is expressed as:

$$D_T = D_W + D_M \tag{10.51}$$

In most of the common fiber designs at longer wavelengths than the ZMD point, the D_W and D_M are of opposite sign. Hence D_W and D_M can be made to cancel each other at some longer wavelengths. Hence the parameters of the fibers may be such modified so as to provide low dispersion and low loss fiber. This may be achieved by a reduction in the fiber core diameter with an accompanying increase in the relative or fractional index difference. This mechanism creates *Dispersion Shifted* (DS) single-mode fibers.

Alternatively the dispersion characteristics of single-mode fibers may be modified by the achievement of a low dispersion window over low loss wavelength region between 1300 nm and 1600 nm. Such fibers are called *Dispersion Flattened* (DF) single-mode fibers. These type of fibers relax the spectral requirements for optical sources and allow flexible WDM. To obtain DF fibers multilayer index profiles are fabricated with increased waveguide dispersion which provides overall dispersion over the entire wavelength range 1300 to 1600 nm. Basically, these fibers exhibit two wavelengths of zero total chromatic dispersion.

10.9.1 Dispersion Shifted Fibers

By altering the material composition of the single mode fiber the zero dispersion wavelength point λ_0 can be shifted to longer wavelength. The normalized frequency V have to be maintained within the range 1.5 μm to 2.4 μm to suitably confine the power of the fundamental mode. The fractional index difference must be increased as a square function while the core diameter has to be linearly reduced to keep V constant. This may be achieved by increasing the level of germanium doping within the core of the fiber. With the simple step index approach to dispersion shifting there is a major disadvantage. The problem is that the fibers produced exhibit relatively high dopant dependent losses at wavelengths around 1550 nm. There are several causes which may result in such excess optical loss. It has been found that such problem is reduced while using graded index approach.

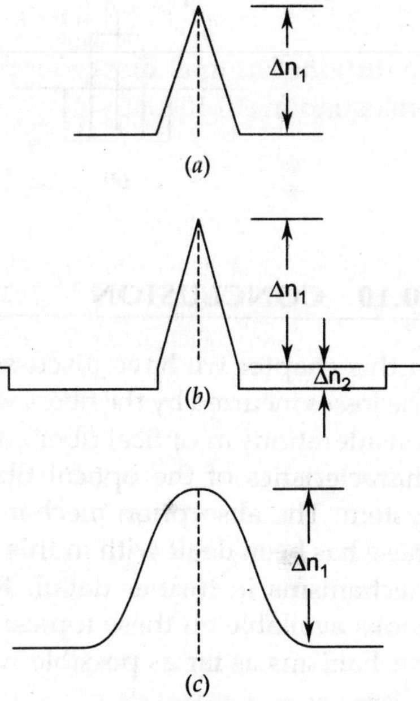

Fig. 10.11

Some graded refractive index profile DS fibers are shown in Fig. 10.11. The triangular profile shown in Fig. 10.11 (*a*) is the simplest and was the first to exhibit low loss (*i.e.*, 0.24 *d*B km^{-1}) which was same to the conventional non-shifted single mode fiber. However such profile caused the LP_{11} mode to cutoff in the wavelength region 850 to 900 nm. Hence the fiber has to be operated far from cutoff which produces sensitivity to bend losses at the 1550 nm wavelength. This drawback may be overcome by employing a triangular index profile combined with a depressed cladding index, as illustrated in Fig. 10.11 (*b*). Low losses and zero dispersion at a wavelength 1550 nm have also been obtained with a Gaussian refractive index profile, as shown in Fig. 10.11 (*c*). This profile produced losses of 0.21 *d*B km^{-1} at the λ_0 wavelength of 1550 nm.

DS single-mode fibers may be alternatively produced by the use of multiple index designs. Doubly clad or W fiber is one such type of fiber which shows dispersion shifting property, but it is more widely used for DF fibers. However the multiple index triangular profile fibers and the segmented-core triangular profile designs which are illustrated in Figs. 10.12 (*a*) and (*b*), respectively have reduced the microbending sensitivity while maintaining a MFD of around 9 μm at a wavelength of 1550 nm. In recent times dual-shaped core DS fibers have come into existence to provide an improvement in bend loss performances over the 1550 nm wavelength region. Such dual-shape core refractive index profile is illustrated in Fig. 10.12 (*c*).

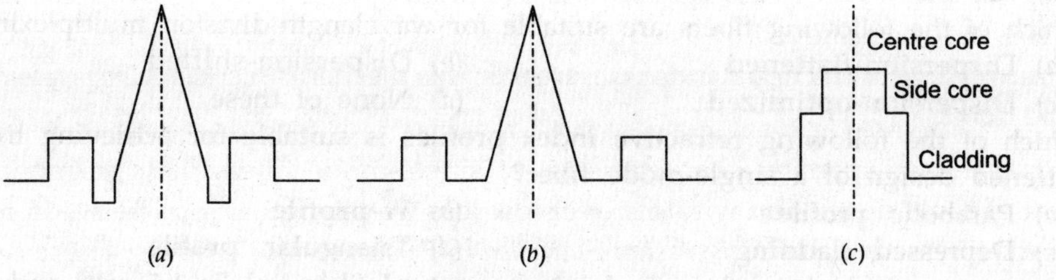

Fig. 10.12

10.9.2 Dispersion Flattened Fibers

The first practical demonstration of dispersion flattening using the W structure was reported in 1981 by T. Miya, K. Okamoto, Y. Ohmori and Y. Sasaki. However, a high degree of dimensional control is to be maintained with the W structural design so as to make reproducible DF fibers. A typical W fiber index profile (double clad) is illustrated in Fig. 10.13 (*a*).

The sensitivity to bend losses associated with the W fiber structure may be reduced by retrapping the light which penetrates into the outer cladding area by introducing a further region of raised index into the structure. This method has given rise to Triple Clad (TC) and Quadruple Clad (QC) structures as illustrated in Fig. 10.13 (*b*) and (*c*) respectively.

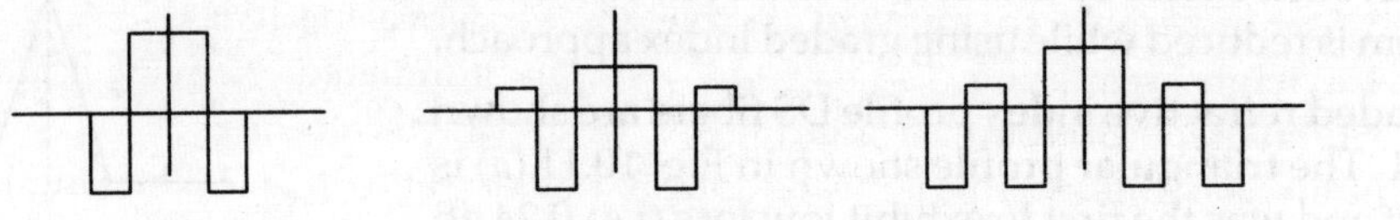
(*a*) (*b*) (*c*)

Fig. 10.13

10.10 CONCLUSION

In this chapter we have discussed on the various transmission characteristics of optical fibers. The losses incurred by the fibers while the optical signal is propagating are one of the most important considerations in optical fiber communication systems. Without detail prior knowledge of these characteristics of the optical fibers an optical system designer will not be able to design the system. The absorption mechanisms, attenuation techniques, the dispersion phenomena – all these has been dealt with in this chapter. It is not within the scope of this book to deal with these mechanisms in further detail. Readers who are interested in these topics can consult further books available on these topics. We have tried to give an overview on all the above mentioned mechanisms as far as possible within the scope of this current text.

REVIEW QUESTIONS

1. An impulse is launched into one end of a 30 km optical fiber with a rated total dispersion of 20 ns/km. What will be the width of the pulse at the other end?
   (*a*) 100 ns (*b*) 300 ns (*c*) 600 ns (*d*) None of these
2. A pulse has half-width of 100 ns. The pulse is transmitted through an optical fiber of length 20 km. The fiber has a rated multipath tine dispersion of 10 ns km^{-1} and a material dispersion of 2 ns km^{-1}. The half-width of the received pulse is
   (*a*) 423 ns (*b*) 332 ns (*c*) 297 ns (*d*) 227 ns
3. Which of the following fibers are suitable for wavelength-division multiplexing of signals?
   (*a*) Dispersion-flattened (*b*) Dispersion-shifted
   (*c*) Dispersion-optimized (*d*) None of these
4. Which of the following refractive index profiles is suitable for achieving the dispersion-flattened design of a single-mode fiber?
   (*a*) Parabolic profile (*b*) W-profile
   (*c*) Depressed cladding (*d*) Triangular profile
5. The mean optical power launched into an optical fiber link is 1.5 mW and the fiber has an attenuation of 0.5 dB km^{-1}. What is the maximum possible link length without repeaters (assuming lossless connectors) when the minimum mean optical power level required at the detector is 2 μW?

(*a*) 57.5 km (*b*) 50.3 km (*c*) 66.9 km (*d*) 21.8 km

6. A glass core optical fiber has an attenuation resulting from Rayleigh scattering of 0.46 dB km^{-1} at a wavelength of 1000 nm. The fictive temperature of the glass is 758 K, isothermal compressibility of 8.4×10^{-11} m^2 N^{-1}, and a photoelastic coefficient of 0.245. What is the value of the refractive index of the glass?

(*a*) 1.31 (*b*) 1.49 (*c*) 1.62 (*d*) 1.55

7. A single-mode step index fiber with a core refractive index of 1.49 has a critical bending radius of 10.4 mm when illuminated with a light at a wavelength of 1300 nm. If the cutoff wavelength for the fiber is 1150 nm estimate its refractive index difference.

(*a*) 0.23% (*b*) 0.32% (*c*) 0.47% (*d*) 0.64%

8. The phenomenon that occurs when the size of the inhomogeneities within the optical fiber are comparable to the size of the guided wavelength is

(*a*) Rayleigh scattering (*b*) Stimulated Raman scattering
(*c*) Stimulated Brillouin scattering (*d*) Mie scattering

9. The numerical input/output mean optical power ratio in a 1 km length of optical fiber is found to be 2.5. What will be the received mean optical power when a mean optical power of 1 mW is launched into a 5 km length of the fiber?

(*a*) 10 mW (*b*) 10 μW (*c*) 15 μW (*d*) None of these

10. A single-mode step-index fiber has a core index of 1.46 and a core radius of 4 μm. The relative refractive index difference is 0.52%. Calculate the cutoff wavelength for the fiber.

(*a*) 1556 nm (*b*) 1230 nm (*c*) 1653 nm (*d*) 1458 nm.

11. What are the intrinsic and extrinsic absorption loss mechanisms in optical fibers?

12. Briefly discuss linear scattering losses in optical fibers with regard to:

(*a*) Mie scattering;
(*b*) Rayleigh scattering.

13. Compare stimulated Raman scattering and stimulated Brillouin scattering in optical fibers, and discuss how they may be avoided in optical fiber communications.

14. What do you mean by the critical bending radius of an optical fiber? Why could bending loss in single-mode fibers be severe? What can be done to minimize this loss?

15. What are the intermodal and intramodal dispersions? Discuss how intramodal dispersion may be minimized within a single-mode fiber.

16. Briefly explain the reasons for pulse broadening due to material dispersion in optical fibers. Derive an expression for the rms pulse broadening due to material dispersion in an optical fiber and define the material dispersion parameter.

17. Discuss the techniques employed and the fiber structures utilized to provide:

(*a*) dispersion flattened single-mode fibers;
(*b*) dispersion shifted single-mode fibers.

18. The threshold optical powers for SRS and SBS in a long 8 μm core diameter single-mode fiber are found to be 1.70 W and 190 mW, respectively. An injection laser source with a bandwidth of 1 GHz is used. Estimate the operating wavelength of the laser and the attenuation in decibels per kilometer of the fiber at this wavelength.

19. A single-mode step index fiber with a core refractive index of 1.49 has a critical bending radius of 10.4 mm when illuminated with light at a wavelength of 1300 nm. If the cutoff wavelength for the fiber is 1150 nm calculate its relative refractive index difference.

20. While being used with a LED source of rms spectral width of 25 nm, a multimode graded index fiber having a parabolic profile has a material dispersion parameter of 30 ps nm^{-1} km^{-1}. The numerical aperture of the fiber is 0.4 and the core axis refractive index is 1.48. Assuming waveguide dispersion to be negligible find out the total rms pulse broadening per kilometer within the fiber. Hence, estimate the bandwidth-length product for the fiber.

11

Optical Transmission Systems, Modulation and Detection Schemes

11.1 INTRODUCTION

The transfer of information in the form of light enables data transmission at high rate in a single communication channel, as it has the potential to be modulated with the carrier frequency as high as 10^{14} Hz, which is much higher than the either radio or microwaves. However successful optical transmission requires a suitable guiding medium. Atmospheric transmission is limited to line of sight communication and is subjected to weather condition. It was the development of laser and cladded dielectric waveguide that became practical proposition for optical communication system. But all the lasers are not equally suitable for launching adequate power into these waveguide. Interestingly, LED in many ways is suitable, though it is an incoherent source. Introduction of optical fiber as guided medium brought the revolution in optical communication system. However, to achieve reliable and secure communication, it is essential that all the components or subsystems within the communication system are compatible, so that overall system performance is enhanced.

The basic components or subsystems of a general optical fiber communication system for either analog or digital transmission are illustrated in Fig. 11.1. The transmitter side consists of an information encoder or signal shaping circuit preceding a modulation or electronic driver stage which operates the optical source. Light emitted from the optical source launches into the optical fiber that constitutes the transmission medium. At the receiver end, the light emerged from the transmission medium is converted into electrical signal by optical detector. The electrical signal is amplified and then decoded or demodulated to obtain the information. To enable the successful incorporation of these components into optical fiber communication system, it is necessary to consider the interaction of one component with another and then to evaluate the overall performance of the system. To optimize the system performance for a specific application, it is often helpful to offset a particular component characteristic by trading it off against the performance of another component, in order to enhance the overall system performance.

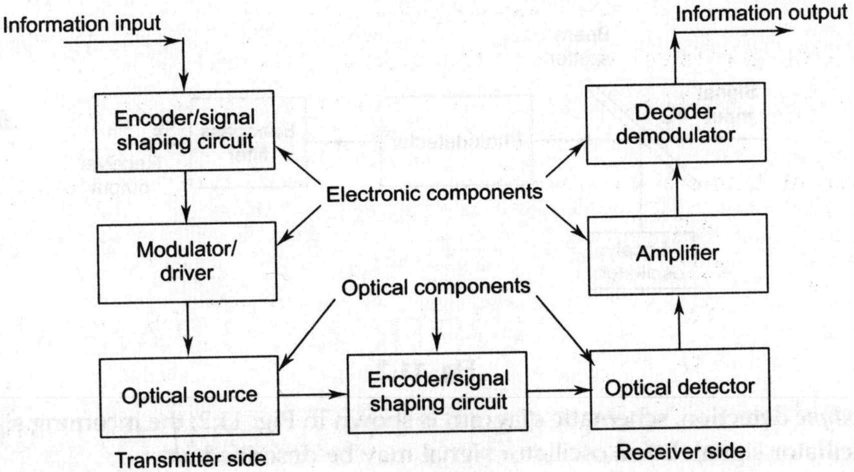

Fig. 11.1

11.2 MODULATION SCHEMES

Modulation is the process of varying one of the parameters associated with a carrier wave to enable it to carry information. Variations in the amplitude, frequency, phase, irradiance or polarization can all be used for this purpose. However, all these possibilities are not suitable when carrier wave oscillates at optical frequencies. Most optical detectors respond only to the irradiance of the light, and so special modulation techniques are employed.

11.2.1 Analog Modulation

With analog modulation, the primary information signal that is a time varying electrical signal continuously varies the appropriate wave parameter. Hence at any time magnitude of the modulated wave is directly related with the original signal amplitude. The magnitude electric field of the carrier may be written as

$$E_c(t) = A_c \cos(\omega_c t + \Phi_c) \tag{11.1}$$

Where A_c ω_c and Φ_c represent the carrier amplitude, angular frequency and phase respectively, all of which may be modulated. If this signal is applied to a detector, the output of the detector O_d which responds only to the irradiance is given by

$$O_d = R\frac{1}{T}\int\left\{A_c \cos\left(\omega_c t + \Phi_c\right)\right\}^2 dt$$

or $$O_d = RA_c^2/2 \tag{11.2}$$

R is the detector responsivity. The equation 11.2 indicates that in direct detection the output depends only upon the signal irradiance or square of the signal amplitude. Therefore modulation of other wave parameters is not possible. There are other techniques which enable to extract further information from the received signal.

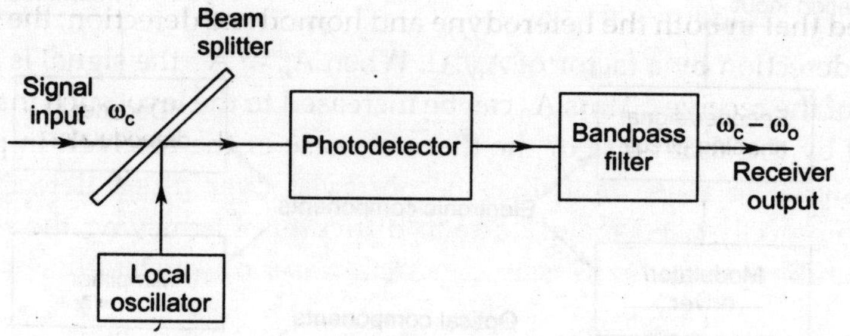

Fig. 11.2

In *heterodyne* detection, schematic diagram is shown in Fig. 11.2, the incoming signal is mixed with local oscillator signal. Local oscillator signal may be described as

$$E_0(t) = A_0 \cos(\omega_0 t + \Phi_0) \tag{11.3}$$

where ω_0 is chosen very closed to ω_c. A_0, in contrast to A_c, is not time dependent. The output of the detector is now expressed as

$$O_d = R\frac{1}{T}\int\left\{E_c^{\,2}(t) + E_0^{\,2}(t)\right\}dt$$

$$= R\frac{1}{T}\int\left\{A_c\cos(\omega_c t + \Phi_c) + A_0\cos(\omega_0 t + \Phi_0)\right\}^2 dt$$

$$= R\frac{1}{T}\int\left[\left\{A_c\cos(\omega_c t + \Phi_c)\right\} + \left\{A_0\cos(\omega_0 t + \Phi_0)\right\}^2\right.$$

$$\left. + 2A_cA_0\cos(\omega_c t + \Phi_c)\cos(\omega_0 t + \Phi_0)\right]dt$$

$$= R\left(\frac{A_c^{\,2}}{2} + \frac{A_0^{\,2}}{2} + \frac{1}{T}\int 2A_cA_0\cos(\omega_c t + \Phi_c)\cos(\omega_0 t + \Phi_0)dt\right) \tag{11.4}$$

If ω_0 is closed to ω_c and a bandpass filter of center frequency $(\omega_c - \omega_0)$ is used, equation (11.4) may be reduced to

$$O_d = R\left(\frac{A_c^{\,2}}{2} + \frac{A_0^{\,2}}{2} + A_cA_0\cos\left[(\omega_c - \omega_c)t + (\Phi_c - \Phi_0)\right]\right) \tag{11.5}$$

Equation (11.5) indicates that O_d is dependent on carrier amplitude, angular frequency and phase. Therefore information may be carried by any of these parameters.

In *homodyne* detection system, the local oscillator is set to the same frequency as carrier. The equation (11.4) may be deduced as

$$O_d = R\left(\frac{A_c^{\,2}}{2} + \frac{A_0^{\,2}}{2} + A_cA_0\cos(\Phi_c - \Phi_0) + \frac{1}{T}A_cA_0\cos\left\{2\omega_c t + (\Phi_c + \Phi_0)\right\}\right) \tag{11.6}$$

In this case, the detected signal is passed through a bandpass filter whose ceter frequency is $\ll \omega_c$, so that it blocks all the terms except $A_cA_0\cos(\Phi_c - \Phi_0)$. If in addition $A_0 \gg A_c$, then equation (11.6) is reduced to

$$O_d = R\,A_cA_0\cos(\Phi_c - \Phi_0) \tag{11.7}$$

Hence both amplitude and phase modulation may be detected by homodyne receiving system.

It may be noticed that in both the heterodyne and homodyne detection, the signal amplitude is larger than direct detection by a factor of A_0/A_c. When $A_0 >> A_c$, the signal is amplified enabling higher sensitivity of the receiver. Thus A_0 can be increased to the level such that the receiver noise is only dominated by the shot noise or the thermal noise of the receiver. In practice 10 to 20 *d*B improvement in signal to noise ratio may be observed over direct detection. However such systems are not commonly used mainly due to difficulty of achieving the stability in required frequency, phase and polarization for semiconductor laser or optical fiber systems.

11.2.2 Digital Modulation

In digital modulation system, the analog signal is first sampled, then quantized and coded. The information is coded into a series of pulses whose temporal positions are fixed. The simplest scheme is called *two level binary*, where pulses are occurred at predetermined time slots having the amplitudes of only one of the two levels—*zero* and *one*. The process is illustrated in Fig. 11.3, where each sample of analog signal is coded to three-bits digital signal. In practice, codification is done with higher number of bits.

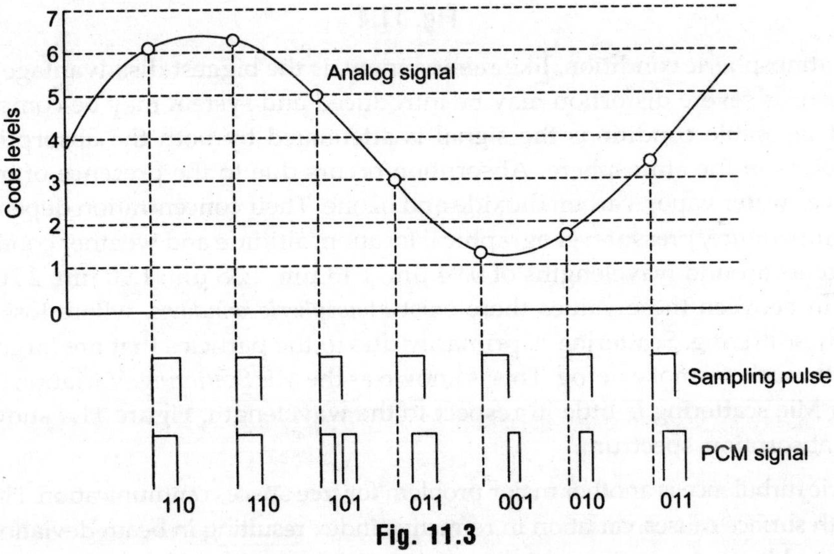

Fig. 11.3

The system frequency requirement in digital modulation is much higher compared to analog modulation. As example, for transmission of a 4 KHz signal the digital system with 8 bits codification the frequency requirement is at least $4 \times 2 \times 8 = 64$ KHz, analog system frequency requirement is only $4 \times 2 = 8$ KHz. But great advantage in digital modulation is its relative freedom from noise or distortion. Digital system is capable of accurately imparting original information from a severely distorted signal. Several schemes are available for codification of quantized signal that includes the error correction facility such that receiving signal can be corrected if any bit error occurs.

11.3 FREE SPACE COMMUNICATION

Although fiber optical communication is the recent trend, free space communication over a distance of few kilometers is a practical possibility for fairly low bandwidth data transmission. Systems can be set up easily and costs are considerably low compared to fiber optic systems. Well-collimated beam of low divergence can provide considerable security advantages over radio communication. Typical applications would be a voice or low data rate link between nearby buildings in urban environment, or at check posts situated at mountain peaks (for military use).

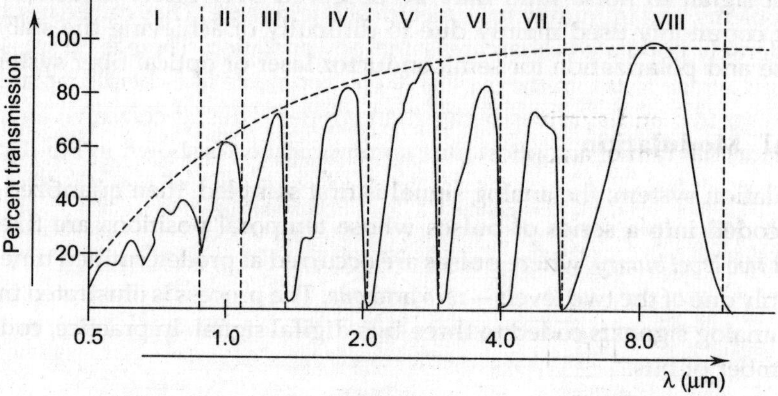

Fig. 11.4

Adverse atmospheric condition, like rain or snow, is the biggest disadvantage for free space communication, as severe distortion may be introduced and system may be completely inoperable. Even at favorable conditions the signal is attenuated by both the absorption as well as scattering process in the atmosphere. Absorption occurs due to the presence of molecular constituents such as water vapor, carbon dioxide and ozone. Their concentration depends on various factors like temperature, pressure, geographical location, altitude and weather conditions. Strong absorption occurs around wavelengths of 0.94 μm, 1.13 μm, 1.38 μm, 1.90 μm, 2.70 μm, 4.30 μm and 6.0 μm. In between these values there exist *atmospheric windows*, where losses are mainly determined by scattering. Scattering is primarily due to the particles that are large compared to the wavelength, such as smoke or fog. This is known as the *Mie Scattering*. Variation of the effective absorption in Mie scattering is little in respect to the wavelength. Figure 11.4 shows the typical atmospheric absorption spectrum.

Atmospheric turbulence is another major problem for free space communication. Heating of air in contact of earth surface causes variation in refractive index resulting in beam deviation and spreading. This gives sudden and pronounced fading of the optical signal. In ideal condition, optical communication is possible up to the range of 150 Km, however this range is limited to only 2 Km to achieve reasonably large bandwidths. Fluctuations in absorption and scattering tend to modulate the signal strength in the range of 1 to 200 Hz and hence amplitude modulation is usually not satisfactory.

For simplest system, LED emitters and silicon photodetectors coupling with suitable lens and mirror system may be used in free space optical communication. Nd-YAG or CO_2 laser sources may be used to form a complex system, as their small beam divergence simplifies the beam collimation. Frequency modulation can be done in CO_2 laser operating in single mode by introducing an electro-optical crystal into the resonant cavity. A signal voltage applied to the crystal changes its length and hence the optical length of the cavity resulting in change in single mode resonance frequency. Nd-YAG laser may be made to emit a train of pulses by mode-locking technique and a signal is impressed on this pulse train using an external modulator. However the use of such system is limited due to high system complexity and cost.

11.4 OPTICAL FIBER COMMUNICATION

With the advent of optical fibers, low loss and high information rate communication became feasible, as the optical fiber has very less attenuation in the range of 1dB/Km and high bandwidth up to several gegaherz. The advantages of optical fiber include small physical size and weight, elimination of ground loop problems, and immunity from electrical interference. They can be used in noisy environments such as densely populated urban areas and power station. There is no spark hazard which is the additional advantage to use in the explosive environments like chemical plants. Optical fiber is relatively corrosion free and survives better than copper wires at corrosive environment like sea water. A schematic diagram of an optical fiber communication is shown in Fig. 11.5.

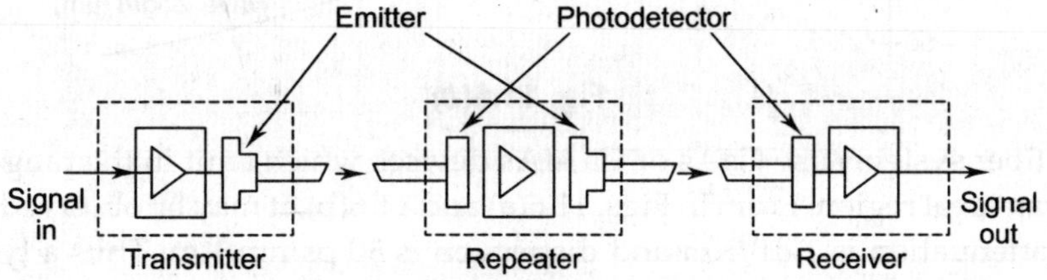

Fig. 11.5

The basic components of transmitter, receiver and repeater are emitters, photodetectors and optical fibers as shown in Fig. 11.5. They all have been discussed in earlier chapters. Here the requirements and criteria for communication system will be discussed.

11.4.1 Operating Wavelength

Two important characteristics of optical fibers depend on wavelength are attenuation and material dispersion. Typical curves for attenuation verses wavelength and material dispersion verses wavelength of a silica based fiber are shown in Figs. 11.6(*a*) and 11.6(*b*) respectively. The pulse spread over $\Delta\tau$ over a fiber length L may be expressed as

$$\Delta\tau = \left| \frac{-L\lambda}{c} \left(\frac{d^2 n}{d\lambda^2} \right) \Delta\lambda \right| \tag{11.8}$$

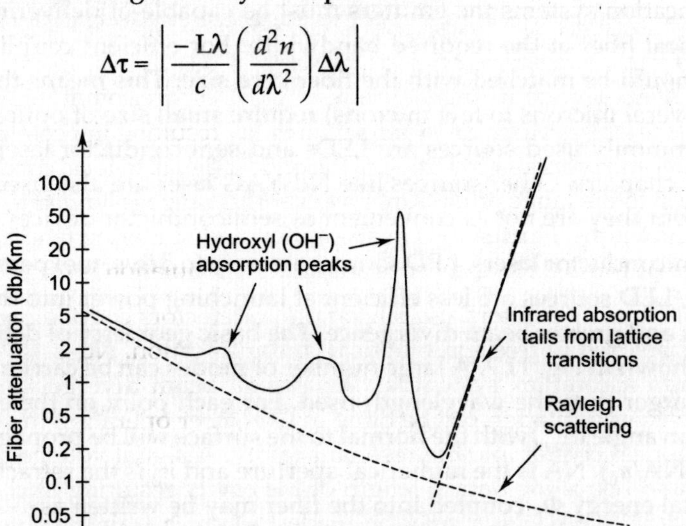

Fig. 11.6 (*a*)

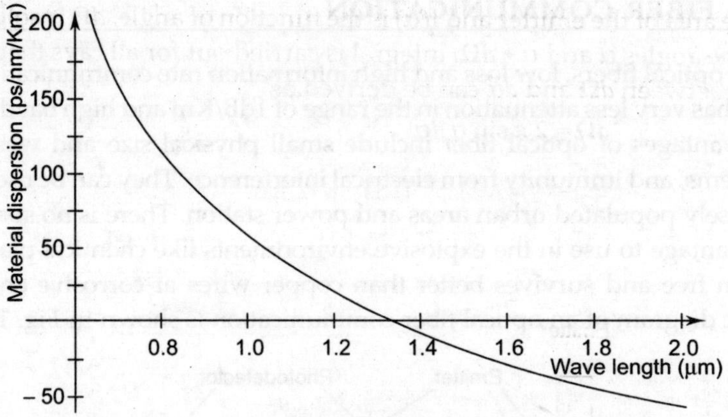

Fig. 11.6(*b*)

Many glass fiber systems use GaAs or GaAlAs devices which emit in the range 0.82 μm to 0.9 μm. This is not an ideal region. From in Figs. 11.6(a) and 11.6(b), it may be observed that at 0.85 μm, the minimum attenuation is 2 dB/Km and dispersion is 80 ps/nm/Km. Thus a typical LED with linewidth of 50 nm would suffer a dispersion of 4 ns/Km and hence the system bandwidth would be limited to 25 MHz for a 10 Km fiber length. However laser has a considerable advantage because of its narrow linewidth and higher bandwidth may be achieved. Fiber attenuation can be considerably reduced by operating at longer wavelengths. At 1.55 μm the attenuation is about 1.55 dB/Km. The material dispersion is also low at the range. Suitable detector like GaInAs based device may be used for longer wavelengths of 1.3 μm to 1.55 μm.

Plastic fibers have extremely high attenuations outside the visible region and usually used in conjunction with red emitting LED.

11.4.2 Emitter Design

In optical communication systems the emitters must be capable of delivering useful amount of power into the optical fiber at the required bandwidth. For efficient coupling of radiation into fiber emitter size should be matched with the fiber core size. This means the small diameter of fiber cores (from several microns to few microns) require small size of optical sources. The most convenient and commonly used sources are LEDs and semiconductor lasers, which have been discussed at earlier chapters. Other sources like Nd-YAG laser are also used matching with the diameter of fibers, but they are not as convenient as semiconductor devices.

Compared to semiconductor lasers, LED sources are easy to drive, inexpensive and have higher lifetimes. However, LED sources are less efficient at launching power into the fibers due to their larger emitting area and greater beam divergence. The basic geometry of delivering power into a step index fiber is shown in Fig. 11.7. A large number of modes can be carrier by fiber if fiber core diameter is much larger than the wavelength used. For each point on the emitter surface, only light emitted up to an angle α_{max} with the normal to the surface will be propagated down the fiber, where $\alpha_{max} = \sin^{-1}(NA/n_0)$. NA is the numerical aperture and n_0 is the refractive index (also refer section 2.5). The total energy Φ_F coupled into the fiber may be written as

$$\Phi_F = A_s \int_{\Omega} I(\alpha)\, d\Omega \tag{11.9}$$

where A_s is surface area of the emitter and $I(\alpha)$ is the function of angle. $d\Omega$ is solid angle subtended by rays between the angles α and $\alpha + d\Omega$. Integral is carried out for all rays that are trapped in the fiber. The relation between $d\Omega$ and $d\alpha$ can be derived as

$$d\Omega = 2\pi \sin\alpha \, d\alpha \tag{11.10}$$

Hence

$$\Phi_F = 2\pi A_s \int_0^{\alpha_{min}} I(\alpha) \sin\alpha \, d\alpha \tag{11.11}$$

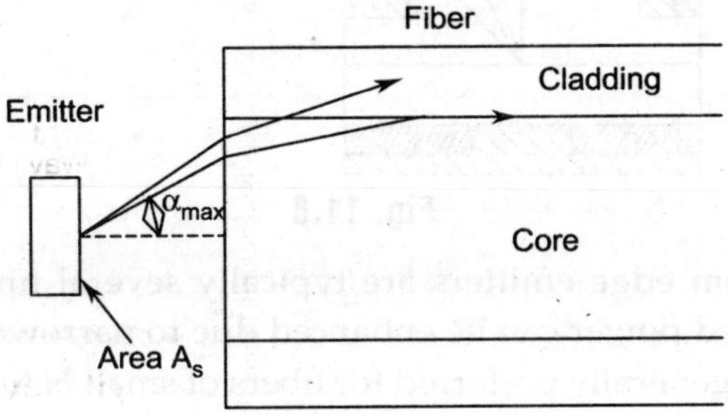

Fig. 11.7

LED sources are generally Lambertian so that we may write $I(\alpha) = I(0) \cos\alpha$. Therefore

$$\Phi_F = 2\pi A_s \int_0^{\alpha_{min}} I(0) \cos\alpha \sin\alpha \, d\alpha$$

$$= \pi A_s I(0) \sin^2 \alpha_{max}$$

So

$$\Phi_F = \pi A_s I(0) (NA/n_0)^2 \tag{11.12}$$

The total output power of a LED is $\pi A_s I(0)$ and hence the coupling efficiency may be derived as

$$\eta_c = \frac{\text{Energy coupled into fiber}}{\text{Energy emitted by LED}} = \frac{\pi A_s I(0)(NA/n_0)^2}{\pi A_s I(0)} = \frac{(NA)^2}{n_0^2} \tag{11.13}$$

For a typical value of NA of 0.3, with air interference only 9% of total emitted power of LED is coupled into the fiber. In the above derivation the Fresnel losses have been ignored. Moreover, if A_s is larger than A_c (area of the core), the effieciency will be further reduced by the factor of A_c/A_s.

For a graded index fiber the calculation is more complicated. As a rough estimate, coupling power is reduced by a factor of 2 at same core diameter, as graded index fiber supports about half number of the modes compared to step index fiber.

Therefore, to maximize the coupling efficiency, the fiber core of large NA and source area smaller than the core area are required. But signal bandwidth decreases with the increase of NA, due increase of mode dispersion. On the other hand reduction in emitter area of the LED tends to reduce the operating lifetime.

The two main types of LED most often used in optical fiber communication system are the surface etched well emitter (Burrus type) and the edge emitter (refer section 2.6.2 and 2.6.3). In Burrus type, a well is etched into the top of a planner LED structure enabling the fiber end to be as close as possible to the light emitting region (*i.e.*, the *p-n* junction). If the emitting area is less than that of fiber core, some form of optical coupling like hemispherical lens on top as in Fig. 11.8 is advantageous. This will magnify the selective surface area, but demagnify the solid angular distribution of the radiation, so that the resulting radiation pattern becomes better compatible to the fiber acceptance pattern.

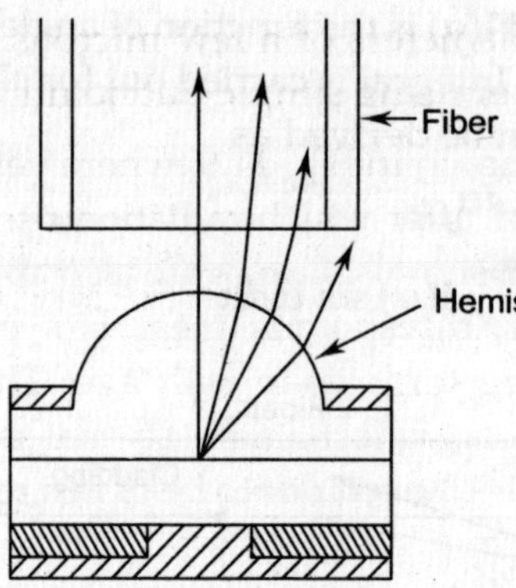

Fig. 11.8

Total optical powers from edge emitters are typically several times smaller than that from surface emitters. But coupled power can be enhanced due to narrower beam divergence at edge emitters. Edge emitters are generally preferred for fibers of small NA (NA = 0.4), whereas surface emitters are more suitable for fibers of larger NA because of their total output power.

Since LEDs have an almost linear relationship between drive current and light output, they are more suitable for amplitude modulation. In case of pulse modulation drive current must be from high value to low value to obtain a wide ratio between ON state and OFF state. This ultimately affects the modulation bandwidth and lifetime. At low current levels junction capacitance limits the bandwidth and at high current levels lifetime of carriers injected into recombination region is affected. If τ is the minority carrier lifetime, the relative response R(f) at frequency f may be written as

$$R(f) = \frac{R(0)}{\left(1 + 4\pi^2 f^2 \tau^2\right)^{1/2}} \tag{11.14}$$

In GaAs the typical value of τ is 10^{-9} s, and the half-power bandwidth is obtained around 300 MHz.

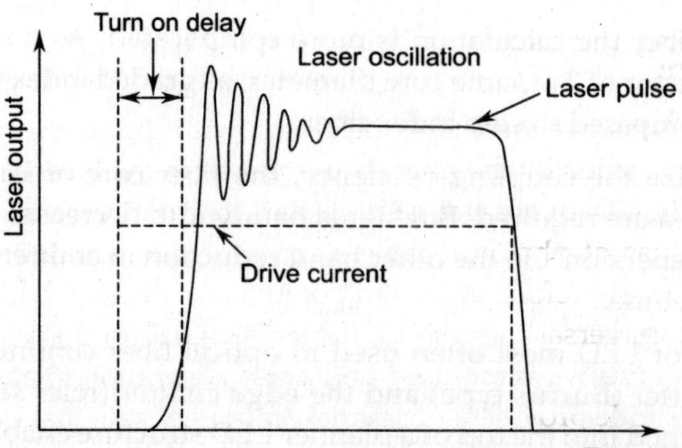

Fig. 11.9

Lasers, on the other hand, exhibit much output power, narrower beam divergence and smaller emitting area. All these factors enable lasers to couple significantly more power into the fibers than LEDs. For fibers of small diameter and low numerical aperture, lasers are especially useful, in particular single mode fibers of core diameters of a few microns. A laser of 10 μm strip can couple about 1 mW into a single mode fiber using simple butt jointing techniques.

The output characteristics of lasers is shown in Fig. 11.9, where very little radiation is emitted until the current reaches a threshold value, after which radiation rises sharply. This characteristic of lasers makes them suitable for digital signals. Lasers are also applicable to the transmission of analog signals as the characteristic above threshold is linear. However the threshold current is temperature sensitive and can show long term drift with age. Therefore good temperature stabilization is required with some feedback mechanism so that a constant power output is maintained. Generally a photodetector (which need not to be of fast response) is mounted next to rear facet of laser to regulate the drive current.

Laser modulation bandwidths are larger compared to that of LEDs, as the carrier recombination time is appreciably low by the action of stipulated emission. However two undesirable features are observed in case of laser pulses. First, the optical output is delayed in respect of the drive current pulse and secondly, some rapid oscillation is superimposed at the beginning of the pulse. These are illustrated in Fig. 11.9. Permanent DC bias current may be incorporated for laser operation to reduce the turn-on-delay, so that the total current never falls below the threshold level. The magnitude of initial oscillation or spiking at laser pulse can be reduced and its frequency is increased with the increase of drive current. Laser construction has also significant effect on the spike magnitude. Some single tranverse mode narrow stripe lasers show little spiking.

The spectral bandwidth also depends on the drive current and wide bandwidth is obtained if the laser is operated close to its threshold. Consequently, narrow bandwidth is achievable if it is biased above the threshold. However noise performance is degraded as the signal will not fall to a low value when indicating a 'zero'. Shot noise is also increased. However, for narrow bandwidth, material dispersion is much less compared to LEDs.

Therefore lasers have considerable advantages over LEDs from the point of view of coupling the maximum amount of power into fibers and minimum material dispersion. But laser transmission is expensive, drive circuitry is complex, and elaborate temperature control and output stabilization are required.

11.4.3 Detector Design

Most detectors have large sensitive areas in comparison with the fiber core areas and can easily detect most of the radiation carried by the fiber core, apart from the Fresnel losses. Detectors of too large sensitive area have the disadvantages of high dark current noise and high detector capacitance. Due to this reason, use of photomultipliers are generally avoided. Thermal detectors are usually slow and insensitive. Therefore, some kind of junction devices like *pin*-photodiode or avalanche photodiode are the universal choice. The detected signal for both the detectors is low and post-detection amplification is always required and noise generation in amplifier is of crucial importance. Avalanche photodiode provides substantial amount of gain that is useful in enhancement of system sensitivity. But it's limited by the noise in amplifier following the detector.

Noise Analysis for PIN Detector

If a signal of optical power of P at wavelength λ falls on a *pin* detector with its quantum efficiency η, then the resulting signal current i_s may be expressed as

$$i_s = \frac{\eta P e \lambda}{hc} \tag{11.15}$$

To analyze the noise performance let us consider the equivalent circuit of pin diode as in Fig. 11.10. If it is assumed that the Johnson noise in amplifier is generated from the load resistor R_L at effective temperature T_{eff} and bandwidth of Δf, then the rms noise voltage ΔV_j and current Δi_j can be expressed as

$$\Delta V_j = (4\,K\,T_{eff}\,R_L\,\Delta f)^{1/2} \tag{11.16}$$

$$\Delta i_j = \frac{\left(4KT_{eff}\,R_L\,\Delta f\right)^{1/2}}{R_L} \tag{11.17}$$

Noise is originated in detector from two sources; shot noise in the signal current $\Delta i_s^{\,shot}$ and dark current $\Delta i_D^{\,shot}$. They may be expressed as

$$\Delta i_s^{\,shot} = (2\,i_s\,e\,\Delta f)^{1/2} \tag{11.18}$$

$$\Delta i_D^{\,shot} = (2\,i_D\,e\,\Delta f)^{1/2} \tag{11.19}$$

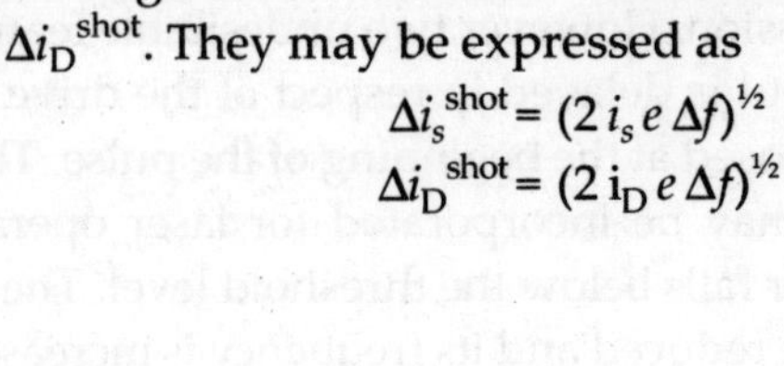

Fig. 11.10

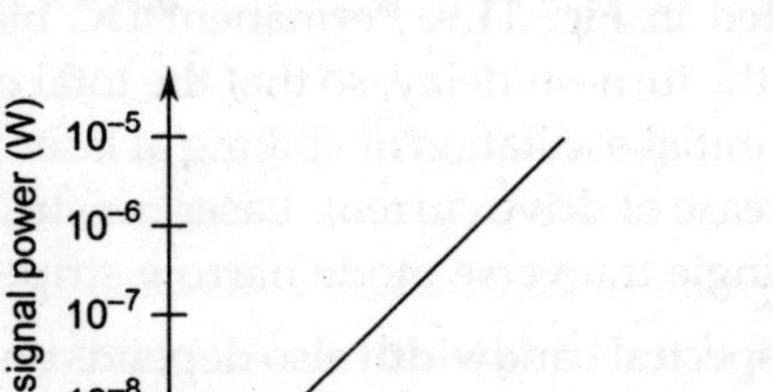

Fig. 11.11

Shot noises are small compared to Δi_j and may be ignored while calculating the signal-to-noise ratio S/N the sensitivity is limited by the amplifier noise alone.

$$\frac{S}{N} = \frac{i_s^{\,2}\,R_L}{\Delta i_j^{\,2}\,R_L} = \frac{i_s^{\,2}\,R_L}{4KT_{eff}\,\Delta f} \tag{11.20}$$

Δf is limited by the detector capacitance C_j and related as $\Delta f = (2\pi\,R_L C_j)^{-1}$. Therefore we may write

$$\frac{S}{N} = \frac{i_s^{\,2}}{8\pi\,K\,T_{eff}\,\Delta f^2\,C_j} \tag{11.21}$$

Assuming the minimum detectable signal $(i_s)_{min}$ and minimum detectable optical power P_{min} are obtained at S/N = 1. Then

$$(i_s)_{min} = 2\Delta f\,(2\pi\,K\,T_{eff} C_j)^{1/2} \tag{11.22}$$

and

$$P_{min} = \frac{2hc}{\eta e \lambda}\,\Delta f\,(2\pi\,K\,T_{eff} C_j)^{1/2} \tag{11.23}$$

A plot of P_{min} verses Δf is shown at figure 11.11 for $\lambda = 0.85\ \mu m$, $C_j = 1$ pF, $\eta = 1$ and $T_{eff} = 1000°$ K.

Noise Analysis For Avalanche Photodiode

If the gain of the avalanche photodiode is M, the signal current will be amplified by the factor of M *i.e.*, Mi_s. Shot noises can not be ignored for APD. Furthermore the avalanche itself introduces some excess noise. The excess noise is assumed as F(M). Hence, with the similarity of the equations (11.18) and (11.19), the shot noise current and dark noise current may be expressed as

$$\Delta i_s^{\,shot} = M\{2\, i_s\, e\, \Delta f\, F(M)\}^{1/2} \tag{11.24}$$

and

$$\Delta i_D^{\,shot} = M\{2\, i_D\, e\, \Delta f\, F(M)\}^{1/2} \tag{11.25}$$

The signal-to-noise ratio may be derived as

$$\frac{S}{N} = \frac{M^2 i_s^2 R_L}{\left(\Delta i_j^2 + \Delta i_s^{\,shot2} + \Delta i_D^{\,shot2}\right) R_L} \tag{11.26}$$

Substituting $R_L = (2\pi\, \Delta f\, C_j)^{-1}$ and equations (11.24) and (11.25),

$$\frac{S}{N} = \frac{M^2 i_s^2}{2M^2 e\, \Delta f\, F(M)\left(i_s + i_D\right) + 8\pi K T_{eff}\, \Delta f^2\, C_j} \tag{11.27}$$

It may be noticed that the factor F(M) is unity the signal to noise ratio would increase with increase of M and attain a maximum asymptotic value of $i_s^2/2\, e\, \Delta f\, (i_s + i_D)$. This is much larger than the corresponding value for a *pin* diode. However, as the factor F(M) increases with the increase of M, the optimum S/N value is some what less than this. It has been observed that the optimum value of M, *i.e.*, M_{opt} may be obtained by solving the equation

$$M_{opt}\left(\frac{M^2_{opt}}{r} - \frac{1}{r} + 1\right) = \frac{2 \times \text{Amplifier noise}}{\text{Shot noise}} = \frac{8\pi K T_{eff}\, \Delta_f\, C_j}{e\left(i_s + i_D\right)} \tag{11.28}$$

where r is the ratio of the electron to hole ionization probabilities. The optimum signal to noise ratio $(S/N)_{opt}$ may be derived from equations (11.27) and (11.28) as

$$\left(\frac{S}{N}\right)_{opt} = \frac{M^2 i_s^2}{\left\{\dfrac{2M_{opt}\, F(M)}{\left(\dfrac{M^2_{opt}}{r} - \dfrac{1}{r} + 1\right)} + 1\right\} 8\pi K T_{eff}\, \Delta f^2\, C_j} \tag{11.29}$$

when M = 1,

$$\left(\frac{S}{N}\right)_{M=1} = \frac{i_s^2}{8\pi K T_{eff}\, \Delta f^2\, C_j} \tag{11.30}$$

It may be noted that if M >> 1 then F(M) = M/r and if $M^2 >> r$, then $M^2/r - 1/r + 1 \approx M^2/r$. The equation (11.30) then reduces to

$$\left(\frac{S}{N}\right)_{opt} = \left(\frac{S}{N}\right)_{M=1} \frac{M^2_{opt}}{3} \tag{11.31}$$

From the above derivation, it may be observed that in respect of gain there may be little or no advantage using APD rather than *pin* detector. In fact signal to noise ratio may be worsen by APD. From equation (11.27), it may be stated that signal to noise ratio in APD is smaller by the factor of F(M) than the *pin* detector.

11.4.4 Limitations on Signal Size

Although in all practical amplifiers often noise is the limiting factor, pulse size also play an important role on the optical amplifier performance. Optical amplifier acts on the emission of photons and probability of detecting photons is governed by Poisson statistics. The probability of detecting n number of photons per unit time interval with mean arrival rate n_m is given by the relation

$$p(n,n_m) = \frac{(n_m)^n}{n!}\exp(-n_m) \tag{11.32}$$

For example, with mean arrival rate of 20 photons per pulse, the probability that a given pulse contains no photon at all is given by

$$p(0,20) = \frac{(20)^0}{0!}\exp(-20) = 2\times 10^{-9} \tag{11.33}$$

Thus if a signal consists of equal numbers of 'ones' and 'zeros', where the 'ones' correspond to the pulses containing on average 20 photons and 'zeros' to the arrival of no photons, then there is a probability of 2×10^{-9} that a 'one' may be mistaken as 'zero'. No error can be made in any of the 'zeros', since there is no fluctuation in zero photon. Therefore considering both 'ones' and 'zeros', a signal containing an average of 10 photons will have the *bit error rate* (BER) of 10^{-9}. With less than 10 photons per bit the BER increases, *i.e.*, BER is degraded. Conversely, with greater than 10 photons per bit the BER decreases, *i.e.*, BER is improved. BER of 10^{-9} is fairly standard for optical communication systems. Figure 11.12 shows the typical example of detector performance of minimum signal power required for a bit error rate of 10^{-9}. Minimum signal power required for APD and *pin* detector at a wavelength of 0.85 μm is also shown in this figure.

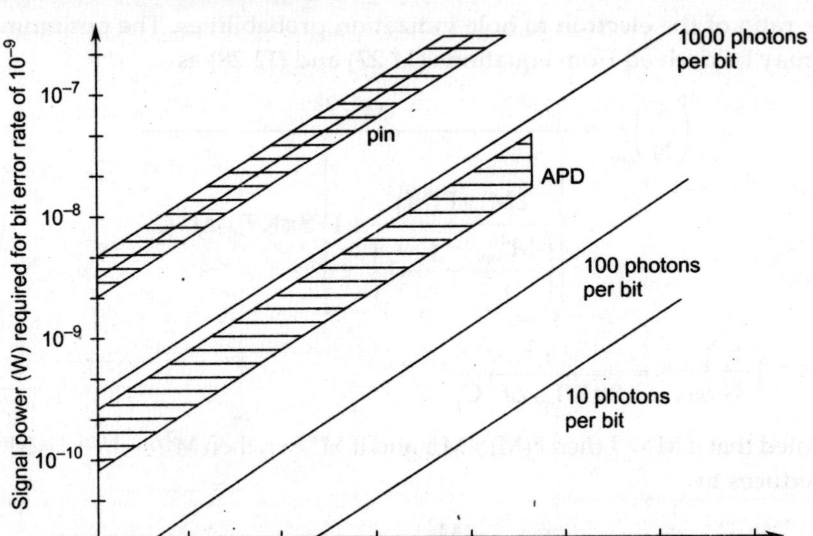

Fig. 11.12

Maximum signal to noise ratio is given by

$$\left(\frac{S}{N}\right)_{max} = \frac{i_s^2}{\left(\Delta i_{s}^{shot}\right)^2} = \frac{i_s}{2e\,\Delta f} = \frac{\eta P\lambda}{2hc\,\Delta f} \quad \text{(substituting } i_s \text{ from equation 11.15),} \qquad (11.34)$$

In digital system, a reasonable value for $(S/N)_{max}$ is 50 *db* for the bit error rate of 10^{-9}. to achieve this the minimum signal strength P_{min} can be derived as

$$P_{min} = \frac{2 \times 10^5\, hc\,\Delta f}{\eta\lambda} \qquad (11.35)$$

Considering $\lambda = 0.85$ μm, $\eta = 1$ and $\Delta f = 6.25$ MHz (equivalent to the digital signal of 6.25×16 Mbs or 100 Mb/sec) we obtain that $P_{min} = 0.3$ μW. It may be observed from figure 11.12 that a digital system of 100 Mb/sec requires less power to achieve a bit error rate of 10^{-9}. This indicates that digital system is superior for low noise long distance optical transmission.

11.4.5 Amplifier Design

It has been observed that amplifier immediately following the detector plays an important factor in the performance of the detector-amplifier combination. Considering the Fig. 11.10, the diode capacitance C_j and load resistor R_L act in parallel to the photogenerated current of the detector. The bias resistor R_a of the amplifier input stage also comes in parallel with R_L. Therefore the effective load resistor $R_e = R_L R_a/(R_L + R_a)$. As the signal bandwidth is inversely proportional to C_j and R_e, both R_L and R_a are to be as small as possible to obtain maximum bandwidth. This design is relatively simple when noise limitation is not a criteria and usually implemented using bipolar transistor as in Fig. 11.13. The output is proportional to the integral of signal current. However major disadvantage in this design is that the system is dominated by Johnson noise and its signal to noise ratio is very poor. The frequency response in this circuit may be improved by incorporating differentiation circuit after the amplifier, but this leads to the reduction of dynamic range *i.e.*, the ratio of maximum to minimum input usable input signal.

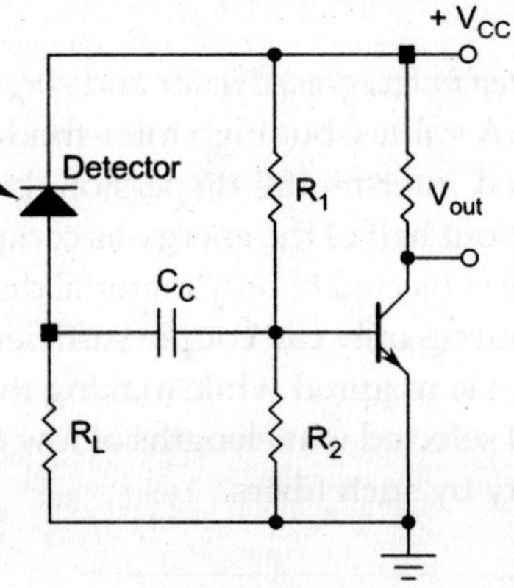

Fig. 11.13

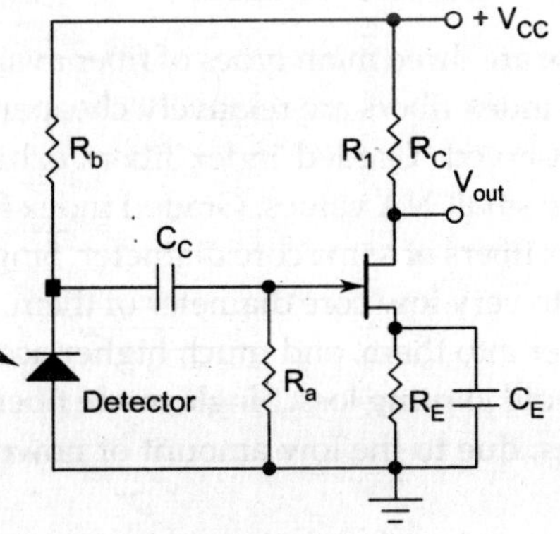

Fig. 11.14

For low noise level, high input impedance amplifier is required. This may be accomplished by introducing Field Effect Transistor (FET) as shown in Fig. 11.14. The silicon FET exhibits very high input impedance as well as low noise performance. The main disadvantage of this circuit is that the gain is limited. Above 25 MHz the gain drops close to unity and therefore above this frequency the bipolar transistor is preferable.

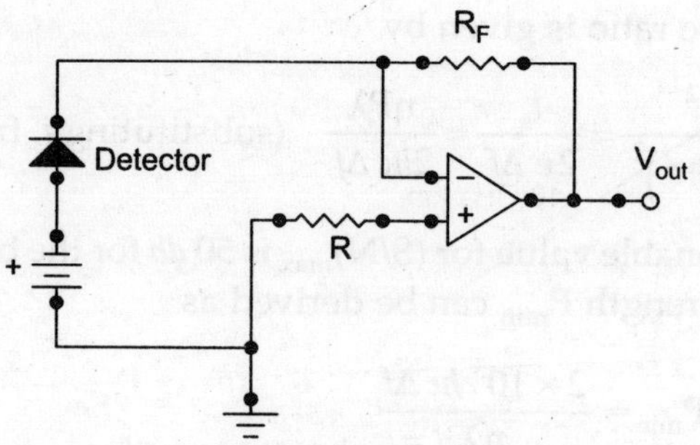

Fig. 11.15

A transimpedance design approach for amplifier front end may be incorporated where a high impedance amplifier is used in conjunction with negative feedback. In this design the frequency response is extended as well as noise performance is improved. A schematic diagram of this type of design is shown in Fig. 11.15. FET operational amplifier in association with high value feedback resistance R_F is used.

11.4.6 Detector Materials

Most silicon detectors operates at the maximum wavelength of 1.1 μm. From section 11.3 it has been observed that higher wavelength gives improved signal to noise ratio as well as less fiber loss. Generally wavelengths around 1.3 μm or 1.55 μm are used for practical purposes. Germanium detectors that give good response characteristics in this range of wavelengths may be used. Unfortunately they suffer from higher dark current in comparison with the silicon detectors. Also optimum gain in germanium detectors is relatively small as its carrier ionization ratio is about unity resulting in without any improvement in signal to noise ratio. Alternatively photodetector with GaInAs and GaInAsP materials may be used, but they also suffer from low carrier ionization ratio and high dark current.

11.4.7 Choice of Fiber

There are three main types of fiber available, known as *step index*, *graded index* and *single mode fibers*. Step index fibers are relatively cheaper and have large NA values, but high inter-modal dispersion is observed. Graded index fibers exhibit greatly reduced inter-modal dispersion, but have relatively small NA values. Graded index fibers carry only about half of the energy as compared to step index fibers of same core diameter. Single mode fibers are subjected to only material dispersion. But due to very low core diameter of them, generally laser sources only can couple sufficient amount of power into them, and much higher accuracy in alignment is required while making the splice joint to avoid jointing loss. Single mode fibers are to be used at selected wavelengths of low transmission losses, due to the low amount of power carrying capacity by such fibers.

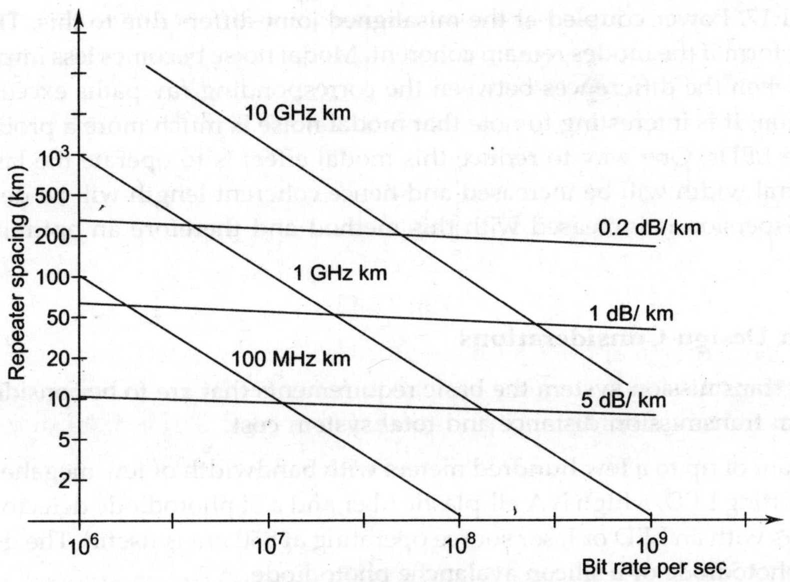

Fig 11.16

The distance between transmitter and receiver or between the repeaters sometimes is the influencing factor for the choice of fibers. This is depended on two key parameters of fibers, namely fiber attenuation and fiber dispersion. For a low bandwidth system, fiber attenuation is the limiting factor whereas dispersion plays as the limiting factor for high bandwidth system. As example, for a fiber cable of attenuation of 5 dB/km, if 40 dB loss is tolerable, the maximum length of fiber is to be limited to 40/5 or 8 km. As another example, if a step index fiber of inter-modal dispersion of 50 ns/km (or 20 MHz km) is used for a signal bandwidth of 10 MHz, then repeater must be provided within 20/10 or 2 km. The effects of these factors on maximum fiber lengths achievable have been illustrated in Fig. 11.16. In this figure a total loss of 50 dB at 1 Gbit/sec is assumed. The attenuation lines are slightly inclined due to reduced receiver sensitivity with increasing bit rate.

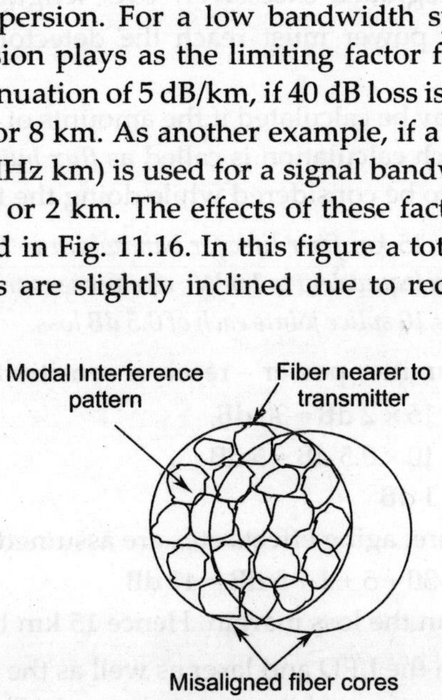

Fig 11.17

In practical systems, an additional source of noise we come across is modal noise which is generated due to flexing of cable containing misaligned fiber joints. An interference pattern is generated. The dark and bright areas in this pattern are roughly proportional to the number of modes propagating down the fiber. Physical movement of the fiber causes slight change in the relative irradiance and phase of the modes, and also the appearance of the interference pattern as shown in Fig. 11.17. Power coupled at the misaligned joint differs due to this. The interference pattern can only form if the modes remain coherent. Modal noise becomes less important for long length of fiber when the differences between the corresponding ray paths exceed the coherent length of radiation. It is interesting to note that modal noise is much more a problem with laser sources than the LEDs. One way to reduce this modal effect is to operate the laser well below threshold. Spectral width will be increased and hence coherent length will be decreased. However material dispersion is increased with this method and therefore an optimization is to be made.

11.4.8 System Design Considerations

For optical fiber transmission system the basic requirements that are to be considered are bandwidth, maximum transmission distance and total system cost.

For short distant of up to a few hundred meters with bandwidth of few megahertz, the combination of red emitting LED, a high NA all-plastic fiber and a Si photodiode detector may be used. Silica based fibers with an LED or laser source operating at 850 nm is useful. The detector may be either a silicon photodiode or a silicon avalanche photodiode.

To ensure satisfactory system performance a numbers of checks must be made.

- Both the emitter and detector must be capable of handling the required signal bandwidth.
- The signal must not be degraded excessively over length due to fiber dispersion.
- Minimum average signal power must reach the detector for a given bit error rate of digital signal.

The maximum fiber length may be calculated if the amounts of power launched into cable and fiber attenuation are known. Such calculation is called as *flux budget calculation*. Splice joint loss and a safety margin of 5 dB are to be considered while doing the flux budget calculation.

Example 11.1. *Verify whether a 15 km fiber link for bandwidth is 100 Mb/s is possible or not, where transmitter throughput power is 0 dBm (equivalent to 1 mW), receiver sensitivity is -50 dBm receiver coupling loss of 1 dB, fiber loss is 2 dB/km. Assume 10 splice joints each of 0.5 dB loss.*

Solution: Loss margin = transmitter power – receiver sensitivity = 0 – (– 50) dBm = 50 dBm

$$\text{Fiber loss} = 15 \times 2 \text{ dB} = 30 \text{ dB},$$

$$\text{Total splice joint loss} = 10 \times 0.5 \text{ dB} = 5 \text{ dB}$$

$$\text{Detector coupling loss} = 1 \text{ dB}$$

Other losses due to temperature, aging effects etc., are assumed as 5 dB.

$$\text{Therefore total loss} = 30 + 5 + 1 + 5 \text{ dB} = 41 \text{ dB}$$

Total estimated loss is less than the loss margin. Hence 15 km link is suitable for operation.

Typical launch power for both the LED and laser as well as the receiver power requirement to achieve 10^{-9} bit error rate as a function of bit rate are shown in Fig. 11.18. It may be noticed that relative advantage of the laser/ APD combination increases with the increase of bit rate.

In case of transmission rate of more than 400 Mb/sec, the usable fiber length is determined by fiber dispersion rather than fiber attenuation (refer Fig. 11.16). Single mode fibers are to be used using operating wavelengths close to minimum material dispersion (*i.e.*, near 1.3 μm in SiO_2 based fibers). With bit rate of a few hundreds Mb/sec unrepeated transmission over the distance of 100 km is achievable.

11.4.9 Local Area Networks

Although the optical fibers have been primarily developed for long distance repeaterless data transmission with high bandwidth, they can make important contribution to local area networks or LANs. LAN is aimed to provide communication among a group of users within a restricted geographical area of short distance that might be of a few kilometers. Availability of wider bandwidth links of optical fiber facilitates other type of communication like high speed data communication or video transmission where data communication is required to be of several megabits within the networks linking up to say one hundred users.

The main disadvantage of using optical communication in LAN is the losses involved in large number of taps or switches. With copper wire technology of coaxial cable these losses are low. However optical fiber system do offer the advantages of high bandwidth, immunity from local interference, higher level of information security, and small size and weight of the cable.

There may be several topologies for interconnection in LANs as shown in Fig. 11.18. The fully connected network is shown in Fig. 11.18(*a*) where every user or node is connected to every other. It is the most reliable topology since if one link is damaged, other links are unaffected and rerouting is possible to restore the connection. However this arrangement complex, expensive and requires considerable rewiring when more nodes are added. Some links may never be used.

The star topology is illustrated in Fig. 11.18(*b*), where each node is connected at a central node. If the central node is active, each signal can be directed to its required destination. If the central node is passive then a particular node sends signal to all the nodes, each of which decides whether to accept the signal or not. The number of links in this topology is minimum but the reliability of whole network depends on the central node. The entire system will break down if the central node fails. The data bus topology is illustrated in Fig. 11.18(*c*). Great flexibility, reliability and efficient use of fiber are achieved from the data bus topology.

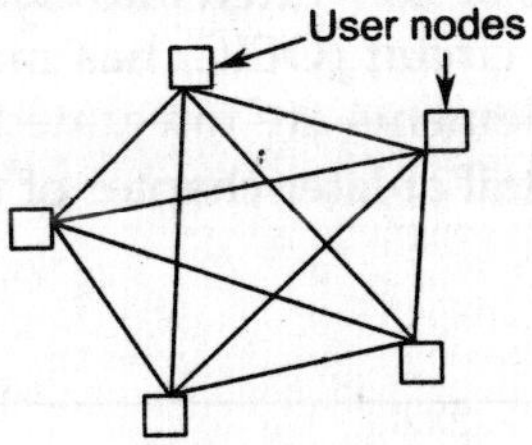

Fig 11.18 (a)

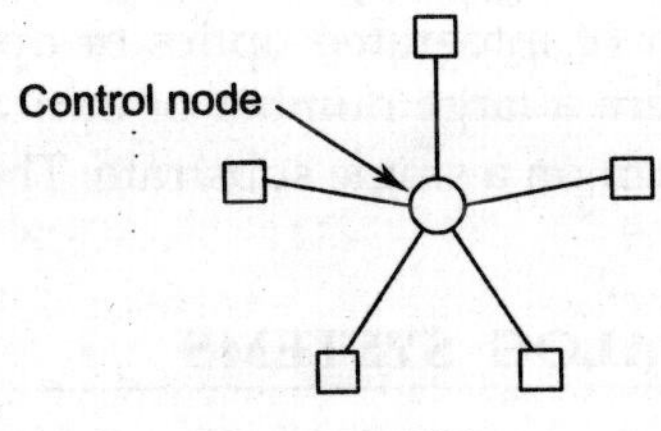

Fig 11.18 (b)

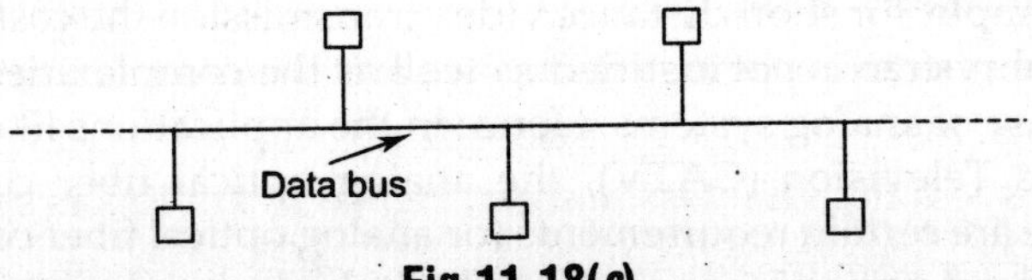

Fig 11.18(*c*)

11.4.10 Coherent Detection System

So far the direct detection aspects are discussed here. There are also considerable advantages in adopting a coherent detection system. With such schemes the semiconductor laser are required to have frequency and phase stability. A normal semiconductor laser might have the linewidth of 500 GHz, whereas for coherent detection the linewidths of 15 MHz or less may be required. The semiconductor lasers of such narrow stable bandwidths can be obtained by using DFB (distributed feedback) structures and external cavities. However, the system complexity increases considerably with the use of such lasers.

A problem for conventional fibers is that they are not good at maintaining polarization and for coherent detection. Both the signal and local oscillator must have the same plane of polarization for mixing properly at the detector. Special fibers of maintaining polarization are available, but they are expensive as well as of higher losses. Therefore the state of polarization of the incoming signal in coherent link is to be monitored, that leads to higher complexity and hence the cost at the detection. However for long haul high bandwidth links, the coherent system is attractive because of its ability to operate without repeaters.

With the incorporation of wavelength multiplexing technique, the bandwidth of an existing optical link can be enhanced without any modification in the fiber. In this technique, several different signals are modulated with carriers of different wavelength and then transmitted simultaneously down the fiber. Separation of signals at the detector end is little complex and inevitably appreciable amount of signal loss as well as crosstalk are involved. There are several methods for signal separation available. One of the method is illustrated in Fig. 11.19, where a miniature prism is incorporated to spatially separate the transmitted signal wavelengths enabling separate fibers to pick up the signals for respective detectors. This approach also enables two way transmissions over a single fiber, different wavelengths for different directions.

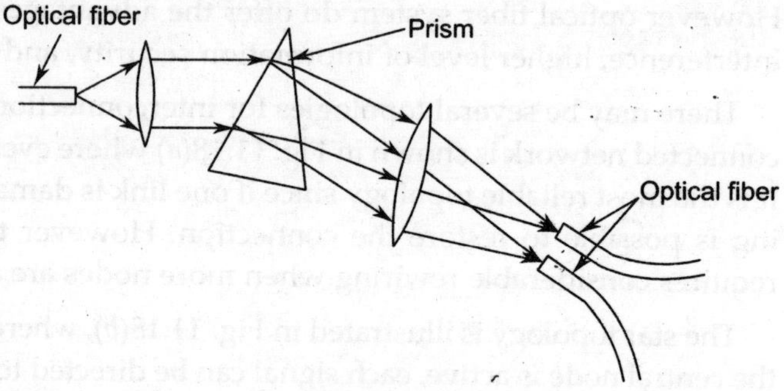

Fig 11.19

After receiving the optical signal at detector, it is required to be converted into electrical signal. The advent of integrated optics or optoelectronic integrated circuit (OEIC) has made this task easier, where a large number of optical and electro-optical elements are integrated in thin film planner form on a single substrate. These are discussed in detail at later chapter of this book.

11.5 ANALOG SYSTEMS

In certain areas of telecommunication application analog system is advantageous over the digital communication. As example, for short distance video transmission the cost of high speed A/D and D/A converter of digital system is not justified as well as the complexities of digital equipments can be avoided by the use of analog systems. Hence in the applications like direct cable television and Common Antenna Television (CATV), the analog optical fiber communication may be utilized. However there are certain requirements for analog optical fiber communication are to be fulfilled. High optical input power is required to obtain high signal to noise ratio at the receiver, and high end to end linearity is necessary to avoid distortion and crosstalk.

11.5.1 Direct Intensity Modulation

The basic principle of analog modulation and detection has been already discussed in section 11.2.1. The simplest form of analog modulation for optical fiber is direct intensity modulation (D-IM) of optical source. In this technique the optical output from the source is modulated by varying the current flowing in the device around a suitable bias or mean level in proportion to the massage. Hence the message signal is transmitted directly in the baseband. A block schematic for direct intensity modulation is shown in Fig. 11.20(*a*). No electrical modulation or demodulation is required with this technique and so it is easy to implement and inexpensive.

The transmitted optical power $P_{opt}(t)$ as the function of time may be expressed as

$$P_{opt}(t) = P_i [1 + P_m(t)] \tag{11.36}$$

P_i is the carrier power and $P_m(t)$ is intensity modulating signal power proportional to message signal. The waveform is illustrated in Fig. 11.20(*b*). For a sinusoidal signal $P_m(t)$ can be written as

$$P_m(t) = m_a \cos \omega_m t \tag{11.37}$$

where m_a is modulation index defined as the ratio of peak excursion from the average power and the average power *i.e.*, $m_a = (P_{max} - P_i)/ P_i$. Therefore,

$$P_{opt}(t) = P_i (1 + m_a \cos \omega_m t) \tag{11.38}$$

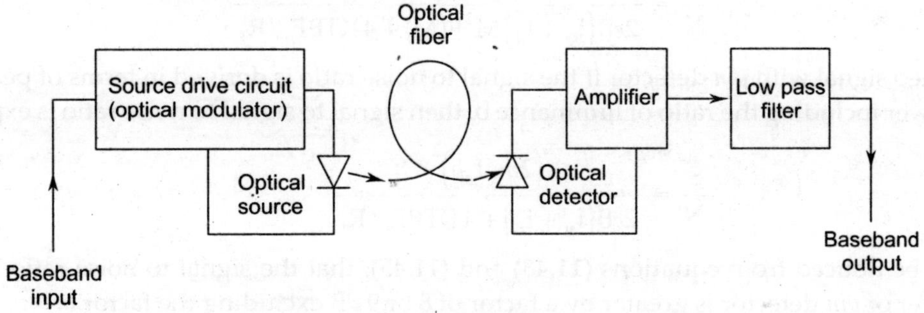

Fig 11.20(*a*)

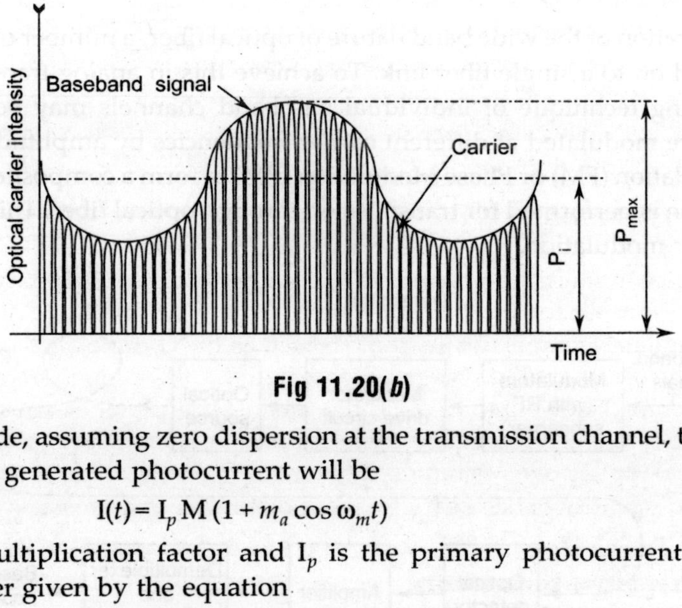

Fig 11.20(*b*)

At the receiver side, assuming zero dispersion at the transmission channel, the average received power is P_o and the generated photocurrent will be

$$I(t) = I_p M (1 + m_a \cos \omega_m t) \tag{11.39}$$

where M is the multiplication factor and I_p is the primary photocurrent obtained with the unmodulated carrier given by the equation

$$I_p = \frac{\eta e}{hf} P_0 \tag{11.40}$$

From equation 11.39 the mean square signal current is

$$i_s^2 = \frac{1}{2}\left(m_a \mathrm{MI}_p\right)^2 \tag{11.41}$$

The total noise in the system consists of due to quantum current, dark current and thermal noise. Thermal noise is equal to $4\mathrm{KTBF}_n/\mathrm{R_L}$ ($\mathrm{R_L}$ is load resistance) and mean square noise current for *pin* detector is given as

$$i_N^2 = [2e\mathrm{B}(\mathrm{I}_p + \mathrm{I}_d) + 4\mathrm{KTBF}_n/\mathrm{R_L}]\,\mathrm{M}^2 \tag{11.42}$$

The signal to noise ratio for *pin* detector is derived as

$$\frac{\mathrm{S}}{\mathrm{N}} = \frac{\frac{1}{2}\left(m_a \mathrm{I}_p\right)^2}{2e\mathrm{B}\left(\mathrm{I}_p + \mathrm{I}_d\right) + 4\mathrm{KTBF}_n/\mathrm{R_L}} \tag{11.43}$$

The signal to noise ratio for APD detector is

$$\frac{\mathrm{S}}{\mathrm{N}} = \frac{\frac{1}{2}\left(m_a \mathrm{MI}_p\right)^2}{2e\mathrm{B}\left(\mathrm{I}_p + \mathrm{I}_d\right)\mathrm{M}^2\mathrm{F(M)} + 4\mathrm{KTBF}_n/\mathrm{R_L}} \tag{11.44}$$

For video signal with *pin* detector if the signal to noise ratio is derived in terms of peak to peak signal power including the ratio of luminance b, then signal to signal to noise ratio is expressed as

$$\frac{\mathrm{S}}{\mathrm{N}} = \frac{\left(2m_a \mathrm{I}_p b\right)^2}{2e\mathrm{B}\left(\mathrm{I}_p + \mathrm{I}_d\right) + 4\mathrm{KTBF}_n/\mathrm{R_L}} \tag{11.45}$$

It may be noticed from equations (11.43) and (11.45), that the signal to noise ratio in peak to peak power of *pin* detector is greater by a factor of 8 or 9 *d*B excluding the factor *b*.

11.5.2 Subcarrier Intensity Modulation

For maximum utilization of the wide band nature of optical fiber, a number of baseband channels may be multiplexed on to a single fiber link. To achieve this in analog transmission, frequency division multiplexing technique of individual baseband channels may be used. Initially the baseband signals are modulated at different carrier frequencies by amplitude modulation (AM) or Frequency Modulation (FM) or Phase Modulation (PM) to form a composite FDM signal as then intensity modulation is performed for transmission through optical fiber. This technique is called subcarrier intensity modulation.

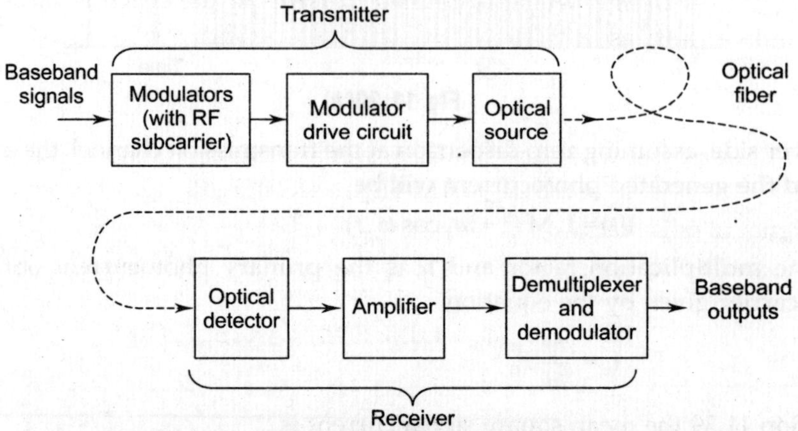

Fig 11.21

The block schematic of subcarrier intensity modulation is illustrated in Fig. 11.21. The baseband signals are modulated at different RF carriers. Depending on the adaptation of modulation procedures they are named as AM-IM, FM-IM or PM-IM. Thereafter the modulated signals are multiplexed to form a composite signal and fed to the drive circuit for optical intensity modulation. At the receiver end the received signal is detected by optical receiver, amplified, demultiplexed and demodulated. In practice, the FM-IM and PM-IM formats are widely used because of their better SNR performance.

Subcarrier Double Sideband Intensity Modulation (DSB-IM)

If the carrier waveform is represented as $A_C \cos \omega_c t$ and the massage signal as $a(t)$, where A_C is amplitude of carrier and ω_c is angular frequency of carrier, then the modulated waveform may be represented as

$$m(t) = A_C\, a(t) \cos \omega_c t \tag{11.46}$$

For a sinusoidal baseband signal of angular frequency ω_m, the subcarrier electric field may be derived as

$$E_m(t) = A_C [\cos(\omega_c + \omega_m)t + \cos(\omega_c - \omega_m)t]/2 \tag{11.47}$$

The modulated signal is shown in Fig. 11.22. It may be observed in the equation 11.47 that there are two frequency components or sidebands as $\omega_c + \omega_m$ and $\omega_c - \omega_m$ or $f_c + f_m$ and $f_c - f_m$. This modulation technique is called as double sideband (DSB) modulation or more specifically double sideband suppressed carrier (DSBSC) modulation. After intensity modulation, the optical power waveform may be written as (assuming A_C equal to unity for simplicity)

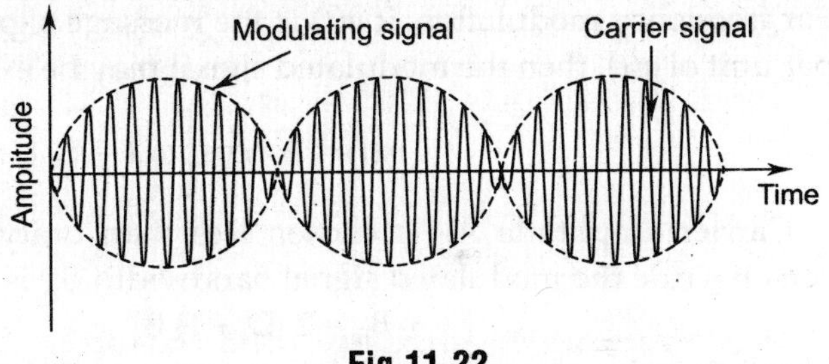

Fig 11.22

$$P_{opt}(t) = P_i [1 + a(t) \cos \omega_m t] \tag{11.48}$$

At the receiver end the photocurrent generated by APD detector as stated at equation 11.39

$$I(t) = I_p M (1 + m_a \cos \omega_m t) \tag{11.49}$$

The mean square signal current is

$$i_s^2 = (MI_p)^2 P_m \tag{11.50}$$

P_m is the total power of $m(t)$ which is defined in terms of the spectral density $S_n(\omega)$ of $m(t)$ occupying a bandwidth B_m as

$$P_m = \frac{1}{2\pi} \int_{-2\pi B_m}^{2\pi B_m} S_n(\omega)\, d\omega \tag{11.51}$$

Mean square noise power with excess noise power as F(M) is

$$i_N^2 = 2eB_m(I_p + I_d) M^2 F(M) + 4KTB_m F_n / R_L \tag{11.52}$$

Therefore signal to noise ratio

$$\frac{S}{N} = \frac{(MI_p)^2 P_m}{2eB_m(I_p + I_d) M^2 F(M) + 4KTB_m F_n / R_L} \tag{11.53}$$

$$= \frac{I_p^2 P_m}{2eB_m(I_p + I_d) F(M) + 4KTB_m F_n / M^2 R_L} = \frac{I_p^2 P_m}{2B_m N_0} \tag{11.54}$$

where $N_0 = 2e(I_p + I_d)F(M) + 4KTF_n/M^2R_L$. If the responsivity of the receiver is R, then referring equation 4.3, $R = I_p/P_{inc}$. Hence

$$\frac{S}{N} = \frac{(RP_{inc})^2 P_m}{2B_m N_0} \tag{11.55}$$

Now for a double sideband system $B_m = 2f_m$ and $P_m = P_a/2$, where P_a is the power of message signal. Therefore the signal to noise ratio at the input of double sideband demodulator system is

$$\frac{S}{N} = \frac{(RP_{inc})^2 P_a/2}{2\times 2f_m N_0} = \frac{(RP_{inc})^2 P_a}{8 f_m N_0} \tag{11.56}$$

However, the detection gain is obtained 2 to 3 dB higher from an ideal DSB demodulator and hence the signal to noise ratio at output of DSB demodulator is improved as

$$\frac{S}{N} = \frac{(RP_{inc})^2 P_a/2}{2\times 2f_m N_0} = \frac{(RP_{inc})^2 P_a}{4 f_m N_0} \tag{11.57}$$

Subcarrier Frequency Modulation (FM-IM)

For frequency modulation, if $a(t)$ is the message signal and k_f is the angular frequency deviation per unit of $a(t)$, then the modulated signal may be expressed as

$$m(t) = A_C \cos\left[\omega_c t + k_f \int_0^t a(\tau)\,d\tau\right] \tag{11.58}$$

Carrier amplitude A_C is chosen less than unity to prevent overmodulation. According to carson's rule the modulated signal bandwidth B_m is given as

$$B_m = 2\,(D_t + 1)\,B_a \tag{11.59}$$

where B_a is the message signal bandwidth and D_f is frequency deviation ratio defined as

$$D_t f = \frac{\text{Peak frequency deviation}}{\text{Bandwitdth of } a(t)} = \frac{f_d}{B_a} \tag{11.60}$$

SNR at the input of subcarrier FM demodulator

$$\frac{S}{N} = \frac{(RP_{inc})^2 (A_C^2/2)}{2B_m N_0} \tag{11.61}$$

SNR at the output of subcarrier FM demodulator

$$\frac{S}{N} = 6D_f^2 (D_f^2 + 1) \frac{(RP_{inc})^2 (A_C^2/2) P_a}{2B_m N_0}$$

or,

$$\frac{S}{N} = \frac{3D_f^2 (RP_{inc})^2 (A_C^2/2) P_a}{2B_m N_0} \tag{11.62}$$

Equation (11.62) indicates a significant improvement in the postdetection SNR at wideband FM-IM format.

Subcarrier Phase Modulation (PM-IM)

For phase modulation, if $a(t)$ is the message signal and k_p is the phase constant deviation per unit of a(t), then the modulated signal may be expressed as

$$m(t) = A_C \cos[\omega_c t + k_p a(t)] \tag{11.63}$$

Carrier amplitude A_C is chosen less than unity to prevent overmodulation. According to carson's rule the modulated signal bandwidth B_m is given as

$$B_m = 2(D_p + 1) B_a \tag{11.64}$$

where B_a is the message signal bandwidth and D_p is frequency deviation ratio defined as

$$D_t f = \frac{\text{Peak frequency deviation}}{\text{Bandwitdth of } a(t)} = \frac{f_d}{B_a} \tag{11.65}$$

SNR at the input of subcarrier PM demodulator

$$\frac{S}{N} = \frac{(RP_{inc})^2 (A_C^2/2)}{2B_m N_0} \tag{11.66}$$

SNR at the output of subcarrier PM demodulator

$$\frac{S}{N} = \frac{D_p^2 (RP_{inc})^2 (A_C^2/2) P_a}{2B_m N_0} \tag{11.67}$$

Again the equation (11.67) indicates higher SNR in the postdetection of PM-IM format over the DSB-IM format. However FM-IM format gives greater SNR performance in comparison with PM-IM.

11.5.3 Pulse Analog Techniques

The most common techniques of pulse modulation for analog transmission are Pulse Amplitude Modulation (PAM), Pulse Width Modulation (PWM), Pulse Position Modulation (PPM) and Pulse Frequency Modulation (PFM). All the pulse analog techniques are to be employed in the electrical regime prior to the intensity modulation. However, PAM-IM is affected by source nonlinearities and is less efficient than D-IM. PWM-IM is also less efficient because a large part of the transmitted energy carries no information as only nominal variations of the pulse width are of interest. PPM-IM and PFM-IM offer distinct advantages over the system and also allow narrow pulses for transmission. If source linearity problem is handled carefully, the SNR performances of both of these formats are similar and they are advantageous for long haul fiber transmission links. However, the terminal equipments for PWM-IM technique is less complex and its wideband FM gain is additive advantage over the other, it is generally for pulse analog transmission.

The block schematic of PFM -IM system is given in Fig. 11.23. Pulse repetition rate is varied acc-ording to the modulating signal in the PFM modulator that consists of a Voltage Controlled Oscillator (VCO). This is fed to the drive circuit of optical source with a fixed pulse width or a fixed duty cycle. At the receiver end, individual pulses are detected in a wideband receiver

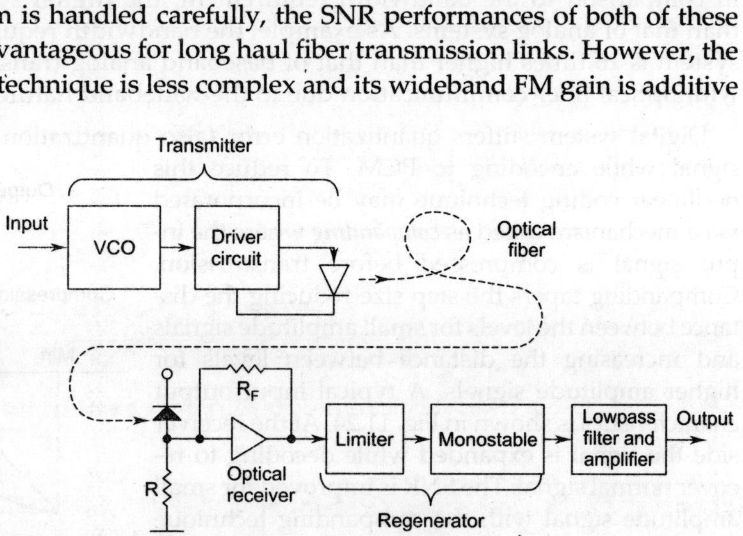

Fig 11.23

before they are regenerated with a limiter and monostable. The desired modulating signal is recovered through a low pass filter. Regenerative baseband recovery gives the best SNR performance and the SNR in terms of peak to peak signal power to rms noise power of a PFM-IM system is given by

$$\frac{S}{N} = \frac{3\left(T_0 f_D MRP_{po}\right)^2}{\left(2\pi T_R B\right)^2 i_N^2} \tag{11.68}$$

where T_0 is the nominal pulse period ($T_0 = 1/f_0$, f_0 is the pulse rate), f_D is peak to peak frequency deviation, R is the responsivity of the photodiode, M is photodiode multiplication factor, P_{po} is peak received optical power, T_R is the pulse rise time at the regenerator circuit input, B is the post detection effective noise bandwidth and i^2_N is the mean square noise current at receiver. It may be noted that improved SNR is obtained if the rise time of detected pulse is short. The pulse rise time is dictated by the overall system rise time T_{syst} (time to attain 10 to 90% of the pulse amplitude) and hence there is no advantage to use a wideband receiver with better rise time. In fact such system would degrade the system performance by passing increased front end noise. For an optimized PFM regenerative receiver design

$$T_R = T_{syst} = 1.1\,(T^2_s + T^2_o + T^2_c + T^2_D)^{1/2} \tag{11.69}$$

where T_s, T_o, T_c and T_D are the rise times for source or the transmitter, the intermodal fiber, intramodal fiber and the detector respectively.

11.6 DIGITAL SYSTEMS

At the present scenario of telecommunication networks digital transmission systems are widely used as digital technology and especially the integrated circuit technology made the transmission of discrete signals both advantageous and economic. Digital transmission systems generally offer superior performance over their analog counterparts in respect of signal to noise ratio as well as compatibility with digital computing techniques. In digital system, as stated earlier in this chapter, the analog signals are sampled, quantized and then suitably encoded in binary form. Pulse code modulation or PCM is one of the widely used digitization techniques. However, in comparison to the bandwidth requirement, the digital systems require greater bandwidth than that of analog systems. As example, the bandwidth requirement of 8 bit PCM transmission system is 16 times higher than that of baseband analog transmission. But this is not a problem with optical fiber communication due to the wideband nature of optical channel.

Digital system suffers quantization error (also quantization noise) on small amplitude analog signal while encoding to PCM. To reduce this nonlinear coding technique may be incorporated via a mechanism called as *companding* where the input signal is compressed before transmission. Companding tapers the step size reducing the distance between the levels for small amplitude signals and increasing the distance between levels for higher amplitude signals. A typical input output characteristics is shown in Fig. 11.24. At the receiver side the signal is expanded while decoding to recover normal signal. The SNR is improved for small amplitude signal with the companding technique resulting in reduction of overall signal degradation caused by quantization process.

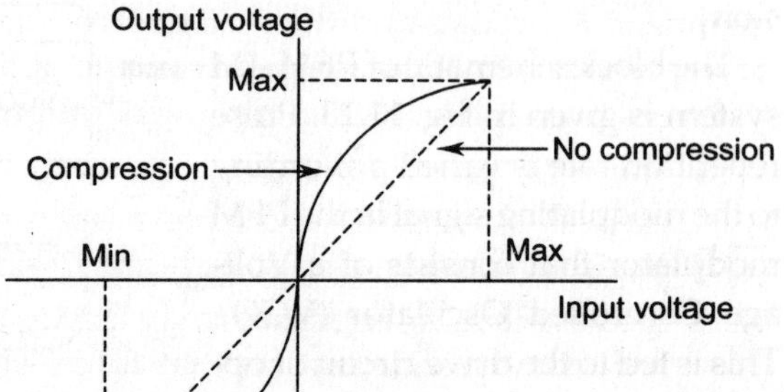

Fig 11.24

PCM signal utilizes the optical fiber communication link after intensity modulation of optical source and usually referred as PCM-IM. For high capacity optical transmission of PCM-IM signal, performance and spacing of regenerative repeaters are the most important parameters. It is generally designed with the same optical components as the terminal equipment. Figure 11.25 illustrates the functional parts of a regenerative repeater. The receiver consists of a photodiode followed by a low noise amplifier for detection and amplification of attenuated and dispersed optical pulse train from transmission channel. The acquired signal is further amplified and passed through a pulse-reshaping circuit or equalizer. Depending on the photodetector used, Automatic Gain Control (AGC) may be incorporated at the amplifier or bias current.

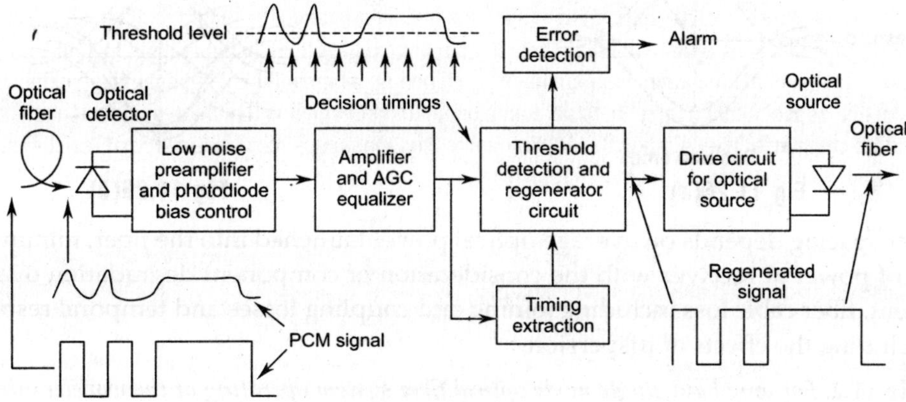

Fig 11.25

Accurate timing/clock information may be obtained from the amplified and equalized waveform using a time extraction circuit such as ringing circuit or phase lock loop. This enables the regenerator circuit for precise operation within the bit intervals of the original pulse train. The regenerator circuit reconstitutes the originally transmitted pulse train. This is achieved with the incorporation of threshold below which a binary zero is recorded while a binary one is recorded for threshold above. Thus binary pulse train is formed and it is rectified with error detection technique in case of presence of any error. The exact replica of transmitted information is formed and then retransmitted through drive circuit and optical source to optical fiber. Analog repeaters can filter equalize and amplify the received signal, but are unable to reconstitute the originally transmitted waveform entirely free from the distortion and noise. Signal degradation in long-houl analog systems is accumulative and distance between the repeaters is less and to be optimized to keep the signal degradation at tolerable level. In contrast to that signal degradation in PCM system is purely a function of the quantization process and the system error rate, and therefore distance between the repeaters may be kept longer. This the major advantage of the digital transmission systems.

Errors may occur during the regeneration process at the following situations:

- Accurate decision making may no be possible due to noise. Binary zero may be considered as binary one and binary one may be granted for binary zero if the noise level at decision-making circuit is higher than tolerable limit.
- Intersymbol interference may occur at the fiber link due to dispersion. This may be reduced by equalization that forces the transmitted binary one to pass through zero at all neighboring decision times.
- Variation of clock rate and phase degradations such as distortion of the zero crossing and static decision time misalignment introduce errors.

Eye pattern or *eye diagram* is one of the important methods for monitoring the performance of the regenerative repeaters. Received waveform is examined on the oscilloscope using a sweep rate as the fraction of the bit rate. The display obtained over the duration of two bits interval resulting in superimposing of all possible pulse sequences is called an eye pattern. An eye pattern of a binary system with little distortion and no additive noise is shown in Fig. 11.26(*a*). It may be observed that the pattern has the shape of human eye with wide open and decision time is corresponding to the centre of openings. To regenerate the pulse sequence without error, the decision time and the threshold (decision crosshair) must be at the open area of the eye pattern. The effect of distortion or intersysmbol interference reduces the size or shape of eye pattern as shown in Fig. 11.26(*b*).

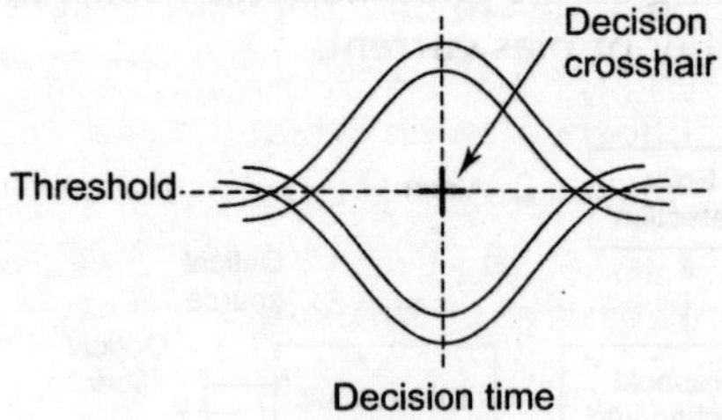

Fig 11.26(*a*)

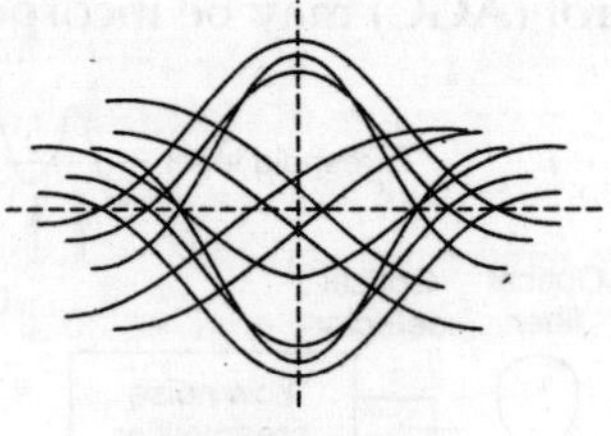

Fig 11.26(*b*)

Repeater spacing depends on average optical power launched into the fiber, minimum acceptable level of power at receiver with the consideration of component degradation over time and environment, fiber cable loss including joining and coupling losses and temporal response of the system including the effects of dispersion.

Example 11.2. *For long haul single mode optical fiber system operating at the wavelength 1.3 μm, the following parameters are established;*

(a) Mean power launched from the laser transmitter-3 dBm
(b) Cabled fiber loss-0.4 dB/km
(c) Splice loss-0.1 dB/km
(d) Connector losses at the transmitter and receiver-1 dB each
(e) Mean powers required at the APD receiver
at 35 Mb/s (BER 10^{-9})-55 dBm
at 400 Mb/s (BER 10^{-9})-44 dBm
(f) Required safety margin-7 dB
Estimate:
(a) The maximum possible link length without repeaters operating at 35 Mb/s (BER 10^{-9}), assuming that there is no dispersion equalization penalty at this bit rate.
(b) The maximum possible link length without repeaters operating at 400 Mb/s (BER 10^{-9}), assuming that there is no dispersion equalization penalty at this bit rate.
(c) The reduction at maximum possible link length without repeaters operating at 400 Mb/s (BER 10^{-9}), if the dispersion equalization penalty is 1.5 dB.

Solution:

(*a*) Transmitted power – received power = (cabled fiber loss + splice loss) × length of the link + connector loss + safety margin.

So, -3 dBm – (-55 dBm) = (0.4 + 0.1) × Length of the link + 2 × 1 + 7 (for 35 Mb/s (BER 10^{-9})

or, 0.5 × Length of the link = 52 – 9 = 43

Therefore,length of the link = 43/0.5 km = 86 km

(*b*) Similarly for 400 Mb/s (BER 10^{-9}),

-3dBm – (-44 dBm) = (0.4 + 0.1) × Length of the link + 2 × 1 + 7

or, length of the link = 32/0.5 km = 64 km

(*c*) Transmitted power – received power = (cabled fiber loss + splice loss) × length of the link + connector loss + dispersion equalization penalty + safety margin.

So, 3 dBm – (– 44 dBm) = (0.4 + 0.1) × Length of the link + 2 × 1 + 1.5 + 7

or, Length of the link = (41 – 2 – 1.5 – 7)/0.5 = 30.5/0.5 = 61 km.

11.7 CONCLUSION

The basic concepts of modulation and detection schemes are presented in this chapter. The operation and characteristics of optical components of general system have been discussed. However to enable the successful incorporation of these components into optical fiber communication it is necessary to consider the compatibility of one component to other and thereafter the overall performance is to be evaluated. It is often helpful to offset the characteristics of certain component by trading off against the performance of another component to optimize the overall system performance.

REVIEW QUESTIONS

1. Which of the type of modulation can be detected by a homodyne receiver?
   (*a*) Amplitude modulation (*b*) Phase modulation
   (*c*) Both of these (*d*) None of these
2. For which of the following situation the source to fiber coupling efficiency is maximum?
   (*a*) Large NA and large source area (*b*) Large NA and small source area
   (*c*) Small NA and large source area (*d*) Small NA and small source area
3. Which of the following characteristics is exhibited by graded index fiber?
   (*a*) Large inter modal dispersion and large NA
   (*b*) Large inter modal dispersion and small NA
   (*c*) Low inter modal dispersion and large NA
   (*d*) Low inter modal dispersion and small NA
4. Which of the following types of intensity modulation exhibits better signal to noise ratio?
   (*a*) AM-IM (*b*) PM-IM (*c*) FM-IM (*d*) DSB-IM
5. In terms of éfficiency and long haul transmission which modulation technique is superior to others?
   (*a*) PAM-IM (*b*) PWM-IM (*c*) PPM-IM (*d*) All of these
6. If analog signal of 4 MHz bandwidth is required to transmit digitally with 8 bit format, what will be minimum bit rate?
   (*a*) 8 Mb/s (*b*) 16 Mb/s (*c*) 32 Mb/s (*d*) 64 Mb/s
7. An analog signal of 1.5 MHz bandwidth is sampled, quantized and encoded to binary code of 512 levels. What will be the minimum bit duration?
   (*a*) 667 ns (*b*) 333 ns (*c*) 74 ns (*d*) 37 ns
8. An analog signal of 4 KHz bandwidth and dynamic range of 200 mV is sampled, quantized and encoded to 8 bit format. What is the maximum quantization error?
   (*a*) 195 μV (*b*) 390 μV (*c*) 780 μV (*d*) 1560 μV
9. Twenty-four 4 KHz speech channels are sampled, quantized, encoded to PCM format and then time division multiplexed for transmission on optical link. The quantizer is linear with 0.5 mV steps over a dynamic range of 2.048 V. Assuming an additional channel time slot for synchronization, what will be minimum frame length?
   (*a*) 25 bits (*b*) 150 bits (*c*) 300 bits (*d*) 600 bits

10. A frequency division multiplexed optical system uses FM-IM with fifty voice channels, each of 3.5 KHz bandwidth and 1 KHz as guard band. The peak frequency deviation is 1.35 MHz. Determine the transmission bandwidth for this FDM system.
    (*a*) 3.15 MHz (*b*) 2.25 MHz (*c*) 1.8 MHz (*d*) 1.575 MHz
11. Why temperature stabilization is required for laser sources and how this can be accomplished?
12. Compare the merits and demerits of LED and Laser as optical sources regarding coupling them to optical fiber.
13. An optical link is designed for signal bandwidth of 6 MHz with LED source of rise time of 10 ns, APD detector of rise time of 3 ns, repeaterless length of 5 Km. Fiber cable intermodal and intramodal risetimes are 9 ns/Km and 2 ns/Km respectively. Calculate the system risetime and determine whether this components combination gives adequate temporal response.
14. Derive the expressions for signal to noise ratio for *p-i-n* detector and APD detector. Which one is superior and how much?
15. An optical fiber PFM-IM system for video transmission uses multimode graded index fiber. It employs regenerative baseband recovery and has the following operational parameters.
    (*i*) Nominal pulse rate : 20 MHz
    (*ii*) Peak to peak frequency deviation : 5 MHz
    (*iii*) APD responsivity : 0.7
    (*iv*) APD multiplication factor : 60
    (*v*) Total system rise time : 12 ns
    (*vi*) Baseband noise bandwidth : 6 MHz
    (*vii*) Receiver mean square noise current : 10^{-17} A^2
    Calculate (*a*) the optimum receiver bandwidth, (*b*) ratio of peak to peak signal power to rms noise power when peak input optical power to the receiver is –40 dBm.
16. A D-IM analog optical fiber system employs *p-i-n* photodiode receiver. Derive an expression for the rms signal power to rms noise power ratio in the quantum limit for the system. The photodiode of the above system has a responsivity of 0.5 at the operating wavelength of 0.85 μm. The system bandwidth is 5 MHz and modulation index is 0.5. The mean square thermal noise current for the receiver is 2×10^{-22} A^2/Hz. Determine the ratio of peak to peak signal power to rms noise power at the receiver, when average incident optical power at receiver is –17.5 dBm. Photodiode dark current is negligible.
17. A 80-channel analog optical fiber system employs subcarrier amplitude modulation (AM-IM) on RF carrier. Forty of these channels have pre-modulation index of 2%, twenty have 3% and twenty have 4% per channel. Calculate the optical modulation of the transmitter. For multichannel frequency division multiplexed system, the optical modulation index m may be related as $m = \left(\sum_{k=1}^{N} m_k^2 \right)^{1/2}$ where m_k is the pre-channel modulation index and N is the number of channels.
18. Describe in brief the governing parameters for free space optical communication. What is Mie Scattering?
19. For optical fiber communication system, discuss the selection criteria for (*i*) operating wavelength, (*ii*) emitter and (*iii*) detector.
20. What is eye pattern and how it is used? Describe the schematic of regenerative repeater and its design consideration.

12

OPTICAL MULTIPLEXING CONCEPTS

12.1 INTRODUCTION

In order to maximize the information transfer over an optical fiber communication link it is usual to multiplex several signals on to a single fiber. Time Division Multiplexing (TDM) scheme prior to intensity modulation of a single optical source permits the transmission of multichannel digital pulse modulated signals under the control of a common clock. Pulses from individual channels are interleaved and transmitted sequentially, enhancing the bandwidth utilization of a single fiber link.

Alternatively, a number of baseband channels may be combined by Frequency Division Multiplexing (FDM). In this technique, optical channel bandwidth is divided into a number of non-overlapping frequency bands and each signal is assigned one of these bands of frequencies. By appropriate electrical filtering at the receiving end the combined FDM signal can be extracted. Frequency division multiplexing is generally performed electrically at the transmitting terminal prior to intensity modulation of a single optical source.

However, a number of optical sources may be utilized, each operating at different wavelength on a single fiber link. The technique is often referred to as wavelength division multiplexing (WDM), where the separation and extraction of the multiplexed signals is performed with optical filters *e.g.*, interference filters, diffraction grating filters or prism filters.

12.2 TIME-DIVISION MULTIPLEXING (TDM)

The basic concept of Time Division Multiplexing (TDM) is illustrated in Fig. 12.1(*a*) and 12.1(*b*). It is a type of digital or (rarely) analog multiplexing in which two or more signals or bit streams are transferred apparently simultaneously as sub-channels in one communication channel, but are physically taking turns on the channel. The time domain is divided into several recurrent timeslots of fixed length, one for each sub-channel. A sample byte or data block of sub-channel 1 is transmitted during timeslot 1, sub-channel 2 during timeslot 2, etc. One TDM frame consists of one timeslot per sub-channel. After the last sub-channel the cycle starts all over again with a new frame, starting with the second sample, byte or data block from sub-channel 1, etc. Figure 12.2 explains time slots for different channels. At the receiver side information is retrieved after demultiplexing with proper synchronization.

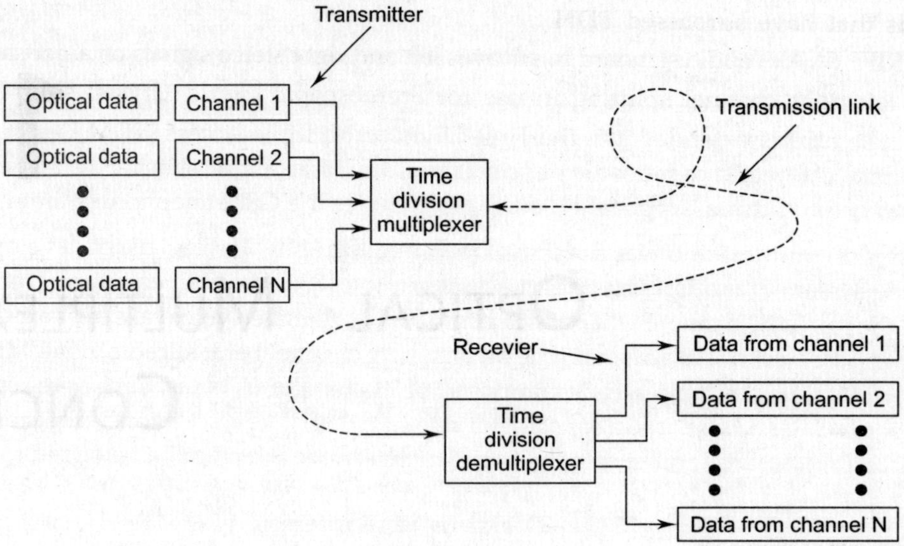

Fig. 12.1(*a*)

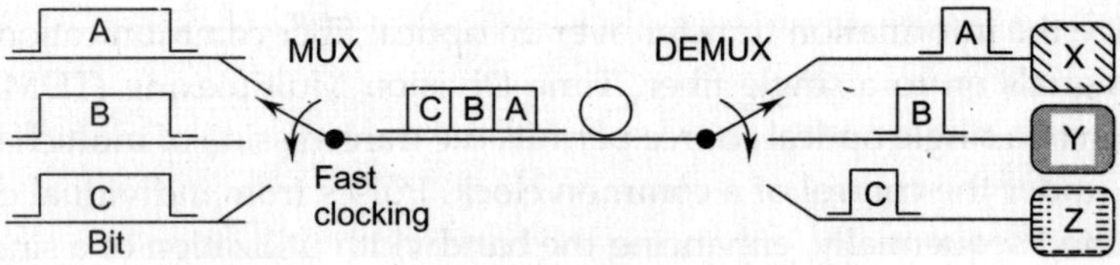

Fig. 12.1(*b*)

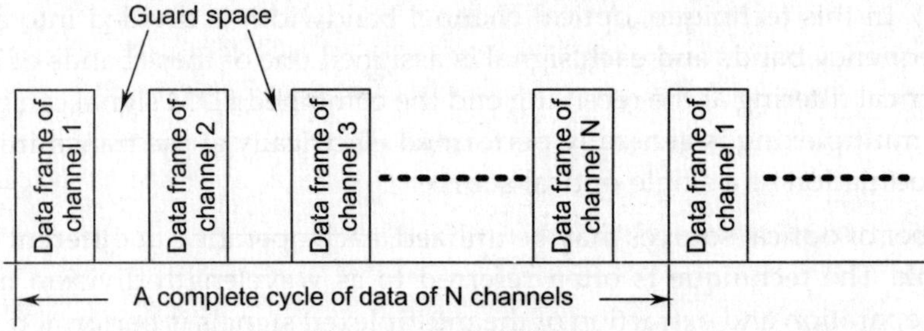

Fig. 12.2

This Time Division Multiplexing (TDM) can be accomplished in the electrical or optical domain, with each lower-speed channel transmitting a bit (or allocation of bits known as a packet) in a given time slot and the waiting its turn to transmit another bit (or packet) after all the other channels have had their opportunity to transmit. TDM is quite popular with today's electrical networks, and is fairly straightforward to implement in an optical network at < 100-Gbps speeds. This scheme by itself cannot hope to utilize the available bandwidth because it is limited by the speed of the time-multiplexing and -demultiplexing components.

The application of TDM technique may be incorporated at the several transmission protocols as indicated below.

- The Plesiochronous Digital Hierarchy (PDH) system, also known as the PCM system, for digital transmission of several telephone calls over the same four-wire copper cable (T-carrier or E-carrier) or fiber cable in the circuit switched digital telephone network.

- The SDH and Synchronous Optical Networking (SONET) network transmission standards that have surpassed PDH.
- The RIFF (WAV) audio standard interleaves left and right stereo signals on a per-sample basis.
- The left-right channel splitting in use for Stereoscopic Liquid Crystal shutter glasses.

TDM can be further extended into the Time Division Multiple Access (TDMA) scheme, where several stations connected to the same physical medium, for example sharing the same frequency channel, can communicate. Application example includes the GSM telephone system.

Time-division multiplexing was first developed in telegraphy. Émile Baudot developed a time-multiplexing system of multiple Hughes machines in the 1870s. In 1962, engineers from Bell Labs developed the first D1 Channel Banks, which combined 24 digitized voice calls over a 4-wire copper trunk between Bell central office analogue switches. A channel bank sliced a 1.544 Mbit/s digital signal into 8,000 separate frames, each composed of 24 contiguous bytes. Each byte represented a single telephone call encoded into a constant bit rate signal of 64 kbit/s. Channel banks used a byte's fixed position (temporal alignment) in the frame to determine which call it belonged to.

12.2.1 Transmission Using Time Division Multiplexing (TDM)

In circuit switched networks such as the Public Switched Telephone Network (PSTN) there exists the need to transmit multiple subscribers' calls along the same transmission medium. To accomplish this, network designers make use of TDM. TDM allows switches to create channels, also known as tributaries, within a transmission stream. A standard DSB voice signal has a data bit rate of 64 kbit/s determined using Nyquist's sampling criterion. TDM takes frames of the voice signals and multiplexes them into a TDM frame which runs at a higher bandwidth. So if the TDM frame consists of n voice frames, the bandwidth will be $n \times 64$ kbit/s.

Each voice sample timeslot in the TDM frame is called a channel. In European systems, TDM frames contain 30 digital voice channels, and in American systems, they contain 24 channels. Both standards also contain extra bits (or bit timeslots) for signaling and synchronization.

Multiplexing more than 24 or 30 digital voice channels is called *higher order multiplexing*. Higher order multiplexing is accomplished by multiplexing the standard TDM frames. For example, a European 120 channel TDM frame is formed by multiplexing four standard 30 channel TDM frames. At each higher order multiplex, four TDM frames from the immediate lower order are combined, creating multiplexes with a bandwidth of $n \times 64$ kbit/s, where $n = 120, 480, 1920$, etc.

12.2.2 Synchronous Time Division Multiplexing (Sync TDM)

There are three types of (Sync TDM): T1, SONET/SDH, and ISDN. *Plesiochronous Digital Hierarchy* (PDH) was developed as a standard for multiplexing higher order frames. PDH created larger numbers of channels by multiplexing the standard Europeans 30 channel TDM frames. This solution worked for a while; however PDH suffered from several inherent drawbacks which ultimately resulted in the development of the *Synchronous Digital Hierarchy* (SDH). The requirements which drove the development of SDH were as follows:

- All clocks in the system must align or synchronize with a reference clock.
- It should be service-oriented, *i.e.*, SDH must route traffic from End Exchange to End Exchange without worrying about exchanges in between, where the bandwidth can be reserved at a fixed level for a fixed period of time.

- Allow frames of any size to be removed or inserted into an SDH frame of any size.
- Easily manageable with the capability of transferring management data across links.
- Provide high levels of recovery from faults.
- Provide high data rates by multiplexing any size frame, limited only by technology.
- Reduces bit rate errors.

SDH Network functions are connected using high-speed optic fiber. SDH has become the primary transmission protocol in most PSTN networks. It was developed to allow streams 1.544 Mbit/s and above to be multiplexed, so as to create larger SDH frames known as *Synchronous Transport Modules* (STM). The STM-1 frame consists of smaller streams that are multiplexed to create a 155.52 Mbit/s frame. SDH can also multiplex packet based frames such as Ethernet, PPP and ATM.

While SDH is considered to be a transmission protocol (Layer 1 in the OSI Reference Model), it also performs same switching functions, as stated in the third bullet point requirement listed above. The most common SDH Networking functions are as follows:

- *SDH Crossconnect*: The SDH Crossconnect is the SDH version of a Time-Space-Time crosspoint switch. It connects any channel on any of its inputs to any channel on any of its outputs. The SDH Crossconnect is used in Transit Exchanges, where all inputs and outputs are connected to other exchanges.
- *SDH Add-Drop Multiplexer*: The SDH Add-Drop Multiplexer (ADM) can add or remove any multiplexed frame down to 1.544Mb. Below this level, standard TDM can be performed. SDH ADMs can also perform the task of an SDH Crossconnect and are used in End Exchanges where the channels from subscribers are connected to the core PSTN network.

12.2.3 Statistical Time-division Multiplexing (STDM)

STDM is an advanced version of TDM in which both the address of the terminal and the data itself are transmitted together for better routing. Using STDM allows bandwidth to be split over one line. Many colleges and corporate campuses use this type of TDM to logically distribute bandwidth.

If there is one 10 MBit line coming into the building, STDM can be used to provide 178 terminals with a dedicated 56 K connection (178×56 K = 9.96 Mb). A more common use however is only to grant the bandwidth when that much is needed. STDM does not reserve a time slot for each terminal, rather it assigns a slot when the terminal is requiring data to be sent or received.

This is also called Asynchronous Time-division Multiplexing (ATDM), in an alternative nomenclature in which *STDM* or *Synchronous Time Division Multiplexing* designates the older method that uses fixed time slots.

12.3 TIME DIVISION MULTIPLE ACCESS (TDMA)

Time Division Multiple Access (TDMA) is a channel access method for shared medium networks. It allows several users to share the same frequency channel by dividing the signal into different time slots. Figure 12.3 illustrates TDMA frame structure showing a data stream divided into frames and those frames divided into time slots. The users transmit in rapid succession, one after the other, each using his own time slot. This allows multiple stations to share the same transmission medium (*e.g.*, radio frequency channel) while using only a part of its channel capacity. However

TDMA is widely used in the digital 2G cellular systems such as Global System for Mobile Communications (GSM), IS-136, Personal Digital Cellular (PDC) and iDEN, and in the Digital Enhanced Cordless Telecommunications (DECT) standard for portable phones. It is also used extensively in satellite systems, and combat-net radio systems.

TDMA is a type of Time-division multiplexing, with the special point that instead of having one transmitter connected to one receiver, there are multiple transmitters. In the case of the *uplink* from a mobile phone to a base station this becomes particularly difficult because the mobile phone can move around and vary the *timing advance* required to make its transmission match the gap in transmission from its peers.

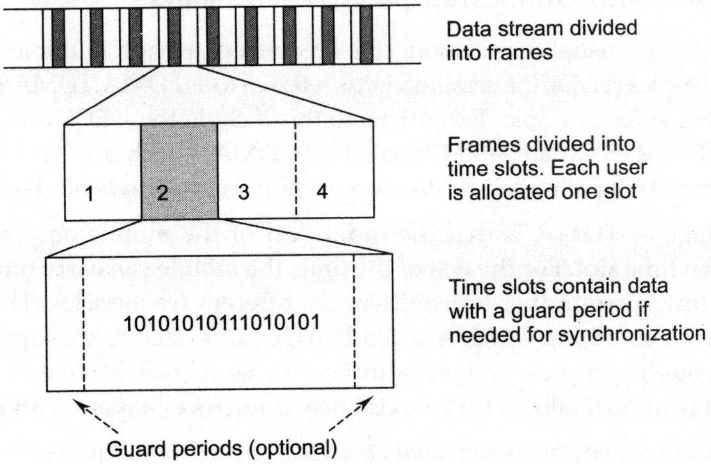

Fig. 12.3

12.3.1 TDMA Characteristics

- Shares single carrier frequency with multiple users.
- Non-continuous transmission makes handoff simpler.
- Slots can be assigned on demand in dynamic TDMA.
- Less stringent power control than CDMA due to reduced intra cell interference.
- Higher synchronization overhead than CDMA.
- Advanced equalization may be necessary for high data rates if the channel is frequency selective and creates intersymbol interference.
- Cell breathing (borrowing resources from adjacent cells) is more complicated than in CDMA.
- Frequency/slot allocation is complex.
- Pulsating power envelop: May be suffered by interference with other devices.

Most cellular systems, with the notable exception of IS-95, are based around TDMA. GSM, D-AMPS, PDC, iDEN, and PHS are examples of TDMA cellular systems. GSM combines TDMA with Frequency Hopping and wideband transmission to reduce interferences. TDMA is combined with CDMA and Time Division Duplexing in two standard UMTS UTRA modes, UTRA TDD-HCR (better known as TD-CDMA), and UTRA TDD-LCR (better known as TD-SCDMA). In each mode, more than one handset may share a single time slot. UTRA TDD-HCR is used most commonly by UMTS-TDD to provide Internet access, whereas UTRA TDD-LCR provides some interoperability with the forthcoming Chinese 3G standard. TDMA is used along with FDMA by the Iridium and Thuraya satellite phone networks.

The ITU-T G.hn standard which provides high-speed local area networking over existing home wiring (power lines, phone lines and coaxial cables) is based on a TDMA scheme. Optical fiber plays important role in this. In G.hn, a "master" device allocates "Contention-Free Transmission Opportunities" (CFTXOP) to other "slave" devices in the network. Only one device can use a CFTXOP at a time, thus avoiding collisions. FlexRay protocol, which is also a wired network used for safety. Critical communication in modern cars uses the TDMA method for data transmission control.

12.3.2 Comparison with Other Multiple-Access Schemes

In radio systems, TDMA is usually used alongside Frequency-division multiple access (FDMA) and Frequency Division Duplex (FDD); the combination is referred to as FDMA/TDMA/FDD. This is the case in both GSM and IS-136 for example. Exceptions to this include the DECT and PHS micro-cellular systems, UMTS-TDD UMTS variant, and China's TD-SCDMA, which use Time Division duplexing, where different time slots are allocated for the base station and handsets on the same frequency.

A major advantage of TDMA is that the radio part of the mobile only needs to listen and broadcast for its own time slot. For the rest of the time, the mobile can carry out measurements on the network, detecting surrounding transmitters on different frequencies. This allows safe inter frequency handovers, something which is difficult in CDMA systems, not supported at all in IS-95 and supported through complex system additions in Universal Mobile Telecommunications System (UMTS). This in turn allows for co-existence of microcell layers with macrocell layers.

CDMA, by comparison, supports *soft hand-off* which allows a mobile phone to be in communication with up to 6 base stations simultaneously, a type of *same-frequency handover*. The incoming packets are compared for quality, and the best one is selected. CDMA's *cell breathing* characteristic, where a terminal on the boundary of two congested cells will be unable to receive a clear signal, can often negate this advantage during peak periods.

A disadvantage of TDMA systems is that they create interference at a frequency which is directly connected to the time slot length. This is the buzz which can sometimes be heard if a TDMA phone is left next to a radio or speakers. Another disadvantage is that the "dead time" between time slots limits the potential bandwidth of a TDMA channel. These are implemented in part because of the difficulty in ensuring that different terminals transmit at exactly the times required. Handsets that are moving will need to constantly adjust their timings to ensure their transmission is received at precisely the right time, because as they move further from the base station, their signal will take longer to arrive. This also means that the major TDMA systems have hard limits on cell sizes in terms of range, though in practice the power levels required to receive and transmit over distances greater than the supported range would be mostly impractical anyway.

12.3.3 Dynamic TDMA

In **dynamic time division multiple access**, a scheduling algorithm dynamically reserves a variable number of time slots in each frame to variable bit-rate data streams, based on the traffic demand of each data stream. Dynamic TDMA is used in

- HIPERLAN/2 broadband radio access network.
- IEEE 802.16aWiMax

- Bluetooth
- The Packet Radio Multiple Access (PRMA) method for combined circuit switched voice communication and packet data.
- TD-SCDMA
- ITU-T G.hn

12.4 FREQUENCY-DIVISION MULTIPLEXING (FDM)

Frequency Division Multiplexing (FDM) is a form of signal multiplexing which involves assigning non-overlapping frequency ranges to different signals or to each user of a medium.

12.4.1 Non Telephone

FDM can also be used to combine multiple signals before final modulation onto a carrier wave. In this case the carrier signals are referred to as subcarriers: an example is stereo FM transmission, where a 38 kHz subcarrier is used to separate the left-right difference signal from the central left-right sum channel, prior to the frequency modulation of the composite signal. A television channel is divided into subcarrier frequencies for video, color, and audio. DSL uses different frequencies for voice and for upstream and downstream data transmission on the same conductors, which is also an example of frequency duplex.

Where frequency-division multiplexing is used as to allow multiple users to share a physical communications channel, it is called *Frequency Division Multiple Access* (FDMA). FDMA is the traditional way of separating radio signals from different transmitters. In the 1860 and 70s, several inventors attempted FDM under the names of Acoustic telegraphy and Harmonic telegraphy. Practical FDM was only achieved in the electronic age. Meanwhile their efforts led to an elementary understanding of electroacoustic technology, resulting in the invention of the telephone.

12.4.2 Telephone

For long distance telephone connections, 20th century telephone companies used L-carrier and similar co-axial cable systems carrying thousands of voice circuits multiplexed in multiple stages by channel banks.

For shorter distances, cheaper balanced pair cables were used for various systems including Bell System K- and N-Carrier. Those cables didn't allow such large bandwidths, so only 12 voice channels (Double Sideband) and later 24 (Single Sideband) were multiplexed into four wires, one pair for each direction with repeaters every several miles, approximately 10 km. By the end of the 20th Century, FDM voice circuits had become rare. Modern telephone systems employ digital transmission, in which Time Division Multiplexing (TDM) is used instead of FDM.

Since the late 20th century Digital Subscriber Lines have used a Discrete multitone (DMT) system to divide their spectrum into frequency channels. The concept corresponding to frequency-division multiplexing in the optical domain is known as wavelength division multiplexing.

12.5 WAVELENGTH-DIVISION MULTIPLEXING (WDM)

In fiber-optic communications, **Wavelength-Division Multiplexing (WDM)** is a technology which multiplexes multiple optical carrier signals on a single optical fiber by using different wavelengths (colours) of laser to carry different signals. This allows for a multiplication in capacity, in addition to enabling bidirectional communications over one strand of fiber. WDM is similar to Frequency Division Multiplexing (FDM). But instead of taking place at Radio Frequencies (RF), WDM is done in the IR portion of the electromagnetic (EM) spectrum. Each IR channel carries several RF signals combined by means of FDM or Time Division Multiplexing (TDM). Each multiplexed IR channel is separated, or demultiplexed, into the original signals at the destination. Using FDM or TDM in each IR channel in combination with WDM of several IR channels, data in different formats and at different speeds can be transmitted simultaneously on a single fiber.

The true potential of optical fiber is fully exploited when multiple beams of light at different frequencies are transmitted on the same fiber. This is a form of Frequency Division Multiplexing (FDM) but is commonly called wavelength division multiplexing. The term wavelength-division multiplexing is commonly applied to an optical carrier (which is typically described by its wavelength), whereas frequency-division multiplexing typically applies to a radio carrier (which is more often described by frequency). However, since wavelength and frequency are inversely proportional, and, since radio and light are both forms of electromagnetic radiation, the two terms are equivalent.

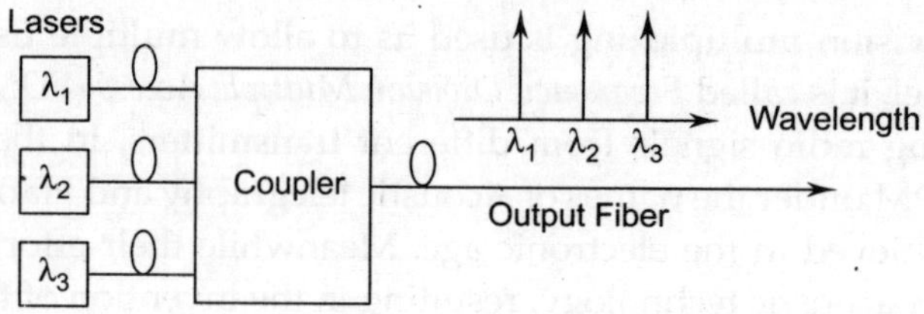

Fig. 12.4

12.5.1 WDM Systems

In WDM several baseband-modulated channels are transmitted along a single fiber but with each channel located at a different wavelength as illustrated in Fig. 12.4. Each of N different wavelength lasers is operating at the slower Gbps speeds, but the aggregate system is transmitting at N times the individual laser speed, providing a significant capacity enhancement. The WDM channels are separated in wavelength to avoid cross-talk when they are de multiplexed by a non-ideal optical fiber. The wavelengths can be individually routed through a network or individually recovered by wavelength-selective components. WDM allows us to use much of the fiber bandwidth, although various device, system, and network issues will limit the utilization of the full fiber bandwidth. Note that each WDM channel may contain a set of even slower time-multiplexed channels.

A WDM system uses a multiplexer at the transmitter to join the signals together and a demultiplexer at the receiver to split them apart. With the right type of fiber it is possible to have a device that does both simultaneously, and can function as an optical add-drop multiplexer. The optical filtering devices used have traditionally been etalons, stable solid-state single-frequency Fabry-Perot interferometers in the form of thin-film-coated optical glass. Figure 12.5 shows Nortel's WDM System as example.

The concept was first conceived in 1970, and by 1978 WDM systems were being realized in the laboratory. The first WDM systems only combined two signals. Modern systems can handle up to 160 signals and can thus expand a basic 10 Gbit/s fiber system to a theoretical total capacity of over 1.6 Tbit/s over a single fiber pair.

Fig. 12.5

WDM systems are popular with telecommunications companies because they allow them to expand the capacity of the network without laying more fiber. By using WDM and optical amplifiers, they can accommodate several generations of technology development in their optical infrastructure without having to overhaul the backbone network. Capacity of a given link can be expanded by simply upgrading the multiplexers and demultiplexers at each end.

This is often done by using optical-to-electrical-to-optical (O/E/O) translation at the very edge of the transport network, thus permitting interoperation with existing equipment with optical interfaces. Most WDM systems operate on single mode fiber optical cables, which have a core diameter of 9 μm. Certain forms of WDM can also be used in multi-mode fiber cables (also known as premises cables) which have core diameters of 50 or 62.5 μm. Early WDM systems were expensive and complicated to run. However, recent standardization and better understanding of the dynamics of WDM systems have made WDM less expensive to deploy. Optical receivers, in contrast to laser sources, tend to be wideband devices. Therefore the demultiplexer must provide the wavelength selectivity of the receiver in the WDM system.

Another method conceptually related to WDM is subcarrier multiplexing (SCM). Instead of directly modulating a terahertz optical carrier wave with 100s Mbps baseband data, the baseband data are impressed on a gigahertz subcarrier wave that is subsequently impressed on the THz optical carrier. Figure 12.6 illustrates the situation in which each channel is located at a different subcarrier frequency, thereby occupying a different portion of the spectrum surrounding the optical carrier. SCM is similar to commercial radio, in which many stations are placed at different RF (Radio Frequency) such that a radio receiver can tune its filter to the appropriate subcarrier RF. The multiplexing and demultiplexing of the SCM channels is accomplished electronically, not optically. The obvious advantage of cost-conscious users is that several channels can share the same expensive optical components; electrical components are typically less expensive than optical ones. Just as with TDM, SCM is limited in maximum subcarrier frequencies and data rates by the available bandwidth of the electrical and optical components. Therefore, SCM must be used in conjunction with WDM if we want to utilize any significant fraction of the fiber bandwidth, but it can be used effectively for lower-speed, lower-cost multiuser systems.

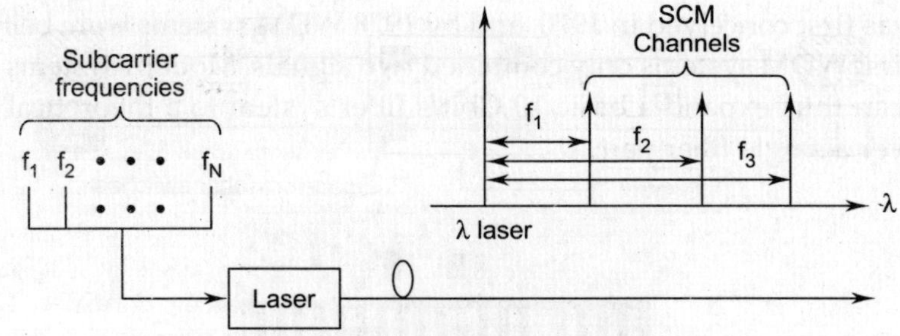

Fig. 12.6

Additional method is the Code-Division Multiplexing (CDM). Instead of each channel occupying a given wavelength, frequency or time slot, each channel transmits its bits as a coded channel-specific sequence of pulses. This coded transmission typically is accomplished by transmitting a unique time-dependent series of short pulses. These short pulses are placed within chip times within the larger bit time. The basic concept of a coded pulse sequence for CDM, with each pulse located in a chip time and the entire code occupying a larger bit time slot is shown at Fig 12.7. All channels, each with a different code, can be transmitted on the same fiber and asynchronously demultiplexd. One effect of coding is that the frequency bandwidth of each channel is broadbanded. If ultra-short (<100 fs) optical pulses can be successfully generated and modulated, then a significant fraction of the fiber bandwidth can be used. Unfortunately, it is difficult for the entire system to operate at these speeds without incurring enormous cost and complexity.

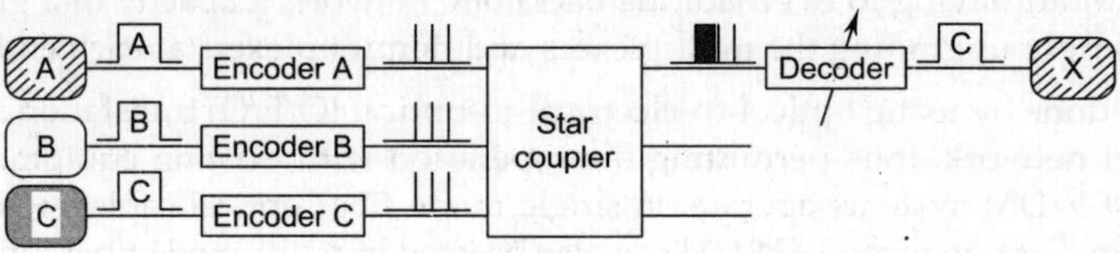

Fig. 12.7

Yet another optical multiplexing scheme is called Space-Division Multiplexing (SDM), in which the channel-routing path is determined by different spatial position (*i.e.*, a different output fiber). A simple example of 1-by-N spatial multiplexing is shown in Fig. 12.8, in which the optical output of a fiber is split into N different and parallel optical beam paths. Each of the N output beams is passed through a light-modulating switch and then coupled to a different output fiber. By controlling the transmissivity of each optical modulator, a signal on the input fiber can be routed to any fiber output port. By extending this scenario, N input fiber ports can be fully interconnected with N output fiber ports by an array of N^2 optical switches. The technology for implementing moderate-speed systems is already commercially available. In contrast to all other methods, however, each channel occupies its own spatial coordinate, and all other channels cannot be transmitted simultaneously on the same fiber. In other words, we are not more fully utilizing the high bandwidth of the fiber, but we are creating a high-bandwidth space-switching matrix, with the result that a high overall switching capacity can be realized.

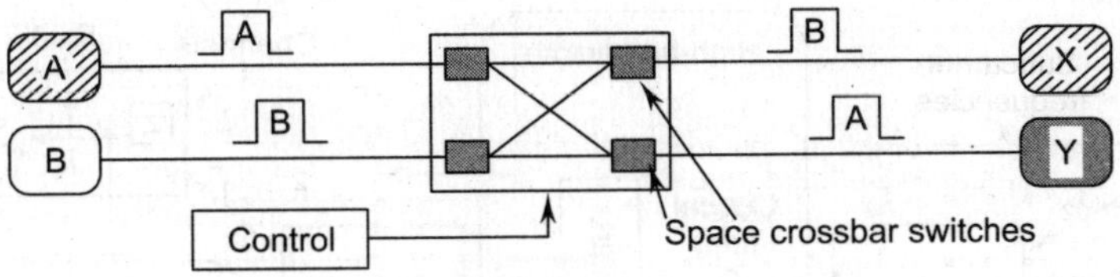

Fig. 12.8

Until the late 1980s, optical fiber communications was mainly confined to transmitting a single optical channel. Because fiber attenuation was involved, this channel required periodic regeneration, which included detection, electronic processing, and optical retransmission. Such regeneration causes a high-speed optoelectronic bottleneck and can handle only a single wavelength. After the new generation amplifiers were developed, it enabled us to accomplish high-speed repeaterless single-channel transmission. We can think of single channel of the bandwidth in the order of Gbps as a single high-speed lane in a highway in which the cars are packets of optical data and the highway is the optical fiber. However, the optical fiber of the capacity of 25 THz can accommodate much more bandwidth than the traffic from a single lane. To increase the system capacity we can transmit several different independent wavelengths simultaneously down a fiber to fully utilize this enormous fiber bandwidth. Therefore, the intent was to develop a multiple-lane highway, with each lane representing data traveling on a different wavelength. Thus, a WDM system enables the fiber to carry more throughputs. By using wavelength-selective devices, independent signal routing also can be accomplished. WDM is one of the methods of choice for future ultra-high bandwidth multichannel systems. Of course, this could be changed as the technology evolves.

12.5.2 Basic Operation

As explained before, WDM enables the utilization of a significant portion of the available fiber bandwidth by allowing many independent signals to be transmitted simultaneously on one fiber, with each signal located at a different wavelength. Routing and detection of these signals can be accomplished independently, with the wavelength determining the communication path by acting as the signature address of the origin, destination or routing. Components are therefore required that are wavelength selective, allowing for the transmission, recovery, or routing of specific wavelengths.

In a simple WDM system as illustrated in Fig. 12.9, each laser must emit light at a different wavelength, with all the lasers light multiplexed together onto a single optical fiber. After being transmitted through a high-bandwidth optical fiber, the combined optical signals must be demultiplexed at the receiving end by distributing the total optical power to each output port and then requiring that each receiver selectively recover only one wavelength by using a tunable optical filter. Each laser is modulated at a given speed, and the total aggregate capacity being transmitted along the high-bandwidth fiber is the sum total of the bit rates of the individual lasers. An example of the system capacity enhancement is the situation in which ten 2.5-Gbps signals can be transmitted on one fiber, producing a system capacity of 25 Gbps. This wavelength-parallelism circumvents the problem of typical optoelectronic devices, which do not have bandwidths exceeding a few gigahertz unless they are exotic and expensive. The speed requirements for the individual optoelectronic components are, therefore, relaxed, even though a significant amount of total fiber bandwidth is still being utilized.

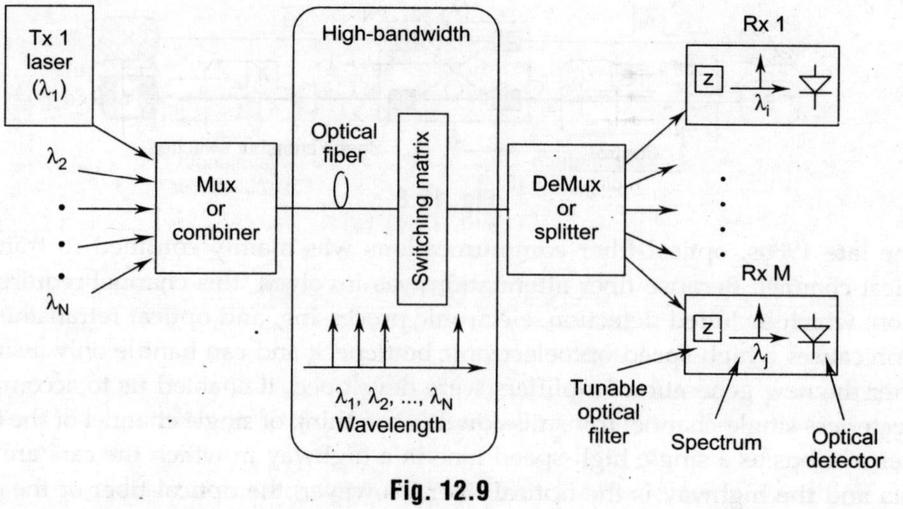

Fig. 12.9

The concept of wavelength demultiplexing using an optical filter is illustrated in Fig. 12.10. In the figure, four channels are input to an optical filter that has a nonideal transmission filtering function. The filter transmission peak is centered over the desired channel, thereby transmitting that channel and blocking all other channels. Because of the nonideal filter transmission function, some optical energy of the neighboring channels leaks through the filter, causing interchannel, interwavelength cross-talk. This cross-talk has the effect of reducing the selected signals contrast ratio and can be minimized by increasing the spectral separation between channels. Although there is no set definition, a nonstandardized convention exists for defining optical WDM as encompassing a system for which the channel spacing is approximately 10 nm.

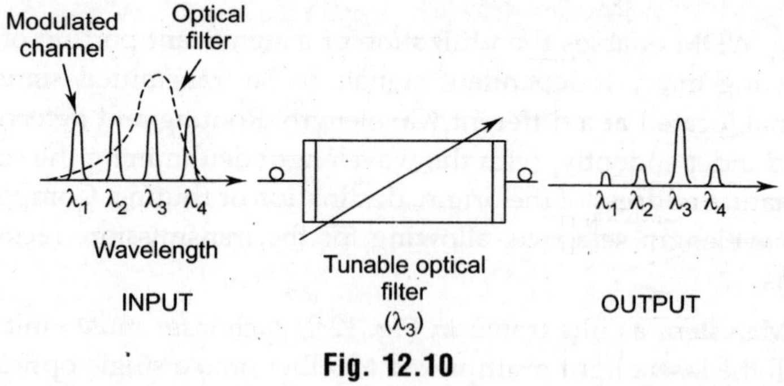

Fig. 12.10

12.5.3 Topologies and Architectures

Let us consider a simple point-to-point WDM system as of Fig. 12.11(*a*), in which several channels are multiplexed at one node, the combined signals are transmitted across through fiber, and the channels are demultiplexed at a destination node. This facilitates high-bandwidth fiber transmission. Additionally, high-bandwidth routing can be facilitated through a multiuser network as illustrated in Fig. 12.11(*b*). The wavelength becomes the signature address for either path through an optical network. Because nodes will want to communicate with each other, either the transmitters or the receivers must be wavelength tunable to facilitate the proper link set-up (in this example, the transmitters are chosen to be tunable).

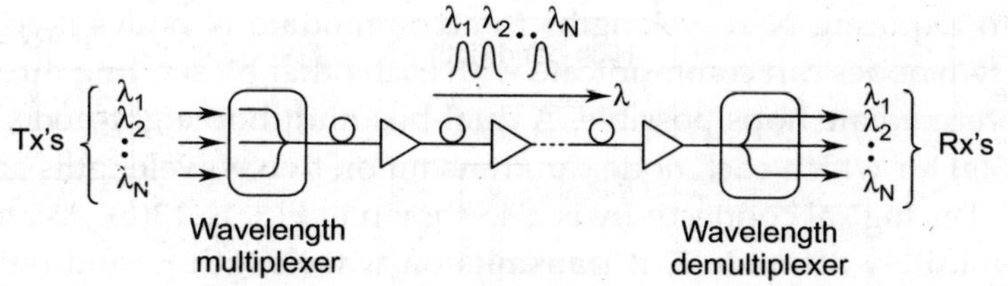

Fig. 12.11 (a)

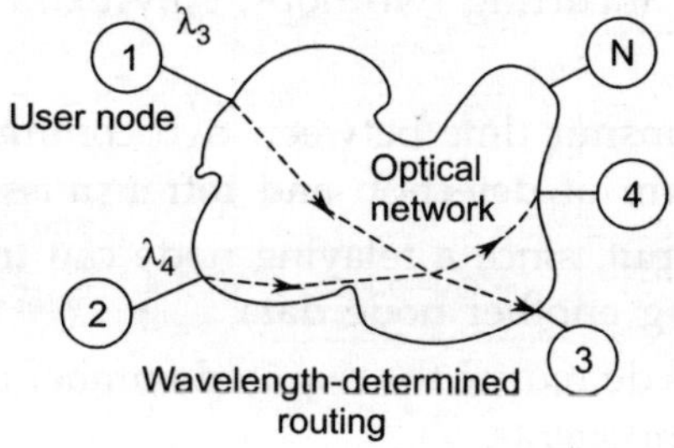

Fig. 12.11(b)

Two common network topologies can use WDM, namely, the star as of Fig. 12.12(*a*) and the ring networks as of Fig. 12.12(*b*). Each node in the star has a transmitter and a receiver, with the transmitter connected to one of the central passive star inputs and the receiver connected to one of the star outputs. WDM networks can also be of the ring variety. Rings are popular because so many electrical networks use this topology and because rings are easy to implement for any network geographical configuration. In this example, each node in the unidirectional ring can transmit on a specific signature wavelength, and each node can recover any other node's wavelength signal by means of a wavelength-tunable receiver.

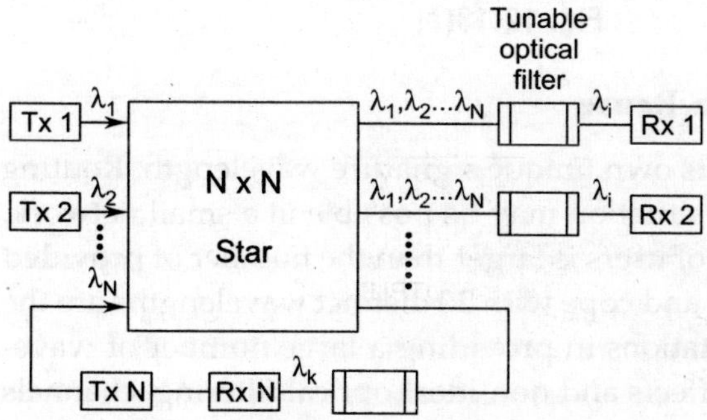

Fig 12.12(a)

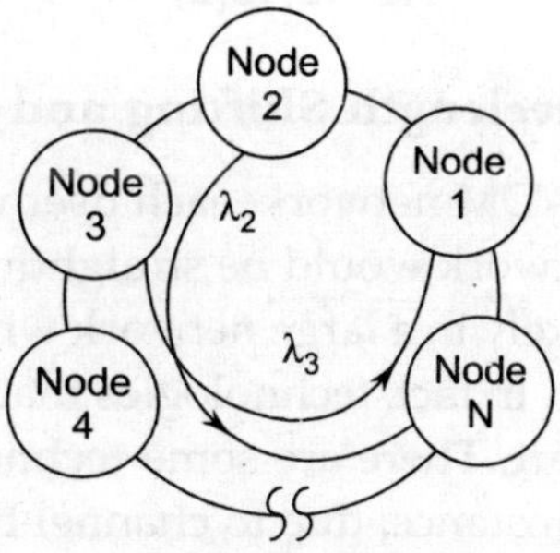

Fig 12.12(b)

In both the star and the ring scenarios, each node has a signature wavelength, and any two nodes can communicate with each other by transmitting on that wavelength. This implies that we require N wavelengths to connect N nodes. The obvious advantage is that data transfer occurs with an uninterrupted optical path between the origin and the destination, known as a single-hop network. The optical data start at the originating node and reach the destination node without stopping at any other intermediate node. A disadvantage of a single-hop WDM network is that the network and all its components must accommodate N wavelengths, which may be difficult and too expensive to achieve in a large network.

An alternative to requiring N wavelengths to accommodate N nodes is to have a multihop network, in which two nodes can communicate with each other by sending through a third node, with many such intermediate hops possible. A dual-bus multihop eight-node WDM network is shown in Fig. 12.13(*a*) for which each node can transmit on two wavelengths and receive on two other wavelengths. The logical connectivity is also shown in Fig. 12.13(*b*). As an example, if node 1 wants to communicate with node 5, it transmits on wavelength λ_1 and only a single hop is required. However, if node 1 wants to communicate with node 2, it first must transmit to node 5, which then transmits to node 2, incurring two hops. Any extra hops are deleterious due to following reasons:

1. Extra hops increase the transmit time between two communicating nodes, since a hop typically requires some form of detection and retransmission
2. They decrease the throughput, since a relaying node can transmit its own data while it is in the process of relaying another node data

However, a multihop networks do reduce the required number of wavelengths and the wavelength tunability range of the components.

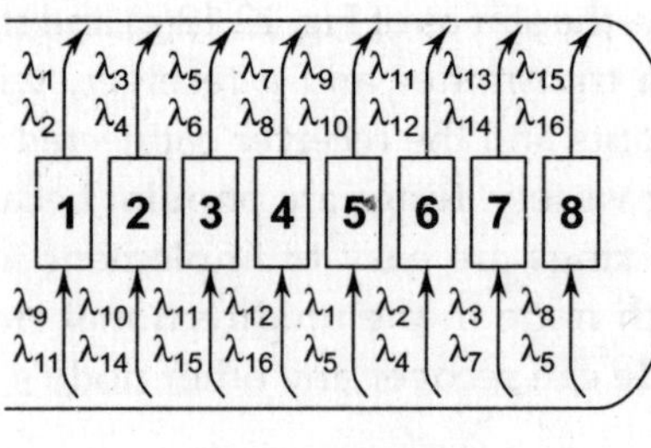

Fig. 12.13(*a*)

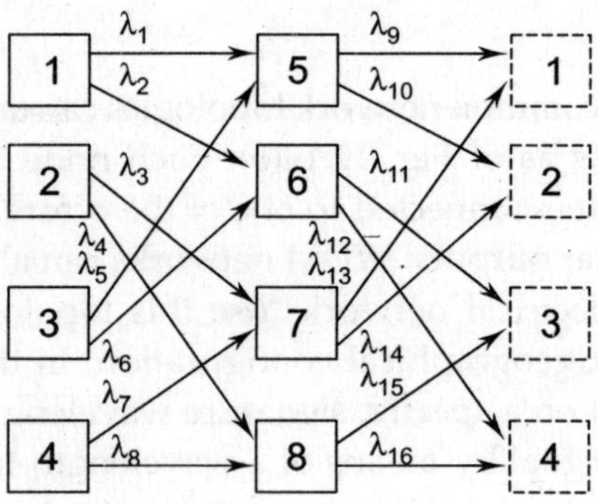

Fig. 12.13(*b*)

12.5.4 Wavelength Shifting and Wavelength Reuse

In an ideal WDM network, each user would have its own unique signature wavelength. Routing in such a network would be straightforward. This situation may be possible in a small network, but it is unlikely in a large network whose number of users is larger than the number of provided wavelengths. In fact, technologies that can provide and cope with 20 distinct wavelengths are the state of the art. There are some technological limitations in providing a large number of wavelengths, for instance: due to channel-broadening effects and non-ideal optical filtering, channels must have minimum wavelength spacing. Wavelength range, accuracy, and stability are extremely difficult to control. Therefore, it is quite possible that a given network may have more users than available wavelengths, which will necessitate the reuse of a given set of wavelengths at different points in the network.

Passive Wavelength Routing

In case of a limited number of available wavelengths, a network can use passive routing of a signal through the network based only on its wavelength. The routing is designed to reuse wavelengths in non-shared links. For example, we can see in Fig. 12.14 that user I can use wavelength λ_1 to establish a link with user II, while simultaneously user V can reuse the same wavelength λ_1, to establish a connection with user III. This functionality is accomplished by the proper arrangement of the cross-connects that route an input signal to a wavelength-determined output. A simple example of the operation of a passive WDM cross-connect is shown in Fig. 12.15. The cross-connect is composed of wavelength demultiplexers for the input stage, wavelength multiplexers for the output stage, and fibers interconnecting the two stages. In the example, although there are only two wavelengths, there are four possible non-interfering routing paths based on both wavelength and origin. In general, instead of N wavelengths and N possible connection paths, now there are N wavelengths and N^2 connections. The same wavelength could be reused by any of the input ports to access a completely different output port and establish an additional connection. This technique increases the capacity of a WDM network.

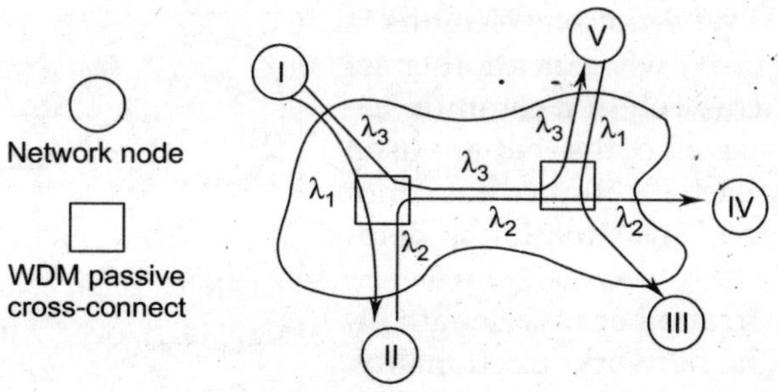

Fig. 12.14

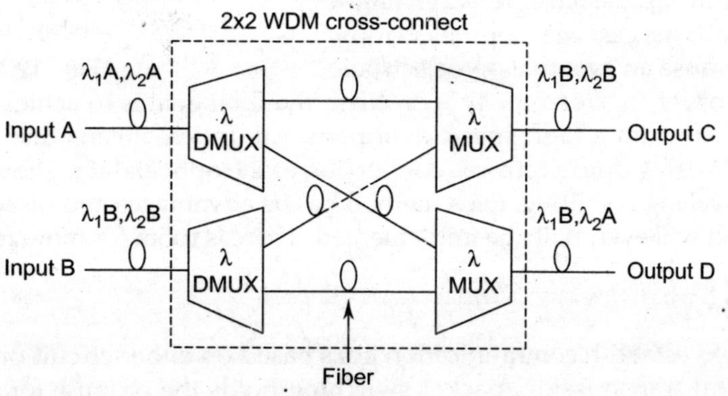

Fig. 12.15

Active Wavelength Shifting

In contrast to passive routing, which is limited to a static network conditions, active wavelength shifting is dynamically deals with changes of the network condition. It does that by changing the routing depending on the available links and wavelengths. This concept of a network requiring active wavelength shifting is illustrated in Fig. 12.16. In the figure there are two small LANs connected to a larger WAN, and each LAN can transmit on only two available wavelengths (λ_a and λ_b). Node I wishes to communicate with node II, the only wavelength available is λ_a. However, when the signal reaches the right LAN, it is revealed that λ_a is already being used by the right LAN. Therefore, the only way for the signal to reach node II is to be actively switched onto the available λ_b.

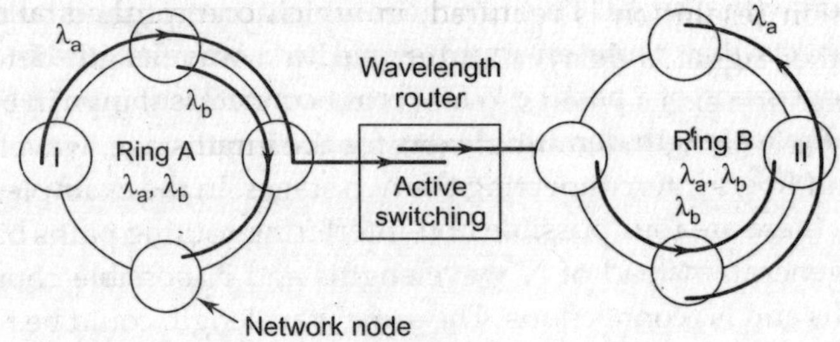

Fig. 12.16

Another scenario that would require active wavelength switching is where one set of wavelengths is used exclusively by each LAN, whereas another set of wavelength is used exclusively for communication between LANs. The wavelengths that are used in a LAN can be reused by each LAN since it will not interfere with another LAN. This situation is demonstrated in Fig. 12.17.

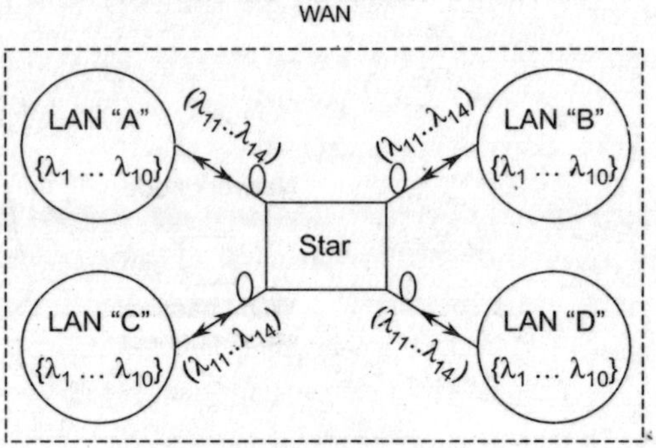

Local wavelengths: $\lambda_1 \ldots \lambda_{10}$
Global wavelengths: $\lambda_{11} \ldots \lambda_{14}$

Fig. 12.17

Shifting one wavelength to another wavelength introduces complexities in the network functionality. One method to perform the active wavelength switching is to employ optoelectronic wavelength shifters. This method necessitates optoelectronic conversions and will cause an eventual optoelectronic speed bottleneck. In order to overcome this problem the final goal is to achieve all-optical active wavelength shifting to retain a high speed data path. All-optical means that all the shifters are purely optical, *i.e.*, not using optoelectronic conversion of the optical data. There are several methods for all-optical wavelength shifting. Each method has its advantages and disadvantages, and it is not clear if any method will eventually be implemented. There is room for more research in this area.

12.5.5 Switching

We know that networks establish communication links based on either circuit or packet switching. For high-speed optical transmission, packet switching holds the promise for more efficient data transfer. Network packet switching can be accomplished in a straightforward manner by requiring a node to optoelectronically detect and transmit each and every incoming optical data packet. As for the routing, all the switching functions can occur in the electrical domain prior to optical retransmission of the signal. Unfortunately, this approach suffers from an optoelectronic speed bottleneck. Alternatively, much research is focused toward maintaining an all-optical data path and performing the switching functions all optically with only some electronic control of the optical components. However, there are many difficulties with optical switching, for instance:

1. A redirection of an optical path is not easy since photons do not have as strong interaction with their environment as electrons do.
2. Switching has to be extremely fast due to the high speed of the incoming signal.
3. Switching nodes cannot easily tap a signal and acquire information about the channel.

12.5.6 Contention Resolution

Consider a situation in which two or more input ports request a communications path with the same output port, known as output-port contention. Since we are dealing with a high-speed system, a rapid contention resolution is required, in which one signal is allowed to reach its destination while the other signal is delayed or rerouted in some fashion. In our multiplexing scheme, the issue of contention exists when signals from two different input ports would request routing to the same output port and contain identical wavelengths.

Several approaches exist for resolving contention. One of them is buffering. The packet is retained locally at the switching node and then it is switched to the appropriate output port when that port is available. The local buffering can be implemented either in electrical or optical form. Electronic buffering is straightforward but requires undesirable optoelectronic conversions and may require very large buffers. On the other hand, optical buffering is difficult because many buffering schemes require updating a priority bit (it is difficult to change a priority bit of an optical data stream), and optical memory is not an advanced art, consisting mostly of using an optical delay line.

12.5.7 Synchronization

A high-speed network transmitting digital signals must have adequate time for synchronization to recover the data stream. Time synchronization is especially required with packet switching, asynchronous packet arrival times, and long-distance transmission. In a WDM network, it is also possible that wavelength synchronization will be required in addition to time synchronization. In such a scenario, a wavelength standard could be broadcast through the network. However, the hope is that the network wavelength stability and accuracy will be robust and will not require its own system overhead and complexity.

12.5.8 Data-Format Conversion

In a large network, it is quite possible that a combination of data formats will be used. This may occur, for instance, if some links may more efficiently use TDM signaling, whereas other links may more effectively use WDM. This explains the need for data-format conversion at network gateways, as illustrated in Fig. 12.18. TDM data is converted to several lower-speed WDM signals at a network gateway.

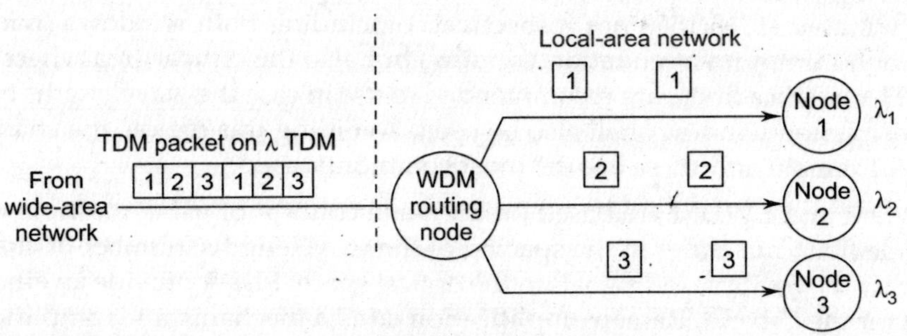

Fig. 12.18

12.5.9 Protocols

A standardized network protocol must be used to ensure that data packets are all formatted with recognizable routing information so that the packet can be switched through the network with full global compatibility. There are two standards that show the most promise of full adoption for a global optical network:

1. SONET - Synchronous Optical Network
2. ATM - Asynchronous Transfer Mode.

These two standards can be combined in one network. Data and header information are bounced into small ATM packets as shown in Fig 12.19(*a*). These packets arrive at a switching node at random times and are grouped together into a large SONET frame as in Fig. 12.19(*b*), which makes its way in predetermined synchronous time slots through the network. The ATM packets are unloaded by the SONET frame when its direction is switched through the network and it can be placed into a different SONET frame. ATM packets can be thought of as people randomly boarding a time-scheduled SONET train.

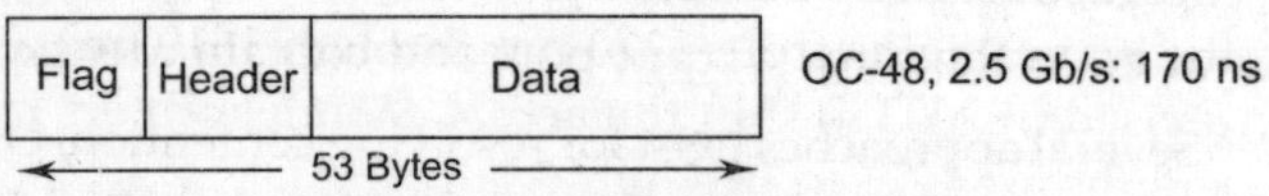

Fig. 12.19(*a*)

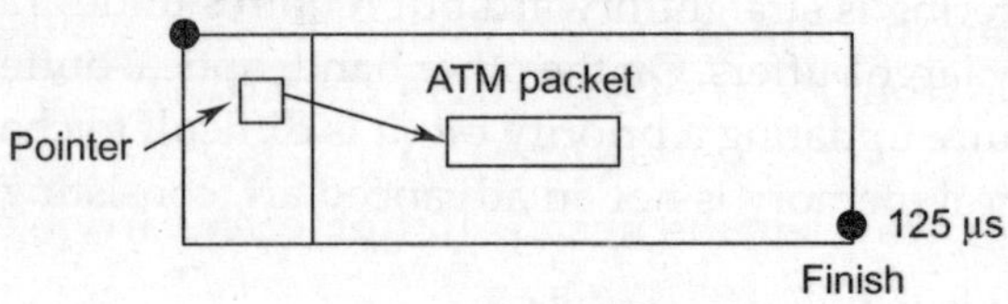

Fig. 12.19(*b*)

12.5.10 Different Wavelength Patterns of WDM

WDM systems are divided in different wavelength patterns, *conventional* or *coarse* and *dense* WDM. Conventional WDM systems provide up to 16 channels in the 3rd transmission window (C-band) of silica fibers around 1550 nm. DWDM Dense WDM uses the same transmission window but with denser channel spacing. Channel plans vary, but a typical system would use 40 channels at 100 GHz spacing or 80 channels with 50 GHz spacing. Some technologies are capable of 25 GHz spacing (sometimes called ultra dense WDM). New amplification options (Raman amplification) enable the extension of the usable wavelengths to the L-band, more or less doubling these numbers.

CWDM (Coarse WDM) in contrast to conventional WDM and DWDM uses increased channel spacing to allow less sophisticated and thus cheaper transceiver designs. To provide additional 16 channels on a single fiber CWDM uses the entire frequency band between second and third transmission window (1310/1550 nm respectively) including both windows (minimum dispersion window and minimum attenuation window) but also the critical area where OH scattering may occur. OH-free silica fibers are recommended to use in case the wavelengths between second and third transmission window shall also be used. Avoiding this region, the channels 31, 49, 51, 53, 55, 57, 59, 61 remain and these are the most commonly used.

WDM, DWDM and CWDM are based on the same concept of using multiple wavelengths of light on a single fiber, but differ in the spacing of the wavelengths, number of channels, and the ability to amplify the multiplexed signals in the optical space. EDFA provide an efficient wideband amplification for the C-band, Raman amplification adds a mechanism for amplification in the L-band. For CWDM wideband optical amplification is not available, limiting the optical spans to several tens of kilometres.

Coarse WDM

Originally, the term "Coarse Wavelength Division Multiplexing" was fairly generic, and meant a number of different things. In general, these things shared the fact that the choice of channel spacings and frequency stability was such that Erbium Doped Fiber Amplifiers (EDFAs) could not be utilized. Prior to the relatively recent ITU standardization of the term, one common meaning for Coarse WDM meant two (or possibly more) signals multiplexed onto a single fiber, where one signal was in the 1550 nm band, and the other in the 1310 nm band.

In 2002 the ITU standardized a channel spacing grid for use with CWDM (ITU-T G.694.2), using the wavelengths from 1270 nm through 1610 nm with a channel spacing of 20 nm. (G.694.2 was revised in 2003 to shift the actual channel centers by 1, so that strictly speaking the center wavelengths are 1271 to 1611 nm. Many CWDM wavelengths below 1470 nm are considered unusable on older G.652 specification fibers, due to the increased attenuation in the 1270–1470 nm bands. Newer fibers which conform to the G.652.C and G.652.D standards, such as Corning SMF-28e and Samsung Widepass nearly eliminate the "water peak" attenuation peak and allow for full operation of all twenty ITU CWDM channels in metropolitan networks.

The Ethernet LX-4 10 Gbit/s physical layer standard is an example of a CWDM system in which four wavelengths near 1310 nm, each carrying a 3.125 Gb/s data stream, are used to carry 10 Gb/s of aggregate data.

The main characteristic of the recent ITU CWDM standard is that the signals are not spaced appropriately for amplification by EDFAs. This therefore limits the total CWDM optical span to somewhere near 60 km for a 2.5 Gbit/s signal, which is suitable for use in metropolitan applications. The relaxed optical frequency stabilization requirements allow the associated costs of CWDM to approach those of non-WDM optical components.

CWDM is also being used in cable television networks, where different wavelengths are used for the *downstream* and *upstream* signals. In these systems, the wavelengths used are often widely separated, for example the downstream signal might be at 1310 nm while the upstream signal is at 1550 nm.

An interesting and relatively recent development relating Coarse WDM is the creation of GBIC and Small Form Factor Pluggable (SFP) transceivers utilizing standardized CWDM wavelengths. GBIC and SFP optics allow for something very close to a seamless upgrade in even legacy systems that support SFP interfaces. Thus, a legacy switch system can be easily "converted" to allow wavelength multiplexed transport over a fiber simply by judicious choice of transceiver wavelengths, combined with an inexpensive passive optical multiplexing device.

Passive CWDM is an implementation of CWDM that uses no electrical power. It separates the wavelengths using passive optical components such as bandpass filters and prisms. Many manufacturers are promoting passive CWDM to deploy fiber to the home.

Dense WDM

Dense Wavelength Division Multiplexing, or DWDM for short, refers originally to optical signals multiplexed within the 1550-nm band so as to leverage the capabilities (and cost) of erbium doped fiber amplifiers (EDFAs), which are effective for wavelengths between approximately 1525–1565 nm (C band), or 1570–1610 nm (L band). EDFAs were originally developed to replace SONET/SDH Optical-Electrical-Optical (OEO) regenerators, which they have made practically obsolete. EDFAs can amplify any optical signal in their operating range, regardless of the modulated bit rate. In terms of multi-wavelength signals, so long as the EDFA has enough pump energy available to it, it can amplify as many optical signals as can be multiplexed into its amplification band (though signal densities are limited by choice of modulation format). EDFAs therefore allow a single-channel optical link to be upgraded in bit rate by replacing only equipment at the ends of the link, while retaining the existing EDFA or series of EDFAs through a long haul route. Furthermore, single-wavelength links using EDFAs can similarly be upgraded to WDM links at reasonable cost. The EDFAs cost is thus leveraged across as many channels as can be multiplexed into the 1550 nm band.

A basic DWDM system contains several main components as described below:

1. *DWDM terminal multiplexer:* The terminal multiplexer actually contains one wavelength converting transponder for each wavelength signal it will carry. The wavelength converting transponders receive the input optical signal (*i.e.*, from a client-layer SONET/SDH or other signal), convert that signal into the electrical domain, and retransmit the signal using a 1550 nm band laser. (Early DWDM systems contained 4 or 8 wavelength converting transponders in the mid 1990s. By 2000 or so, commercial systems capable of carrying 128 signals were available.) The terminal multiplexer also contains an optical multiplexer, which takes the various 1550 nm band signals and places them onto a single SMF-28 fiber. The terminal multiplexer may or may not also support a local EDFA for power amplification of the multi-wavelength optical signal.
2. *Intermediate optical terminal, or Optical Add-drop multiplexer*: This is a remote amplification site which amplifies the multi-wavelength signal that may have traversed up to 140 km or more before reaching the remote site. Optical diagnostics and telemetry are often extracted or inserted at such a site, to allow for localization of any fiber breaks or signal impairments. In more sophisticated systems (which are no longer point-to-point), several signals out of the multiwavelength signal may be removed and dropped locally.
3. *DWDM terminal demultiplexer*: The terminal demultiplexer breaks the multi-wavelength signal back into individual signals and outputs them on separate fibers for client-layer systems (such as SONET/SDH) to detect. Originally, this demultiplexing was performed entirely passively, except for some telemetry, as most SONET systems can receive 1550 nm signals. However, in order to allow for transmission to remote client-layer systems (and to allow for digital domain signal integrity determination) such demultiplexed signals are usually sent to O/E/O output transponders prior to being relayed to their client-layer systems. Often, the functionality of output transponder has been integrated into that of input transponder, so that most commercial systems have transponders that support bi-directional interfaces on both their 1550 nm (*i.e.*, internal) side, and external (*i.e.*, client-facing) side. Transponders in some systems supporting 40 GHz nominal operation may also perform Forward Error Correction (FEC) via 'digital wrapper' technology, as described in the ITU-T G.709 standard.
4. *Optical Supervisory Channel (OSC):* This is an additional wavelength usually outside the EDFA amplification band (at 1510nm, 1620nm, 1310nm or another proprietary wavelength). The OSC carries information about the multi-wavelength optical signal as well as remote conditions at the optical terminal or EDFA site. It is also normally used for remote software upgrades and user (*i.e.*, network operator) Network Management information. It is the multi-wavelength analog to SONET's DCC (or supervisory channel). ITU standards suggest that the OSC should utilize an OC-3 signal structure, though some vendors have opted to use 100 megabit Ethernet or another signal format. Unlike the 1550 nm band client signal-carrying wavelengths, the OSC is always terminated at intermediate amplifier sites, where it receives local information before retransmission.

The introduction of the ITU-T G.694.1 frequency grid in 2002 has made it easier to integrate WDM with older but more standard SONET/SDH systems. WDM wavelengths are positioned in a grid having exactly 100 GHz (about 0.8nm) spacing in optical frequency, with a reference frequency fixed at 193.10 THz (1552.52nm). The main grid is placed inside the optical fiber amplifier bandwidth, but can be extended to wider bandwidths. Today's DWDM systems use 50 GHz or even 25 GHz channel spacing for up to 160 channel operation.

DWDM systems have to maintain more stable wavelength or frequency than those needed for CWDM because of the closer spacing of the wavelengths. Precision temperature control of laser transmitter is required in DWDM systems to prevent drift off a very narrow frequency window of the order of a few GHz. In addition, since DWDM provides greater maximum capacity it tends to be used at a higher level in the communications hierarchy than CWDM, for example on the Internet backbone and is therefore associated with higher modulation rates, thus creating a smaller market for DWDM devices with very high performance levels. These factors of smaller volume and higher performance result in DWDM systems typically being more expensive than CWDM.

Recent innovations in DWDM transport systems include pluggable and software-tunable transceiver modules capable of operating on 40 or 80 channels. This dramatically reduces the need for discrete spare pluggable modules, when a handful of pluggable devices can handle the full range of wavelengths.

12.5.11 Wavelength Converting Transponders

Wavelength converting transponders serve originally to translate the transmitted wavelength of a client-layer signal into one of the DWDM system's internal wavelengths in the 1550 nm band. It may be noted that even external wavelengths in the 1550 nm will most likely need to be translated, as they will almost certainly not have the required frequency stability tolerances nor will it have the optical power necessary for the system's EDFA.

In the mid-1990s, however, wavelength converting transponders rapidly took on the additional function of signal regeneration. Signal regeneration in transponders quickly evolved through 1R to 2R to 3R and into overhead-monitoring multi-bitrate 3R regenerators. These differences are outlined below.

1R-Retransmission — Basically, early transponders were "garbage in garbage out" in that their output was nearly an analog copy of the received optical signal, with little signal cleanup occurring. This limited the reach of early DWDM systems because the signal had to be handed off to a client-layer receiver (likely from a different vendor) before the signal deteriorated too far. Signal monitoring was basically confined to optical domain parameters such as received power.

2R-Re-time and re-transmit — Transponders of this type were not very common and utilized a quasi-digital Schmidt-triggering method for signal clean-up. Some rudimentary signal quality monitoring was done by such transmitters that basically looked at analog parameters.

3R-Re-time, re-transmit, re-shape — 3R Transponders were fully digital and normally able to view SONET/SDH section layer overhead bytes such as A1 and A2 to determine signal quality. Many systems will offer 2.5 Gb transponders, which will normally mean the transponder is able to perform 3R regeneration on OC-3/12/48 signals, and possibly gigabit Ethernet, and reporting on signal health by monitoring SONET/SDH section layer overhead bytes. Many transponders will be able to perform full multi-rate 3R in both directions. Some vendors offer 10 Gb transponders, which will perform Section layer overhead monitoring to all rates up to and including OC-192.

12.5.12 Muxponder

The muxponder (from multiplexed transponder) has different names depending on vendor. It essentially performs some relatively simple time division multiplexing of lower rate signals into a higher rate carrier within the system (a common example is the ability to accept four OC-48s and then output a single OC-192 in the 1550 nm band). More recent muxponder designs have absorbed more and more TDM functionality, in some cases obviating the need for traditional SONET/SDH transport equipment.

12.5.13 Reconfigurable Optical Add-Drop Multiplexer (ROADM)

As mentioned above, intermediate optical amplification sites in DWDM systems may allow for the dropping and adding of certain wavelength channels. Adding or dropping of wavelengths requires manually inserting or replacing wavelength-selective cards. This is costly, and in some systems requires that all active traffic be removed from the DWDM system, because inserting or removing the wavelength-specific cards interrupts the multi-wavelength optical signal.

With a ROADM, network operators can remotely reconfigure the multiplexer by sending soft commands. The architecture of the ROADM is such that dropping or adding wavelengths does not interrupt the 'pass-through' channels. Numerous technological approaches are utilized for various commercial ROADMs, the trade off being between cost, optical power, and flexibility.

12.5.14 Optical Cross Connects (OXCs)

When the network topology is a mesh, where nodes are interconnected by fibers to form an arbitrary graph, an additional fiber interconnection device is needed to route the signals from an input port to the desired output port. These devices are called optical crossconnects (OCXs).Various categories of OXCs include electronic, optical, and wavelength selective devices.

12.6 CONCLUSION

Experimental results on WDM point-to-point links can be divided into two groups based on whether the transmission distance is 100 km or exceeds 1000 km. Since thé in the order of experiment in 1995 in which ten 2-Gbps channels were transmitted over 68.3 km, both the number of channels and the bit rate of individual channels have increased considerably. By 1995, a capacity of 340 Gbps was demonstrated by transmitting 17 channels, each operating at 20 Gbps, over 150 km. This record was broken within a year by three experiments that used WDM to realize the total bit rate of 1 Tbps or more. By the end of 1996, a bit rate of 2.64 Tbps was demonstrated in a 132-channel WDM experiment using 0.27 nm channel spacing. The following table lists several record-setting WDM transmission experiments performed after 1995:

The second group of WDM experiments worked on a transmission distance of more than 1000 km. A 1994 experiment realized transmission of 40 Gbps over 1420 km by multiplexing sixteen 2.5 Gbps channels while maintaining an amplifier spacing of about 100 km. It was followed by many experiments that increased either the transmission distance or the bit rate. In one test-bed experiment, a transmission distance of 6000 km at 20 Gbps (8 channels at 2.5 Gbps) has been realized with an amplifier spacing of 75 km. On the high-bit-rate end, a 1996 experiment multiplexed sixteen 10 Gbps channels to realize transmission at 160 Gbps, but the link length was only 531 km. Using very sophisticated techniques, 160 Gbps transmission over a transoceanic distance of 9100 km has been realized.

The development of WDM fiber links has led to the advent of the fourth generation of lightwave systems, which make use of the WDM technology to increase the bit rate and in-line optical amplifiers to increase the transmission distance. Needless to say that WDM network is the fastest network in the world. The use of WDM can multiply the effective bandwidth of a fiber optic communications system by a large factor. But its cost must be weighed against the alternative of using multiple fibers bundled into a cable. A fiber optic repeater device called the erbium amplifier promises to make WDM a cost-effective long-term solution to the bandwidth exhaustion problem.

REVIEW QUESTIONS

1. Which of the followings is similar to the scheme of WDM?
   (*a*) FDM in RF transmission (*b*) TDM
   (*c*) OTDM (*d*) SDM
2. What is the function of WDM?
   (*a*) Change the transmission speed of the input signal
   (*b*) Combine signals of different wavelengths to pass through a single fiber
   (*c*) Separate signals of different wavelengths and couple them to different detectors
   (*d*) Tap off part of energy of the incoming signal
3. What is the channel spacing specified by ITU-T G.694.1recommendation for DWDM?
   (*a*) 25 GHz (*b*) 50 GHz (*c*) 100 GHz (*d*) 200 GHz
4. What is the channel spacing in wavelengths specified by ITU-T G.694.1recommendation for DWDM?
   (*a*) 0.4 nm (*b*) 0.8 nm (*c*) 1.2 nm (*d*) 1.6 nm
5. The power available at output ports of a 2 × 2 directional coupler are 45 μW and 45 μW, when input power level is 100 μW. What is the coupling ratio?
   (*a*) 45% (*b*) 50% (*c*) 90% (*d*) 100%
6. For the coupler as mentioned at question 12.5, what the insertion loss for the path from input port 1 to output port 2?
   (*a*) 3.46 dB (*b*) 5.23 dB (*c*) 6.92 dB (*d*) 10 dB
7. For the coupler as mentioned at question 12.5, what is the excess loss?
   (*a*) 0.46 dB (*b*) 0.5 dB (*c*) 1 dB (*d*) 3 dB
8. What is the power level of each of the output port for a 1 × 10 coupler when the input signal is 0 dBm?
   (*a*) –10 dBm (*b*) –3 dBm (*c*) 1 dBm (*d*) 0 dBm
9. Which of the following tunable filters is most suitable for DWDM?
   (*a*) Fabry-Perot filters (*b*) Fiber Bragg gratings
   (*c*) Acousto-optic tunable filters (*d*) Mach-Zehnder interferometer
10. Which of the following schemes is most suitable for DWDM?
    (*a*) Arrayed waveguide grating multiplexer
    (*b*) Blazed reflection gratings
    (*c*) Fiber Bragg gratings
    (*d*) Mach-Zehnder interferometer
11. Explain the principle of operation of TDM system. How STDM differs from TDM? What will be the minimum bandwidth of a 24 channel TDM system of 64 bits frame-size?
12. Describe the characteristics of TDMA. Compare this with CDMA.
13. Describe the basic operations of WDM, CWDM and DWDM. Explain how wavelength shifting and reuse can be achieved. What is the base frequency and channel spacing specified by ITU for DWDM?
14. Explain Maxponder, ROADM and OXC.
15. What is FWM? What are the negative effects of EDM in WDM systems? How can it be used in beneficial way?
16. Illustrate SCM (Subcarrier multiplexing), CDM (Code division multiplexing) and SDM (Space division multiplexing) with schematic diagrams. Mention the advantages of these systems.
17. How wavelength shifting and wavelength reuse are incorporated in WDM network?
18. Discuss the Switching, Contention resolution, Synchronization, Data format conversion and Protocols in high speed WDM system.
19. Illustrate CWDM and DWDM. What are main components for DWDM system?
20. Write short notes on (*i*) Wavelength converting transponders, (*ii*) Muxponders, (*iii*) ROADM, (*iv*) OXCs.

13

FIBER OPTIC INTERCONNECTION DEVICES

13.1 INTRODUCTION

Optical fiber links, which is the integral part of recent communication systems, have the requirements for both jointing and termination of transmission medium. The number of intermediate fiber connections or joints is dependent on the link length between the repeaters. The continuous length of fiber cables that may produced has been described in chapter 8. Although by preparation process continuous single mode fiber may be produced up to the length of 200 km, from practical point of view such fiber spans cannot be readily installed. However, continuous cable lengths of tens of kilometers have already been deployed, in particular within submarine systems where continuous cable laying presents fewer problems.

Repeater spacing on optical communication link is one of the critical parameters. Currently installed digital systems are operating over spacing in the range of 40 to 60 km at transmission rates of between 400 Mbits/sec and 1.7 Gbits/sec. Spacings in excess of 100 km are also readily achievable with practical systems at such transmission rates. A fully operational 2.4 Gbits/sec system operating with 100 km repeater spacing using dispersion shifted single mode fiber has already reported, together with the experimental transmission at 10 Gbits/sec over a similar unrepeated distance.

It is obvious that the higher spacing between the repeaters at optical fiber communication can be achieved by low loss and minimum distortion *i.e.*, modal noise. These factors are also related to the number of terminal connections permissible within the developing the optical fiber networks. Although the jointing techniques of fiber appear to lag behind the technology associated with the other components required in optical fiber communication systems (sources, detectors etc.), significant developments have been made in the recent years. In this chapter the theoretical and practical aspects of fiber to fiber connection with regard to both single mode and multimode systems will be discussed. There are two major categories of fiber joints currently both in use and development.

- Semipermanent or permanent joints called as fiber splices find major use in most optical fiber telecommunication systems. It is analogous to electrical soldered joints.

- Demountable fiber connectors or simple connectors which are removable joints allow easy, fast, manual coupling and uncoupling of fibers. These are analogous to electrical plugs and sockets.

The fiber joints are designed ideally to couple all the light propagating in one fiber to the adjoining fiber. Fiber couplers are the branching devices that split all the light from one main fiber to two or more fibers, or alternatively it may be said that they couple a proportion of the light propagating in the main fiber into a branch fiber. These devices are usually bidirectional providing for the combination of light from two or more branch fibers to a main fiber. The requirements of these devices have been substantially increased over the recent years in order to facilitate the widespread deployment of optical fiber within communication networks. They became the essential components for optical fiber communication and there is growing demand for more sophistication of these components.

A critical aspect of fiber jointing concerns the optical loss associated with the connection. The joint loss is dependent upon the alignment of the two fibers and very little misalignment causes large optical loss, loss of bandwidth, crosstalk and bit error. The basic techniques and technology associated with both fiber joints and couplers are therefore must be considered which will be outlined in this chapter.

13.2 FIBER ALIGNMENT AND JOINT LOSS

A major consideration with all types of fiber to fiber connection is the optical loss encountered at the interface. Even when the two jointed fiber ends are smooth and perpendicular to the fiber axes, and the two fiber axes are perfectly aligned, a small proportion of light may be reflected back into transmitting fiber resulting in attenuation at the joint. This phenomenon is called as *Fresnel reflection*. This associated with the step changes in refractive index at the jointed interface (*i.e.*, glass-air-glass). The amount of this partial reflection of the transmitted light through the interface may be estimated using the classical Fresnel formula for light of normal incidence as given by

$$r = \left(\frac{n_1 - n}{n_1 + n} \right)^2 \tag{13.1}$$

where r is the fraction of the light reflected at a single interface, n_1 is the refractive index of the fiber core and n is the refractive index of the medium between the two jointed fibers. In this case the medium is air, where $n = 1$. The loss in decibels due to Fresnel reflection at single interface is given by

$$\text{Loss}_{\text{Fres}} = -10 \log_{10} (1 - r) \tag{13.2}$$

Using the relationship above, it is possible to determine the optical attenuation due to Fresnel reflection at a fiber to fiber joint. However, Fresnel reflection at both the fiber interfaces must be considered in order to estimate the amount of light reflected at fiber joint.

It may be noticed that the Frenel reflection may cause a significant loss at a fiber joint, even when all other aspects of the connection are ideal. However, Fresnel reflection at a fiber to fiber connection can be significantly reduced with the use of index matching fluid in the gap between the jointed fibers. When index matching fluid has the same refractive index as the fiber core, the losses due to Fresnel reflection may be eliminated.

Unfortunately, the Fresnel reflection is not the only source of optical loss at the fiber joint. Misalignment of the two joint fibers is another potentially greater source of loss in fiber to fiber connection and proper jointing technique must be adopted to minimize this. Any deviation in the geometrical and optical parameters, as indicted below, of the two optical fibers which are to jointed will affect the optical attenuation or insertion loss.

(*a*) Different core and/or cladding diameters;
(*b*) Different numerical apertures and/or relative refractive index;
(*c*) Different refractive index profiles;
(*d*) Fiber faults – core ellipticity, core connectivity, etc.

These aspects cause inherent connection problems while making fiber joints and it is not possible to avoid all these variation within any particular connection. The losses caused by these factors together with the Fresnel reflection loss are generally referred to as *intrinsic loss*. The best result may therefore be achieved with compatible or similar fibers which are manufactured with the lowest tolerance. Still problem may be persisted due to imperfect quality of the fiber alignment provided by the jointing mechanism. Figure 13.1 illustrates the examples of misalignment. Misalignment may occur in three dimensions. Separation between the fibers is referred as longitudinal misalignment, offset perpendicular to the fiber core axes is lateral or radial or axial misalignment, while the angle between the fiber cores is referred as angular misalignment. Longitudinal, lateral and angular misalignments are described by Figs. 13.1(*a*), 13.1(*b*) and 13.1(*c*) respectively.

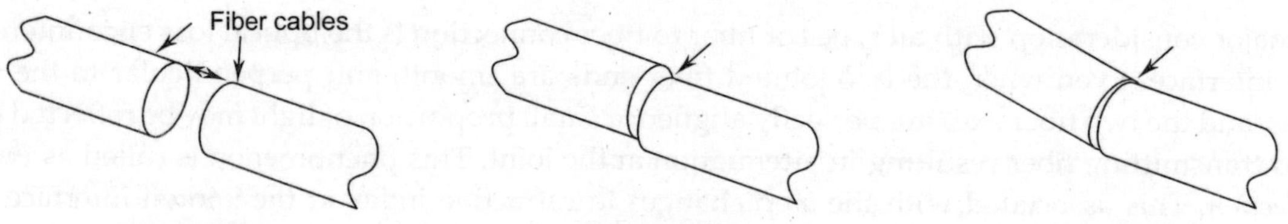

Fig. 13.1(*a*) **Fig. 13.1(*b*)** **Fig. 13.1(*c*)**

Optical losses due to these three types of misalignments are also influenced by the fiber type, core diameter and distribution of optical power between the propagating modes. Typical graphs for losses with respect to geometrical and structural parameters of fiber cores are shown in Fig. 13.2. Figure 13.2(*a*) illustrates the losses for longitudinal and lateral misalignments for a 50 μm core diameter graded index fiber. It may be noted that loss due to lateral misalignment has greater effect with comparison to the longitudinal misalignment. As example, for 10 μm lateral displacement the insertion loss is 1 dB, whereas insertion loss is 0.1 dB in case of longitudinal displacement of similar amount. The Fig. 13.2(*b*) demonstrates the attenuation due to angular misalignment of two multimode step index fibers with Numerical Apertures (NA) of 0.22 and 0.3. An insertion loss of 1 dB is observed for angular misalignment of 4° and 5° with numerical apertures 0.22 and 0.3 respectively. It may be noticed that the use of index matching fluid at angular misalignment results in the increase in attenuation, which is shown in Fig. 13.2(*b*). Therefore, it is clear that relatively small levels of lateral and/or angular misalignment can cause significant attenuation at a fiber joint. This is especially for the fibers of small core diameter (less than 150 μm), which are currently employed for most telecommunication systems.

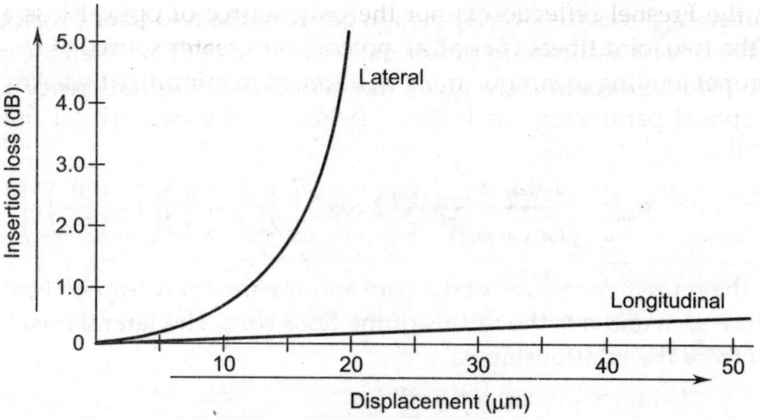

Fig. 13.2(*a*)

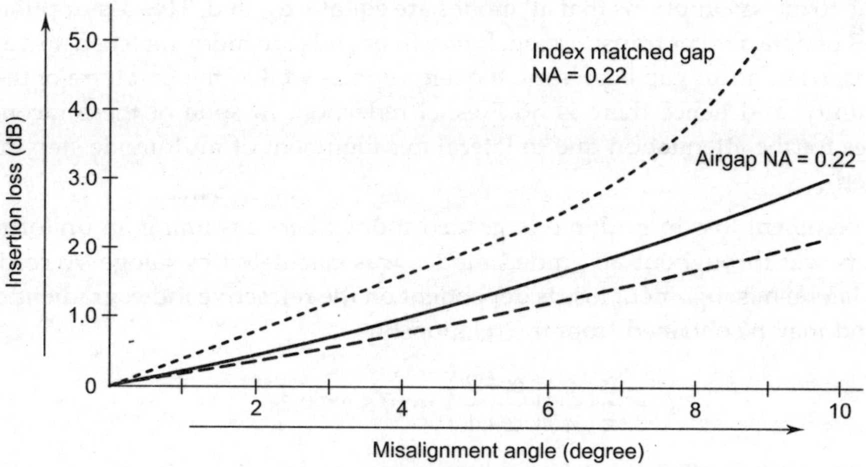

Fig. 13.2(*b*)

13.2.1 Multimode Fiber Joints

Approximate determination of losses encountered with various misalignments of different fiber types have been established by theoretical and experimental studies of fiber misalignment in optical fiber connections. Mathematical expressions have been developed to calculate the losses. Longitudinal misalignment has the least important effect and may be largely avoided if proper care is taken while making the fiber connection. There is some disagreement over the magnitude of the losses due to longitudinal misalignment when it is calculated theoretically between Miyazaki *et al*. and Ţanchiya *et al*. due to consideration of number of variables involved in the measurement. However, lower losses predicted by Tanchiya *et al*. are agreeable other researchers. Also higher losses are predicted for fibers with larger numerical apertures. The larger is the numerical aperture, the grater is the spread of the output light and higher the optical loss at a longitudinally misaligned joint.

Theoretical expressions for determination of lateral and angular misalignment losses are by no means definitive, although they are close to experimental results. Experimental results from different sources tend to vary due to difficulties of measurements, especially for angular misalignment losses. It is therefore implied that the expressions as would be presented here are not essentially the most accurate.

Lateral misalignment reduces the overlap region between the two fiber cores. Uniform excitation of all the optical modes is assumed in a multimode step index fiber. Then the overlapped area between both fiber cores approximately gives the lateral coupling efficiency η_{lat} that can be expressed as

$$\eta_{lat} = \frac{16(n_1/n)^2}{\{1+(n_1/n)\}^4}\,\frac{1}{\pi}\left\{2\cos^{-1}\left(\frac{y}{2a}\right)-\left(\frac{y}{a}\right)\left(1-\left(\frac{y}{2a}\right)^2\right)^{1/2}\right\} \tag{13.3}$$

where n_1 and n are the refractive indices of the core and the medium respectively, y is the lateral offset of the fiber core axes and a is the radius of the fiber core. The lateral misalignment loss in decibel determined from the relationship as

$$\text{Loss}_{lat} = -10\log_{10}\eta_{lat}\ \text{dB} \tag{13.4}$$

The loss calculated from the equations 13.3 and 13.4 is generally slightly higher than the measured values due to the assumptions that all modes are equally excited. This assumption is correct for certain cases of optical fiber transmission. If two fiber ends are index matched and assumed to be of close contact, *i.e.*, no air gap is present, the term n_1/n as well as the first term of the equation 13.3 becomes unity, and hence there is no Fresnel reflection. In spite of these inconsistencies, useful estimates for the attenuation due to lateral misalignment of multimode step index fibers may be obtained.

Lateral misalignment loss in multimode graded index fibers assuming an uniform distribution of optical power throughout all guided modes was calculated by Gloge. According to his estimation the lateral misalignment loss is dependent on the refractive index gradient α for small lateral offset and may be obtained from the relationship

$$L_t = \frac{2}{\pi}\left(\frac{y}{a}\right)\left(\frac{\alpha+2}{\alpha+1}\right) \quad \text{for } 0 \le y \le 0.2a \tag{13.5}$$

where the lateral coupling efficiency is given by

$$\eta_{lat} = 1 - L_t \tag{13.6}$$

Above equation may be utilized to calculate the lateral misalignment loss in decibels. As example, for a parabolic refractive index profile of $\alpha = 2$,

$$L_t = \frac{8}{3\pi}\left(\frac{y}{a}\right) = 0.85\left(\frac{y}{a}\right) \tag{13.7}$$

Revised expression for lateral misalignment loss of equation 13.5 may be obtained considering further estimation including the leaky modes. This analysis may be extended to step index fibers, where α = infinity, lateral misalignment losses become 0.64 (y/a) for guided modes and 0.5 (y/a) for both guided plus leaky modes.

It may be noticed from the Fig. 13.2(*a*) that the measured loss due to lateral misalignment for a 50 μm diameter graded index fiber is very pessimistic. Miller and Mettler have proposed a model for lateral misalignment loss measurement where predicted insertion loss is more accurate. In this model the power distribution is assumed to be of Guassian form. The analysis is too complex and involves complex integrations using numerical techniques.

Angular misalignment losses at joints in multimode step index fibers may be determined with reasonable accuracy using the expression for coupling efficiency

$$\eta_{ang} = \frac{16(n_1/n)^2}{\{1+(n_1/n)\}^4}\left(1-\frac{n\theta}{\pi n_1 (2\Delta)^{1/2}}\right) \quad (13.8)$$

where θ is the angular displacement in radians and Δ is the relative refractive index difference for the fiber. Angular loss may be calculated from coupling efficiency in the similar manner.

$$\text{Loss}_{ang} = -10 \log_{10} \eta_{ang} \quad (13.9)$$

From above equations it may be observed that the smaller the values of Δ the larger the insertion loss due to angular misalignment. The Numerical Aperture (NA) is related to the refractive index difference as

$$\text{NA} = n_1 (2\Delta)^{1/2} \quad (13.10)$$

Therefore the insertion loss due to angular misalignment is reduced by using fibers with larger numerical apertures. This is a reverse trend of the longitudinal misalignment, where longitudinal misalignment loss at a joint is increased higher numerical apertures.

The major factors causing fiber to fiber intrinsic loss comprising of mismatch in fiber core diameters (Fig. 13.3(*a*)), mismatch in fiber numerical apertures (Fig. 13.3(*b*)) and mismatch in the refractive index profiles (Fig. 13.3(*c*)), are quite common while making the joints, particularly when a pigtailed optical source is used. The fiber pigtail possesses different characteristics from the main transmission fiber. Moreover, diameter variations can occur with same fiber type.

Assuming all modes are equally excited in a multimode step or graded index fiber, and that the numerical apertures and index profiles are the same, then the loss resulting from a mismatch of core diameters may be determined by the relationship,

$$\text{Loss}_{CD} = \begin{cases} = -10\log_{10}(a_2/a_1)^2 \text{ dB} & \text{for } a_2 < a_1 \\ = 0 \text{ dB} & \text{for } a_2 \geq a_1 \end{cases} \quad (13.11)$$

where a_1 and a_2 are the core radii of the transmitting and receiving fiber respectively. It may be noticed that if the core diameter of the receiving fiber is larger than the transmitting fiber, there is no loss due to the mismatch of the core diameters. Also it may be noted that only a relatively small amount of loss (approximately 0.09 dB) is incurred if the receiving fiber core diameter is 1% smaller than the transmitting fiber.

When the transmitting fiber has a higher numerical aperture than the receiving fiber, then some portion of emitted light will fall outside the acceptance angle of the receiving fiber and therefore coupling joint will not be perfect Fig. 13.3(*b*). Assuming an uniform modal power distribution, similar refractive index profiles and equal core diameters, the loss due to mismatch of numerical core apertures can be obtained from the relationship,

$$\text{Loss}_{NA} = \begin{cases} = -10\log_{10}(\text{NA}_2/\text{NA}_1)^2 \text{ dB} & \text{for NA}_2 < \text{NA}_1 \\ = 0 \text{ dB} & \text{for NA}_2 \geq \text{NA}_1 \end{cases} \quad (13.12)$$

NA_1 and NA_2 are the numerical apertures of the transmitting and receiving fibers respectively. Equation 13.12 is valid for both step and graded index fibers. Here also it may be noticed that no loss is incurred if numerical aperture of receiving fiber is greater than that of transmitting fiber.

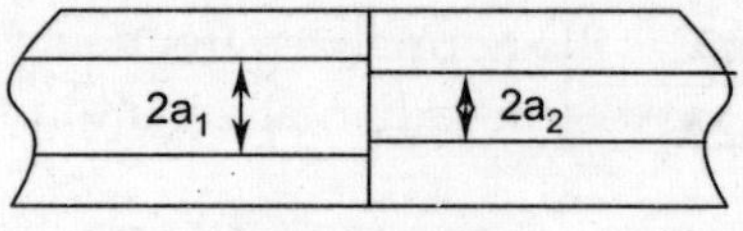

Fig. 13.3(*a*)

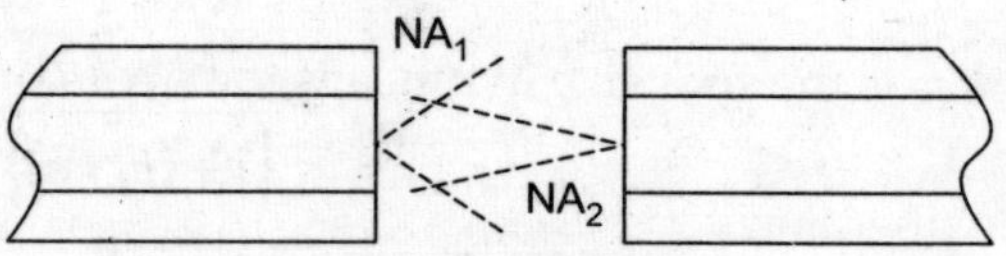

Fig. 13.3(*b*)

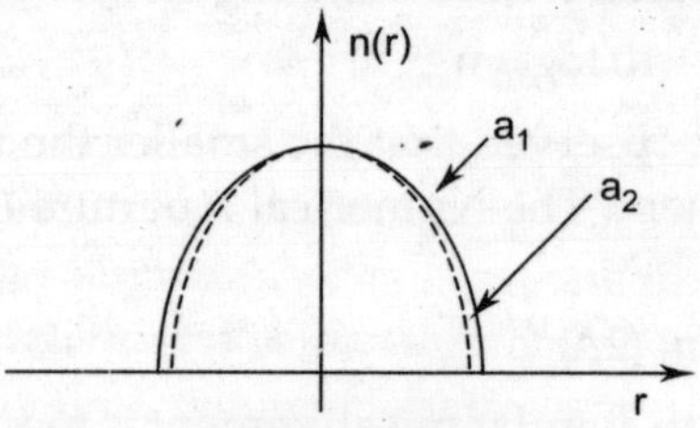

Fig. 13.3(*c*)

Finally, loss due to mismatch in refractive index profiles, while other parameters remain similar, may be determined from the relationship,

$$\text{Loss}_{\text{RI}} = \begin{cases} = -10\log_{10} \dfrac{\alpha_2(\alpha_1+2)}{\alpha_1(\alpha_2+2)} \text{ dB} & \text{for } \alpha_2 < \alpha_1 \\ = 0 \text{ dB} & \text{for } \alpha_2 \geq \alpha_1 \end{cases} \tag{13.13}$$

α_1 and α_2 are the refractive index profiles of the transmitting and receiving fibers respectively. The equation 13.13 suggests that a loss of 3 dB is produced when transmitting fiber of step index profile of α_1 is connected to a fiber of parabolic profile graded index of $\alpha_2 = 2$, while other parameters are similar. However, on the other hand, for reverse connection no loss is observed due to mismatch of refractive index profiles.

Assuming that the three mismatches are occurred together, the total intrinsic loss at multimode fiber joint may be derived, combining by equations (13.11), (13.12) and (13.13), by the relationship,

$$\text{Loss}_{\text{int}} = \begin{cases} = -10\log_{10} \dfrac{(a_2\text{NA}_2)^2(\alpha_1+2)\alpha_2}{(a_1\text{NA}_1)^2(\alpha_2+2)\alpha_1} \; dB & \text{for } a_2 < a_1, \text{NA}_2 < \text{NA}_1, \alpha_2 < \alpha_1 \\ = 0 \text{ dB} & \text{for } a_2 \geq a_1, \text{NA}_2 \geq \text{NA}_1, \alpha_2 \geq \alpha_1 \end{cases} \tag{13.14}$$

13.2.2 Single-Mode Fiber Joints

Theoretically the misalignment losses in single mode fiber are analyzed by Marcuse and Gambling *et al*. Theoretical analysis by Marcuse is based on the Guassian or near Gaussian shape of the modes propagating in single mode fibers regardless of the fiber type (*i.e.*, step index or graded index). Gambling *et al* provides further simplification of the formulae for both the lateral and angular misalignment losses at joints in single mode fibers. According to Gambling *et al*, the loss T_l due to lateral offset y in the absence of angular misalignment is given by

$$T_l = 2.17\left(\frac{y}{w}\right)^2 \text{ dB} \tag{13.15}$$

w is normalized spot size of the fundamental mode. (Spot size is defined as mode field radius which is the nominal half width of the input excitation). Normalized spot size for LP_{01} mode that corresponds to HE mode may be obtained from the formula,

$$w = a\,(0.65 + 1.62\,V^{-3/2} + 2.88\,V^{-6})/2^{1/2} \qquad (13.16)$$

where w is the spot size in μm, a is the fiber core radius and V is the normalized frequency for the fiber. The insertion loss T_a caused by an angular misalignment θ (in radian)at a joint in a single mode fiber is given by

$$T_a = 2.17\left(\frac{\theta\, w\, n_1 V}{a\,\text{NA}}\right)^2 \text{dB} \qquad (13.17)$$

where n_1 is the refractive index of the fiber core and NA is the numerical aperture of the fiber. It must be noted that equations (13.16) and (13.17) are based on the assumption that the spot sizes of the modes in the two coupled fibers are same. A formula with good approximation for the combined losses due to both lateral and angular misalignment at a fiber joint has been given by Gambling *et al*, but it is somewhat complicated. However both the calculation processes indicate that for small losses, less than 0.75 dB, combination of the equations (13.16) and (13.17) delivers a reasonable approximation.

Loss due to angular misalignment is generally larger than that due to lateral misalignment. It may be noticed that the insertion losses incurred are also strongly dependent upon the normalized frequency of the fiber. This is especially the case with angular misalignment at a single mode fiber joint where insertion losses of less than than 0.3 dB may be obtained when the angular misalignment is 1° with appropriate normalized frequency V value. To achieve low loss single mode fiber joints it is important that angular alignment is less than 1°.

Nemota and Makimoto utilized the theoretical model developed by Marcuse to obtain a general equation for determining the coupling loss between single mode fibers, considering all the extrinsic factors (lateral, angular and longitudinal misalignment and Fresnel reflection), as well as the intrinsic factors associated with the connection of fibers with unequal mode-field diameters. This generalized equation is confirmed to a great degree of approximation by various experimental investigations. However, a reduced expression will be employed here, which allows calculation of the most commonly occurred intrinsic factors in the interconnection of single mode fibers. With the assumption of the absence of any loss due to extrinsic factors, the intrinsic coupling loss is determined from the relationship,

$$\text{Loss}_{\text{int}} = -\,10\log_{10}\left(4\left\{\frac{w_{02}}{w_{01}} + \frac{w_{01}}{w_{02}}\right\}^{-2}\right) \text{dB} \qquad (13.18)$$

where w_{01} and w_{02} are the spot sizes of the transmitting and receiving fibers respectively. Equation 13.18 enables to derive the additional coupling loss resulting from mode-field diameter mismatch between two single mode fibers.

It may be noted from the equation (13.18) that same loss intrinsic loss is incurred at the coupling joint irrespective of which fiber is transmitting or receiving. This is in contrast to the multimode fiber joint. This suggests that intrinsic loss through a single mode fiber joint is independent of the direction of propagation.

So far some detail on optical attenuation at fiber to fiber connections has been discussed. Another possible loss at the fiber to fiber joint is due to distortion of transmitted signal. An increased interest has been generated with the use of highly coherent sources (injection lasers) and very low dispersion fibers. It is apparent that the signal transmission is strongly affected by fiber connections causing modal noise and nonlinear distortion when a coherent light source is utilized with a multimode fiber. It has been also reported that the transmission loss of a connection in coherent multimode system is extremely wavelength dependent, exhibiting a possible 10% change in the transmitted optical wavelength for a very small change in the order of 0.001 nm in the laser emission wavelength. It has been observed that these problems may be reduced by the use of single mode fiber. A theoretical model for wavelength dependence of joint losses in the single mode fiber has been developed which predicts that the width of the fundamental mode field increases with the increase in wavelength and hence the joint loss decreases for a given lateral offset or angular tilt at the joint. As example, the lateral offset loss at a wavelength of 1.5 μm is calculated to be around 80% of the loss at a wavelength of 1.3 μm.

The above modal effects become negligible when an incoherent light source (light emitting diode or LED) is used in multimode fiber. However, at the fiber joint some mode conversion often takes place where the connection effectively acts as a modal mixer or filter. This phenomenon is related to the type of fiber joint, also called as fiber splice, -mechanical type or fusion type.

13.3 FIBER SPLICES

A permanent joint formed between two individual optical splices in the field or factory is known as a *fiber splice*. Smaller lengths of fibers needed to be joined with the use of fiber splices to establish longer length of optical fiber. This is done if there is no requirement of repeated connections and disconnections. Depending on the use of splicing technique, splices may be divided into broad categories, —mechanical type and fusion or welding type.

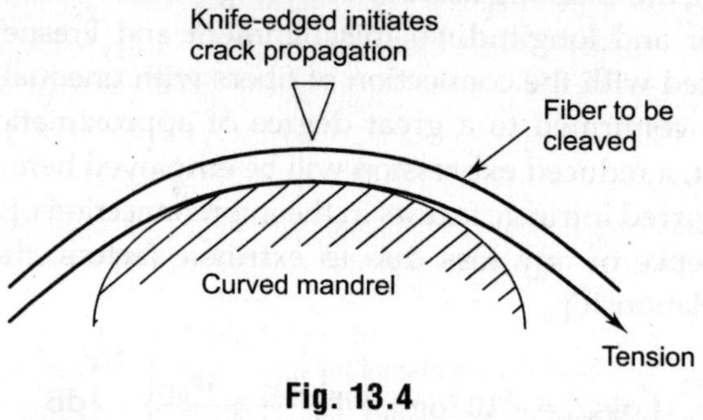

Fig. 13.4

In mechanical splicing, fibers are held in alignment by some mechanical means and fiber joint is made with the use of tubes around the fiber ends (tube splices) or V-grooves into which the butted fibers are placed (groove splices). On the hand, fusion splicing is accomplished by applying localized heating by flame or electric arc at the interface between two butted fibers causing them to soften and fuse, after proper alignment of them. All these techniques are required the fiber end preparation and alignment for optimization of the splice performance to reduce the insertion loss at the joint. Typical average insertion loss in multimode fibers at the spliced joint is in the range of 0.1 to 0.2 dB, which is generally a better performance than that exhibited by demountable connections. It may be noted that the insertion loss is generally much less than the possible Fresnel reflection loss at a butted fiber to fiber joint. This is because with fusion splice there is no large step change in refractive index and it forms a continuous fiber connection. For mechanical splices, some method of index matching (*e.g.*, a fluid) is utilized minimizing the insertion loss.

However the process of fiber splicing, especially the fusion splicing is little difficult at field environment and suffers from practical problems in the development of field usable tools.

The fibers intended for splicing are required to have smooth and square end faces. Suitable tool is used for fiber-end preparation which cleaves the fiber as illustrated in the Fig. 13.4. This process is often referred to as *scribe and break* or *score and break* as it involves the scoring of the fiber surface under tension with a cutting tool with sapphire or diamond or tungsten carbide blade. The surface scoring may create failure as fiber is tensioned, but reasonable square fiber ends can be produced. Figure 13.4 demonstrates this process with fiber tensioned around a curved mandrel. Straight pull scribe and break tools are also utilized deliver better results. An alternative technique involves circumferential scoring providing controlled method of lightly scoring around the fiber circumference. In this method the score can be smooth and uniform and large diameter fiber may be prepared by simple straight null with end angle less than 1°.

13.3.1 Fusion Splices

The fusion splicing of single fibers involves the heating of two prepared fiber ends to their fusing point with the application of sufficient axial pressure between the two optical fibers. The ends of the fibers are stripped of cabling and buffer coating and it is essential that they are perfectly positioned and aligned to achieve good continuity of transmission medium at the junction point. Therefore inspection microscope is used to check the precision placement of the fiber ends.

Microplasma torches (argon and hydrogen) and oxydric microbusters (oxygen, hydrogen and alcoholic vapors) are utilized as the flame heating sources for fusion splicing. However, the widely used heating source is an electric arc. This technique has the advantages of consistent, easily controllable heat with adaptability for use under field conditions. A schematic diagram of the basic arc fusion method is shown in Fig. 13.5(*a*). Figure 13.5(*b*) illustrates the development of the basic fusion process which involves the rounding of the fiber ends with a low energy discharge before pressing the fibers together and fusing with a stronger arc. This technique is known as prefusion that removes the requirement of fiber end preparation. This has the distinct advantage in field environment and is generally utilized with the multimode fibers yielding average splice losses of 0.09 dB.

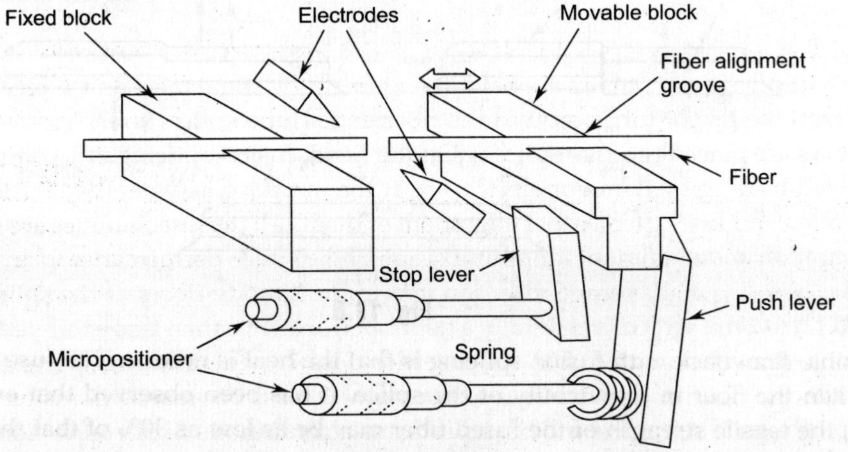

Fig. 13.5(*a*)

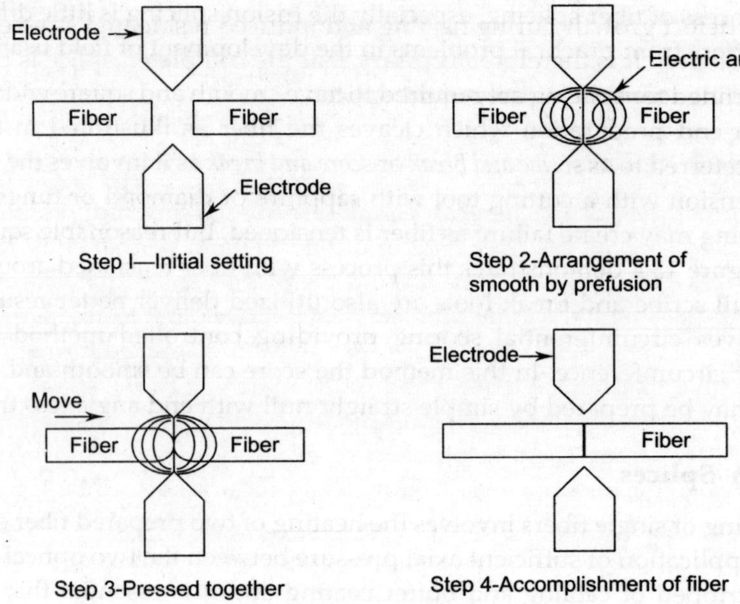

Fig. 13.5(*b*)

More critical fiber alignment is required in fusion splicing of single mode fibers with typical core diameter between 5 to 10 μm. The lateral offset must be less than 1 μm to obtain low joint loss. However, splice insertion losses below 0.3 dB may be achieved due to a self alignment phenomenon which partially compensates for any lateral offset.

Self alignment is achieved due to surface tension effects between the two fiber ends during fusing. This has been illustrated in Fig. 13.6. An early field trial of single mode fiber fusion splicing over 31.6 km link observed splice insertion losses of 0.18 dB and 0.12 dB at wavelengths of 1.3 μm and 1.55 μm respectively. Mean splice losses of only of 0.06 dB have also been obtained with a fully automatic single mode fiber fusion splicing machine.

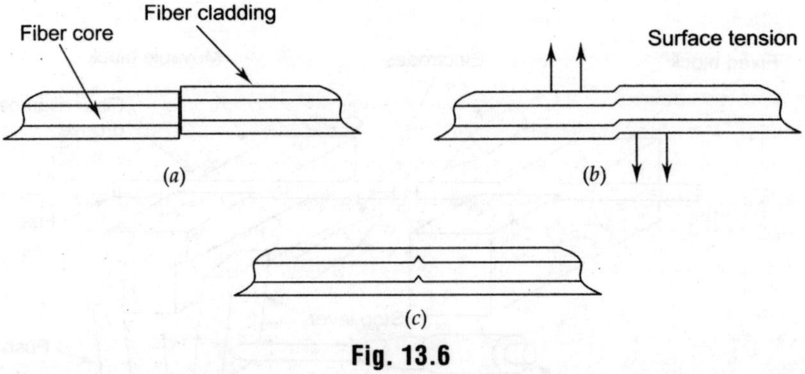

Fig. 13.6

A possible drawback with fusion splicing is that the heat is necessary to fuse the fibers which may weaken the fiber in the vicinity of the splice. It has been observed that even with careful handling, the tensile strength of the fused fiber may be as low as 30% of that the uncoated fiber before fusion. In the heat-affected zone adjacent to the fused joint the fiber fracture generally occurs. The reduced tensile strength results in the combined effects of surface damage caused by handling, surface defect growth during heating and induced residential stresses due to changes in chemical composition. It is therefore necessary that the completed splice is packaged so as to reduce tensile loading upon the vicinity of the splice.

13.3.2 Mechanical Splices

There are various methods of mechanical techniques for splicing individual optical fibers. A common method is to prepare the fiber ends and then to bond permanently with the use of an accurately produced rigid alignment tube around them. This snug tube splice may be made of glass or ceramic capillary with the inner diameter just large enough to accept the optical fibers as illustrated in Fig. 13.7. Transparent adhesive, *e.g.*, epoxy resin, is injected through a transverse bore in the capillary to give mechanical sealing and index matching of the splice. Average insertion loss as low as 0.1 dB may be obtained with multimode graded index and single mode fibers using ceramic capillaries. However, there is problem of capillary diameter mismatch in snug tube splices, which increases the insertion loss. Commercial product may exhibit losses of up to 0.5 dB.

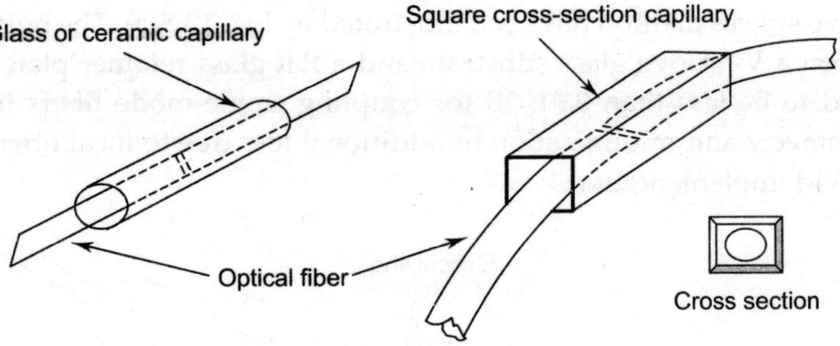

Fig. 13.7

Oversized square cross-section metal tube which easily accepts the prepared fibers may be used instead of snug capillary tubes to overcome the critical tolerance requirements of capillary diameter. This is shown in Fig. 13.7. Transparent adhesive is first inserted into the tube followed by the fibers. The splicing is self-aligning when the fibers are curved in the same plane, forcing the fiber ends simultaneously into same corner of the tube. Mean splice insertion losses of 0.073 dB may be achieved in this method using multimode graded index fibers.

Another common method of mechanical splicing techniques is to use of grooves to secure the fibers to be jointed. A simple method is to utilize a V-groove into which the two prepared fiber ends are pressed. The schematic arrangement is described in Fig. 13.8(*a*), where the jig with V-groove gives alignment of the prepared fibers through insertion in the groove. The splice is made permanent by securing the fibers in the V-groove with epoxy resin. This method is quite successful as the insertion loss may be achieved within 0.1 dB.

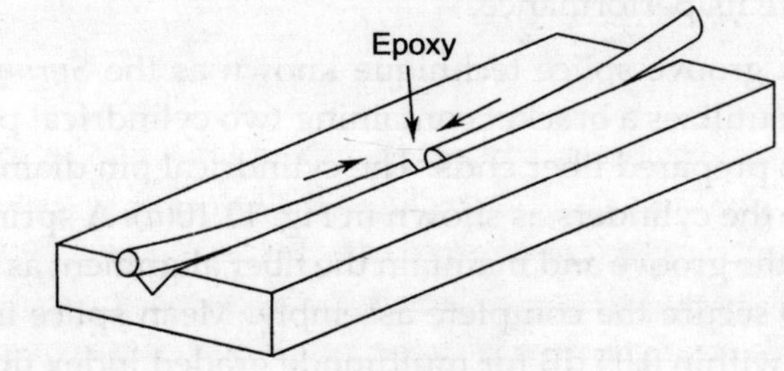

Fig. 13.8(*a*)

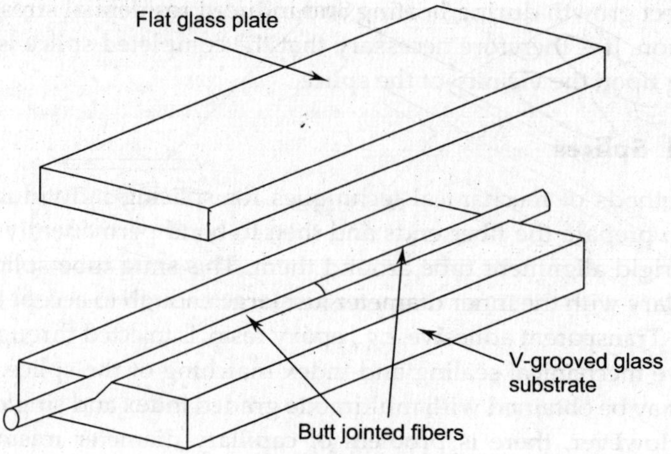

Fig. 13.8(*b*)

Another V-groove splices method has been illustrated in Fig. 13.8(*b*). The butted fiber ends are sandwiched between a V-groove glass substrate and a flat glass retainer plate. Splice insertion losses are observed to be less than 0.01 dB for coupling single-mode fibers in this technique. However, fiber geometry and minimization of additional loss due to local fiber bending restrict this technique at field implementation.

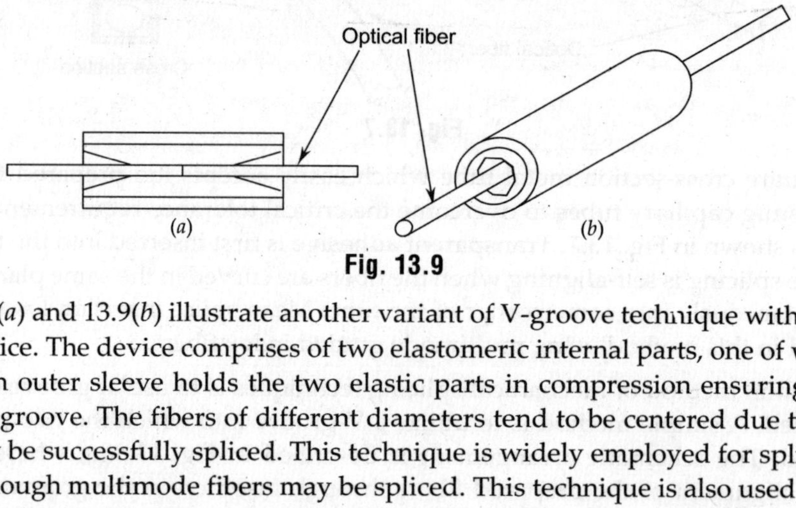

Fig. 13.9

Figures 13.9(*a*) and 13.9(*b*) illustrate another variant of V-groove technique with elastic tube or elastomeric splice. The device comprises of two elastomeric internal parts, one of which contains a V-groove. An outer sleeve holds the two elastic parts in compression ensuring alignment of fibers in the V-groove. The fibers of different diameters tend to be centered due to compression and hence may be successfully spliced. This technique is widely employed for splicing of single-mode fibers, though multimode fibers may be spliced. This technique is also used for temporary splicing especially at laboratory investigations. The splice loss observed in this splicing technique is less than 0.25 dB for commercial products. Index matching gel is normally used within the device to improve its performance.

A slightly more complex groove splice technique known as the *Springroove* is illustrated in Figs. 13.10(*a*) and 13.10(*b*). It utilizes a bracket containing two cylindrical pins which serves as an alignment guide for the two prepared fiber ends. The cylindrical pin diameter is chosen to allow the fibers to protrude above the cylinders as shown in Fig. 13.10(*a*). A spring or elastic element is used to press the fibers into the groove and maintain the fiber alignment as shown in Fig. 13.10(*b*). Epoxy resin is employed to secure the complete assembly. Mean splice insertion losses by this technique is observed to be within 0.05 dB for multimode graded index fibers.

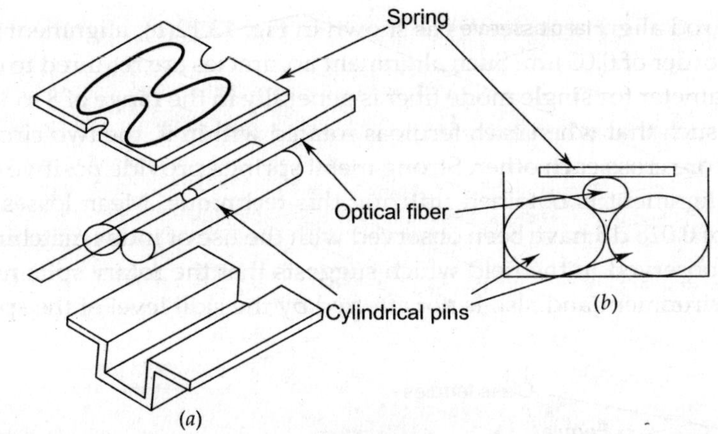

Fig. 13.10

The above-mentioned mechanical splicing methods employ alignment of the bare fibers. However, alignment of secondary elements around the bare fibers also gained favor because of its ruggedness. Secondary alignment technique also provides a structure that can be ground and polished for fiber end preparation. Furthermore, with a good design the fiber coating can be terminated within the secondary element leaving only the fiber end face exposed. Hence when the fiber end face is polished flat to the secondary element, a very rugged termination is produced. The technique is particularly advantageous for the use in the fiber remountable connectors. The main disadvantages of this technique are longer time to make termination and the often increased splice losses resulting from tolerances on the secondary elements which tend to contribute to the fiber misalignment.

A secondary aligned mechanical splice for multimode fiber is shown in Fig. 13.11. Precision glass capillary tubes called ferules are employed as the secondary elements with an alignment sleeve of metal or plastic into which the glass tubed fibers are inserted. An average loss of around 0.2 dB is obtained with normal assembly of the splice using 50 μm core diameter fibers.

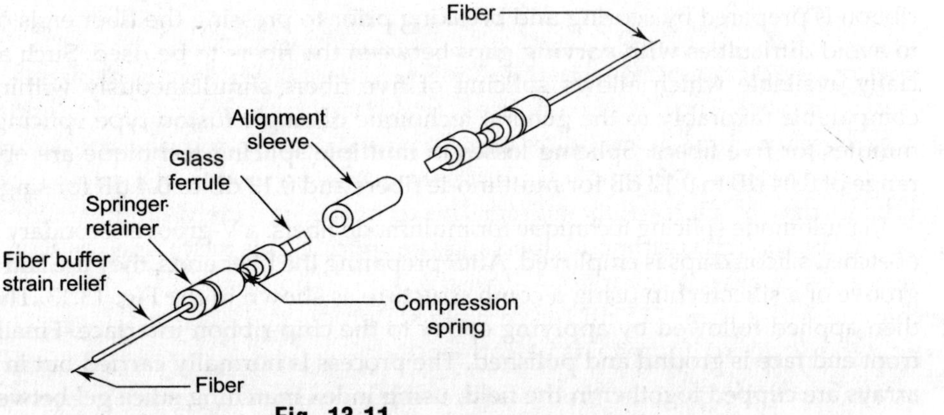

Fig. 13.11

The secondary alignment technique has also been employed in the realization of a low loss, single mode fiber mechanical splice. This device known as a single mode rotary splice is illustrated in Figs. 13.12(*a*) and 13.12(*b*). The fibers to be spliced are initially terminated in precision glass capillary tubes which are designed to make use of the small eccentricity that is present. The fiber terminations are prepared with a simple grinding and polishing operation and an ultraviolet curable adhesive is employed to cement the fibers in the glass tubes.

Using three glass rod alignment sleeves as shown in Fig. 13.12(*b*), alignment accuracy may be obtained within the order of 0.05 μm. Such alignment accuracies are required to obtain low losses as the mode-field diameter for single mode fiber is generally in the range of 8 to 10 μm. The sleeve has a built-in offset such that when each ferule is rotated within it, the two circular paths of the centre of each fiber core cross each other. Strong metal springs provide positive alignment retention and excellent alignment is obtained utilizing this technique. Mean losses of 0.03 dB with standard deviation of 0.018 dB have been observed with the use of index matching gel. Moreover this result has been observed in the field which suggests that the rotary splicing technique may be used in harsh environment and also is not affected by the skill level of the splicer.

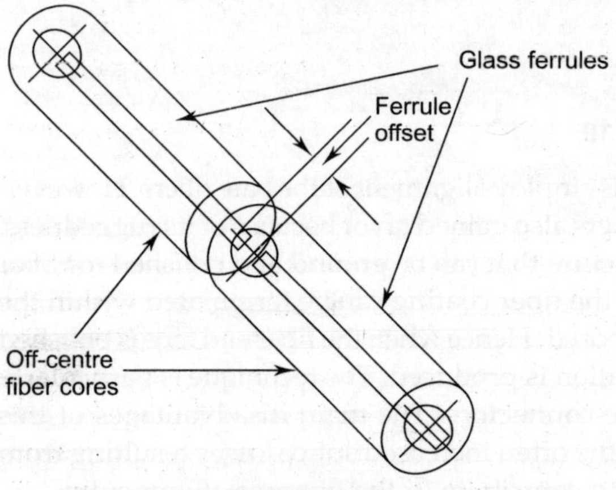

Fig. 13.12(*a*)

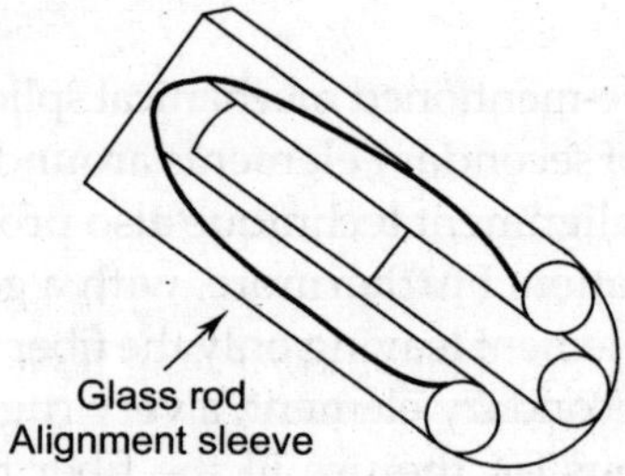

Fig. 13.12(*b*)

13.3.3 Multiple Splices

Multiple simultaneous fusion splicing of an array of fibers in a ribbon cable for both multimode and single mode fibers has been schematically illustrated in Fig. 13.13. in both the cases fiber ribbon is prepared by scoring and breaking prior to pressing the fiber ends on to a contact plate to avoid difficulties with varying gaps between the fibers to be used. Such a device is commercially available which allows splicing of five fibers simultaneously within 5 minutes that is comparable favorably to the general technique of single fusion type splicing, where it takes 15 minutes for five fibers. Splicing losses in multiple splicing technique are observed to be in the range of 0.04 dB to 0.12 dB for multimode fibers and 0.13 dB to 0.4 dB for single mode fibers.

In multimode splicing technique for multimode fibers, a V-groove secondary element comprising of etched silicon chips is employed. After preparing the fiber ends, they are laid into the trapezoidal groove of a silicon chip using a comb structure as shown in the Fig. 13.13. The top silicon chip is then applied followed by applying epoxy to the chip-ribbon interface. Finally, after curing, the front end face is ground and polished. The process is normally carried out in the factory and the arrays are clipped together in the field, using index matching silica gel between the fiber ends.

The major advantages of this method are the substantial reduction of splicing time per fiber (by more than a factor of 10) and the increased robustness of the final connection. The major splicing loss in this technique is due to intrinsic fiber mismatch. Early splicing investigation using silicon chip has demonstrated the feasibility of connecting 12 × 12 fiber arrays. However,in practice only single twelve fiber array is used for splicing at one time to minimize the splicing tolerances and cross interferences from telecommunication channels that are present in large numbers in multidimensional array of fibers.

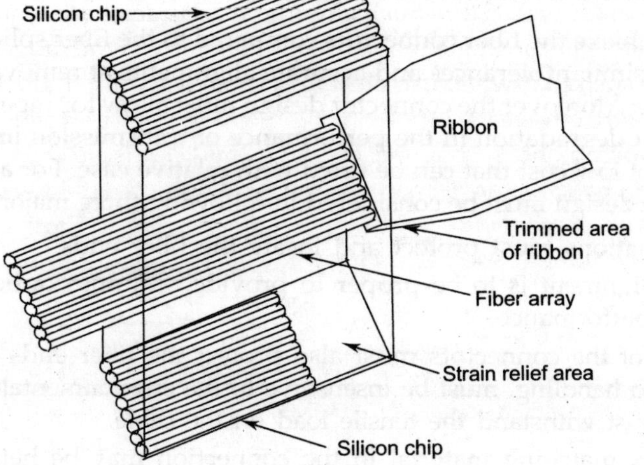

Fig. 13.13

An alternative V-groove flat chip moulded from glass filled polymer resin may be employed, where simultaneous preparation of all fibers of either end is achieved using ribbon grinding and polishing procedure. The fiber ribbon is laid in guides and all the fibers are positioned in groove in the glass filled polymer substrate as shown in Fig. 13.14. A vacuum technique is used to hold the fibers in position whilst cover plate is applied and spring clips holds the fiber assembly together. Index matching gel is applied through a hole in the cover plate giving average splicing losses of 0.1 dB with multimode fiber.

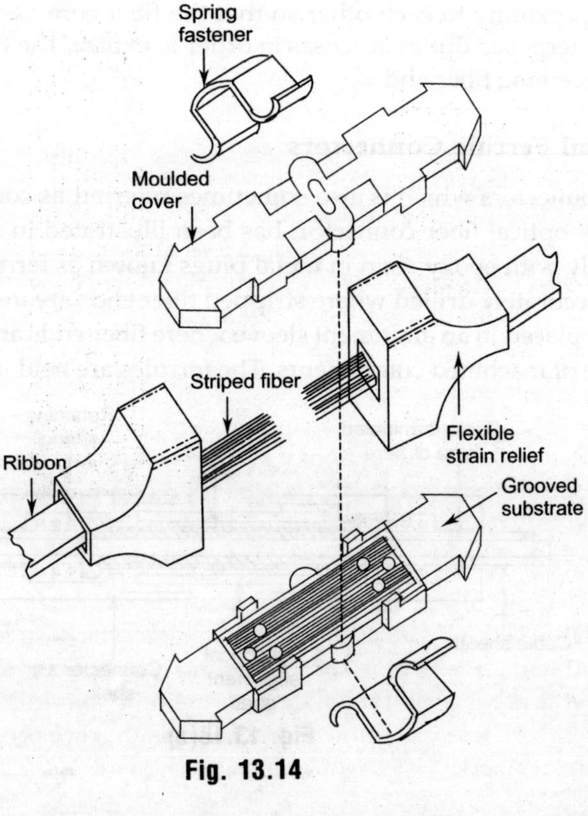

Fig. 13.14

13.4 FIBER CONNECTORS

It is more difficult to achieve the fiber connectors compared to the fiber splices. Connectors must be accomplished with similar tolerances and accurate alignments at removable fashion which is very difficult to achieve. Moreover the connector design must allow for repeated connections and disconnections without degradation in the performance of transmission line at the joints. Fiber connectors should be of low cost that can be fitted with relative ease. For an optimum performance at connectors, the design must be considered in following three major areas.

(*a*) The fiber terminations must protect and locate the fiber ends.

(*b*) The fiber end alignment is to be proper to provide optimum optical coupling without degradation of performance.

(*c*) The outer shell of the connectors must also protect the fiber ends from damage which may occur due to handling, must be insensitive to the environmental factors like dust and moisture, and must withstand the tensile load on the cable.

The use of an index matching material in the connection may be helpful in two ways. It increases the light transmission through the connections as well as it keeps the dust and moisture away from the connections. However, this type of design aspect is not always practical for demountable connectors, as problem exists for sealing and replacement of index matching fluid, when the connector is disconnected and reconnected. In addition to that liquids may have detrimental effect, attracting dust and dirt to the connection.

There are a large number of commercially available demountable single fiber connectors, which have insertion losses within the range of 0.2 dB to 3 dB. Fiber connectors may be classified into two broad categories -*butt jointed connectors* and *expanded beam connectors*. Butt jointed connectors are the most widely used type of connectors which rely upon alignment of the two prepared fiber ends in close proximity to each other so that the fiber core axes coincide. Expanded beam connectors utilize interposed optics or lenses in order to expand the transmitting beam and again reducing it at the receiving fiber end.

13.4.1 Cylindrical Ferrule Connectors

The basic ferrule connectors which is also sometimes referred as concentric sleeve connector, is the simplest type of optical fiber connector, has been illustrated in Fig. 13.15(*a*). Two fibers are bonded permanently with epoxy resin in metal plugs known as ferrules. The centers of the ends of the ferrules are accurately drilled where stripped fiber ends are inserted. Within the connector the two ferrules are placed in an alignment sleeve where fiber ends are placed accurately and butt jointed with the use of machined components. The ferrules are held in place with help of springs.

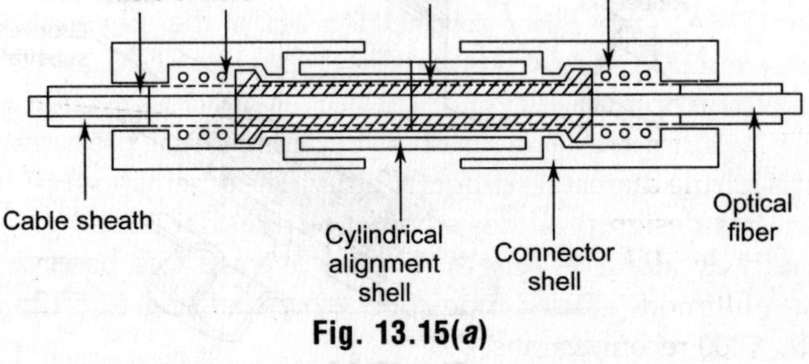

Fig. 13.15(*a*)

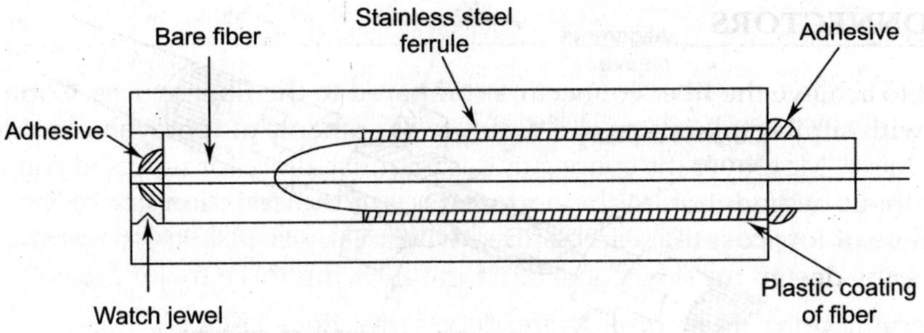

Fig. 13.15(*b*)

In this type of connection mechanism, it is essential that the fiber faces are smooth and square, *i.e.*, perpendicular to the fiber axis that can be achievable by the following procedures:

(*a*) Cleaving the fiber before insertion into the ferrule,

(*b*) Inserting and bonding before cleaving the fiber close to ferrule end face,

(*c*) Using either (*a*) and (*b*) and polishing the fiber end face until it is flush with the end of the ferrule.

Polishing the fiber end faces after insertion and bonding provides the best results, but it is time-consuming especially in the field.

The fiber alignment accuracy of the basic ferrule is largely dependent upon the ferrule hole into which fiber is inserted. Hence some connectors are incorporated with a watch jewel in the ferrule end face (called as *jewelled ferrule connector*), which is illustrated in Fig. 13.15(*b*). In this case the fiber is centered with respect to the ferrule through watch jewel hole. Close diameter and tolerance requirements of the ferrule end face hole are easily obtained with use of watch jewel. Typical concentricity error between the fiber core and the outside diameter of the jeweled ferrule are in the range of 2 to 6 μm yielding insertion losses in the range 1 to 2 dB with multimode step index fibers.

Ceramic materials have found widespread application for manufacturing of precision capillary ferrule connectors due to their outstanding thermal, mechanical and chemical resistance characteristics in comparison to metals and plastics. Such capillary ferrules have a precision bore which is accurately centered in the ferrule. Final assembly of the connector requires the use of adhesive prior to grinding and polishing for end preparation. The ceramic ferrule material is harder than the optical fiber and is unaffected by grinding and polishing processes, thus assisting the production of low loss connectors. Typical average losses are 0.2 dB for multimode graded index fibers (core/cladding: 50/125 μm) and 0.3 dB for single-mode fiber (core/cladding: 9/125 μm).

Numerous cylindrical sleeve ferrules are commercially available for both multimode and single-mode fiber termination. The most common design types are the Straight Tip (ST), the subminiature assembly (SMA), the Fiber Connection (FC), the Physical Contact (PC), the Subscriber Connector (SC) and D3/D4. Also combination types like ST-PC and FC-PC are available. ST series connectors exhibits an optimized cylindrical sleeve with cross-section designed to expand uniformly when ferrules are inserted. The constant circumferential pressure provides accurate alignment, even when the ferrule diameters differs slightly. The schematic of ST connector is shown in Fig. 13.16. SMA connector design provides stepped ferrule that is a single step ferrule which reduces the diameter midway along its length. Typical average loss observed in these type of connectors is 0.22 dB for multimode graded index fiber (core/cladding: 62.5/125 μm) with less than 0.1dB change in loss after 1000 reconnections.

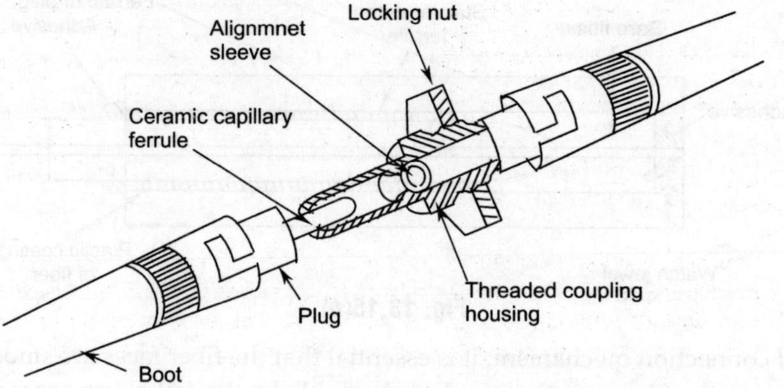

Fig. 13.16

13.4.2 Biconical Ferrule Connectors

Biconical ferrule connectors are also referred as *biconic connectors* which are also widely used as part of jumper cable in variety of applications. This ferrule type biconical plug connectors are either transfer moulded directly on to the fiber or cast around the fiber using a silica loaded resin ensuring concentricity to within 5 μm. The fiber ends are polished before the plugs are inserted and then plug is attached with the alignment to the biconical moulded sleeve. The schematic of the biconic connector arrangement is shown in Fig. 13.17. Using this connector with 50 μm core diameter graded index fiber, the mean insertion losses are observed to be as low as 0.21 dB. In the original design transparent silicon resin pads were placed over the fiber end faces to provide index matching. However, preparation process has been modified where the polished fiber end faces are butted directly and the gap as well as parallelism of end faces being controlled to a degree that improves insertion losses better than the level normally exhibited by Fresnel reflection. This connector may also be used with single mode fibers by reducing the eccentricity of the tapered cone and the fiber core to 0.33 μm, while limiting the tilt angle to 35° or less. The average losses in case of single mode fiber with this arrangement may be achieved up to 0.28 dB whereas the maximum loss is observed to be 0.7 dB.

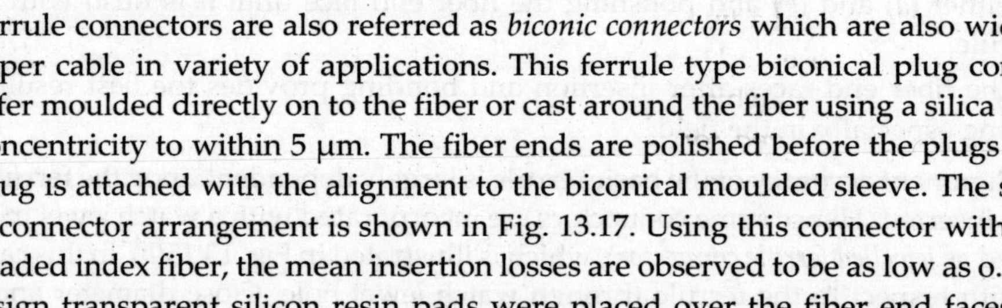

Fig. 13.17

13.4.3 Double Eccentric Connectors

The double eccentric connector is an active adjustable assembly allowing close alignment of fiber axes and does not rely upon on a concentric fixed sleeve. The connector mechanism as illustrated in Fig. 10.18 consists of two eccentric cylinders within the outer plug.

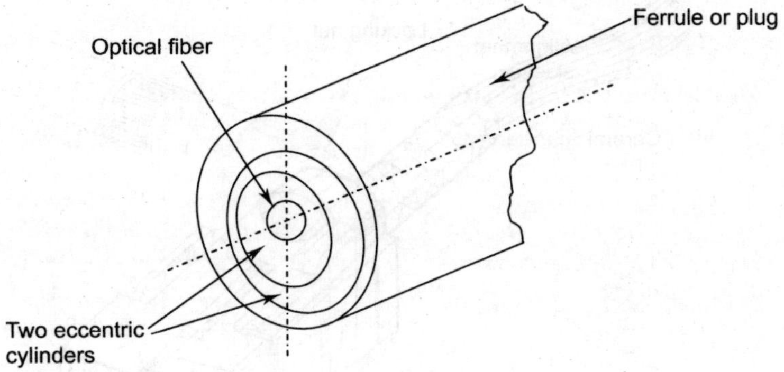

Fig. 13.18

The optical fiber is mounted eccentrically within the cylinder. When the two connector halves are mated, the fiber core axes coincide through rotation operation using either an inspection microscope or a peak optical adjustment. Finally they are locked to make permanent alignment. The insertion losses are observed to be within 0.48 dB for multimode graded index fibers with the connectors. Use of index matching liquid reduces the loss within 0.2 dB. The double eccentric connector may also be utilized for single mode fibers where its adjustable nature has proved advantageous for alignment of small core diameter fibers yielding losses of 0.46 dB without index matching.

13.4.4 Duplex and Multiple Fiber Connectors

Duplex fiber connector provides the facility for two way communication. As well as fiber distributed data interface. Figure 13.19 shows the schematic of this type connector, which comprises two ST ferrules housed in a protective moulded shroud and exhibits a typical loss of 0.6 dB. For interconnections of large number of fiber, multiple fiber connector is useful. Multiple fiber configuration is formed by assembling of both cylindrical and biconical ferrule connectors. Single ferrule connectors generally allow the alignment sleeve to float within the housing, and therefore, any requirement of high tolerance on ferrule positioning within the multiple ferrule versions is avoided. However, insertion force of multiple cylindrical ferrules may be large when ferrules are involved. In this regard multiple biconical ferrule connector is advantageous due to low insertion force of biconical configuration.

There exits other type of multiple fiber connectors apart from assembly of a number of single fiber connector as discussed above. Silicon chip arrays have been suggested for the jointing of fiber ribbon cable. Figure 13.20 shows such type of multiple fiber connector which utilizes V-grooved silicon chips. In this connector ribbon fibers are mounted and bounded, and bonded into V-grooves in order to form a plug together with precision metal guiding rods and coil springs. The fiber connections are then accomplished by butt jointing the two pairs of guiding rods in the slitted sleeves located in the adaptor. This type of multiple fiber connectors gives insertion losses of 0.8 dB which may be reduced to 0.4 dB using index matching fluid.

Improved loss characteristics of mean loss and standard deviation of 0.45 dB and 0.12 dB respectively has also been observed in this type of connector, while terminating of 50 μm core multimode fibers.

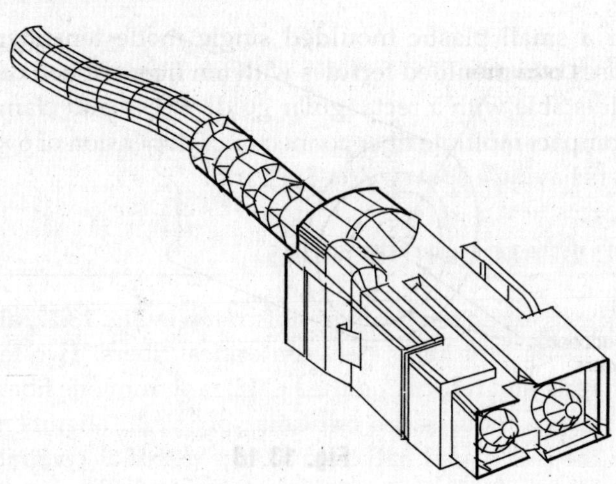

Fig. 13.19

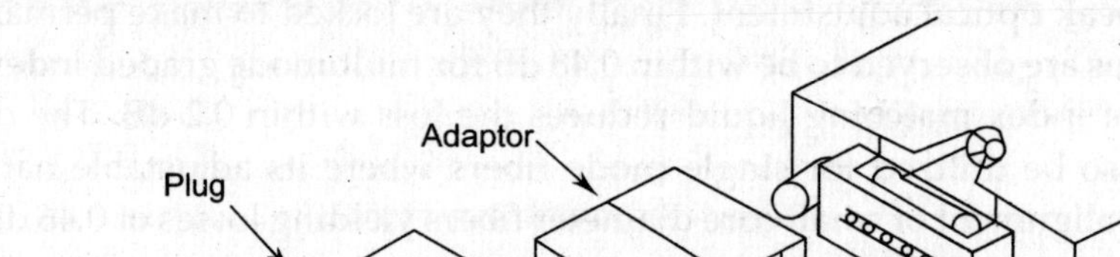

Fig. 13.20

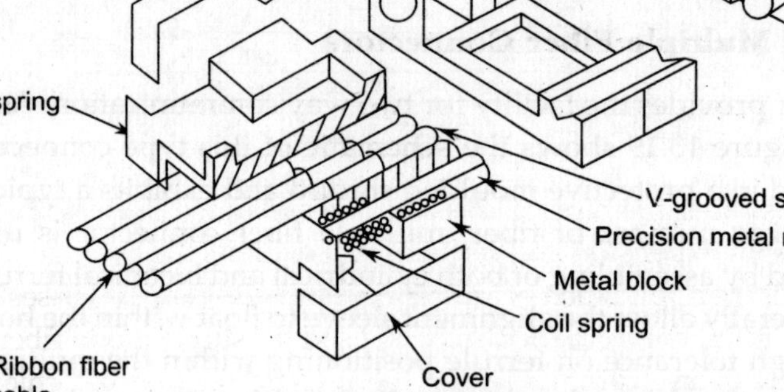

Fig. 13.21

The structure of a small plastic moulded single mode ten fibers connector is shown in Fig. 13.21. It comprises two moulded ferrules with ten fibers ribbon cables accurately aligned by guide pins. It is held stable with a rectangular guide sleeve and cramp spring. Typical average insertion loss of a compact multiple fiber connector of dimension of 6 × 4 mm is observed 0.43 dB for single mode fibers having a spot size (ω_0) of 5 μm.

13.5 EXPANDED BEAM CONNECTORS

Based on the principle of expanded beam as illustrated in Fig. 13.22, alternate type of connectors have been designed using butt joints between optical fibers. Two lenses are provided in this connector for collimating and refocusing the light beam from one fiber to the other. Such type of interposed optics concept reduces the criticality of lateral alignment as well as longitudinal separation between the two mated halves to a large extent as compared to normal butt jointed fiber connectors. However, the angular alignment is much more critical in this case. In spite of this disadvantage, expanded beam connectors are useful for multiple fiber connection and edge connection for printed circuit boards where lateral and longitudinal alignment are frequently difficult to achieve.

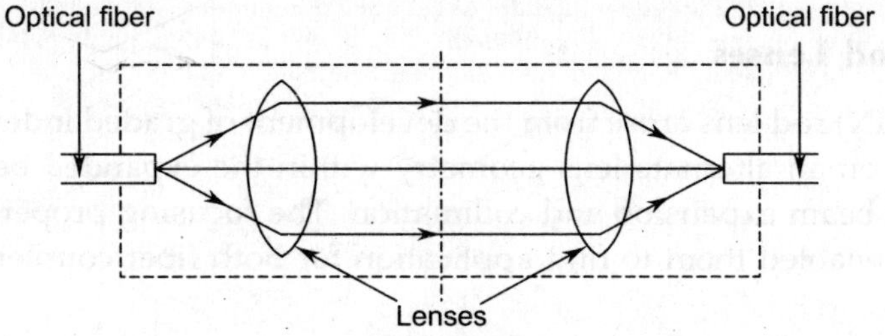

Fig. 13.22

Figure 13.23(*a*) illustrates one type of lens coupled expanded beam connectors. It utilizes spherical microlenses for beam expansion and reduction. It exhibits the average loss of 1 dB which can be reduced to 0.7 dB with the application of antireflection coating on the lenses for 50 μm core diameter graded index fiber. Similar configuration may be used for single mode fiber using lenses of 2.5 mm diameter. With antireflection coating on lenses, the average loss may be achieved around 0.7 dB using 8 μm core diameter single mode fibers. Much smaller lenses of 250 μm diameter sapphire ball can be used for single mode fiber connections to restrict the losses within the range of 0.4 dB to 0.7 dB for over 1000 connections.

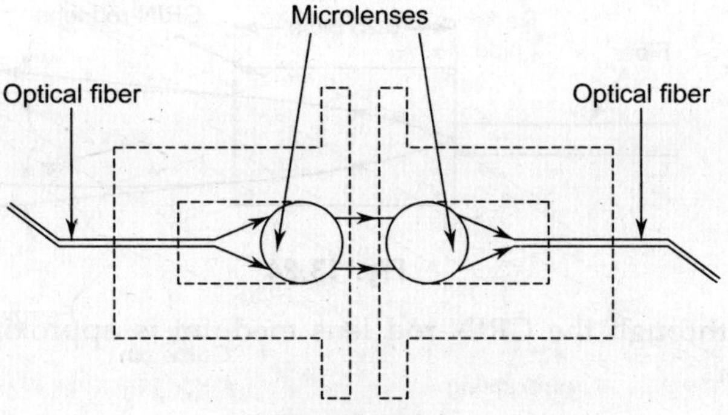

Fig. 13.23(*a*)

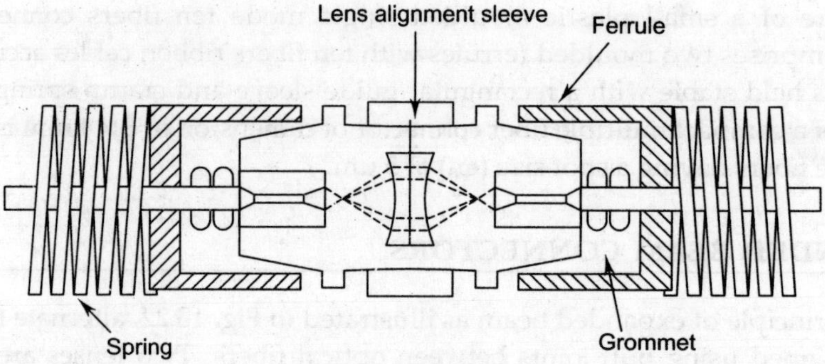

Fig. 13.23(*b*)

An expanded beam connector employing a moulded spherical lens is shown in Fig. 13.23(*b*). In order to obtain a collimated beam, the fiber is placed approximately at the focal length of the lens. Thus lens to lens longitudinal misalignment effects are minimized. A lens alignment sleeve is provided to minimize the effects of angular misalignment. The complete connector structure is provided with ferrule, grommet, spring and external housing. The repeatability of this straightforward lens design is found to be good, incurring losses of around 0.7 dB.

13.5.1 GRIN-Rod Lenses

Graded index (GRIN) rod lens arose from the development of graded index fiber waveguides. Its principle is based on an alternate lens geometry within the expanded beam connectors which facilitates efficient beam expansion and collimation. The focusing properties of such microlens devices have also enabled them to find application for both fiber couplers as well as source to fiber coupling.

GRIN-rod lens comprises of a cylindrical glass rod typically of 0.5 mm to 2 mm diameter exhibiting a parabolic refractive index profile with a maximum at the axis similar to graded index fiber. Light propagation through the lens is determined by the lens dimension and wavelength of the light, as refractive index is a wavelength dependent parameter. The GRIN-rod lens can produce a collimated light output within the divergent angle α of 1° to 5° at the other end of the light source, as illustrated in Fig. 13.24. Conversely, incoming light beam within above angle α can be focused at small area at the other end of the lens. Typically, using GRIN-rod lens, light emerging from a 50 μm diameter fiber core results in a collimated output beam of between 0.5 to 1 mm.

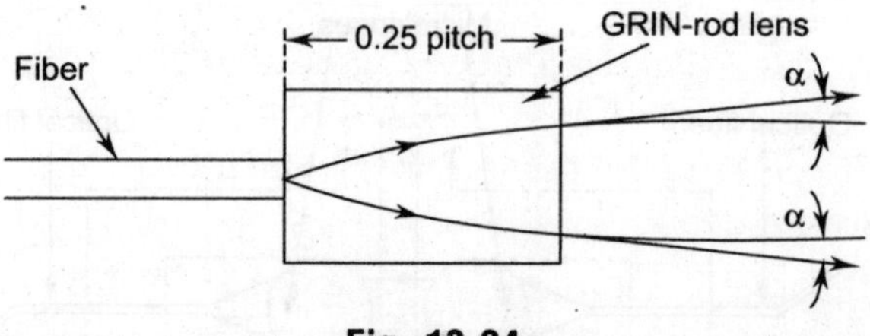

Fig. 13.24

Ray propagation through the GRIN-rod lens medium is approximately governed by the paraxial ray equation.

$$\frac{d^2r}{dz^2} = \frac{1}{n}\frac{dn}{dr} \tag{13.19A}$$

where r is the radial coordinate, z is the distance along the optical axis and n is the refractive index at a point.

Furthermore, the refractive index at r in a graded index medium can be expressed as the following relationship:

$$n(r) = n_1\left(1 - \frac{Ar^2}{2}\right) \tag{13.19B}$$

where n_1 is the refractive index at the optical axis and A is a positive constant. Therefore above two relations the position r of the ray is given by

$$\frac{d^2r}{dz^2} = -Ar \tag{13.20}$$

The general solution of the equation 13.20 is given by

$$r = K_1 \cos A^{-1}r + K_2 \sin A^{-1}r \tag{13.21}$$

K_1 and K_2 are constants.

Therefore, all the input rays follow a sinusoidal path through the lens medium with refractive index variation. The transversion of one sinusoidal period is termed as one full pitch. The GRIN-rod lenses are manufactured with several pitch lengths considering different aspects and applications as follows:

1. The quarter pitch or 0.25 pitch lens produces a perfectively collimated output beam at opposite lens face, when the input light emerges from a point source. Conversely, light beam from spread source converges at the centre of lens at opposite face, as illustrated in Fig. 13.25(*a*). The focal point of the quarter pitch GRIN-rod lens is coincident with the lens face providing efficient direct butted connection to the optical fiber.
2. The focal point of 0.23 pitch GRIN-rod lens lies outside the lens providing a collimated beam projection at the other side as shown in Fig. 13.25(*b*). This type of lens is employed to convert the diverging beam from a fiber or a laser diode into a collimated beam.
3. The focal points of 0.29 pitch GRIN-rod lens are designed to be at the outside of both the sides of the lens faces. It is generally used for coupling the output of laser diode into an optical fiber which convert a diverging beam into a converging beam. The ray diagram is shown in Fig. 13.25(*c*).

Apart from the pitch lengths as discussed above, various pitch of 0.5 pitch and 0.75 pitch lenses are also available with numerical apertures 0.37, 0.46 and 0.6. The GRIN-rod lenses are generally manufactured with the diameter of 0.5 mm to 2 mm which may be employed for either single mode or multimode step or graded index fiber.

The divergence of collimated beam from GRIN-rod lenses is affected by a number of factors which include errors in the lens cut length, the finite size of fiber core and chromatic aberration. Losses are observed around 1 dB for divergence angle within 1°. In contrast to the butt jointed multimode fiber connectors, GRIN-rod connectors performance characteristics are independent of the modal power distribution in the fiber.

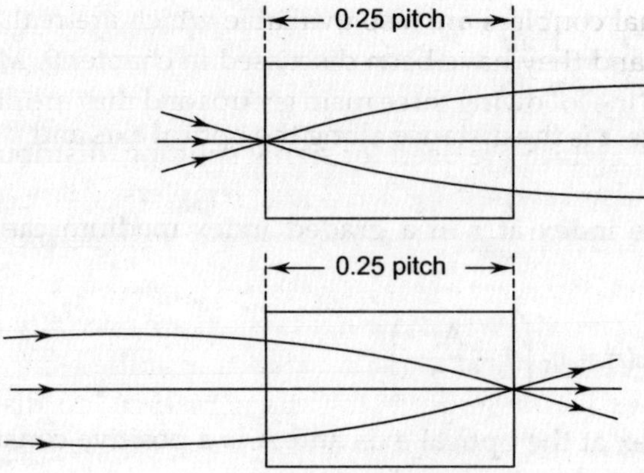

Fig. 13.25(*a*)

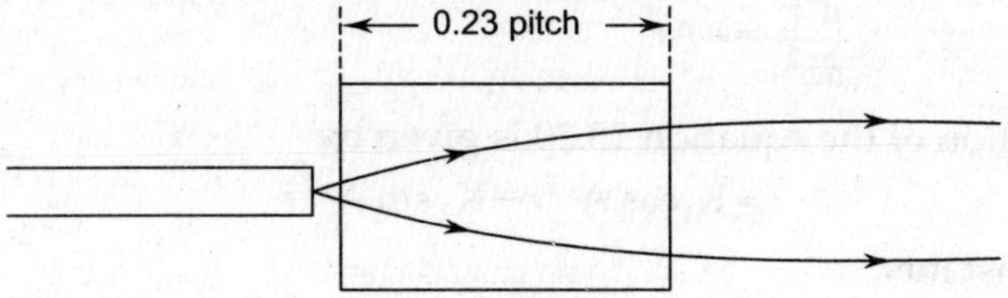

Fig. 13.25(*b*)

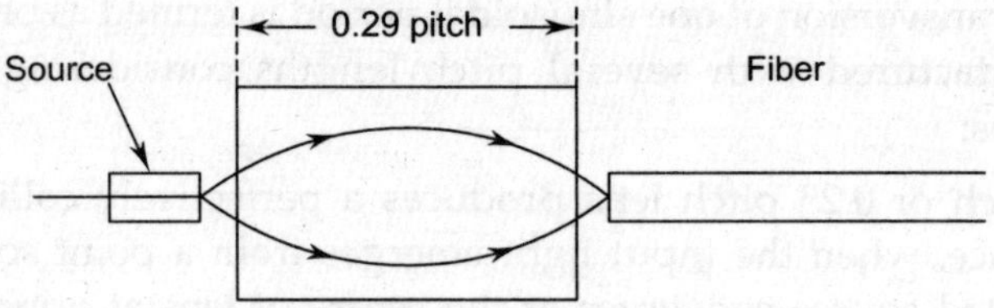

Fig. 13.25(*c*)

13.6 FIBER COUPLERS

An optical coupler, also referred as the direction coupler is the device that distributes light from a main fiber into one or more branch fibers. General use of optical couplers is to distribute or combine light to or from a number of branch fibers for the application within optical fiber information distribution systems including data buses, local area networks, computer networks and telecommunication access networks. Such devices are often passive type and also known as multiport fiber couplers. The power transfer in optical couplers are taken place by one of the following two methods:

(*a*) **Core interaction type:** Through the fiber core cross-section by butt jointing the fibers or by using some form of imaging optics between the fibers as shown in Fig. 13.26(*a*).

(*b*) **Surface interaction type:** Through the fiber surface and normal to its axis by converting the guided core modes to both cladding and refracted modes which then enable the power sharing mechanism as shown in Fig. 13.26(*b*)

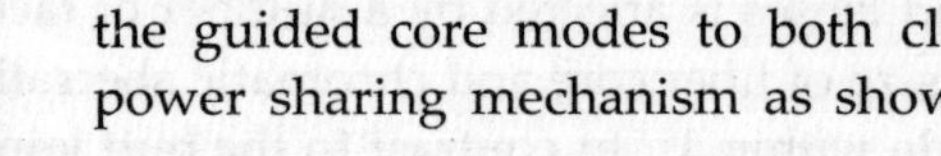

Fig. 13.26(*a*)

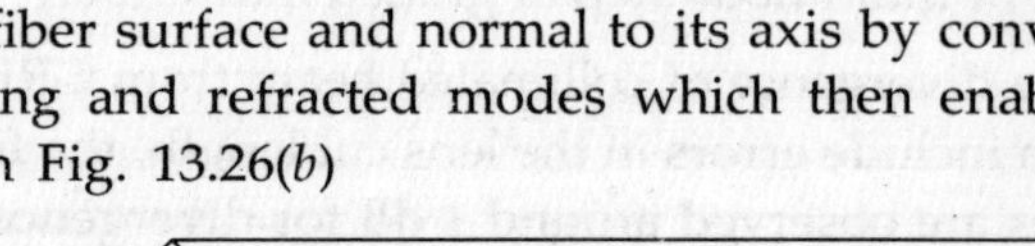

Fig. 13.26(*b*)

Active waveguide directional couplers are also available which are realized using integrated optical fabrication techniques and they have been discussed in chapter 9. Multiport optical fiber couplers can be subdivided into the following three main groups and they are illustrated in Fig. 13.27.

1. *Three and four port couplers* which are used for signal splitting, distribution and combining. Four port couplers are also referred as 2 × 2 star couplers.
2. *Star couplers* which are generally used for distributing a single input signal to multiple outputs.
3. *Wavelength Division Multiplexing* (WDM) devices which are specially designed to permit a number of peak wavelength optical signals to be transmitted on a single fiber. These devices are also designed for demultiplexing the signal, *i.e.,* to distribute a number of signals into different fibers with respect to their peak wavelength from a single fiber.

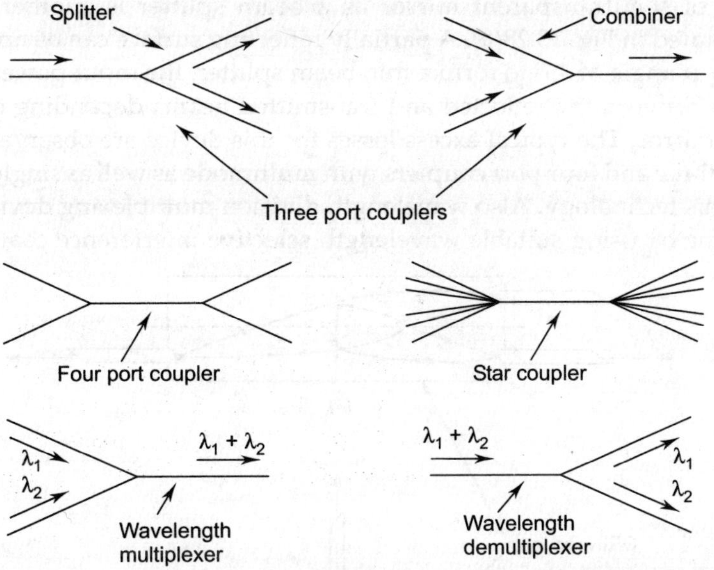

Fig. 13.27

Ideal fiber couplers should distribute light signals to the branch fibers without any scattering loss (also referred as the excess loss) or generation of any noise. They should operate with complete insensitivity to factors including the distribution of light between the fiber modes as well as the state of polarization of the light. However, in practice, the passive couplers do not maintain all these properties and hence performance of the optical fiber networks is affected by the characteristics of the coupling devices. The finite scattering loss at the couplers limits the number of terminals that can be connected or the span of the network. On the other hand, generation of noise and modal effects can cause problems in the specification of the network performance. Therefore, the couplers are to be judiciously used in the network with certain compromises in their application.

13.6.1 Three and Four Port Couplers

There are several methods of fabrication of three and four port couplers. In Fig. 13.28(*a*), the lateral offset method is illustrated, which relies on the overlapping of fibers end faces. Light from input fiber is coupled to the output fibers according to the degree of overlap. Hence the input power can be distributed in a well defined proportion by appropriate control of the amount of lateral offset between the fibers. This technique is well suited for multimode step index fibers as it can provide a bidirectional coupling capability. But high excess loss or scattering loss may be observed in this technique as all input light cannot be perfectly coupled to the output branches.

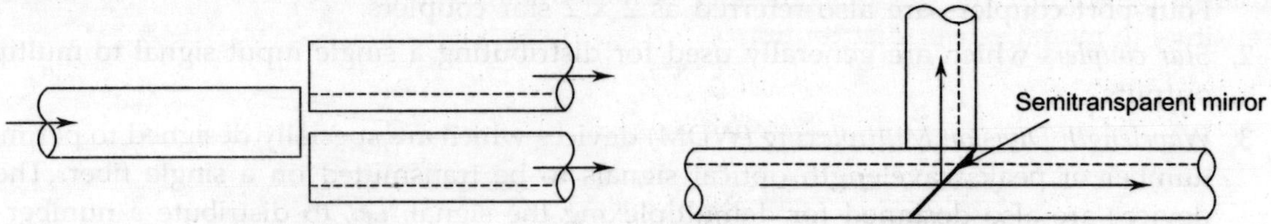

Fig. 13.28(*a*) **Fig. 13.28(*b*)**

Incorporation of semitransparent mirror as a beam splitter is another technique of fiber coupling as illustrated in Fig. 13.28(*b*). A partially reflecting surface can be applied directly to the fiber end face cut at angle of 45° to form a thin beam splitter. the input power may be divided in any desired ratio between the reflected and transmitted beams depending on the properties of semitransparent mirror. The typical excess losses for this device are observed to be in the range of 1 to 2 dB. Both three and four port couplers with multimode as well as single mode fibers can be fabricated with this technology. Also wavelength division multiplexing device can be fabricated with this technique by using suitable wavelength selective interference coatings.

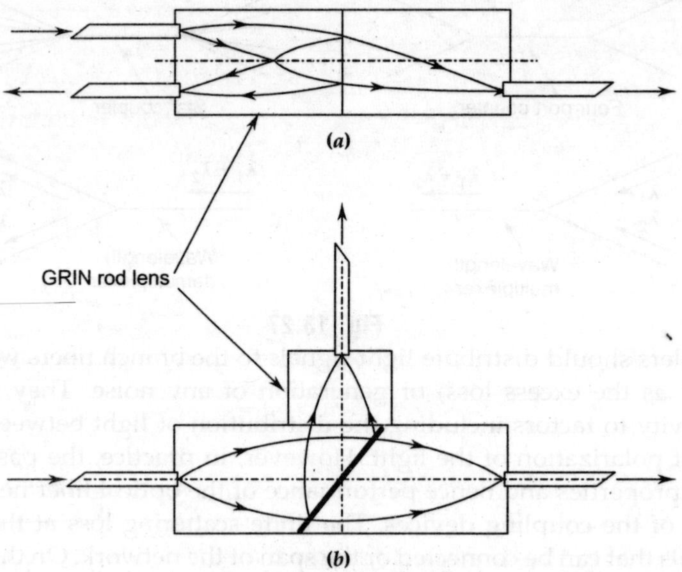

Fig. 13.29

A wide range of couplers are available using micro-optic components. With the utilization of beam expansion and collimation properties of graded index (GRIN) rod lens combined with spherical retro-reflecting mirrors these couplers are fabricated. Such devices are accomplished with miniature optical assemblies of compact construction. Figure 13.29(*a*) shows the structure of a parallel surface type GRIN-rod lens three port coupler that comprises of two quarter pitch lenses with a semitransparent mirror in between. Light rays from input fiber collimate in the first lens before they are incident on the mirror. A portion of the incident beam is reflected back and while the transmitted portion is focused in the second lens. Figure 13.29(*b*) shows the slant surface version of the similar coupler. The parallel surface type however is the most attractive due to its simplicity, compactness, ease of fabrication and relatively low insertion loss. WDM application may be achieved in this device with the use of interference filter instead of mirror. Interference filter transmit only certain wavelength band and reflects all other signal of longer or shorter wavelengths. The devices generally exhibit low insertion loss of less than 1 dB and are insensitive to modal power distribution.

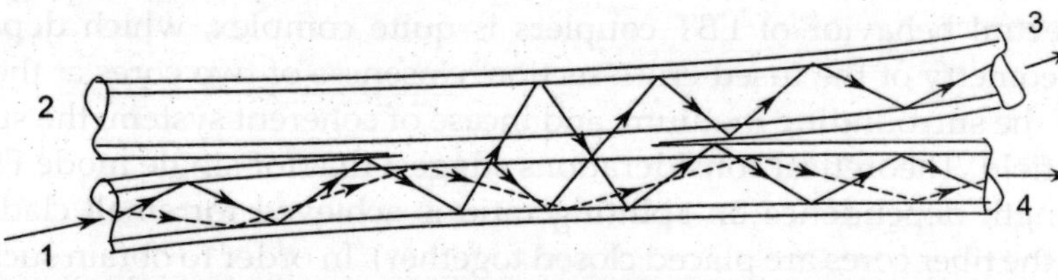

Fig. 13.30

One of the most common methods for manufacturing couplers is the *Fused Biconical Taper* (FBT) technique. Its basic structure and principle of operation is illustrated in Fig. 13.30. In this method the fibers are twisted together and then spot-fused under tension such that the fused portion is elongated to form a biconical taper structure. Optical power of the input fiber propagates in the form of guided core modes. The higher order modes, however, leave the fiber core because of its reduced size in the tapered-down region and therefore guided as cladding modes. These modes are transferred back to guided core modes in the tapered-up region of the output fiber with an approximately even distribution between the two fibers.

As only the higher order modes take part in the coupling process, generally a portion of total power is coupled between the two fibers and lower order modes remain within the main fiber. Therefore, mode dependent or wavelength dependent coupling ratio is obtained. However, if the tapered region is made sufficiently narrow, then the entire mode volume can be participated in the coupling process and larger proportion of input power can be shared between the output fibers. With this principle, improvement in both the power and modal uniformity of the coupler may be made.

With reference to Fig. 13.30 of four port coupler, various parameters may be calculated.

$$\textit{Excess loss} \text{ in four port coupler} = 10 \log_{10} \frac{P_1}{P_3 + P_4} \text{ dB} \tag{13.22}$$

$$\textit{Insertion loss} \text{ for port 1 to 4} = 10 \log_{10} \frac{P_1}{P_4} \text{ dB} \tag{13.23}$$

$$\textit{Crosstalk} \text{ in four port coupler} = 10 \log_{10} \frac{P_2}{P_1} \text{ dB} \tag{13.24}$$

$$\textit{Coupling ratio} \text{ or } \textit{split ratio} = \frac{P_3}{P_3 + P_4} \times 100\% \tag{13.25}$$

$$= \left(1 - \frac{P_4}{P_3 + P_4}\right) \times 100\% \tag{13.26}$$

Coupling ratio or split ratio indicates the proportion of power available at the output ports. This is also dependent on the relative cross sections of the fibers and the mode coupling mechanism in both single mode and multimode fibers. The major advantages of the FBT structure are relatively low excess loss (within 0.5 dB), low crosstalk (usually better than – 50 dB) and capability of coupling any fiber and geometry. Hence this type of couplers can be adopted for specific requirements of a system or network. However, a major disadvantage is that the mode dependent splitting can result in differing losses through the coupler. Therefore, a wavelength dependent performance as well as generation of modal noise is observed when coherent light sources are employed.

The precise spectral behavior of FBT couplers is quite complex, which depends upon the dimensions and geometry of the fused cross-section, closeness of two cores at the fused section, refractive index of the surrounding medium, and incase of coherent system, the state of polarization of the optical field. Theoretical considerations suggest that for single mode FBT coupler, the minimum wavelength dependence on splitting ratio is achieved for small cladding radii and strong fusing (*i.e.*, the fiber cores are placed closed together). In order to obtain such performance, it is necessary to taper the fibers down to radius of around 1.5 μm or less, and to ensure that the rate of taper is such that the major proportion of coupling occurs in the neck region. The wavelength dependent behavior associated with single mode FBT couplers follow an approximately sinusoidal pattern over the wavelength range 0.8 μm to 1.5 μm as a result of single mode coupling length between two fibers. This mechanism is also useful in the manufacturing of WDM multiplexer/ de-multiplexer couplers which will be discussed at later section. With polarization maintaining fiber that preserves the polarization of input signals, it is also possible to fabricate single mode fiber couplers (also called as hi-birefringence couplers) as well as polarization sensitive couplers effectively functioning as beamsplitters.

Optical waveguide coupler technique is an alternative technology for both the single mode and multimode fibers. The production of such couplers involves two basic processes. Firstly, a mask of the desired branching function is deposited on to a glass substrate using photolithographic process. The substrate is then subjected to a two stage ion exchange creating virtually circular waveguides embedded within the surface of the substrate on which the mask was deposited. Figure 13.31 shows a schematic example of a three-port integrated waveguide coupler using this process. Multimode fibers are bonded to the structure using etched V-groove. Typical excess losses are observed to be 0.5 dB for three port couplers and 0.6 dB for 1×8 star couplers. These waveguide couplers are attractive because of the flexibility at the masking stage. The same technique can also be used for fabrication of WDM multiplexer/ de-multiplexer devices.

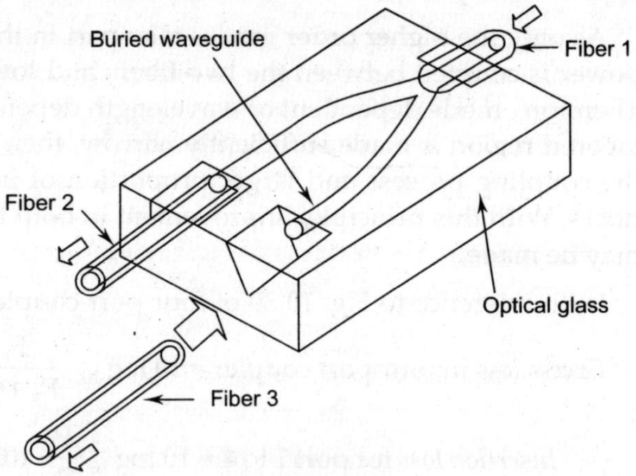

Fig. 13.31

When a periodic deformation is applied to the fiber, mode coupling can take place between guided and radiation modes. Based on this idea directional couplers may be fabricated with the application of microbend as illustrated in Fig. 13.32. Mode coupling may be obtained by pressing the fiber in close contact with a transparent mechanical grating. The radiated optical power is collected by a lens or a shaped curved glass plate. Variable coupling ratio of wide range can be obtained by altering the pressure on the fiber. Also low light can be extracted from fiber with very little excess loss (typically 0.05 dB).

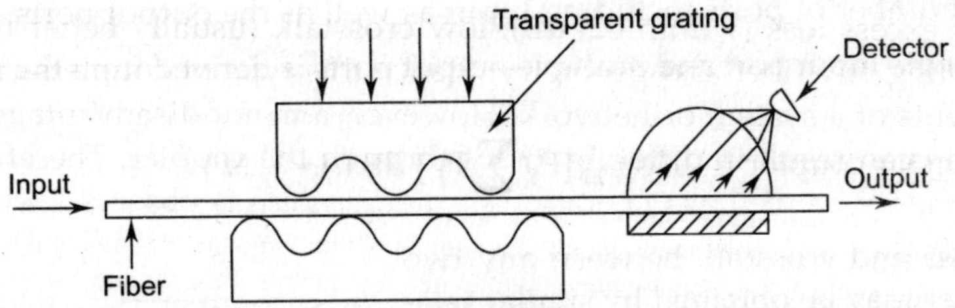

Fig. 13.32

13.6.2 Star Couplers

Star couplers are used for distribution of the optical signal from a single input fiber to multiple output fibers. Two basic principles are adopted for fabrication of star couplers, —mixer rod method and Fused Biconical Taper (FBT) method. In the mixer rod technique, thin glass platelet is employed which effectively mixes the light from one fiber and distributing among output fibers. Transmissive star couplers or reflective star coupler can be manufactured with mixer rod method as illustrated in Figs. 13.33(*a*) and 13.33(*b*) respectively.

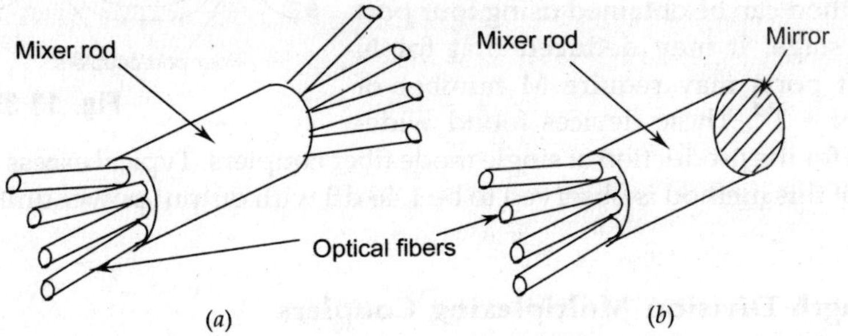

Fig. 13.33

The typical insertion loss for an 8 × 8 mixer rod star coupler with fiber pigtails is 12.5 dB with port to port uniformity within 0.7 dB.

The manufacturing process for fused biconical taper or FBT star coupler is similar to the process as discussed in section 13.6.1 for three and four port FBT couplers. The fibers are bundled, twisted, heated and pulled to form the FBT star couplers as shown in Fig. 13.34. With multimode fiber, this method relies upon the coupling of higher order modes between different fibers. Therefore, FBT couplers are highly mode dependent resulting in wide port to port variation in comparison with mixer rod star couplers.

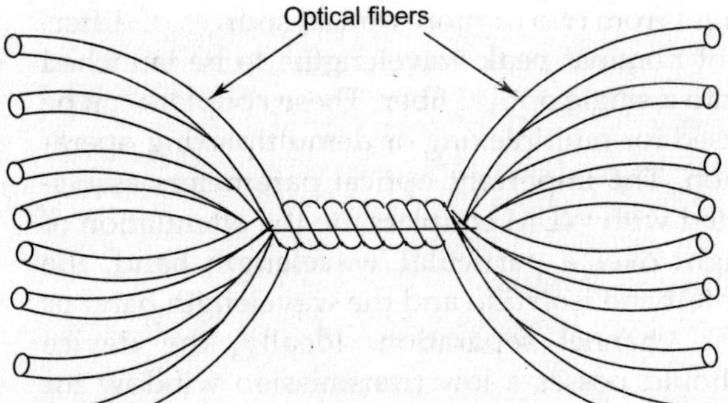

Fig. 13.34

In an ideal star coupler the optical power from input fiber is evenly distributed among the output fibers. The total loss associated with the star coupler comprises of its theoretical splitting loss together with the excess loss. The splitting loss may be calculated as per following relation.

$$\text{Splitting loss for star coupler} = 10 \log_{10} N \text{ dB} \quad (13.27)$$

Where N is the number of ports including Input as well as the output ports.

Excess loss for single input port and multiple output ports is derived from the relation as below.

$$\text{Excess loss for star coupler} = 10 \log_{10} [P_i / \sum_{1}^{N} P_j] \text{ dB } (j = 1, N) \quad (13.28)$$

The insertion loss and crosstalk between any two ports on star coupler may be obtained by similar relations to four port coupler as given at equations 13.23 and 13.24 respectively.

An alternative technique for realization of star coupler is to construct a ladder coupler by cascading multiple stages of three or port FBT couplers as illustrated in Fig. 13.35 (8 × 8 star coupler has been shown). It may be noted that using three port couplers at each stage, it do not form symmetrical star configuration (as example, it may provide 1 × N couplers rather than N × N couplers). However symmetrical configuration of star coupler in this method can be obtained using four port couplers at each stage. It may deduced that for N number of output ports may require M number of stages such that $N = 2^M$. These devices found widespeard application for the production of single mode fiber couplers. Typical excess loss observed for 9 × 9 star coupler of this method is observed to be 1.46 dB with output power uniformity of within 1.5 dB.

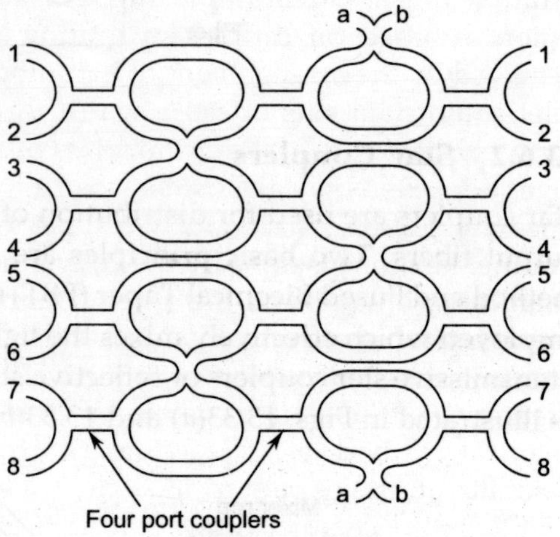

Fig. 13.35

13.6.3 Wavelength Division Multiplexing Couplers

Wavelength Division Multiplexing (WDM) couplers are the special type of couplers that enable light from two or more optical sources of different nominal peak wavelengths to be launched into a single optical fiber. These couplers can be used for multiplexing or demultiplexing operation. The important optical parameters associated with WDM couplers are the attenuation of light over a particular wavelength band, the interband isolation and the wavelength band or the channel separation. Ideally, the device should posses a low transmission window for each wavelength band, giving a low insertion loss. On the other hand the device should exhibit high interband isolation, thus minimizing crosstalk. Furthermore, the channel separation should be as small as may be permitted by the light source availability together with crosstalk consideration. However, in practice, high interband isolation is required at the receiver end (*i.e.*, for demultiplexing operation), or at both end in case of bidirectional system. The spectral performance characteristic for a typical WDM coupler having flat passband output characteristic is shown in Fig. 13.36.

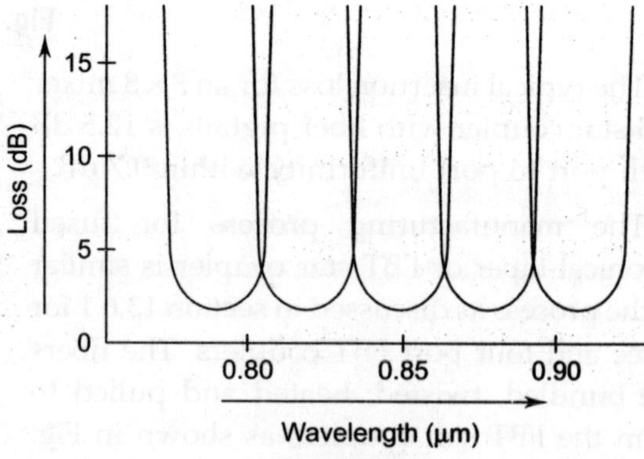

Fig. 13.36

Various techniques have been developed for the implementation of WDM couplers. Passive devices may be classified into three major categories. Two of them – angularly dispersive (usually diffraction grating) coupler and filter, are the core insertion type, while the third one is surface interaction type that may be employed with single mode fiber in the form of a directional coupler. Hybrid combinations of the two core interaction types form other types of implementations.

A glass prism may be utilized as an angularly dispersive element to facilitate wavelength multiplexing and demultiplexing. The implementation is based on diffraction grating which is achieved by the arrangement of a number of parallel equidistant slits of same width. A common form of diffraction grating comprises of an epoxy layer deposited on a glass substrate, on which the lines are blazed. There are two types of blazed grating. First one is done by the conventional mechanical technique, and the other is fabricated by the anisotropic etching of single crystal silicon, hence it is called as silicon grating. The silicon grating is considered to be superior to the mechanical method, as it provides greater design flexibility in choice of blazing angle θ_B (refer Fig. 13.37) and grating constant (number of lines per unit length). Also the silicon grating is efficient and environmentally stable.

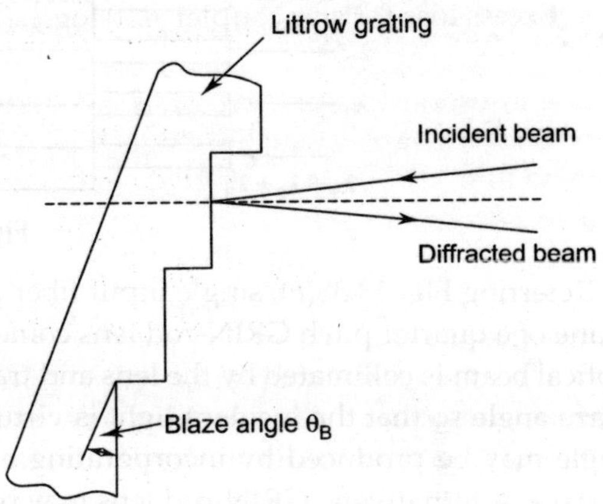

Fig. 13.37

In diffraction grating, the light is reflected at particular directions according to the grating constant, the angle at which the light is incident on grating and the optical wavelength. To manufacture the WDM couplers, two main structural types are used. First one is the Littrow device that employs a single lens and separate plane of grating, and the second one is the concave grating that performs both focusing and diffraction functions without utilizing any lens.

In Littrow mounted grating as illustrated in Fig. 13.37, the blaze angle of the grating is such that the incident and reflected light beams follow virtually same path. Thus maximizes the grating efficiency and minimizes the lens astigmatism. For a given wavelength λ and line spacing of grating $\times$, the blaze angle θ_B is set by the following relation.

$$\theta_B = \sin^{-1}(\lambda/2\times) \tag{13.29}$$

Schematic diagrams of Littrow type grating demultiplexers employing a conventional lens and a GRIN-rod lens are shown in Figs. 13.38(*a*) and 13.38(*b*) respectively. Spherical ball microlens may be used as conventional lens. However, GRIN-rod lens configuration is found to be advantageous because of its compactness and ease of alignment.

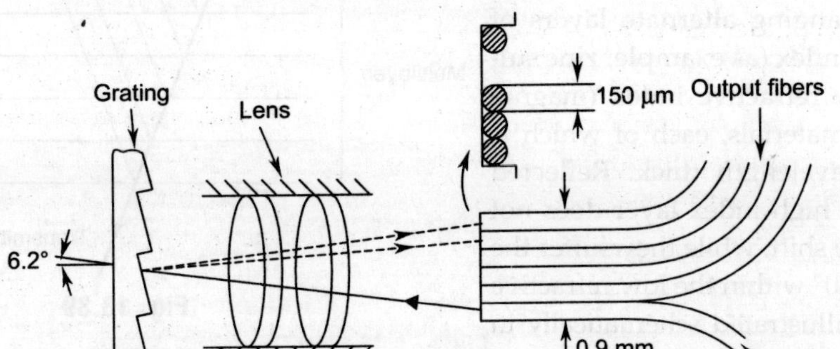

Fig. 13.38(*a*)

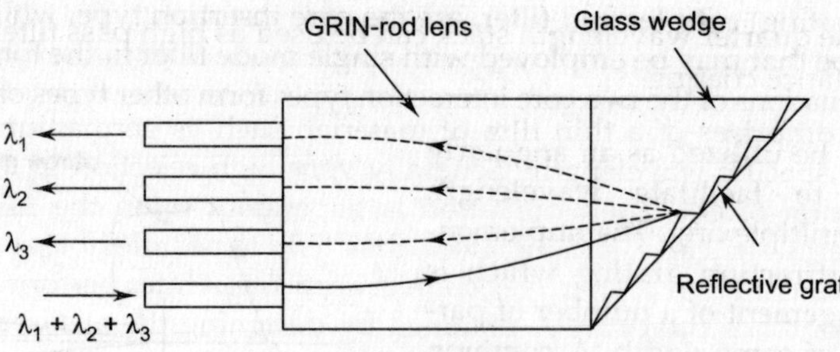

Fig. 13.38(*b*)

Referring Fig. 13.38(*b*), single input fiber and multiple output fibers are arranged on the focal plane of a quarter pitch GRIN-rod lens coinciding with the fiber end face. The multiplexed input optical beam is collimated by the lens and transmitted to the diffraction grating, which is offset at blaze angle so that the incident light is virtually normal to the groove faces. The required offset angle may be produced by incorporating a prism (glass wedge) between the lens and grating surface. Alternatively, GRIN-rod lens may be processed with grating on its end face after cutting and polishing. The former method exhibits superior performance as the optical properties of the GRIN-rod lens are unaltered. On reflection from grating, the light is angularly diffracted according their optical wavelengths and travel through the lens and thereafter focused onto the different collecting output fibers. These type of device exhibits channel insertion losses of less than 2 dB and channel spacing of 18 nm with low crosstalk.

Single mode wavelength multiplexer and demultiplexer pairs are also fabricated by plannar diffraction technique with a lithium niobate waveguide structure. Typical performance characteristics for a six channels multiplexer/ demultiplexer, three over the wavelength region of 1275 nm to 1335 nm and three over the wavelength range of 1510 nm to 1570 nm, with this fabrication method has been observed of less than – 25 dB crosstalk, and insertion losses of 5 dB to 8 dB for multiplexing and 1 dB to 2.2 dB for demultiplexing.

The other major core interaction type WDM devices employ optical filter technology. There are two main categories of optical spectral filters — interference filters and absorption filters. Dielctric thin film (DTP) absorption filters may be constructed by arranging alternate layers of high refractive index (as example, zinc sulphide) and low refractive index (magnesium fluoride) materials, each of which is one quarter wavelength thick. Reflected light within the high index layer does not suffer any phase shift, while they suffer the phase shift of 180° within the low refractive index layer, as illustrated schematically in Fig. 13.39. The successive reflected beams recombine constructively at the filter front face, producing a high reflectance over a limited wavelength region that is dependent on the ratio between the high and low refractive indices. The reflectance changes abruptly to low value at outside this

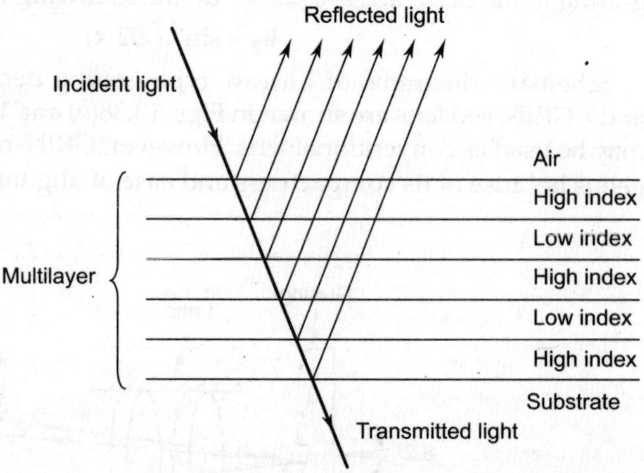

Fig. 13.39

reflectance region. The quarter wavelength stack can be used as high pass filter or low pass filter as well as high reflectance coating.

Absorption filter comprises of a thin film of material, such as germanium, that exhibits an absorption edge at a specific wavelength. Generally very high rejection in the cutoff region is observed for absorption filters. Their operation is dependent upon the fundamental optical properties of the material structure, and therefore, they tend to be inflexible as the edge positions are fixed. However, combination of sharp rejection of absorption filter performance as well as the flexibility of the interference filter can be obtained by designing the absorption filter with the fabrication of interference filters on to an absorption layer substrate. High performance edge filters utilize such combined structures.

Edge filters are generally used for the requirement of separation of two wavelengths differed by 10% or more of median wavelength. One of the filter configuration adopt the technique of interposing the edge filter between two fiber ends, when fiber is cleaved at a specific angle, as shown in Fig. 13.40(*a*). in a demultiplexer structure, light of one wavelength is reflected by the filter while light beam other optical wavelength is transmitted through the filter and then propagates down the cleaved fiber. In such device, the insertion losses are observed to be 2 dB to 3 dB with crosstalk levels less than – 60 dB using LED sources of centre wavelengths of 755 nm and 825 nm. In alternative arrangement WDM device, a cascaded bandpass filter may be incorporated between two GRIN-rod lenses as shown in Fig. 13.40(*b*). This design is capable of two channels multiplexer/demultiplexer operation of wavelengths of 1.2 μm and 1.3 μm at either side, with low insertion loss of around 1.5 dB and crosstalk level less than – 58 dB. This device also exhibits environmentally stable performance with the variation of less than 0.3 dB in insertion loss.

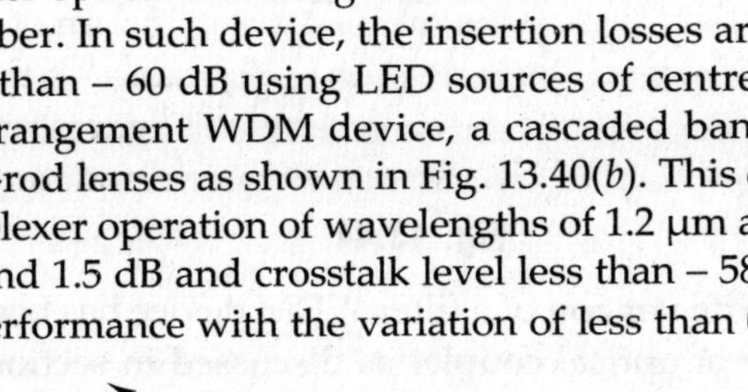

Fig. 13.40(*a*)

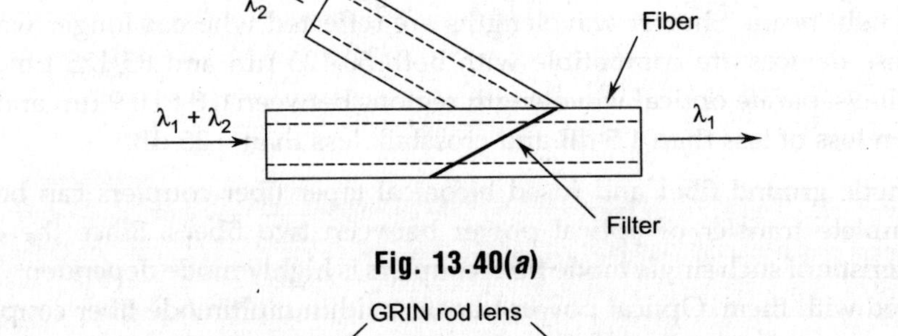

Fig. 13.40(*b*)

Multiple wavelength multiplexer/demultiplexer devices employing DTF interference filters may be fabricated from a suitably aligned series of bandpass filters. These bandpass filers are of different passband wavelength regions, cascaded in such a way that each filter transmits a particular wavelength, but reflects all others. Figure 13.41 illustrates such type multiplexer device. This structure has the disadvantage that with the increase of number of multiplexed channels the insertion losses increase linearly, as losses are incurred at each successive reflection due to filter imperfections and the difficulties of maintaining good alignment.

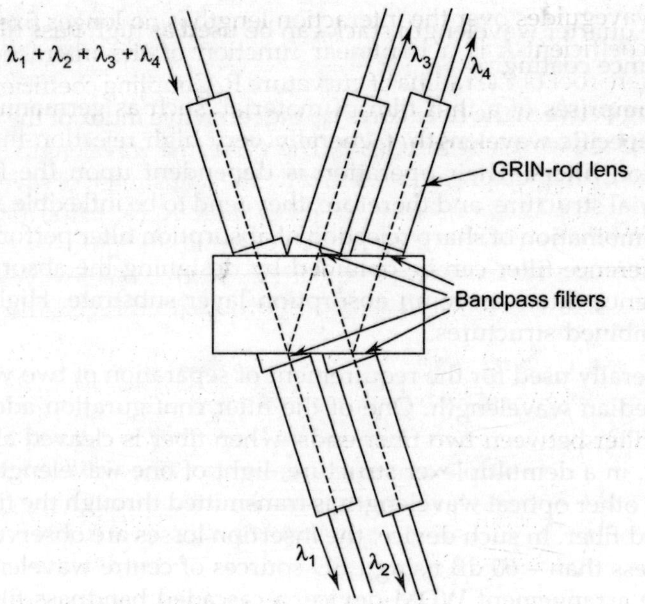

Fig. 13.41

A two channel slab waveguide version of a filter WDM device has been developed by Corning based on the same technology of optical coupler as discussed in section 13.6.1. The wavelength separation is accomplished within the waveguide using a diochroic filter that interacts the path of the incoming light beam. Shorter wavelengths are reflected whereas longer wavelengths are transmitted. These devices are compatible with both 50/125 μm and 85/125 μm graded index fibers and combine/separate optical wavelength regions between 0.8 to 0.9 μm and 1.2 to 1.4 μm with the insertion loss of less than 1.5 dB and crosstalk less than – 25 dB.

Both single mode ground fiber and fused biconical taper fiber couplers can be fabricated to provide the complete transfer of optical power between two fibers. Since the optical power coupling characteristic of such single mode fiber couplers is highly mode dependent, WDM devices can be constructed with them. Optical power transfer within multimode fiber couplers is a mode dependent phenomenon that generally takes place between the higher order modes propagating in the outer portions of the fiber cores as well as in the cladding regions. These higher order modes couple more easily with other fibers of close proximity. Consequently, the spectral dependence of light transfer within the multimode fiber couplers is far less pronounced and predictable than that of single mode fiber structures. Therefore, using fused biconical taper or ground fiber techniques, multimode fiber WDM devices are difficult to be fabricated.

Optical fibers may be coupled between two single mode fibers when they are brought close together over a region that is known as interaction length. In one of the techniques, the fibers are bend and fixed into two blocks of suitable material. (as example, quartz) followed by grinding of two blocks down so that a proportion of each of the fiber cladding regions worn away. Finally two blocks are brought together as shown in the Fig. 13.42.

Parallel single mode fiber waveguides exchange optical energy with a spatial period (coupling length) $L = 2\pi/k$, where k is the coupling coefficient (k has the unit of inverse length), for the two interacting waveguide modes. This relation holds good at the curved regions also, where the spacing between the waveguides over the interaction length is no longer fixed. Hencee, at curved region the coupling coefficient k is a nonlinear function of the interaction length, and it is proportional to the square root of the radius of curvature R. Coupling coefficient is also the function of the minimum spacing between the fiber cores d_0, the refractive index of the intervening material, the fiber parameters and the wavelength of light. Therefore, the wavelength dependent properties of a single mode ground fiber coupler can be altered by the adjustment of several different parameters. As example, referring Fig. 13.42, offsetting the cores laterally, *i.e.*, in the direction z or altering d_0, the spectral characteristics of coupler can be changed. The coupling ratio (optical power transferred between two fibers) follows a typical sinusoidal pattern. Interchannel wavelength spacing of 140 μm with insertion losses as low as 0.1 dB may be achieved with this construction of couplers.

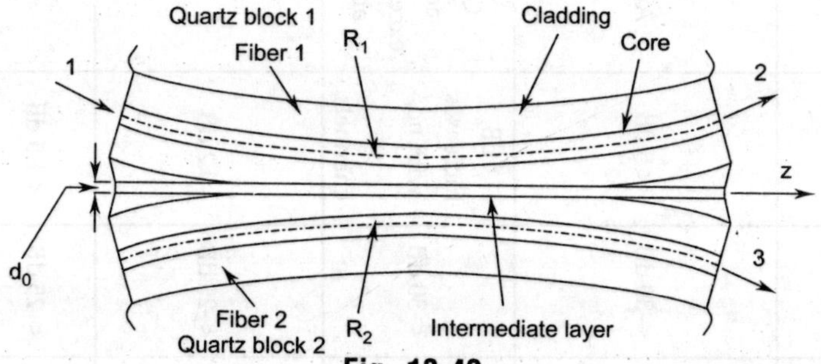

Fig. 13.42

Since the propagation constants of wavelength selective ground fiber directional couplers constructed from single mode fibers of different core diameters and refractive indices are matched at only one wavelength, they can be used as true bandpass filters. The fiber parameters determine the central wavelength and spectral bandwidth of these couplers. Typical insertion losses in longer wavelength operation are measured between 0.5 dB to 0.6 dB with crosstalk levels less than – 22 dB.

Fused Biconical Taper (FBT) technique is another method of fabricating a single mode fiber WDM coupler. Single mode fibers are fused together at around 1500°C before pulled while the heat is still applied. The pulling process decreases the fiber core size causing the evanescent field of the transmitted optical signal to spread out further from its core, thus enabling light to couple into the adjacent fiber. This manufacturing process requires monitoring of the optical power output from two fibers, and it is discontinued when the required coupling ratio is reached. Careful fabricated fused couplers in this process exhibit very low insertion losses and a high degree of environmental stability.

The coupling ratio of FBT couplers as a function of wavelength is sinusoidal with a period dependent on the dimensions and geometry of the fused cross section, and on the refractive index of the surrounding medium. Closeness of fibers at fused region is also a parameter to determine the coupling ratio. If the fiber cores are too close, it is said as strongly fused, and for relatively far apart cores it is weakly fused. Extension of the interaction region of the fibers at the time of fusing is one of the popular methods of varying the periodic coupling function. Such two channel devices exhibit insertion losses of 0.25 dB to 0.37 dB with crosstalk levels lower than – 22 dB. It may be noted that these WDM couplers are not well suited for the provision of closely spaced or multiple channels.

13.7 CONCLUSION

The generalized performance characteristics of various passive wavelength division multiplexing couplers is provided in table 13.1. It may be noted that data given in the table signifies the characteristics of typical WDM devices which is guide to the reader and not a definite statement.

Table 13.1 Generalized characteristics of passive wavelength division miltiplexer/demultiplexer

Device Type	*Mechanism*	*Implemen-tation*	*Channel Bandwidth*	*Channel Separation*	*No. of Channels*	*Crosstalk Level*	*Insertion Loss*	*Environmental Stability*	*Drawbacks*
Prism	Angularly dispersive	Bulk optic, difficult/ expensive	10–20 nm	20 nm	Multiple		< 10 dB	Acceptable	Optical material produce low values of angular dispersion and bulky
Diffraction grating	Angularly dispersive	Bulk optic, Straight forward Compact	Min 10 nm Dependent on fiber type and dimension, grating line constant	< 40 nm	Multiple, but usually in one wavelength region	<–30 dB	< 5 dB	Acceptable	High insertion loss for equal fibers. To be used with particular filters.
Filter	Interference of absorption	Bulk optic, Straight forward Compact	Min 2.5 nm at 850 nm, 4 nm at 1300 nm	10 nm or more	Two for single fiber, Cascading requires for multiple	<–30 dB	1 dB, Increases with no. of channels	Generally acceptable except moisture absorption	Cascaded filters increases insertion loss
Fiber directional coupler	Waveguide dependent power transfer between adjacent fibers	Fused biconical Ground fiber Compact	Generally periodic, large at around 40 nm	100 nm	Two for single fiber, Cascading requires for multiple	<–20 dB	0.2 dB	Good	Only single mode fiber type device. Ground fiber type requires careful fabrication.
Integrated waveguide	Slab	Integrated Very compact	100 nm	Wide 350/1300 nm	Two	<–25 dB	< 1.5 dB	Good	Commercially available
	Multimode	Do not require lens		5 nm for grating type	Multiple	<–20 dB	2 – 5 dB	Good	Generally commercially not available

REVIEW QUESTIONS

1. What type of optical fibers can be drawn from a solid perform (formed by collapsing a solid rod or hollow tube deposited by vapour-phase oxidation method)?
   (a) Multimode SI fibers (b) Multimode GI fibers
   (c) Single mode fibers (d) All of these
2. What is effect in refractive index (RI) for the increase in concentration of GeO_2 in SiO_2 ?
   (a) Decrease in RI (b) Increase in RI
   (c) No change in RI (d) Change in RI randomly
3. What is the placement of strength number in a multifiber cable?
   (a) Along the central axis of the cable
   (b) In coaxial cylindrical configuration
   (c) Anywhere within the cable
   (d) Not required
4. If an air gap is introduced while splicing two compatible fibers with core indices of 1.46, what is the loss at the joint due to Fresnel reflection?
   (a) Zero (b) 0.154 dB (c) 0.309 dB (d) 0.36 dB
5. If two optical fibers with numerical apertures of 0.17 and 0.20 are spliced, what will be the loss at the joint at forward direction?
   (a) Zero (b) 1.41 dB (c) 1.82 dB (d) 2.50 dB
6. When two Si fibers of 62.5/125 μm and 50/125 are spliced (both have the core indices of 1.5), what will be the joint loss at forward direction?
   (a) Zero (b) 0.97 dB (c) 1.94 dB (d) 2.45 dB
7. When two optical fibers with numerical apertures of 0.17 and 0.20 are spliced, what will be the loss at the joint at backward direction?
   (a) Zero (b) 1.41 dB (c) 1.82 dB (d) 2.50 dB
8. Two identical multimode SI fibers with core RI of 1.50 are spliced together. If the splice loss is 0.7 dB mainly due to 5°angular misalignment of the fiber axes, what is the numerical aperture?
   (a) 0.17 (b) 0.21 (c) 0.28 (d) 0.30
9. Which of the following mechanical splices will be most suitable to splice two optical fibers of different diameters?
   (a) Snug tube splice (b) Loose tube splice
   (c) Spring groove splice (d) V-groove splice
10. What is the use of OTDR?
    (a) To know the location dependence attenuation
    (b) To know the overall link length
    (c) To detect splice and connector losses
    (d) all of the above
11. (a) Distinguish between a splice and a connector.
    (b) How the Fresnel reflection loss at joint can be avoided?
    (c) Distinguish between fusion and mechanical splicing of optical fibers. Discuss the advantages and drawbacks of these techniques.
12. A broken multimode step index fiber is spliced with fusion process. The splice shows a loss of 0.36 db which is mainly due to an air gap. Calculate the refractive index of the fiber core.

13. A multimode step index fiber with core refractive index of 1.46 is spliced with fusion process. Assuming the entire joint loss of 0.6 dB is due to 4° misalignment of core axes, calculate the numerical aperture.
14. Discuss the design of multifiber cables that employ (*a*) a central strength member, (*b*) a structural member that also act as a strength member (*c*) fiber ribbons.
15. Discuss, with suitable diagrams, the three types of fiber to fiber misalignment which may contribute to insertion loss at a joint.
16. Discuss, with suitable diagrams, the following techniques of mechanical splicing (*a*) snug tube splice, (*b*) spring groove splice, (*c*) elastomeric splice.
17. (*a*) Describe the double crucible method for producing optical fibers. What are the limitations of this method?

    (*b*) Distinguish between the outside vapour phase oxidation method and inside vapour phase oxidation for manufacturing optical fibers. Compare the salient features of both the methods.
18. A mechanical splice of multimode step index fiber has a lateral offset of 12% of the fiber core diameter. The refractive index of the core is 1.5 and index matching fluid of refractive index of 1.47 is inserted in the splice between the fiber ends. Assuming there is no other losses, determine the insertion loss of the splice.
19. Discuss, with suitable diagrams (*a*) ferrule connector, (*b*) expanded beam connector.
20. How can the electro-optic effect be used for coupling between two parallel waveguides in an integrated-optical device?

14

Optoelectronic Integrated Circuits

14.1 INTRODUCTION

Optoelectronic technology and devices have made tremendous progress in the recent past. In the previous chapters, the materials used for optoelectronic devices and their properties as well as the working principles have been discussed. The role of optoelectronic devices have been increased in manifolds, — from sensors to display to active components in information handling systems at communication, data transfer, modulation, computing systems, etc. optical systems provide the inherent advantages of large bandwidth, parallelism and reconfigurable configurations. However, like electronic devices, optics does not provide the input-output isolation and it can be difficult to focus multiple beams in a parallel system. Therefore, combining of electronic with photonic devices is the logical approach, resulting in optoelectronic integration.

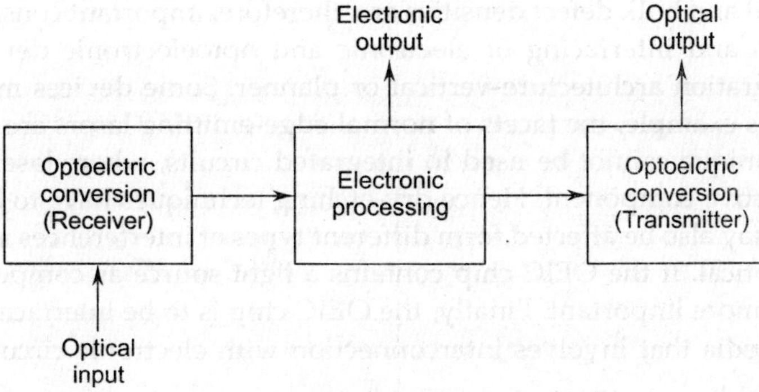

Fig. 14.1

Interconnection medium is one of the important aspects of both optical and computing systems. The conventional electrical medium for interconnection performs adversely, as it is affected by the increases in reactance and reflections due to impedance mismatch at higher frequencies. Though chip-to-chip communication may be achieved by the use of multilevel board technology through via-holes in the wafers, they fail to perform at higher operating frequencies in the range of gigahertz (GHz). An optical interconnect medium that may be of the form of free space or integrated optical waveguides or optical fibers is the alternative to enhance system performance.

Optical interconnects and transmission media provide large bandwidth and high speed data transmission, as well as provide immunity to mutual interference, crosstalk and capacitive loading effects. Reduced system size, reduced system power and increased fan out capabilities are achieved through optical integration.

Integration of electronic and optical components as well as the interconnect medium are therefore involved in optoelectronic integrated circuits or OEIC. The essential elements of an OEIC are shown as block diagram in Fig. 14.1. It is expected that the monolithic integration of electronic and optical components on the same chip will deliver high speed, high sensitivity, compactness and reliability at low cost. Many factors are to be considered for successful integration architectures. First, the different layer structures of the device, unlike the electronic components, must be of high quality. Secondly, compatibility and impedance matching between the devices and the devices with interconnect media must be considered. And lastly is the cost factor, which must not be unnoticed at all.

14.2 APPLICATIONS OF OPTOELECTRONIC INTEGRATED CIRCUITS

Silicon based microelectronics technology provides advantages of low cost, large scale integration and ruggedness for the complex electronic circuitries like signal processing functions in communication and computing applications. Similarly photonic devices and circuits can serve unique functions that are complementary to those electronic counterparts. Integration of electronic and photonic devices and circuits can also provide new functional capabilities.

For successful integration of hybrid approach, first and foremost requirement is to develop high performance discrete devices with minimized parasitic effects ensuring desired frequency performance. Such high performance devices, both electronic and optoelectronic, like transistors, lasers, detectors, etc. are available. However, it is challenging task to obtain monolithic integration due to dissimilar material requirements. The individual devices are to be fabricated with non-optical materials and lattice-mismatched heteroepitaxy, or may be realized by epitaxial regrowth. Interfacial and bulk defect densities are, therefore, important considerations, as well as the interconnection and interfacing of electronic and optoelectronic devices, which depend closely on the integration architecture-vertical or planner. Some devices may have to be fabricated differently. As example, the facets of normal edge-emitting lasers are created by cleaving. Obviously this technique cannot be used in integrated circuits, where laser is coupled on-chip with a passive or active component. Hence dry-etching techniques have to be used to create the laser facets. OEIC may also be affected form different types of interferences and cross talk-electrical, thermal and optical. If the OEIC chip contains a light source as component device, optical cross talk becomes more important. Finally, the OEIC chip is to be interfaced with other chip or the transmission media that involves interconnection with electronic circuitry.

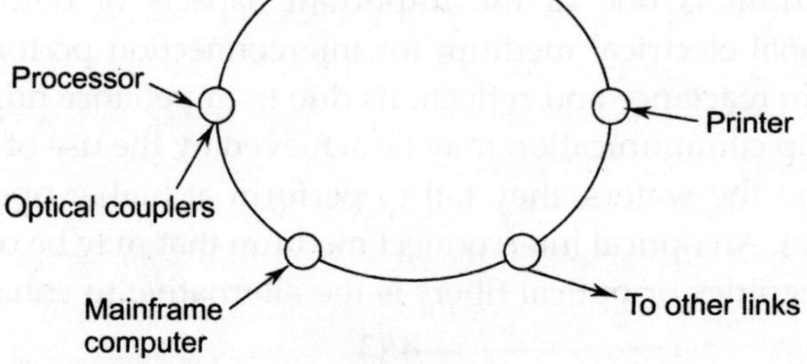

Fig. 14.2

Optoelectronic devices and OEIC have made a great impact in the field of telecommunication, where large bandwidth and lightweight optical fibers are necessary. Optical fiber systems and optoelectronic technologies have extended its application for the domestic and individual subscribers in the form of telephone links and broadcast cable television. These systems require lasers with precise frequency control, tenability and wavelength selective detectors and receivers. Data transmission rate for these systems and circuit is as high as several tens of gigabits/second. Similar technology is applicable for optical computing systems to interface several input/output ports as shown in Fig. 14.2. However, due to shorter links used at the optical computing system, GaAs-based technology may be used without incurring much of loss in the optical fibers. Since such systems are essentially digital systems, Bit Error Rate (BER) must be very low (less than 10^{-13}) even in most noisy environment, and they should have high reliability and redundancy in critical paths with transmission error correcting capability. Architecture and interfacing of active and passive components are also critical issues in this application.

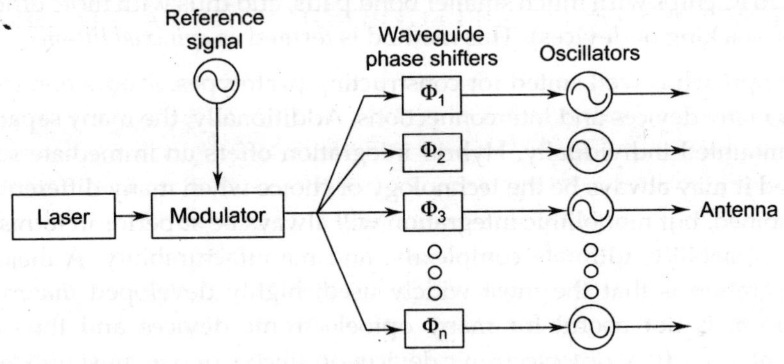

Fig. 14.3

Optoelectronic integration also finds its application in the field of radar systems. Figure 14.3 represents the schematic of a microwave or millimeter wave phased-array radar system. Phased-array radar system involves the elements of phase-shifted and modulated high frequency optical signals that are injection locked to free-running microwave or millimeter-wave oscillator. Beam steering of such radars is achieved by incorporation of a progressive phase shift between successive elements. It is more efficient to use optical phase shifters than to use conventional phase shifters at microwave frequencies that are bulky and cumbersome. Each element of the phased array consists of a source, a modulator, a phase shifter, waveguide and a oscillator, and they can be combined with hybrid integration for compactness, space reduction and ruggedness. However, it is more desirable to realize by monolithic integration, resulting in development of OEIC chips, though it is extremely complex, and they can be used on the ground, on aircrafts and in satellites.

14.3 OPTOELECTRONIC INTEGRATION TECHNOLOGIES

In optoelectronic devices, the electron and photons interact effectively complementary and collaboratively in direct bandgap semiconductor to produce optoelectrnic conversion. The idea of optical integrated circuit was suggested from the analogy of electronic integrated circuit, but with some optical devices connected by optical wiring on the same substrate or chip. Optoelectronic integrated circuit or OEIC combines the basic optoelectronic device with some electronic functions in same package to enhance speed and bandwidth as well as to minimize loss and interference. Technologies for integrating optoelectronic devices and electronic circuitry can be classified as either hybrid or monolithic. Hybrid integration, as its name suggests, involves combining optoelectronic devices and integrated circuits in the same package or substrate. This can be done using traditional hybrid techniques for simply combining packaged devices on a ceramic substrate. An advantage of such hybrid integration is the possibility of using high performance discrete decives as components. But the device density of the resulting optoelectronic integrated circuit (OEIC) is very low and many of the advantages of using optics are lost. Consequently, alternative hybrid techniques have been proposed to increase device density while maintaining manufacturability. Currently the most successful such techniques involve flip-chip/ solder-ball or -bump integration of discrete optoelectronic devices on multi-chip modules or directly on silicon IC chips, but even these techniques consume an inordinate amount of area and even higher densities are needed. A possible solution is the use of specially thinned discrete devices (and, ultimately, moderate sized arrays of devices) which can be flip-bonded to IC chips with much smaller bond pads, and thus with more efficient use of space (and even some stacking of devices). This method is termed as *epitaxial lift-off*.

The hybrid approach is well suited for constructing prototypes. Production can be costly if the circuit contains many devices and interconnections. Additionally, the many separate devices must be tested and mounted individually. Hybrid integration offers an immediate solution to certain OEIC needs, and it may always be the technology of choice when many different kinds of devices need to be combined, but monolithic integration will always be superior in terms of speed, device density, system reliability, ultimate complexity, and manufacturability. A major difficulty with monolithic integration is that the most widely used, highly developed material for integrated electronics, silicon, is not useful for many optoelectronic devices and thus one must either monolithically integrate III-V optoelectronic devices on silicon, or one must utilize III-V electronics.

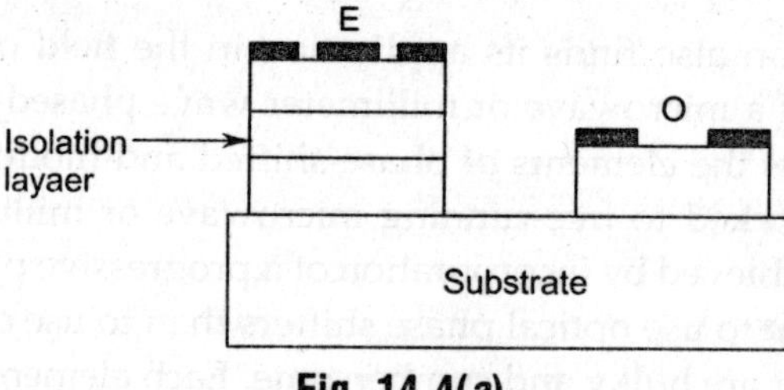

Fig. 14.4(*a*)

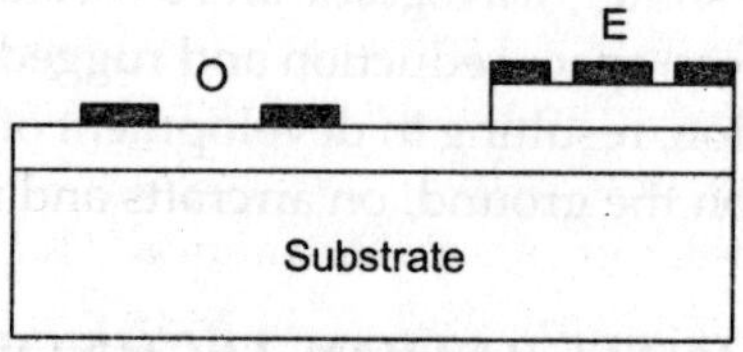

Fig. 14.4(*b*)

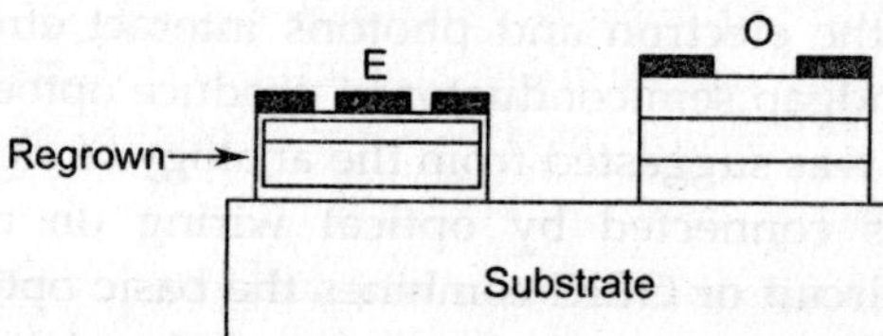

Fig. 14.4(*c*)

Monolithic integration can be achieved in either a vertical or a horizontal configuration. In the vertical configuration, electronic and optical device structures are epitaxially grown sequentially with an isolation layer in between. This is schematically shown in Fig. 14.4(*a*). Lack of planarity is the obvious disadvantage of this scheme. The horizontal scheme of monolithic integration may be classified into the planner compatible type, as shown in Fig. 14.4(*b*), where both devices are developed from same heterostructure, or planner regrown as shown in Fig. 14.4(*c*), where one of the device is selectively regrown after the growth of first device. This technique provides a large freedom in the choice of heterostuctures. However, a large density of traps and other electrically active defects may be produced in the regrown interface, which can affect the performance of the regrown device.

The choice of materials and the epitaxial techniques used to synthesize the OEICs are governed by same requirements as those for high performance discrete devices. These are operating wavelength, lattice matching considerations and the choice of component devices. GaAs based technologies and related materials are matured enough and they are widely used at local area networks, computer interconnects and optical information processing. InP based materials are more suitable for the application of long distance communication. Another approach to use of materials for OEIC is heteroepitaxy, or the use of mismatched materials that include III-VI or II-VI compounds on similar semiconductors, or GaAs and InP based compounds on Si. There are potential advantages and disadvantages in this type of heteroepitaxy. The greatest problem in these mismatched materials is the large density of misfit dislocations that are created at the heterointerface and are propagated upward as threading dislocations into active regions of the device. In spite of this shortcoming, because of low cost and large density Si electronic circuits and devices, this technique is used to realize high performance OEICs, with plausible configuration having pockets of III-VI optoelectronic devices in the comparatively larger area of Si VLSI. Similar reasoning is also appropriate for heteroepitaxy of GaAs and InP substrates, and encouraging results are obtained from the devices made with these materials.

Monolithic integration of InGaAlAs and InGaAsP heterostructures on silicon (a process generically called GaAs-on-Si is attractive because it would allow one to make use of the wealth of silicon BiCMOS electronics integrated circuit technology. Unfortunately, III-V epitaxy on silicon (or vice versa) is made difficult by the facts that the lattice constants of GaAs and Si differ by 4%, and that the thermal expansion coefficients differ by almost 50%. The lattice mismatch can be overcome by special growth initiation techniques and interfacial buffer layers, but the thermal mismatch is more troublesome. Significant numbers of defects (and even cracks) are introduced as GaAs-on-Si structures are cooled from the growth temperature to room temperature, and the III-V layers are under severe tension, which reduces device lifetimes dramatically, reduces reliability, and in general adversely effects performance. The most promising solution to this incompatibility involves introducing an AlAs layer between the growth-initiation layers and the active device heterostructures, and subsequently etching this AlAs layer away to physically separate the III-V device structure from the Si, but otherwise leave it monolithically integrated on the chip. The separated structure can be thermally annealed to remove the thermally induced strain and defects (significantly improving the material quality and device performance), and more research work in this area is under process.

The difficulty of monolithic integration on silicon, and steady advances in the levels of electronic integration demonstrated on III-V substrates have led to a shift of emphasis to monolithic integration on either gallium arsenide (GaAs) or indium phosphide (InP). Two approaches are being pursued. The first family of techniques, which we will call "epitaxy-first" or "stacked-heterostructure" methods, involve first growing all of the epitaxial layers needed for both the optoelectronic devices and electronic devices on a suitable substrate. The top layers are then removed selectively in areas where devices are to be fabricated on lower layers and each of the various device types is formed in the appropriate layers. In variations on this idea, some of the epitaxial layers are grown, they are etched away where they will not be needed, and more epitaxial layers are grown in newly cleared areas; this approach results in a more planar surface but the growth and processing are much more complex. An important problem with the epitaxy-first approach is that pursuing it involves simultaneous development of sophisticated technologies for optoelectronic devices and integrated circuits. The task is enormous and as a practical matter only small scale OEICs have been demonstrated following this approach.

While the integration levels of epitaxy-first III-V OEICs have remained low, the degree of integration in commercial GaAs electronic integrated circuits has reached VLSI levels in recent years, and the technologies used in this integration use only refractory metals in contact with the semiconductor, and use Si IC upper-level interconnect and passivation technologies. These advances offer a route to achieving much higher levels of optoelectronic integration through epitaxial growth of III-V heterostructures on GaAs-based VLSI electronics. This is essentially the GaAs-on-Si idea without the myriad of materials problems inherent in growing the III-Vs on silicon, and with all the advantages of building on a commercial integrated electronics foundation. This approach is called *epitaxy-on-electronics* (*epi-on-electronics*, for short) that is most promising for realizing high density, high performance optoelectronic integrated circuits.

14.4 INTEGRATED TRANSMITTERS AND RECEIVERIES

The key element of an optical communication system is receiver. The basic purpose of the receiver is to detect the incident light and to convert it into electrical signal containing information impressed on the light at the transmitting end. The important performance characteristics of a photoreceiver are operating bandwidth and sensitivity. Sensitivity plays an important role in deciding the number of repeaters needed in a long distance communication system. The receiver sensitivity is defined as the minimum amount of optical power level needed at the receiver input so that the signal to noise ratio is greater than a given value. In digital communication system, this translates as the bit error rate that is to be less than a given value. For some specific applications, other features such as dynamic range become vital. Block diagram of a photoreceiver circuit is shown in Fig. 14.5, which consists of photodetector, a low noise amplifier, equalizer circuit and lastly pulse shaping, gain control and other decision-making circuits.

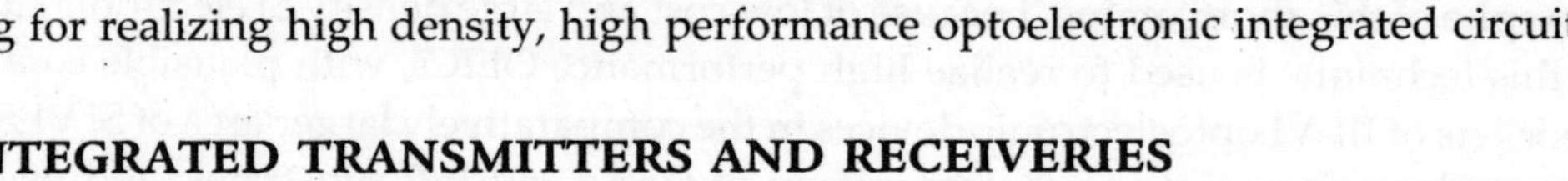

Fig. 14.5

In most photoreceiver circuits, multiple stages of amplification are included to increase the gain of the signal. The overall performance of the circuit is mostly dependent on the photodetector and the pre-amplifier at the input stage. The desirable features of the photodetector are high quantum efficiency, low capacitance, small response time and low dark current. Avalanche photodiodes, PIN photodiodes and Metal-Semiconductor-Metal (MSM) photodetectors are among the most commonly used photodetectors for this application. Low noise and high output gain are essential for the preamplifier sections, as well as the leakage current and parasitic capacitance are to be minimized. Both field-effect transistors and bipolar transistors are used in the preamplifier circuits.

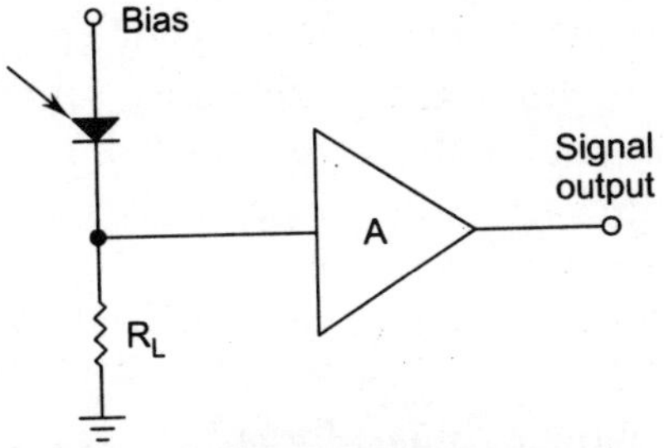

Fig. 14.6(*a*)

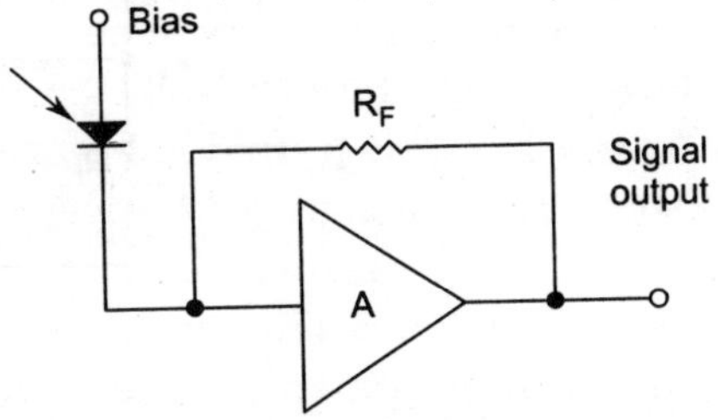

Fig. 14.6(*b*)

The selection of the integration scheme for the photodetector and the low-noise preamplifier circuit is also very important in determining the overall performance of the photoreceiver. Three types of integration are commonly used-low input impedance type, high input impedance type and transimpedance type, and the latter two are mostly used. The schematics of low input impedance type and high input impedance type are similar, and shown in Fig. 14.6(*a*), whereas Fig. 14.6(*b*) represents the schematic of transimpedance design. The high input impedance design requires additional equalization circuit to extend the value of the bandwidth due to the large RC constant. Transimpedance design is most popular as no equalization is generally required. A field-effect transistor may also be employed as active feedback element, replacing the passive resistor and enhancing the impedance value to as high as 80 KΩ, resulting in significant reduction of thermal noise.

Figure 14.7(*a*) represents a typical schematic diagram for a monolithic integration technique of $In_{0.53}Ga_{0.47}As$ photodiode with a $In_{0.53}Ga_{0.47}As/In_{0.52}Al_{0.48}As$ modulation doped FET (MODFET) by regrown on InP, and its equivalent circuit is represented in Fig. 14.7(*b*). In this integration scheme, the PIN diode is grown in the first epitaxial step and MODFET is regrown. The MODFET consists of a layer of undoped low-bandgap material forming a heterojunction with highly doped high-bandgap material. Due to difference in the electron affinities of the two layers, electrons move from high bandgap material to low bandgap material forming a quasi-two-dimensional electron gas (2DEG).

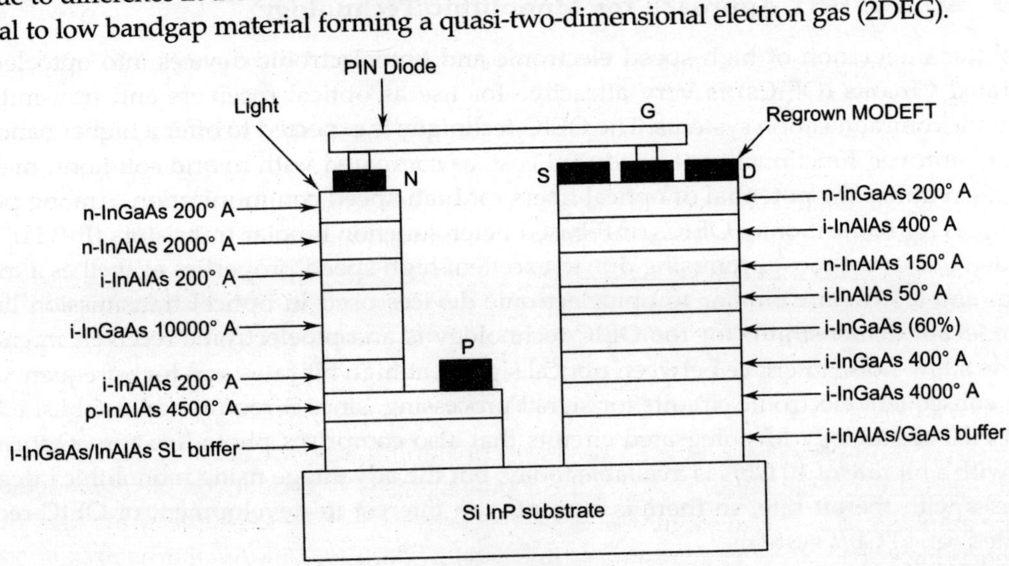

Fig. 14.7(*a*)

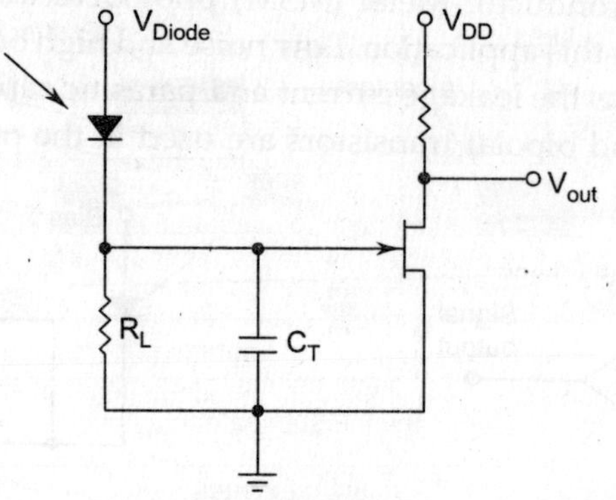

Fig. 14.7(*b*)

The main advantage of such structure is that the electrons are separated from their parent donors and the coulombic scattering is greatly reduced, resulting in higher carrier mobility and drift velocity. A high bandgap undoped spacer layer is typically added between the highly doped high bandgap layer and the undoped low bandgap layer to further separation of electrons from parent donors. The MODFET structure has an advantage in the proximity of the channel to the gate, typically less than 500°A, which becomes important as the gate length is shrunk to submicron dimensions. Also MODFET exhibits lower noise figure and noise temperature compared to other FETs. The InGaAs/InAlAs/InP structure is considered superior to GaAs/AlGaAs structure for MODFET due to large conduction band discontinuity (ΔE_c = 0.5 eV) enabling increased confinement of carriers in the quasi two-dimensional electron gas, higher mobility (greater than 10000cm^2/Vsec) and peak velocity in InGaAs layer. Device performance may be further improved using $In_xGa_{1-x}As$ channel layer, resulting in even higher mobility and peak velocity (approximately 15% increase for $x = 0.65$). Pseudomorphic $In_xGa_{1-x}As/In_{0.52}Al_{0.48}As/InP$ MODFETs have demonstrated extrapolated frequencies greater than 200 GHz and maximum oscillation frequencies greater than 400 GHz, with noise figure as low as 1.4 dB and associated gain of 6.6 dB at W-band.

14.4.1 An INP-HBT Approach for Monolithic Technology

Monolithic integration of high-speed electronic and optoelectronic devices into optoelectronic Integrated Circuits (OEICs) is very attractive for use as optical receivers and transmitters in fibre-optic communication systems. The OEIC-technique is expected to offer a higher bandwidth and an improved functionality at a reduced cost, as compared with hybrid solutions, making it possible to utilize the potential of optical fibers for high-speed communication. Among possible techniques for realization of OEICs, InP-based heterojunction bipolar transistors (InP-HBTs) are regarded one of the most promising due to excellent high-speed properties as well as a material system and a structure similar to optoelectronic devices used in optical transmission links. A very relevant demonstrator for the OEIC-technology is an optoelectronic receiver, intended to provide a low-noise interface between optical signals at high bit rates and high-frequency input of the subsequent electronic circuits for signal processing, since it requires an established high-speed HBT-technology for integrated circuits that also comprises photodetectors. Optical fiber links with a bit rate of 10 Gb/s is available today, but the advantage using monolithic integration increases with the bit rate, so there is currently an interest in development of OEIC-receivers intended for 40 Gb/s systems.

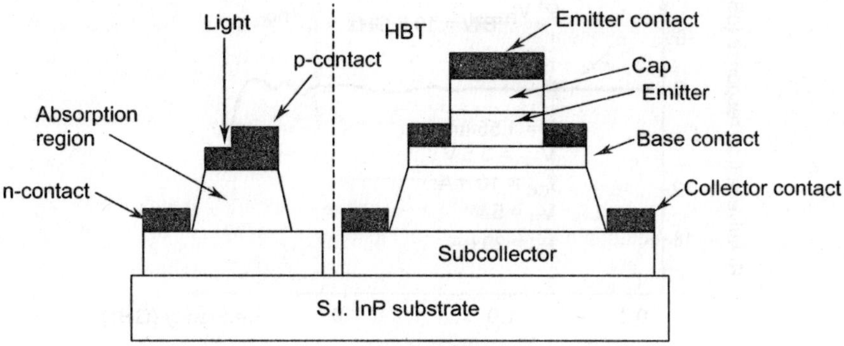

Fig. 14.8(*a*)

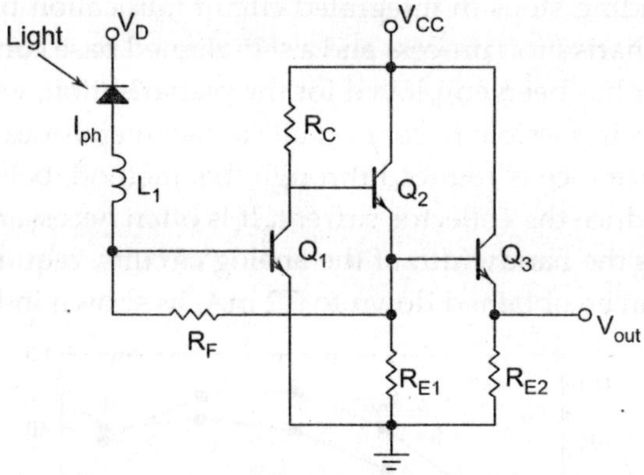

Fig. 14.8(*b*)

Heterojucntion Bipolar Transistor (HBT) can be monolithically integrated with *p-i-n* diode to form the integrated photoreceiver. The important point to realize is that the integration involves a single step epitaxy of the HBT serving as the preamplifier, from which the *p-i-n* diode as photodetector is selectively defined by processing. The collector region of the HBT also serves as the i-region of the diode. The two devices are then monolithically integrated with the addition of the required passive components. The other important reasons for using a PIN-HBT combination are the noise reduction and sensitivity improvement. It has been also noticed that the FET based photoreceiver further improves the sensitivity at higher frequencies. The sensitivity of PIN-FET receiver is proportional to B^3, where B is the bit rate, while the sensitivity of PIN-HBT is proportional to B^2.

The epitaxial layer structure of a transimpedance amplifier design is shown in Fig. 14.8(*a*), and its equivalent circuit with active and passive elements is given in Fig. 14.8(*b*). A valuable technique to enhance the photoreceiver response at high frequencies is inductive peaking, where an inductor is placed in series after the photodiode at the input of the amplifier. The resistors and inductors are usually formed with evaporated single or multilayered metals in the monolithic circuits. Typical values of peaking inductors are in the range of 2 nH to 5 nH. Figure 14.9 represents a typical graph for optical to electrical small signal frequency response of the photoreceiver circuit. A – 3 dB modulation bandwidth of 19.5 GHz is suitable for 24 Gbits/sec bit rate in digital application and better than hybrid circuits.

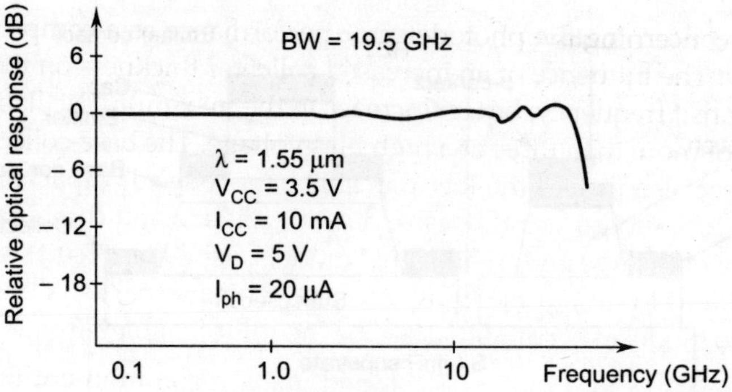

Fig. 14.9

Two of the most demanding steps in integrated circuit fabrication using high-speed InP-HBTs are to achieve a reliable planarisation process and a self-aligned base contact for small-area devices. A silicon nitride (SiN) layer has been employed for the planarisation, with a thickness equal to the device height, which makes it possible to connect all contact metals outside the active device area. The external parasitic capacitance is reduced through this method, being important when scaling down the device area to reduce the collector current. It is often necessary to use cutoff frequencies that are at least three times the bandwidth of the analog circuits, requiring more than 80 GHz for 40 Gb/s operations. This can be obtained down to 0.2 mA, as shown in Fig14.10.

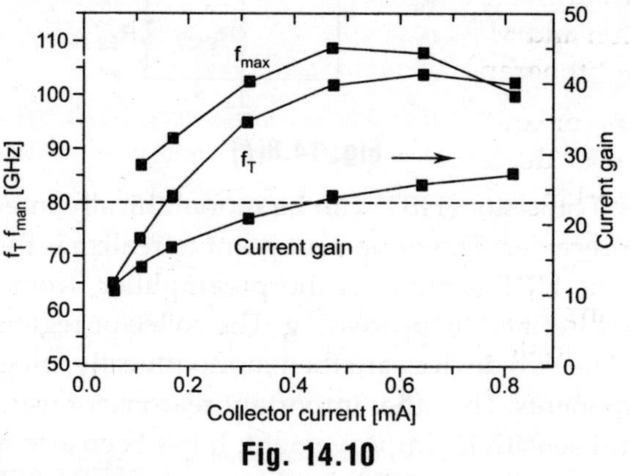

Fig. 14.10

A self-aligned base contact metal reduces both base resistance and base-collector capacitance, and is also very important for uniform values. The critical step is the required undercutting of the emitter metal, since the etch-rate is strongly crystal orientation dependent and requires excessive etching to produce sufficient undercutting in both directions. The SiN-planarisation technique makes it possible to protect both short sides of the rectangular emitter with an overlapping emitter metal during mesa formation. Complete self-alignment is then possible due to uniform undercutting of the emitter metal outside the active area, since the etch-rate of SiN is isotropic. The resulting cutoff frequencies have been demonstrated as $f_T = 94.7 \pm 2.6$ GHz, and $f_{max} = 95.1 \pm 4.4$ GHz, at base current of 15 μA and a collector-to-emitter voltage of 1.2 V.

Photodetectors can be fabricated from the PIN-structure formed by the base-collector-subcollector region of an HBT. This technique only requires that the emitter layer be removed, which can be done in the same process step as formation of the emitter mesas, so no extra steps are required to include the PIN-diodes in the fabrication process. The thickness of the absorbing *i*-layer must be chosen concerning the photodetector performance as a compromise between speed and quantum efficiency. The influence of an increased collector thickness on the HBT properties is a slight decrease in the transit frequency, but an increase in the maximum oscillation frequency, resulting in an f_{max} to f_T ratio of more than three at a high bias voltage. The base-collector delay $\tau_{BC} = R_B C_{BC}$ that approaches 0.05 psec also proves the low parasitic base-collector capacitance. S-parameters for both the HBTs and the PIN-diodes were measured up to 60 GHz and combined with DC-measurements in optimization of both large and small-signal models, using the circuit simulators HSPICE and EEsof Touchstone, required to design the OEIC-receiver. The internal serial resistance had to be included in the PIN-diode model to accurately describe the roll-off at high frequencies.

A transimpedance preamplifier may be designed using a common-emitter stage, two emitter followers, and a feedback resistor. The emitter areas are $2.5 \times 5.0\ \mu m^2$ for the common-emitter stage and the first emitter follower, and $2.5 \times 10\ \mu m^2$ for the second emitter follower. Both f_T and f_{max} were at least 100 GHz for the transistors in the common-emitter stage, at the chosen bias point. The photodiode is internally reverse-biased by the circuit to about 1.5 V. Necessary level-shifting diodes consisted of HBTs with the base shorted to the collector, and with $2 \times 2.5 \times 10\ \mu m^2$ using two emitter fingers. They were employed to prevent transistor avalanche breakdowns when using a single power supply. The size of the layout was $0.35 \times 0.5\ mm^2$ including a 50 pF decoupling capacitance, a DC-pad, and HF-probe pads for the output signal.

Epitaxial HBT-structures were grown on a SI-InP substrate, using Solid-Source Molecular Beam Epitaxy (SSMBE) and the HBTs were fabricated in our in-house process. Two metal layers are used for interconnection and MIM-capacitors, and a sputtered NiCr-layer for thin film resistors. Conventional optical lithography was used in all steps.

Optoelectronic receivers for several bitrates ranging from 10 to 40 Gb/s have been fabricated. The epitaxial structure was designed for 40 Gb/s, meaning that the sensitivity would be improved for receivers working at lower bitrates if a photodetector were employed that were optimised for the actual bitrate. The results prove the possibility to produce OEIC-receivers that provide a compact interface between high-speed modulated light and standard 50 Ω impedance high frequency electronic input, and the obtained bandwidth may be large enough for 40 Gb/s transmission, depending on system requirements. The 40 Gb/s receivers was not optimized for maximum speed, so a higher bandwidth is expected through a reduction of the feedback resistor value, on cost of an increased noise level.

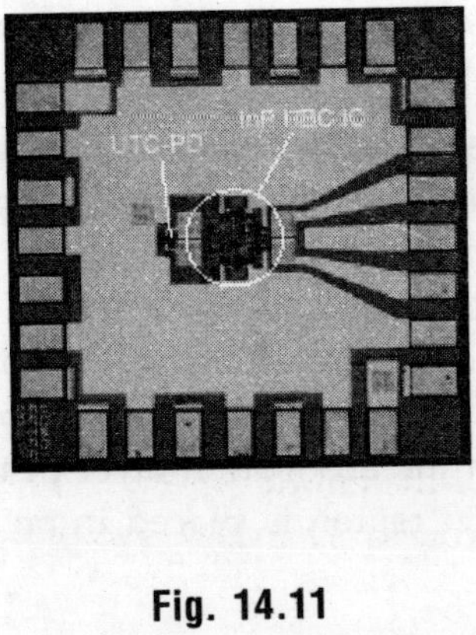

Fig. 14.11

(*a*)

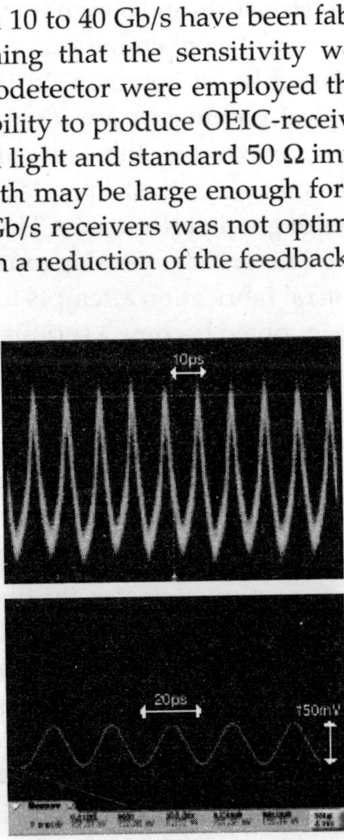

(*b*)

Fig. 14.12

To improve the OEIC performance in InP-HBT method, OEIC structures have been developed with non-selective regrowth technique and an encouraging result has been obtained. Figure 14.11 shows a chip photograph of optical input frequency divider OEIC using InP-HBTs with unitravelling-carrier photodiode (UTC-PD). The regrowth technique makes it possible to optimize the layer design for HBTs and UTC-PDs. The fabricated HBT has 0.8 μm wide emitter and 200 μm thick collectors. It has f_T of 260 GHz and f_{max} of 320 GHz at a collector current density of 1.5 mA/μm^2. The UTC-PD exhibits an output voltage of 1.0V, maintaining 3dB bandwidth at 100 GHz. Figure 14.12(*b*) shows the output of OEIC with an optical clock input as shown at Fig. 14.12(*a*). This result shows that non-selective regrowth technique is promising for the application over 100 Gbits/sec OEICs.

In recent years, there has been tremendous interest in the optoelectronic integrated circuit (OEIC) receivers for high performance optical communication systems. The performance of a photoreceiver is mainly controlled by its front-end components—a photodetector and a preamplifier. A monolithically integrated photoreceiver has advantages of low parasitics, compact size and low cost. Commonly used photodetectors are Metal-Semiconductor-Metal (MSM) photo-detector, PIN photodiode and avalanche photodiode. Among these, however, the first two are important for monolithic integration with amplifiers to form an integrated photoreceiver. State-of-the-art amplifiers used are based on the High Electron Mobility Transistors (HEMTs) and Heterojunction Bipolar Transistors (HBTs). Between MSM and PIN photodectors the former has an added advantage of high intrinsic bandwidth, ultra-low capacitance and easy planar integratibility with HEMTs and HBTs.

Various structures and designs of photo detectors and transistors are used to get high performance photoreceivers. To obtain the best possible photoreceiver performance, it is very essential that the design should be an optimum. In the design of integrated photoreceivers, various device and technological parameters are involved. For optimization purposes, these parameters are to be tailored by exploiting the fabrication technology and device design. In fact, experimental researchers may use their knowledge and insights to tailor the parameters directly to produce devices with good performance. However, this becomes a difficult task in most cases because the device parameters are inter-dependent and hence cannot be controlled independently. This might mean several fabrication attempts to produce optimized structures and devices, but now the cost and time involved become a serious concern. However, both the cost and time can be significantly reduced and the task simplified if model equations and inter-relations among the parameters are known. Then analytical and/or computational techniques can be used in an efficient manner to show how the device parameters can be tailored for structures with optimum perform ance. This approach can be a powerful guide to experimental researchers and device manufacturers in creating monolithically integrated photoreceivers for desired optimum performance.

To get a good design, it is always desirable that the devices are modeled as accurately as possible. For a photodetector, the two important parameters are the Quantum Efficiency (QE) and bandwidth (BW). In conventional surface-illuminated photodetectors, there is a trade-off between these two parameters-large thick-ness of the absorption layer means large transit time limited bandwidth whereas small thickness reduces the quantum efficiency. Resonant Cavity Enhanced (RCE) structures with a very thin active layer is a possible solution. The thin layer gives rise to large BW, and the multiple passes of light in the absorption layer in the resonant cavity increases the quantum efficiency. A thin absorption region is placed in an asymmetric

Fabry-Perot cavity. The top and bottom reflectors that can be fabricated by Distributed Bragg Reflectors (DBR) form the cavity. Another way of improving the QE as well as transit time limited BW is to use Edge-Coupled (EC) structures. Here, the requirements for high efficiency and high bandwidth are decoupled by illuminating the PD from the side of the absorption layer. Now, efficiency is a function of the length of the absorption layer, not its thickness. So, a thin absorption layer is good enough to get high efficiency at the same time increasing the transit time limited bandwidth. In the resonant cavity structure, the construction of a DBR mirror is a complex process while the present one is a simpler one. EC structure uses the attributes of the waveguides. These are also suitable structures for coplanar optical interconnects. The incorporation of Ge into the base of a silicon bipolar transistor can significantly increase the bipolar transistor performance. This makes it possible to consider the use of Si-Ge electronics to produce performance compatible with long-haul telecommunications requirements. It also provides access to a material with, meaningful absorption at much longer wavelengths than native Si. In particular it becomes possible to consider detector diodes with both increased absorption and longer wavelength sensitivity. Because of the 4% lattice mismatch between Si and Ge, the implementation of the receiver becomes more challenging at longer wavelength since the detector requires strained-layer epitaxial Ge_xSi_{1-x} films with higher Ge fraction. To get the best performance, optimum design of the OEICs is based on the representative models of its components is required.

14.4.2 OEIC Transmitters

An optical transmitter circuit includes a light source in form of high power LED or laser. Integration of the laser with the associated electronics, like driver circuitry consisting of transistor, is more complicated compared to the fabrication of a photoreceiver. This is because the laser has more stringent material and processing requirements than a photodetector. First, since the laser structure is nearly 4 μm high, the integration processing steps with compatible heterostructure for the electronic device is very difficult. Secondly, the optical cavity in an edge-emitting laser needs to be defined by two end mirrors. Third, electrical and optical confinement is required to be achieved in the lateral dimension. And finally, heat sink is necessary with the chip for the operation of laser. However, in spite of these disadvantages, it is worthwhile to integrate the transmitter device to ensure a higher modulation bandwidth, which is achieved by the close proximity of the devices and associated reduction of parasitics from the devices and interconnects. An OEIC transmitter using the same heterostructure for the laser and the driver transistor is shown in Fig. 14.13 However, to obtain the optimized performance of both devices, it is desirable to use separate optimized structures.

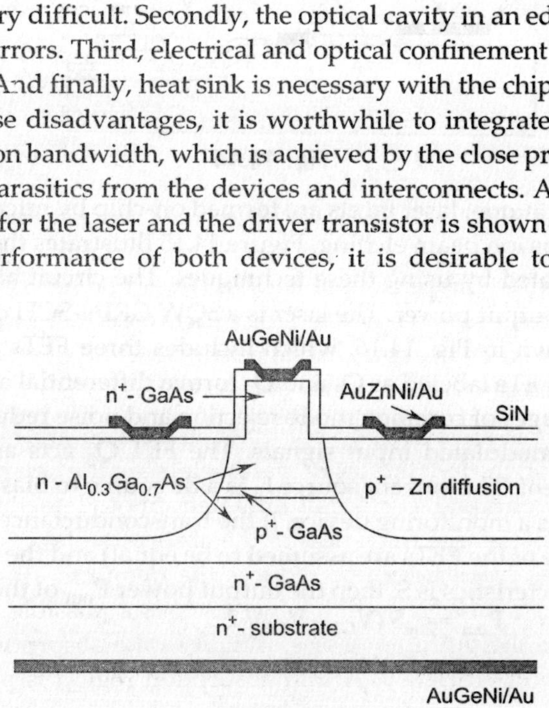

Fig. 14.13

Early generation transmitter circuits consisted of a single edge-emitting laser, whose facet was created by cleaving, integrated with a single transistor. The cross-section of a GaAs based laser-MESFET OEIC and steps involved in fabrication are shown in Fig. 14.14. At first step, n^+-GaAs layer is formed on SiGaAs substrate by groove-etching process. MQW structure is then fabricated, which is followed by the formation of n-layer of FET is with n-GaAs. At Final step, the ridge formation and contact formation are done.

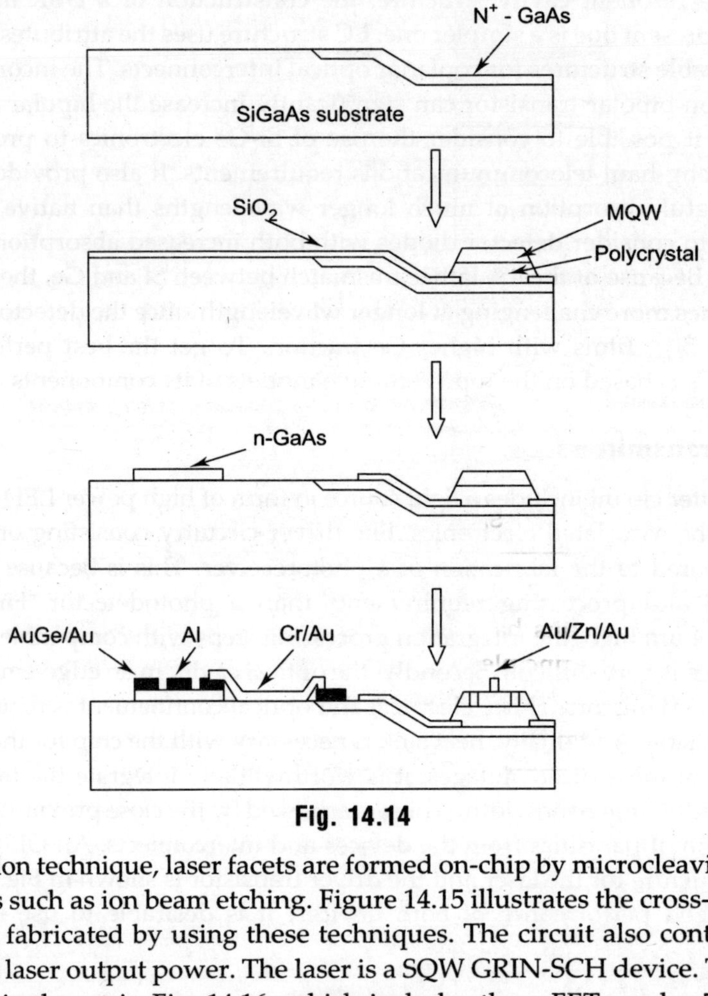

Fig. 14.14

At next generation technique, laser facets are formed on-chip by microcleaving or by using dry etching techniques such as ion beam etching. Figure 14.15 illustrates the cross-section of a simple transmitter circuit fabricated by using these techniques. The circuit also contains a photodiode for monitoring the laser output power. The laser is a SQW GRIN-SCH device. The corresponding equivalent circuit is shown in Fig. 14.16, which includes three FETs and a 50 ohm input load resistance. The identical FETs labeled as Q_1 and Q_2 form a differential amplifier or current source that provides the advantages of common mode rejection and noise reduction. V_{G1} and V_{G2} are the respective small signal modulated input signals. The FET Q_3 acts as constant current source providing the drive current. The current source I_B is a dc source to bias the laser at threshold and the photodiode PD acts as a monitoring device. If the transconductance of the FETs Q_1 and Q_2 are g_m (the transconductance of the FETs are assumed to be equal) and the slope in the lasing portion of the light-current characteristics is S, then the output power P_{out} of the laser can be expressed as

$$P_{out} = g_m S (V_{G1} - V_{G2})/2$$

It may be noted that the detector will be almost transparent to the laser output, since the emission of a laser occurs at energies slightly below the absorption edge of the material.

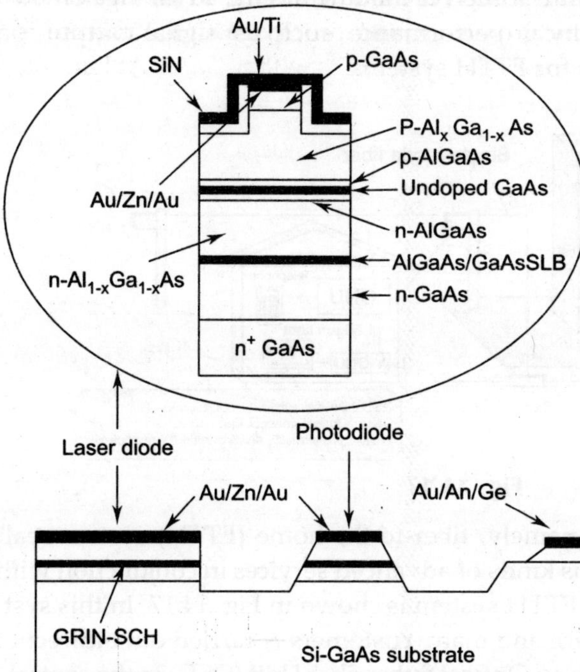

Fig. 14.15

Simple GaAs based circuits have been described here to illustrate the basic techniques of fabrication and operating principles. However more complex circuits and techniques are used utilizing both GaAs and InP based materials.

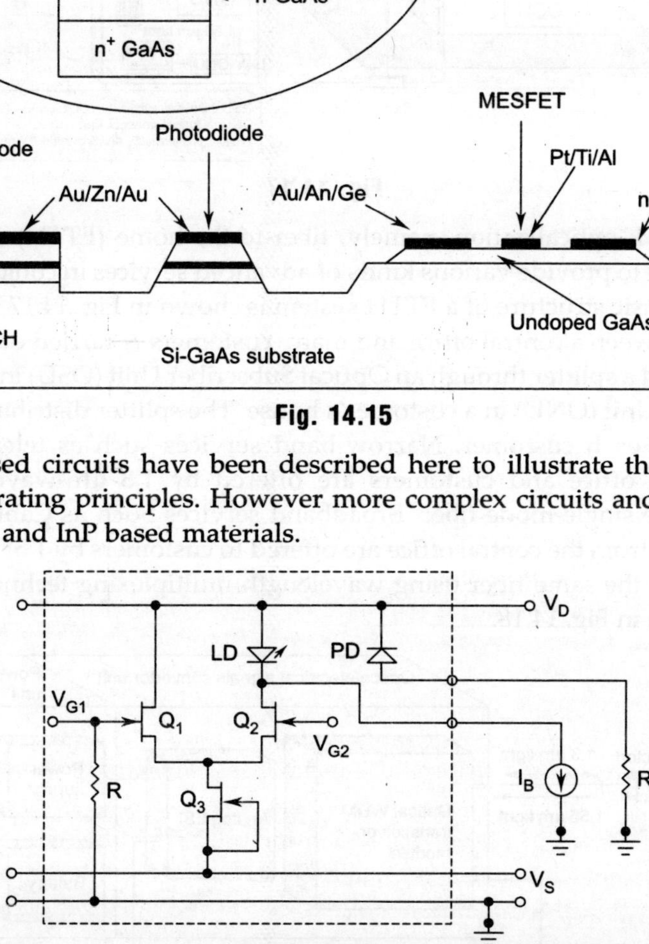

Fig. 14.16

14.4.3 Hybrid Integrated Optical WDM Transreceiver Module

A hybrid integrated optical wavelength-division-multiplexer/ demultiplexer (WDM) transceiver module, which is one of the key components for constructing fiber-to-the home (FTTH) systems, has been developed on a Planar-Lightwave-Circuit (PLC) platform assembled by passive alignment. The optical WDM transceiver module provides the interface between the optical transmission world and the electrical transmission world through the use of opto-electronic devices (OE-devices). This optical WDM transceiver module has the advantage of low cost, high performance, and compact size. Optical performance such as signal output power and responsivity satisfies the requirements for FTTH systems.

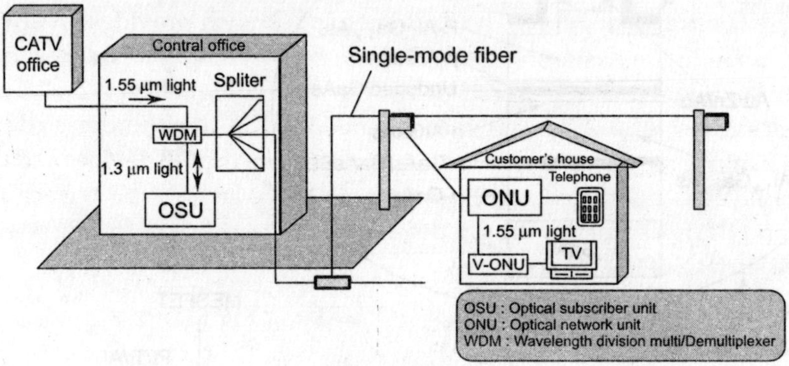

Fig. 14.17

Full-access network opticalization, namely, fiber-to-the-home (FTTH) system realization, is considered as the key to provide various kinds of advanced services in conjunction with personal computer use. The basic structure of a FTTH system is shown in Fig. 14.17. In this system, information exchange between a central office and many customers is carried out bidirectionally on a single-mode fiber and a splitter through an Optical Subscriber Unit (OSU) in the central office and an Optical Network Unit (ONU) in a customer's house. The splitter distributes information from the central office to each customer. Narrow-band services such as telephone and facsimile between the central office and customers are offered by 1.3 μm-wavelength bidirectional transmissions on one single-mode fiber. Broadband services Such as Cable Television (CATV) and video telephony from the central office are offered to customers by 1.55 μm wavelength light which is overlaid on the same fiber using wavelength multiplexing technology. The ONU discussed here is shown in Fig. 14.18.

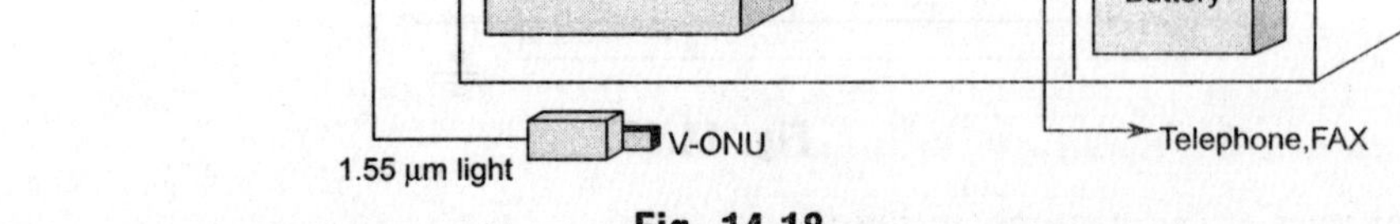

Fig. 14.18

The cost of fabrication process of the conventional optical module using micro-optics technology is quite high, as this consists of several individual parts: a Laser Diode (LD) module, a photodiode (PD) module, and a WDM module, and they all are required to be mechanically assembled, with less than a micrometer tolerance with respect to optical axes, with lenses and fibers.

Also this assembly requires use of high-precision materials and components as well as time-consuming manual labor. Therefore, hybrid-integration-type module with a PLC-platform has been proposed, in which the structure of the WDM transceiver modules is shown in Fig. 14.19. A spot-size-converted laser (SSC-LD), a waveguide photodiode (WG-PD) and monitor WG-PD are flip-chip mounted onto a PLC-platform, which has a 1.3/1.5 μm WDM dielectric filter and an asymmetric Y-branch silica waveguide. The asymmetric Y-branch provides an arbitrary splitting ratio, resulting in flexibility in designing the WDM transceiver module. For the fabricated optical WDM transceiver módule, The PLC consists of a silica-waveguide with a 7.2 × 7.2 μm core with a refractive-index difference of 0.45%. The core is embedded in thick under-cladding and over-cladding layers. To minimize the optical coupling loss between the OE devices and the waveguide core, a thin glass layer is deposited on the under-cladding layer. A Si terrace is used as the alignment plane and the heat-sink for the OE devices. And index marks for the passive alignment are formed on the surface of the Si terrace.

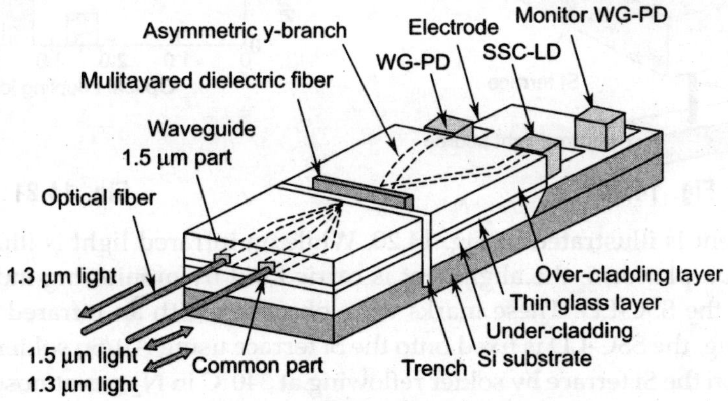

Fig. 14.19

In the process for fabricating the PLC-platform, the Si terrace is fabricated by using standard bulk micromachining techniques. A 4-inch-diameter *p*-type (100) Si wafer is used as a substrate. The wafer is anisotropically etched with a solution of potassium hydroxide to a depth of 30 μm. An under-cladding layer is deposited on the Si substrate using flame hydrolysis deposition (FHD), followed by the flattening of the deposited surface by mechanical polishing. A thin glass layer for adjusting OE-device height is deposited on the polished surface using the FHD method. A 7.2 μm thick Ti-doped core layer is formed on the thin glass layer by electron-beam vapor deposition. Then, the core ridge is formed by processing the core layer through photolithography and reactive ion etching (RIE). The core ridge is then embedded in an over-cladding layer by FHD. The index marks on the Si terrace are formed at the same time as the waveguide-core pattern. The surface of the Si terrace for OE-device assembly is revealed by RIE. Finally, a 0.50 μm thick Au/Pt/Ti electrode is formed on a 1.0 μm-thick SiO2 layer, which is deposited on the Si terrace and the over-cladding layer through a lift-off technique.

To implement passive alignment, index marks need to be accurately formed simultaneously with the waveguide pattern. The formation process is carried out in two steps. First, the index mark pattern, together with the waveguide pattern, is formed by RIE, by which the Si terrace is revealed. Then, while a WSi film (used as a glass etching mask) is removed by RIE, the Si is slightly etched, and thus index marks are formed on the Si terrace. This is a very simple and stable process. The position accuracy of an index mark measured with a laser microscope is within 0.20 μm. The main cause for its error is considered to be the low accuracy of a photomask.

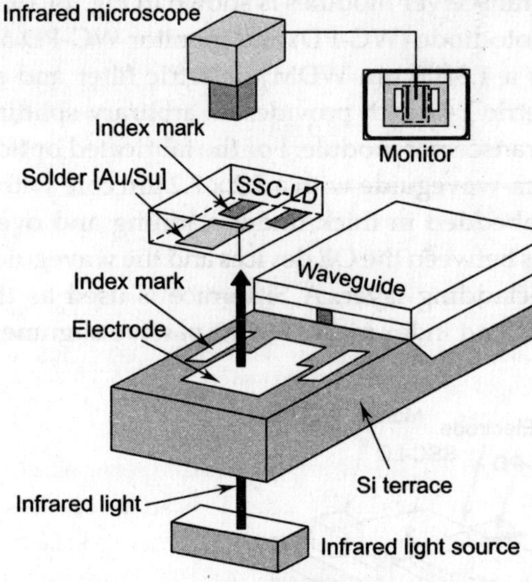

Fig. 14.20

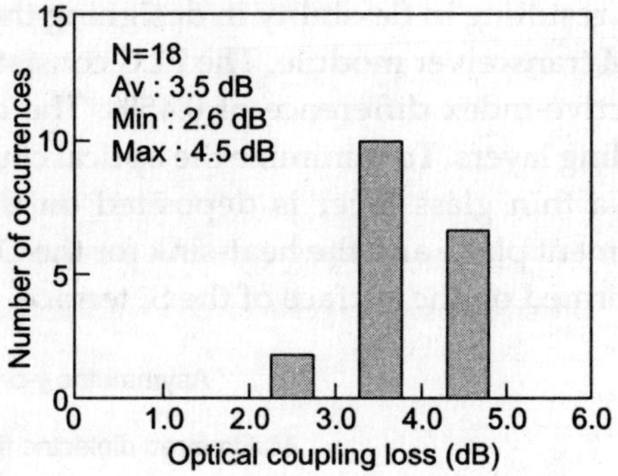

Fig. 14.21

Passive alignment is illustrated in Fig. 14.20. While an infrared light is illuminated from the backside of the PLC-platform, the alignment is carried out by monitoring index marks on both the Si terrace and the SSC-LD. These marks were observed with an infrared microscope. After accurate positioning, the SSC-LD is fixed onto the Si terrace using Au/Sn solder. The SSC-LD chip was then bonded on the Si terrace by solder reflowing at 340°C in N_2-gas atmosphere. Figure 14.21 shows the optical coupling loss distribution for bonded SSC-LD chips. The distance from waveguide core to the SSC-LD facet was 20 μm, resulting in an intrinsic loss of 3.2 dB. A high average coupling efficiency with a loss of 3.5 dB may be obtained.

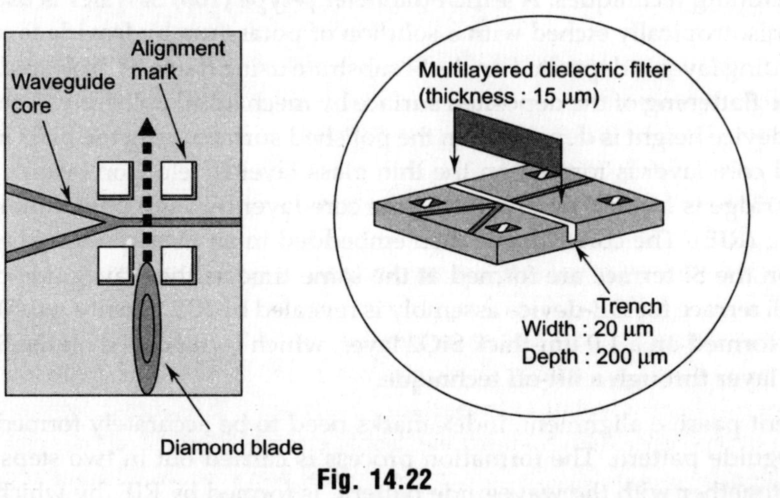

Fig. 14.22

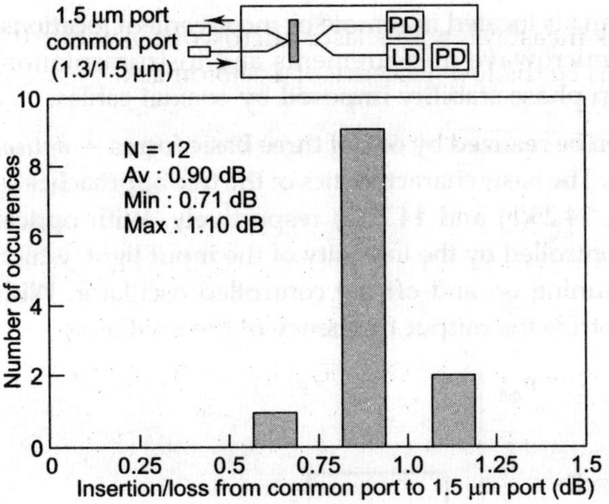

Fig. 14.23

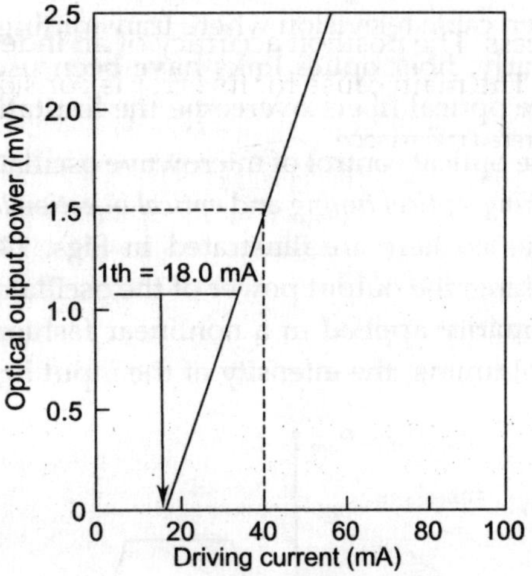

Fig. 14.24

To fit more devices on a single wafer, the device structure has adopted a multilayered dielectric filter for the 1.3/1.5 μm WDM function, resulting in a smaller device size. This structure has the advantage that 1.3/1.5 μm and 1.5 μm output waveguides can be arranged on the same side of the PLC-platform, but it has the drawback that the trench into which the multilayered dielectric filter is inserted needs to be set in the right position without any angle deflection to the waveguide. Some deviation in position and/or angle causes an impermissible increase in the insertion loss from the common port to the 1.5 μm port. To prevent excess loss, a 20 μm wide and 200 μm-deep trench needs to be formed by a cutting machine for insertion of a 15 μm thick multilayered dielectric filter plate. A diamond blade is set in the correct position by monitoring the alignment marks on the Si terrace, which are formed in the same way as the index marks described above. The precise positioning method for the diamond blade is shown in Fig. 14.22. This method ensures high-precision processing of the trench, with the positional deviation being within 0.5 μm. The distribution of the insertion loss from the common port to the 1.5 μm port for PLC chips is shown in Fig. 14.23 and the average loss is 0.9 dB.

The optical output power of the module with respect to driving current is plotted in Fig. 14.24. For the 1.3 μm transmitter port, the output power from the pigtail fiber is 1.5 mW when the SSC-LD driving current is 40.0 mA. For the 1.3 μm receiver port, the module responsivity is 0.40 A/W. The insertion loss from the common port to the 1.5 μm port is 1.0 dB, including fiber coupling losses at the input and output ports. The 1.3/1.5 μm isolation for the 1.5 μm port is 45.0 dB.

14.5 OPTICAL CONTROL OF MICROWAVE OSCILLATORS

The optical phased array antenna as shown in the Fig. 14.3 is a typical example of the application of optical control of high-speed electronic circuits. Due to the wide bandwidth of optical control signals and their inherent isolation from Radio Frequency (RF) signals, it is advantageous to use of optical signals to control high-speed electronic circuits. Optical signals can be routed through lightweight fibers or monolithically integrated optical waveguides without affecting signals transmitted on microwave waveguides. Microwave optical links can be used for remote control application of antenna, which allows expensive control equipments to be located at the command center, rather than in the antenna. This scheme is extremely useful in cellular telephone communication or even cable television where transmitting antenna is located in remote or inconvenient locations. Similarly, fiber optics links have been used in microwave measurements and instrumentation, where optical fibers overcome the limitations on phase stability imposed by coaxial cables.

The optical control of microwave oscillators can be realized by one of three basic forms — *optical switching, optical tuning* and *optical injection locking*. The basic characteristics of the tree approaches as mentioned here are illustrated in Figs. 14.25(*a*), 14.25(*b*) and 14.25(*c*) respectively. With optical switching, the output power of the oscillator is controlled by the intensity of the input light, which is primarily applied in a nonlinear fashion for tuning on and off the controlled oscillator. With optical tuning, the intensity of the input light controls the output frequency of the oscillator.

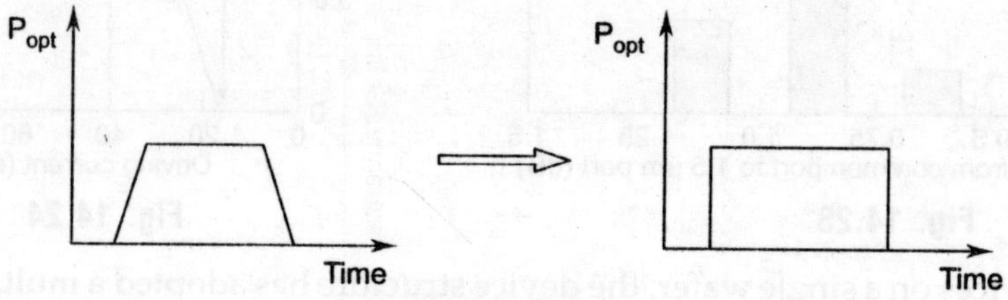

Fig. 14.25(*a*)

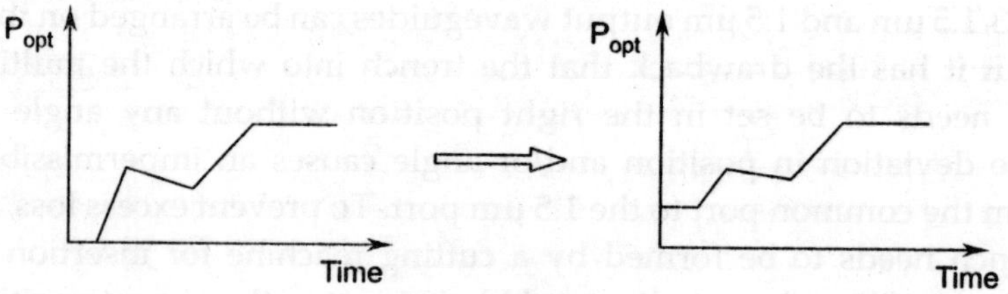

Fig. 14.25(*b*)

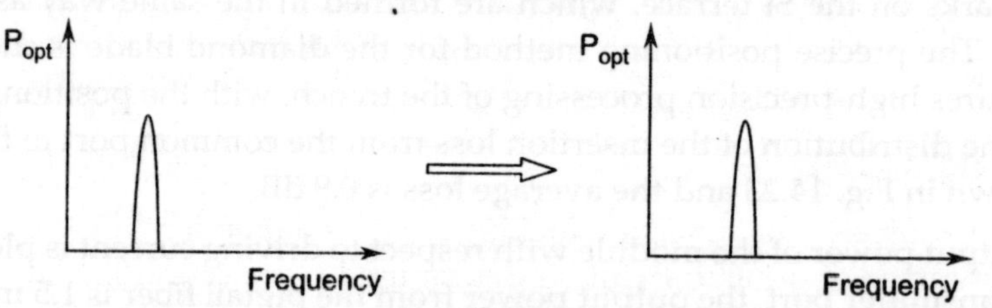

Fig. 14.25(*c*)

In optical injection locking, the optical control signal is intensity modulated at a frequency close to the free running frequency of the oscillator (fundamental locking), one of its harmonics (harmonic locking), or one of its subharmonic (subharmonic locking). The modulated optical signal absorbed in the device active region gives rise to a current flow at the modulating frequency, and this acts in a very similar way to direct microwave signal injection. When in locked condition, the oscillator will oscillate at the same frequency as the injection-locking signal. The variation of phase offset Φ between the oscillator output signal and the injected signal is limited within ± 90° over the injection locking bandwidth ω_{max}. The injection locking bandwidth and phase error are expressed as

$$\omega_{max} = \frac{\omega_0}{2Q}\sqrt{\frac{P_{inj}}{P_0}}$$

and $\Phi = \sin^{-1}\left(\dfrac{\omega_{lock} - \omega_0}{\omega_{max}}\right)$

where ω_0 is the free running oscillation frequency, ω_{lock} is the injection locking frequency, P_{inj} and P_0 are the injected power and oscillator output power. Q is the quality factor of the oscillator, which provides a measure for loss in the resonant circuit and is defined by

$$Q = \omega\, x \,\frac{\text{average energy stored}}{\text{energy lose/second}}$$

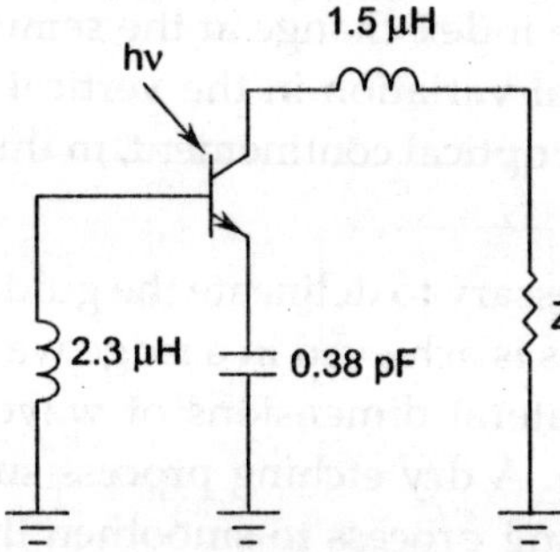

Fig. 14.26(*a*)

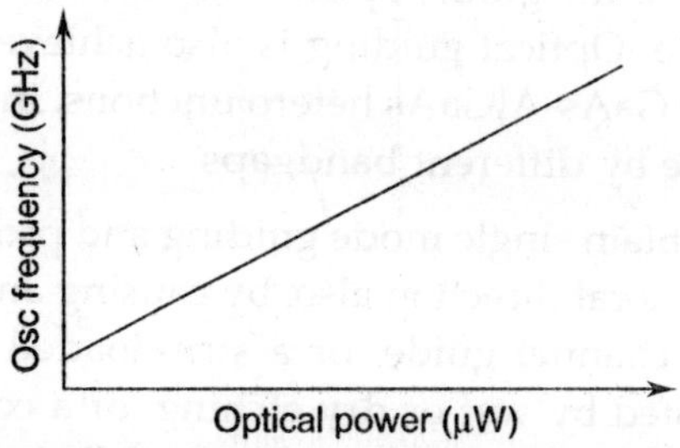

Fig. 14.26(*b*)

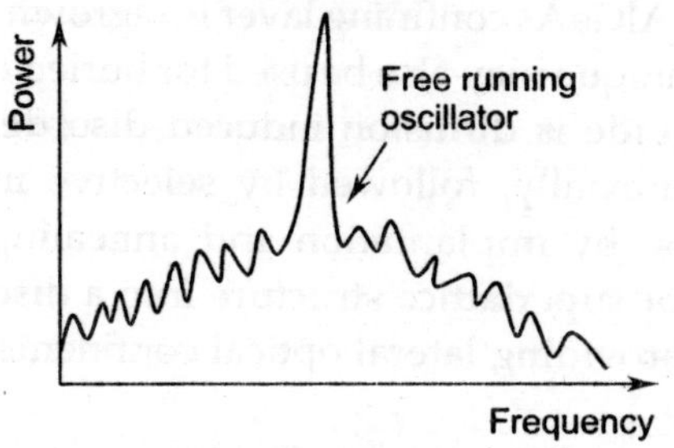

Fig. 14.26(*c*)

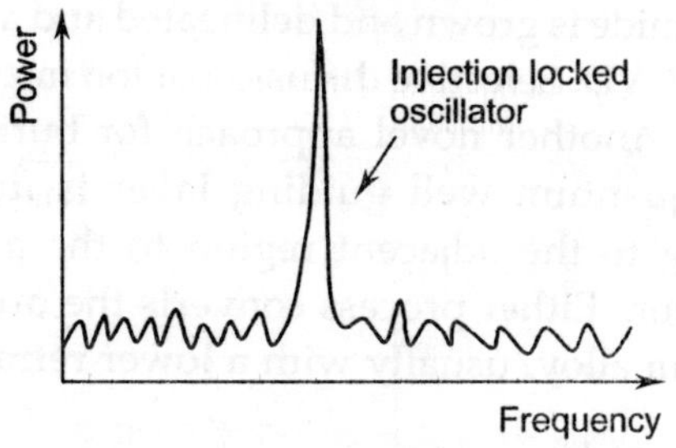

Fig. 14.26(*d*)

Above equations indicates that injection locking range is dependent on square root of injected power and quality factor of oscillator can be determined by injection locking experiment. The tuning and injection locking characteristics of an HBT based oscillator is shown in Fig. 14.26.

14.6 GUIDED WAVE DEVICES

Guided wave components are required for routing signals on a chip and also for the functions of directional coupling, filtering and modulation. In complex QEIC it would be advantageous to have optical waveguides and other passive or quasi-passive components monolithically integrated with active components, sources, detectors, modulators and other electronic devices. Optical fibers may be used, but these are limited from both the point of view of integration and functionality. Lithium niobate ($LiNdO_3$) and similar dielectric materials that have very a large electro-optic coefficient are very suitable for most of these applications, but integration of such components with active semiconductor devices is extremely difficult.

Waveguide is a region of dielectric through which light is propagated surrounded by dielectric regions or air having a smaller dielectric constant. Therefore, to form a guide, it is essential to employ techniques that will effectively and selectively create regions of varying refractive index. The simplest technique of delineating a guiding region is by the incorporation of free carriers. When a semiconductor material is doped with a large density of free carriers, its refractive index reduces with comparison to pure material, due to the negative contribution of the free carrier plasma to the dielectric constant. This change in refractive index is adequate for light confinement. Based on this principle, waveguides can be produced either by growth of an undoped epitaxial layer on a highly doped substrate, or by implantation damage. In the implantation damage technique, the waveguide-cladding layer interface is not well defined, since the implantation profile is not rectangular, but Guassian. In the epitaxial technique the interface is more abrupt. At the top of the guide, optical confinement is provided by the index change at the semiconductor interface. Optical guiding is also achieved by compositional variation in the vertical direction, such as GaAs-AlGaAs heterojunctions. Index difference and optical confinement, in this case, are possible by different bandgaps.

To obtain single mode guiding and propagation, it is necessary to delineate the guiding region in the lateral direction also, by causing an index change. This is achieved in a ridge waveguide, a buried channel guide, or a strip-loaded guide. First the lateral dimensions of waveguide are delineated by wet or dry etching, or a combination of both. A dry etching process such as ion-milling or reactive ion-etching is followed by the wet etching process to smoothen the surface. Buried channel waveguides can be produced by regrowth technique, where, for example, GaAs waveguide is grown and delineated and a higher index AlGaAs confining layer is regrown by LPE or MOCVD. Selective diffusion or ion implantation technique may also be used for buried channel guide. Another novel approach for buried channel guide is diffusion induced disordering. A multiquantum well guiding layer is first grown epitaxially, followed by selective masking doping to the adjacent region to the guiding portion by implantation and annealing or by diffusion. Either process converts the ordered MQW or superlattice structure into a disordered random alloy, usually with a lower refractive index, providing lateral optical confinement.

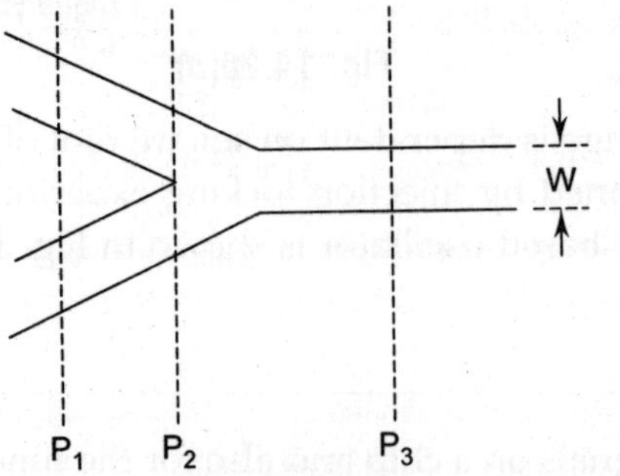

Fig. 14.27(*a*)

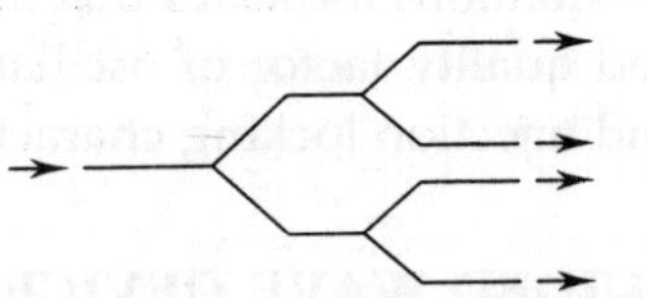

Fig. 14.27(*b*)

Passive semiconductor waveguides can be used in branching networks for integrated optics. Branching networks, the schematic of which is shown in Fig. 14.27, divide optical power among two or more outputs (*beam splitters*) or combine power from two or more inputs. Important parameters of branching networks are *power division ratio* that indicates the ratio of output power at the output ports, and the *excess* or *insertion* loss, which reflects the loss of the input power due to material losses. A commonly used branching network is the Y-structure waveguide as shown in Fig. 14.27(*a*), which can be used as a symmetric power divider or combiner. At point P_1 the two arms are uncoupled and behave as independent single mode guides, suporting the lowest order mode. Near the taper at P_2, the structure transitions from a single waveguide of width 2W to a waveguide of width W. Therefore at P_2, both the symmetric and astmmetric modes are supported. While propagating towards P_3, the asymmetric mode becomes cut off and its energy is radiated into substrate to restore the single mode charateristics. For equal input intensities in the two arms, the Y-combiner is required to behave like the 3 dB directional coupler. More complex variations of directional couplers and Y-structures are shown in Fig. 14.27(*b*). Apart from the straight type of waveguide, various types of guided devices are required in integrated optics, which are lenses mirrors, grating for the purpose of beam focusing, reflection and filtering.

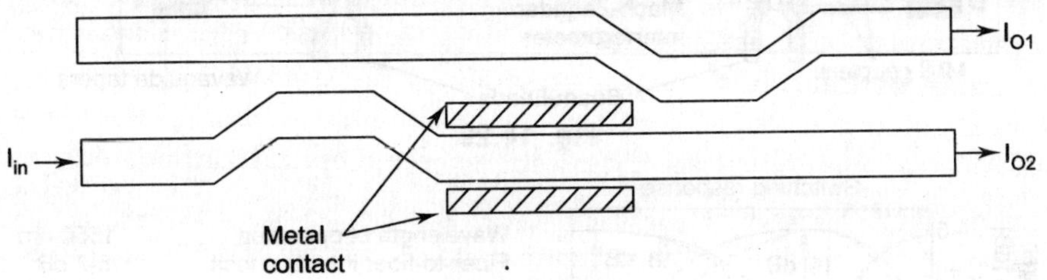

Fig. 14.28

Active or quassi-passive guided wave components can also be integrated in OEIC with active optoelectronic devices. A simple guided wave modulation/ switching device based on the electro-optic effects is the Mach-Zehnder interferometer, schematic of which is shown in Fig. 14.28. The incoming optical beam is split equally between the two branches of the input coupler and then recombined at the coupler at the other end. Care must be taken in material growth and processing such that the tapers are very gradual to reduce bend losses, as well as the spatial uniformity is maintained. With no applied bias, the phase in the two arms is equal and at the output coupler the two waves interfere constructively and all the power (less the insertion loss of the device) appears at the output. If electric field is applied in one arm, there will be phase difference at the output ends due to propagation over the wave guide. Thus the output power can be controlled in a tunable manner by varying the applied bias and its phase. Eventually very complex integrated optical guided wave devices can be realized with GaAs or InP based semiconductors.

14.6.1 Optically Space Switches for Fiber Optic Communication

Electrically controlled optical space switches are key components for use in practical optical communication networks, where they can perform fast switching, rerouting and channel add-drop functions. Typical example of InGaAsP/InP optical waveguide switch matrices, configured as four switch units in a two-stage architecture with four optical inputs and four optical outputs is shown in Fig. 14.29. These switch matrices consist of four 2 × 2 Mach-Zehnder interferometer type space switches, exploiting the Franz-Keldysh electro-optic effect in a specially oriented waveguide configuration to achieve polarization insensitivity. To minimize the overall losses, tapered input and output waveguides are integrated with the switches and combined with self-aligned lensed fibers on a V-grooved silicon motherboard. For high-speed operation the switch matrix is interconnected with high-speed InP-HBT driver electronics. The best fully packaged switch module in a two-by-two cascaded configuration with four fiber inputs and four fiber outputs reach record low fiber-to-fiber losses of less than 6 dB in best and 7 dB in all channels. On-state polarization sensitivity is less than 0.5 dB and on-off ratios greater than 15 dB in regular and greater than 30 dB in dilated arrangements, both maintained over

30 nm throughout the 1550 nm wavelength range. Full turn-on and turn-off times of these switches are below 200 *ps*. The performance characteristics are given in Fig. 14.30.

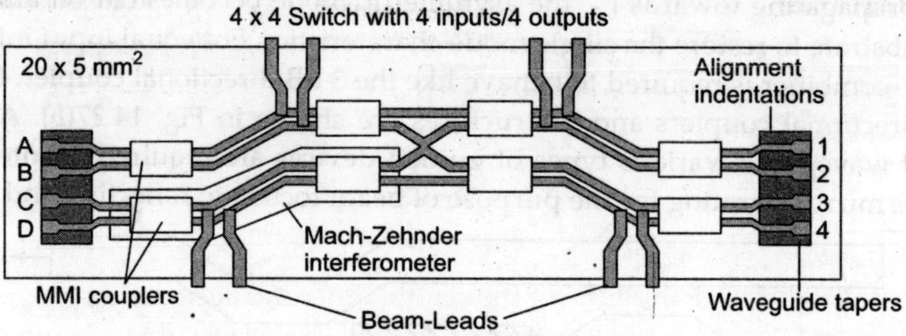

Fig. 14.29

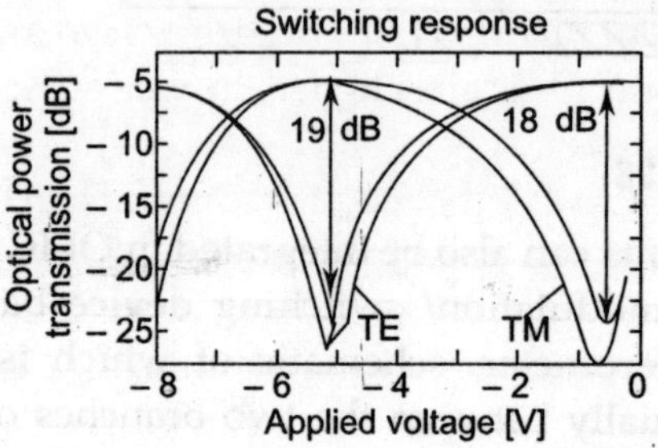

Wavelength of operation		1550 nm
Fiber-to-fiber insertion loss		5-7 dB
Polarization sensitivity		± 0.5 dB
Optical on-off ratios	non dilated	> 15 dB
	dilated	> 30 dB
Spectral bandwidth		30 nm
Switch rearrangement time		< 300 ps
Operational voltage		5 Volt

Fig. 14.30

14.6.2 Multichannel Duplexer Module for Fiber In The Loop Application

This multi channel duplexer is an optical combining/splitting element with integrated photodiode in an array configuration for multiple subscriber applications. The Fiber in the loop (FITL) concept pursued at Ascom Tech provides a bidirectional service via one single mode fiber for both, downstream and upstream communication. Laser diodes at the central office distribute the information to several (4, 8 or 12) subscribers and low cost reflective modulators are implemented at the subscriber side. Each channel has to be terminated at the central office with its own dedicated receiver [Fig. 14.31(*a*) shows the schematic diagram and photograph].

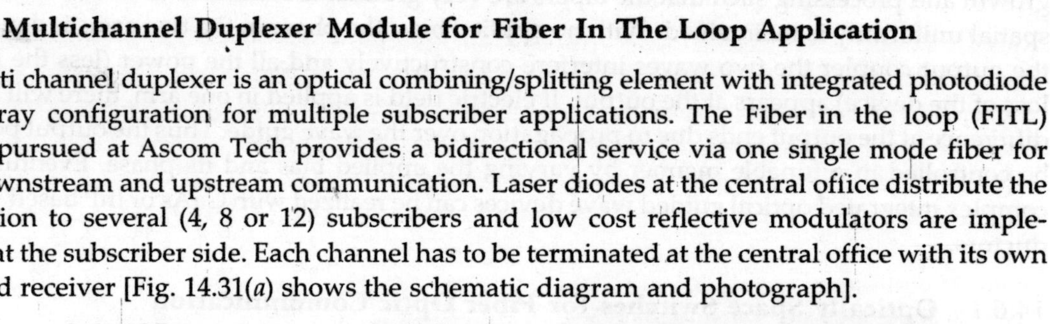

Fig. 14.31(*a*)

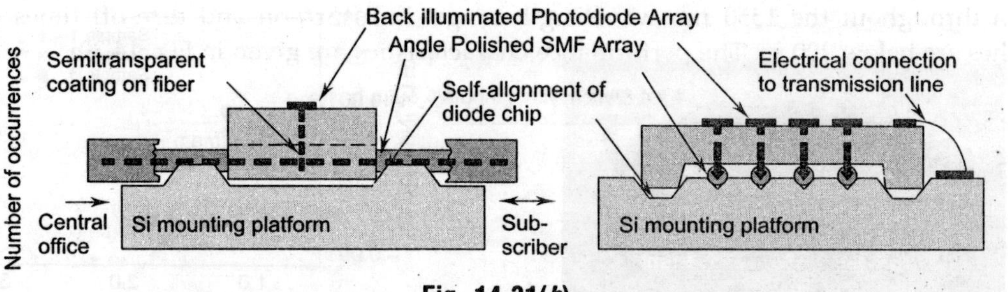

Fig. 14.31(*b*)

A low cost parallel duplexer module for the central office has been developed, providing access to the back reflected light of the upstream channel. As packaging dominates the cost of optoelectronic modules, the realized device is based on only three components that are passive aligned on a silicon mounting board with alignment features [Fig. 14.31(*b*)]. Two angle polished fiber ribbons, one coated with a beam splitter thin film, are aligned in wet etched V-grooves realizing the transmission for the downstream channel. A photodiode array chip with backside processing fits into etched grooves on the same Si motherboard providing self-alignment with the beam splitters for extraction and detection of the upstream information. The realized 4 channel modules (center part of the photograph as shown in Fig. 14.30(*a*) show low polarization sensitivities of less than 1 dB, low beam splitter and alignment losses within 1dB, low crosstalks of less than–35 dB and good thermal stability in a – 40°C to + 80°C cycling test.

14.6.3 Buried Optical Waveguides in Indium Phosphide

Ridge waveguide structures usually show a polarization dependency not in losses but in effective refractive index. In order to realize polarization insensitive optical devices based on the concept of phased arrays, such as waveguide routers, wavelength demultiplexers, add-drop multiplexers, polarization insensitive optical single mode waveguides in both loss and effective refractive index are required. In order to obtain the same effective refractive index for both TE and TM modes, an almost quadratic InGaAsP waveguide core embedded in InP is utilized. It is chosen a low index contrast between core and cladding, leading to a large mode size and thus low waveguide-to-fibre coupling losses. The fabrication of the waveguides is based on two MOVPE growth steps and dry etching. Waveguides with different widths and core compositions were realized. A SEM picture of an example is shown in Fig. 14.32(*a*). The sample was stain etched to enhance the contrast between core and claddings. Optical characterization of the waveguides at λ = 1.55 μm show propagation losses of about 2 dB/cm. The difference in loss between the polarizations is below 0.3 dB/cm and the birefringence is less than 0.001 for a reasonable wide range of waveguide widths. Typical birefringence of the ridge waveguides is 0.003. The buried waveguides are being implemented in phased array devices. Measured birefringence of buried waveguides as a function of the mask width at λ = 1.55 μm. Figure 14.32(*b*) represents its performance characteristics.

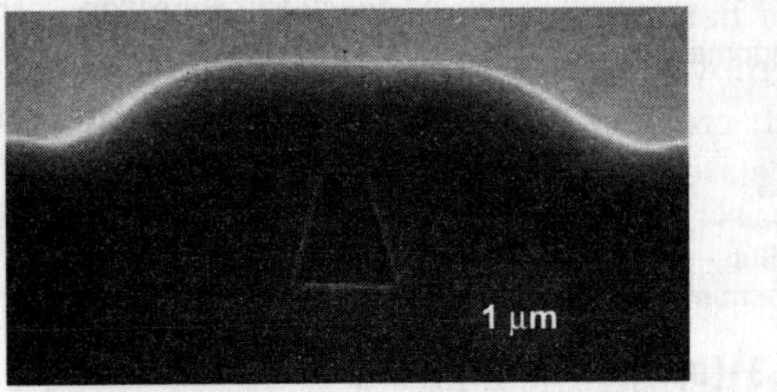

Fig. 14.32(*a*)

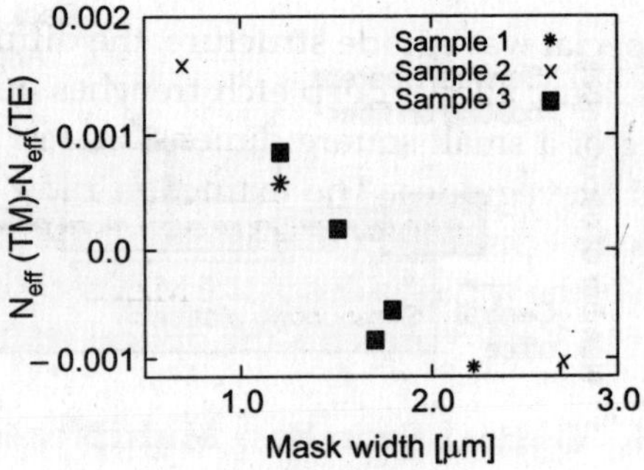

Fig. 14.32(*b*)

14.6.4 Low Power Waveguide Switch with Fast Response

For low speed applications as bypass or protection switches, thermo-optic silica on silicon waveguide switches are advantageous because of their low loss, polarization independence and long term stability. Besides the switching speed, the power consumption and needed space per switch are important parameters. The scheme for 2 × 2 thermo-optic silica on silicon waveguide switch is mention here that combines compactness, low polarization independent losses with low heating powers of 110 mW for π phase shift and response speeds as fast as 150 µs rise and 180 µs fall times.

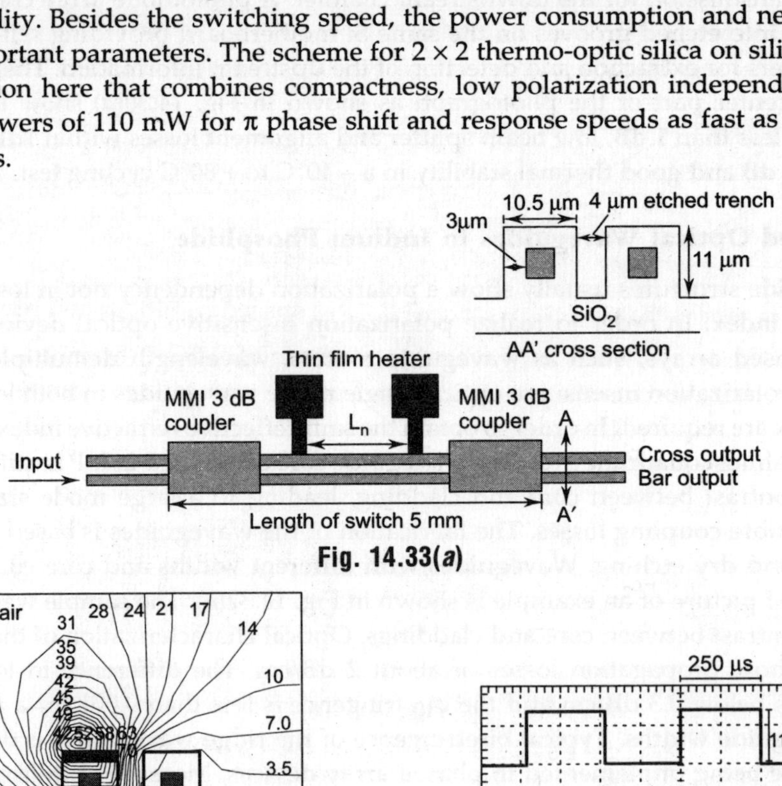

Fig. 14.33(*a*)

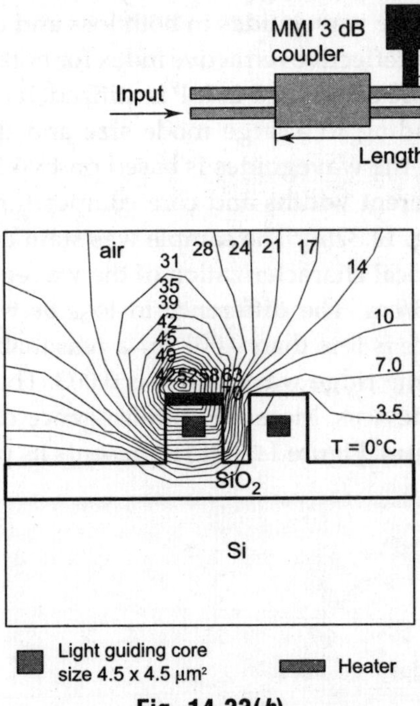

Fig. 14.33(*b*)

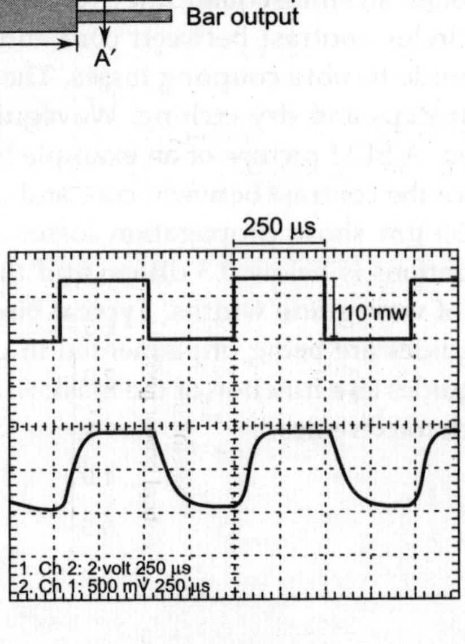

Fig. 14.33(*c*)

A special waveguide structure, the diffused square waveguides shown as AA′ cross section in Fig. 14.33(*a*), allows deep etch trenches outside or between waveguides. This provides efficient heating of a small square dimension, fast heat removal and good thermal separation of closely spaced waveguides. The extinction ratio for both cross and bar output exceeds 21dB and the fiber-to-fiber insertion loss is about 1dB. Steady state thermal analysis and temperature distribution of the waveguides in the MZI section is shown in Fig. 14.33(*b*) for heating power of 110mW. Figure 14.33(*b*) represents the optical response of thermo-optic switch to current switching.

14.6.5 Optical Switches: Multiplexers and Demultiplexers

All-Optical Mach-Zehnder Interferometer Switches and Demultiplexers: Optical switches that are controlled by optical pulses hold promise for extensions of switching speeds beyond the limits of current electronics. Mach-Zehnder interferometers (MZI) type indium-phosphide switches with semiconductor optical amplifiers (SOAs) have been realized. If symmetrically biased, input signals (P_s) are directed to the cross port (P_x). Upon application of a short (P_s) control pulse P_{c_1}, SOA_1 is disturbed, changing its gain and its refractive index, causing signal pulses to switch from cross port to bar port ($P_=$). A second control pulse (P_{c_2}), applied shortly (t_c) thereafter, switches the signals back to the cross port (P_x). Through superposition of two closely spaced control pulses, the signals passing through the MZI-SOA are, thus, switched from port to port. These optically controlled MZI-SOAs are suitable for high-speed optical demultiplexers, optical sampling gates and wavelength converters. A schematic of 10 ps gating obtained with an MZI-SOA is shown in Fig. 14.34(*a*) and the waveforms are shown in Fig. 14.34(*b*).

Two hundred femtosecond control pulses (P_{c_1}, P_{c_2}) spaced (t_c =10 *ps*) and repeated at 80 MHz from a 1.53 mm Spectra Physics laser system control the MZI-SOA. By means of a pump probe technique the switching from the cross (P_x) to the bar ($P_=$) and gating is demonstrated.

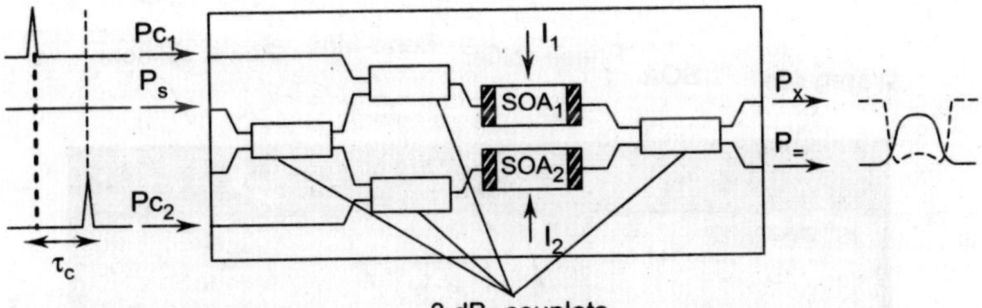

Fig. 14.34(*a*)

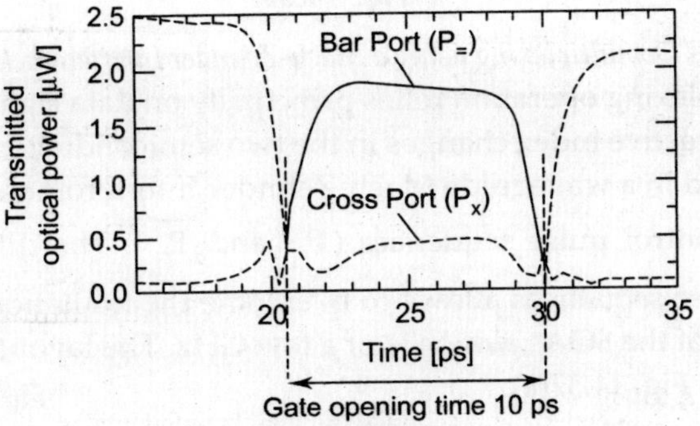

Fig. 14.34(*b*)

Monolithically Integrated Ultra-High Speed All-Optical Indium Phosphide Waveguide Switches and Demultiplexers: All-optical switches use optical controls to switch the light-path of signals between the bar and cross state. Major applications for these components are optical demultiplexing of high bit-rate signals with add-drop function and wavelength conversion for wavelength division multiplexed (WDM) networks.

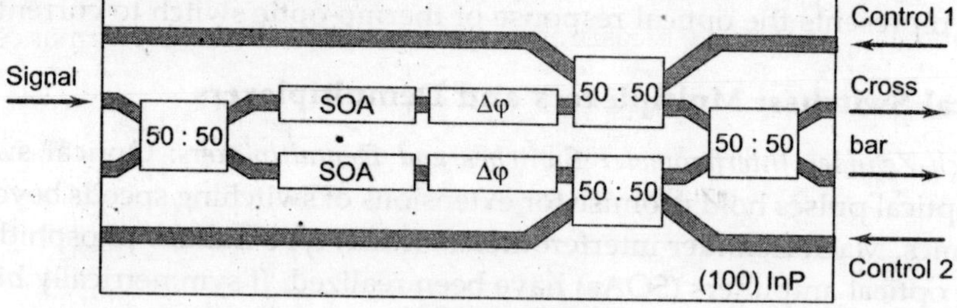

Fig. 14.35

Switching is done by introducing optical control-pulses into the arms of a Mach-Zehnder Interferometer (MZI) that contain SOAs and phase-shifting elements. The control pulses generate refractive index changes that alter the optical length of the arms in the MZI, leading to constructive interference either in the bar or in the cross output waveguide. The InP device is built from a number of different components: Semiconductor Optical Amplifiers (SOA), phase shifters (Δφ), passive single mode waveguides, and power splitters (50:50). A schematic view of the device is shown in Fig. 14.35.

The monolithic integration of these components have been developed using multiple-step low pressure metal-organic vapor phase epitaxy, in combination with wet and dry etching techniques. Pyralin is used for insulation of the electron-beam evaporated and electroplated electrical contacts. The photograph in Fig. 14.36 shows a mounted and bonded all-optical switch with two SOAs followed by two-phase shifters.

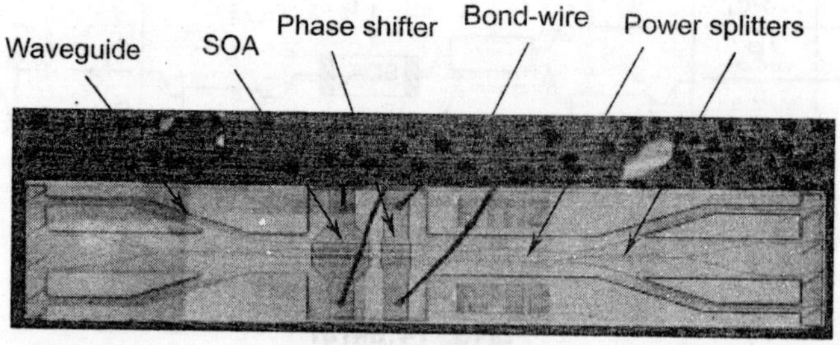

Fig. 14.36

All Optical 80 to 10 Gbit/s Demultiplexing using a Mach-Zehnder Interferometer with Semiconductor Optical Amplifiers: The demultiplexing operation relies principally on data signal phase and gain changes induced by gain and refractive index changes in the two semiconductor optical amplifiers (SOAs) monolithically integrated in a waveguide Mach-Zehnder Interferometer (MZI) through properly synchronized optical control pulse sequences (P_{c_1} and P_{c_2}). The differential injection of two delayed (Δ_t) control pulse sequences allows to overcome the limitations imposed by the optical modulation bandwidth of the SOAs, which is of a few GHz. The layout of the device used for the experiments is shown in Fig. 14.37(*a*).

The excellent performances of the MZI-SOA device operating at 1.55 μm are demonstrated by error free all optical demultiplexing of 80 Gbit/s random data signals. The bit-rate flexibility of the MZI-SOA is demonstrated by penalty-free alloptical demultiplexing of 40 Gbit/s data streams into 10 Gbit/s streams. Bit-error curves are shown on the right. The same bit error rate performances are obtained for an input power of the data signal ranging between –15 and +2 dBm (17 dB dynamic range). Injection currents for the SOAs are 234 and 247 mA respectively and the control powers for P_{c_1} and P_{c_2} are +5 dBm, measured in the fibers. The device performance is indicated in Fig. 14.37(*b*).

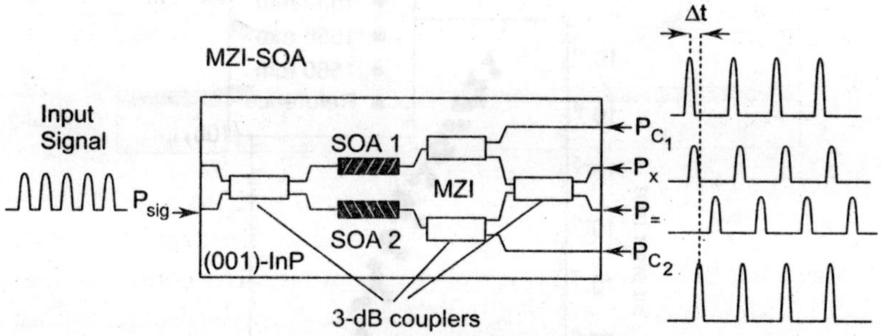

Fig. 14.37(*a*)

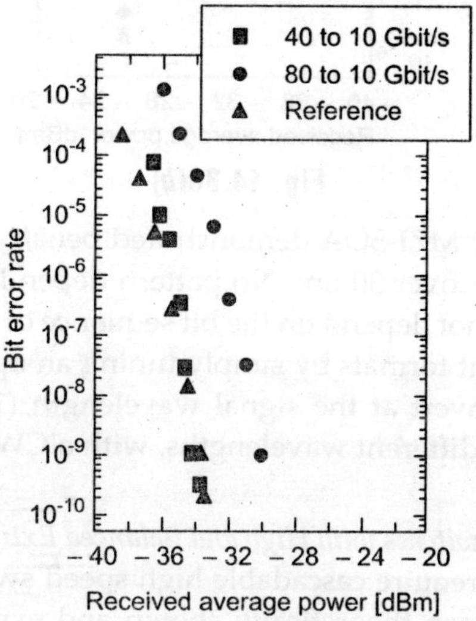

Fig. 14.37(*b*)

All Optical Wavelength Conversion for RZ-Signals using high-performance Mach-Zehnder Interferometer: The operation principle of the all-optical wavelength converter relies on the differential control scheme, which allows to overcome the bandwidth limitations of the semiconductor optical amplifiers (SOAs). The 40 Gbit/s RZ-signal is split in two fibers of approximately the same length, one of them including an optical delay (τ) to fine adjust the synchronization of the bit pattern sequence. The device used for the experiments is shown in Fig. 14.38(*a*). It consists of single-mode quaternary InP-waveguide Mach-Zehnder interferometer (MZI) with polarization insensitive and low-reflection SOAs and 50 : 50 % multi-mode interference couplers (MMI) as 3 dB couplers. Device dimensions are 6.5 × 1.5 mm.

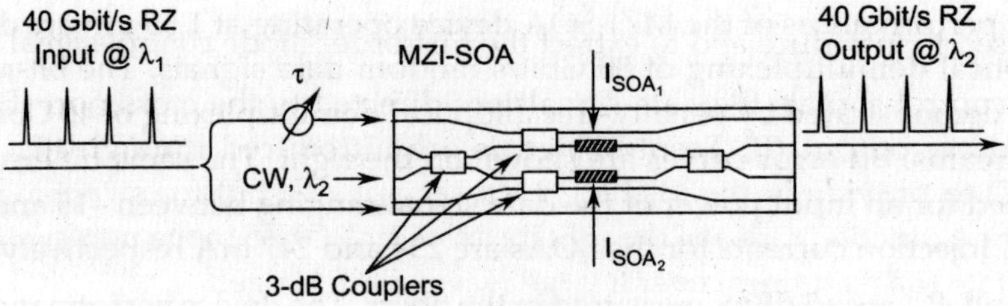

Fig. 14.38(*a*)

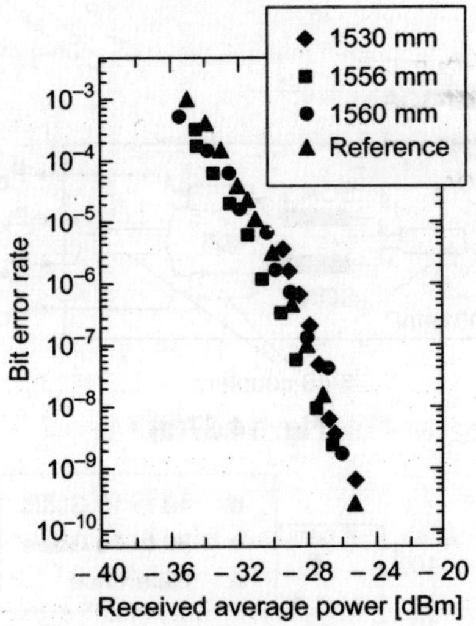

Fig. 14.38(*b*)

System experiments with the MZI-SOA demonstrated penalty free wavelength conversion of 40 Gbit/s data streams, possible over 30 nm. No pattern dependant effects are presents, *i.e.*, the Bit-Error Rate (BER) values do not depend on the bit sequence of the input signal. It is possible to obtain both RZ and NRZ output formats by simply tuning an optical delay. No blocking effects present *i.e.*, the device can convert at the signal wavelength (1556 nm for the measurements reported here). BER curves for different wavelengths, with a CW input in 1540 nm are shown in Fig. 14.38(*b*).

Cascadable All-Optical Space Switches with High and Balanced Extinction Ratios: Future optical time division multiplexing systems require cascadable high speed switches and demultiplexers with high extinction ratios. It has been theoretically shown and experimentally demonstrated that asymmetries are necessary to improve the extinction ratios of all-optical Mach-Zehnder Interferometer space switches. With the incorporation of new MMI-couplers, the control-signal is possible to convert into a higher order mode while coupling it into the signal path of the fundamental mode data-signal. With that trick the control and data signals can be easily separated after signal processing rendering the devices cascadable. The large optical bandwidths of these so-called MMI-converter-combiners were theoretically predicted and experimentally testified. The all-optical space switch, shown in the Fig. 14.39(*a*), delivers high and balanced extinction ratios over a large wavelength range. The device comprises two asymmetric Multi-Mode Interferometer (MMI) couplers for data signal splitting of (P_s), four Semiconductor Optical Amplifiers (SOAs) on the MZI-arms providing the nonlinearities for switching and four MMI-Converter-Combiners

(CC), that are used to introduce and to extract the first order mode control-signal (P_{C_1} , P_{C_2}). In the absence of the control-signals the data-signals are directed to the cross-port (P_x). To switch the signals into the bar output ($P_=$), control-signals are introduced through the MMI-Converter-Combiners (CC) to provide in the MZI the necessary phase-shift for switching. Figure 14.39(*b*) demonstrates the extinction ratios of the switch, exceeding 20 dB in the non-switched (solid lines) and switched state (dashed lines) over a large wavelength range.

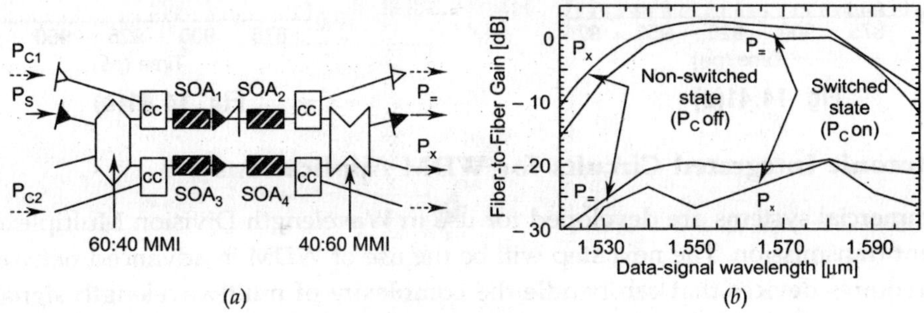

Fig. 14.39

14.6.6 Dynamic Analysis of Mach-Zehnder SOA

All optical switches are expected to be key devices in future high capacity optical time division multiplex systems. Monolithically integrated Mach-Zehnder interferometer type all optical switches with Semiconductor Optical Amplifiers (SOAs) in their arms have been developed for high speed switching. 40 to 10 Gbits/s demultiplexer (DEMUX) experiments were demonstrated using a differential control scheme. However, the gain in SOA brings about complicated switching characteristics because transmission characteristic of these integrated interferometric switches with SOAs is determined not only by phase but also by gain. As an example, unbalanced extinction ratios for both switching ports were observed. To improve the unbalanced extinction ratio, asymmetrically biasing the two SOAs was proposed and the balanced extinction ratios were demonstrated in the experiment with a CW data signal. Dynamic switching characteristics of MZI-SOA type all optical switches under high speed operation in a time domain large signal analysis have been examined. Uniform data output powers and high extinction ratios were observed for all-optical switches with asymmetrically biased SOAs.

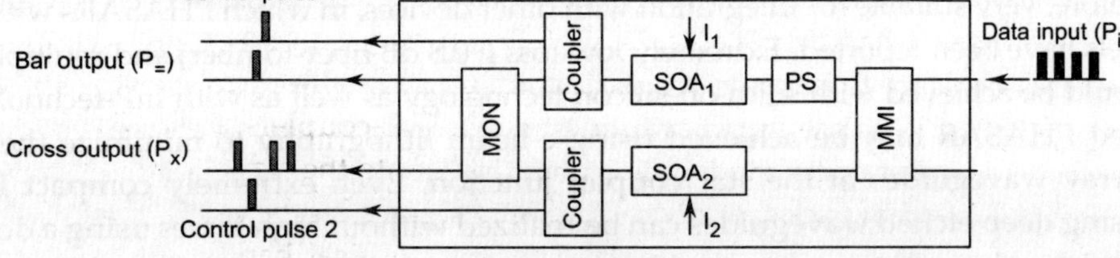

Fig. 14.40

The schematic configuration of MZI-SOA all optical switches is illustrated in Fig. 14.40. The data signals injected from the input port are directed to the cross-port or the bar-port depending on the phase difference between the two SOAs. The phases in the MZI configuration can be changed by properly adjusting the injection times between the two control pulses. Switching characteristics for the cross and bar port under 40 to 10 Gbits/s demultiplexer operation are shown in Figs. 14.41(*a*) and 14.41(*b*). Almost uniform output powers as well as balanced high extinction ratios are obtained by asymmetrically biasing the SOAs.

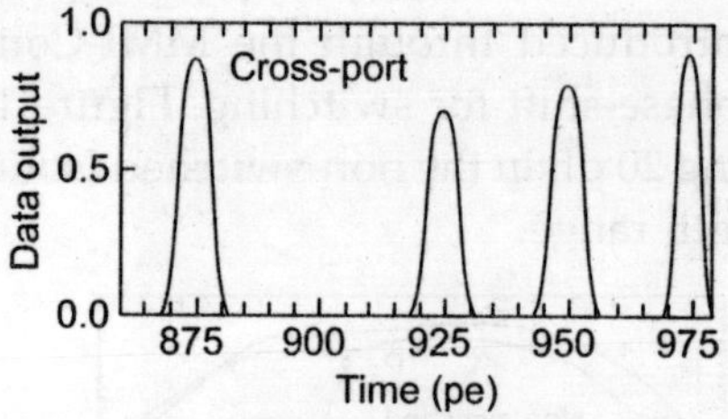

Fig. 14.41(*a*)

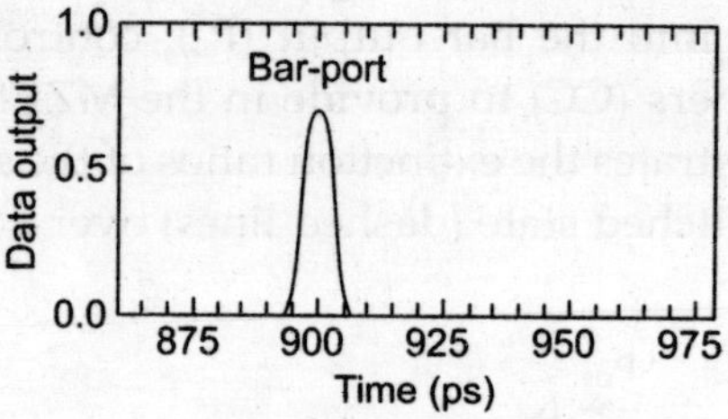

Fig. 14.41(*b*)

14.6.7 Photonic Integrated Circuits for WDM Applications

Today's commercial systems are developed for use in Wavelength Division Multiplexed (WDM) point-to-point transmission. The next step will be the use of WDM in advanced network applications. This requires devices that can handle the complexity of multiwavelength signals, such as tunable or multiwavelength transmitters and receivers, and WDM add-drops and cross-connects. Most of these subsystems are assembled from discrete components which are coupled to each other with fibers. Photonic integration of these devices has the potential to reduce packaging cost by eliminating most of the interconnection fibers. At the same time it leads to a drastic reduction in volume of complex devices.

The key component in Photonic Integrated Circuits (PICs) for WDM applications is the phased-array Multiplexer/Demultiplexer or PHASAR (also called Arrayed Waveguide Grating, AWG, or Waveguide Grating Router, WGR). Most devices required in advanced WDM systems can be realized using a combination of PHASARs, Semiconductor Optical Amplifiers (SOAs) and optical switches. InP-based technology is potentially capable of integrating these three basic components and thus suitable for realization of a broad class of optical devices. A variety of functions have already been demonstrated. Since InP-based devices are smaller by one or two orders they are especially suitable for complex circuits. The integration of SOAs and fast electro-optic switches will help InP-technology to provide competitive solutions.

PHASAR Demultiplexers

Phased array Multiplexers/Demultiplexers have been realized in almost any waveguide technology and are, therefore, very suitable for integration with other devices, in which PHASARs with as many as 256 channels have been reported. Extremely low loss (~0.8 dB fiber-to-fiber) and multiple channel PHASARs could be achieved with silica-on-silicon technology as well as with InP-technology. Low loss InP-based PHASAR may be achieved using e-beam lithography to minimize the gaps between the array waveguides at the star coupler junction. Even extremely compact InP-based PHASARs using deep etched waveguides can be realized without high losses using a double etch process in which a low contrast waveguide structure is used in the start coupler region and combined with a high contrast waveguide structure in the bends.

Important issues in PHASAR Multiplexer/Demultiplexer are the central wavelength, crosstalk and passband shape. Especially when large numbers of wavelength filters are cascaded, spectral misalignment and passband narrowing will result in severe power penalties. By controlling the phase independently in each array waveguide, it is possible to affect all of these issues. Thermo optic heaters in silica-technology or electro-optic effect in an InP-based device may be used. Figure 14.42 shows a photograph of the widely tunable PHASAR using electro optic phase shifters.

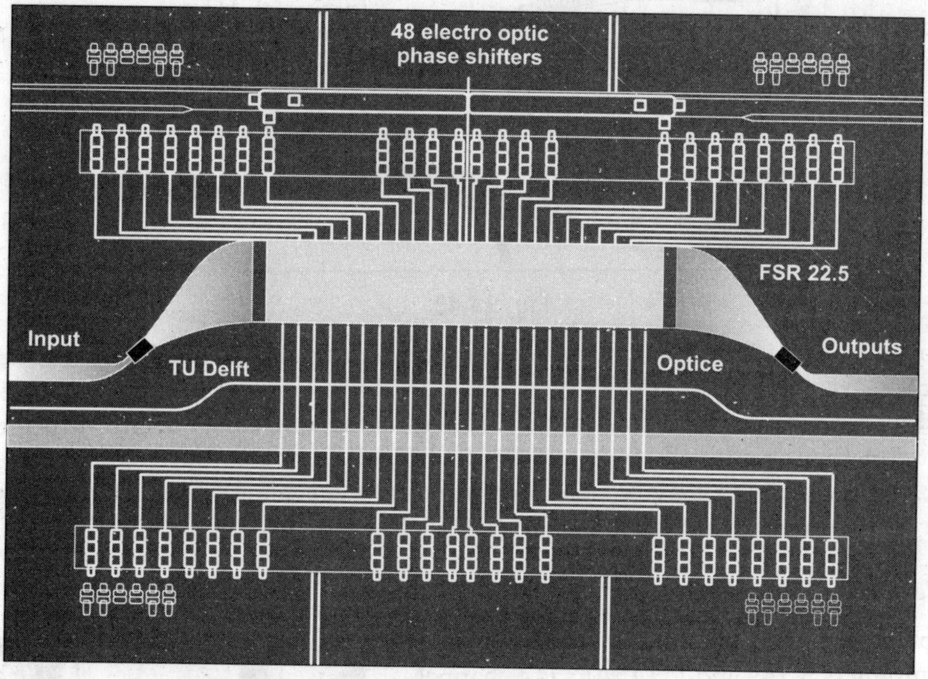

Fig. 14.42

WDM Sources

Today's WDM systems use wavelength -selected or tunable lasers as sources multiplexed into one fiber using a power combiner or wavelength multiplexer. A disadvantage of this solution is the large number of individual lasers required, each of which has to be wavelength controlled individually. Integrated multiple wavelength lasers have been realized by combining an array of lasers operating at the correct wavelengths with a power combiner. Even though the use of a power combiner inherently introduces a loss of 10 log N dB, with N the number of wavelength channels, a high fiber coupled power of +10 dBm was demonstrated for an 8-channel device. However, for large numbers of wavelength channels the combiner losses will become a serious problem.

The combiner loss can be drastically reduced by using a PHASAR multiplexer to combine the wavelength signals. Since the multiplexer has a wavelength selective response, broadband semi-conductor amplifiers can be used. Lasing will occur at the passband wavelengths of the PHASAR, which have a well defined spacing. Figure 14.43 shows the layout of this device, which uses only 5 + 8 SOAs cleverly combined with a chirped PHASAR which enables laser operation at $5 \times 8 = 40$ discrete wavelengths. Figure 14.43 represents the layout of the 40-channel digitally tunable laser.

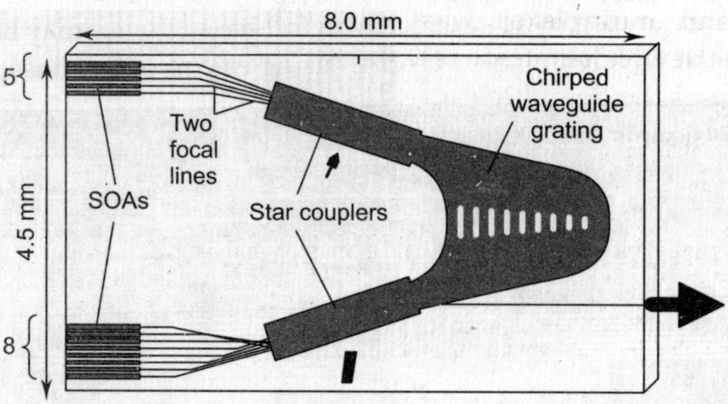

Fig. 14.43

A way to avoid the difficult monolithic integration is hybridly coupled chip with an SOA array to an InP chip containing a PHASAR demultiplexer. The disadvantage of the multiwavelength laser formed by the combination SOAs and a PHASAR is the long cavity length, which restricts the direct modulation speed, and mode stability is difficult to obtain due to the closely spaced cavity modes. Mode stability may be obtained from multiwavelength source by integration of mode stable DBR lasers with a PHASAR.

Channel Selectors and Equalisers

Another application of the combination of semiconductor optical amplifiers SOAs and PHASARs is a channel selector. Monolithically integrated channel selectors on InP have been realized which can pass or block each of the 8 and 16 wavelengths applied to the input, respectively. These channel selectors exhibit high extinction ratios of more than 40 dB and zero loss between fibers. In both devices a deep etched ridge waveguide structure was used in the passive PHASARs for compact dimensions. The use of low bandgap material passive regions enables the use of polarization independent waveguides. Figure 14.44 shows a photograph 16-wavelength channel selector realized monolithically integrated on InP with a channel spacing as small as 100 GHz indicating a good control of the polarization behavior of the waveguides. Besides monolithically integrated devices, hybrid channel selectors are reported combining silica-technology and InP-SOA arrays using PLC technology.

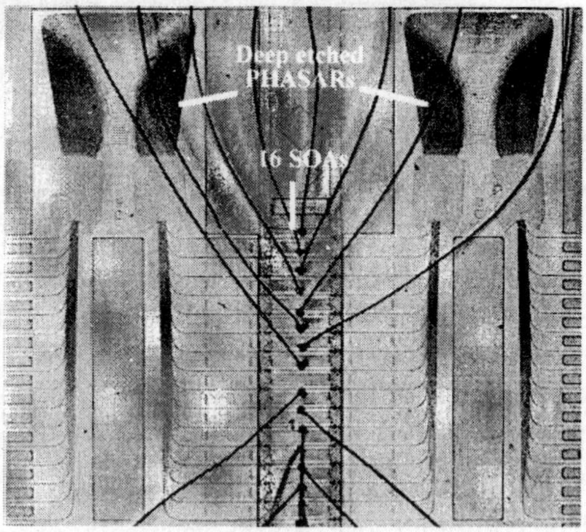

Fig. 14.44

WDM Switching Devices

A more common application of the integration of PHASAR multiplexers/demultiplexers and phase shifters are WDM switch devices. Monolithic integration of wavelength selective switches or WDM cross-connects was first reported in silica technology using a combination of wavelength multiplexers/demultiplexers or routers (PHASARs) combined with thermo-optic space switches.

Silica-based WDM switching device with add-drop multiplexer capable of handling 40 channels at 100 GHz spacing and of low crosstalk has been realized. Spectral misalignment between the demultiplexer and multiplexer was eliminated because the device operates in reflective mode and reuses the demultiplexer for remultiplexing the signals to the output. The device has a flat through spectrum, which eliminates passband narrowing effects when cascading multiple nodes. WDM cross-connects with InP-technology using different arrangements of PHASARs and electro optic Mach-Zehnder Interferometer switches has equal potential to obtain good crosstalk values as silica-based devices. InP-technology, however, offers much more compact devices (less than 1 cm^2), much lower power consumption (less than 1 mW) and much higher switching speeds (typically ns). Disadvantages of InP are the losses, both the fiber coupling and on-chip losses, which are still too high, and therefore much research work is still in process. A Photograph of InP WDM cross-connect with dilated switches is given in Fig. 14.45.

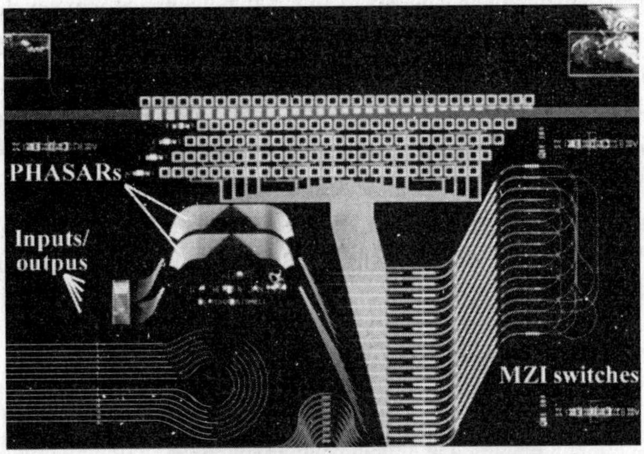

Fig. 14.45

Semiconductor-based Photonic Integrated Circuits are rapidly gaining performance. In the coming years they will play an increasingly important role in providing the functionality required in advanced WDM networks in a compact form and at steadily decreasing costs. Most of the key components for WDM applications can be realized with a combination of PHASARs, SOAs and optical switches. Performance and fabrication costs of InP-based devices are still prohibitive for broad application, but they show steady improvement. The huge market for WDM lasers will accelerate the InP technology development and pave the way for integration processes which will provide compact and cost-effective solutions for many functions in WDM applications.

14.7 COMPLEX CIRCUITS AND ARRAYS

An array consists of an assemblage of identical devices monolithically integrated on the same chip. It is increasingly necessary, where a number of identical devices are used. As example, high data rate local area networks are being installed in an increasing number of locations due to the merging of a high density of fibers in each of these locations. Therefore, it is desirable to use multiple optoelectronic devices, or arrays, for reduction of component cost, space and weight.

14.7.1 Monolithic Two-Dimensional PbTe Infrared Sensor Array

Infrared focal plane arrays for thermal imaging with the highest sensitivity use semiconductor infrared sensor pixels and are presently fabricated in a hybrid manner: The chip containing the 2-D array is fabricated in a suitable narrow gap material (or a III-V superlattice structure), while the read-out multiplexer is constructed with Si-technology. Hybrid connections are usually formed by Indium bumps for each individual pixel. This makes the technique rather expensive for sensor arrays containing thousands of pixels. Due to the easy molecular beam epitaxy of narrow gap PbTe and $Pb_{1-x}Sn_xSe$ layers on Si-substrates, a monolithic design becomes possible by growing the infrared sensor material as a thin layer directly onto the Si read-out chip. The read-out chip is designed in CMOS technology on double side polished Si. It contains a switching transistor for each individual pixel, and a shift register to access the lines serially. The columns are fed out in parallel to an amplifier chip. The array may contain 128 × 96 pixels with the epitaxial growth of the material, each pixel dedicated for the infrared sensors. A special low temperature cleaning technique is to be used in order not to destroy the chip structure including the Al-interconnections. The maximum temperature used is about 450°C for 30 min (for wafer cleaning and growth of the CaF_2 buffer layer), followed by about 3 hrs at 400°C for the growth of PbTe. Figure 14.46(*a*) shows a photograph of PbTe infrared sensor array and individual pixel is shown in Fig. 14.46(*b*).

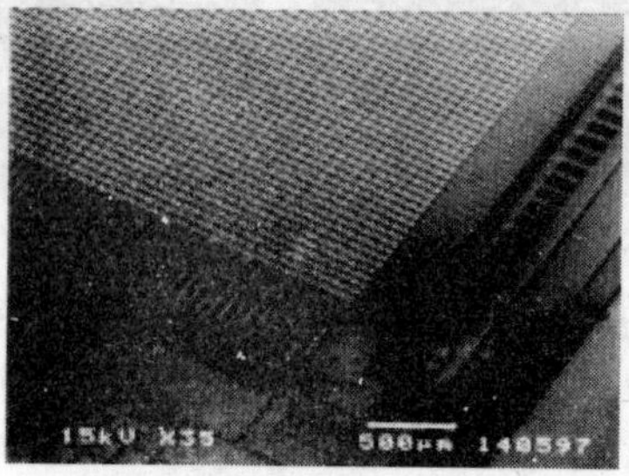

Fig. 14.46(*a*)

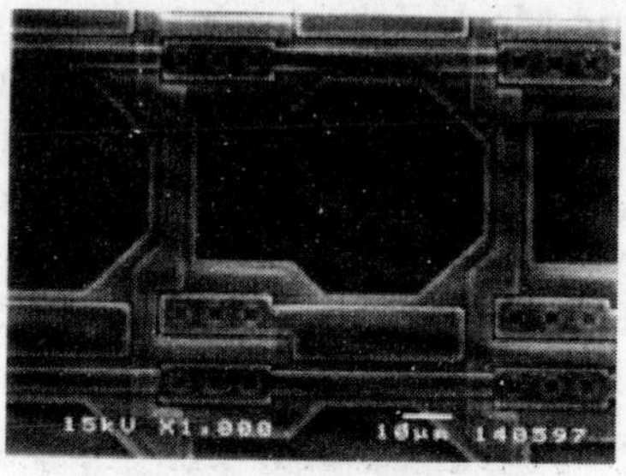

Fig. 14.46(*b*)

14.7.2 Packaging of Thermoelectrocally Cooled Infrared Sensor Array

While sensitive semiconductor infrared sensor arrays have to be cooled to below 130°K, thermo-electric cooling can be employed for certain applications for sensors. A compact, completely sealed package is designed. The sensor arrays (linear arrays with 128 elements or 2-D arrays with 128 × 96 elements) are soldered on 4-stage thermoelectric coolers. Two amplifier chips each with 64 channels are bonded on a ceramic substrate and operate at room temperature inside the evacuated package. The electrical connections with the cooled sensor chip are done with a novel low thermal conductivity flexible kapton foil coated with thin film Au stripes. The vacuum feed throughs are constructed again with a novel thin film technique. The Au-pads on the ceramic substrate crossing the encapsulation are covered with a second ceramic ring and soldered using a special glass solder. The rest of the package as well as the sapphire window are again soldered, brazed or welded. Figure 14.47 shows a cross section of the encapsulation. A non-evaporable getter tube is built in since the maximum out-gassing temperature is below 170°C, therefore, reconditioning of the vacuum is possible.

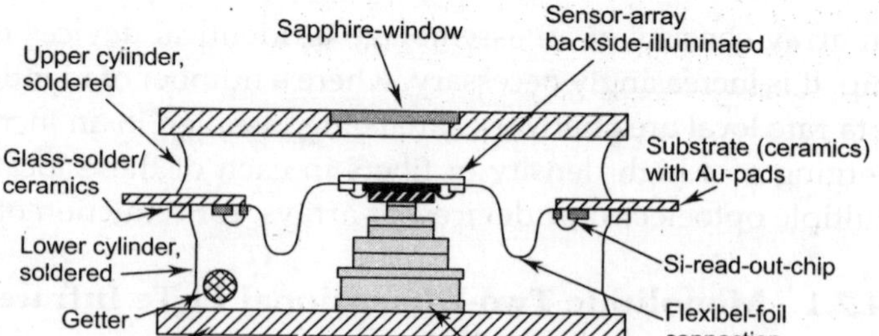

Fig. 14.47

14.7.3 Infrared Sensor Arrays On Si-Substrate

Narrow bandgap lead-chalcogenides are of use for fabrication of infrared detectors for the thermal wavelength (3-12 m)range. Layers such as PbSnSe and PbTe with band gaps of 0.1–0.2 eV are grown by Molecular Beam Epitaxy (MBE). An intermediate CaF_2 buffer layer serves for compatibility. The quality of infrared devices critically depends on the quality of the layers. Contrary to narrow gap HgCdTe, dislocations in IV-VI layers can move easily along their main glide system at high and even at low temperatures. This behavior can be used to improve the structural quality of the layers. Photograph in Fig. 14.48(*a*) shows how dislocations moved towards the edges of a sample. The movement is caused by the thermal mismatch strain induced on temperature changes. A dislocation density as low as 1×10^6 cm^{-2} and below were obtained after such anneals, and even on samples of several cm size. Keeping in mind that these values are obtained in 3 m layers on heavily lattice mismatched substrates, these values compare favorably to HgCdTe layers on lattice matched CdZnTe substrates where dislocation densities of 10^5 cm^{-2} are observed. The layers are used to fabricate photovoltaic infrared sensor arrays for thermal imaging and spectroscopic applications. Figure 14.48(*b*) shows the resistance-area product (essentially the inverse of the noise current density) as a function of temperature. The theoretical values as predicted by the Schottky theory is obeyed down to about 100 K. Below this temperature, the change in slope is due to inhomogeneities of the Schottky barrier height.

Fig. 14.48(*a*)

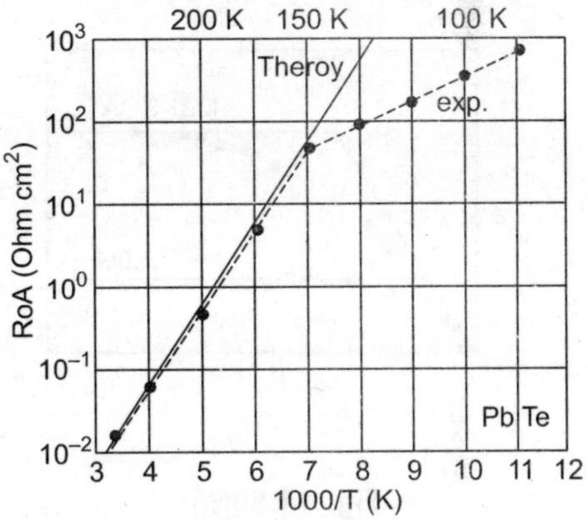

Fig. 14.48(*b*)

14.7.4 InGaAs/InP Photodiode Array for Optical Interconnect

Optical interconnects offer the possibility to remove the electronic interconnect bottleneck which arise from ever increasing IC complexity, speed and interconnect requirements. Two-dimensional flip-chip mounted InGaAs/InP photodiode arrays are key components for high-throughput optical intra-chip and inter-chip interconnects (as opposed to edge). In order to obtain an efficient, high speed optoelectronic interface between the silicon integrated VLSI electronics and optical fibers, an InGaAs/InP pin photodiode array is developed that is flipchip mounted on the silicon IC and that allows simultaneous illumination with a wavelength of 980 nm through the substrate. A cross-section through the photodiode chip, illustrating its concept is shown in the Fig. 14.49 below. The photodiode array chip consists of 4 × 8 back-illuminated photodiodes with the active regions on a 250 × 250 grid. The light sensitive region is made accessible through a hole etched in the back-side of the wafer. The bottom of the hole is coated with an antireflection coating optimized for a wavelength of 980 nm, whereas the sidewalls are metallized in

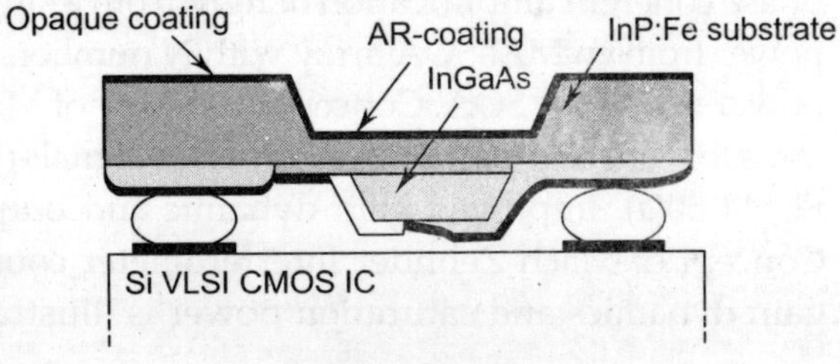

Fig. 14.49

order to collect a maximum of the divergent light from the optical fiber and to minimize crosstalk due to light coupled to neighbor photodiodes in the array. In order to prepare the photodiode array chip for flip-chip solder bonding, the solder is deposited on the photodiode InP wafer.

14.7.5 Semiconductor Optical Amplifier Arrays

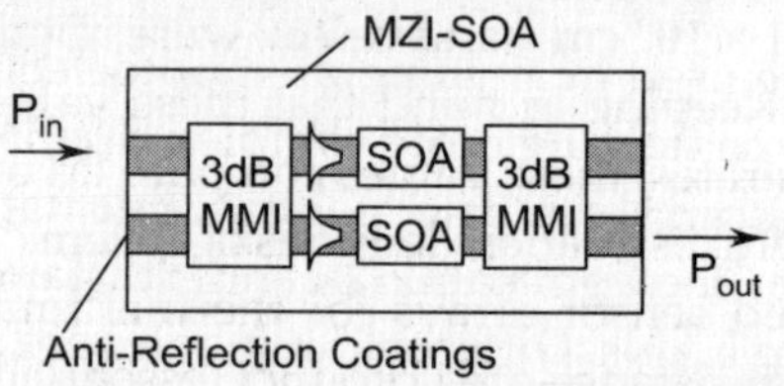

Fig. 14.50(*a*)

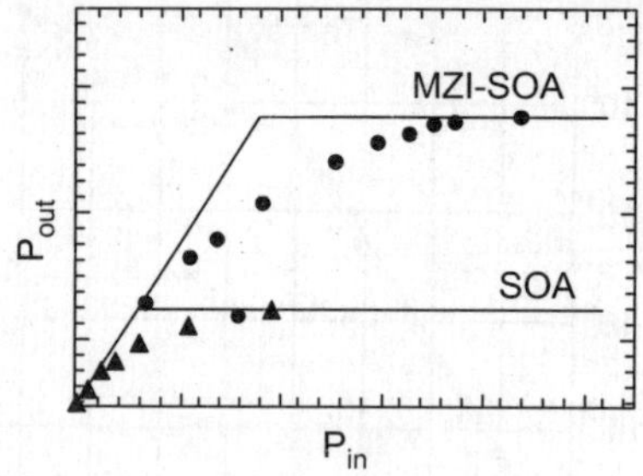

Fig. 14.50(*b*)

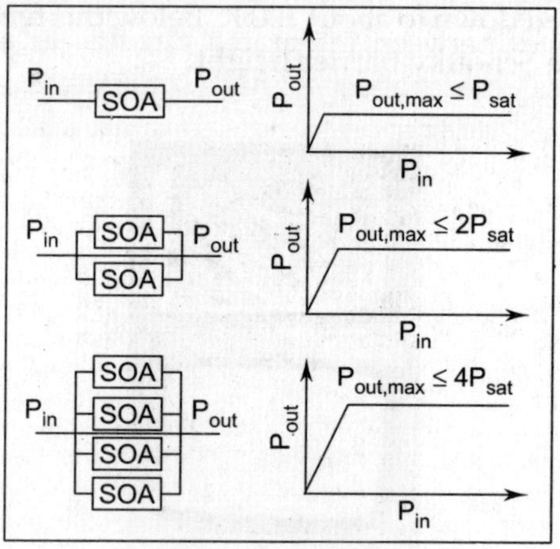

Fig. 14.50(*c*)

High single-mode optical powers coupled into single-mode fibers are a key for many applications in meteorology, medical surgery, laser machining and optical communications. To achieve this, the coherent couplings of laser diode or Semiconductor Optical Amplifier (SOA) arrays have been investigated. In one such an arrangement semiconductor optical amplifiers are monolithically embedded into monolithic Mach-Zehnder interferometer structures and used for phase coherent amplification of light from a single longitudinal mode laser. The maximum output power from an MZI-SOA array with N numbers of SOAs is expected to reach N times the saturation power of a single SOA. Conceptional view of Mach-Zehnder Interferometer (MZI) with SOAs in its arms for coherent coupling of amplified signals (MMI = Multi Mode Interference Coupler) is given in Fig. 14.50(*a*). Improved gain dynamic and output saturation powers are shown in Fig. 14.50(*b*). Concept of Mach-Zehnder interferometer coupled with SOA arrays for improved overall SOA gain dynamics and saturation power is illustrated in Fig. 14.50(*c*).

14.8 OPTICAL INTERCONNECTS

With the advent and development of low-loss and low-dispersion silica fibers, long distance optical fiber communication has received a significant boost for operation in the spectral range of 1.1 to 1.6 μm. Therefore, most existing fiber optic links are for high data rate, point-to-point, long distance communication. There has been increasing awareness of the fact that optical fibers may also be used in short distance, high connectivity applications. As example, the overall speed of VLSI chips is not only limited by the speed of individual devices, but also by the interconnect delays. The time taken for transmission of signal across a conducting wire is determined by the RC time constant that is proportional to the square of the wire length. Therefore, there has been growing interest in replacing such interconnects with optical interconnects to obtain higher transmission speed — the speed of light.

Optical interconnects may be used in an integrated optoelectronic chip, between two chips or between printed circuit boards containing more complex circuits. Interconnects are also required between a group of devices that can share data, perform calculations or can perform input-output operation. These are usually for the application of short distance and high connectivity and are termed as *Local Area Networks* or LANs. Different topologies are used for the connectivity scheme. The star and ring network scheme are shown in the Figs. 14.51(*a*) and 14.51(*b*) respectively. Massively parallel interconnect network schemes are required for computing systems with fast data transfer between integrated circuits. In addition to the long distance communication systems, optical interconnects of high connectivity, parallelism, high speed, low loss and low crosstalk are desired for short distance communication.

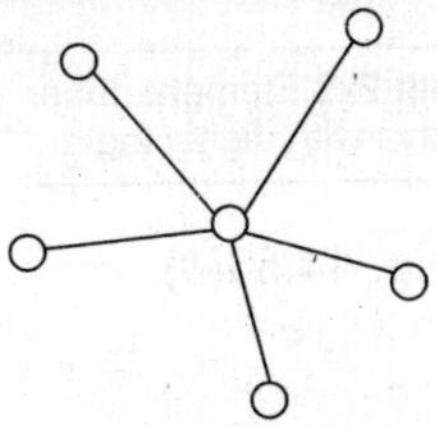

Fig. 14.51(*a*)

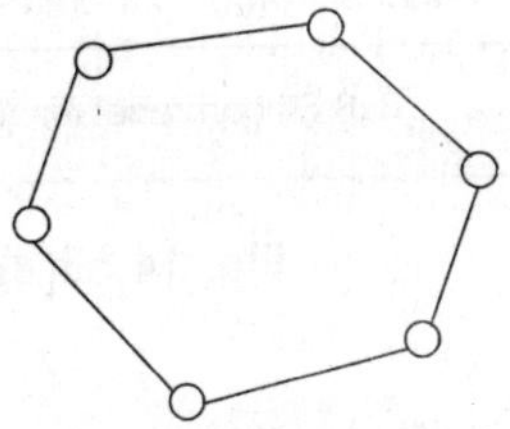

Fig. 14.51(*b*)

Three types of optical interconnects can basically be used for inter-chip, or intra-chip, or interboard connections. They are for optical waveguides, which are already described, free space communication and optical fibers. Free space communication includes holographic techniques and is used in optical cross-bars for imaging and pattern recognition.

In high speed LAN network, optical interconnects principally interface various functional modules as shown in Fig. 14.2. Coupling is accomplished by the use of optical fiber backplanes and backplane couplers. A common fiber is an optical guide characterized by an inner high index core and outer low index cladding. Fiber properties are largely determined by core and cladding diameters and refractive indices. Fiber insertion loss, coupling loss and intermodal distortion are the important parameters for the optical fibers. There are several difficult technological hurdles in fiber interconnect technology. First is the interfacing or coupling with active or passive devices like detectors, lasers, and waveguides. The second aspect is related to the fact that bidirectional transmission, or the fabrication of fiber couplers and power splitters/combiners may require splitters and isolators. Low coupling and insertion losses are to be provided for these components. The efficiency of optical fiber links is governed by power budget and time budget calculation. These are related to the important parameters of the fiber - sensitivity and bandwidth respectively. Both the properties are also dependent on the length of the fiber. For long distance application of spectral range of 1.1 to 1.6 μm exceeding several kilometers fiber loss and dispersion become the important issues, and InP based alloys and hetrostructures such as InGaAs/InGaAsP/Inp are used. For short distance communication and LAN application, GaAs based devices are suitanle where dispersion loss is not of much importance.

14.8.1 Silicon Integrated Circuits for Optical Interconnects

In the near future the data throughput of inter-chip and intra-chip communication will exceed the capabilities of standard electrical interconnections. A processing architectures allowing the introduction of high density two dimensional interconnections with aggregate capacities in the Gbit/s range is mentioned here. Figures 14.52(*a*) and 14.52(*b*) show the set of CMOS driver and receiver arrays respectively for bit rates up to 600 Mbit/s with 150 Mbit/s per channel.

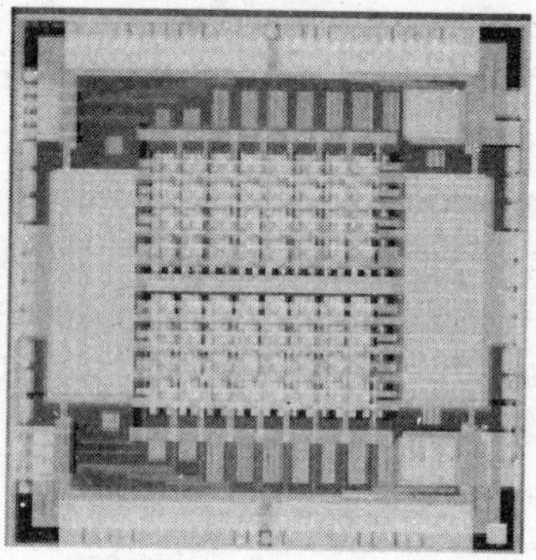

8x8 Silicon receiver array

Fig. 14.53(*a*)

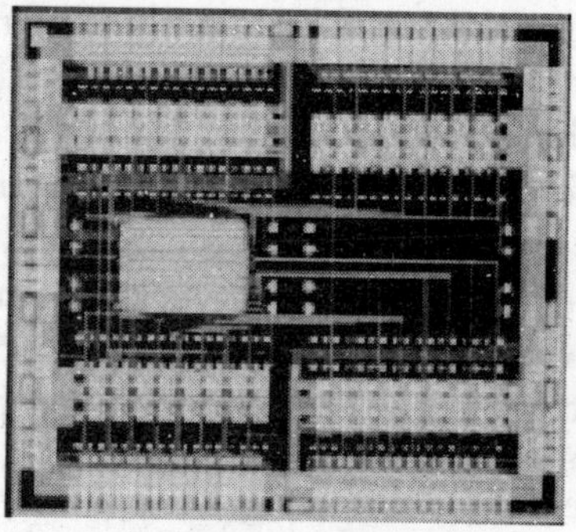

Demonstrator with 2x8 Element driver and receiver arrays with digital logic

Fig. 14.53(*b*)

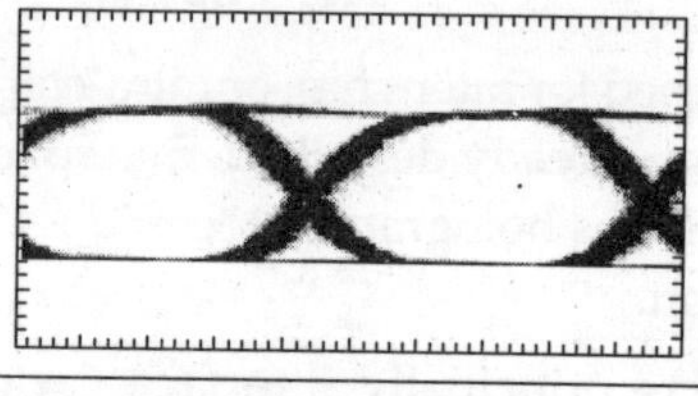

RX: 100 Mn/s, I_{PD} = 20 μAPP
V_{SUPPLY} = 3.3 V, PRBS = 2^{15}–1
500mV/Div, 2ns/Div

Fig. 14.53(*a*)

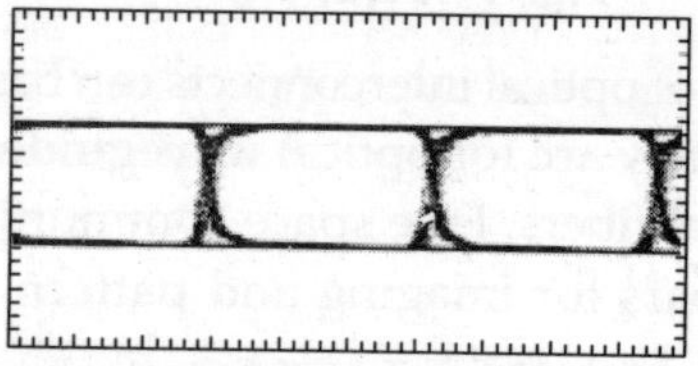

TX: 600 Mb/s, I_{MOD} = 1.4 mA, I_{BIAS} = 3.4 mA.
V_{SUPPLY} = 3.3 V, PRBS = 2^{31}–1
0.5ns/Div

Fig. 14.53(*b*)

The arrays are of 2 × 8 and 8 × 8 driver/receiver elements. The pitch of the single interconnection channels is 250 μm in each direction. A photodetector array with 4 × 8 photodetectors has been designed and fabricated. The detector array will be flipchip mounted on the silicon receiver ICs. Figures 14.53(*a*) and 14.53(*b*) represent the eye-diagrams of receiver and driver.

14.8.2 CMOS Driver and Receiver Circuits with Surface Emitting Lasers

Optical interconnects combining parallel optical paths consisting of transmitter, optical fiber or waveguide and receiver-arrays are of interest for high throughput interconnections between electronic systems. For such interconnects, that use vertical cavity surface emitting laser diode (VCSEL) arrays, electronic drivers and receiver circuits are developed. A set of driver and receiver demonstrator chips, which allow comparing the propagation delay and the power dissipation of optical and electrical links have been designed.

	Measurement setup with demonstrator chip containing two independent VCSEL, drivers and equivalent electrical output buffers in 0.8mm CMOS. Both electrical and optical output signals are generated either from the on chip VCO or from an external applied input signal.
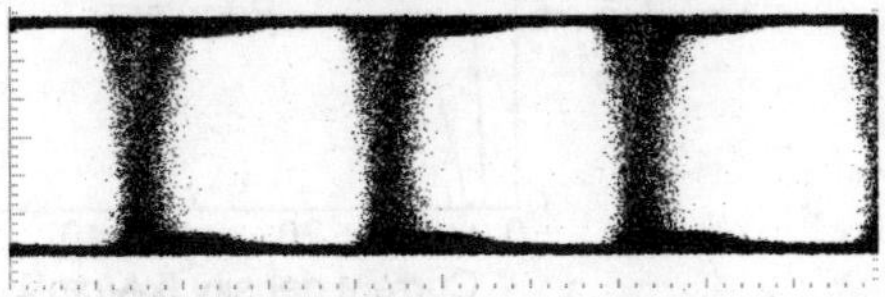	Eye-diagram of transmitter demonstrator at 700 Mbit/s; Supply voltage 3.3V; I_{bias} = 2.5 mA; I_{mod} = 4 mA PRBS = 2^{15}–1; Scale: 0.5ns /dv; 50 mV/div
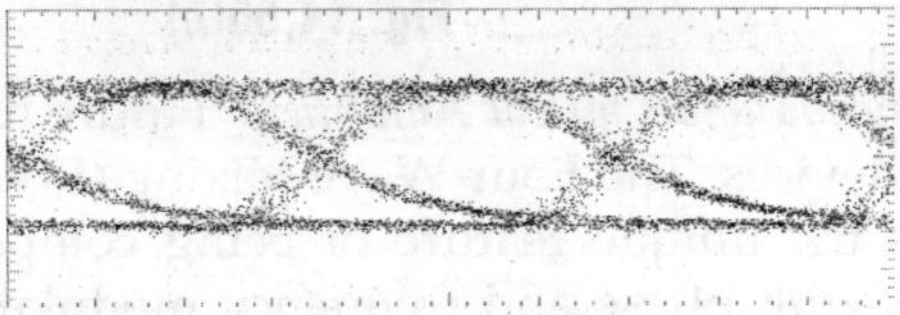	Eye-diagram of receiver setup at 300 Mbit/s; Supply voltage 3.3 V; $I_{Detectro}$ = 15 μ A_{ppx}; PRBS = 2^7–1; Scale: 1 ns/div; 100mV/div

Fig. 14.54

The driver works from DC up to 800 Mbit/s. In this driver either the on chip Voltage Controlled Oscillator (VCO) or an external applied signal can be selected as input signal. This input signal can be directed either to the optical or the equivalent electrical output. The receiver works up to 300 Mbit/s. An on chip phase measurement circuit allows to measure the phase difference between the optical and the equivalent electrical link. The integrated circuits have been fabricated on a commercial 0.8 μm CMOS technology. Figure 14.54 illustrates the measurement set up with demonstrator chip containing two independent VCSEL drivers and equivalent electrical output buffers in 0.8 μm CMOS, eye-diagram of transmitter and receiver. The supply voltage for both circuits is 3.3 V. Special care has been taken in the design to minimize the power dissipation.

14.9 OTHER OEIC DEVICES

Single Transverse Mode Active Multi-Mode Interferometer Laser Diode for High Power Applications: High power Laser Diode (LD) is attractive for many applications, for instance, long distance telecommunication system, compact disk pick-up, and others. One of the ways to obtain high power LD is to utilize multi mode waveguide structure since it provides wider pumping area than one of regular single mode LD. On the other hand, single transverse mode output, which is contrary to widening the pumping area, is one of the essential requirements for those above applications.

To realize single transverse mode LD by utilizing multi-mode waveguide, a novel active multi-mode interferometer (active MMI) LD has been developed, which incorporates 1 × 1 MMI in between single mode waveguides. By exploiting it to regular infrared LDs, efficiency and single transverse output power have been increased significantly compared with regular LD. Moreover, significant threshold current density reduction has also been obtained. The schematic diagram and the performance characteristics of this are shown in Figs. 14.55(*a*) and 14.55(*b*). These results predict the drastic improvements in every LDs characteristic.

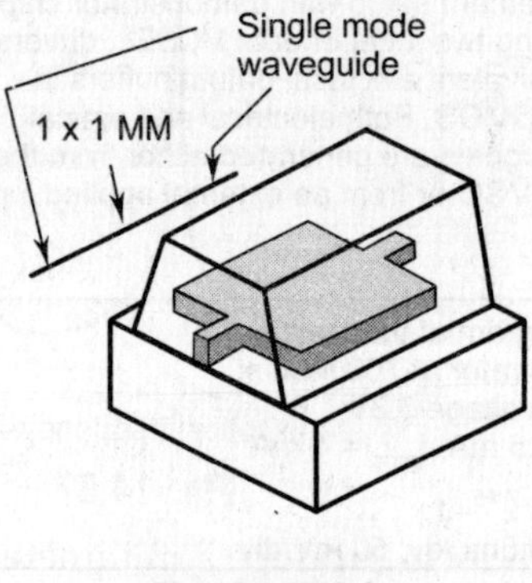

Fig. 14.55(*a*)

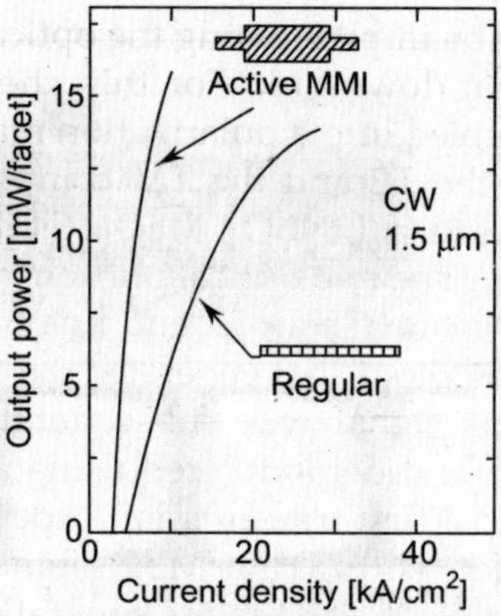

Fig. 14.55(*b*)

Highly Efficient Four-Wave Mixing in Bulk SemiconductorOptical Amplifiers: Future high bit-rate optical networks call for high-speed all-optical devices. The Four-Wave Mixing (FWM) is a key contender for optical networking, since it has the unique feature of being compatible with wavelength division multiplexed data streams, with phase and frequency modulated signals, and with analog signals.

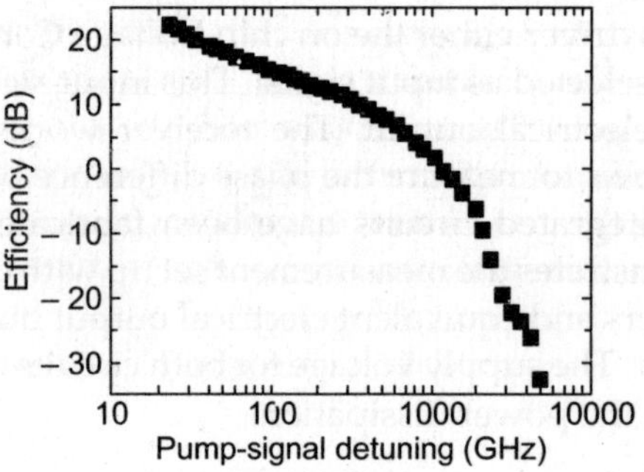

Fig. 14.56(*a*)

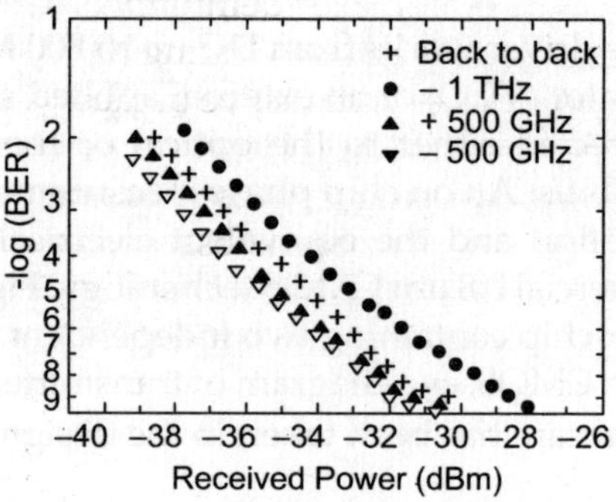

Fig. 14.56(*b*)

The FWM effect in Semiconductor Optical Amplifiers (SOA) was however thought to be a rather inefficient process. But the increase of the length has been shown to lead to a higher efficiency, due to the increase of the unsaturated gain. It has also been shown that the limitation of the gain due to amplified spontaneous emission does not limit the FWM performance. Record efficiency of +5 dB at 1 THz pump-signal detuning together with a high signal to background ratio have been obtained in 1.5 and 2 mm long bulk SOAs. The use of pigtailed devices in sub-system experiments has lead to a successful signal frequency conversion up to 2 THz at 2.5 Gbit/s. Improvements of the subsystem setup are in progress in order to measure the performance of the devices at higher bit-rate by using FWM and cross gain modulation schemes, which are shown at Figs. 14:56(*a*) and 14.56(*b*).

Antenna Fiberoptic Remcte Feeding Using a Dual-Polarization External Cavity Diode Laser: The optical generation and distribution of millimeter-wave antenna signals to be employed broadband mobile communications networks has recently attracted much interest. A 1.55 µm dual-polarization emission external cavity laser (DP-ECDL) of which both, the TE and the TM mode can be tuned independently in wavelength over approximately 60 nm has been demonstrated. In a fast photodetector, the beating of the two modes creates the millimeter-wave of desired frequency according to the wavelength separation of the optical modes. It is feasible of data transmission using this

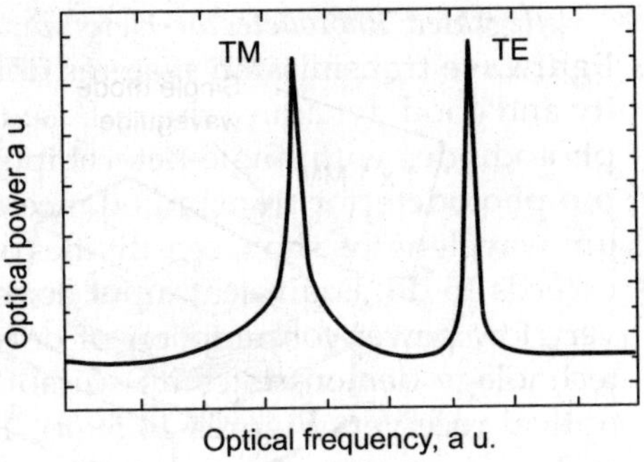

Fig. 14.57

technique by implementing the optical transmission part of a virtual down-link. For this, the output of the DP-ECDL coupled into a polarization maintaining fiber was split into the TE and the TM component, of which the latter passed a $LiNbO_3$ Mach-Zehnder interferometer modulator. The two optical portions were recombined in a fiber directional coupler and finally detected in a high-speed photodiode that generated the electrical millimeter-wave. This signal was subsequently down-converted, amplified, and demodulated using a self-multiplication scheme. In a first transmission experiment, the modulator was fed with a 140 Mbit/s binary amplitude shift keyed signal, and the mode separation was adjusted such as to produce a millimeter-wave carrier at 58 GHz. Figure 14.57 shows Fabry-Perot spectrum of the DP-ECDL under modulation of the TM mode. The modulation sidebands are apparent around the TM mode. The actual mode separation of 58 GHz cannot be deduced from the figure.

Wavelength Multiplexer/Demultiplexer Filters in Glass on Silicon: Wavelength division multiplexing (WDM) is becoming increasingly important in fiber optical communications, since it offers full exploitation of the almost unlimited bandwidths of optical fibers. Key elements of WDM are wavelength multiplexer/demultiplexer filters. For an optical packet switching and routing network the Arrayed Waveguide Grating (AWG) concept is used to realize wavelength filters. The silica on silicon technology is attractive because it offers low insertion loss, low crosstalk, good reproducibility, and long term stability. However it holds undesirable polarization dependence owing to the strain birefringence. Using a stress releasing technique, significant reduction in the polarization dependence of the filter has been observed while maintaining the good filter performance in crosstalk and loss. Figures 14.58(*a*), 14.58(*b*) and 14.58(*c*) illustrate the schematic and performance characteristics of WDM filter in glass on silicon.

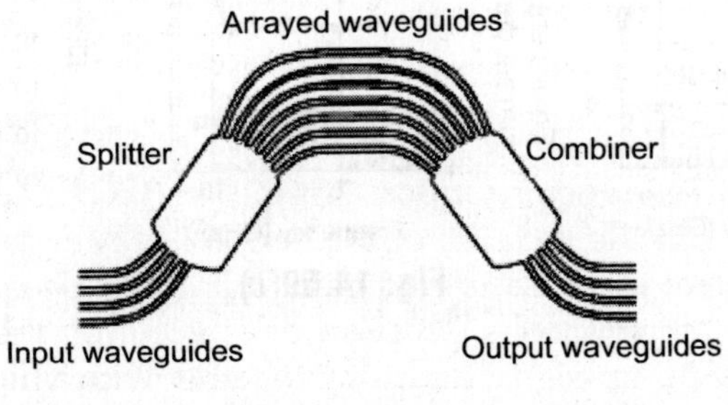

Fig. 14.58(*a*)

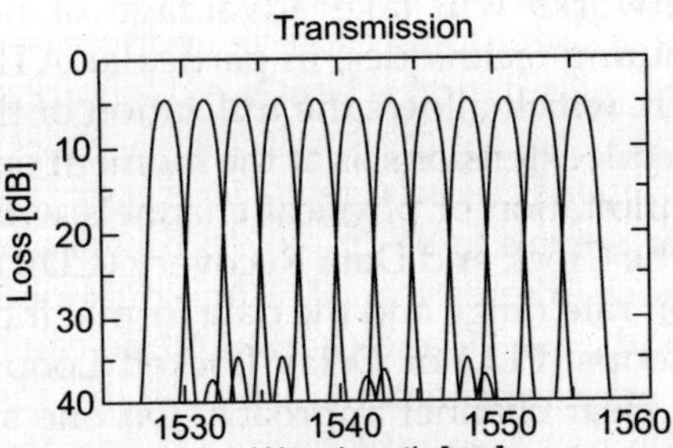

Fig. 14.58(*b*)

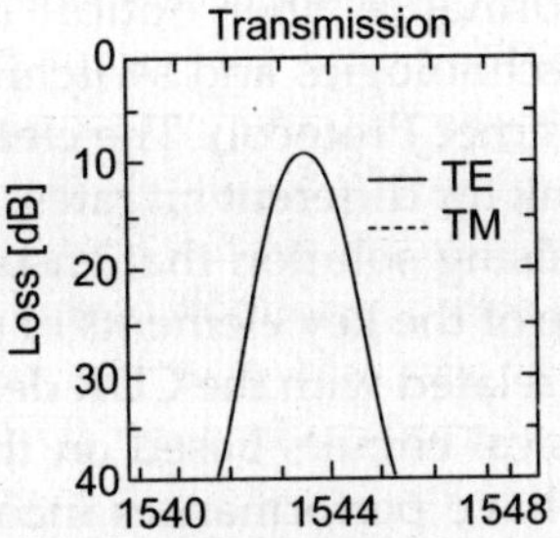

Fig. 14.58(*c*)

Integrated Photodetector-Heterobipolar Transistor Optical Receivers: Increasing data rates of lightwave transmission systems demand for high speed optical receivers with high sensitivity and good dynamic ranges. Based on InP/InGaAs technology, monolithically integrated pin photodiodes with single-heterobipolar transistors has been realized. The packaged monolithic pin photodetector transimpedance amplifiers designed for 10 and 20 Gb/s operation at 1.55 μm wavelengths show sensitivities of –19.7 dBm and –15 dBm, respectively. Their dynamic range exceeds 15 dB. Equivalent input noise current density is as low as 6.5 pA/Hz$^{1/2}$. Together with the very low power consumption of only 36 mW of the receiver chips this InP/InGaAs pin/SHBT technology demonstrates its suitability for high-speed, low-noise monolithically integrated optical receivers Figures 14.59(*a*), 14.59(*b*) and 14.59(*c*) describe the schematic, photograph and performance characteristics.

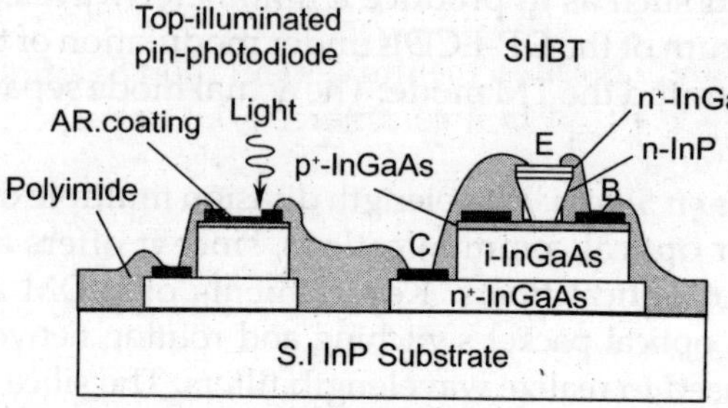

Fig. 14.59(*a*)

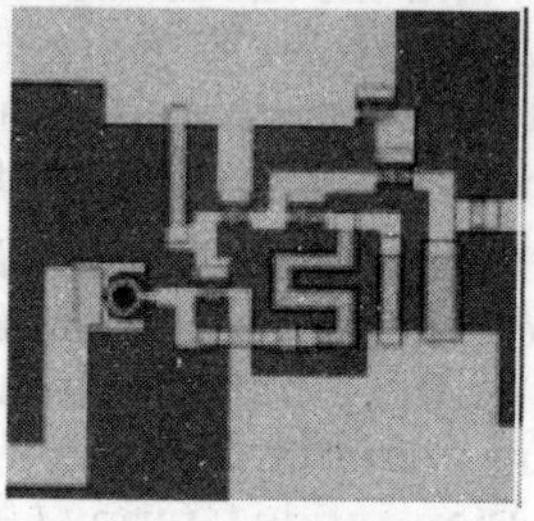

Fig. 14.59(*b*)

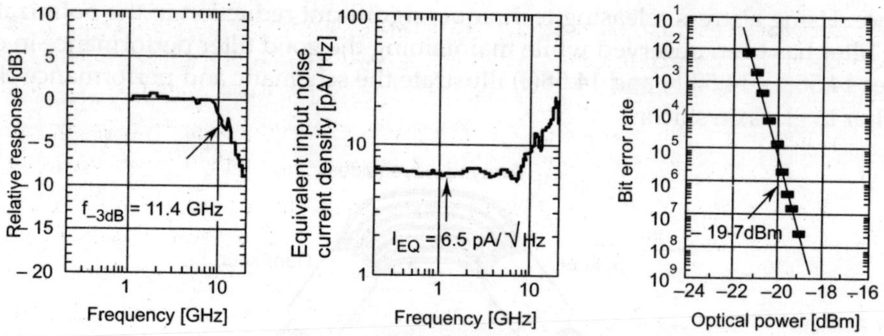

Fig. 14.59(*c*)

14.10 CONCLUSION

High performance fiber optical communication networks will take advantage of new and evolving technologies and switching/multiplexing/routing techniques, in particular ATM, SDH and IP (Internet Protocol). The clear channel approach, which allows the utilization of the same physical link for different bit rates, protocols and network extensions, is at the moment one of the most promising solution that ensures the optimal exploitation of physical channels within networks. One of the key elements of such a network is the Clock and Data Recovery (CDR) circuit. The issues related with the CDR design are the wide bit rate range and the data format independence. Classical circuits based on the Phase Locked Loops (PLL) or Delay Locked Loops (DLL) structures have performances incompatible with the clear channel approach. On one side, the PLLs work with different protocols but only within a restricted bit rate range, while on the other side DLLs allow a wide bit rate range operation at cost of the protocol independence.

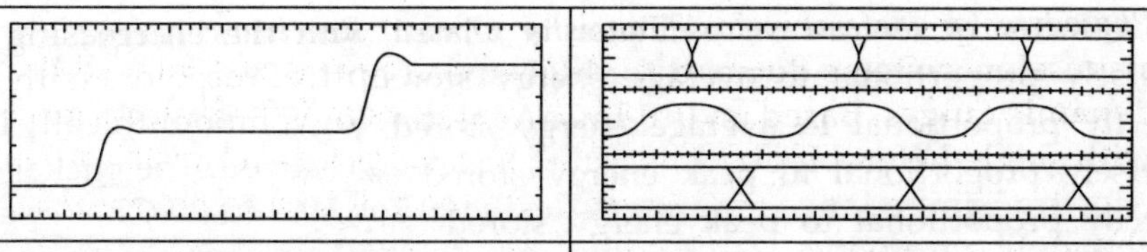

PLL loop filter voltage showing the circuit step response with small overshoots. The picture has been taken while switching the three different data bit rates 600, 700 and 800Mbit/s. The three voltage levels shown above correspond to these three bit rates. (20 ms Horizontal.; 200 m Vertical)

Eye diagram at 1062Mbit/s with a PRBS $2^{31}-1$ input. The top trace shows the input data, the middle trace the extracted clock signal and the bottom trace displays the extracted data. (3 ns Horizontal.; 400 mV Vertical)

Fig. 14.60

A modification of the CDR based on the PLL structure has fed to the implementation of a first prototype suitable for clear channel links. The circuit, realized in a commercial 0.8 µm silicon BiCMOS technology, works at the bit rates of 622 Mbit/s, 737 Mbit/s and 1.06 Gbit/s requiring only a small preamble to synchronize the incoming data. The required power supply ranges from 3 V to 5.5 V at the low power dissipation of 100 mW to 190 mW. The output jitter is lower than 20 *ps* at 622 Mbit/s with PRBS $2^{31}-1$ random sequence, decreasing further at higher bit rates. Figure 14.60 illustrates the step response and eye-diagram of the circuit with a PRBS $2^{31}-1$ random input data sequence at 3.3 V power supply. The CDR loop filter step response has been shown when changing the data bit rate from 600 to 700 and to 800 Mbit/s, some eye diagrams at 1062 Mbit/s and the output clock jitter at 1062 Mbit/s.

Optical system has made a large impact on communication system because of its several advantages. Modern day communication system can not dispense with the optical system and continuous upgradation of technology is being carried out. Scientists and researchers are always looking for new and upgraded integration technology and integrated circuits, which might be suitable for application. In this chapter, only a few is described. As technology advances, it is sure many more to come.

REVIEW QUESTIONS

1. Which technique is used to create in-chip laser facets?
(*a*) Epitaxial regrowth (*b*) Planner
(*c*) Cleaving (*d*) Dry-etching

2. What is resulted using the MODFET structure of photoreceiver?
(*a*) Higher carrier mobility (*b*) Higher drift velocity
(*c*) Both of these (*d*) None of these

3. If B is the bit rate, then how sensitivity of photoreceiver with PIN-HBT structure behaves?
(*a*) Proportional to B (*b*) Proportional to B^2
(*c*) Proportional to B^3 (*d*) None of these

4. What is observed if the thickness of absorption layer is increased for conventional surface-illuminated photodetector?
(*a*) Higher transit time and higher qauantum efficiency
(*b*) Higher transit time and lower qauantum efficiency
(*c*) Lower transit time and higher qauantum efficiency
(*d*) Lower transit time and lower qauantum efficiency

5. How the Q-factor of microwave oscillator is related with the energy?
   (*a*) Inversely proportional to average energy stored
   (*b*) Directly proportional to average energy stored
   (*c*) Inversely proportional to peak energy stored
   (*d*) Directly proportional to peak energy stored
6. Fiber in the loop (FITL) concept used for
   (*a*) Unidirectional communication in single mode fiber
   (*b*) Unidirectional communication in multimode fiber
   (*c*) Bidirectional communication in single mode fiber
   (*d*) Bidirectional communication in multimode fiber
7. InP based buried optical waveguides exhibit
   (*a*) Low mode size and low waveguide-to-fiber coupling losses
   (*b*) Low mode size and large waveguide-to-fiber coupling losses
   (*c*) Large mode size and large waveguide-to-fiber coupling losses
   (*d*) Large mode size and low waveguide-to-fiber coupling losses
8. For practical epitaxy-first method is most suitable for fabrication of
   (*a*) Small scale OEIC (*b*) Medium scale OEIC
   (*c*) Large scale OEIC (*d*) Very large scale OEIC
9. In injection locking for microwave oscillator, how the locking range is dependent with injected power?
   (*a*) Directly proportional to injected power
   (*b*) Proportional to square of injected power
   (*c*) Proportional to square root of injected power
   (*d*) None of these
10. When a semiconductor material is doped with a large density of free carriers, how its refractive index changes with comparison to pure material?
    (*a*) Decreases (*b*) Increases
    (*c*) No change (*d*) Changes depending on doping material
11. What are the fields of application for optoelectronic integrated circuits?
12. What are the advantages and disadvantages of monolithic and hybrid methods of fabrication of OEIC? What is epitaxial lift off?
13. Discuss the choice of materials and governing parameters for fabrication of OEIC.
14. What is MODFET scheme? Discuss the monolithic process of fabrication of photoreceiver. What are the two most demanding steps of InP-HBT process for photoreceiver and why?
15. Describe the process of fabrication of QEIC with laser in-chip.
16. Describe the processes to develop integrated optical transmitter and transreceiver.
17. Explain the terms optical switching, optical tuning and optical injection locking. Where are these used? What is Q-factor?
18. Explain the operation of integrated Mach-Zehnder interferometer. What are applications of it?
19. What are PHASAR and its application? Discuss the its basic structure and fabrication. Describe its advantages.
20. Write short notes on
    (*i*) WDM switching devices (*ii*) Infrared sensor array (*iii*) optical interconnect (*iv*) Photodiode array for optical interconnect (*v*) CMOS Driver and Receiver Circuits with Surface Emitting Lasers (*vi*) Single transverse mode LD

15

FIBER OPTICAL NETWORK

15.1 INTRODUCTION

Optoelectronic technology and optical fiber system have made tremendous impact on transmission and communication systems. The improvement of communication systems has been further enhanced by the development of superior light sources and detectors as well as low loss optical fibers. The optical fiber transmission offers several advantages as summarized below:

- Enormous potential bandwidth
- Small size and weight
- Ruggedness and flexibility
- Immunity to interference and crosstalk
- Electrical isolation
- Low transmission loss and dispersion
- Signal security
- System reliability
- Reduced power consumption
- Low cost.

The optical fiber communication system offers very attractive proposition for use of short distance as well as long distance communication within national and international telecommunication networks. In this chapter some network application and topologies will be discussed.

15.2 NETWORK TOPOLOGIES AND CONFIGURATIONS

Optical networks interconnect nodes which are in the form of terminals, switches, telephones, computers or any other equipment. Different types of communication systems are shown in Fig. 15.1. Figure 15.1(*a*) illustrates point to point communication link which may be unidirectional or bidirectional. Figure 15.1(*b*) explains multipoint network, but all nodes are connected to a master node. This type of network is useful for broadcasting system and eventually as fiber-to-home concept. Figure 15.1(*c*) is the combination of first two where each of multiple nodes is inter-connected to more than one node and they are used for medium to long distance communication systems.

Fig 15.1(*a*)

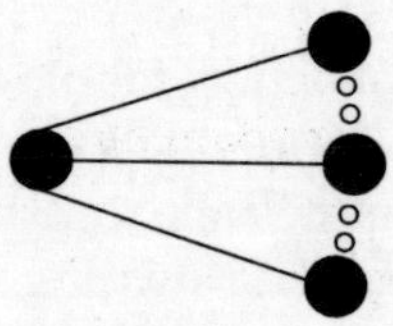

Fig 15.1(*b*)

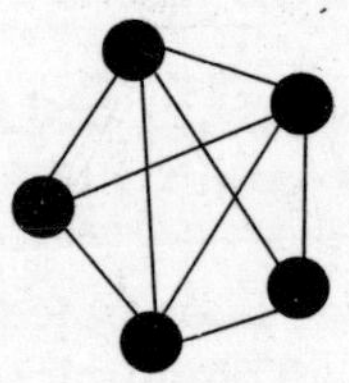

Fig 15.1(*c*)

Any optical link is characterizes by its bandwidth, error rate and propagation delay. Depending on the span of networks, they are classified as *Local Area Network* (LAN), *Metropolitan Area Network* (MAN) and *Wide Area Network* (WAN). The spans are 1 to 2 km, 2 to 1000 km and more than 1000 km for LAN, MAN and WAN respectively.

The most common network topologies that exploit the properties of the fiber and route the information with minimal optoelectronic conversions are *star*, *bus* and *ring* topologies. In star topology a passive central star coupler receives all inputs and distributes the sum to each output. In bus topology any transmission can be directly attributed to any receiver. In ring topology consecutive nodes are arranged around a fiber ring by point to point links. Each interfacing node is an active device that can act according to information packet and instruction bits received. Based on the instruction bits or header bits the node can listen, transmit or bypass. Nodes can also direct to other station that the network serves. The star, bus and ring topologies are shown in Figs. 15.2(*a*), 15.2(*b*), 15.2(*c*) respectively. The combination of these topologies as shown in Fig. 15.2(*d*) may also be used which is called *tree* topology. To utilize the network bandwidth, each node can have more than one transmitter and receiver.

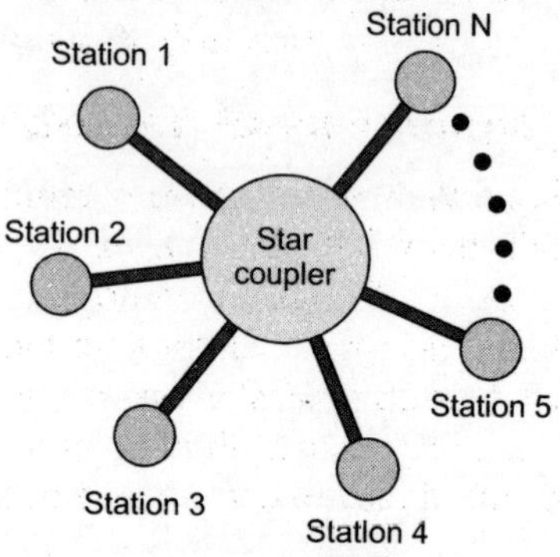

Fig 15.2(*a*)

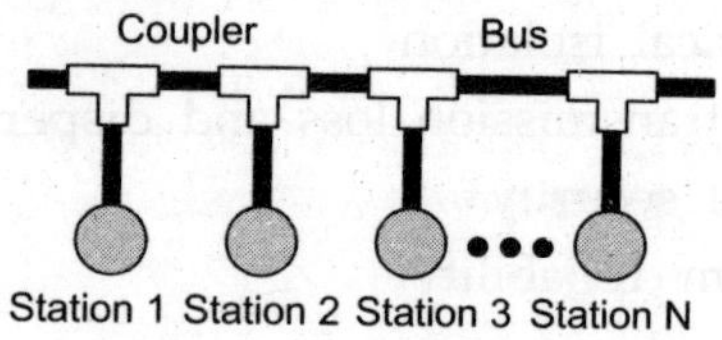

Fig 15.2(*b*)

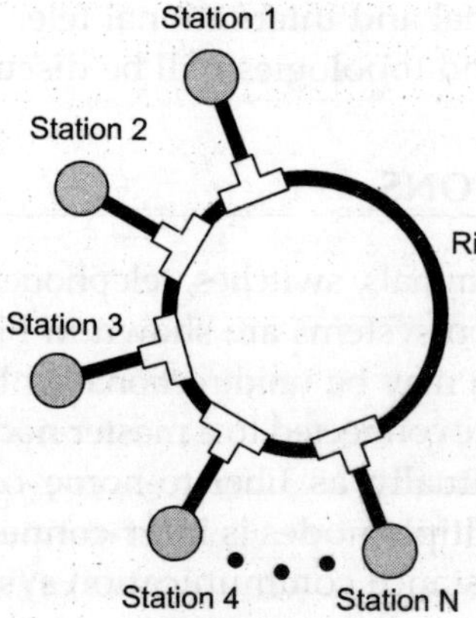

Fig 15.2(*c*)

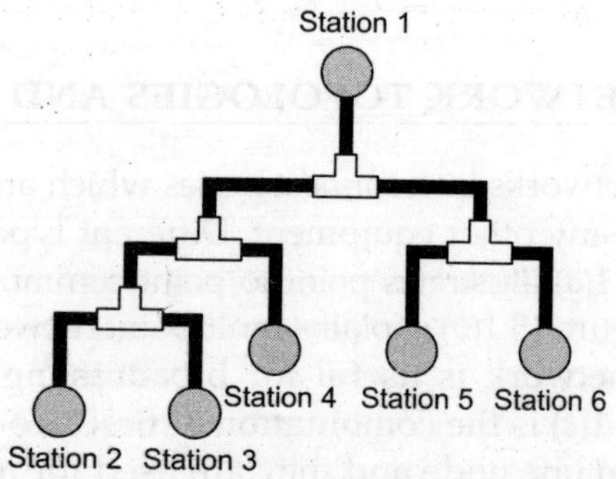

Fig 15.2(*d*)

15.3 LOCAL AREA NETWORK

Local area network or LAN is an interconnection topology usually confined to either a single building or group of buildings within a confined site or establishment (*e.g.*, industrial, educational, military etc.). The LAN typically operates at a transmission rate of between 100 Kbit/sec to 100 Mbit/sec is operated and controlled by the owning body rather than a common carrier. Optical fiber communication technology is being utilized within LANs to meet the on-site communication requirements of large commercial organizations and to enable access to distributed or centralized computing resources.

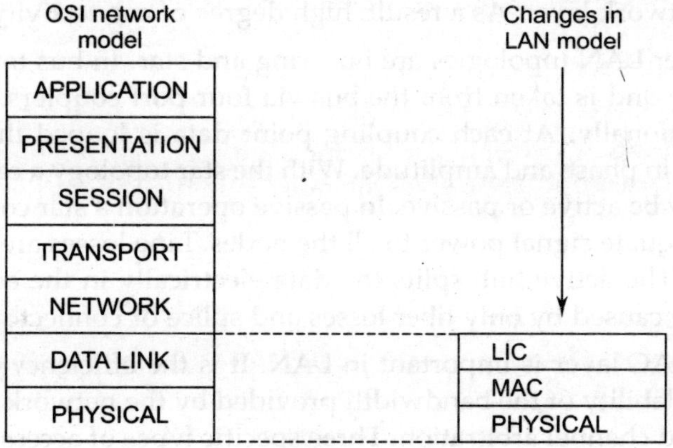

Fig 15.3

Table 15.1

Principal Functions for the Seven Level OSI Model	
Physical	This is related to transmission of unstructured bit stream over the physical link and interfaces the physical transmission medium.
Data link	Provides for reliable data transfer over the physical link and controls the transmission and reception of blocks with necessary synchronization, error and flow control.
Network	For interworking with the underlying telecommunication network. Provides independence to the upper layers from data transmission technologies.
Transport	Provides reliable transparent transfer between end points, end to end synchronization, flow control and error recovery.
Session	Engages terminals to applications during a session. Establishes, manages and terminates connections (sessions).
Presentation	Performs useful manipulation or transformation on data structure (*e.g.*, encryption, text compression, reformatting).
Application	Provides services to the users of the OSI environment, *e.g.*, data base management, network management.

Open Systems Interconnection (OSI) seven layer network as developed by the International Organization for Standardization (ISO) to provide standard architecture defining communication tasks into a set of hierarchical layers within networks is shown in Fig. 15.3. Each layer performs certain set of function in order to communicate one system (nominally a heterogeneous computer) to another as illustrated in table 15.1. This was primarily developed for Wide Area Networks (WAN).

For LAN the bottom two layers of OSI model are modified to three layers. The physical layer as in WAN, is responsible for the physical transmission of information.

In LAN the functions of data link layer are distributed over two layers, namely the Logical Link Control (LLC) and Medium Access Control (MAC). The LLC layer assembles/ disassembles data frames or packets and provides appropriate address and error checking fields. MAC layer organizes communication over the link. The MAC layer provides the set of logical rules or the access protocol which allow nodes to access the common communication channel. Therefore several MAC options may be provided for LLC layer. Another major difference between WANs and LANs is that each node on the network can be directly connected to all others, eliminating the need for routing via network layer. As a result, high degree of connectivity is achieved.

The basic optical fiber LAN topologies are bus, ring and star. In bus topology data generally circulate bidirectionally and is taken from the bus via four port couplers. In ring configuration data circulate unidirectionally. At each coupling point data is looped through the nodes and repeatedly regenerated in phase and amplitude. With the star topology a central hub is formed in the network which may be active or passive. In passive operation a star coupler at the hub splits the data to provide adequate signal power to all the nodes. Line losses are therefore determined by degree of splitting. The active hub splits the data electrically in the active star coupler and therefore line losses are caused by only fiber losses and splice or connector losses.

The design of the MAC layer is important in LAN. It is the efficiency of the access protocol which governs the availability of the bandwidth provided by the network for the dual functions of data transmission and channel arbitration. Three specific types of access protocol are accepted for their simplicity. They are - (1) random access protocol, the most popular example of which is carrier sense multiple access with collision detection (CSMA/CD) used on the Ethernet LAN, (2) token passing protocols, and (3) time division multiple access (TDMA) protocols.

With CSMA/CD protocols, data is allowed to be transmitted by the nodes as soon as the communication channel is found to be idle. The physical layer hardware detects if the transmission from two or more nodes collide and the nodes involved terminate transmission before attempting retransmission after a random time interval. Nodes obtain permission sequentially to use the channel from token passing protocol which behaves like a distributed polling system. The channel arbitration is determined by the possession of a small distinctive bit packet, known as token that can only be held by one node permitted to use the channel at any one time. After completion of data transfer from that node the token packet is passed to the next node in the logical sequence. Finally, with basic or fixed assignment TDMA protocols, a fixed length time slot is assigned to each node allowing for data transmission. Thus the protocol operates in time multiplexing manner on a point to point link.

Optical fiber regenerative repeaters may be used to restore the attenuated signal power in an optical fiber LAN as well as an active tap to allow the interconnection of numerous nodes within ring or bus topologies. Passive components like four-port tap or three-port tap may also be used to reduce the number of repeaters, in turn the cost. However these passive components incur excess transmission losses of the order of 0.1 dB to 1 dB and upset the power budget. Excessive use of taps (more than thirty) requires active components like optical amplifiers to provide gain.

A major problem associated with the tapping methods in optical fiber buses is uneven distribution of the transmitted optical power between the connected receivers. This results in two distinct drawbacks. First, a large dynamic range is required for the optical receivers and secondly, during detection of collision events CSMA/CD protocol becomes difficult. Use of passive star couplers overcomes this problem. Each node in the star topology is provided with a single input and single output fiber connected to the hub of the star, so that optical power within the input fiber can be more evenly distributed between the output fibers. Thus more uniform reception levels is achieved. Typical port to port insertion losses are around 19 dB and 23 dB for 32 × 32 and 64 × 64 port transmissive passive stars respectively with multimode fiber. Use of active star hub and cascaded passive stars are the variations on the passive star topology.

Increasing demand for information technology within commerce and industry has intensified the activity in the LAN area. Optical fiber LAN or mixed-media LAN products (fiber and twisted wire pairs) are available from several equipment suppliers and a good number of networks are at varying stages of development.

15.4 PUBLIC NETWORK APPLICATIONS

Optical fiber communication systems provide a variety of applications in public telecommunication network. The suitability of optical fibers for transmission first made an impact in this area. In worldwide telecommunication applications optical fiber links are the alternative to coaxial and high frequency pair cable systems. Even millimetric waveguide transmission is abandoned in favor of optical fiber system.

15.4.1 Trunk Network

The trunk or toll network is used to handle telephonic traffic between major trunk terminals situate or urban stations. The primary requirement of this is high capacity transmission system with minimum cost per circuit. The transmission distance for trunk system can vary from 20 km to over 100 km and occasionally as much as 1000 km. Therefore low attenuation at transmission link is desirable to minimize the number of repeater stations as well as cost. The optical fiber system is distinctly advantageous for this because of its increased bandwidth and repeater spacings. Also the optical fiber system is most economic compared to other transmission systems. Furthermore digital transmission through optical links is more economic with low error and higher signal to noise ratio. This aspect of cost reduction leads to the use of optical links for interoffice or junction network. The speed of operation has also enhanced to a great extent with the digital transmission and optical links. Table 15.2 indicates the typical optical power budget for a 140 Mbit/sec trunk system operating over 8 Km of multimode graded index fiber at the wavelength of 0.85 μm. Table 15.3 shows the typical optical power budget for high performance 565 Mbit/sec single mode fiber system operating at 1.3 μm and 1.55 μm transmission wavelengths.

Table 15.2

Mean power launched from the laser transmitter (butt Coupling)	–4.5 dB
APD receiver sensitivity at 140 Mbit/sec (BER10^{-9})	48.0 dB
Total system margin	43.5 dB
Cabled fiber loss (8 × 3 dB/Km)	24.0 dB
Splice losses (9 × 0.3 dB each)	2.7 dB
Connector loss (2 × 1 dB each)	2.0 dB
Dispersion equalization penalty	6.0 dB
Safety margin	7.0 dB
Total system loss	41.7 dB
Excess power margin	1.8 dB

Table 15.3

Transmitter type	*F-P laser 1.3 μm*		*DFB Laser 1.55 μm*	
Receiver type	PIN	APD	PIN	APD
Mean power launched from the laser transmitter (dBm)	0	0	– 3	– 3
Receiver sensitivity (dBm)	– 32	– 39	– 33	– 40
Total system margin (dB)	32	39	30	37
Connector loss (2 × 1 dB each)	2	2	2	2
Penalty (dB)	3	3.5	3	3.5
Ageing and temperature margin (dB)	2	2.5	3	3.5
Design margin (dB)	3	3	3	3
Safety margin (dB)	5	5.5	6	6.5
Maximum repeater spacing (Km)	55	70	95	125
Cabled fiber loss (dB/Km)	0.4	0.4	0.2	0.2
Total cable loss (dB)	22	28	19	25
Total system loss (dB)	32	39	30	37

15.4.2 Junction Network

The junction or interoffice network usually consists of routes within major urban stations over a distance of typically 5 to 20 km. The distribution of distances between switching centers (telephone exchanges) or offices in the junction networks of large urban areas varies considerably for various countries. The use of optical fiber in junction network may not be apparently benefited due to shorter link lengths, but it is advantageous because of high bandwidth, high channel capacity and repeaterless operation.

Typical transmission rate of the junction network is 8 Mbit/sec with both the injection laser and LED sources via multimode graded index fiber to APD detectors. Optical power budget of such system at wavelength of 0.88 μm over 12 km is indicated in table 15.4. It may be noted that the mean power launched from the laser is reduced below the level obtained with multimode graded index fiber of similar dimension in the optical power budget as in indicated in table 15.2. This is due to lower duty factor for the use of 2B3B code fiber on the 8 Mbit/sec systems in comparison with 7B8B code fiber for 140 Mbit/sec systems.

Table 15.4

Mean power launched from the laser transmitter (butt Coupling)	– 6.0 dB
Receiver sensitivity at 8 Mbit/sec (BER10^{-9}) at wavelength of 0.88 μm	– 63.0 dB
Total system margin	57.0 dB
Cabled fiber loss (12 × 3.5 dB/Km)	42.0 dB
Splice losses (13 × 0.3 dB each)	3.9 dB
Connector loss (2 × 1 dB each)	2.0 dB
Dispersion equalization penalty	0 dB
Safety margin	7.0 dB
Total system loss	54.9 dB
Excess power margin	2.1dB

15.4.3 Local Area Network

The local access network or subscriber loop connects telephone subscribers to the local switching centre or office. Possible network topologies are ring, tree or star types as discussed in section 15.2. In a ring network a number of transmission channel must be provided, as any information fed by the subscriber into the network passes through all the network nodes. Time multiplex system may be incorporated utilizing a broadband transmission medium so that only information addressed to a particular subscriber is taken from the network at that subscriber node. The tree network consists of several branches and may be considered as a multiple bus network in which access to the local switching centre from a node requires the use of a common bus link. Bus topology is sometimes referred to as the base configuration. A number of transmission channels on its common links must also be provided in tree network. In comparison with the ring network, the tree topology has the advantage of greater flexibility at topology enlargement. However in both the configuration the number of subscriber is limited by the transmission capacity of the links.

Star topology provides separate link to every subscriber to the local switching centre. In comparison to other topologies, star configuration is reliable and simple subscriber equipment is adequate (*i.e.*, no TDM is required) and network expansion is straightforward. However the amount of cable required is considerably high. Star configuration may be used to provide wideband services like videophone, television, stereo hi-fi, facsimile, data transmission etc together with narrowband speech transmission.

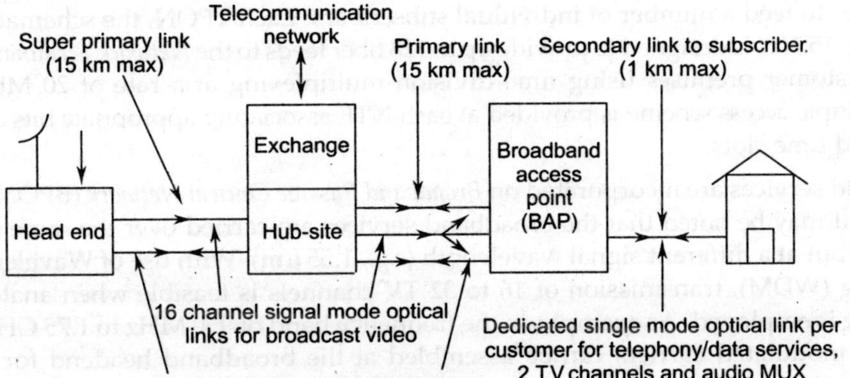

Fig 15.4

Different strategies may be adopted for efficient use of optical fiber network. A schematic for *Broadband Integrated Distributed Star* (BIDS) is shown in Fig. 15.4, which is a modified switched star network incorporating telephony services to take single mode fiber through local access network to subscriber premises. At the headend, broadband signals are collected and then transmitted to a number of hub-sites, generally located at the local telephone exchange, on 1.3 μm single mode fiber link. Each fiber is capable of carrying up to 16 television channels using FM modulation. The scheme provides the interface of telephony services from telecommunication network at 140 Mbit/sec as well as broadband services through single mode primary links to the *Broadband Access Point* (BAP). The telephony and data services plus two TV channels are provided to the subscriber or customer premises by employing secondary single mode fiber links.

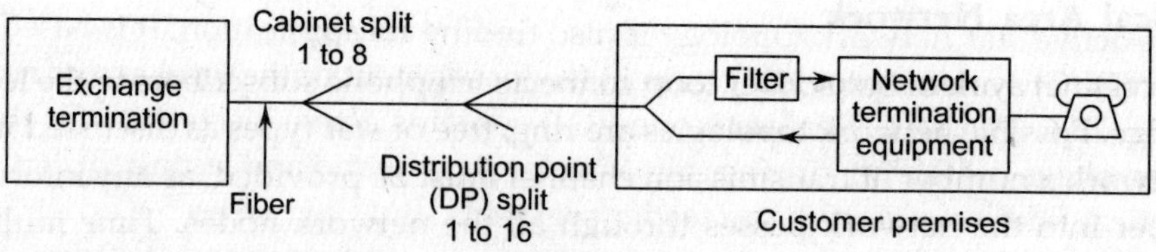

Fig. 15.5(*a*)

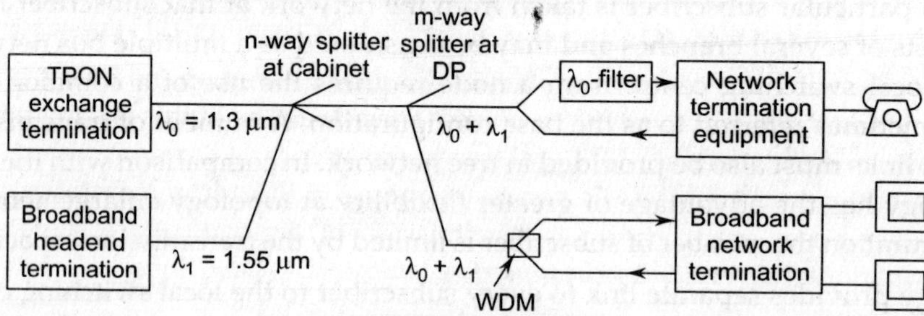

Fig. 15.5(*b*)

An alternate strategy has been developed for initial provision of telephony and other low bit rate services while allowing broadband services. *Telephony on Passive Optical Network* (TPON) is employed by one single mode fiber operating at 1.3 μm fed from the local telephone exchange and fanned out via passive optical splitters or tree couplers at cabinet and *Distribution Point* (DP) positions in order to feed a number of individual subscribers. Each TPON, the schematic of which is shown in Fig. 15.5(*a*), is designed to provide up to 128 fiber feeds to the *Network Terminating Equipment* (NTE) in customer premises using time division multiplexing at a rate of 20 Mbits/sec. Time division multiple access scheme is provided at each NTE associating appropriate bits of information into assigned time slots.

Broadband services are incorporated on *Broadband Passive Optical Network* (BPON) as shown in Fig. 15.5(*b*). It may be noted that the broadband services are carried over the same fiber network architecture but at a different signal wavelength (*e.g.*, 1.55 μm). With use of Wavelength Division Multiplexing (WDM), transmission of 16 to 32 TV channels is feasible when analog subcarrier multiplexing is employed. As example, in the frequency band of 950 MHz to 1.75 GHz a multiplex of 32 video modulated carriers can be assembled at the broadband headend for downstream transmission to the customer premises on the 1.55 μm wavelength signal.

Basic *Integrated Services Digital Network* (ISDN) together with CATV is also feasible on single mode fiber. End to end digital connectivity with standard set of interfaces for integrated services delivery is described by the ISDN terminology. The basic or narrowband ISDN service is intended to provide the subscribers with access of two 64 Kbit/sec digital switched channels and 16 Kbit/sec packet switched channel. Broadband ISDN (BISDN) that can offer transmission rates in the hundreds of Mbit/sec may facilitate the worldwide compatibility of broadband services of switched access TV, videophone, high quality audio and multimedia desktop teleconferencing within local access network.

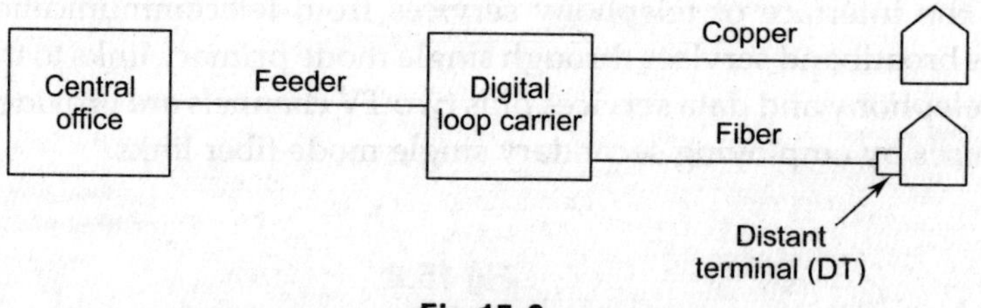

Fig 15.6

A double active star network topology is also finding its application. It is based on the *Digital Loop Carrier* (DLC) systems providing gain in feeder segment of the local access network as illustrated in Fig. 15.6. The DLC Remote Node (RN) contains active electronics for the provision of battery, overvoltage protection, ringing, signaling, coding, hybrid and testing (BURSCHT) functions. Initially optical fiber was used in the feeder segments. But SLC series 5 Carrier System as marketed by AT&T uses physical/logical star topology in both feeder and distribution segments to extend the fiber to the subscriber premises. In this case the remote node electronics have been modified to provide the BURSCHT functions at the Distant Terminal (DT) instead of remote node.

Alternate local access network as supplied by Raynet is based on the bus/star topology in the distribution segments using active service points. The schematic of this is shown in Fig. 15.7. The system employs an *Office Interface Unit* (OIU) at the central office or remote node that interfaces with eight digital lines of 1.5 Mbit/sec and two protection lines from central office switch. Two dimensional fiber buses are utilized for transmission of digitized voice signals together with signaling information to several *Subscriber Interface Units* (SIU). Each Subscriber Interface Unit (SIU) supports eight subscriber lines. SIUs are designed to be placed near the subscriber premises such that copper conductors may be used for short distances to the subscribers. This architecture requires less quantity of fiber at feeder and distribution segments, but increases number of locations containing active electronics.

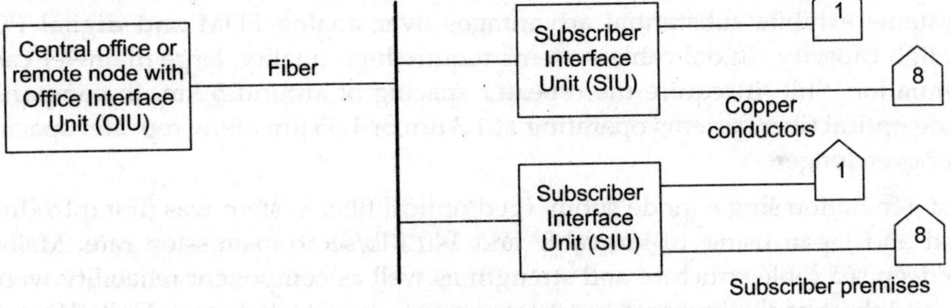

Fig 15.7

Double star topology may also be employed with passive fiber access network. It is called *Passive Photonic Loop* (PPL) and based on very high capacity multiwavelength optical network with LAMBDANET system. Schematic of PLL architecture is illustrated in Fig. 15.7. Wavelength Division Multiplexing (WDM) technique is utilized at central office as well as remote node or service access point to support over twenty wavelength channels. A single feed fiber is used from central office to remote node and dedicated optical fiber links connect the subscribers from remote node. Hence the PLL comprises of a physical star/logical star in both feeder and distribution segments of local access network.

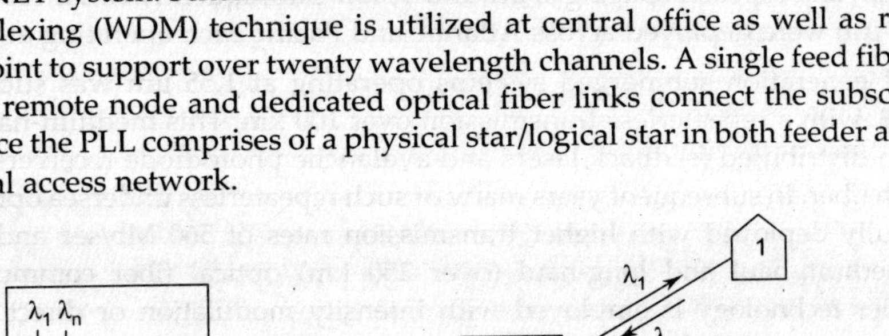

Fig 15.8

In the LABMDANET concept each subscriber is assigned two unique wavelengths, one for upstream transmission and one for downstream transmission. Information signal for a particular subscriber is modulated with assigned downstream wavelength at central office and multiplexed to feeder fiber. This is demultiplexed at remote node and passed down to the subscriber. Therefore, no filtering is necessary at the subscriber premise. Feeder fiber length is 10 km approximately and distribution fiber length varies between 2.2 km and 3.3 km with bit error rate of less than 10^{-9} at 600 Mbit/sec or 1200 Mbit/sec for bidirectional transmission.

Synchronous optical network (SONET) is generally adopted for broadband services. However some broadband services, particularly bursty traffic are not easily accommodated by SONET alone. Hence alternate packetized multiplexing and switching technique called. *Asynchronous Transfer Mode* (ATM) is employed to carry the information within the SONET payload. Local access network must be compatible for handling ATM data system and therefore the APON system allows 155 Mbit/sec ATM data stream to be distributed over the shared PON architecture.

15.4.4 Submerged Systems

Undersea cable systems are integral part of international telecommunication network. Undersea cable systems of shorter routes are utilized especially at Europe. Longer routes, such as across the Atlantic, provide route diversity in conjunction with satellite links. The requirements of submerged systems and their capacities are steadily increasing. In this context digital optical fiber communication systems exhibits substantial advantages over analog FDM and digital PCM coaxial systems. High capacity coaxial cable systems require high quality, large diameter cable to overcome attenuation. Still it require the repeater spacing of around 5 km. In comparison to this, single mode optical fiber systems operating at 1.3 μm or 1.55 μm allow repeater spacing of 25 km to 30 km or even longer.

The first generation single mode submerged optical fiber system was first introduced in 1980 at Scotland and Japan using 10 km cable and 140 Mb/sec transmission rate. Major problems related to deep sea cable structure and strength as well as component reliability were gradually overcome, and short-haul submerged systems were operative at UK, France, United States, Japan and European countries using transmission rate of 280 Mb/sec and 400 Mb/sec. In 1988 repeater service was introduced in transoceanic optical fiber systems. This single mode fiber system operates at the wavelength of 1.3 μm with the transmission capacity of 560 Mb/sec (two fibers at 280 Mb/sec each) and repeater spacing of around 50 km. Subsequently two more single mode fiber systems of 1.3 μm were deployed across Atlantic and Pacific each operating at 420 Mb/sec.

The second generation submerged systems operating at 1.55 μm was successfully demonstrated in 1988, with a repeaterless transmission over 100 km, This medium-haul fiber system is equipped with distributed feedback lasers and avalanche photodiode receivers operating at 417 Mb/sec on each fiber. In subsequent years many of such repeaterless undersea optical cable systems were successfully deployed with higher transmission rates of 560 Mb/sec and 565 Mb/sec. For submerged medium-haul and long-haul (over 250 km) optical fiber communication systems optical amplifier technology is employed with intensity modulation or direct modulation. It is possible to achieve up to 1000 km of distance using ten optical amplifiers without the need for optical isolators or filters. Coherent optical system is important perspective for long-haul transmission systems over the distance of 2000 km. The main limitations, however, is the spontaneous emission at each amplifier which is both random and signal related degrading signal to noise ratio. Furthermore dispersion and nonlinear effects build up due to long interaction length resulting in distortion.

15.5 COMPUTER APPLICATIONS

Modern computers systems consist of large number of interconnections. These range from few micrometers (on-chip interconnections in VLSI chips) to thousands of kilometers for terrestrial links in computers networks. A wide range of transmission rate of around 100 b/sec for some teletype terminals to several hundred Mb/sec for on chip connections is covered over these interconnections. Optical fibers are being utilized in such types of interconnections where secure and interference free transmission is required. optical techniques are being utilized in integrated circuits, hybrid integrated circuits and data processing equipments. Digital data buses use optical fibers for their small size, low loss, low radiation properties and freedom from ground loops.

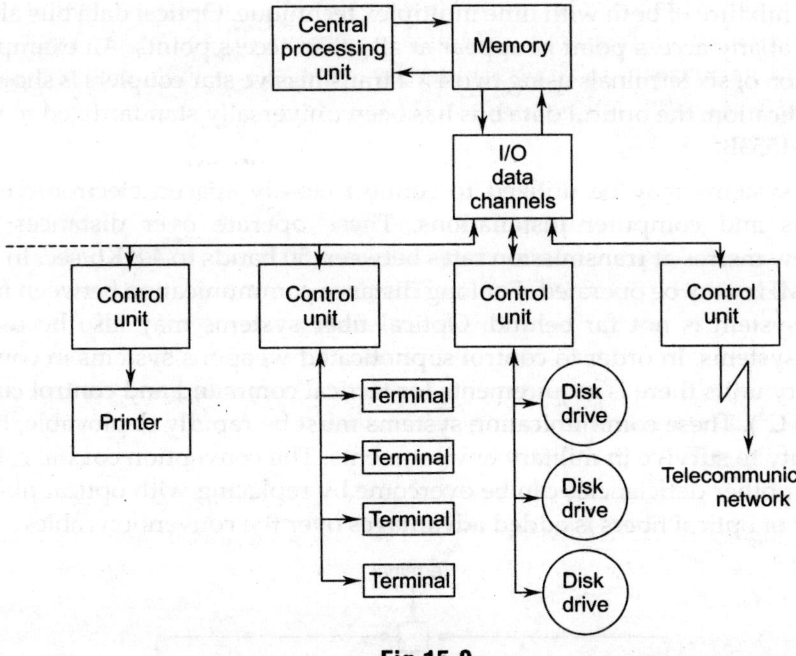

Fig 15.9

Interequipement connections topology for a typical mainframe computer is shown in Fig. 15.8. This comprises of several I/O devices in form of teletypes, printers, magnetic disk etc., generally handled by a central processing unit and attached to main storage device. I/O requirements concurrently work through control units. A large number of parallel interconnections is required where optical fiber may be utilized to replace the coaxial cables. Multiplexed operation with optical fiber results in significant reduction of cable and connectors, improving connection reliability and communication security. High speed transmission of 500 Mb/sec up to the radius of 30 km is achievable.

15.6 MILITARY APPLICATIONS

The design considerations of military equipments are size, weight, ruggedness, deployability, survivability in conventional as well as nuclear attack and above all the security. All these special features are attributed to optical fiber communication systems. The small size and weight of optical fibers provide an attractive solution to space problems in the military mobiles such as aircraft, ships and tanks, which are equipped with sophisticated electronics. The wideband nature of optical fiber transmission allows the multiplexing of a number of signals on a common bus. Furthermore, the electromagnetic interference (EMI) generated from noisy environment of military mobiles cannot make any effect on optical signal due to its inherent nature of immunity to EMI noise. Other advantage to use the optical fibers is the immunity to lightning and electromagnetic pulses (EMP) especially within avionics. The electrical isolation and so the safety aspect of optical fiber communication makes it attractive to be used in military application even routing through fuel tanks and magazines.

The design approaches include multiterminal data systems with single fibers and use of an optical data bus. The multiterminal single fiber system network is realized by ring or star configuration or mixture of both with time multiplex technique. Optical data bus allows injection of optical signal at any access point to appear at all other access points. An example illustrating the interconnection of six terminals using two 4 × 4 transmissive star couplers is shown in Fig. 15.9. For military application, the optical data bus has been universally standardized to the guidelines set by MIL-STD-1553B.

Optical fiber systems may be utilized to connect closely spaced electronic equipments in operation rooms and computer installations. These operate over distances from several centimeters to few metres at transmission rates between 50 bauds to 4.8 Kb/sec. In addition to it, video links of 7 MHz may be operated. For long distance communication between military installations, optical system is not far behind. Optical fiber systems may also be used in tactical communication systems. In order to control sophoticated weapons systems in conjunction with dispersed military units there is requirements for tactical command and control communication (often termed as C^3). These communication systems must be rapidly deployable, highly mobile, reliable and ability to survive in military environments. The convention coaxial cables are bulky enough and their other deficiencies can be overcome by replacing with optical fibers. Immunity to EMI and EMP of optical fibers is added advantages over the convention cables.

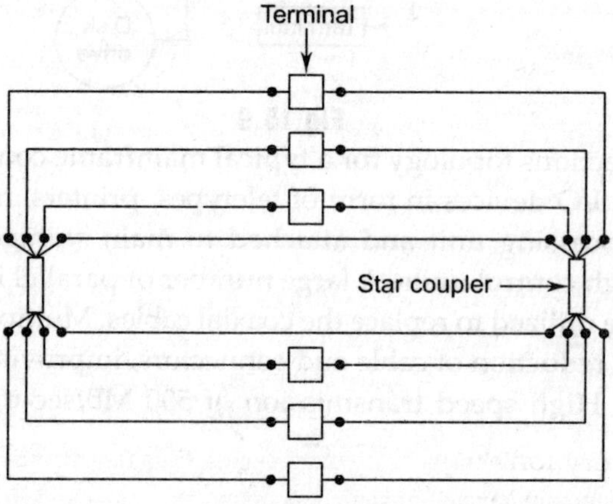

Fig 15.10

15.7 CIVIL, CONSUMER AND INDUSTRIAL APPLICATIONS

The introduction of optical fiber communication systems into public network has stimulated its application in public utility organizations that provide their own communication facilities over moderately long distances. These transmission techniques are utilized in signaling, information and communication services at railways, and along pipe and electrical power lines. Although high capacity transmission is not required in these applications, optical fibers provide low cost implementation and enhanced protection in harsh environments, especially from EMI and EMP.

Optical systems are suitable for video transmission and hence commercial television signal transmission is incorporated with fiber networks. These applications include short distance links between studios and outside broadcast vans or receiving aerials, and close circuit television (CCTV) links for security and traffic surveillance. Optical fiber communication is also incorporated at larger networks for cable and common antenna television (CATV) providing the advantages of bandwidth and unrepeatered transmission distance over conventional video links. Various techniques may be used for video transmission including baseband intensity modulation, subcarrier intensity modulation (FM-IM), pulse analog technique (PFM-IM) and digital pulse code modulation (PCM-IM). Generally on larger CATV networks digital transmissions preferred as it allows time division multiplexing as well as greater repeaterless transmission. It avoids problems of nonlinearities of optical sources. However digital transmission is not always economic due to the cost of terminal equipments.

A major consumer application of optical fiber systems is at automobile electronics. Microprocessors and microcontrollers are now widely used in automobile electronics to provide the features of convenience such as power windows, seat controls and control panels. Optical fiber links are of low cost and weight, and immuned to EMI noise, and hence may be routed through fuel tanks and magazines similar to military mobiles as explained in earlier section. Other consumer applications where optical systems may be useful are the home appliances based on microprocessors and microcontrollers technology.

A variety of on-premises industrial operation within a single operational site may use optical fiber communications links. Especially at harsh environments of electromagnetic interference optical links are advantageous. These system offers reliable telemetry within reasonable cost. Furthermore optical fiber systems provide a far safer solution than conventional electrical wirings at the environments of explosive and corrosive gases, as example, at chemical process plant and petroleum refineries. Nevertheless to mention that LANs are widely used in industry for interoffice data communication and information services which are already described earlier.

15.8 CONCLUSION

The growing utilization of optical fiber systems has stimulated tremendous research efforts towards enhanced fiber design. This has resulted in improvement of the associated optoelectronics as well as investigation of passive optics to provide advanced state of art technology to optical fiber communications together with the expansion in the areas of use. Development of fluoride glass fibers for very long haul mid-infrared transmission or soliton (very narrow optical pulse with high peak power) propagation are the examples. Utilization of nonlinear pulse propagation in optical fibers is investigated to provide greatly increased channel capacity exhibiting compatibility with integrated optical systems in a nonlinear environment. Advancement of ôptical transmission systems will not stop here and many more to come.

REVIEW QUESTIONS

1. A fiber optic link of length 50 Km has two splices each exhibiting a loss of 1 dB. Fiber loss is 0.2 dB/Km. What power must be supplied by the source if the minimum power required by the photodetector is 20 nW?
   (a) 20 nW (b) 317 nW (c) 317 μW (d) 500 μW
2. A fiber used in atypical link of length 8 Km has intermodal dispersion time as 3 ns/Km and intramodal dispersion time as 1 ns/Km. The transmitter and receiver rise times are 12 ns and 11 ns respectively. Find the overall rise time of the system.
   (a) 23 ns (b) 27 ns (c) 30 ns (d) 35 ns
   [Hint: $t_f = (t^2_{\text{intermodal}} + t^2_{\text{intramodal}})^{1/2}$ and $t_r = (t_{tr}^2 + t_f^2 + t_{re}^2)^{1/2}$]
3. What is the bit duration of a 2.5 Gb/s signal?
   (a) 0.1 ns (b) 0.4 ns (c) 1 ns (d) 2.5 ns
4. For a passive star network the total optical power is supplied by the central node is 1 μW and each receiver requires 0.1 μW. If fractional insertion loss at each coupler is 0.05 μW, what is the number of subscribers (nodes)?
   (a) 50 (b) 100 (c) 250 (d) 500
5. Which of the following is a nonlinear inelastic process?
   (a) SRS (b) SPM (c) CPM (d) FWM
6. Which of the following nonlinear effects arise from the intensity dependent variation of the refractive index of a fiber?
   (a) SPM (b) CPM (d) FWM (d) All of these
7. Assuming the fiber link span link to be large, what will be the effective length if its attenuation is 0.2 dB/Km?
   (a) 95.6 Km (b) 50.2 Km (c) 22.2 Km (d) 10 Km
8. The repeaterless dispersion shifted fiber length is 50 Km with dispersion parameter of 16 ps/nm/Km. What should be the dispersion parameter of a DCF of length of 2 Km in order to compensate for the accumulated dispersion of the DSF for a repeaterless distance?
   (a) 80 ps/nm/Km (b) – 80 ps/nm/Km
   (c) – 200 ps/nm/Km (d) – 400 ps/nm/Km
9. The number of wavelength channels in a WDM system is 10. What will be the number of frequencies crated by four-waving mixing?
   (a) 30 (b) 100 (c) 450 (d) 900
10. Which of the following process is used to compensate for GVD-induced dispersion in a soliton?
    (a) FWD (b) SPM (c) CPM (d) All of these
11. What are the link power budget and rise time budget analyses? Perform these analyses for a fiber optic link that uses a conventional single mode fiber that has been upgraded by DCF loops and optical amplifiers.
12. Discuss different types of network topologies and their applications
13. What are the functions of logical link transfer (LLC) and Medium Access Control (MAC)? Discuss the different layers and their functions in LAN and WAN.
14. Discuss various parts of public network system Define TPON, BPON, DP and ISDN. Describe the schemes how they are useful to optical communication.
15. In a typical scheme of distribution of video signals in CATV system using optical bus, the transmitter of 0 dBm is coupled to optical bus. Each receiver has the sensitivity is

of –42 dBm, each optical tap couples 5% of optical power to the subscriber and has insertion loss of 0.3 dB, optical fiber loss is 0.01 dB/m and minimum distance between two subscribers is 50 m. How many subscribers may be added to the bus without any use of line amplifier?

16. A typical digital fiber optic link uses GaAs LED as transmitter and Si-pin as receiver. It is designed for 12 Km for operation of 0.85 μm at 100 Mb/s. The link uses step index fiber of n_1 of 1.46 with relative refractive index difference Δ of 0.01 and material dispersion parameter of 80 ps/Km/nm. The rise times of transmitter and receiver are 10 ns and 12 ns respectively. The spectral half width of the LED is 30 nm. Prepare a rise time budget of the link. Is it possible to operate the system with NRZ format?

17. A single mode fiber link is to be designed for operation of 1.55 μm at 500 Mb/s. A single mode InGaAsP laser of average optical power of –13 dBm is coupled to the link. Fiber is available with 5 Km pieces with attenuation of 0.5 dB/Km and each splice loss of 0.5 dB. InGaAs APD is used as receiver with the sensitivity of –40 dBm and coupling loss of 0.5 dB. Set up a link power budget keeping a safety margin of 7 dB and calculate the repeaterless distance.

18. What are the advantages and disadvantages of optical transmission? What are the parameters to characterize the optical link? How the optical links are classified and what are they?

19. What are the protocols used for LAN system? Discuss them in detail.

20. What are junction network and trunk network? Indicate their applications. Explain BIDS with its schematic diagram. Illustrate the LABMDANET concept and SONET.

21. How the optical transmission is adopted for submerged links and military applications?

INDEX

D

E

F

G